Air Quality
Control
Handbook

ALDRICH • *Pollution Prevention Economics: Financial Impacts on Business and Industry*

AMERICAN WATER WORKS ASSOCIATION • *Water Quality and Treatment*

BAKER, HERSON • *Bioremediation*

BRUNNER • *Hazardous Waste Incineration, Second Edition*

CASCIO, WOODSIDE, MITCHELL • *ISO 14000 Guide: The New International Environmental Management Standard*

CHOPEY • *Environmental Engineering in the Process Plant*

COOKSON • *Bioremediation Engineering: Design and Application*

CORBITT • *Standard Handbook of Environmental Engineering*

CURRAN • *Environmental Life-Cycle Assessment*

FIKSEL • *Design for Environment: Creating Eco-efficient Products and Processes*

FREEMAN • *Hazardous Waste Minimization*

FREEMAN • *Standard Handbook of Hazardous Waste Treatment and Disposal, Second Edition*

FREEMAN • *Industrial Pollution Prevention Handbook*

HARRISON • *Environmental, Health and Safety Auditing Handbook, Second Edition*

HAYS, GOBBELL, GENICK • *Indoor Air Quality: Solution and Strategies*

HEUMANN • *Industrial Air Pollution Control*

JAIN, URBAN, STACEY, BALBACH • *Environmental Impact Assessment*

JOHNSON • *ISO 14000 Road Map*

KALETSKY • *OSHA Inspections: Preparation and Response*

KOLLURU • *Environmental Strategies Handbook*

KOLLURU • *Risk Assessment and Management Handbook for Environmental, Health and Safety Professionals*

KREITH • *Handbook of Solid Waste Management*

LEE • *Dictionary of Environmental Legal Terms*

LEVIN, GEALT • *Biotreatment of Industrial and Hazardous Waste*

LUND • *The McGraw-Hill Recycling Handbook*

MARRIOTT • *Environmental Impact Assessment: A Practical Guide*

ROSSITER • *Waste Minimization Through Process Design*

SELDNER, COETHRAL • *Environmental Decision Making for Engineering and Business Managers*

SMALLWOOD • *Solvent Recovery Handbook*

WILLIG • *Environmental TQM, Second Edition*

Air Quality Control Handbook

E. Roberts Alley & Associates, Inc.

McGraw-Hill

New York   San Francisco   Washington, D.C.   Auckland   Bogotá
Caracas   Lisbon   London   Madrid   Mexico City   Milan
Montreal   New Delhi   San Juan   Singapore
Sydney   Tokyo   Toronto

Library of Congress Cataloging-in-Publication Data

Air quality control handbook / E. Roberts Alley & Associates, Inc.
   p.   cm.
   Includes index.
   ISBN 0-07-001411-6
   1. Air—Pollution.   2. Air quality management.   I. E. Roberts
   Alley & Associates.
   TD883.A576   1998
   363.739'26—dc21   98-9163
                                                                CIP

McGraw-Hill

A Division of The McGraw·Hill Companies

1 2 3 4 5 6 7 8 9 0   DOC/DOC   9 0 3 2 1 0 9 8

ISBN 0-07-001411-6

The sponsoring editor for this book was Robert Esposito, the editing supervisor was David E. Fogarty, and the production supervisor was Sherri Souffrance. It was set in Century Schoolbook by Estelita F. Green of McGraw-Hill's Professional Book Group composition unit.

Printed and bound by R. R. Donnelley & Sons Company.

 This book is printed on recycled, acid-free paper containing a minimum of 50% recycled, de-inked fiber.

McGraw-Hill books are available at special quantity discounts to use as premiums and sales promotions, or for use in corporate training programs. For more information, please write to the Director of Special Sales, McGraw-Hill, 11 West 19th Street, New York, NY 10011. Or contact your local bookstore.

Contents

Part 3 Air Pollution Regulation

Part 4 Pollutant Control Systems

Preface

Introduction

Global warming, the ozone hole, acid rain, Los Angeles smog, and ultraviolet radiation all have something in common. Even though these hackneyed expressions have been biased with emotionalism, there is no question that air pollution has caused potential or actual damage to the environment, which makes each one of these terms an interest and perhaps a concern for those in the environmental field. Without this concern and the improvements resulting from this concern, environmental deterioration can very well eliminate sustainable development locally, regionally, or globally. We have inherited our planet from previous generations along with a legacy of associated pollution. The question crying out from this inheritance is whether we can maintain sustainable growth or whether the technical progress which has brought us prosperity will destroy the health of future generations. There is an escalating conflict between "environmentalists," who seek to prevent the discharge of the most minute pollutants into the atmosphere, and "industrialists," whose experience and familiarity with their product convinces them that their pollutants do no damage to the atmosphere. The public, meanwhile, is confused by reports of global warming, the ozone hole, and acid rain and continually supports environmental control. This Handbook addresses these concerns and explains the chemistry and the meteorology which can cause damage to our sensitive ecosystem.

There exists in our society a need among industrial, government, and consulting engineers for a handbook to allow quick reference to the major areas of air pollution control—theory, characterization, regulations, management, and implementation. This Handbook fills that need and relates the five areas to one another so that a rational approach can be followed in solving air pollution control issues. The Handbook first explains the types of pollutants which can be discharged into the air and how they are dispersed into the atmosphere. A detailed discussion of methods for monitoring and testing for the pollutants is then present-ed. At this point a discharger should be able to characterize its waste in terms of quality and quantity and estimate how the waste enters and is carried through the atmosphere. The Handbook next reviews and rec-ommends air control management programs which can be used to implement a control policy. Air regulations including the 1990 Clean Air Act Amendments are discussed in detail, and suggestions are made for filing applications and meeting the requirements of the regulations. The last part reviews the currently used methods for controlling air pollu-tion including design criteria.

Summary

This Handbook examines the use of the air as a realm for the dis-charge of pollutants and proposes practical methods for controlling air pollution by presenting recommendations in the following areas:

1. The theory of air pollution
2. Modeling the diffusion of pollution in the atmosphere
3. Monitoring air emissions
4. Testing air pollution characteristics
5. Setting a management policy to control air pollution
6. Establishing management programs to control air pollution
7. Preparing for an air pollution audit
8. History of air pollution
9. The Clean Air Act requirements of the U.S. Environmental Protection Agency
10. Collecting and transporting of air pollutants
11. Control of particulate (solid) air pollutants
12. Control of gaseous organic air pollutants
13. Control of gaseous inorganic air pollutants

Handbook Use

The following paragraphs suggest ways to use this Handbook in solv-ing air pollution control problems.

Pollutant characterization

As stated above, air pollutants have a wide range of characteristics. These are summarized in the chapters of this Handbook as follows:

State (gas, liquid, or solid in air)	Chapter 2
Organic or inorganic	Chapter 2
Toxicity to the environment	Chapter 3
Regulatory controls	Chapters 11 through 17
Discharge rate	Chapter 3
Discharge concentration	Chapters 5 through 7
Variability of discharge rate and concentration	Chapter 5

The Handbook makes specific recommendations for monitoring and testing methods for determining these characteristics.

Atmospheric assimilation

The atmosphere can serve to transfer pollutants so that they are dispersed to the extent that they can do no environmental harm, but it can also transfer pollutants in such a way that humans or other parts of the environment can be negatively affected. The following chapters of this Handbook address the atmosphere assimilation issues:

Atmosphere diffusion models	Chapter 3
Air monitoring (ambient and stack)	Chapters 4 and 5
Pollutant testing	Chapter 6
Fugitive emission modeling	Chapter 7

Air pollution control management

Knowledge of the character of air pollutants and the character of the atmosphere into which these pollutants will be discharged is only academic if a system is not developed to manage the discharge of these pollutants. Top management should set a policy of air pollution control and conformance standards to ensure that this policy is followed. These standards will typically be in the form of documentation and tests of the effectiveness of the documentation. The following chapters of this Handbook recommend management techniques for air pollution control:

Management policy	Chapter 8
Management methods	Chapter 9
Management to meet regulations	Chapter 9
Management programs	Chapter 9
How to prepare for an audit	Chapter 10

Environmental regulations

Even though ideally the control of environmental air pollution should be limited to the influences of the pollutant characterization and the assimilative capacity of the atmosphere, governments have been forced to develop regulations in order to protect their environments and establish equality between the potential polluters of the environment. In the United States, these regulations have been the result of a series of laws, the latest of which is called the Clean Air Act Amendments (CAAA) of 1990. The Environmental Protection Agency has had the responsibility of developing the rules and regulations to enforce this Act and its Amendments.

This Handbook present a practical summary of the Act and its Amendments and makes recommendations for meeting these regulatory requirements in the following chapters:

History of air pollution	Chapter 1
Definitions and acronyms	Appendixes A and B
Attainment and nonattainment areas	Chapter 11
National emissions standards	Chapter 11
Air pollution from moving vehicles	Chapter 12
Hazardous air pollutants	Chapter 13
CAAA, Title IV, acid rain	Chapter 14
CAAA, Title V, operating permits	Chapter 15
CAAA, Title VI, ozone protection	Chapter 16
CAAA, Title VII, enforcement and administration	Chapter 17

Air pollution control

Once the air pollutant characteristics, the atmospheric character, the management policy, and the environmental regulations are known, the air pollution control technology must be addressed. The following chapters of this Handbook make specific practical recommendations for controlling various forms of air pollution:

Hood and ductwork design for collecting and transporting air pollutants	Chapter 18
Control of particulate (solid) air pollutants	Chapter 19
Control of gaseous organic air pollutants	Chapters 20 through 24
Control of gaseous inorganic air pollutants	Chapters 25 and 26

These chapters should enable the Handbook user to compare the characteristics of the air pollutants with alternative treatment technologies and to design and select the most efficient and feasible control method.

Acknowledgments

The genesis of this Handbook is a course in air pollution control taught by E. Roberts Alley and William L. Cleland through George Washington University. The authors would like to thank the Continuing Engineering Education Program of the University for the opportunity to develop the course and the notes which have been expanded and organized into this Handbook.

The employees of E. Roberts Alley & Associates, Inc., have compiled this additional information through their experience and research, and we would like to thank them for their efforts.

The following employees authored the chapters of the Handbook:

1.	Introduction	E. Roberts Alley
2.	Basic Air Pollution Theory	Mike Ayers
3.	Atmospheric Dispersion Models	Andrew T. Allen
4.	Ambient Air Monitoring	Steven Marquardt
5.	Stack Sampling and Monitoring	Lem B. Stevens, III
6.	Air Pollution Testing	E. Roberts Alley, Jr.
7.	Fugitive Emissions	Taylor H. Wilkerson
8.	Air Quality Management Policy	E. Roberts Alley
9.	Air Quality Management Program	Taylor H. Wilkerson
10.	Air Quality Audit	John Coulter
11.	Air Quality	William L. Cleland
12.	Mobil Sources	Charles Gallagher
13.	Hazardous Air Pollutants	William L. Cleland
14.	Acid Rain	Charles Gallagher
15.	Operating Permits	William L. Cleland
16.	Stratospheric Ozone Protection	Charles Gallagher
17.	Enforcement and Administration	Charles Gallagher
18.	Ventilation	Barney Fullington

The readers are invited to contact these authors for additional information on their subject.

Introduction

E. Roberts Alley

1.1 Background

Whenever energy of any kind is used on our planet, there is a possibility of residual pollution. This pollution can take the form of a gas, a liquid, or a solid, and it can come from inorganic or organic sources. Much of the pollution we experience is caused by natural processes not associated with human activity. Natural occurrences can both cause pollution and affect the distribution of pollution into the environment. It is therefore imperative that we understand both pollution sources and pollution assimilation in order to control pollution and prevent environmental deterioration.

Humans, in their activities, emit gaseous, solid, and liquid pollutants, which must be assimilated into the air, land, or bodies of water. Through pollution control, these emissions can be transferred among these three realms; i.e., a gaseous pollutant can be absorbed into a liquid and discharged to a stream and therefore can change the realm into which it is discharged. When we choose air, the discharges are merely dispersed into the environment; they are not normally diminished in quantity. Air chemistry can even increase the toxicity of a pollutant. The air, because of natural forces, can distribute pollution over large geographic areas and is therefore of global concern.

Bodies of water act as dispersers of pollution but can also provide minor treatment to organic pollution because of the oxygen present in the water. Water is typically of regional environmental concern in scope.

The land is the only one of the three discharge possibilities which will provide significant treatment, because of biological activity. Land is normally of only local environmental concern.

Since with any activity on this planet, pollution of some kind will occur, the decision must be made, in order to protect the environment, as to whether the air, the water, or the land can best assimilate the particular pollutants involved. Before this can be discerned, the pollutants must be characterized as to quality and quantity. A pollutant can be gaseous, liquid, or solid, which may dictate its final disposition into the environment. But it can also be organic or inorganic; microscopic or macroscopic; corrosive; toxic; or it can have any combination of other characteristics. Each of these characteristics can be critical to the pollutant's effects on the environment, and this demonstrates the importance of accurately determining the characteristics of the pollutants as to concentration and variability.

To control pollution of the environment, therefore, we must consider two main factors: the characteristics of the pollutants and the natural characteristics of the realm of the environment into which the pollutant might be discharged. Most countries have regulated the discharge of pollution into the environment for protection purposes. These environmental regulations may also affect the decision for discharging pollutants.

1.2 History of Air Pollution

We think of air pollution as something that has developed in the last half of the 20th century, but this is not the case. The types of pollution to which people have been exposed go back to at least the 14th century, when coal was first used to heat homes. In England in the early part of that century, with sea coal that was particularly smoky when burned, there were recorded protests, men were actually tortured for producing a "pestilent odor," and coal was restricted through taxation. Queen Elizabeth I passed a law early in her reign forbidding the use of coal while Parliament was in session. As cities grew and the industrial revolution developed, the use of coal and smoke pollution increased.

In the United States, the first recognition of the air pollution problem was also due to coal smoke. Large cities such as Chicago, St. Louis, and Cincinnati passed smoke ordinances. It was established in the early years that responsibility rested with the state and local governments. Even so, air pollution was not considered a health hazard as it is known to be today.

During World War II, there was no time or materials to control air pollution. Smoke concentrations rose to new highs as the national effort to win the war proceeded. When the war was over, once again attention was turned to action to address the problem. In Pittsburgh, an ordinance was enacted to control the type of coal used.

In October 1948, disaster struck in Donora, Pennsylvania. This industrial town, located about 20 mi from Pittsburgh, had a population of around 12,000. Steel, zinc, and sulfuric acid plants had been spewing smoke into the air for decades. Normally, winds of sufficient velocity carried away the pollutants in the smoke and caused dilution in a fairly deep layer of the atmosphere. Over a 5-day period from October 26 to 31, 1948, air pollution accumulated in stagnant air at a concentration that caused several thousand people to become ill. Many were hospitalized and 20 died. The health effects of air pollution are sometimes slow and cumulative, but it seems to take a catastrophe such as this to get the attention of the public and the government.

Table 1.1 shows a list of severe instances in the past. In November 1950, some 350 people were hospitalized and 22 were killed when concentrated hydrogen sulfide from a plant recovering sulfur from natural gas escaped into the air during a thermal inversion. In early December 1952, after most of the British Isles were covered by a fog and temperature inversion, many persons became ill with shortness of breath from a condition in which the body turns blue because of

TABLE 1.1   Instances of Severe Air Pollution

Location	Date	Deaths	Reported illness	Conditions
Meuse Valley, Belgium	12/1/30	63	6000	Low atmospheric dilution
Donora, Pennsylvania	10/25/48	18	5900 (43%)	Fog and gaseous materials
London, England	11/26/48	700–800		
Pozo Rica, Mexico	11/21/50	22	>350	
London, England	12/5/52	3500–4000	Unknown	
New York, New York	11/22/53	175–260	Unknown	
London, England	11/56	1000	Unknown	
London, England	12/2/57	700–800		
London, England	1/26/59	200–250		
London, England	12/5/62	700	Unknown	
London, England	1/7/63	700		
New York, New York	1/9/63	200–400		
New York, New York	11/23/66	170		

insufficient oxygen in the blood. Almost 4000 people were killed by the polluted air. The most severe disaster in the United States occurred in late November 1953, when for 11 days New York City was engulfed by a windless thermal inversion condition. This huge stagnant air mass accumulated over regions of Ohio, Pennsylvania, New Jersey, and New England and resulted in between 175 and 260 deaths above what was considered the norm for the area.

Around the time of World War II, another type of air pollution was being noticed with increasing frequency, especially in the atmosphere in and around the Los Angeles area. As a result of what was being recognized as photochemical smog, eye and skin irritation and plant damage were occurring. At first it was thought that this smog was caused primarily by oil refineries and facilities for storing oil and gasoline. Later it was found that much of it, if not all, was being caused by the internal combustion engine.

As the population increased and industrial growth continued, the problems from air pollution have become increasingly worse. Besides the increased use of the automobile, industrial processes were causing more complex types of air pollution than that from coal smoke and smog.

Beginning in 1962, several articles and books were published on environmental issues. Since that time, the general population has been increasingly aware of the effect of using chemicals which eventually affect humanity, either directly or indirectly, since we are at the end of the food chain.

Meanwhile, increased power requirements due to the rise in population have led to the use of high-sulfur fuels burned in utility boilers. This in turn has led to the poisoning of lakes and forests from acid rain.

Depletion of ozone in the stratosphere has also been brought to the attention of the public, to environmentalists, and to the national governments, especially after issuance of the Montreal Protocol in 1986.

1.3   Development of Air Quality Regulations

1.3.1   Federal laws and regulations from 1955 to 1970

"Air Pollution Control Research and Technical Assistance," enacted in 1955, provided money for state and local agencies and authorized federal technical assistance. This set the pattern for federal-state cooperation and interaction which still exists today.

The Clean Air Act (CAA) in 1963 gave the U.S. Public Health Service, Department of Health, Education, and Welfare (HEW), a role in handling air pollution control and provided grants-in-aid to states, but let HEW itself intervene directly in instances of "health and welfare" endangerment. The 1965 CAA amendment gave the federal government even more direct involvement, letting HEW set automobile emission standards without involving the states. The Air Quality Act of 1967 allowed the Secretary of HEW to issue mandatory air quality criteria and designate air quality control regions. Performance standards were set within regions, leaving it up to states to enforce.

1.3.2   Federal laws and regulations from 1970 to 1990

The Clean Air Act Amendments (CAAA) of 1970 resulted in the establishment of the Environmental Protection Agency (EPA). The EPA was charged with setting national ambient air quality standards (NAAQSs), primarily to protect health and secondarily to protect welfare, for "criteria" pollutants. Attainment to the standards was mandated to occur no later than 1977. States were required to submit state implementation plans (SIPs) to accomplish attainment. If a state failed to adopt a plan acceptable to the EPA, then the EPA could adopt a Federal Implementation Plan (FIP) for the state. New-source performance standards (NSPSs), by category, were set for new specific sources, and national emission standards were set for seven hazardous air pollutants (NESHAPs). Most states instituted an operating permit program.

The 1977 CAAA and subsequent regulations provided schemes for maintaining the national ambient air quality standards in areas already in attainment and for reaching attainment in areas that were not reaching attainment. A prevention of significant deterioration (PSD) preconstruction review was required for major sources in attainment areas. PSDs required best-available control technology (BACT). In nonattainment areas, new-source review (NSR) was required for major sources, and it was necessary to meet the lowest-achievable emission rate (LAER) in order for the area to maintain reasonable further progress (RFP) toward meeting the standards. While stationary sources were primary in attempting to maintain or attain the standards, automotive pollution was also to be addressed in a state implementation plan.

In 1980, amendments to the CAA modified PSD requirements, basically redefining major status in some categories.

1.3.3   The 1990 CAAA—Brief overview

The Clean Air Act Amendments of 1990 expanded the CAA and made it one of the longest environmental laws on the books. It is extremely complex and seeks to deal with subjects left out of previous versions of the act. The general theme is one of distrust—of industry and of the EPA.

Title I—Nonattainment Provisions. Title I still has the basic structure for setting ambient air quality standards and SIPs to obtain them. Standards are significantly more stringent than before, requiring new SIPs and new deadlines (from 3 to 20 years) for obtaining attainment depending on the severity of the nonattainment status. New offset requirements and sanctions for failure to achieve attainment were included, as well as excess-emission fees.

Title II—Mobile Sources. Title II imposes more-stringent standards for mobile sources. New standards for 1994 models continue to tighten through model year 2004. Vehicles must meet standards over a longer life span (10 years and 100,000 mi). Fuel standards are tight, requiring reformulated and alternative fuels and less-polluting engines.

Title III—Air Toxics. Rather than the seven hazardous air pollutants that have been regulated under the old rules, 189 compounds are now regulated. Standards based on maximum achievable control technology (MACT) are imposed. These are determined by the EPA or by states on a case-by-case basis. Also, a Chemical Safety and Hazard Identification Board is created to investigate and recommend regarding accidental releases of 100 chemical substances.

Title IV—Acid Rain. Acid deposition or the acid rain problem is addressed, targeting (in phase I) 111 power generating plants in 22 states permitted by the EPA in 1995. Remaining power plants will be permitted by 2000 in phase II. A trading policy through the Chicago Board of Trade allows the buying of SO_2 emission allowances to be used after the year 2000. The allowances are based on reduction beyond what is required and may be used or sold. The scheme is for SO_2 emissions in 2000 to be 10 million tons below 1980 levels and NO_x emissions to be down 2 million tons.

Title V—Operating Permits. A new operating permit program is established that requires sources to obtain a federally enforceable operating permit. Fees are required to fund the state-operated program. States were required to submit a plan by November 15, 1993, for approval by the EPA. It parallels provisions of NPDES requirements, except it is to be operated by states.

Title VI—Stratospheric Ozone Protection. Title VI phases out many ozone-destructive chemicals by 2000 and requires recycling of refrigerants.

Title VII—Enforcement. Title VII strengthens CAA enforcement provisions and allows civilian suits against violators. It significantly enhances criminal penalties for "knowing" violations and establishes the definition of the designated official of companies and his/her responsibilities. It allows minor penalties on the basis of "field citations."

Titles VIII and IX. They provide for a variety of authorities relating to investigations and research of different causes of air pollution and of different air pollution impacts.

Title X. This title seeks to secure fair treatment for disadvantaged business concerns in providing funding for environmental research.

Title XI. Title XI provides for "clean air employment transition assistance." It amends the Job Training Partnership Act to provide funding for persons who have been dislocated or displaced by new Clean Air Act requirements.

The Theory and Quantification of Air Pollution

Basic Air Pollution Theory

Mike Ayers

Some knowledge of the properties of gases, vapors, and aerosols is necessary to perform air pollution engineering calculations, whether such calculations deal with sampling and analysis or with control. Obtaining properties of gases and vapors is relatively easy at standard temperature and pressure (STP), typically 20°C and 1 atmosphere (atm), but it is often difficult to ascertain how the property varies with temperature and/or pressure.

Air, as well as all its gaseous and vaporous components (see Table 2.1), is considered to be perfect or ideal if its properties are similar to what they would be at standard temperature and pressure. This is

TABLE 2.1 Chemical Composition of Standard Air

Substance	Percent by volume in dry air
N_2	78.09
O_2	20.94
A	0.93
CO_2	0.03
Ne	0.0018
He	0.00052
CH_4	0.00022
Kr	0.00010
N_2O	0.00010
Xe	0.00008
H_2	0.00005

the case for the vast majority of air pollution engineering applications. The water vapor in air often varies from ideal conditions somewhat more than gases, but the errors in using ideal gas laws for water vapor are usually negligible.

Most practical problems are worked on the assumption that the air, or gas, is incompressible. The errors in doing so are negligible up to velocities of about one-third the speed of sound [Mach 0.33 or 250 miles per hour (mi/h)].

Some things are counterintuitive and therefore require careful attention. For example, the viscosity of gases increases with temperature, a fact that can be explained logically; but since this is counter to the case for liquids, acceptance is often difficult. Furthermore, gas viscosity is independent of pressure up to several atmospheres.

Aerosol dynamics are based on spherical particles, a premise which almost never exists in practice. However, if there is consistency in handling the aerosol dynamics calculations, the aerodynamic diameter that is measured gives fairly accurate predictions of the aerodynamic behavior. As a result, the difference between the real shape and size of the particles and the aerodynamic shape and size is unimportant for most practical purposes.

Another intriguing concept is the behavior of clouds of aerosols— whether air will blow around the clouds or through them, i.e., whether the cloud acts as many distinct, discrete individual particles or whether the whole mass acts as a unit. The answer depends upon the proximity of the particles to one another. If the settling velocity for the cloud as a whole is much greater than that for the individual particles, then the air goes around the particles. This is especially important in considering the possible dilution of a fog containing a highly toxic gas or mist.

Air pollution is mainly a phenomenon of urban living when the capacity of the air to dilute pollutants is overburdened. Population, industrial growth, and the dependence on internal combustion engines cause increased concentrations of gaseous and particulate emissions. Vehicular transportation accounts for 42 percent of all pollutants by weight, the majority of which is carbon monoxide that dissipates quickly. Fuel combustion in stationary sources accounts for only 21 percent of all pollutants by weight, but produces about 25 percent of total emissions of sulfur oxides. Therefore, the relative danger of the pollutant emitted must be taken into account.

2.1   Main Classes of Air Pollutants

The Environmental Protection Agency has, in the Clean Air Act of 1972, listed five main classes of air pollutants, referred to as *criteria pollutants.* They are particulate matter, sulfur dioxide, carbon monoxide, nitrogen oxides, and hydrocarbons. Lead was added to this list in 1976.

Particulate matter consists of solid or liquid substances that are visible as well as invisible. The particles affect visibility and can be transported over long distances by winds. The small particles, PM_{10} [particulate matter smaller than 10 micrometers (μm)], are particularly dangerous to human health because their small size makes it possible for them to pass through nostril hairs and enter the lungs.

Sulfur oxides (SO_x) are acrid, corrosive, and poisonous gases which are produced when fuel containing sulfur is burned. Power plants burning coal produce about 60 percent of the total SO_x in the atmosphere, oil burning about 14 percent, and industrial processes approximately 22 percent.

Carbon monoxide (CO) is a colorless, odorless, and poisonous gas that is produced by the incomplete burning of carbon in fuels. Approximately two-thirds of the CO in the atmosphere is produced by internal combustion engines, overwhelmingly from gasoline-powered motor vehicles.

Nitrogen oxides (NO_x) are produced when fuel is burned at very high temperatures. Stationary sources produce approximately 49 percent of the NO_x in the atmosphere, motor vehicles 39 percent, and other sources 12 percent. Under the influence of sunlight, NO_x combines with gaseous hydrocarbons to form photochemical oxidants, primarily ozone (O_3). Other harmful nitrogen compounds include peroxyacyl nitrates (PANs), aldehydes, and acrolein. They cause eye and lung irritation, damage to vegetation, offensive odors, and thick haze.

Hydrocarbons can be produced by unburned and wasted fuel. Hydrocarbons are important because of their role in forming photochemical smog. The most reactive compounds, called *volatile organic compounds* (VOCs), are produced by evaporation from industrial processes, mainly solvent evaporation from painting processes.

Lead has an adverse effect on health and welfare. Lead poisoning has a destructive effect on the human central nervous system. It comes mainly from the combustion of lead in fuel and from lead-based painting operations.

Additional information on the criteria pollutants is shown in Table 2.2.

2.2 Effects of Air Pollution

The primary impact of air pollution is on human health. Weather can play a significant role in how pollution affects us. Wind speed and the depth of the atmosphere determine how pollutants are dispersed. Natural low-lying gathering places such as basins, rivers, bays, and

TABLE 2.2   Criteria Pollutants

Name:	**Particulate matter**
Definition:	Any finely divided solid or liquid material other than uncombined water as measured by the federal reference methods (40 CFR 53)
Examples:	Dust, smoke, fumes, oil droplets, beryllium, asbestos
Sources:	Kilns, crushers, mills, grinders, dryers, furnaces, calciners, boilers, incinerators, conveyors, textile finishing, mixers and hoppers, cupolas, chemical processing equipment, spray booths, digesters, forest fires
Effects:	Decreased visibility; smoke and dust effect on human health; chronic diseases of the respiratory tract; asbestosis; lead poisoning; soiling of homes and clothing; destruction of plant life and agriculture; effects on climate

NAAQS:

	TSP	PM_{10}
Primary—annual	75 $\mu g/m^3$	50 $\mu g/m^3$
24-h	266 $\mu g/m^3$	150 $\mu g/m^3$
Secondary—annual	60 $\mu g/m^3$	50 $\mu g/m^3$
24-h	150 $\mu g/m^3$	120 $\mu g/m^3$

Name:	**Sulfur dioxides** SO_x (especially SO_2)
Definition:	Acrid, corrosive, poisonous gases produced when fuel containing sulfur is burned.
Examples:	SO_2 in gaseous form or sulfurous acid (H_2SO_3) and sulfuric acid (H_2SO_4) in liquid form, a component of acid rain
Sources:	Electric utilities, industrial boilers, copper smelters, petroleum refineries, mobile sources, residential and commercial heating
Effects:	Breathing difficulty when dissolved in the nose and upper air passages; chronic coughing and mucous secretion; acid rain destroys plant life, fish, especially in pristine areas such as national parks and national forests; contributes to lowered visibility as the sulfate portions of suspended particulates.

NAAQS:

	TSP	PM_{10}
Primary—annual	80 $\mu g/m^3$	0.03 ppm
24-h	365 $\mu g/m^3$	0.14 ppm
Secondary—3-h	1300 $\mu g/m^3$	0.5 ppm

Name:	**Carbon monoxide**
Definition:	A colorless, odorless, poisonous gas, lighter than air, produced by incomplete combustion of carbon in fuels
Sources:	Stationary fuel burning sources; mobile fuel burning sources (internal combustion engines, primarily gasoline engines)
Effects:	Can be fatal in a short time in an enclosed area; reacts with the hemoglobin in blood to prevent oxygen transfer (equilibrium coefficient for CO in hemoglobin is 210 times that of oxygen)

NAAQS:

	NAAQS	Increment
Primary—8-h	10,000 $\mu g/m^3$	9 ppm
Secondary—1-h	40,000 $\mu g/m^3$	35 ppm

TABLE 2.2   Criteria Pollutants (*Continued*)

Name:	**Nitrogen oxides**
Definition:	Seven oxides of nitrogen: NO, NO_2, NO_3, N_2O, N_2O_3, N_2O_4, N_2O_5. Air pollution work generally refers to NO and NO_2, nitric oxide and nitrogen dioxide. Both are colorless gases. Excessive concentrations in air cause a brownish color due to light absorption in the blue-green area of the spectrum.
Sources:	Produced by burning fuel at very high temperatures from nitrogen in the air. Also produced from organic nitrogen in coal and heavy oils: large electric power generators, large industrial boilers, internal combustion engines, nitric acid plants.
Effects:	Reduced visibility; nose and eye irritation, pulmonary edema, bronchitis and pneumonia; react with VOCs under the influence of sunlight to form ozone, PAN, and smog. Ozone and PAN are powerful oxidants that are severe eye, nose, and throat irritants and that cause the cracking of rubber, paint, textiles, etc., and damage plant life.
NAAQS:	Primary and secondary—annual 100 $\mu g/m^3$, 0.05 ppm
Name:	**Hydrocarbons—VOCs**
Definition:	Any compound of carbon, excluding carbon monoxide, carbon dioxide, carbonic acid, metallic carbides or carbonates, ammonium carbonate, and acetone which precipitates in atmospheric photochemical reactions. EPA has a list of light chloro-, fluoro-, and heavy compounds specifically exempted by name from the VOC definition.
Examples:	Propane, toluene, methyl ethyl ketone, di-isocyanates, xylene, gasoline
Sources:	VOCs are sometimes produced from incomplete burning, but are primarily emitted from evaporative sources such as surface coating, printing, and solvent cleaning operations: surface coating, graphic arts, petroleum refineries and tanks, gasoline storage and transfer, vegetable oil manufacturers, tire production, dry cleaning, synthetic organic chemical manufacturers, plastic manufacturers, and boat and toilet component manufacturers
Effects:	Primary pollutant in forming ozone and photochemical smog. (See NO_x for ozone effect.)
NAAQS:	Primary and secondary—1-h 235 $\mu g/m^3$, 0.12 ppm
Name:	**Hydrocarbons—Non-VOCs**
Definition:	All compounds of carbon and hydrogen, in liquid or gaseous form, specifically exempted as VOCs. Most are on the list of EPA HAPs. CFCs and HCFCs are included.
Examples:	Freons (CFCs, HCFCs), carbon tetrachloride, methylene chloride, methyl chloroform
Sources:	Air conditioning systems, solvent cleaning and degreasing, dry cleaning, foam product manufacturing
Effects:	Most are stratospheric ozone depleters. Ozone layer in the stratosphere protects the earth from excessive uv radiation.

TABLE 2.2   Criteria Pollutants (*Continued*)

Name:	**Lead**
Definition:	A heavy metal with a molecular weight of 207.
Sources:	Combustion of lead in fuels, lead-based paint, lead-containing piping, storage batteries
Effects:	Attacks the central nervous system, with subsequent neurological damage. Cannot be removed from the body easily.
NAAQS:	Primary and secondary 1.5 μg/m^3 (calendar quarter)

areas around mountains trap pollutants during temperature inversions. The atmosphere is also taxed over cities with industry and heavy traffic, even in favorable locations.

Dramatic examples of hazards to human health are the 1948 incident in Donora, Pennsylvania, and the 1952 killer smog in London. The long-range lasting effect on human health is even more pervasive. Chronic diseases such as emphysema and bronchitis have doubled every 5 years since World War II. Cancer and arteriosclerotic heart disease have been linked to pollution, as have asthma, acute respiratory infections, allergies, and other ailments in children.

Steel corrodes 2 to 4 times faster in industrial areas than in rural areas because of particulate and sulfur pollution. Particulate pollution speeds deterioration of statuary and buildings but also soils clothing, cars, and houses. Ozone damages textiles, dyes, and rubber.

Sulfur dioxide damages timberlands and vegetation. Smog has caused the decline in citrus grove production in the Los Angeles Basin, and fluorine and sulfur oxides have blighted pines and citrus orchards in Florida. Livestock suffer from fluorosis (from fluorine in vegetation). At high SO_2 and NO_x levels, plants suffer "early aging." Ozone is a significant threat to leafy vegetables, field crops, shrubs, fruit, and forest trees.

Carbon monoxide replaces oxygen in the blood and slows reactions. This causes a heavier burden on the heart, lungs, thyroid, and mental capacity. Sulfur oxides cause temporary and permanent injury to the respiratory system, and produce sulfuric acid, which causes further damage. Photochemical oxidants, O_3, and PAN cause eye irritation and asthma attacks and affect healthy people during exercise. Studied little until recently, NO_x contributes to the formation of O_3. Particulates cause lung lesions and respiratory diseases, and some contribute to the damage of other organs such as those containing beryllium and lead.

Measurable costs of air pollution to health, vegetation, materials, and property values are estimated to be at least $16 billion per year ($80 per person in the United States). In addition, there are the immeasurable costs such as aesthetics, pain, and discomfort from illness.

Many experts believe that the effects of air pollution on our climate are causing CO_2 to increase. Carbon dioxide causes surface warming and absorbs incoming radiation to raise the temperature. This action is known as the *greenhouse effect* and may be augmented by other natural circumstances such as fog, clouds, and dust from volcanoes.

Studies on the causes and effects of air pollution have documented the need for emission inventories, emission factors, auto emission modeling, land use control, studies of impact on health, economical and environmental strategies for pollution prevention, and alternative fuels.

2.3   Ideal Gas Law

Under ideal conditions, standard temperature and pressure, dry atmospheric air contains 78.09 volume percent nitrogen, 20.94 volume percent oxygen, 0.93 volume percent argon, and the rest carbon dioxide and other gases (see Table 2.1). Volume percent is the same as mole percent for an ideal gas. The ideal gas law states that

$$PV = nRT \qquad (2.1)$$

where P = absolute pressure
V = volume
n = number of moles
R = ideal gas constant
T = absolute temperature

All units must be consistent. For example, for P in atmospheres (atm), V in liters (L), n in gmol, and T in kelvins (K), R is 0.0821 L•atm/(gmol•K).

The following modification of the ideal gas law is important because it is accurate at low temperatures and can be used to find mass and molecular weight as well:

$$PV = \frac{M}{\text{MW}} RT \qquad (2.2)$$

where M = mass of the sample and MW = molecular weight of gas. Or to find the density of the gas, we can solve for ρ, or M/V:

$$\rho = \frac{M}{V} = \frac{P \cdot \text{MW}}{RT} \qquad (2.3)$$

In the above units, ρ is in grams per liter (g/L).

Most air quality measurements are given in parts per million (ppm), which can be converted to micrograms per cubic meter ($\mu g/m^3$) by using the ideal gas law. Parts per million equals the mole (volume) fraction of the pollutant in the air times 1,000,000.

$$\text{ppm} = \frac{\text{volume of pollutant gas}}{\text{volume of gas mixture} \times 10^6} \tag{2.4}$$

This is true only for gases at standard temperature and pressure because the gas volume can change significantly with variations in pressure and temperature. Typically, standard temperature and pressure are 20°C (68°F) and 1.0 atm.

To find the pollutant concentration at nonstandard conditions, we can use the volume fraction ratio and substitute the ideal gas law:

$$\frac{V_p}{V_t} = \frac{n_p RT/P}{n_t RT/P} \tag{2.5}$$

where p = pollutant properties and t = total gas properties. Since $n_p = M_p/MW_p$:

$$M_p = n_t \frac{V_p}{V_t} MW_p = \frac{PV_t}{RT} \frac{V_p}{V_t} MW_p \tag{2.6}$$

The mass pollutant concentration M_p/V_t is

$$\frac{M_p}{V_t} = \frac{P}{RT} \frac{V_p}{V_t} MW_p \tag{2.7}$$

At standard conditions [P = 1.0 atm, T = 293 K, R = 0.082057 L·atm/(gmol·K)], $P/(RT)$ is 0.0416 gmol/L, or the reciprocal of molar volume.

This mass concentration is in units of grams per liter. To convert to the more common units of $\mu g/m^3$, multiply by 10^9:

$$1.0 \text{ g/L} \times 10^6 \text{ }\mu g/g \times 1000 \text{ L/m}^3 = 10^9 \text{ }\mu g/m^3 \tag{2.8}$$

At standard conditions, 1.0 g/L = 10^3 ppm = 10^9 $\mu g/m^3$ and

$$C_{\text{mass}} = 41.6 C_{\text{ppm}} (MW_p) \tag{2.9}$$

where C_{mass} is the mass concentration ($\mu g/m^3$) and C_{ppm} is the volume concentration (ppm). See Table 2.3.

As stated by the ideal gas law, a specific amount of gas occupies a certain volume at a given temperature and pressure. If the tempera-

TABLE 2.3   Values for the Ideal Gas
Law Constant

0.08205	atm•L/(gmol•K)
82.057	atm•cm³/(gmol•K)
62.361	mmHg•L/(gmol•K)
0.08314	bar•L/(gmol•K)
0.08478	[(kg/cm³)•L]/(gmol•K)
1.314	atm•ft³/(lbmol•K)
998.9	mmHg•ft³/(lbmol•K)
0.7302	atm•ft³/(lbmol•R)
21.85	inHg•ft³/(lbmol•R)
555	mmHg•ft³/(lbmol•R)
10.73	psia•ft³/(lbmol•R)
1545	psia•ft³/(lbmol•R)
8.314	Pa•m³/(gmol•K)
1.987	cal/(gmol•K)
8314	J/(kgmol•K)

ture and/or pressure is changed, the new volume can be calculated from the following form of the ideal gas law:

$$\frac{PV}{T} = nR \tag{2.10}$$

Because the ideal gas constant R and the number of moles of the gas n are constant, the quantity nR is the same before and after a pressure or temperature change. Thus,

$$\frac{P_{orig}V_{orig}}{T_{orig}} = nR = \frac{P_{new}V_{new}}{T_{new}} \tag{2.11}$$

This equation can be rearranged to find the new volume after such a pressure or temperature change:

$$V_{new} = V_{orig}\frac{P_{orig}}{P_{new}}\frac{T_{new}}{T_{orig}} \tag{2.12}$$

This equation is typically used to calculate the volume that a specific amount of gas occupies at a nonstandard temperature or pressure from the known volume at standard conditions.

In order to calculate the dry molar gas flow rate if the dry volumetric flow rate V_{flow} is known, use the following equation, derived from the ideal gas law:

$$n_t = \frac{PV_{flow}}{RK} \tag{2.13}$$

where V_{flow} has units of volume per time and n_t is in moles per time.

2.3.1 Gas flow rate measurement

A number of different devices are used to measure the flow rate of gases, including venturi meters, orifice meters, and rotameters. The flow through these devices is described by the Bernoulli equation. Because gases are compressible and occupy different volumes at different temperatures and pressures, the ideal gas law must also be incorporated into flow measurement. A more rigorous explanation of measurement is presented in Chap. 5.

The orifice and venturi meter equation is

$$Q = k\sqrt{\Delta P}\sqrt{\frac{T_c}{P_c \cdot MW_c}} \tag{2.14}$$

where Q = volumetric flow rate
k = calibration constant
ΔP = pressure drop
T_c, P_c = absolute temperature and pressure of calibration
MW_c = molecular weight of calibration gas

This equation must only be used for turbulent gas with a constant density through the meter.

The equation for rotameters is

$$Q = k'RR\sqrt{\frac{T_c}{P_c \cdot MW_c}} \tag{2.15}$$

where RR = rotameter reading. This equation also can only be used in turbulent flow of constant density.

If the conditions of measurement vary in temperature or pressure from the calibration conditions, then ideal gas law considerations must be added to the formula

$$Q_a = Q_i\sqrt{\frac{T_a}{P_a \cdot MW_a} \cdot \frac{P_c \cdot MW_c}{T_c}} \tag{2.16}$$

where a = actual conditions
i = indicated conditions
c = calibration conditions

2.4 Volative Organic Compounds

The difference between gas and vapors can be subtle, but important. Gas properties are quite accurately, at ambient conditions, shown with the ideal gas law, but vapor is more complex and different from the gas law.

The reason for making the distinction is that gases and vapors interact differently with control devices, and systems should therefore be designed accordingly. Both gases and vapors are made up of molecules that have individual motions and are considerably distant from one another. They expand to fill the size and shape of the vessel in which they are stored, and they exert pressure in all directions.

The internal molecular energy between gases and vapors is different. A vapor is in the gaseous state, but close to the liquid state, and it can be easily condensed into that liquid state. The gas is also in the gaseous state, but much farther away from the liquid state from a temperature or pressure standpoint. A gas is considered to be at a temperature that is much higher than the critical point (highest temperature at which it can be condensed to liquid) for that substance. A vapor is typically closer to the dew point temperature.

SO_2, NO, NO_2, and CO are gases, but VOCs are vapors (except for methane, ethane, ethylene, and other low-boiling-point chemicals).

2.4.1 Vapor pressure

The vapor pressure of a liquid is the force exerted by its pure component in vapor form above the liquid. All liquids have unique boiling points; and as the temperature and pressure approach this condition for each liquid, the molecules of liquid have a greater tendency to reach the required energy to escape the liquid and move into the vapor form. At any given temperature and pressure, a pure component reaches an equilibrium in which a certain number of molecules transcend to vapor. These vapor molecules exert a certain pressure, called *vapor pressure*. In general, as the temperature increases and approaches the boiling point, the vapor pressure increases.

In the practice of air quality control, conditions are typically at atmospheric pressure. Therefore, temperature is normally the only factor affecting the vapor pressure of a pure component. The *Antoine equation* relates vapor pressure to temperature change as follows:

$$\log P_{vi} = A_i + \frac{B_i}{C_i} + T \tag{2.17}$$

where P_{vi} = vapor pressure of pure component liquid i
    T = temperature
A_i, B_i, C_i = curve-fit constants for component i

When two or more compounds exist in a liquid, which is often the case, the vapor pressures must be found for each individual compound once equilibrium has been established. Once the system comes to equilibrium with the vapor phases, the vapor is saturated and the vapor pressures of each pure component are equal to the partial pressures for that mixture of gases.

When one pure liquid component is placed in a closed system with air, evaporation or volatilization of some of the liquid molecules will occur until the equilibrium between the vapor and liquid is established. This is the point at which the pressure exerted by the volatilized liquid is equal to the vapor pressure:

$$\overline{P}_i = y_i P = P_{vi} \qquad (2.18)$$

where $\overline{P}_i$ = partial pressure of component i, atm
y_i = mole fraction of component i in gas phase
P = total pressure, atm

The above equation can be used with vapor-liquid equilibrium (VLE) data to determine the temperature to cool a VOC airflow in order to condense a certain amount of the VOCs. Partial pressure is less than or equal to vapor pressure. For a normal airstream, the air is not saturated. Therefore, when it is cooled, the vapor pressure starts out above the partial pressure. As the temperature is reduced, so is the vapor pressure, but the partial pressure stays constant. When the vapor pressure is equal to the partial pressure, equilibrium is reached and condensation begins.

The following equation can be used to determine the gas- or liquid-phase compositions of vapor-liquid systems with two or more components at equilibrium (Smith and Van Ness):

$$\phi_i y_i P = \gamma_i x_i P_{vi} \qquad (2.19)$$

where ϕ_i = vapor-phase activity coefficient for component i
γ_i = liquid-phase activity coefficient for component i
x_i = mole fraction of component i in liquid

The activity coefficients for ideal systems are equal to 1.0. From a practical standpoint, they are heavily related to mixture composition and concentration in organic mixtures and can range from 1 to more than 10.

2.4.2 Diffusivities

Matter spontaneously moves from a highly concentrated region to a less concentrated region, according to the second law of thermodynamics. This diffusion is known through Ficks law as the proportionality constant between the matter flux and the concentration gradient:

$$\frac{\dot{M}}{A} = (-)D\,\frac{dC}{dx} \qquad (2.20)$$

where $\dot{M}$ = mass transfer rate, mol/s
      A = area normal to direction of diffusion, cm^2
      D = diffusivity, cm^2/s
  dC/dx = concentration gradient, mol/cm^4

The minus sign shows that the concentration gradient decreases in the direction of diffusion. Diffusivity is a function of the specific substance and the medium through which it is diffusing. Diffusion is particularly important in absorption, adsorption, chemical reactions, and catalytic incineration. Diffusion can typically be the rate-limiting step in pollution control. Even in turbulent flow, there is a boundary layer through which the pollutant must molecularly diffuse in order to pass to the other phase, such as onto the surface of carbon grains in the case of adsorption.

2.4.3 Gas-liquid and gas-solid equilibria

Solubility is an important concept in pollution control. For example, in absorption processes, once the pollutant molecule has diffused through the boundary layer, as mentioned above, it must be absorbed into the liquid. The rate of absorption is essentially instantaneous, therefore it is not the rate-limiting step, but the extent of absorption, i.e., the solubility, is very important to the mass-transfer process.

As mentioned above, pollution control systems are generally operated near atmospheric pressure, such that pressure is not a major factor. As a result, an equation known as *Henry's law* is often used to illustrate how the gas and liquid concentrations are related.

$$\overline{P}_i = H_i x_i \qquad (2.21)$$

where H_i is Henry's law constant and x_i = mole fraction of component i in the liquid. Henry's law constants are typically in units of atmospheres per mole fraction. Henry's law varies considerably with temperature. The relationship is valid for small mole fractions. At higher concentrations, the relation is nonlinear.

Adsorption is the process by which contaminants diffuse from air or water onto the surface of a solid. The equilibrium relationships with adsorption are not linear as with solubility. The adsorption process is governed by particular vapor, partial pressure of the vapor, the type of adsorbent (the substance onto which the vapor adsorbs), the temperature of the process, and the surface area of the adsorbing material.

There are two kinds of adsorption: physical and chemical. The process of physical adsorption gives off heat of the same order of magnitude as the heat of condensation. Chemical adsorption, or *chemisorption,* releases quite a bit more energy because chemical bonds are broken and reformed in the process. Physical adsorption is reversible. This is particularly useful in cleaning the adsorbent, typically activated carbon in air systems.

2.4.4 Chemical reactions

Once a pollutant has changed phases, from air to liquid or solid, it can react with another compound to produce less harmful substances. It is important to understand the chemical reactions that produce the harmful compounds as well as the ones that eliminate them.

There are two essential principles of chemical reactions: kinetics and thermodynamics. Kinetics affects the rate at which a chemical reaction proceeds. Thermodynamics affects the heat requirements and equilibrium of a reaction.

A *reaction* can be defined as the rate of disappearance of the reactants, or rate of formation of the products.

$$\text{Reaction rate} = r_P = -r_R \tag{2.22}$$

where r_P = rate of generation of products P, mol/(L•s), and r_R = rate of generation of reactants R, mol/(L•s).

The concentration of the reactants, or frequency of collisions of molecules, and the reaction temperature, which is actually the energy of the reaction, are the two major factors that determine the rate of the reaction. So the reaction

$$A + B \rightarrow C + D \tag{2.23}$$

can be described by

$$r_P = kC_A^x C_B^y \tag{2.24}$$

where k = reaction rate constant
C_A, C_B = concentrations of reactants, mol/L
x, y = exponents—order of reaction

The exponents x and y define the *order* of the reaction. The reaction is x order in reactant A and y order in reactant B and $x + y$ order for the whole reaction. The reaction orders are generally whole numbers that must be calculated from experimental data.

The Arrhenius equation is often used to find the rate constant k

$$k = Ae^{-E/(RT)} \tag{2.25}$$

where A = frequency factor
E = activation energy
R = universal gas constant
T = absolute temperature

There are two models typically used for reacting vessels: continuous stirred tank reactor (CSTR) and plug flow reactor (PFR). In a CSTR, a continuous flow of reacting material is sent into the reactor, and the contents are stirred and reacted such that the mixture in the reactor is homogeneous and is the same as the exit stream. The volume in is equal to the volume out, so that the balance is zero.

Material in + material generated = material out

$$0 = Q_{in}C_{in} - Q_{out}C_{out} + r_iV \qquad (2.26)$$

where Q = volumetric flow rate, L/s
$\quad C$ = Concentrations of reactants, mol/L
$\quad r$ = rate of generation of product i, mol/(L•s)
$\quad V$ = volume of reactor, L

In a PFR, however, the reaction takes place in a long tube in which the mixture approximates the concentrations of the reactants at the beginning of the tube and the concentration of the products at the end of the tube. The material balance for this system describes a differential slice of the reactor ΔV:

$$0 = Q_V C_{iv} - Q_{V+\Delta V}\, C_{iv+\Delta V} + r_i\, \Delta V \qquad (2.27)$$

This can be reduced to

$$\frac{dC_i}{r_i} = \frac{1}{Q}\, dV \qquad (2.28)$$

for the conditions in which the flow rate is constant and r_i is not a function of position. Mass balance may be better because some volumetric inconsistencies may occur in the reactor with position.

Reactions can have very different thermodynamics. Some are exothermic and release large amounts of energy, usually heat. Others are endothermic and absorb energy. This can affect the rate of reaction as well as the concentrations and volumetric flow rates.

Combustion reactions, especially of some organics, are some of the most energetic of chemical reactions.

Because chemical reactions rarely go to completion (i.e., where 100 percent of the reactants are converted to the products), it is very important to understand chemical equilibrium. For the reaction above, the *equilibrium constant K_p* is defined as

$$K_p = \frac{P_c P_d}{P_a P_b} \qquad (2.29)$$

where P = molar concentration at equilibrium for reactants a and b and products c and d. At a given temperature, if K_p is large, more of products c and d will be formed. Conversely, if K_p is small, the equilibrium shifts more toward the reactants and less c and less d are formed.

For exothermic reactions, the equilibrium constant generally decreases as the temperature increases, but the kinetic rate constant always increases. Therefore, there is some optimal temperature and pressure at which the reaction will proceed rapidly and to an acceptable amount of completion.

TABLE 2.4   Properties of Air and Factors for Air Calculations*

Property	Symbol	Value	Remarks
Molecular weight	MW	28.966	Dry
		28.84	At 50% relative humidity
Molar volume		24.053 m³/kgmol	At STP
		385 ft³/lbmol	At STP
		22.4146 m³/kgmol	At 0°C
		359 ft³/lbmol	At 0°C
Density	ρ	1.204 kg/m³	Dry
		0.0752 lb/ft³	Dry
		1.199 kg/m³	At 50% relative humidity
		0.0749 lb/ft³	At 50% relative humidity
		1.292 kg/m³	At STP
Liquid density	γ	0.92 g/cm³	At -147°C
Universal gas constant	R	8.314×10^7 erg/(K•mol)	
		1.987 cal/(°C•mol)	
		53.3 ft/(°F•mol)	
		1716 ft•lb/(°R•slug)	
		287 J/(K•kg)	
Avogadro's number	N_a	6.023×10^{23}	Molecules per gmol
Refractive index	m	1.00029	Sodium light ($c = 2.998 \times 10^{10}$ cm/s)
Velocity of sound	c	1129 ft/s	
		344 m/s	
		331.4 m/s	STP $c = c_o \times [(T(°C) + 273/273]^{0.5}$
Dynamic viscosity	μ	1.83×10^{-4} P	P = poise = g/(s•cm)
		170.8 μP	STP $m = (11.554 \times T^{0.5425} - 70)$ $\times 10^{-6}$
Kinematic viscosity	υ	15.2 cSt	St = cm²/s
			cSt = centistoke $\upsilon = \mu/\rho$
Thermal conductivity	k	5.68×10^{-5} cal/(cm²•s•°C/cm)	
Specific heat	C_p	0.24 cal/(°C•g)	Constant pressure
		0.24 Btu/(°F•lb)	
	C_v	0.17 cal/(°C•g)	Constant volume $C_p/C_v = \gamma = 1.4$
Relative humidity	RH	$100 p_v/p_s$	p_v = water vapor pressure in air
			p_s = saturated water vapor pressure

*All properties are at STP, 20°C, and 1 atm, unless otherwise noted.

2.4.5 Reactions in the atmosphere

1. Types of reactions in the atmosphere (all types):

   - *Thermal.* Reactions that are caused by heat absorption
   - *Photochemical.* Reactions that are caused by light absorption
   - *Heterogeneous.* Reactions with particulates

     There are no clear-cut boundaries between the three types. All cause pollution.

2. Examples:

   - *Thermal reaction:*

$$2NO + O_2 \rightarrow 2NO_2 \tag{2.30}$$

   - *Photochemical:*

$$NO_2 + h\nu \rightarrow NO + O \tag{2.31}$$

$$O_3 + h\nu \rightarrow O_2 + O \tag{2.32}$$

where $h\nu$ is a photon of sunlight. The reactive oxygen atom then reacts to form ozone:

$$O + O_2 \rightarrow O_3 \tag{2.33}$$

with SO_2

$$O + SO_2 \rightarrow SO_3 \tag{2.34}$$

with olefins

$$O + \text{olefin} \rightarrow \text{several fragments}$$

   - *Heterogeneous.* Reactions on the surfaces of particulate matter or in solution in the droplets create a catalytic effect:

$$ZnO + H_2O \text{ (vapor)} + h\nu \rightarrow H_2O_2 + Zn \tag{2.35}$$

$$SO_2 + H_2O \text{ (drop)} \rightarrow H_2SO_3 \tag{2.36}$$

$$H_2SO_3 + O \rightarrow H_2SO_4 \tag{2.37}$$

$$H_2SO_4 + CaO \text{ (limestone dust)} \rightarrow CaSO_4 + H_2O \tag{2.38}$$

2.5 Particulate Matter

Particulate matter has a large variance in shape and size, ranging from large drops of liquid to microscopic dust particles, each with its own set of physical and chemical properties. Particulate matter is emitted from a wide range of industrial and natural sources, including combustion, mining, and construction as well as windstorms, forest fires, and volcanoes.

Size, size distribution, shape, density, stickiness, corrosiveness, reactivity, and toxicity are common characteristics of particles that are useful to the environmental engineer.

Shown in Table 2.5 is the size distribution characteristic. Particulates cover five orders of magnitude from micrometers to meters. Most collection systems work better for specific ranges of particulates. The design engineer must choose the appropriate process for the specific particulate.

The control efficiency of a particulate removal system is determined by the following equation, which is simply the difference between the inlet and emission rates per inlet rate:

TABLE 2.5   Particulate Sizes, μm

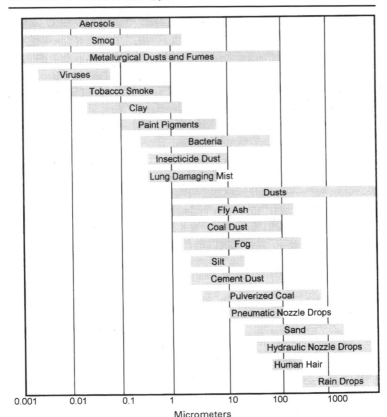

$$\eta = \frac{M_i - M_e}{M_i} = \frac{L_i - L_e}{L_i} \qquad (2.39)$$

where η = overall collection efficiency (as a fraction)
M_i = total mass input rate, g/s
M_e = total mass emission rate, g/s
L_i = particulate loading in inlet gas to system, g/m^3
L_e = particulate loading in exit gas stream, g/m^3

If the size distribution of the particles is known, as well as the system efficiency as a function of particle size, the overall efficiency can be found from

$$\eta = \Sigma \eta_j m_j \qquad (2.40)$$

where η_j = efficiency of collection for jth size range and m_j = mass percent of particles in the jth size range. Because particulate mass variance can often be quite large, the mass efficiency can vary significantly.

A major assumption is that the above relationship assumes that all the particles are spherical. This may be a valid assumption in most cases, but in reality, particles come in a variety of nonspherical shapes. When one is determining a characteristic diameter for an irregularly shaped particle, the decision is ultimately based on how the particle acts in a flowing gas stream as opposed to some detailed inspection of the individual particle.

Therefore, a standardized diameter and density must be selected as a basis for comparing different particles. The aerodynamic diameter serves this purpose. According to Friedlander (1977), *the aerodynamic diameter of a specific particle is the diameter of a sphere with unit density (density of water) that will settle in still air at the same rate as the particle in question.* This is the diameter used to select a control technique.

It is imperative to have accurate information on the characteristics of the emission stream. To determine the size distribution in the stream, a cascade impactor is used. This device separates the particles by forcing the stream through varying sizes of sieves. The impactor actually has slots and impaction plates that separate the particles based on their aerodynamic diameter in the gas stream. The first stage contains the widest slot and largest distance to the impaction plate, and each successive stage has a smaller slot and distance. Thus progressively smaller particles are filtered out in each stage.

The resulting mass of particles at each stage gives the particle size distribution for that stream. Size distribution and statistics, gaussian distributions, etc., are covered in Chap. 5.

2.5.1   Behavior of particulates in fluids

In water, particles are removed by sedimentation tanks and filters. When the fluid is air, typical removal systems consist of gravity settlers, centrifugal settlers, fabric filters, electrostatic precipitators, and wet scrubbers. These devices are all based on utilizing a force (gravity, inertia, centrifugal acceleration, or static electricity) to separate particles from the airstream by accelerating them in a direction opposite the stream flow.

To determine the most efficient way to separate particulates, an analysis should be completed to determine how the specific particles in question interact in the airstream.

The drag force of a particle in a fluid (air) describes how that particle moves relative to the fluid. If a particle is in motion (because of gravity or some other force) in a fluid, that fluid exerts a force, known as the *drag force,* in the opposite direction.

$$F_D = \frac{C_D A_p \rho_F V_r^2}{2}$$

(2.41)

where F_D = drag force, N
C_D = drag coefficient
A_p = projected area of particle, m^2
ρ_F = density of fluid, kg/m^3
V_r = relative velocity, m/s

For best results, the drag force should be determined experimentally with each specific particle and fluid, but it can be estimated with the *Reynolds number* Re and Fig. 2.1:

$$\text{Re} = \frac{d_p V_r \rho_F}{\mu}$$

(2.42)

where d_p = particle diameter, m, and μ = fluid viscosity, kg/(m•s). The Reynolds number for a fluid tells how the fluid is behaving—whether it is turbulent, in which case there is little to no variation across the area of the flow, or laminar, in which case the particles closest to the wall essentially are stagnant and there is no slip.

The flow regime that includes Reynolds numbers less than 1 is known as the *Stokes regime* because particles in this regime follow Stokes' law:

$$F_D = 3\pi\mu d_p V_r$$

(2.43)

or
$$C_D = \frac{24}{\text{Re}}$$

(2.44a)

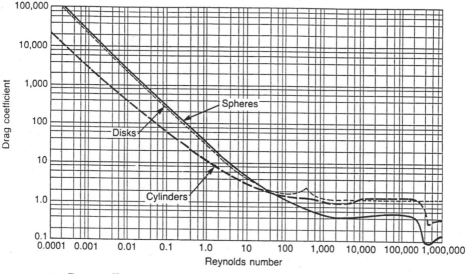

Figure 2.1   Drag coefficients.

$$C_D = \alpha Re^{-\beta} \qquad (2.44b)$$

where α and β are constants according to the following table for each section:

Reynolds number	α	β
<2.0	24.0	1.0
2–500	18.5	0.6
500–200,000	0.44	0.0

Stokes' law is only valid when the particle diameter is significantly larger than the mean free path λ. In a system in which the mean free path of the fluid is approximately the same magnitude as the particle diameter, the fluid acts on the particle as individual molecules instead of as a continuous fluid. The drag coefficient, in this case, is reduced, and the *Cunningham correction factor* C must be used to compensate as follows:

$$C = 1 + 2.0\,\frac{\lambda}{d_p}(1.257 + 0.40e^{-0.55d_p/\lambda}) \qquad (2.45)$$

The mean free path is defined from the kinetic theory of gases as

$$\lambda = \frac{\mu}{0.499P\,\sqrt{8\,MW/(\pi RT)}} \qquad (2.46)$$

where λ = mean free path, m
  P = absolute pressure, Pa
  R = universal gas constant, 8314 J/(kgmol•K)
  MW = molecular weight, kg/kgmol
  T = absolute temperature, K
  μ = absolute viscosity, kg/(m•s)

The Cunningham correction factor varies with particle size. In general, the drag slip for particles 1 μm or smaller is always significant, and it becomes less significant as the particle size increases. Dividing the Stokes' law drag coefficient by the Cunningham correction factor, we get the corrected drag coefficient

$$C_D' = \frac{C_D}{C} = \frac{24}{C(\mathrm{Re})} \tag{2.47}$$

2.5.2 External forces

For a particle to have its own motion relative to the carrier fluid, there must be at least one additional force acting on it besides that of the fluid. Gravity is always acting on it and is sometimes the major force, but occasionally the major force is centrifugal acceleration or static electricity.

Assuming one additional force, in the opposite direction of the drag force, Newton's second law says

$$F_e - F_D = M_p \frac{dv_r}{dt} \tag{2.48}$$

where F_e = net external force and M_p = mass of the particle. Assuming a spherical particle and Stokes' law,

$$\frac{dv_r}{dt} + \frac{18\mu}{\rho_p d_p^2} v_r = \frac{F_e}{M_p} \tag{2.49}$$

where ρ_p = particle density. The term $\rho_p d_p^2/(18\mu)$ is a very important characteristic of particle fluids called the *characteristic time* τ, and it has units of time. This term corrected for slip is τ', which equals τC. If this term is the same for two apparently different systems, they will behave in like fashion.

Therefore, to describe the general motion of a particle in a fluid in the Stokes regime, use the following differential equation:

$$\frac{dv_r}{dt} + \frac{v_r}{\tau'} = \frac{F_e}{M_p} \tag{2.50}$$

2.5.3   Gravitational settling

This is the case in which gravity is the only additional force acting on the particle. The equation is

$$\frac{dv_r}{dt} + \frac{v_r}{\tau'} = \frac{\rho_p - \rho_F}{\rho_p g} \tag{2.51}$$

where g = gravitational constant. The buoyancy of a particle in this fluid is described by the term $(\rho_p - \rho_F)/\rho_p$. This term approaches 1 as the particle becomes more solid. At that point we have

$$v_r = v_t\left[1 - \exp\left(\frac{-t}{\tau'}\right)\right] \tag{2.52}$$

where v_t is the terminal settling velocity:

$$v_t = \tau' g = \frac{C\rho_p d_p^2}{18\mu} g \tag{2.53}$$

where τ' is in units of time.

Given a graph of V_r/V_t versus t/τ', find the time when the terminal velocity is achieved. After four characteristic times, the velocity of the particle is approximately equal to the terminal velocity for that particle. The time to reach terminal velocity is typically very small, on the order of a few milliseconds, and is usually ignored.

Outside of Stokes' regime (10 to 20 µm+), the Reynolds number is too high and these parameters must be determined by experiment. Empirical models have been developed to assist with this process. Theodore and Buonicore (1976) say that for 2 < Re < 500,

$$v_t = \frac{0.153 d_p^{1.14}\, \rho_p^{0.71}\, g^{0.71}}{\mu^{0.43}\, \rho_F^{0.29}} \tag{2.54}$$

and for 500 < Re < 200,000,

$$v_t = 1.74\left(\frac{\rho_p d_p}{\rho_F}\right)^{0.5} g^{0.5} \tag{2.55}$$

It is usually more convenient and accurate to use an experimental chart such as Fig. 2.1.

2.5.4   Aerodynamic diameter

The motion of a particle in a fluid under Stokes' conditions varies according to the value of τ', which is the ratio of particle properties to fluid properties, such that any system with similar τ' will act the same.

$$d_a = \frac{18\mu v_t}{C\rho_w g} \qquad (2.56)$$

where d_a = aerodynamic diameter, m
$\quad \mu$ = gas viscosity, kg/(m•s)
$\quad v_t$ = settling velocity, m/s
$\quad \rho_w$ = density of water, kg/m^3
$\quad g$ = gravitational acceleration, m/s^2

2.5.5 Collection of particles by impaction, interception, and diffusion

When fluid flows around an object, diverging stream lines are created that converge on the other side of the object. However, particles flowing around an object do not follow stream lines. Their inertia causes them to maintain their present course (see Fig. 2.2). Thus, while the carrier fluid will flow around such objects, particles in the fluid will collide and can be collected by the objects.

When the center of mass of the particle in motion directly strikes an object, we call this *impaction*. When the center of mass of the moving particle passes close to the object such that part of the particle strikes the object, the particle is *intercepted*.

If a particle is added to a fluid stream, it has some initial velocity v_0, and its velocity at some later time is given by

$$v = v_0 e^{-t/\tau} \qquad (2.57)$$

At some point after being injected into the stream, the particle will cease its relative motion and will be carried with the stream. The distance traveled by a particle prior to this point is shown by

$$x_{stop} = \int_0^\infty v\,dt = v_0\tau' \qquad (2.58)$$

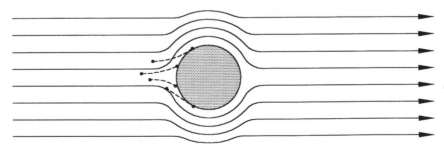

Figure 2.2 Flow of particles around an object.

If the particle comes to rest prior to contact with the object, the particle is entrained in the fluid stream and may flow around the object. If τ' is small, then x_{stop} is also small and the particle will stop in a very short distance.

An impaction parameter N_I is used to determine what fraction of particles will impact the object and what fraction will flow around it.

$$N_I = \frac{x_{stop}}{d_0} \qquad (2.59)$$

Impaction occurs when N_I is large, and the particles flow around the object when N_I is small.

2.5.6 Overview of particulate control equipment

Many different kinds of equipment are used to control particulate emissions. Chapter 20 discusses in detail the design of particulate control equipment.

In a gravity settler, the gas flow is slowed down so that the particles are forced by gravity to settle out in accordance with Stokes' law.

A cyclone swirls the gas around and causes the larger particles to accelerate centrifugally toward the outside wall, where they are collected, and the clean gas flows out the top. Stokes' law also controls this effect.

A baghouse, or fabric filter, is like a large vacuum cleaner. The airstream moves through the cloth filter, trapping the particles on one side. Occasionally the filter must be shaken or the airflow reversed to clean the filter.

An electrostatic precipitator (ESP) uses electricity to attract particles for collection. The gas stream passes through an electric field that charges the particles. The charged particles are then attracted to and collected on an oppositely charged plate. These plates must also be cleaned or replaced periodically.

A wet scrubber employs a mist stream that flows countercurrently to the gas stream. The particles are impacted and intercepted on water drops which are then collected by gravity for reuse or disposal.

In general, the mechanical control devices are less expensive, but not as efficient as the mass-transfer devices. Each air system should be treated individually and a control device designed specifically for it.

2.6 Summary of Terms and Relationships for Particle Dynamics

Acceleration/motion inducing forces. Often shown with suffix of *phoresis*—act of carrying; transmission.

gravitational. Usually taken as 9.8067 m/s² or 32.174 ft/s²—varies with latitude and altitude (9.77 to 9.83)—toward center of the earth

inertial. Tendency to continue rest or motion; motion in a straight line

diffusive. Brownian motion; movement from bombardment by gas molecules

electrical. Electric field moves particle toward opposite charge

thermal. Bombardment and/or pinch effect moves particles from hot to cold

radiative. Movement from unequal bombardment with photons

Aerodynamic drag

$$F_D = \frac{C_D \rho_f A_p \mu^2}{2}$$

Usually work with spheres or spherical equivalent, approximate coefficients:

$$
C_D =
\begin{cases}
\dfrac{24}{N_{\text{Re}}} & \text{for } N_{\text{Re}} < 3 & F_D = 3 \bullet \pi \bullet \mu \bullet \mu \bullet D_P \\[2ex]
\dfrac{14}{(N_{\text{Re}})}^{1/2} & \text{for } 3 \leq N_{\text{Re}} \leq 10^3 & F_D = 1.75 \bullet \pi \bullet (\rho \bullet \mu \bullet \mu^3 \bullet D_P^3)^{1/2} \\[2ex]
0.44 & \text{for } N_{\text{Re}} > 10^3 & F_D = 0.055 \bullet \pi \bullet \rho \bullet \mu^2 \bullet D_P^2
\end{cases}
$$

Acceleration of particles. Generally considered to reach u_t in negligible time and distance—may need to consider for large particles

   laminar region

$$\frac{u}{u_t} = 1 - \exp\left[\frac{-(gh + u_t u)}{u_t^2}\right] \qquad h = \text{distance fallen}$$

References

Benitez, J., *Process Engineering and Design for Air Pollution Control*, PTR Prentice-Hall, Englewood Cliffs, N.J., 1993.

Cooper, C. D., Alley, F. C., *Air Pollution Control—A Design Approach*, Waveland Press, Inc., Prospect Heights, Ill., 1990.

Finlayson-Pitts, B. J., Pitts, J. N., Jr., *Atmospheric Chemistry of Tropospheric Ozone Formation: Scientific and Regulatory Implications,* Department of Chemistry and Biochemistry, California State University, Fullerton, 1993.

Friedlander, S. K., *Smoke, Dust and Haze,* John Wiley & Sons, New York, 1977.

Lapple, C. E., *Stanford Research Journal,* **5**(95), 1961.

Lapple, C. E., Shepherd, C. B., *Industrial and Engineering Chemistry,* American Chemical Society, Washington, 1940.

Smith, J. M., Van Ness, H. C., *Introduction to Chemical Engineering Thermodynamics,* 4th ed., McGraw-Hill, New York, 1987.

Solomons, T. W., *Organic Chemistry,* 4th ed., Graham, John Wiley & Sons, New York, 1988.

Theodore, L., Buonicore, A. J., *Industrial Air Pollution Control Equipment for Particulates,* CRC Press, Cleveland, Ohio, 1976.

Chapter

3

Atmospheric Dispersion Models

Andrew T. Allen

The field of air quality control primarily addresses the impact of the harmful effects of airborne substances on the health of living organisms. Substances which lower the quality of health are classified as *pollutants,* and the presence of pollutants in the atmosphere is termed *air pollution.* Entities which release pollutants into the environment are pollutant *sources.* Living organisms which interact with pollutants are *receptors.*

The movement of the pollutant from the source to the nearby atmosphere is termed *release.* The movement of the pollutant from the location near the source to the location of the receptor is *transport,* and the interaction of the pollutant with the receptor is *exposure.*

An *air quality model* is a physical or mathematical representation of an air pollution event. Its purpose is to predict the ambient air concentration of a pollutant at a given point, typically at the location of a receptor, using information about the entities involved—the source, the atmosphere, and any other objects which may affect the event—and about the processes involved—release and transport.

3.1 Important Characteristics of the Atmosphere

3.1.1 Wind

The wind dilutes the pollutants as they are emitted and carries them away from the source. The dilution of the pollutants is typically assumed to be in direct proportion to the mean wind speed through the plume. The wind also acts to create eddies at the surface of the earth, which in turn increase the dispersion of the plume. In general, the wind dictates the speed and direction in which the bulk of the plume moves, and wind also affects the amount of dispersion that takes place.

Global wind patterns are caused by temperature differences between the poles and the equator, and between the continents and the oceans. Local wind patterns can also be affected by surface features, such as mountains and buildings, and by the presence of lakes and rivers.

Wind speed near the ground is generally slower, due to surface features such as hills, trees, and buildings. The region of the atmosphere affected is called the *planetary boundary layer* and can vary from hundreds of meters to several kilometers. Within this region, the relationship between altitude and wind speed can be represented by the following equation:

$$u = u_0 \left(\frac{z}{z_0} \right)^P \tag{3.1}$$

where u = wind speed at altitude of interest
u_0 = known wind speed at altitude z_0
z = altitude of interest
z_0 = altitude at known wind speed
P = exponent with value between 0 and 1, a function of stability

In the absence of wind data at the source, the estimation of mean wind speed is typically based on data taken at weather stations. Care should be taken in estimating wind speed, as topographical differences may make weather station data unrepresentative.

3.1.2   Stability

Atmospheric stability is the absence of vertical mixing. Vertical mixing is caused primarily by solar insolation and wind shear. Instability due to solar insolation can be determined by use of a temperature profile. A comparison of the slope of the temperature profile, called the *lapse rate,* to the adiabatic lapse rate indicates whether the atmosphere is stable, neutral, or unstable. If the actual lapse rate is greater or flatter than the adiabatic lapse rate (see looping on Fig. 3.3), then a small volume of air forced upward will be warmer, and therefore less dense, than the surrounding air and therefore will have a tendency to move upward even farther. This condition results in atmospheric instability. If the actual lapse rate is close to the adiabatic lapse rate (see coning on Fig. 3.3), then vertical motions are unaffected by thermal forces, and the atmosphere is neutral. If the actual lapse rate is significantly less or steeper (see fanning on Fig. 3.3) than the adiabatic lapse rate, then vertical motions will be inhibited, resulting in atmospheric stability.

3.1.3 Mixing height

The *mixing height* is the height of the region containing turbulent mixing and is therefore the height of the region available for dispersion. Because turbulence is caused by both thermal and mechanical forces, it varies with the seasons and topography. Mixing height data have been collected by the Environmental Protection Agency, using the Miller-Holzworth method, and are available via the Internet from the Office of Air and Radiation of the U.S. Environmental Protection Agency.

This method assumes steady-state conditions and requires a temperature profile, usually based on soundings at various altitudes. The Miller-Holzworth method consists of the following steps, as exemplified in Fig. 3.1:

1. Add 5°C (rural areas) or 3°C (urban areas) to the minimum morning surface temperature ($z = 0$) to determine the starting point A.

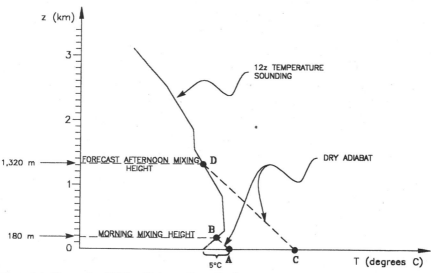

Figure 3.1 Example of Miller-Holzworth method.

2. From point A, follow the dry adiabat upward to its intersection with the profile, located at point B.

3. Assign the z value at this intersection as the morning mixing height.

The afternoon mixing height is determined in a similar manner, as indicated by points C and D, using the maximum afternoon temperature (do not add 5°C) as the starting point C. The mixing height calculated is used to establish the effects of the associated inversion on plume behavior. If mixing height is small enough, an elevated inversion will inhibit the vertical dispersion of pollutants.

3.2   Dispersion Modeling

As stated at the beginning of this chapter, an air quality model is a physical or mathematical representation of an air pollution event. The process of correctly analyzing the effects of a source requires one to identify the meteorological, topographical, and emission conditions present; select an appropriate model; acquire the necessary data; run the model; and evaluate the results. The factors to be considered in selecting a model include the meteorological and topographical complexities of the area, the level of detail and accuracy needed, the technical competence of the modelers, the available resources, and the detail and accuracy of the available data. It is cautioned that the best model cannot completely duplicate field conditions, and environmental conditions and resulting values may vary considerably from actual air pollution measurements. The models most widely used, and discussed in detail in this text, are gaussian-based models. They are relatively simple to understand and use, and they have provided reasonable fits to experimental data.

3.2.1   Plume characterization

Once a plume is released from a stack into the atmosphere, its behavior follows certain general patterns, represented schematically in Fig. 3.2. The plume first rises and then, while continuing to rise, begins to move horizontally under the influence of the wind. As the plume continues to move horizontally, it rises less and less, until its motion is observed to be virtually horizontal. As it travels, the plume also spreads out, both vertically and laterally.

Plume rise is attributed to *momentum* and *buoyancy*. Momentum is reflected in the tendency of the plume to rise due to its release velocity. Buoyancy is due to the difference between plume density and atmospheric density. The plume's movement downwind, also called

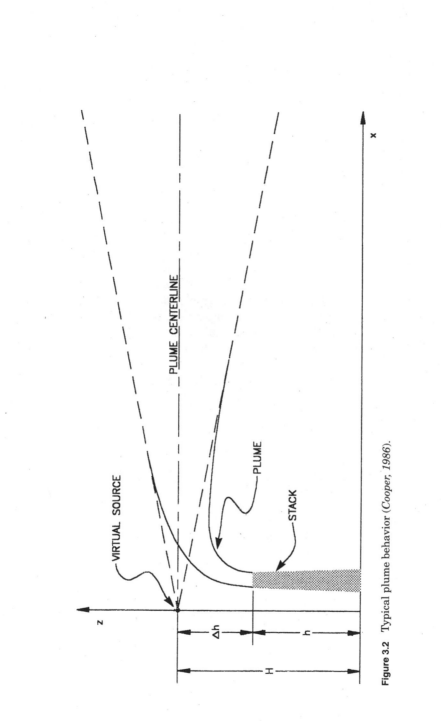

Figure 3.2 Typical plume behavior (*Cooper, 1986*).

3.5

bulk transport, is caused by wind, and will follow the average direction of the wind. Plume spread, or dispersion, is caused primarily by atmospheric turbulence and the random shifting of the wind.

Turbulent flow contains eddies, which are swirling currents that move at variance with the mean wind. Eddies will displace a portion of a plume, forcing the displaced pollutant to move some distance from the plume. Eddies in the atmosphere are caused by *thermal* and *mechanical* processes. Thermal processes involve the conversion of sunlight to heat as it strikes the surface of the earth. This heat is then transferred to the lower atmosphere by convection and conduction, resulting in thermal eddies. Mechanical processes involve the creation of shear forces as wind blows over rough surfaces, which form mechanical eddies.

Plume behavior varies with the degrees of thermal and mechanical turbulence present. Six general classifications of plume behavior have been identified, as tabulated in Fig. 3.3. A *looping* plume occurs in the presence of dominating convective (thermal) turbulence, as indicated by a superadiabatic lapse rate. A *coning* plume occurs when the atmosphere has neutral stability, and therefore small-scale mechanical turbulence is most influential. Extremely stable conditions, as indicated by a highly subadiabatic lapse rate, result in a *fanning* plume. Under extremely stable conditions, ground-level concentrations are nearly zero. *Fumigation* results from the breakup of an inversion from unstable air below. A *lofting* plume is caused by a stable layer of air below a plume contained in an unstable layer. A *trapped* plume is in an unstable layer bounded above and below by stable layers.

3.2.2   Gaussian dispersion modeling

Because of the random nature of dispersion, the distribution of the pollutant in both the vertical and lateral directions can be represented by a gaussian model, in many cases with good results. Concentration estimates are time-averaged due to the presence of eddies and the shifting of the wind. Figure 3.4 illustrates the three-dimensional shape of a time-averaged plume as well as typical vertical and lateral concentration profiles downwind of the stack.

The gaussian equation for average concentration downwind of an elevated continuous source is

$$
C = \frac{Q}{2\pi u \sigma_y \sigma_z} \exp\left(-\frac{1}{2}\frac{y^2}{\sigma_y^2}\right)\left\{\exp\left[-\frac{1}{2}\frac{(z-H)^2}{\sigma_z^2}\right]\right.
$$

$$
\left. + \exp\left[-\frac{1}{2}\frac{(z+H)^2}{\sigma_z^2}\right]\right\} \quad (3.2)
$$

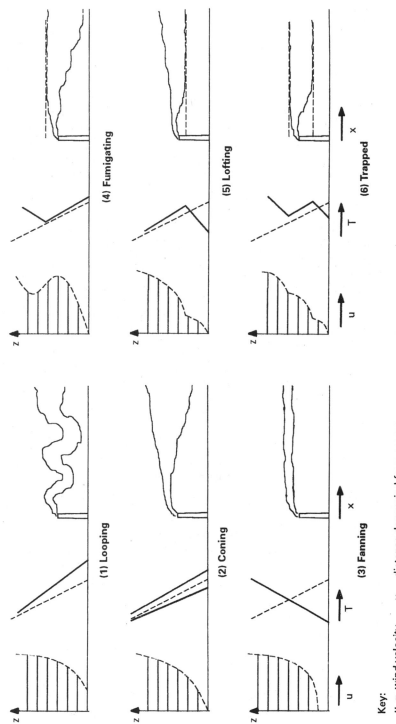

Key:

u = wind velocity   x = distance downwind from source
T = temperature   z = height above ground

Figure 3.3 Plume classifications (*Wark*, *1981*).

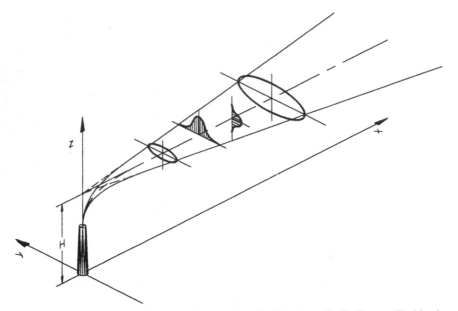

Figure 3.4 Time-averaged plume with gaussian distributions. (*D. B. Turner,* Workbook of Atmospheric Dispersion Estimates, *Environmental Protection Agency, Washington, 1970.*)

where C = time-averaged steady-state pollutant concentration at a point, $\mu g/m^3$

$\quad\quad Q$ = emissions rate, $\mu g/s$

$\quad \sigma_y, \sigma_z$ = horizontal and vertical dispersion parameters, respectively

$\quad\quad u$ = average wind speed at effective stack height, m/s

$\quad\quad y$ = horizontal distance from plume centerline, m

$\quad\quad z$ = vertical distance from ground level, m

$\quad\quad H$ = effective stack height, m, equal to $h + \Delta h$ (see Fig. 3.2)

The presence of the ground is accounted for by the inclusion of concentrations from an image source located at $z = -H$. This is accomplished in Eq. 3.2 by the exponential term containing $z + H$. The effect of the image source is to include in the region of actual dispersion the plume mass which would be underground. Due to the use of an image source, this effect is typically referred to as *ground reflection.*

In the case of a plume's dispersing under an elevated inversion, the inversion is also treated mathematically as a reflecting surface, located a distance L from the ground surface. The mathematical interaction of the two surfaces requires several image sources in the model.

The gaussian equations for continuous point sources are presented in their general form in Fig. 3.5, along with the dispersive conditions for which they are applicable.

Plume Types	General Equation
Looping, coning	$$C(x,y,z) = \frac{Q}{2\pi u \sigma_y \sigma_z} \exp\left(\frac{-y^2}{2\sigma_y^2}\right)\left\{\exp\left[\frac{-(z-H)^2}{2\sigma_z^2}\right] + \exp\left[\frac{-(z+H)^2}{2\sigma_z^2}\right]\right\} \quad (3.3)$$
Trapped	$$C(x,y,z) = \frac{Q}{2\pi u \sigma_y \sigma_z} \exp\left(\frac{-y^2}{2\sigma_y^2}\right)$$ $$\times \left\{\exp\left[-\frac{(z-H)^2}{2\sigma_z^2}\right] + \exp\left[\frac{-(z+H)^2}{2\sigma_z^2}\right] + E_T\right\} \quad (3.4)$$ where $$E_T = \sum_{n=1}^{n=\infty}\left\{\exp\left[-\frac{1}{2}\left(\frac{z-H-2nL}{\sigma_z}\right)^2\right] + \exp\left[-\frac{1}{2}\left(\frac{z+H-2nL}{\sigma_z}\right)^2\right]\right.$$ $$\left. + \exp\left[-\frac{1}{2}\left(\frac{z-H+2nL}{\sigma_z}\right)^2\right] + \exp\left[-\frac{1}{2}\left(\frac{z+H+2nL}{\sigma_z}\right)^2\right]\right\} \quad (3.5)$$
Trapped $(x>2x_L)$, fumigating	$$C(x,y,z) = Q\backslash(2\pi)^{1/2}\,\sigma_y L u \exp\left[-\frac{1}{2}\left(\frac{y}{\sigma_y}\right)^2\right] \quad (3.6)$$

Figure 3.5 Gaussian equations for different types of plumes.

3.2.3 Estimation of dispersion parameters

Dispersion parameters (σ_y, σ_z) can be estimated using data available in the form of graphs or curve-fit formulas. Values must be selected based on meteorological conditions and averaging time as well as the downwind distance from the source. Meteorological data required are the mean wind speed at plume height and stability classification. Atmospheric stability is assumed to be a function of surface wind speed and net radiation. During the day, net radiation is indicated by two factors: the sun's angle with the earth's surface and the presence of clouds. At night, net radiation is assumed to be affected by cloud cover only. Site-specific meteorological data are the most desirable, but in most cases wind speed data are unavailable. The first alternative is typically data taken at weather stations.

3.2.4 Plume rise estimation

As discussed in Sec. 3.2.1, plume rise is attributed to momentum and buoyancy. Momentum is exhibited by the tendency of the plume to rise due to its release velocity. Buoyancy is due to the difference between plume density and atmospheric density. Density differences may be due to the temperature or composition of the plume at release. The surrounding air quickly depletes a plume's initial momentum, but the rising effect of buoyancy can continue for a much greater distance. Due to the number of parameters involved, plume rise estimation can be complex, and several methods have been developed, with varying results. Briggs' method is a widely used method, and it has been recommended by the Environmental Protection Agency.

The following are plume rise models for various conditions:

1. Momentum-dominated plume (neutral/unstable atmosphere)

$$\Delta h = 3.78 \left[\frac{v^2}{u(v + 3u)} \right]^{2/3} \left(\frac{Xr^2}{2} \right)^{1/3} \tag{3.7}$$

or when $v/u \geq 4$

$$\Delta h_{max} = \frac{6vr}{u} \tag{3.8}$$

where X = downwind distance, m
r = stack radius, m
v = gas exit velocity, m/s
u = wind speed, m/s

2. Momentum-dominated plume (stable atmosphere). Use the lowest value of the following three equations:

$$\Delta h = 1.5 \left(\frac{F_m}{u} \right)^{1/3} S^{-1/6} \tag{3.9}$$

$$\Delta h = 4 \left(\frac{F_m}{S} \right)^{1/4} \tag{3.10}$$

$$\Delta h = \frac{6vr}{u} \tag{3.8}$$

where

$$F_m = v^2 r^2 \tag{3.11}$$

3. Momentum and buoyancy mixed (neutral/unstable atmosphere).

$$\Delta h = \frac{vD}{u} \left[1.5 + \frac{2.68E - 3PD(T_s - T_a)}{T_s} \right] \tag{3.12}$$

where T_s = stack gas temperature, K, and T_a = ambient temperature, K. For different stability classes,

$$H = h + \Delta h(1.4 - 0.1N_c) \tag{3.13}$$

where N_c is a class number corresponding to the atmospheric stability classification

$$N_c = \begin{cases} 1 \text{ for class A} \\ 2 \text{ for class B} \\ 3 \text{ for class C} \\ 4 \text{ for class D} \end{cases}$$

Note:

$$T(\text{K}) = T(°\text{C}) + 273$$

4. Buoyancy-dominated plume (neutral/unstable atmosphere). Rising stage:

$$\Delta h = \frac{1.6 F^{1/3} X^{2/3}}{u} \qquad \text{for } X \leq 3.5 X^* \qquad (3.14)$$

Final stage:

$$\Delta h = \frac{1.6 F^{1/3} (3.5 X^*)^{2/3}}{u} \qquad \text{for } X \geq 3.5 X^* \qquad (3.15)$$

where

$$X^* = \begin{cases} 14 F^{5/8} & \text{for } F < 55 \\ 34 F^{2/5} & \text{for } F \geq 55 \end{cases}$$

and

$$F = \frac{g V r^2 (T_s - T_a)}{T_s}$$

5. Buoyancy-dominated plume (stable atmosphere):

$$\Delta h = 2.6 \left(\frac{F}{uS} \right)^{1/3} \qquad \text{for } u \geq 1 \text{ m/s} \qquad (3.16)$$

The following items refer to the class numbers in Table 3.1.

1. Clear skies, solar altitude greater than 60° above the horizon, typical of a sunny summer afternoon. Very convective atmosphere.

2. Summer day with a few broken clouds.

3. Typical of a sunny fall afternoon, summer day with clear skies and solar altitude from only 15° to 35° above horizontal.

4. Can also be used for a winter day.

TABLE 3.1   Stability Classifications

Surface wind speed at 10 m (m/s)	Day			Night	
	Incoming solar radiation			Cloud cover	
	Strong	Moderate	Slight	Mostly overcast	Mostly clear
Class*	1	2	3	4	5
<2	A	A–B	B	E	F
2–3	A–B	B	C	E	F
3–5	B	B–C	C	D	E
5–6	C	C–D	D	D	D
>6	C	D	D	D	D

*The neutral class, D, should be assumed for overcast conditions during day or night. Class A is the most unstable, and class F is the most stable, with class B moderately unstable and class E slightly stable.

SOURCE: D. B. Turner, *Workbook of Atmospheric Dispersion Estimates,* HEW, Washington, 1969.

3.2.5   Available models

Models can be categorized by their ability to account for characteristics of the source, the pollutant, the terrain, and atmospheric conditions. In addition to simple gaussian models for continuous point sources, models have been developed to account for mobile sources, area sources, line sources, instantaneous releases, complex terrain, reactive pollutants, dense pollutants, long-range transport, and other particulars. Several models are available via the Internet from the Office of Air and Radiation of the U.S. Environmental Protection Agency.

References

Cooper, C. David, and F. C. Alley, *Air Pollution Control: A Design Approach,* Waveland Press, Inc., Prospect Heights, Ill., 1986.
Guidelines on Air Quality Models (Revised), Office of Air Quality Planning and Standards, Environmental Protection Agency, Research Triangle Park, N.C., July 1986.
Turner, D. B., *Workbook of Atmospheric Dispersion Estimates,* Environmental Protection Agency, Washington, 1970.
Wark, Kenneth, and Cecil F. Warner, *Air Pollution: Its Origin and Control,* 2d ed., HarperCollins Publishers, New York, 1981.

Ambient Air Monitoring

Steven Marquardt

4.1 NAAQSs and PSD Monitoring

The importance of clean air to human health and the health of the environment goes without argument. Just how clean this air must be is of considerable debate. Regardless of the level of pollution deemed acceptable, a means must be established to measure the concentration of pollutants in the ambient air to show compliance with regulatory guidelines, track pollution trends, and provide a background level against which to predict the outcome of proposed emission increases. While many different types of pollutants are discharged into the air, this discussion is limited to the EPA-designated criteria pollutants: particulate, sulfur dioxide, nitrogen oxides, carbon monoxide, ozone, and lead. However, the basic principles of ambient air monitoring of the criteria pollutants could be applied to other pollutants, such as the various hazardous air pollutants (HAPs).

The Clean Air Act Amendments (CAAA) of 1970 required the EPA to set national ambient air quality standards (NAAQSs) to protect the health of both the environment and its human inhabitants. The act also required the states to submit state implementation plans (SIPs) to attain and maintain the NAAQSs. The Clean Air Act was modified further by the CAAA of 1977 which updated the NAAQSs directive. An important addition to the act was Prevention of Significant Deterioration (PSD), the purpose of which is to limit the increase in ambient concentrations of pollutants to prevent degradation of air quality in areas meeting the NAAQSs or in pristine natural areas. Currently, NAAQSs have been established for six pollutant groups: particulates, sulfur dioxides, nitrogen oxides, carbon monoxide, ozone, and lead. The standards are of two types: primary and secondary. Primary standards are designed to protect human health, and secondary standards are set to guard the public welfare. The NAAQSs can be found in 40 CFR 50 and are summarized here in Table 4.1. Because concentrations in air vary with temperature and pressure, the EPA has referenced the NAAQSs to a temperature of 25°C and a pressure of 760 mmHg. The current standards based on other than annual average are not to be exceeded more than once per year. The EPA is required to periodically review the effectiveness of the NAAQSs, and users of this handbook are cautioned to consult the latest local, state, and federal regulations to be sure of the standards

TABLE 4.1   National Ambient Air Quality Standards (NAAQSs)

Pollutant	Averaging time	Primary standard		Secondary standard	
Sulfur dioxide	1 year	80 μg/m^3	0.03 ppm	—	—
	24 h	365 μg/m^3	0.14 ppm	—	—
	3 h	—	—	1300 μg/m^3	0.5 ppm
Particulate matter (PM$_{10}$—particles of 10-μm aerodynamic diameter or less)	1 year	50 μg/m^3		Same	
	24 h	150 μg/m^3		Same	
Carbon monoxide	8 h	10,000 μg/m^3	9 ppm	Dropped	
	1 h	40,000 μg/m^3	35 ppm	Dropped	
Nitrogen dioxide	1 year	100 μg/m^3	0.053 ppm	Same	
Ozone	1 h	235 μg/m^3	0.12 ppm	Same	
Lead	Calendar quarter arithmetic average	1.5 μg/m^3		Same	
Proposed 1997 changes:					
Particulate matter (PM$_{2.5}$—particles of 2.5-μm aerodynamic diameter or less. The present PM$_{10}$ standard will be retained with revision of monitoring network requirements.)	1 year	15 μg/m^3		Same	
	24 h	50 μg/m^3		Same	
	1 year	50 μg/m^3		Same	
	24 h	150 μg/m^3		Same	
Ozone	8 h	155 μg/m^3	0.08 ppm	Same	

NAAQSs are referenced to 25°C and 760 mmHg.

before beginning monitoring for the purpose of demonstrating attainment.

As of the writing of this handbook, the EPA was under a court order to issue a proposal as to whether to retain or revise the particulate and ozone standards. The proposed ozone standard would be met when the 3-year average of the annual third-highest daily maximum 8-h average concentration is less than or equal to 0.08 ppm. The EPA has proposed to retain the PM_{10} standard with reporting and network changes and to add a new standard for particle fines smaller than 2.5 μm in diameter. The new $PM_{2.5}$ standard would be met when the 3-year average of the annual arithmetic mean concentrations, spatially averaged across an area, is less than or equal to 15 μg/m^3. The new 24-h $PM_{2.5}$ standard would be met when the 3-year average of the 98th percentile of the 24-h concentrations at each monitor within an area is less than or equal to 50 μg/m^3. The PM_{10} standard would be met when the 3-year average of the annual arithmetic PM_{10} concentrations at each monitor within an area is less than or equal to 50 μg/m^3. The current 24-h PM_{10} standard will be maintained at the level of 150 μg/m^3, but revised such that the standard would be met when the 3-year average of the 98th percentile of the monitored concentrations at the highest monitor in an area is less than or equal to 150 μg/m^3.

To ascertain whether a specific area meets the NAAQSs, each state must develop and maintain an ambient air monitoring network for each of the six groups of pollutants. Additionally, facilities desiring to locate or expand in an area may be required to perform ambient air monitoring to establish a baseline from which to model proposed emission increases under PSD. The Clean Air Act stipulates that an air quality analysis be conducted before construction or significant modification of a major source is permitted. The study can be conducted with the use of modeling and/or monitoring of air quality. The EPA has directed that continuous monitoring be employed for establishing existing air quality concentrations for the criteria pollutants SO, CO, and NO_x. Postconstruction monitoring is generally not required; however, the EPA has the authority to require postconstruction monitoring in situations where the NAAQSs are threatened or in cases in which the effect of increased pollution is uncertain. Generally, the purpose of PSD monitoring is to ascertain the effect of a proposed increase in emissions on air quality. The data gathered during monitoring are used to establish background air quality in the area surrounding the source and to validate and refine predictive models. Regardless of whether ambient monitoring is being done for a source requiring a PSD review or for state monitoring of NAAQSs, the basic operation of the monitoring station is the same. The main difference in operating the networks lies in the system auditing and reporting.

4.2 Reference Method Descriptions

Each group of pollutants has a specific monitoring methodology which must be followed. The EPA has established standard procedures for sampling and analyzing the ambient air for each of the six pollutant groups, referred to as *reference methods*. The reference methods are just that—a means of establishing reference methods or benchmarks to ensure that all organizations performing sampling produce comparable results. Because several of the reference methods are complicated and do not readily lend themselves to field use, the EPA also recognizes equivalent methods which produce results as good as, or better than, those of the reference methods. The equivalent methods are carefully tested against the reference methods and must have an EPA equivalent certification number indicating the EPA's approval. When considering the purchase of equipment or analyzers for ambient air sampling, the operator must be certain to choose instruments which carry EPA equivalent approval if they differ from the reference method. A short explanation of each reference method is given below. For a complete discussion of the reference methods, see the appendices of 40 CFR Part 50.

Sulfur dioxide (SO_2).   Ambient air is drawn through a solution of potassium tetrachloromercurate (TCM). The sulfur dioxide present in the airstream reacts with the solution to produce a monochlorosulfonatomercurate complex which is resistant to oxidation. The volume of air drawn through the solution is measured with a flowmeter and must be corrected to standard conditions. The sample is analyzed by reacting the complex with acid-bleached pararosaniline dye and formaldehyde. The result of this reaction is a solution with an optical density directly related to the amount of SO_2 collected. The concentration is computed and expressed in micrograms per dry standard cubic meter.

Total suspended particulate (high-volume method).   This method provides a measure of the mass concentration of the total amount of suspended particulate. The mass concentration is determined by drawing the sample through a flowmeter and a particulate filter which is 99 percent efficient for particles 0.3 μm in diameter. The filter is weighed before and after sampling to determine the mass of the particulate collected. The sample air volume must be corrected to the reference standard temperature and pressure of 25°C and 760 mmHg. For purposes of determining attainment of the primary and secondary standards, particulate matter must be measured in the ambient air as PM_{10}. The high-volume method can be used to show compliance with PM_{10} requirements provided the total mass per unit volume does not exceed the PM_{10} guidelines. See Fig. 4.1.

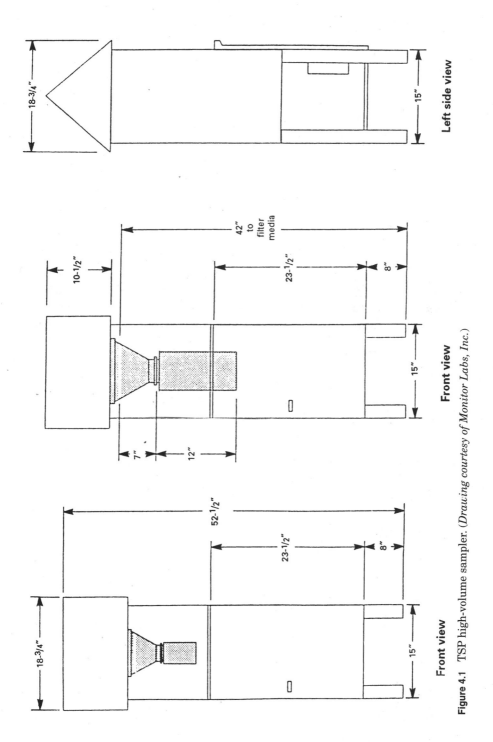

Left side view

Front view

Front view

Figure 4.1 TSP high-volume sampler. (*Drawing courtesy of Monitor Labs, Inc.*)

PM₁₀. This method relies on inertial separation to divide the particles into one or more diameter classes within the 10-µm-diameter cutoff. Each size division is captured on a separate filter. Each filter is weighed before and after each test to determine the mass of PM_{10} collected. The airflow through the collector must be corrected to standard conditions. The mass concentration of PM_{10} in the ambient air is computed as the total mass of collected particles in the PM_{10} size range divided by the volume of air sampled, and it is expressed in micrograms per standard cubic meter. See Fig. 4.2.

Proposed PM₂.₅. The requirements for reference method designation for $PM_{2.5}$ are to be performance-based specifications for the operational aspects of a reference method sampler; therefore, it is expected that various manufacturers will design and build different samplers to meet the required specifications. Sampler designs would have to meet flow rate, flow rate regulation, flow rate measurement accuracy, ambient air temperature and barometric pressure measurement accuracy, filter temperature control and measurement accuracy, and sampling time accuracy requirements. Other tests would be of the units leak test procedure, flow rate cutoff function, and field operational precision. Three classes of equivalent method are also proposed. Class I would include methods based on samplers that are very similar to a reference method sampler. The primary difference from reference method samplers is one or more modifications necessary to provide capability for collection of several sequential samples automatically without operator service. Class II equipment would include all other $PM_{2.5}$ methods based on a 24-h integrated filter sample. Class III would include both continuous and semicontinuous methods having other than a 24-h sample collection interval followed by moisture equilibration and gravimetric mass. Class III instruments would include beta attenuation, harmonic oscillating element and other complete in situ monitor types, as well as non-filter-based methods such as nephelometry or other optical instruments.

Carbon monoxide (CO) (nondispersive infrared photometry). This reference method is based on carbon monoxide's ability to absorb infrared energy. In a photometer analyzer, infrared energy is passed through the gas sample, and the sample's absorptive capacity at a wavelength of 4.6 µm is compared to that of "zero air," free of CO. Because infrared absorption by CO is a nonlinear relationship, the raw signal must be converted electronically to provide a linear output. The analyzer is calibrated against CO gas standards traceable to National Bureau of Standards (NBS) CO in air standard reference material (SRM) or to an NBS/EPA-approved commercially available certified reference material (CRM). Calibration gases are available

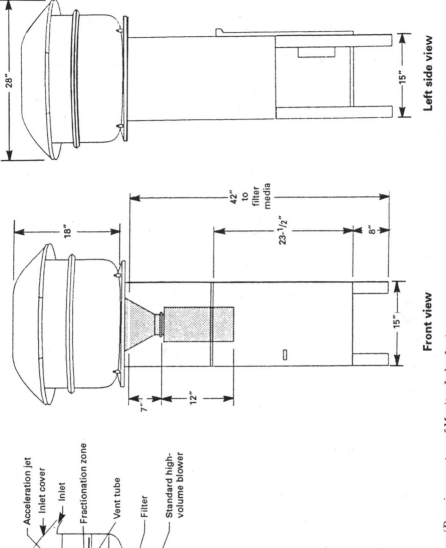

Flow diagram

Front view

Left side view

Figure 4.2 PM₁₀ high-volume sampler. (*Drawing courtesy of Monitor Labs, Inc.*)

from specialty gas suppliers in various concentrations and bottle sizes.

Ozone. The EPA reference method specifies that the amount of ozone in a sample of air be measured by detecting the light energy emitted when the ozone in the sample is mixed with ethylene. The amount of light emitted is detected with a photomultiplier tube. The resultant signal is amplified and displayed directly or converted to reflect the concentration.

Nitrogen dioxide (NO_2). The ambient concentration of NO_2 in air is determined by measuring the intensity of light at wavelengths greater than 600 nm resulting from the chemiluminescent reaction of nitric oxide (NO) and ozone (O_3). The sample is split into two portions. The NO_2 in the first portion is reduced to NO by passing the sample through a converter. The second portion of the sample is bypassed around the converter, and the ambient level of NO in the air is exposed to ozone. The level of NO_2 in the ambient air is determined as the difference between the converted and nonconverted readings. See Fig. 4.3.

Lead. Particulate matter is collected using the high-volume TSP method and then analyzed for lead using atomic absorption spectrometry following extraction of the lead using nitric acid.

4.3 General Considerations Common to All Methods

All the methods share several common considerations in the design and operation of the monitoring network. Among these considerations are the purpose of the monitoring, site selection, sampling considerations such as environmental control and manifold design, data han-

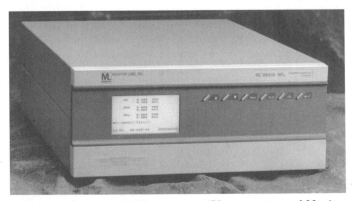

Figure 4.3   Automated NO_x monitor. (*Photo courtesy of Monitor Labs, Inc.*)

dling and reporting, testing equipment selection, and quality assurance. The EPA has summarized these considerations in its *Quality Assurance Handbook for Air Pollution Measurement Systems,* vol. 2, *Ambient Air Specific Methods.* The user of this handbook is encouraged to read and understand both volumes 1 and 2 of this series. [Reproductions of these references are available from National Technical Information Service (NTIS) and can be ordered by phone.]

4.3.1 Monitoring objective

The first monitoring network design problem to be addressed is to state the purpose for which monitoring is to be conducted. Monitoring is most often conducted to judge compliance or measure progress made toward compliance with air quality standards, and to provide a baseline for pollution dispersion modeling such as preconstruction monitoring for PSD. Other monitoring purposes include trend evaluation and evaluation of the effectiveness of control measures. The purpose of monitoring can impact the number and placement of monitoring stations. A single station with only one analyzer may suffice for a PSD review, while design and operation of a state or local air monitoring stations (SLAMS) network as provided for in a state implementation plan would obviously require a much more extensive layout. It is strongly recommended that the monitor operator contact the state or federal agency to which the monitoring data will be submitted prior to beginning work, to ensure that any special conditions or requirements will be met so that the data produced are of acceptable quality and suit the purpose for which they are to be gathered.

4.3.2 Site selection

Site selection is the next step in network design. For NAAQS compliance monitoring, the sites selected should be in the area or zone at which the highest concentrations are expected to be encountered. Sites should also be placed at the edge of the region or area being evaluated, to gain an idea of the pollutant concentrations entering and leaving the area. Several stations should be placed based on long-term considerations, such as providing a means to track trends and to measure the impact of source additions and abatement measures through time. Guidelines for the number and placement of monitoring sites required for a network can be found in 40 CFR Part 58, Appendix D. Although no longer in force, 40 CFR Part 51.17, July 1975, contained a chart of the minimum acceptable number of monitoring sites for the criteria pollutants, and it would serve as a good starting point for determining the number of sites required. This chart is shown in Table 4.2.

TABLE 4.2 Recommended SIP Monitoring Requirements by Pollutant

Classification of region	Pollutant	Measurement methods or principle[i]	Minimum frequency of sampling	Region population	Minimum number of air quality monitoring sites[h]
I	Suspended particulates	High-volume sampler	One 24-h sample every 6 days[a]	Less than 100,000 100,000–1,000,000 1,000,000–5,000,000	4 4 + 0.6 per 100,000 population[c] 7.5 + 0.25 per 100,000 population[c]
		Tape sampler	One sample every 2 h	Above 5,000,000	12 + 0.16 per 100,000 population[c] One per 250,000 population[c] up to eight sites
	Sulfur dioxide	Pararosaniline or equivalent[d]	One 24-h sample every 6 days (gas bubbler)[a]	Less than 100,000 100,000–1,000,000 1,000,000–5,000,000	3 2.5 + 0.5 per 100,000 population[c] 6 + 0.15 per 100,000 population[c]
			Continuous	Above 5,000,000 Less than 100,000 100,000–5,000,000 Above 5,000,000	11 + 0.05 per 100,000 population 1 1 + 0.15 per 100,000 population[c] 6 + 0.05 per 100,000 population[c]
	Carbon monoxide	Nondispersive infrared or equivalent[e]	Continuous	Less than 100,000 100,000–5,000,000 Above 5,000,000	1 1 + 0.15 per 100,000 population[c] 6 + 0.05 per 100,000 population[c]
	Photochemical oxidants	Gas-phase chemiluminescence or equivalent[f]	Continuous	Less than 100,000 100,000–5,000,000 Above 5,000,000	1 1 + 0.15 per 100,000 population[c] 6 + 0.05 per 100,000 population[c]
	Nitrogen dioxide	24-h sampling method (Jacobs-Hochheiser method)	One 24-h sample every 14 days (gas bubbler)[b]	Less than 100,000 100,000–1,000,000 Above 1,000,000	3 4 + 0.6 per 100,000 population[c] 10

II	Suspended particulates	High-volume sampler	One 24-h sample every 6 days[a]	3
	Sulfur dioxide	Tape sampler	One sample every 2 h	1
		Pararosaniline or equivalent[d]	One 24-h sample every 6 days (gas bubbler)[a]	3
			Continuous	1
III	Suspended particulates	High-volume sampler	One 24-h sample every 6 days[a]	1
	Sulfur dioxide	Pararosaniline or equivalent[d]	One 24-h sample every 6 days (gas bubbler)[a]	1

[a]Equivalent to 61 random samples per year.

[b]Equivalent to 26 random samples per year.

[c]Total population of a region. When required number of samplers includes a fraction, round to nearest whole number.

[d-f][Reserved, 40 FR 7042, February 18, 1975]

[g]It is assumed that the federal motor vehicle emission standards will achieve and maintain the national standards for carbon monoxide, nitrogen dioxide, and photochemical oxidants; therefore, no monitoring sites are required for these pollutants.

[h]In interstate regions, the number of sites required should be prorated to each state on a population basis.

[i]Named methods and principles, except the tape sampler method, are described in Part 50 of this chapter. The tape sampler method is described in Hemeon, W. C. L., Haines, G. F., Jr., and Ide, H. M., "Determination of Haze and Smoke Concentrations by Filter Paper Samplers," *J. Air Pollution Control Association*, vol. 3, pp. 22–28, 1953. Use of these and other methods shall be as specified in § 51.17a.

SOURCE: CFR, Part 51.17, Air Quality Surveillance, July 1975, p. 43.

The proposed $PM_{2.5}$ standards are intended to protect against exposure to fine particulate, while the PM_{10} standards are designed to protect against exposure to coarse particles. The new standard will require spatial averaging. The new $PM_{2.5}$ network will consist of core population-oriented monitors and supplementary population-oriented monitors. The core monitors are required in all the largest metropolitan areas. There will be at least one supplementary monitor for each 250,000 people.

A number of factors impact pollutant concentration and hence site selection, including weather, topography, population densities, known source locations, and pollutant interactions. The single largest weather consideration is the effect of the wind. Wind speed and direction determine the pollutant movement and the time required to move from a source to the monitoring station. Wind speed also plays an important part in pollutant concentration; the higher the wind speed, the greater the dispersion and, therefore, the lower the concentration.

Topographic features, such as valleys which are infamous for inversions, must also be factored into site selection decisions. Large bodies of water and their effects on wind currents due to their temperature differences with the land at certain times of the year are yet another example of the effects of topography. On a microscale, buildings and trees can block the flow of wind or create eddies and swirls which can influence sampling concentrations. In general, the greater the variation of the terrain, the larger the number of monitors required and the greater the care needed in placing the monitors to produce a true representation of area concentrations. The final sites selected should also be free of sources of dust and nearby pollutant sources and still provide ready access to electrical service and allow easy ingress for the required station servicing.

4.3.3   Control of sampling environment

To ensure that the data collected are representative of actual conditions, steps must be taken to preserve the integrity of the sample and the conditions under which the sample is evaluated. Some of the more important conditions to be controlled include the physical conditions under which the monitor will be maintained. For instance, many of the continuous analyzers must be operated within specified temperature ranges in order to satisfy the conditions of method equivalency. Obviously, dust, humidity, and vibration are deleterious to the reliability and service life of the sensitive electronics and optics contained in most monitors. Light, natural or artificial, especially certain spectra, can affect analytical chemicals and internal monitor components.

The easiest way to control these parameters is by placing the monitors in an enclosed shelter with a sampling probe to the ambient air.

Often, the shelter can be a permanent or temporary building and should be equipped with an HVAC system. One convenient method of shelter is a purpose-built trailer similar to a construction site office trailer. Such a trailer has the advantages of being mobile and relatively inexpensive. See Fig. 4.4.

As a final note on monitor shelters, the shelter in all likelihood will require electrical service. The value of a reliable electrical supply cannot be overstated in terms of prevention of lost data or possible equipment damage. For these reasons, the shelter should be equipped with protected circuits, surge protectors, and backup power supplies for the analyzers and any computers which will allow the station to continue to operate during short-term electric power outages.

4.3.4 Sampling probes and manifolds

The sampling probes and manifolds should be constructed of nonreactive materials such as glass or Teflon. The probe and manifold should be easy to disassemble for cleaning and maintenance. The most widely used probe and manifold design is the "conventional" system which consists of a vertical probe with a U-shaped section near the probe opening, designed to keep out rain and to avoid accumulating dirt in the line. The manifold consists of a horizontal section connected to the vertical probe section with a tee connector and a knockout chamber. The length of the probe and manifold should be as short as possible to reduce sample residence time and sample degradation through conta-

Figure 4.4 Trailer serving as a monitor shelter.

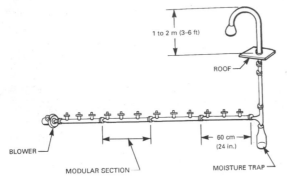

Figure 4.5 Conventional manifold system.

mination or breakdown. The probe and manifold should be insulated to avoid condensation in the line. An illustration of a conventional manifold system is shown in Fig. 4.5.

4.3.5 Probe siting

Ambient air monitoring probes must be located in such a manner that they sample the maximum concentration while not introducing errors as the result of placement too close to sources of pollutant, such as roadways and exhaust vents. Each pollutant has somewhat different requirements, and the horizontal and vertical placement requirements are summarized in Table 4.3.

4.3.6 Monitoring equipment selection

The analyzer selected for use must carry either an EPA reference method or equivalent method designation. Whether the analyzer is of one designation or the other is not of great importance; either can be used interchangeably. What is important is that the monitor have a valid EPA designation number. Without the EPA designation, the data collected are going to be regarded with suspicion and are likely to be rejected by the regulatory authority. Beyond EPA acceptance, what is important in an analyzer? Reliability, vendor support, ease of use, and cost are easily the most important factors. An unreliable, difficult-to-use machine which suffers regular breakdowns and is backed by an unresponsive technology-help line and parts program holds countless frustrations, including the loss or invalidation of irreplaceable data.

Additional required equipment will likely include an analog strip chart recorder, a backup power source for each analyzer, and possibly a personal computer equipped with a modem and communication

TABLE 4.3 Summary of Probe Siting Criteria

Pollutant	Height above ground, m*	Distance from supporting structure, m†		Other spacing criteria
		Vertical	Horizontal	
TSP	2–15	—	>2	1. Should be >20 m from the drip line and must be 10 m from the drip line when the trees act as an obstruction. 2. Distance from sampler to obstacle, such as buildings, must be at least twice the height that the obstacle protrudes above the sampler. 3. Must have unrestricted airflow 270° arc around the sampler. 4. No furnace or incineration flues should be nearby.‡ 5. Must have minimum spacing from roads. This varies with height of monitor.
PM₁₀ (impact near major roadway and/or ground-level sources)	2–7	—	>2	1. Should be >20 m from the drip line and must be 10 m from the drip line when the trees act as an obstruction. 2. Distance from sampler to obstacle, such as buildings, must be at least twice the height that the obstacle protrudes above the sampler. 3. Must have unrestricted airflow 270° arc around the sampler. 4. No furnace or incineration flues which emit particulate matter should be nearby.‡ 5. Must be 5 to 15 m from roads.
PM₁₀	2–15	—	>2	1. Should be >20 m from the drip line and must be 10 m from the drip line when the trees act as an obstruction. 2. Distance from sampler to obstacle, such as buildings, must be at least twice the height that the obstacle protrudes above the sampler. 3. Must have unrestricted airflow 270° arc around the sampler. 4. No furnace or incineration flues which emit particulate matter should be nearby.‡
SO₂	3–15	>1	>1	1. Should be >20 m from the drip line and must be 10 m from the drip line when the trees act as an obstruction. 2. Distance from inlet probe to obstacle, such as buildings, must be at least twice the height that the obstacle protrudes above the inlet probe. 3. Must have unrestricted airflow 270° arc around the inlet probe, or 180° if probe is on the side of a building. 4. No furnace or incineration flues should be nearby.‡

TABLE 4.3 Summary of Probe Siting Criteria (Continued)

Pollutant	Height above ground, m*	Distance from supporting structure, m		Other spacing criteria
		Vertical	Horizontal†	
CO (street and canyon)	$3 \pm \frac{1}{2}$	>1	>1	1. Must be >10 m from intersection and should be at a midblock location. 2. Must be 2 to 10 m from edge of nearest traffic lane. 3. Must have unrestricted airflow 180° around the inlet probe.
CO (nonstreet canyon and corridor)	3–15	>1	>1	Must have unrestricted airflow 270° around the inlet probe, or 180° if probe is on the side of a building.
O_3	3–15	>1	>1	1. Should be >20 m from the drip line and must be 10 m from the drip line when the trees act as an obstruction. 2. Distance from inlet probe to obstacle, such as buildings, must be at least twice the height that the obstacle protrudes above the inlet probe. 3. Must have unrestricted airflow 270° arc around the inlet probe, or 180° if probe is on the side of a building. 4. Spacing from roads varies with traffic.
NO_2	3–15	>1	>1	1. Should be >20 m from the drip line and must be 10 m from the drip line when the trees act as an obstruction. 2. Distance from inlet probe to obstacle, such as buildings, must be at least twice the height that the obstacle protrudes above the inlet probe. 3. Must have unrestricted airflow 270° arc around the inlet probe, or 180° if probe is on the side of a building.
Pb (impact near major roadway and/or ground-level sources)	2–7	—	>2	1. Should be >20 m from the drip line and must be 10 m from the drip line when the trees act as an obstruction. 2. Distance from sampler to obstacle, such as buildings, must be at least twice the height that the obstacle protrudes above the sampler. 3. Must have unrestricted airflow 270° arc around the sampler. 4. No furnace or incineration flues which emit lead should be nearby.‡ 5. Must be 15 to 30 m from major roadways.

TABLE 4.3 Summary of Probe Siting Criteria (Continued)

Pollutant	Height above ground, m*	Distance from supporting structure, m† Vertical	Distance from supporting structure, m† Horizontal†	Other spacing criteria
Pb	2–15	—	>2	1. Should be >20 m from the drip line and must be 10 m from the drip line when the trees act as an obstruction. 2. Distance from sampler to obstacle, such as buildings, must be at least twice the height that the obstacle protrudes above the sampler. 3. Must have unrestricted airflow 270° arc around the sampler. 4. No furnace or incineration flues which emit lead should be nearby.‡
Particulate noncriteria pollutants	2–7 for ground-level sources; 2–15 for elevated sources	—	>2	1. Should be >20 m from the drip line and must be 10 m from the drip line when the trees act as an obstruction. 2. Distance from sampler to obstacle, such as buildings, must be at least twice the height that the obstacle protrudes above the sampler. 3. Must have unrestricted airflow 270° arc around the sampler. 4. No furnace or incineration flues which emit the noncriteria pollutant should be nearby.‡
Gaseous noncriteria pollutants	3–15	>1	>1	1. Should be >20 m from the drip line and must be 10 m from the drip line when the trees act as an obstruction. 2. Distance from inlet probe to obstacle, such as buildings, must be at least twice the height that the obstacle protrudes above the inlet probe. 3. Must have unrestricted airflow 270° arc around the inlet probe, or 180° if the probe is on the side of a building. 4. No furnace or incineration flues which emit the noncriteria pollutant should be nearby.‡

*For ground-level sources, monitors and inlet probes should be placed as close to the breathing zone as possible.
†When probe is located on rooftop, this separation distance is in reference to walls, parapets, or penthouses located on the roof.
‡Distance is dependent on height of furnace or incineration flue, type of fuel or waste burned, and quality of fuel. This is to avoid influences from minor pollutant sources.

software to allow remote access to station control and data retrieval. The type of computer selected is obviously dependent on the tasks required and the level of complexity of the operation. Many monitors are fully programmable and have internal hard and floppy drives for data reduction and storage.

4.3.7 Monitoring duration

Monitoring is usually conducted for at least a 1-year period for PSD purposes. However, in order to be judged representative, the EPA has established that a minimum of 4 months of monitoring data is acceptable provided that they are collected during the months during which the historical high concentrations of a pollutant exist. Ozone monitoring is an exception to this rule. Because ozone formation is affected by temperature and sunlight, monitoring is required for the four warmest months of the year, June through September. In addition, the 4 months of highest historical concentration must also be included if they are different from the four warmest months of the year. If a time gap exists between the warmest months of the year and the historically highest months or month, then data must also be gathered during this time period. For purposes of demonstrating attainment with the NAAQSs, the monitoring program must obviously consist of data for more than 1 year for certain pollutants. To ensure completeness, at least 80 percent of the individual hourly averages for continuous monitors and 80 percent of the individual 24-h values for manual methods should be reported.

4.4   Monitoring Plan Preparation

Although not required prior to the start of data collection, a monitoring plan must be assembled and approved before the data will be accepted and utilized by the regulatory authority. At a minimum, the plan must contain a source environment description, sampling program description, monitoring site description, monitor description, data-reporting description, and a quality assurance program description.

4.4.1 Source site description

The source site description should include topography surrounding the source for a 2-km radius. The description should include the typical land use and include a map detailing the existence of other stationary sources. A climatological description including quarterly wind roses is also required.

4.4.2   Sampling program description

This portion of the plan must detail the time period for which the data will be gathered and the rationale for the number and location of monitors. The location and number of monitors are dependent on the complexity of the environment in which the source and monitors will be located. Factors to be considered include topography, wind direction, and number and location of other nearby stationary sources. A single source located on relatively flat terrain in a rural or semirural area can be satisfactorily served by a single monitoring station placed directly downwind at a distance from the source where the concentration of a given pollutant will be at its maximum level. A plant located in a large urban area with many industries or in mountainous terrain may require a much more complex arrangement. The regulatory agency in charge of permitting can provide recommendations for the number and location of monitors needed.

4.4.3   Site and monitor description

The description of the monitoring site includes such items as the location's universal transverse mercator (UTM) coordinates, the height of the probe above the ground, a description of any obstructions, distances to nearby roads, other sources, and photographs of the site. This portion of the plan includes the name of the manufacturer of the monitor, a description of the calibration procedure to be followed, and a description of any related recording equipment and controls. The analyzer used must carry an EPA equivalent method designation number to ensure that the data generated meet the EPA standards. The equivalent method designation number is issued to the equipment manufacturer following testing according to 40 CFR Part 53.

4.4.4   Probe and shelter description

The monitor must be placed in a stable environment where the temperature must not be allowed to vary outside the limits set by the equipment manufacturer. Care must also be taken to ensure that the sensitive internal working components of the monitor are protected from dust and moisture. A temporary building or trailer with heating and cooling provides the required attributes. A proposed monitoring site will generally require electrical and phone service, which are important factors to consider in choosing a monitoring site. The probe must be made of a nonreactive material such as glass or Teflon and located according to the guidelines given in Table 4.3.

4.4.5   Data reporting

Monthly monitoring reports are required and must be prepared for the month just ended by the 15th of the following month. The report must be submitted in a standardized computer data file arrangement used by the EPA to compile a nationwide database. The report must be submitted in magnetic media form as well as hard copy.

4.4.6   Quality assurance

Quality assurance is arguably the most important segment of the monitoring plan. Data which are rejected due to questions of its validity or accuracy can cause expensive delays in the granting of a construction permit. Included in this section is a complete examination of how the data are to be generated and analyzed. The quality assurance requirements for ambient air monitoring for PSD can be found in Appendix B of 40 CFR Part 58, Ambient Air Quality Surveillance.

4.4.7   Maintenance and data recording

A schedule of routine maintenance procedures should be supplied by the monitor manufacturer. A stockpile of frequently needed spare parts should be kept at the monitor site for use as necessary. Following maintenance of automated analyzers, a zero and span check should be conducted and the monitor recalibrated if indicated. Continuous monitors are available which can be linked to a personal computer (PC) to store the data gathered. Alternatively, monitors are available with internal floppy drives. Observation data are collected on a continuous basis and are recorded both digitally and via an analog strip chart, which maintains a continuous record of activity at the site. The analog record also creates a backup in case the computer-stored data are lost. The computer or floppy drive internal to the analyzer can be configured to record the data at specified intervals, typically 10 seconds. These time intervals are then averaged and recorded on the floppy disk in periods of up to 1 hour. At the end of each month, the data are downloaded and arranged in the standardized electronic reporting format. Generally, the data for each month must be submitted to the regulator by the 15th of the following month; however, the site operator should verify the reporting deadlines.

4.4.8   Documentation of quality control methods

The documentation of quality control methods is achieved by keeping accurate records of all procedures carried out during the monitoring period. In addition to developing and following the monitoring plan, records should be kept of calculations used for calibrations and data checks. Further, a logbook should be kept of all adjustments and maintenance procedures carried out. The logbook should be a bound volume with numbered pages, and each entry should include the date and time, the activity and reason, and the name and signature of the technician.

4.5 Quality Assurance Plan and Procedures

The purpose of quality assurance is to provide data of adequate quality to ensure that monitoring objectives are met and to minimize the loss of data due to malfunctions and out-of-control conditions. These objectives are met by implementing a two-part system of operational procedures and data analysis. The purpose of operational procedures is to control the measurement process through the use of specific procedures and requirements. The data analysis is designed to assess the quality of the data produced following the procedures and requirements of the first part of the program. These two activities are interconnected and form a feedback loop, thereby indicating when either segment of the loop needs correction or adjustment. The operational procedures should include calibration frequency and procedures, independent auditing conducted quarterly by the regulatory agency, an annual audit by the EPA, internal quality assurance control procedures such as zero and span checks designed to track instrument drift, and data precision and accuracy calculation procedures.

4.5.1 Calibration of automated analyzers

The calibration procedure to be followed should be spelled out in the quality assurance plan. The calibration of newer automated analyzers for CO, SO_2, and NO_x is usually accomplished using an automated two-point procedure built into the analyzer software. The first calibration point is the zero concentration and is set by introducing "zero gas" scrubbed of the pollutant. After the analyzer response is allowed to stabilize, the zero calibration mode is activated and the analyzer response is corrected to display zero. The upper calibration point is set by returning the analyzer to sample mode and supplying a source of calibration gas containing a known concentration of pollutant corresponding to 80 percent of the analyzer's full-scale range. Allow the analyzer to sample the calibration gas in the normal operational mode until a stable response is attained. The calibration is completed by entering the range calibration mode and adjusting the analyzer response to display the concentration of the calibration gas. Following calibration, ensure that the analyzer is returned to the normal sampling mode, and connect the analyzer to the sampling manifold. The calibration gas must be an EPA traceable protocol standard. The pollutant gas is contained in a noninterfering, diluting, carrier gas; and the zero air must be made up of the same components as the calibration carrier gas. Manual or multipoint calibrations can also be done if required. Multipoint calibrations are made by challenging the analyzer with several different known concentrations, noting the response, and changing the offset and range of the analyzer to show linearity across the spectrum of concentrations introduced. To avoid the expense of having several bottles of gases of different concentrations, a gas divider can be used with a single high concentration which is diluted with zero gas.

The calibration procedure for ozone differs somewhat from the other automated methods and is found in Appendix D of 40 CFR, Part 50. Regulations require that the ozone analyzer be calibrated using either a primary ozone standard utilizing the uv photometry method of measurement or a certified transfer standard. See Fig. 4.6. (The photometric assay of ozone concentration in an absorption cell is determined from a measurement of the amount of 254-nm light absorbed by the sample.) A multipoint calibration should be performed at least quarterly during the ozone season and following any major repair or maintenance work on the monitor. Additionally, daily zero and span checks are performed to track the performance of the

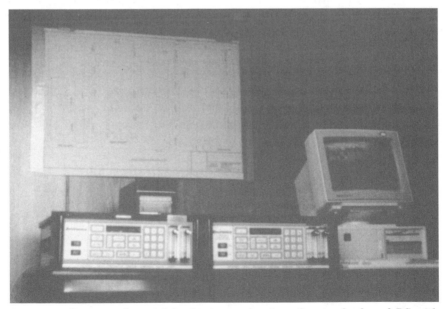

Figure 4.6 Ozone analyzer, strip chart recorder, transfer standard, and PC with modem for remote communication.

monitor. When the analyzer checks deviate significantly from the previous results, a calibration of the analyzer is in order. Most commonly, a primary standard is created from which transfer standards are developed, which are then in turn used to calibrate the site analyzers. The transfer standard is usually physically similar to the monitoring analyzer, but is capable of generating ozone and must be used exclusively for the purpose of auditing or calibrating the monitors in the network. The transfer standard should be calibrated against the primary standard at least once each monitoring season.

4.5.2 Zero and span checks

Zero and span checks are performed to track the "drift" or change of the response of automated analyzers over time. Zero checks are made by supplying the analyzer with a source of air scrubbed of the pollutant of interest and then recording the response of the analyzer. Span checks are performed in a similar way by supplying the monitor with a source of air containing a known concentration of the pollutant. The span concentration should be approximately 80 percent of the instrument's operational range. The results of the zero and span checks are compared to previous test results to ensure that the drift is not too great. One method commonly used to determine if the result of a zero or span check is outside an acceptable range uses control charts. Monitors are available on the market that are capable of performing remote zero and span checks via a computer modem hookup, and this allows the analyzer to be placed far from the operator. Control charts provide a tool for distinguishing whether an observation is due to random variation or is the result of some assignable cause. If the zero and span checks exceed the control limits, the data generated prior to the last acceptable zero and span check should be invalidated, and the cause of the problem found and corrected.

4.5.3 Single automated analyzer precision

A precision check is to be conducted at least once every 2 weeks by challenging the analyzer with a known concentration of pollutant between 0.008 and 0.10 ppm. Carbon monoxide is an exception and is evaluated between 8.0 and 10.0 ppm. At the end of each quarter, the precision probability interval for each analyzer must be reported. The precision probability interval for the 95th percentile is calculated as follows:

$$\text{Upper 95\% probability limit} = d_j + 1.96S_j \qquad (4.1)$$

$$\text{Lower 95\% probability limit} = d_j - 1.96S_j \qquad (4.2)$$

where d_j = quarterly average
   S_j = standard deviation
   j = quarter of year, 1 to 4

The quarterly average d_j and the standard deviation S_j are calculated as follows:

$$\overline{d}_j = \frac{1}{n} \sum_{i=1}^{n} d_i \qquad (4.3)$$

$$S_j = \sqrt{\frac{1}{n-1} \left[\sum_{i=1}^{n} d_i^2 - \frac{1}{n} \left(\sum_{i=1}^{n} d_i \right)^2 \right]} \qquad (4.4)$$

The percentage difference d_i for each precision check is

$$d_i = \frac{Y_i - X_i}{X_i} \times 100 \qquad (4.5)$$

where Y_i = analyzer's indicated concentration from the ith precision check and X_i = known concentration of the test gas used for the ith precision check.

  Precision checks and zero and span checks *must* be conducted prior to making any adjustments to the analyzer calibration settings. Records must be kept of all the precision and zero and span checks and analyzer calibrations and must be submitted with the final monitoring report. Automated method precision—zero and span check and two-point calibration forms are shown in Figs. 4.7 and 4.8.

4.5.4   Independent auditing

Once a quarter during the monitoring season, the regulating agency performs a performance audit of the monitor. The audit is conducted by challenging the monitor with four known concentrations of pollutant from the regulatory agency's transfer standard. The monitor in question passes the audit if it reads within 15 percent of the standard at all four points. In the event the tested monitor fails, then the data gathered previous to the last successful audit are invalidated. In addition, agencies operating state or local networks must participate in the EPA's National Performance Audit Program (NPAP). Under NPAP, the EPA sends each company a cylinder containing a concentration of pollutant known to the EPA. The agency must then analyze the sample and report the response of its monitor.

4.5.5   Single automated analyzer accuracy

At the end of each quarter, the percentage difference for each audit concentration must be reported. The calculation is made using Eq. (4.5) for the percentage difference for each precision check, except that

```
                     ONE-POINT PRECISION CHECK
         AND LEVEL ONE ZERO/SPAN CHECK WORKSHEET
                       FOR AUTOMATED METHODS


  Name of Client and Site: _____

  Name of Analyst: _____

  Date: _____

  Time: Beginning: _____  Ending: _____



                     ONE-POINT PRECISION CHECK

  Standard Concentration       Concentration Generated     Analyzer Response

  (0.008< CONC. <0.100 ppm)
  (CO 8< CONC. < 10 ppm)


  _____             _____            _____



               Percentage Difference: _____



                     LEVEL 1 ZERO/SPAN CHECK

  Standard Concentration       Concentration Generated     Analyzer Response

  Zero Air                     _____            _____

  Span - 80% of Scale

  (_____)            _____            _____


       Zero Check Percentage Difference: _____


       Span Check Percentage Difference: _____


  Signature of Analyst: _____
```

Figure 4.7 Automated method precision—zero and span check worksheet.

d_i = percentage difference for each audit concentration

Y_i = analyzer's indicated concentration from ith audit check

X_i = known concentration of audit gas used for ith audit check

```
                        AUTOMATED METHODS CALIBRATION WORKSHEET
     Name of Client and Site: _____

     Name of Analyst: _____

     Date: _____

     Time: Beginning: _____    Ending: _____

     Date of Last Calibration: _____

     Last Calibration Settings:    Instrument Zero: _____


                                   Instrument Gain: _____

                              TWO POINT CALIBRATION

     Standard Concentration    Concentration Generated      Analyzer Response

     Zero Air                      _____          _____


                       Analyzer Zero Set to: _____


     80% of Range ( _____ )    _____          _____


                       Analyzer Span Set to: _____

     2 - Pt Calibration Settings:      Instrument Zero: _____

                                       Instrument Gain: _____

              VERIFICATION OF TWO POINT CALIBRATION USING MULTIPOINT CHECK

     Standard Concentration     Concentration Generated      Analyzer Response

     Zero Air                      _____          _____

     10% of Range _____        _____          _____

     20% of Range _____        _____          _____

     40% of Range _____        _____          _____

     60% of Range _____        _____          _____

     80% of Range _____        _____          _____

     90% of Range _____        _____          _____



     FINAL CALIBRATION SETTINGS:

             Instrument Zero: _____

             Instrument Gain: _____

             Signature of Analyst : _____
```

Figure 4.8   Automated methods calibration worksheet.

4.5.6 Single instrument precision for manual methods

For sampling networks incorporating manual methods, one sampling site must have collocated samplers. One monitor is designated as the reporting monitor, and one monitor is designated as the duplicate sampler. The site selected for the duplicate monitors must be the site projected to have the highest 24-hour pollutant concentration. The two monitors must be within 4 m of each other, but at least 2 m apart to avoid airflow disruption. Calibration and sampling must be the same for both monitors. The difference in measured concentration between the two samplers is used to calculate the precision at the end of each quarter. The percentage difference for each pair of weekly concentrations is

$$d_i = \frac{Y_i - X_i}{(Y_i + X_i)/2} \times 100 \tag{4.6}$$

where Y_i = concentration of pollutant measured by duplicate monitor and X_i = concentration of pollutant measured by reporting monitor. The quarterly average percentage difference and standard deviation are calculated using Eqs. (4.5) and Eq. (4.6). Finally, the upper and lower 95 percent probability limits for precision are

$$\text{Upper 95\% probability limit} = d_j + \frac{1.96S_j}{\sqrt{2}} \tag{4.7}$$

$$\text{Lower 95\% probability limit} = d_j - \frac{1.96S_j}{\sqrt{2}} \tag{4.8}$$

4.5.7 Single-instrument accuracy for manual TSP, PM$_{10}$ and lead methods

The manual methods include the reference methods for TSP, PM$_{10}$ and lead. The flow rate of each sampler must be audited at least once each quarter. The audit must be performed at the normal flow rate and must be conducted using a certified flow transfer standard. The flow transfer standard must not be the same one used to calibrate the flow of the sampler being audited. The difference between the transfer standard and the indicated flow of the sampling monitor is used to calculate the accuracy. The percentage difference d_i for each precision check is calculated by using Eq. (4.5), where Y_i = indicated flow rate of the sampling monitor and X_i = known flow rate of the audit transfer standard. For the lead reference method, the lead analysis is audited by using filter strips containing a known quantity of lead. The difference between the audit concentration and the measured concentration is used to calculate the analysis accuracy. The percentage difference d_i for each precision check is calculated from Eq. (4.5), where Y_i = indicated value of lead and X_i = known value of the audit sample.

4.6   Control Charts

Control charts are used to indicate when observed variations in normally distributed testing results are greater than expected variations due to chance, signaling the need for corrective action. To distinguish between chance causes of variation which occur in a stable pattern and assignable causes which have large variations that lie outside the stable pattern, control limits are established based on statistics.

Control charts are generally constructed based on subgroups of several observations made at the same time. Often, however, only one observation, such as span and zero checks, is made per day. Happily, it is still possible to construct charts based on the results of only one observation per sample period. This is accomplished through the charting of the moving average and range of two successive pairs of observations. The control limits are generally established as the mean, or range ± 3 standard deviations of the stable pattern. Initial control limits can be established based on manufacturers' recommendations and then refined after 15 to 20 test values have been collected. Every 3 to 6 months, the control limits should be recalculated.

4.6.1   Steps in constructing a control chart for single observations

1. Determine which data are to be tracked. For a continuous monitor, daily zero and span checks would be of importance.

2. Decide what statistic to plot. For single observations of daily zero and span checks and biweekly precision checks, the appropriate statistics would be the moving average and range.

3. Evaluate the form of the distribution of the observations to evaluate whether the data fit a normal distribution. If the distribution is not normal, either the data must be modified to fit or an alternative method of statistical process control must be found.

4. Examine the data from the preceding monitoring quarter, and eliminate any outliers and out-of-control points.

5. Determine the control limits (± 3 deviations). For control charts based on moving average and range of two consecutive measurements:

$$\text{Upper control limit for moving range (UCL}_R) = D_4\overline{R} \qquad (4.9)$$

$$\text{Lower control limit for moving range (LCL}_R) = D_3\overline{R} \qquad (4.10)$$

$$\text{Upper control limit for moving average (UCL}_{\overline{\overline{x}}}) = \overline{\overline{X}} + A_2\overline{R} \quad (4.11)$$

$$\text{Lower control limit for moving average (LCL}_{\overline{\overline{x}}}) = \overline{\overline{X}} - A_2\overline{R} \quad (4.12)$$

where D_4, D_3, A_2 = control chart factors for normal distributions. If desired, warning control limits can also be set for ± 2 deviations:

$$\text{Upper warning limit for moving range (UWL}_R) = D_5\overline{R} \qquad (4.13)$$

$$\text{Lower warning limit for moving range (LWL}_R) = D_6\overline{R} \qquad (4.14)$$

$$\text{Upper warning limit for moving average (UWL}_{\overline{x}}) = \overline{\overline{X}} + \frac{2}{3} A_2\overline{R}$$
$$(4.15)$$

$$\text{Lower control limit for moving average (LCL}_{\overline{x}}) = \overline{\overline{X}} - \frac{2}{3} A_2\overline{R}$$
$$(4.16)$$

Factors for computing control chart lines can be found in quality control and statistics texts or in Appendix H of the EPA's *Quality Assurance Handbook for Air Pollution Measurement Systems,* vol. 1. The values corresponding to two averaged observations used in moving-average and range control charts are

$$D_4 = 3.267 \qquad D_3 = 0 \qquad A_2 = 1.880$$
$$D_5 = \frac{1 + 2D_4}{3} = 2.51 \qquad D_6 = \frac{5 - 2D_4}{3} = 0$$

6. Draw control chart indicating central line, and warning and control limits.

7. Post the result of each consecutive pair of observations, and join the points with a straight line.

8. Study the chart for any visible trend. Highlight any out-of-control points.

9. Take corrective action needed to restore control.

10. Revise control limits periodically. Once per quarter is good practice.

11. Maintain a historical file of all control charts. These may be of help in spotting a recurring problem or of interest for instrument

diagnostic work. An example of a control chart and the supporting calculations for tracking the moving average and range are given in Fig. 4.9 and Table 4.4.

4.6.2 Interpretation of control charts

The criteria to be used in judging whether or not a process is in control include the following:

1. Point falls outside control limits. This indicates that the point has only a 1 percent probability of being within the normal distribution, and action should be taken to correct any problem. The data gathered after the last point in control are highly suspect and should be invalidated.

2. Two points fall outside the warning limits. While these points may not require that the preceding data be invalidated, they strongly suggest that a problem may lie ahead.

3. There are runs or cycles of points. Runs of points up or down or recurring patterns indicate a systematic effect which should be investigated.

4.7 Meteorological Monitoring

The transport and dispersion of primary pollutants and photochemical pollution formation are all dependent on climatological influences. Because the effect of meteorology on ambient air quality is complex, meteorological monitoring requirements will vary with the type of monitor and network in question. In many cases, a review of available historical data will be sufficient to make an analysis of conditions affecting ambient air pollutant concentrations and movements. This analysis could be utilized to predict areas of high concentration and the speed and direction of air movement in the design of monitoring networks or to infer the source of a pollutant. When historical data are not available or when more up-to-date and accurate information is needed, such as to validate model calibrations, then a project may require meteorological measurements within the study area. Automated analyzers often contain provisions for the selection of self-correcting temperature and pressure subroutines. Often, however, wind speed and direction will be of interest, or an independent means will be required to confirm the analyzer assessment of pressure and temperature.

Weather monitoring instrumentation should obviously be collocated with the pollutant analyzers. Meteorological stations should also be sited near sources of pollutant and scattered throughout the study

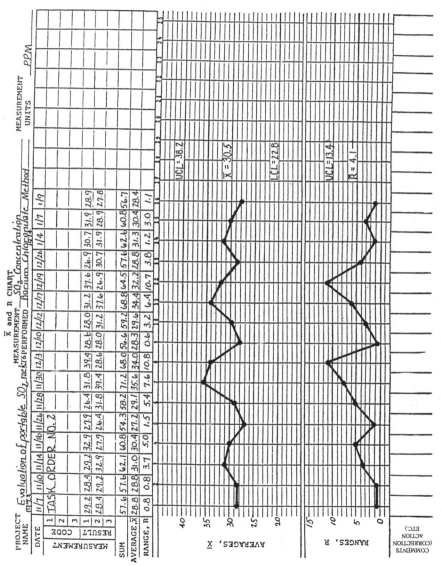

Figure 4.9 Control chart for moving average and range.

TABLE 4.4   Moving Average and Range Table

Sample no.	Value	Moving average of 2	Range
1	29.2	—	—
2	28.4	28.80	0.8
3	29.2	28.80	0.8
4	32.9	31.05	3.7
5	27.9	30.40	5.0
6	26.4	27.15	1.5
7	31.8	29.10	5.4
8	39.4	35.60	7.6
9	28.6	34.00	10.8
10	28.0	28.30	0.6
11	31.2	29.60	3.2
12	37.6	34.40	6.4
13	26.9	32.25	10.7
14	30.7	28.80	3.8
15	31.9	31.30	1.2
16	28.9	30.40	3.0
17	27.8	28.35	1.1
	Totals:	488.30	65.6
		$\overline{\overline{X}} = 30.52$	$\overline{R} = 4.1$

$$\text{UCL}_R = D_4\overline{R} = 3.27 \times 4.1 = 13.4$$
$$\text{LCL}_R = D_3\overline{R} = 0 \times 4.1 = 0$$
$$\text{UCL}_{\overline{X}} = \overline{\overline{X}} + A_2\overline{R} = 30.5 + 1.88 \times 4.1 = 38.2$$
$$\text{LCL}_{\overline{X}} = \overline{\overline{X}} - A_2\overline{R} = 30.5 - 1.88 \times 4.1 = 22.8$$

Warning limits could be computed in a similar manner.

area if dispersion modeling is to be conducted. At a minimum, the parameters being measured include wind speed and direction, average atmospheric stability, surface temperature, and precipitation amounts. An estimation of the average mixing height extrapolated from National Weather Service radiosonde measurements may also be required. Variations in topography must be considered in determining the number and placement of meteorological stations. The Meteorology and Assessment Division of the EPA has published recommendations for meteorological station placement in *Guidelines for Siting and Exposure of Meteorological Instruments for Environmental Purposes.* Generally, at least 1 year of meteorological data is required, and multiple years of data are preferred.

The equipment used for meteorological data gathering must be calibrated and maintained according to a quality assurance program resembling that used for the pollutant-monitoring network. *Quality*

Assurance Handbook for Air Pollution Measurement Systems, vol. 4: *Meteorological Measurements* should be consulted when one is developing the meteorological monitoring network and the related quality plan.

4.8 Conclusions

Whether performed as part of a state's implementation plan, a research study, or a requirement of a PSD review, ambient air monitoring follows a similar path for each application. The monitoring objective must be determined, appropriate monitoring sites found, EPA-approved equipment selected, and a monitoring and quality assurance plan established, prior to the gathering of monitoring data. The data must be gathered according to the monitoring plan, and the monitoring conducted with sufficient care to ensure that no questions arise about the validity of the information obtained. The basic principles of ambient air monitoring are fairly simple, but great care must be exercised in the day-to-day operation of the monitoring network to see that an accumulation of small mistakes does not result in the generation of suspect data which are of little value. The best way to avoid trouble is by developing and following a comprehensive quality control program created with input from the regulatory agency responsible for reviewing the final result.

References

Cooper, C. David, and Alley, F. C., *Air Pollution Control: A Design Approach,* Waveland Press, Inc., Prospect Heights, Ill., 1986.

Ambient Monitoring Guidelines for Prevention of Significant Deterioration (PSD), Environmental Protection Agency, Office of Air Quality Planning and Standards, Research Triangle Park, N.C., publication no. EPA-450/4-87-007, May 1987.

CFR Title 40, Part 50, Appendix D.

CFR Title 40, Part 58, Appendices A, B, and C.

Environics Series 300 Computerized Ozone Analyzer Operating and Service Manual, Environics, West Wilmington, Conn., 1990.

Lodge, James P., Jr. (Ed.), *Methods of Air Sampling and Analysis,* 3d ed., Lewis Publishers, Chelsea, Mich., 1989.

Noll, Kenneth E., and Miller, Terry L., *Air Monitoring Survey Design,* Ann Arbor Science, Ann Arbor, Mich., 1977.

Quality Assurance Handbook for Air Pollution Measurement Systems, vol. 1: *Principles,* Environmental Protection Agency (MD-77), Research Triangle Park, N.C., EPA publication no. EPA-600/9-75-005, March 1976.

Quality Assurance Handbook for Air Pollution Measurement Systems, vol. 2: *Ambient Air Specific Methods,* Environmental Protection Agency (MD-77), Research Triangle Park, N.C., EPA publication no. EPA-66/4-77-027a, May 1977.

Rossano, August T., Jr. (Ed.), *Air Pollution Control, Guidebook for Management.* McGraw-Hill, New York, 1974.

Stern, Arthur C. (Ed.), *Air Pollution,* vol. 3: *Measuring, Monitoring, and Surveillance of Air Pollution,* 3d ed., Academic Press, New York, 1976.

5

Stack Sampling and Monitoring

Lem B. Stevens III

5.1 Goal of Stack Testing

Stack testing is used to determine the actual emissions of a pollutant from a source. It is the most precise method of characterizing emissions available. Stack testing is often performed to show compliance with permit conditions at the request of regulatory agencies. Another benefit of stack testing is that the information from the testing can be used to fine-tune some processes. For example, the concentrations of oxygen, carbon monoxide, and carbon dioxide in stack gas can be used to determine the excess air and combustion efficiency from a boiler. This information is commonly used to optimize combustion. Stack testing is also relatively expensive. In many cases, other options for characterizing emissions, such as engineering calculations, are employed instead of stack testing to avoid the expense.

5.2 General Parts of a Stack Test

A stack test can be divided into three main parts: determination of stack gas flow rate, determination of constituent concentration, and calculation of mass emission rate. Each of these parts is explained below.

5.2.1 Flow rate

The flow rate of stack gas is calculated from the continuity equation $Q = VA$, where Q is the flow rate, V is the velocity of the stack gas, and A is the cross-sectional area of the stack. The velocity of the stack gas is calculated from the average velocity pressure using Bernoulli's equation in the form $VP = \rho V^2/(2g)$, where VP is the velocity pressure, ρ is the stack gas density, and V is the velocity. The velocity pressure is measured with a Pitot tube at several locations in the cross section of the stack. These measurements are used to determine the average velocity pressure. The EPA has developed standard methods for stack testing. The methods discussed in this chapter are found in Appendix A of chapter 40, subpart 60, of the Code of Federal Regulations (CFR). EPA method 1 is used to determine the location of sampling sites for stack testing. Method 2 is used to determine the velocity pressures in the stack. The stack gas density is calculated from the volumetric percentage of oxygen, carbon monoxide, carbon dioxide, and nitrogen present in the stack gas. These constituents make up the vast majority of the volume of most stack gases, and they can therefore be used to determine a weighted-average molecular weight. The molecular weight of the stack gas is converted to density by using the perfect gas relationship that 1 mole of any gas occupies the same volume at standard conditions. EPA method 3 governs the determination of the stack gas molecular weight. The velocity pressure and density are then substituted into the above equation to calculate the velocity of the stack gas. The velocity and cross-sectional area are then substituted into the continuity equation to obtain the actual volumetric flow rate.

The actual volumetric flow rate is reduced to the dry standard volumetric flow rate by subtracting the volume fraction of water vapor in the stack gas and correcting the result to standard temperature and pressure with the perfect gas law. The results of stack tests are usually reported as dry standard conditions. The fraction of water vapor in the stack gas is determined by drawing a known volume of stack gas through a condenser and dessicant to remove all water vapor. The weight of water vapor collected is then transformed to a volume, using the perfect gas law, and compared to the volume of stack gas sampled at the corresponding temperature and pressure to determine the volumetric fraction of water vapor in the stack gas. This procedure is governed by EPA method 4.

5.2.2   Concentration of constituent

The concentration of the constituent can be determined by three main methods: the analyzer method, the particulate method, or the wet method. The analyzer method is relatively straightforward in terms of explanation. A sample of the stack gas is drawn through the analyzer, and the analyzer produces a volumetric concentration of the constituent. These results are averaged over the proper time period to obtain a reportable result. Most analyzer methods reference EPA method 6C for the details of how the test must be performed. Examples of constituents whose concentration can be determined using an analyzer are nitrogen oxides, sulfur dioxide, carbon monoxide, and total hydrocarbons.

The particulate method is used to sample all kinds of particulates, which are filtered out of the sample gas stream on a filter. The type of filter used is determined by the type of particulate being sampled. Particulate sampling is fairly involved because it must be performed isokinetically. Isokinetic sampling is explained later in this chapter. Particulate sampling is described in EPA method 5.

Wet methods are characterized by scrubbing constituent gases out of a known volume of stack gas. The mass of the constituent collected is then measured or determined by chemical analysis. This mass divided by the volume of stack gas sampled is the mass concentration of the constituent in the stack gas. Constituent gases are scrubbed out of the sample gas using impingers filled with solutions known to scrub the constituent gas. Wet methods have been written for most gaseous constituents.

5.2.3   Mass emission rate

Most standards are written with limits for mass emission rates. The mass emission rate of a constituent is simply calculated by multiplying the flow rate by the constituent concentration. Care must be taken to preserve the correct units and to use dry standard conditions for both the flow rate and the constituent concentration.

5.2.4   Isokinetic sampling

Any kind of particulate sampling, whether it be solids or mist, must be sampled isokinetically. Simply put, isokinetic sampling means that the velocity of stack gas drawn into the sampling nozzle is equal to the velocity of the stack gas which flows past the nozzle.

Isokinetic sampling is important for particulates because the velocity at which gas is sampled affects the result of the testing. If the sampling is performed superisokinetically, then the concentration of particulate will be an understatement of the true concentration of particulate in the stack gas. If the sampling is performed subisokinetically, then the concentration of particulate will be an overstatement of the true concentration.

The physical reason why isokinetic sampling is important in particulate sampling lies in the inertia of the particles as they move through the stack and the effect of the sampling on the local flow patterns in the immediate area of the nozzle. If sampling is performed isokinetically, then the particle feels no change in forces as it moves through the stack and into the nozzle. If sampling is performed superisokinetically, then a column of gas with a diameter larger than the diameter of the nozzle is drawn into the nozzle. Not all particles in this column will be drawn into the nozzle. Particles in this column whose inertial path is directed past the nozzle will force their way past the nozzle. The resulting mass of particles collected in the sample gas will be lower than the true mass concentration of particles in the stack gas. This concept is most easily seen in Fig. 5.1.

5.2.5   Number of runs

Generally, three independent tests are required to determine an acceptable average for a stack test result. Three runs are necessary because the accuracy of a single run is suspect. Realistically, the result of any single run is within ±10 percent of the true emission rate. If three runs with such an accuracy are averaged, then the confidence that the average is close to the true emission rate is much greater than that for a single run. Three results within 5 percent of the average of the results are a strong indication of a good test. The best method of determining the quality of the test results is to perform a Student t test on data. The Student t test is a statistical analysis of small data sets for the level of confidence for which the data set represents the true average. Most regulators accept data with a 95 percent confidence interval of the true average as calculated by the Student t test.

5.3   Overseeing a Stack Test

In many cases when a stack test is required, a plant engineer will contract with a testing firm to perform the test. It is very important that the plant engineer actively oversee the test to ensure that the results are accurate and acceptable for the intended purpose of the test. Without an active role played by the plant engineer, it is unlikely that exactly the requested test will be performed. There are a multitude of mistakes that stack testers unfamiliar with the plant can make. For example, it is possible, and has probably occurred, that a test for stack A has been requested and the test was actually performed on stack B. The following section contains guidance to help avoid this type of error.

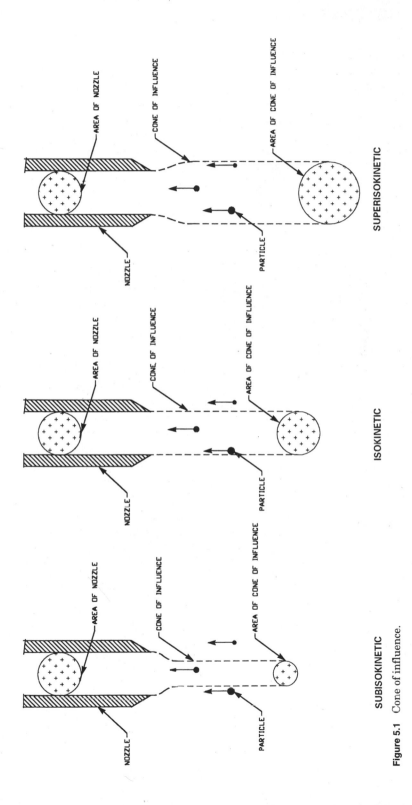

Figure 5.1 Cone of influence.

5.3.1   Preparing a site for a test

Unless prior agreement is made, it is the responsibility of the plant engineer to prepare the stack for the test. Preparation for the test includes providing safe access to the stack and testing ports that meet the method required for the test. EPA methods 1 and 2 provide the requirements for the location of test ports. Simply put, these methods require 10 diameters' worth of straight stack for a stack test. The ports should be installed eight diameters downstream of a flow disturbance and at least two diameters upstream of the stack exit. For example, a 10-ft-diameter stack requires 80 ft of clean straight stack before the test ports and 20 ft of stack above the ports before the stack exit. From this example, it is clear that stack testing is often performed at elevated locations. The equipment necessary to test a stack is quite cumbersome and heavy. The chances for a successful test are greatly increased by safe access to the sampling ports and a sturdy platform from which to conduct the test. Unsafe conditions only complicate an already complex procedure. Please take safety considerations seriously!

5.3.2   Preparing a process for a test

Before a test is performed, the process should be running smoothly and consistently. If process upsets occur during the test, then that part of the test may be nullified or emissions overstated. The plant engineer should make sure that the process will be running on the day of the test and that the process can be run for the required time to obtain the necessary test runs. It requires about 5 h to perform three good 1-h runs. Furthermore, the process should be run as consistently as possible. It is much more likely that accurate results will be obtained if the process is consistent. Inconsistency simply scatters the data points and calls the test results into question. The process operators should be made aware of the testing and the testing schedule so that they can work to ensure consistent process operation during the test. In many cases, permit conditions will specify the process operating conditions for testing. Therefore, it is important to review the permit to make sure any conditions of this type are met for the test. An example of this type of condition is a minimum operating temperature for a thermal oxidizer. If the permit specifies a minimum operating temperature of 1400°F, then do not attempt to run a test with a thermal oxidizer operating at 1350°F.

5.3.3   Reviewing a stack test report

The plant engineer should review the stack test report and accept it before sending it to the regulatory authorities. Some regulatory agencies require that the official report be sent by the company and do not accept a report directly from a stack testing firm, so it is recommended that the plant engineer send in the official report. A good report includes the following main items:

Field data sheets.   Determination of the emissions requires field data sheets such as shown in Fig. 5.2 for the flow rate and conditions, stack gas molecular weight, moisture content, and constituent concentration. The information from these sheets is used in the calculations to obtain the results.

Sample calculations.   A complete sample calculation which details the results of one of the three test runs should be included (see Table 5.1).

Particulate Field Data

Very Important — Fill in all Blanks

Plant _Manufacturing Inc._
Run no. _1_
Location _EAF Baghouse_
Date _3/20/97_
Operator _L. Stevens, T. Wilkerson_
Sample box no. _01_
Meter box no. _01_
Nomograph ID no. _NA_
Orsat no. ____ Date rebuilt ____
Fyrite no. ____ Date rebuilt ____

C_p _0.84_
$\Delta H_@$ _2.34_
P_m, in. Hg _28.60_
P_s, in. Hg _28.65_
B_{ws} (assumed) _0.02_
M_d _29.0_
M_s _28.78_
T_m, °R _555_
T_s, °R _560_
ΔP_{avg}, in. H_2O _0.326_

Test start time _10:17_
Stop time _11:17_
D_n calculated (in.) _0.26_
D_n used (in.) _0.25_
Ambient temp., °F _60_
Bar. pressure, in. Hg _28.6_
Heater box setting, °F _250_
Probe heater setting, °F _250_
Average ΔH _1.47_
Leak rate @15 in. Hg Pre-test _0_ Post-test _0_

Point	Clock time (min)	Dry gas meter CF	Pitot in H_2O Δp	Orifice ΔH in H_2O Desired	Orifice ΔH in H_2O Actual	Dry gas temp. °F Inlet	Dry gas temp. °F Outlet	Pump vacuum in. Hg gauge	Box temp. °F	Impinger temp. °F	Stack press. in. Hg	Stack temp. °F	Fyrite %CO_2
1	0	725.681	0.35		1.63	77	77	3.0	180	67	28.6	128	
2	10	731.7	0.41		1.85	80	77	3.0	248	59	28.6	146	
3	20	737.8	0.44		2.06	84	79	3.5	254	61	28.6	155	
4	30	743.9	0.35		1.57	86	80	3.0	252	61	28.6	144	
5	40	750.0	0.26		1.16	87	81	2.5	251	61	28.6	143	
6	50	756.1	0.23		1.05	88	82	2.5	251	61	28.6	138	
7	60	762.2	0.21		0.96	89	83	2.0	251	61	28.6	139	
8	70	768.3	0.21		0.96	89	84	2.0	250	63	28.6	139	
9	80	774.4	0.26		1.19	90	85	2.5	249	65	28.6	142	
10	90	780.4	0.35		1.56	91	86	3.0	249	66	28.6	154	
11	100	786.5	0.41		1.81	93	87	3.0	250	66	28.6	160	
12	110	792.6	0.41		1.85	94	89	3.0	250	66	28.6	160	
	120	798.686											

Comments: _____

Test observers: _____

Figure 5.2   Field data sheet.

TABLE 5.1 Manufacturing Inc. Method 5 Test

Point	Clock time, min	Dry gas meter, CF	Pitot (inH$_2$O) dP	Orifice dH, inH$_2$O Desired	Orifice dH, inH$_2$O Actual	Dry gas temperature, °F Inlet	Dry gas temperature, °F Outlet	Pump vacuum, inHg gauge	Box temp., °F	Impinger temp., °F	Stack pressure, inHg	Stack temp., °F	Square root dP
1	0	725.681	0.35		1.63	77.0	77.0	3.0	180.0	67.0	28.60	128.0	0.592
2	10		0.41		1.85	80.0	77.0	3.0	248.0	59.0	28.60	146.0	0.640
3	20		0.46		2.06	84.0	79.0	3.5	252.0	61.0	28.60	155.0	0.678
4	30		0.35		1.57	86.0	80.0	3.0	251.0	61.0	28.60	146.0	0.592
5	40		0.26		1.16	87.0	81.0	2.5	251.0	61.0	28.60	143.0	0.510
6	50		0.23		1.05	88.0	82.0	2.5	250.0	61.0	28.60	138.0	0.480
7	60		0.21		0.96	89.0	83.0	2.0	249.0	61.0	28.60	139.0	0.458
8	70		0.21		0.96	89.0	84.0	2.0	249.0	63.0	28.60	139.0	0.458
9	80		0.26		1.19	90.0	85.0	2.5	249.0	65.0	28.60	142.0	0.510
10	90		0.35		1.56	91.0	86.0	3.0	250.0	66.0	28.60	154.0	0.592
11	100		0.41		1.81	93.0	87.0	3.0	250.0	66.0	28.60	160.0	0.640
12	120	798.686	0.41		1.85	94.0	89.0	3.0	248.0	66.0	28.60	160.0	0.640
13													
14													
15													
16													
17													
18													
19													
20													
21													
22													
23													
24													
25													
26													
27													
28													
29													
30													
31													
Average	120	73.01	0.33		1.47	87.33	82.50	2.75	244.08	63.08	28.60	145.83	0.57
Total													

	Before	After	Difference
Filter	0.3329	0.3500	0.0171 g
Impinger 1	482.30	597.60	115.30 g
Impinger 2	618.50	653.70	35.20 g
Impinger 3	616.80	621.00	4.20 g
Impinger 4	695.30	695.20	−0.10 g

Extra particulate	0.002 g	C_p	0.84
Stack diameter	3.25 ft	$dH_@$	2.34
Stack area	8.30 ft^2	P_m	28.60 inHg
Total particulate weighed	0.0191 g	P_s	28.65 inHg
Total water collected	154.60 g	M_s	29.00
Nozzle area	0.00034 ft^2	T_m	545.0°R
Average meter temp.	544.9°R	T_s	605.8°R
Gas meter correction	0.998	D_n (calc.)	0.26 in
Standard volume	67.74 ft^3	D_n (used)	0.25 in
Water collected	7.2894 ft^3	Ambient temp.	40.0°F
Gas moisture content	0.0972 fraction	Bar pressure	28.60 inHg
Average gas velocity	34.95 ft/s	Heater box setting	250.0°F
Average gas volume rate	786802 dscf/h	Probe heat setting	250.0°F
Pollutant mass rate	0.49 lb/h (0.00436 gr/dscf)	Average dH	1.47
Isokinetic variation	104.81%		

The reviewer should check this calculation against the results reported for that run to ensure equivalence.

Process information.   Process variables which quantify the operation during the test should be documented. For example, a strip chart of the chamber temperature is good documentation of the operating temperature of a thermal oxidizer and is useful information to include in a test report. Other examples are input rates and production rates.

Tester's information.   The testing firm, telephone number, and the individual testers should be identified in the test report so that any questions that arise from the report can be directed to them.

Purpose.   The purpose of the test should be clearly stated in the report. As time goes by and employees change companies, it is difficult to remember exactly why some tests were performed. The purpose of the test can be useful information in determining whether the data from the test are applicable to future situations.

Test methods.   The methods used in the test should be stated and any deviations from the methods explained fully. If at all possible, deviations from the test method should be discussed with regulators before the test to obtain their acceptance of the deviation. In any event, it is suggested that deviations from the method be discussed with the regulatory agency before the report is submitted. In this way, answers to any questions the regulators may have about the method used for testing can be addressed in the first submission of the report, and revisions will not be necessary.

5.4   Performing a Stack Test

The successful completion of a stack test involves many tasks and the correct use of many pieces of equipment. It is strongly suggested that stack testers develop a checklist which covers the equipment necessary for a test. Such a checklist can be used to ensure that no equipment is left behind at the home office when the testing crew leaves for the field.

5.4.1   Pretest calibration of equipment

Calibration of equipment is required before a compliance test is performed. Failure to perform the required calibrations can compromise test results and may invalidate the test. In general, the following items must be calibrated:

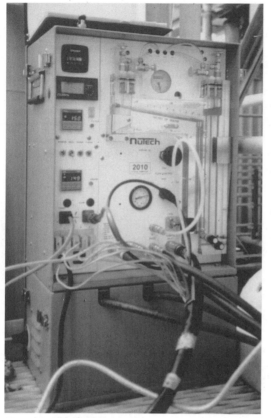

Figure 5.3 Method 5 console.

Method 5 Console (Fig. 5.3)
- Dry gas meter
- Thermocouples
- Differential-pressure sensors

Probe
- Nozzle
- Pitot tube
- Thermocouple

There are several accepted methods of calibrating the dry gas meter. The most common methods are to compare the console meter to a wet test meter, another certified dry gas meter, or critical orifices. Each of these methods of calibration is explained in method 5. It is important to note that the device used to calibrate the console meter must also be certified on a periodic basis. Usually, certifications are good for 1 year.

The thermocouples are calibrated using three known temperatures. Usually, an ice water bath, boiling water, and hot oil are used to create the calibrating temperatures. A laboratory-grade thermometer can be used to determine the actual temperature of these baths.

Differential pressure sensors can be a manometer, a magnehelic gage, or a pressure transducer. Magnehelic gages and transducers must be calibrated against a manometer. If your console has manometers to sense differential pressure, then no calibration of the differential pressure sensors is required.

Nozzles are calibrated by measuring three different inside diameters of the nozzle and averaging the individual results. Calipers must be used for these measurements. For a nozzle to be acceptable, no individual diameter can be more than 2 percent different from any of the other diameters. The nozzle should have some form of identification, and its average diameter should be recorded. If a nozzle fails to meet these criteria, then a plumb bob can sometimes be used to restore the nozzle to a round shape. Press the sharp, round end of the plumb bob into the nozzle to restore its roundness.

Pitot tubes are calibrated by measuring important dimensions of the tip and comparing these dimensions to the values in method 2. Calipers are recommended for these measurements. Pitot tubes can be expected to stay in calibration as long as care is taken to avoid damaging them.

5.4.2   Preparation of the impingers

It is helpful to prepare a set of impingers for each run before going to the field. This involves labeling the impingers, introducing the appropriate volume and type of solution to the impingers, and weighing the impingers. Many stack sampling equipment suppliers sell an impinger case which protects the impingers and keeps them in an upright position. The author uses these cases and considers them essential equipment for efficient stack testing.

5.4.3   Choosing measurement sites

When choosing a site to perform measurements for a test, the tester should find the longest straight, uninterrupted run of duct or stack available and measure the inside diameter of that run. If possible, the measurement sites should have a straight length of flow equivalent to eight diameters before the test site and a straight length of flow equivalent to two diameters after the test site. If these criteria can be met, then the fewest traverse points can be used during the test. The shortest length of straight stack allowed for testing is an equivalent length of $2\frac{1}{2}$ diameters. In this case, choose a site with a straight length of flow equivalent to 2 diameters before the test site and a straight length of flow equivalent to $\frac{1}{2}$ diameter of stack after the test site. This case requires the largest number of traverse points. When the optimal test location is determined, install test ports. The test ports must be located at 90° from each other on the stack. One of the test ports must be on the diameter where the highest concentration of constituent is expected. For example, if a 90° bend exists before the straight run of stack upstream of the sample ports, then one port must be located on a diameter parallel to the duct before the bend.

5.4.4   Determination of traverse points

There are two figures from method 1 which are used to determine the number of traverse points. Figure 5.4 is used for particulate tests, and Fig. 5.5 is used for nonparticulate tests. The figures are used by drawing an imaginary line to the center of the figure from the upstream number of diameters at the bottom of the figure and from the downstream number of diameters at the top of the figure. Write down the number of traverse points that result from each imaginary line, and choose the higher number of traverse points for the test.

The location of the traverse points is determined by using Table 5.2, reproduced from method 1. The number of traverse points on a diameter is equal to $\frac{1}{2}$ the total number of traverse points required. Use

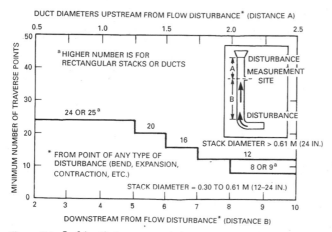

Figure 5.4 Isokinetic traverse points.

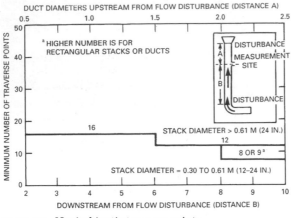

Figure 5.5 Nonisokinetic traverse points.

this value to locate the column in the table corresponding to the test at hand. The numbers in the column represent the percentage of diameter for each traverse point. For example, if the number of traverse points required for a test is 12, then the number of traverse points on a diameter is 6. The first percentage under the column for 6 points is 4.4 percent. Multiply the stack inside diameter by 0.044. Measure this distance from the stack wall toward the center along a diameter containing a sample port to obtain the location of the first traverse point. This procedure is repeated for the other five traverse points on that diameter. The same locations are used for the traverse points on the diameter containing the second sample port.

5.4.5 Preliminary stack gas data

The stack gas velocity is determined by using a Pitot tube. Individual differential pressure readings are taken at each of the traverse points. The same type of Pitot tube should be used for both the preliminary test and the actual test.

If possible, it is also best to determine the temperature at each traverse point. Many probes have both a Pitot tube and a thermocouple at the same relative location, making this task easy. In most cases, a single temperature reading taken about one-third of a diameter toward the center of the stack is sufficient for the preliminary temperature determination.

Stack gas dry molecular weight (M_d) must be estimated in some fashion. Knowledge of the process can be used to estimate the molecular weight. For example, if the tester knows that the stack gas is

TABLE 5.2 Location of Traverse Points in Circular Stacks

[Percentage of stack diameter from inside wall to traverse point]

Traverse point number on a diameter	Number of traverse points on a diameter											
	2	4	6	8	10	12	14	16	18	20	22	24
1	14.6	6.7	4.4	3.2	2.6	2.1	1.8	1.6	1.4	1.3	1.1	1.1
2	85.4	25.0	14.6	10.5	8.2	6.7	5.7	4.9	4.4	3.9	3.5	3.2
3		75.0	29.6	19.4	14.6	11.8	9.9	8.5	7.5	6.7	6.0	5.5
4		93.3	70.4	32.3	22.6	17.7	14.6	12.5	10.9	9.7	8.7	7.9
5			85.4	67.7	34.2	25.0	20.1	16.9	14.6	12.9	11.6	10.5
6			95.6	80.6	65.8	35.6	26.9	22.0	18.8	16.5	14.6	13.2
7				89.5	77.4	64.4	36.6	28.3	23.6	20.4	18.0	16.1
8				96.8	85.4	75.0	63.4	37.5	29.6	25.0	21.8	19.4
9					91.8	82.3	73.1	62.5	38.2	30.6	26.2	23.0
10					97.4	88.2	79.9	71.7	61.8	38.8	31.5	27.2
11						93.3	85.4	78.0	70.4	61.2	39.3	32.3
12						97.9	90.1	83.1	76.4	69.4	60.7	39.8
13							94.3	87.5	81.2	75.0	68.5	60.2
14							98.2	91.5	85.4	79.6	73.8	67.7
15								95.1	89.1	83.5	78.2	72.8
16								98.4	92.5	87.1	82.0	77.0
17									95.6	90.3	85.4	80.6
18									98.6	93.3	88.4	83.9
19										96.1	91.3	86.8
20										98.7	94.0	89.5

mostly air, then a preliminary determination that the dry molecular weight is 29.0 g/mol is acceptable. For processes which involve the combustion of fossil fuels, a dry molecular weight of 30.0 g/mol can be used according to method 3. If the tester wants to obtain a better estimate of the molecular weight, then an analysis of the stack gas must be performed. Typically an ORSAT or Fyrite analyzer is used to make this determination. Using these analyzers, the volume percentages of nitrogen, carbon dioxide, oxygen, and carbon monoxide are determined. From these percentages, the dry molecular weight can be calculated according to the following formula:

$$M_d = \Sigma M_x B_x = 0.44(\%CO_2) + 0.32(\%O_2) + 0.28(\%N_2 + \%CO) \quad (5.1)$$

The volume fraction of moisture in the stack gas (B_{ws}) must also be estimated. But B_{ws} is the most difficult value to estimate, because it has a wide range. The author has tested stacks with a B_{ws} of almost 0 percent and stacks with a B_{ws} of up to 20 percent. There are several methods of obtaining a good estimate of B_{ws}. If the stack temperature is less than 160°F, then a sling psychrometer can be used. A sling psychrometer determines the dry-bulb temperature of the stack gas with a standard thermometer and the wet-bulb temperature of the stack gas with a thermometer covered in a wet sock. The dry- and wet-bulb temperatures designate a specific point on the psychrometric chart. From the psychrometric chart, B_{ws} can be determined. This method cannot be used at higher temperatures because the water wetting the sock of the wet-bulb thermometer will boil and a false reading of 212°F will result until all the water has evaporated. Another useful instrument for estimating B_{ws} is a relative-humidity probe. With the temperature and relative humidity, one can determine the B_{ws} of the stack gas by using a psychrometric chart. Like the sling psychrometer, the relative-humidity probes usually have a temperature limitation. If the stack gas is too hot for these first two methods, then a simple moisture run in accordance with EPA method 4 can be performed. This method requires 30 min of run time, but it is the best method of estimating B_{ws} because it is the method used to determine B_{ws} during the actual testing.

The absolute pressure at the dry gas meter P_m can be assumed equal to the barometric pressure. The barometric pressure should be determined with a precise portable barometer which has been recently calibrated. The author cautions against the use of barometric pressure from a local weather station or airport, because these pressures are often corrected to sea level. The stack tester is interested in the barometric pressure at the elevation of the meter box. Corrections can be made to barometric pressures from a local station, but the elevation of that station and the elevation of the meter box must be known. A correction factor of 0.1 inHg is used for every 100 ft of elevation difference.

The absolute pressure of the stack P_s can be determined by using the manometer on the meter box. Simply disconnect the low-pressure line from the Pitot tube, and zero the manometer. Then insert the Pitot tube at a right angle to the stack gas flow, and read the manometer. Take care to set up the manometer to read negative pressure if the stack has an induced-draft fan. Now P_s is calculated according to the following equation:

$$P_s = P_b + \frac{\Delta P}{13.6} \tag{5.2}$$

If there is very little flow into or out of the sample port in the stack, then P_s can be assumed equal to the barometric pressure.

The temperature of the dry gas meter T_m can be estimated by experience. Generally, T_m will be 30 to 50°F above ambient temperature. After several runs with a specific meter box, calculate the average temperature above ambient that the dry gas meter ran, and use this figure to estimate T_m in the future.

Calculation of the optimal nozzle diameter.   The preliminary stack gas data are used to calculate the optimal nozzle diameter in accordance with the following formula:

$$D_n = \sqrt{\frac{0.0358Q_m P_m}{T_m C_p (1 - B_{ws})}} \sqrt{\frac{T_s M_s}{P_s(\Delta p)_{av}}} \tag{5.3}$$

For the flow rate in this equation Q_m, use 0.75 std ft³/min. This is the flow rate at which $dH_@$ is defined. Therefore, at the calculated nozzle diameter, the average dH during the testing will be approximately $dH_@$. This is advantageous, because $dH_@$ is usually around 2 inH$_2$O, which is a relatively easy value to set with accuracy during the test. If a nozzle is used which requires a dH of 0.2 inH$_2$O, then it is much more difficult to set dH accurately. Once the optimal nozzle diameter has been calculated, then the actual nozzle can be chosen from the tester's stock of nozzles. Choose the nozzle with a diameter closest to the optimal nozzle diameter. In the author's experience, most stacks can be effectively tested with nozzle diameters from 0.25 to 0.5 in.

5.4.6   Assembly of the sampling train

At this point, the sampling train can be assembled (see Fig. 5.6). The nozzle chosen must be connected to the probe with an airtight seal. The probe must be connected to the hot box containing the filter holder (Fig. 5.7). The

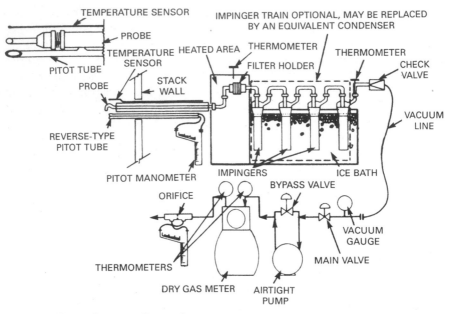

Figure 5.6 Particulate-sampling train.

Figure 5.7 Heated filter holder and impinger case.

glassware must be assembled and the umbilicals hooked up. Commercially available testing trains usually have compression fittings or quick-connect fittings to accomplish an airtight seal. The instructions supplied with the testing equipment being used should be followed to ensure proper operation of the equipment and seals. The following discussion covers some recurring pitfalls to avoid when assembling the testing train. Avoid connecting the Pitot tube manometer backward by labeling the tubes on the umbilicals for high and low pressure. These labels must be affixed at both the sampling console and the probe ends of the umbilicals. Glassware leaks are usually the result of dirty mating surfaces or improper alignment. Clean all mating surfaces before taking the equipment into the field, and take special precautions to prevent particulate from settling on the surfaces in the field. Try to position the impingers so that all the U connectors are horizontal. With impingers that have a compression seal between the stem and the bottle, the stem can be raised or lowered to make U connectors horizontal. When installing the filter holder, make sure that there is a filter in the holder and that the stack gas will be drawn through the filter first and the filter support second. The console should be as level as possible to facilitate leveling the manometer. Most consoles have some mechanism to level the manometer, which must be level to operate properly. Once leveled, the manometers must be zeroed so that the meniscus of the fluid is right on the zero mark when read from a position perpendicular to the manometer. If the manometer is not read from a perpendicular position, then a parallax error will be introduced.

5.4.7 Leak-testing the sampling train

Once assembled, the train should be leak-tested. This test is done by bringing the train up to sampling temperature, capping the nozzle, and drawing a 15-inHg vacuum on the entire train. The test should not be started with a train that has a leak at this vacuum. Instead, the leak should be located and fixed.

There is danger, during the leak test, of drawing water from the impingers into the filter holder. When this happens, the train must be disassembled, a new filter installed, and the impingers reweighed. To avoid this mishap, the following procedure is suggested in EPA method 5 for performing the leak check: Start the pump with the bypass valve fully open and the coarse adjust valve fully closed. Partially open the coarse adjust valve and slowly close the bypass valve until a vacuum of 15 inHg is reached. Do not turn the bypass valve backward. When the leak check is finished, remove the cap from the probe nozzle and immediately turn off the pump.

5.4.8 Calculation of isokinetic sampling rate

The rate of sampling is determined from the simple proportional equation $dH = K\, dP$. During sampling, each time the probe is set at a new position, dP is read from the manometer on the Pitot tube and a dH is calculated. Then the sampling pump is regulated with valves to obtain the calculated dH on the manometer on the orifice downstream of the dry gas meter.

This equation is deceptively simple, since K is determined as follows:

$$K = 846.72 D_n^4\, \Delta H_@\, C_p^2 (1 - B_{ws})^2\, \frac{M_d T_m P_s}{M_s T_s P_m} \tag{5.4}$$

From this equation, it is clear that K is affected by many variables in the stack test. However, most of these variables do not change by a significant percentage and therefore do not radically affect the value of K. On most stacks, the temperature of the stack gas T_s is the only variable that changes enough to significantly affect K. Therefore, it is advisable to recalculate K at each point if T_s changes. If T_s does not change, then the tester can often use the same value of K for the entire test.

5.4.9 Running the test

At this point, the test is ready to be run. Read the exact value on the dry gas meter, and record it on the field data sheet. Position the probe at the first traverse point, and read dP. Calculate the desired dH. Turn on the pump and the timer simultaneously. As quickly as possible, adjust the pump valves to obtain the desired dH. Then fill in the rest of the field data sheet for that traverse point. Maintain operation of the sampling train at these conditions for the duration of the sampling period at the traverse point. In the case of a 60-min run where there are 12 traverse points, the sampling period at each traverse point is 5 min. When the time at the first point is finished, move the probe to the second point and repeat the procedure. The pump and timer do not have to be turned off between sample points if the new dH can be set quickly. When the sample period is complete at the last traverse point on the first diameter, stop the timer and the pump. Then position the probe at the first location at the second sampling port. In the above example, this would be the seventh traverse point. Read dP and calculate dH. Turn on the pump and the timer, and set dH. Repeat this procedure for the remaining traverse points. If the stack temperature is high, the tester may find it advantageous to start sampling at the traverse point farthest from the sampling port.

Then the probe can cool down as it is withdrawn from the stack to sample at the remaining traverse points on that diameter. The cooler probe is easier to move to the second diameter in the middle of the run. At the end of the sampling period for the last traverse point, turn off the pump and the timer. Record the final readings for the timer and the dry gas meter on the field data sheet.

5.4.10   Posttest leak check

Once the probe is out of the stack, cap the nozzle and perform a posttest leak check. The procedure for performing the leak check is identical to that for the pretest leak check, except that the tester only need pull a vacuum as great as the largest vacuum created during the run. Generally, if there is no leak greater than 0.02 dcfm, then the train passes the posttest leak check.

5.4.11   Disassembly of the sampling train

After a successful leak check, the sampling train must be disassembled and the sample collected. The first part of disassembly is to remove the probe and nozzle. A cap should be placed over the nozzle and the probe end if it has to be moved any distance. The probe should be relocated to a staging area out of the wind where the nozzle and probe liner can be cleaned. The outside of the probe and nozzle should be wiped clean if there is any particulate which may fall into the sample jar. The nozzle is disconnected from the probe and cleaned with a nylon brush and research-grade acetone. All the acetone is collected in a sample jar for later analysis. The probe liner is cleaned in the same manner, and the acetone is collected into the sample jar. Care should be taken to clean the probe and nozzle well. For some runs, as much as 90 percent of the particulate is collected in the probe and nozzle.

The filter holder should also be taken to a sheltered staging area before it is opened. The filter should be removed from the holder with tweezers and placed in the appropriate petri dish, dirty side up. Usually, some of the filter paper sticks to the filter support around the edges. This residue should be scraped off the filter support and transferred to the filter in the petri dish. The petri dish should be taped closed and kept upright in a safe place. If the petri dish is not labeled, it should be labeled now to identify the run.

The glassware in contact with the sample upstream of the filter must also be washed with acetone to collect any particulate. This acetone can be drained into the previous sample jar or a new sample jar depending upon the level of acetone in the sampling jars. Before a sample jar is filled, start a new jar. The jars should each be labeled to identify the run.

The impingers are removed from the impinger case, and their exterior is wiped dry with an absorbent cloth. Then the tester should weigh each impinger and record its posttest weight on the impinger field data sheet.

This concludes the field work for the first run. Generally, two more runs are performed for a total of three runs to complete a test. Runs 2 and 3 should be completed in the same manner as run 1.

5.5 Calculations

The following calculations are performed using the field data to determine actual emissions.

5.5.1 Standard volume sampled

The first calculation made is the standard volume of stack gas sampled. This calculation uses the perfect gas law to standardize the volume of stack gas measured directly by the dry gas meter. The calculation is of the form

$$
V_{m(\text{std})} = V_m Y \frac{T_{\text{std}}}{P_{\text{std}}} \frac{P_b + \Delta H/13.6}{T_m}
\tag{5.5}
$$

In this equation, Y is a correction factor for the dry gas meter. Each dry gas meter must be calibrated in accordance with EPA method 5. The result of this calibration is a Y specific to the dry gas meter. It is this Y that must be used in the above equation. And T_{std} and P_{std} are the standard temperature and pressure of 528°R and 29.92 inHg, respectively; T_m is the average of the inlet and outlet temperatures at the dry gas meter during the test, in degrees Rankine. The term P_b + $dH/13.6$ is the pressure in the dry gas meter with P_b being barometric pressure and $dH/13.6$ being the difference between the dry gas meter pressure and barometric pressure. Also dH is the pressure difference across the orifice after the dry gas meter, in inches of H_2O. Therefore, the dry gas meter pressure is different from barometric pressure by dH. So dH is divided by 13.6 to convert it from inches of water to inches of mercury.

5.5.2 Moisture content of stack gas

The moisture content B_{ws} is the volume fraction of moisture vapor in the stack gas and is calculated in the following equation:

$$B_{ws} = \frac{V_w}{V_w + V_{mstd}} \qquad (5.6)$$

where V_w is the volume of water vapor at standard conditions and is determined by the amount of water collected in the impingers and silica gel. The weight of water collected in grams is multiplied by the factor stated in method 5 to yield V_w.

5.5.3 Average stack gas velocity

The average stack gas velocity is computed in feet per second from the following form of Bernoulli's equation:

$$\overline{v}_s = K_p C_p \sqrt{\frac{T_s}{P_s M_s}} \ (\sqrt{\Delta p})_{av} \qquad (5.7)$$

In this equation, K_p is a conversion factor stated in EPA method 5 for English and metric units; C_p is the Pitot tube coefficient. For a Stauschiebe Pitot tube, which meets the physical tolerances in EPA method 5, $C_p = 0.84$. And T_s and P_s are the average stack gas absolute temperature and pressure, respectively; M_s is the stack gas molecular weight. These three terms, grouped under the radical, convert $(\sqrt{\Delta p})_{av}$ to standard conditions and air equivalent molecular weight. This conversion must be performed to use K_p, which is determined at standard conditions. An excellent derivation of K_p is included in the first reference at the end of this chapter.

5.5.4 Average stack gas volumetric flow rate

The average actual stack gas volumetric flow rate $\overline{Q}_a$ is determined from the continuity equation:

$$\overline{Q}_a = \overline{v}_s A_s (3600) \qquad (5.8)$$

This equation yields the flow in actual cubic feet per hour. The dry standard volumetric flow rate Q_s is obtained by applying the following relationship:

$$\overline{Q}_s = 3600(1 - B_{ws})\overline{v}_s A_s \frac{T_{std}}{P_{std}} \frac{P_s}{T_s} \qquad (5.9)$$

In this equation, the temperature and pressure terms are used to convert to standard conditions in accordance with the perfect gas law.

The term $1 - B_{ws}$ is used to subtract the volume of gas resulting from moisture in the stack gas.

5.5.5 Pollutant mass rate

The pollutant mass rate (PMR) is determined from the product of the stack gas flow rate and the concentration of the pollutant. The specific form of the equation is

$$\text{PMR} = \overline{Q}_s \, \frac{M_p}{V_{m(\text{std})}} \tag{5.10}$$

In this equation, the mass of pollutant collected M_p, divided by the standard volume sampled, represents the concentration of the pollutant in the stack gas.

5.5.6 Isokinetic variation

The isokinetic variation describes how close to isokinetic conditions the sampling was performed. The goal for sampling is to be perfectly isokinetic; however, a range of 90 to 110 percent is generally accepted. Isokinetic variation is calculated according to the following relationship:

$$\%I = 100 \, \frac{T_s V_{m(\text{std})} P_{\text{std}}}{T_{\text{std}} \overline{v}_s \theta A_n P_s 60 (1 - B_{ws})} \tag{5.11}$$

This equation compares the velocity in the stack to the velocity in the nozzle. The velocity in the stack was previously calculated in Eq. (5.7). The velocity in the nozzle is calculated by using a form of the continuity equation. The volume sampled $V_{m(\text{std})}$ is divided by the time of sampling and the area of the nozzle to yield a velocity at standard conditions. This velocity is then corrected to wet stack conditions to yield the actual average velocity in the nozzle during the test.

References

Environmental Protection Agency, *Source Sampling for Particulate Pollutants,* EPA 450/2-79-006, December 1979.
Code of Federal Regulations, 40 CFR 60, Appendix A, GPO, Washington, 1992.

6

Air Pollution Testing

E. Roberts Alley, Jr.

A variety of sources exist for analytical methods for air samples, ranging from experimental methods to the official, promulgated EPA emission test methods. In certain situations, such as for general in-house testing, a promulgated and certifiable method may not be necessary. However, for compliance purposes, the promulgated EPA methods may be the only acceptable methods available. The air control boards that are responsible for oversight of a particular facility should always be consulted for specific analytical requirements.

6.1 Environmental Protection Agency

The U.S. Environmental Protection Agency (EPA) has approved a variety of acceptable emission test methods for use in demonstrating compliance with the federal emission standards. The majority of the methods are published in Title 40 of the Code of Federal Regulations, section 60, Appendix A. Revisions, updates, corrections, and new methods are always published in the Federal Register before promulgation, first in draft form and then in final form. Table 6.1 includes a listing of the promulgated and many of the proposed emission test methods, including the Federal Register reference and the date published.

6.2 Emission Measurement Technical Information Center (EMTIC)

The Emission Measurement Technical Information Center is a technical information exchange network for emission test methods.

TABLE 6.1   Summary of EPA Emission Test Methods

Title 40, Office of Air Quality Planning and Standards

Method	Reference	Date	Description
1-8	42 FR 41754	08/18/77	Velocity, Orsat, particulate matter (PM), SO_2, NO_x, etc.
	43 FR 11984	03/23/78	Corr. and amend. to M-1 through 8
1/24	52 FR 34639	09/14/87	Technical corrections
	52 FR 42061	11/02/87	Corrections
2-25	55 FR 47471	11/14/90	Technical amendments
1	48 FR 45034	09/30/83	Reduction of number of traverse points
1	51 FR 20286	06/04/86	Alternative procedure for site selection
1A	54 FR 12621	03/28/89	Traverse points in small ducts
2A	48 FR 37592	08/18/83	Flow rate in small ducts—vol. meters
2B	48 FR 37594	08/18/83	Flow rate—stoichiometry
2C	54 FR 12621	03/28/89	Flow rate in small ducts—std. Pitot
2D	54 FR 12621	03/28/89	Flow rate in small ducts—rate meters
2E P	56 FR 24468	05/30/91	Flow rate from landfill wells
2F		Tentative	3D Pitot for velocity
3	55 FR 05211	02/14/90	Molecular weight
3/3B	55 FR 18876	05/07/90	Method 3B applicability
3A	51 FR 21164	06/11/86	Instrumental method for O_2 and CO_2
3B	55 FR 05211	02/14/90	Orsat for correction factors and excess air
3C P	56 FR 24468	05/30/91	
3	48 FR 49458	10/25/83	Addition of QA/QC
4	48 FR 55670	12/14/83	Addition of QA/QC
5	48 FR 55670	12/14/83	Addition of QA/QC
5	45 FR 66752	10/07/80	Filter specification change
5	48 FR 39010	08/26/83	DGM revision
5	50 FR 01164	01/09/85	Incorp. DGM and probe cal. procedures
5	52 FR 09657	03/26/87	Use of critical orifices as cal. stds.
5	52 FR 22888	06/16/87	Corrections
5A	47 FR 34137	08/06/82	PM from asphalt roofing (Prop. as M-26)
5A	51 FR 32454	09/12/86	Addition of QA/QC
5B	51 FR 42839	11/26/86	Nonsulfuric acid PM
5C		Tentative	PM from small ducts
5D	49 FR 43847	10/31/84	PM from fabric filters
5D	51 FR 32454	09/12/86	Addition of QA/QC
5E	50 FR 07701	02/25/85	PM from fiberglass plants
5F	51 FR 42839	11/26/86	PM from FCCU
5F	53 FR 29681	08/08/88	Barium titration procedure
5G	53 FR 05860	02/26/88	PM from wood stove—dilution tunnel
5H	53 FR 05860	02/26/88	PM from wood stove—stack
6	49 FR 26522	06/27/84	Addition of QA/QC
6	48 FR 39010	08/26/83	DGM revision
6	52 FR 41423	10/28/87	Use of critical orifices for FR/vol. meas.
6A	47 FR 54073	12/01/82	SO_2/CO_2
6B	47 FR 54073	12/01/82	Auto SO_2/CO_2
6A/B	49 FR 09684	03/14/84	Incorp. coll. test changes
6A/B	51 FR 32454	09/12/86	Addition of QA/QC

TABLE 6.1   Summary of EPA Emission Test Methods (*Continued*)

Method	Reference	Date	Description
6C	51 FR 21164	06/11/86	Instrumental method for SO_2
6C	52 FR 18797	05/27/87	Corrections
7	49 FR 26522	06/27/84	Addition of QA/QC
7A	48 FR 55072	12/08/83	Ion chromatograph NO_x analysis
7A	53 FR 20139	06/02/88	ANPRM
7A	55 FR 21752	05/29/90	Revisions
7B	50 FR 15893	04/23/85	UV NO_x analysis for nitric acid plants
7A/B		Tentative	High SO_2 interference
7C	49 FR 38232	09/27/84	Alkaline permanganate/colorimetric for NO_x
7D	49 FR 38232	09/27/84	Alkaline permanganate/IC for NO_x
7E	51 FR 21164	06/11/86	Instrumental method for NO_x
8	36 FR 24876	12/23/71	Sulfuric acid mist and SO_2
8	42 FR 41754	08/18/77	Addition of particulate and moisture
8	43 FR 11984	03/23/78	Miscellaneous corrections
9	39 FR 39872	11/12/74	Opacity
9A	46 FR 53144	10/28/81	Lidar opacity called Alternative 1
10	39 FR 09319	03/08/78	CO
10	53 FR 41333	10/21/88	Alternative trap
10A	52 FR 30674	08/17/87	Colorimetric method for PS-4
10A	52 FR 33316	09/02/87	Correction notice
10B	53 FR 41333	10/21/88	GC method for PS-4
11	43 FR 01494	01/10/78	H_2S
12	47 FR 16564	04/16/82	Pb
12	49 FR 33842	08/24/84	Incorp. method of additions
13A	45 FR 41852	06/20/80	F—colorimetric method
13B	45 FR 41852	06/20/80	F—SIE method
13A/B	45 FR 85016	12/24/80	Corr. to M-13A and 13B
14	45 FR 44202	06/30/80	F from roof monitors
15	43 FR 10866	03/15/78	TRS from petroleum refineries
15	54 FR 46236	11/02/89	Revisions
15	54 FR 51550	12/15/89	Correction notice
15A	52 FR 20391	06/01/87	TRS alternative/oxidation
16	43 FR 07568	02/23/78	TRS from kraft pulp mills
16	43 FR 34784	08/07/78	Amend. to M-16, H_2S loss after filters
16	44 FR 02578	01/12/79	Amend. to M-16, SO_2 scrubber added
16	54 FR 46236	11/02/89	Revisions
16	55 FR 21752	05/29/90	Correction of figure (~10%)
16A	50 FR 09578	03/08/85	TRS alternative
16A	52 FR 36408	09/29/87	Cylinder gas analysis alternative method
16B	52 FR 36408	09/29/87	TRS alternative/GC analysis of SO_2
16A/B	53 FR 02914	02/02/88	Correction 16A/B
17	43 FR 07568	02/23/78	PM, in-stack
18	48 FR 48344	10/18/83	Volatile organic compound (VOC), general GC method
18	49 FR 22608	05/30/84	Corrections to M-18
18	52 FR 51105	02/19/87	Revisions to improve method
18	52 FR 10852	04/03/87	Corrections
18	59 FR 19308	04/22/94	Revisions to improve QA/QC

TABLE 6.1 Summary of EPA Emission Test Methods (*Continued*)

Method	Reference	Date	Description
19	44 FR 33580	06/11/79	F factor, coal sampling
19	52 FR 47826	12/16/87	M-19A incorp. into M-19
19	48 FR 49460	10/25/83	Corr. to F factor equations and F_c value
20	44 FR 52792	09/10/79	NO_x from gas turbines
20	47 FR 30480	07/14/82	Corr. and amend.
20	51 FR 32454	09/12/86	Clarifications
21	48 FR 37598	08/18/83	VOC leaks
21	49 FR 56580	12/22/83	Corrections to Method 21
21	55 FR 25602	06/22/90	Clarifying revisions
22	47 FR 34137	08/06/82	Fugitive VE
22	48 FR 48360	10/18/83	Add smoke emission from flares
23	56 FR 5758	02/13/91	Dioxin/dibenzo furan
23R	60 FR 28378	05/31/95	Revisions and corrections
24	45 FR 65956	10/03/80	Solvent in surface coatings
24A	47 FR 50644	11/08/82	Solvent in ink (prop. as M-29)
24		Tentative	Solvent in water-borne coatings
24	57 FR 30654	07/10/92	Multicomponent coatings
24	60 FR 47095	09/11/95	Radiation-cured coatings
25	45 FR 65956	10/03/80	TGNMO
25	53 FR 04140	02/12/88	Revisions to improve method
25	53 FR 11590	04/07/88	Correction notice
25A	48 FR 37595	08/18/83	TOC/FID
25B	48 FR 37597	08/18/83	TOC/NDIR
25C P	56 FR 24468	05/30/91	VOC from landfills
25D	59 FR 19311	04/22/94	VO from TSDF—purge procedure
25E	59 FR 62896	12/06/94	VO from TSDF—vapor pressure procedure
26	56 FR 5758	02/13/91	HCl
26	57 FR 24550	06/10/92	Corrections to method 26
26	59 FR 19309	04/22/94	Add 26 HCl, halogens, other hydrogen halides
26A	59 FR 19309	04/22/94	Isokinetic HCl, halogens, hydrogen halides
27	48 FR 37597	08/18/83	Tank truck leaks
28	53 FR 05860	02/26/88	Wood stove certification
28A	53 FR 05860	02/26/88	Air-to-fuel ratio
29 P	59 FR 48259	09/20/94	Multiple metals
Part 60, Appendix B			
PS-1	48 FR 13322	03/30/83	Opacity
PS-1 P	59 FR 60585	11/25/94	Revisions
PS-2	48 FR 23608	05/25/83	SO_2 and NO_x
PS 1-5	55 FR 47471	11/14/91	Technical amendments
PS-3	48 FR 23608	05/25/83	CO_2 and O_2
PS-4	50 FR 31700	08/05/85	CO
PS-4A	56 FR 5526	02/11/91	CO for MWC
PS-5	48 FR 32984	07/20/83	TRS
PS-6	53 FR 07514	03/09/88	Velocity and mass emission rate
PS-7	55 FR 40171	10/02/90	H_2S
PS-8	59 FR 64580	12/15/94	VOC CEMS performance specifications
PS-9	59 FR 64580	12/15/94	GC CEMS performance specifications

TABLE 6.1   Summary of EPA Emission Test Methods (*Continued*)

Method	Reference	Date	Description
Part 60, Appendix B			
PS-10		Tentative	Ammonia CEMS
PS-11		Tentative	PM CEMS
PS-12		Tentative	Hg CEMS
Part 60, Appendix F			
Prc 152			
FR 21003		06/04/87	Quality assurance for CEMS
Prc 154			
FR 52207		12/20/89	Revision
Part 60, Appendix J			
App-J55 FR 33925		08/20/90	Wood stove thermal efficiency
Alternative Procedures and Miscellaneous			
	48 FR 44700	09/29/83	S factor method for sulfuric acid plants
	48 FR 48669	10/20/83	Corrections to S factor publication
	49 FR 30672	07/31/84	Add fuel analysis procedures for gas turbines
	51 FR 21762	06/16/86	Alternative PST for low-level concentrations
	54 FR 46234	11/02/89	Misc. revisions to Appendix A, 40 CFR, Part 60
	55 FR 40171	10/02/90	Monitoring revisions to subpart J (Petr. Ref.), Part 60
	54 FR 06660	02/14/89	Test methods and procedures rev. (40 CFR 60)
	54 FR 21344	05/17/89	Correction notice
	54 FR 27015	06/27/89	Correction notice
Part 61, Appendix B			
101	47 FR 24703	06/08/82	Hg in airstreams
101A	47 FR 24703	06/08/82	Hg in sewage sludge incinerators
101A P	59 FR 48259	09/20/94	Revisions—consistency with Method 29
101	49 FR 35768	09/12/84	Corrections to M-101 and 101A
102	47 FR 24703	06/08/82	Hg in H_2 streams
103	48 FR 55266	12/09/83	Revised Be screening method
104	48 FR 55268	12/09/83	Revised beryllium method
105	40 FR 48299	10/14/75	Hg in sewage sludge
105	49 FR 35768	09/12/84	Revised Hg in sewage sludge
106	47 FR 39168	09/07/82	Vinyl chloride
107	47 FR 39168	09/07/82	VC in process streams
107	52 FR 20397	06/01/87	Alternative calibration procedure
107A	47 FR 39485	09/08/82	VC in process streams
108	51 FR 28035	08/04/86	Inorganic arsenic
108A	51 FR 28035	08/04/86	Arsenic in ore samples
108B	55 FR 22026	05/31/90	Arsenic in ore alternative
108C	55 FR 22026	05/31/90	Arsenic in ore alternative
108B/C	55 FR 32913	08/13/90	Correction notice
111	50 FR 05197	02/06/85	Polonium 210
114	54 FR 09612	03/07/89 P	Monitoring of radio nuclides
115	54 FR 09612	03/07/89 P	Radon 222

TABLE 6.1   Summary of EPA Emission Test Methods (*Continued*)

Method	Reference	Date	Description
Part 61			
	53 FR 36972	09/23/88	Corrections
Part 51, Appendix M			
201	55 FR 14246	04/17/90	PM_{10} (EGR procedure)
201A	55 FR 14246	04/17/90	PM_{10} (CSR procedure)
201/A	55 FR 24687	06/18/90	Correction of equations
201	55 FR 37606	09/12/90	Correction of equations
202	56 FR 65433	12/17/91	Condensible PM
203 P	57 FR 46114	10/07/92	Transmissometer for compliance
203A P	58 FR 61640	11/22/93	Visible emissions—2–6 min avg.
203B P	58 FR 61640	11/22/93	Visible emissions—time exception
203C P	58 FR 61640	11/22/93	Visible emissions—instantaneous
204		Tentative	VOC capture efficiency
204A		Tentative	VOC capture efficiency
204B		Tentative	VOC capture efficiency
204C		Tentative	VOC capture efficiency
204D		Tentative	VOC capture efficiency
204E		Tentative	VOC capture efficiency
204F		Tentative	VOC capture efficiency
205		05/30/94	Dilution calibration verification
206		Tentative	Ammonia (NH_3)
207		Tentative	Isocyanates
Part 63, Appendix A			
301	57 FR 61970	12/29/92	Field data validation protocol
302	(Reserved)		
303	58 FR 57898	10/27/93	Coke oven door emissions
304A	59 FR 19590	04/22/94	Biodegradation rate (vented)
304B	59 FR 19590	04/22/94	Biodegradation rate (enclosed)
305	59 FR 19590	04/22/94	Compound specific liquid waste
306	60 FR 4948	01/25/95	Hexavalent chromium
306A	60 FR 4948	01/25/95	Simplified chromium sampling
306B	60 FR 4948	01/25/95	Surface tension of chromium suppressors
307	59 FR 61801	12/02/94	Solvent degreaser VOC
308 P	58 FR 66079	12/17/93	Methanol
309 P		06/06/94	Aerospace solvent recovery material balance
310		Tentative	Residual hexane in EPDM rubber
311 P	59 FR 62652	12/06/94	VOC HAPS in furniture coatings
312		Tentative	Residual styrene in SBR rubber
313		Tentative	Residual styrene in PBR rubber
314		Tentative	Halogenated compounds in solvents
315		Tentative	MeCl extractable organic matter
316		Tentative	Formaldehyde—manual method
317		Tentative	Phenol—manual method
318		Tentative	Formaldehyde, phenol, methanol with FTIR
319		Tentative	Filter efficiency, paint over spray

P = proposal

Tentative = under evaluation

EMTIC's purpose is to promote consistent and accurate test method application in the development and enforcement of emission prevention and control programs.

EMTIC's primary goal is the transmission of test methods, procedures, and other information from government agencies to the sources and testing contractors. The transfer of information includes developing a test methods reference system, conducting emission measurement workshops around the United States, developing a computer bulletin board and Web site for information distribution, making information available that describes recent test method advancements, and implementing special emission testing projects to address specific measurement problems encountered by government agencies, sources, or testing contractors.

The Internet Web Page is perhaps the most powerful method EMTIC has available for the distribution of emissions test information. The EMTIC Web site, located at http://134.67.104.12/html/emtic/, contains information regarding promulgated test methods, proposed test methods, revisions to test methods, emissions testing software, guideline documents, and many general and background information documents. These sources are being continuously updated and should be checked regularly.

Another valuable source of information available from EMTIC is a library of test methods and other procedures that have not been promulgated or officially proposed for adoption. These methods include alternative testing methods and experimental methods for new equipment or regulated compounds.

6.3 Listing of CFR Promulgated Methods

Promulgated test methods have been published in the Federal Register and adopted in the Code of Federal Regulations (CFR) as final rules. These are the official, legal Federal Register versions. Performance specifications have also been published in the Federal Register as final rules. EMTIC has made available for download on a computer bulletin board the official, legal Federal Register versions. These methods can also be downloaded from the EMTIC Web site, http://134.67.104.12/html/emtic/. Table 6.2 shows a typical listing of the promulgated methods available for download. The CFR is available at many public libraries for review and on the Internet. The CFR can be accessed at http://www.access.gpo.gov/nara/cfr/cfr-table-search.html. The Federal Register can be accessed at http://www.epa.gov/docs/fedrgstr/.

TABLE 6.2   Typical Promulgated CFR Methods Available from EMTIC

File name	Date	Bytes	Description
README.TXT	3/18/93	1,920	Read before downloading
M-01.WPF	1/9/97	367,612	Method 1: Traverse Points
M-01A.WPF	9/25/96	100,260	Method 1A: Small Ducts
M-02.ZIP	1/10/97	868,379	Method 2: Velocity—S-type Pitot
M-02A.WPF	9/25/96	33,900	Method 2A: Volume Meters
M-02B.WPF	4/1/96	21,883	Method 2B: Exhaust Volume Flow Rate
M-02C.WPF	9/25/96	121,858	Method 2C: Standard Pitot
M-02D.WPF	1/9/97	27,577	Method 2D: Rate Meters
M-02E.WPF	1/9/97	116,731	Method 2E: Landfill Gas Production Flow Rate
M-03.WPF	9/25/96	311,859	Method 3: Molecular Weight
M-03FIG.WPF	6/13/90	3,103	Method 3: Figure 3.3
M-03A.WPF	6/13/90	14,038	Method 3A: CO_2, O_2—Instrumental
M-03B.WPF	6/13/90	22,991	Method 3B: CO_2, O_2—Orsat
M-03C.WPF	6/20/96	21,543	Method 3C: CO_2, CH_4, N_2, O_2—TCD
M-04.WPF	1/9/97	464,259	Method 4: Moisture Content
M-05.WPF	1/14/97	199,890	Method 5: Particulate Matter (PM)
M-05A.WPF	6/13/90	24,460	Method 5A: PM Asphalt Roofing (Particulate Matter)
M-05B.WPF	6/13/90	5,522	Method 5B: PM Nonsulfuric Acid (Particulate Matter)
M-05D.WPF	8/20/96	223,893	Method 5D: PM Baghouses (Particulate Matter)
M-05E.WPF	8/20/96	26,866	Method 5E: PM Fiberglass Plants (Particulate Matter)
M-05F.WPF	9/25/96	31,728	Method 5F: PM Fluid Catalytic Cracking Unit
M-05G.WPF	8/20/96	62,366	Method 5G: PM Wood Heaters from a Dilution Tunnel
M-05H.WPF	8/20/96	67,300	Method 5H: PM Wood Heaters from a Stack
M-06.WPF	9/25/96	487,136	Method 6: Sulfur Dioxide (SO_2)
M-06FIG.WPF	6/13/90	1,955	Method 6: Figure 6.4
M-06A.WPF	9/25/96	504,307	Method 6A: SO_2, CO_2
M-06B.WPF	8/20/96	18,258	Method 6B:- SO_2, CO_2—Long-Term Integrated
M-06C.WPF	9/25/96	290,996	Method 6C: SO_2—Instrumental
M-07.WPF	3/12/96	38,351	Method 7: Nitrogen Oxide (NO_x)
M-07A.WPF	1/28/91	17,588	Method 7A: NO_x—Ion Chromatographic Method
M-07B.WPF	8/20/96	18,486	Method 7B: NO_x—Ultraviolet Spectrophotometry
M-07C.WPF	9/25/96	316,617	Method 7C: NO_x—Colorimetric Method
M-07D.WPF	9/25/96	128,784	Method 7D: NO_x—Ion Chromatographic
M-07E.WPF	8/14/90	9,787	Method 7E: NO_x—Instrumental
M-08.WPF	10/29/96	549,697	Method 8: Sulfuric Acid Mist
M-09.WPF	8/20/96	53,177	Method 9: Visual Opacity
M-10.WPF	9/25/96	496,761	Method 10: Carbon Monoxide—NDIR
M-10A.ZIP	9/25/96	410,456	Method 10A: CO for Certifying CEMS

TABLE 6.2   Typical Promulgated CFR Methods Available from EMTIC (*Continued*)

File name	Date	Bytes	Description
M-10B.WPF	11/8/94	13,634	Method 10B: CO from Stationary Sources
M-11.ZIP	1/16/97	333,145	Method 11: H_2S Content of Fuel
M-12.WPF	9/25/96	551,702	Method 12: Inorganic Lead
M-13A.ZIP	1/16/97	331,453	Method 13A: Total Fluoride (SPADNS Zirconium Lake)
M-13B.WPF	8/20/96	15,534	Method 13B: Total Fluoride (Specific Ion Electrode)
M-14.ZIP	1/16/97	332,152	Method 14: Fluoride for Primary Aluminum Plants
M-15.WPF	10/10/96	41,642	Method 15: Hydrogen Sulfide, Carbonyl Sulfide, and Carbon Disulfide
M-15A.ZIP	1/16/97	480,179	Method 15A: Total Reduced Sulfur (TRS Alt.)
M-16.WPF	10/10/96	55,030	Method 16: Sulfur (Semicontinuous Determination)
M-16A.ZIP	9/25/96	427,370	Method 16A: Total Reduced Sulfur (Impinger)
M-16B.WPF	4/1/96	35,317	Method 16B: Total Reduced Sulfur (GC Analysis)
M-17.ZIP	1/16/97	414,182	Method 17: In-Stack Particulate (PM)
M-18.ZIP	9/25/96	632,510	Method 18: VOC by GC
REVM-18.WPF	3/3/94	19,023	Recent Revisions to Method 18 (3/2/94)
M-19.WPF	4/15/96	74,490	Method 19: SO_2 Removal and PM, SO_2, NO_x Rates from Electric Utility Steam Generators
M-20.WPF	9/25/96	58,956	Method 20: NO_x from Stationary Gas Turbines
M-21.WPF	4/27/95	19,771	Method 21: VOC Leaks (Corrected on 4/26/95)
M-22.WPF	5/21/93	17,937	Method 22: Fugitive Opacity
M-23.WPF	5/25/95	86,545	Method 23: Dioxin and Furan (02/91 FR Copy)
M-24.WPF	9/7/95	14,872	Method 24: Surface Coatings 9/11/95 FR copy
M-24A.WPF	8/6/93	10,354	Method 24A: Printing Inks and Related Coatings
M-25.WPF	9/25/96	119,616	Method 25: Gaseous Nonmethane Organic Emissions
M-25A.WPF	9/25/96	106,158	Method 25A: Gaseous Organic Concentration (Flame Ionization)
M-25B.WPF	11/8/94	5,476	Method 25B: Gaseous Organic Concentration (Infrared Analyzer)
M-25C.WPF	6/20/96	36,756	Method 25C: NMOC in Landfill Gases
M-25D.WPF	8/20/96	69,675	Method 25D: VOC of Waste Samples (revised —figures added)
M-25E.WPF	9/25/96	30,544	Method 25E: Vapor Phase Organic Concentration in Waste Samples
REVM-26.WPF	3/3/94	34,779	Recent Revisions to Method 26 (3/3/94)
M-26.WPF	12/3/96	57,281	Method 26: Hydrogen Chloride, Halides, Halogens

TABLE 6.2   Typical Promulgated CFR Methods Available from EMTIC (*Continued*)

File name	Date	Bytes	Description
M-26A.WPF	3/3/94	55,936	Method 26A: Hydrogen Halide and Halogen—Isokinetic
M-27.WPF	1/21/94	13,889	Method 27: Vapor Tightness of Gasoline Tank—Pressure Vacuum
M-28.WPF	9/25/96	67,354	Method 28: Certification and Auditing—Wood Heaters
M-28A.WPF	9/25/96	24,716	Method 28A: Air-to-Fuel Ratio, Burn Rate—Wood-Fired Appliances
M-29.ZIP	1/16/97	164,262	Method 29: Metals Emissions from Stationary Sources
M-101.ZIP	9/25/96	479,274	Method 101: Mercury from Chlor-Alkali Plants (Air)
M-101FIGS.WPF	11/8/95	21,577	Figures for M-101
M-101A.WPF	5/20/91	38,589	Method 101A: Mercury from Sewage Sludge Incinerators
M101AREV.WPF	6/25/96	512,028	Revisions to Method 101A (4/25/96)
M-102.WPF	5/20/91	12,678	Method 102: Mercury from Chlor-Alkali Plants (Hydrogen Streams)
M-103.WPF	9/25/96	30,887	Method 103: Beryllium Screening Method
M-104.WPF	1/18/95	21,215	Method 104: Beryllium Emissions Determination
M-105.WPF	9/25/96	32,406	Method 105: Mercury in Wastewater Treatment Plant Sewage Sludge
M-106.WPF	5/23/95	30,433	Method 106: Determination of Vinyl Chloride
M-107.WPF	8/20/96	40,886	Method 107: Vinyl Chloride Content of In-Process Wastewater Samples
M-107A.WPF	9/25/96	32,752	Method 107A: Vinyl Choride Content of Solvents
M-108.WPF	9/25/96	56,080	Method 108: Particulate and Gaseous Arsenic Emissions
M-108A.WPF	8/20/96	17,456	Determination of Arsenic Content in Ore Samples from Nonferrous Smelters
M-108B.WPF	2/1/96	7,619	Method 108B: Arsenic
M-108C.WPF	2/1/96	14,820	Method 108C: Arsenic
M-111.WPF	4/1/96	35,891	Method 111: Polonium 210 Emissions
M-114.WPF	4/1/96	59,874	Method 114: Radionuclide Emissions
M-115.WPF	4/1/96	29,109	Method 115: Radon 222 Emissions
M-201.ZIP	9/25/96	569,836	Method 201: PM_{10} (In-stack, CRS)
M-201A.WPF	1/15/97	579,127	Method 201A: PM_{10} (In-stack, CRS) (revised with figures)
M-202.WPF	9/25/96	284,788	Method 202: Condensible Particulate Matter
M202FIGS.WPF	11/8/95	11,538	Figures for M-202
M-205.WPF	9/25/96	23,225	Method 205: Gas Dilution Calibration
M-301.WPF	12/30/92	54,912	Method 301: Validation Protocol
M-303.WPF	9/25/96	79,843	Method 303: By-product Coke Oven Batteries
M-303A.WPF	10/27/93	19,111	Method 303A: Nonrecovery Coke Oven Batteries

TABLE 6.2   Typical Promulgated CFR Methods Available from EMTIC (*Continued*)

File name	Date	Bytes	Description
M-304A.WPF	3/3/94	47,795	Method 304A: Biodegradation Rates—Vent Option
M-304B.WPF	3/3/94	50,010	Method 304B: Biodegradation Rates—Scrubber Option
M-305.WPF	9/25/96	53,770	Method 305: Potential VOC in Waste
M-306.ZIP	2/8/96	251,088	Method 306: Chromium Emissions Electroplating/Anodizing
M-306A.WPF	9/25/96	372,925	Method 306A: Chromium Emissions Electroplating/Anodizing (Mason Jar Method)
M-306B.WPF	2/8/96	12,480	Method 306B: Surface Tension for Tanks Electroplating/Anodizing
M-307.WPF	9/25/96	38,080	Method 307: Emissions from Solvent Vapor Cleaners (Dec. 2, 1994)
M-311.WPF	1/17/96	84,020	Method 311: HAPS in Paints and Coatings
PS-2.WPF	10/23/91	53,634	CEMS Performance Specification 2 for SO_2 and NO_x
PS-3.WPF	2/11/92	13,987	CEMS Performance Specification 3 for O_2 and CO_2
PS-4.WPF	2/11/92	6,481	CEMS Performance Specification 4 for CO
PS-4A.WPF	10/30/91	6,702	CEMS Performance Specification 4A for CO
PS-5.WPF	5/12/92	5,248	CEMS Performance Specification 5 for TRS
PS-6.WPF	5/12/92	8,832	CEMS Performance Specification 6 for Flow Rate
PS-7.WPF	7/1/91	5,504	CEMS Performance Specification 7 for H_2S
PS-8.WPF	3/3/95	18,928	CEMS Performance Specification 8 for VOC CEMS
PS-9.WPF	3/19/96	21,085	CEMS Performance Specification 9 for GC CEMS (promulgated Dec. 1994)
PT60APPF.WPF	5/17/95	33,240	Appendix F: Quality Assurance Procedures

6.4   Other Sources

A variety of other sources exist for new and modified emission test methods. Among these are the *Journal of the Air and Waste Management Association (JAWMA)*. A good reference manual is *Methods of Air Sampling and Analysis,* 3d ed., edited by James P. Lodge, Jr. The book describes and evaluates many of the traditional emission test methods as well as several of the newer methods. The work is particularly useful in the description of the analytical equipment used in the different methods, providing a good introduction for the tester.

6.5   Analytical Parameters

The following sections include portions of the promulgated test methods as published in the Code of Federal Regulations or updated in the Federal Register. This information is intended to allow readers to familiarize themselves with the analytical procedures and requirements of different methods. This is not intended to be a complete detailing of the methods. In fact, many of the methods require that the analytical work be performed by persons trained in and familiar with specific disciplines beyond the scope of many engineers or scientists.

In general, the following sections include discussions of the method's principles and applicability, apparatus required (for analysis), and procedure to be followed (again, for analysis). Sections pertaining to the sample collection procedures, calibrations, and calculations have been omitted. The text that follows is in large part taken directly from the promulgated versions of the methods, with some minor grammatical and formatting changes for clarity. The actual methods should be reviewed for specific details prior to performing any sampling or analysis.

6.5.1   Physical parameters

Method 2: Determination of stack gas velocity and volumetric flow rate (type S Pitot tube)

Principle. The average gas velocity in a stack is determined from the gas density and from measurement of the average velocity head with a type S (Stausscheibe or reverse-type) Pitot tube.

Applicability. This method is applicable for measurement of the average velocity of a gas stream and for quantifying gas flow.

This procedure is not applicable at measurement sites that fail to meet the criteria of method 1, section 2.1. Also, the method cannot be used for direct measurement in cyclonic or swirling gas streams; section 2.4 of method 1 shows how to determine cyclonic or swirling flow conditions. When unacceptable conditions exist, alternative procedures, subject to the approval of the EPA Administrator, must be employed to make accurate flow rate determinations; examples of such alternative procedures are (1) to install straightening vanes, (2) to calculate the total volumetric flow rate stoichiometrically, or (3) to move to another measurement site where the flow is acceptable.

Apparatus. Specifications for the apparatus are given below. Any other apparatus that has been demonstrated (subject to approval of the Administrator) to be capable of meeting the specifications will be considered acceptable.

Type S Pitot tube.   This Pitot tube is made of metal tubing (e.g., stainless steel). It is recommended that the external tubing diameter (dimension D_t) be between 0.48 and 0.95 cm ($\frac{3}{16}$ and $\frac{3}{8}$ in). There shall be an equal distance from the base of each leg of the Pitot tube to its face-opening plane (dimensions P_A and P_B); it is recommended that this distance be between 1.05 and 1.50 times the external tubing diameter. The face openings of the Pitot tube shall be aligned preferably; however, slight misalignments of the openings are permissible.

The type S Pitot tube shall have a known coefficient, determined as outlined in section 4. An identification number shall be assigned to the Pitot tube; this number shall be permanently marked or engraved on the body of the tube. A standard Pitot tube may be used instead of a type S, provided that it meets the specifications of sections 2.7 and 4.2; note, however, that the static and impact pressure holes of standard Pitot tubes are susceptible to plugging in particulate-laden gas streams. Therefore, whenever a standard Pitot tube is used to perform a traverse, adequate proof must be furnished that the openings of the Pitot tube have not plugged up during the traverse period; this can be done by taking a velocity head (D_p) reading at the final traverse point, cleaning out the impact and static holes of the standard Pitot tube by "back-purging" with pressurized air, and then taking another D_p reading. If the D_p readings made before and after the air purge are the same (± 5 percent), the traverse is acceptable. Otherwise, reject the run. Note that if D_p at the final traverse point is unsuitably low, another point may be selected. If back-purging at regular intervals is part of the procedure, then comparative D_p readings shall be taken, as above, for the last two back purges at which suitably high D_p readings are observed.

Differential pressure gauge.   This is an inclined manometer or equivalent device. Most sampling trains are equipped with a 10-in (water column) inclined-vertical manometer, having 0.01-in H_2O divisions on the 0- to 1-in inclined scale, and 0.1-in H_2O divisions on the 1- to 10-in vertical scale. This type of manometer (or other gauge of equivalent sensitivity) is satisfactory for the measurement of D_p values as low as 1.3 mm (0.05 in) H_2O. However, a differential pressure gauge of greater sensitivity shall be used (subject to the approval of the Administrator) if any of the following is found to be true: (1) The arithmetic average of all D_p readings at the traverse points in the stack is less than 1.3 mm (0.05 in) H_2O; (2) for traverses of 12 or more points, more than 10 percent of the individual D_p readings are below 1.3 mm (0.05 in) H_2O; (3) for traverses of fewer than 12 points, more than one D_p reading is below 1.3 mm (0.05 in) H_2O. Citation 18 in the Bibliography describes commercially available instrumentation for the measurement of low-range gas velocities.

As an alternative to criteria 1 through 3 above, the following calculation may be performed to determine the necessity of using a more sensitive differential pressure gauge:

$$T = \frac{\sum\limits_{i=1}^{n} \sqrt{\Delta p_i + K}}{\sum\limits_{i=1}^{n} \sqrt{\Delta p_i}}$$

where Δp_i = individual velocity head reading at a traverse point, mm (in) H_2O
n = total number of traverse points
$K = 0.13$ mmH_2O in metric units and 0.005 inH_2O in English units

If T is greater than 1.05, the velocity head data are unacceptable and a more sensitive differential pressure gauge must be used.

Note: If differential pressure gauges other than inclined manometers are used (e.g., magnehelic gauges), their calibration must be checked after each test series. To check the calibration of a differential pressure gauge, compare D_p readings of the gauge with those of a gauge-oil manometer at a minimum of three points, approximately representing the range of D_p values in the stack. If, at each point, the values of D_p as read by the differential pressure gauge and gauge-oil manometer agree to within 5 percent, the differential pressure gauge shall be considered to be in proper calibration. Otherwise, either the test series shall be voided, or procedures to adjust the measured D_p values and final results shall be used, subject to the approval of the Administrator.

Temperature gauge.   This is a thermocouple, liquid-filled bulb thermometer, bimetallic thermometer, mercury-in-glass thermometer, or other gauge capable of measuring temperature to within 1.5 percent of the minimum absolute stack temperature. The temperature gauge shall be attached to the Pitot tube such that the sensor tip does not touch any metal; the gauge shall be in an interferencefree arrangement with respect to the Pitot tube face openings (see fig. 2.1 and also fig. 2.7 in section 4). Alternative positions may be used if the Pitot tube-temperature gauge system is calibrated according to the procedure of section 4. Provided that a difference of not more than 1 percent in the average velocity measurement is introduced, the temperature gauge need not be attached to the Pitot tube; this alternative is subject to the approval of the Administrator.

Pressure probe and gauge.   This is a piezometer tube and mercury- or water-filled U-tube manometer capable of measuring stack pressure to within 2.5 mm (0.1 in) Hg. The static tap of a standard-type Pitot tube or one leg of a type S Pitot tube with the face opening planes positioned parallel to the gas flow may also be used as the pressure probe.

Barometer.   This is a mercury, aneroid, or other barometer capable of measuring atmospheric pressure to within 2.5 mm (0.1 in) Hg. See Note in method 5, section 2.1.9.

Gas density determination equipment.   This includes method 3 equipment, if needed (see section 3.6), to determine the stack gas dry molecular weight, and reference method 4 or method 5 equipment for moisture content determination; other methods may be used subject to approval of the Administrator.

Procedure. Measure the velocity head and temperature at the traverse points specified by method 1. Ensure that the proper differential pressure gauge is being used for the range of D_p values encountered. If it is necessary to change to a more sensitive gauge, do so, and remeasure the D_p and temperature readings at each traverse point. Conduct a posttest leak check (mandatory), as described above, to validate the traverse run.

Measure the static pressure in the stack. One reading is usually adequate.

Determine the atmospheric pressure.

Determine the stack gas dry molecular weight. For combustion processes or processes that emit essentially CO_2, O_2, CO, and N_2 use method 3. For processes emitting essentially air, an analysis need not be conducted; use a dry molecular weight of 29.0. For other processes, other methods, subject to the approval of the Administrator, must be used.

Obtain the moisture content from reference method 4 (or equivalent) or from method 5.

Determine the cross-sectional area of the stack or duct at the sampling location. Whenever possible, physically measure the stack dimensions rather than use blueprints.

Method 2A: Direct measurement of gas volume through pipes and small ducts

Applicability. This method applies to the measurement of gas flow rates in pipes and small ducts, either in-line or at exhaust positions, within the temperature range of 0 to 50°C.

Principle. A gas volume meter is used to measure gas volume directly. Temperature and pressure measurements are made to correct the volume to standard conditions.

Apparatus. Specifications for the apparatus are given below. Any other apparatus that has been demonstrated (subject to approval of the Administrator) to be capable of meeting the specifications will be considered acceptable.

Gas volume meter. This is a positive displacement meter, turbine meter, or other direct measuring device capable of measuring volume to within 2 percent. The meter shall be equipped with a temperature sensor (accurate to within ±2 percent of the minimum absolute temperature) and a pressure gauge (accurate to within ±2.5 mmHg). The manufacturer's recommended capacity of the meter shall be sufficient for the expected maximum and minimum flow rates at the sampling conditions. Temperature, pressure, corrosive characteristics, and pipe size are factors to consider in choosing a suitable gas meter.

Barometer. This is a mercury, aneroid, or other barometer capable of measuring atmospheric pressure to within 2.5 mmHg. See also Note in method 5.

Stopwatch. It must be capable of measurement to within 1 s.

Procedure. For sources with continuous, steady emission flow rates, record the initial meter volume reading, meter temperature(s), and meter pressure; and start the stopwatch. Throughout the test period, record the meter temperatures and pressure so that average values can be determined. At the end of the test, stop the timer, and record the elapsed time, final volume reading, meter temperatures, and pressure. Record the barometric pressure at the beginning and end of the test run. Record the data in a table.

For sources with noncontinuous, nonsteady emission flow rates, add the following to the procedure: Record all the meter parameters and the start and stop times corresponding to each process cyclical or noncontinuous event.

Method 2B: Determination of exhaust gas volume flow rate from gasoline vapor incinerators

Applicability. This method applies to the measurement of exhaust volume flow rate from incinerators that process gasoline vapors consisting primarily of alkanes, alkenes, and/or arenes (aromatic hydrocarbons). It is assumed that the amount of auxiliary fuel is negligible.

Principle. The incinerator exhaust flow rate is determined by carbon balance. Organic carbon concentration and volume flow rate are measured at the incinerator inlet. Organic carbon, carbon dioxide (CO_2), and carbon monoxide (CO) concentrations are measured at the outlet. Then the ratio of total carbon at the incinerator inlet and outlet is multiplied by the inlet volume to determine the exhaust volume and volume flow rate.

Apparatus

Volume meter—equipment described in method 2A.

Organic analyzers (2)—equipment described in method 25A or 25B.

CO analyzer—equipment described in method 10.

CO_2 analyzer—a nondispersive infrared (NDIR) CO_2 analyzer and supporting equipment with comparable specifications as CO analyzer described in method 10.

Procedure. Install a volume meter in the vapor line to incinerator inlet according to the procedure in method 2A. At the volume meter inlet, install a sample probe as described in method 25A. Connect to the probe a leaktight, heated (if necessary to prevent condensation) sample line (stainless steel or equivalent) and an organic analyzer system as described in method 25A or 25B. Three sample analyzers are required for the incinerator exhaust: CO_2, CO, and organic analyzers. A sample manifold with a single sample probe may be used. Install a sample probe as described in method 25A. Connect a leaktight, heated sample line to the sample probe. Heat the sample line sufficiently to prevent any condensation. The incinerator exhaust flow rate is determined by carbon balance. Organic carbon concentration and volume flow rate are measured at the incinerator inlet. Organic carbon, carbon dioxide, and carbon monoxide concentrations are measured at the outlet. Then the ratio of total carbon at the incinerator inlet and outlet is multiplied by the inlet volume to determine the exhaust volume and volume flow rate.

The output of each analyzer must be permanently recorded on an analog strip chart, digital recorder, or other recording device. The chart speed or number of readings per time unit must be similar for all analyzers so that data can be correlated. The minimum data recording requirement for each analyzer is one measurement value per minute.

Method 2C: Determination of stack gas velocity and volumetric flow rate from small stacks or ducts (standard Pitot tube)

Applicability. The applicability of this method is identical to that of method 2, except it is limited to stationary source stacks or ducts less than about 0.30 m (12 in) in diameter, or 0.071 m^2 (113 in^2) in cross-sectional area, but equal to or greater than about 0.10 meter (4 in) in diameter, or 0.0081 m^2 (12.57 in^2) in cross-sectional area.

The apparatus, procedure, calibration, calculations, and bibliography are the same as in method 2, except as noted in the following sections.

Principle. The average gas velocity in a stack or duct is determined from the gas density and from measurement of velocity heads with a standard Pitot tube.

Apparatus

Standard Pitot tube (instead of type S)—a standard Pitot tube which meets the specifications of method 2. Use a coefficient of 0.99 unless it is calibrated against another standard Pitot tube with an NBS-traceable coefficient.

Alternative Pitot tube—a modified hemispherical-nose Pitot tube (see fig. 2C.1), which features a shortened stem and enlarged impact and static pressure holes. Use a coefficient of 0.99 unless it is calibrated as mentioned in section 2.1 above. This Pitot tube is useful in particulate liquid droplet-laden gas streams when a back-purge is ineffective.

Procedure. Follow the general procedures of method 2, except conduct the measurements at the traverse points specified in method 1A. The static and impact pressure holes of standard Pitot tubes are susceptible to plugging in particulate-laden gas streams. Therefore, the tester must furnish adequate proof that the openings of the Pitot tube have not plugged during the traverse period; this can be done by taking the velocity head (Δp) reading at the final traverse point, cleaning out the impact and static holes of the standard Pitot tube by back-purging with pressurized air, and then taking another Δp reading. If the Δp readings made before and after the air purge are the same (± 5 percent), the traverse is acceptable. Otherwise, reject the run. Note that if the Δp at the final traverse point is unsuitably low, another point may be selected. If back-purging at regular intervals is part of the procedure, then take comparative Δp readings, as above, for the last two back-purges at which suitably high Δp readings are observed.

Method 2D: Measurement of gas volume flow rates in small pipes and ducts

Applicability. This method applies to the measurement of gas flow rates in small pipes and ducts. It can only be applied to intermittent or variable gas flows with particular caution.

Principle. All the gas flow in the pipe or duct is directed through a rotameter, orifice plate, or similar device to measure flow rate or pressure drop. The device has been previously calibrated in a manner that ensures its proper calibration for the gas being measured. Absolute temperature and pressure measurements are made to adjust volume flow rates to standard conditions.

Apparatus. Specifications for the apparatus are given below. Any other apparatus that has been demonstrated (subject to approval of the Administrator) to be capable of meeting the specifications will be considered acceptable.

Gas metering rate or flow element device—a rotameter, orifice plate, or other volume rate or pressure drop measuring device capable of measuring the stack flow rate to within 5 percent. The metering device shall be equipped with a temperature gauge accurate to within 2 percent of the minimum absolute stack temperature and a pressure gauge (accurate to within 5 mmHg). The capacity of the metering device shall be sufficient for the expected maximum and minimum flow rates at the stack gas conditions. The magnitude and variability of stack gas flow rate, molecular weight, temperature, pressure, dew point, corrosive characteristics, and pipe or duct size are factors to consider in choosing a suitable metering device.

Barometer—same as method 2.

Stopwatch—capable of measurement to within 1 s.

Procedure—volume rate measurement
Continuous, steady flow. At least once an hour, record the metering device flow rate or pressure drop reading, and the metering device temperature and pressure. Make a minimum of 12 equally spaced readings of each parameter during the test period. Record the barometric pressure at the beginning and end of the test period.

Noncontinuous and nonsteady flow. Use volume rate devices with particular caution. Calibration will be affected by variation in stack gas temperature, pressure, and molecular weight. Use the procedure above with the addition of the following: Record all the metering device parameters on a time interval frequency sufficient to adequately profile each process cyclical or noncontinuous event. A multichannel continuous recorder may be used.

Method 2E: Determination of landfill gas production flow rate

Applicability. This method applies to the measurement of landfill gas (LFG) production flow rate from municipal solid waste (MSW) landfills and is used to calculate the flow rate of nonmethane organic compounds (NMOCs) from landfills. This method also applies to calculating a site-specific k value as provided in §60.754(a)(4). It is unlikely that a site-specific k value obtained through method 2E testing will lower the annual emission estimate below 50 Mg/yr NMOC unless the tier 2 emission estimate is only slightly higher than 50 Mg/yr NMOC.

Dry regions may show a more significant difference between the default and calculated k values than wet regions.

Principle. Extraction wells are installed either in a cluster of three or at five locations dispersed throughout the landfill. A blower is used to extract LFG from the landfill. LFG composition, landfill pressures near the extraction well, and volumetric flow rate of LFG extracted from the wells are measured, and the landfill gas production flow rate is calculated.

Apparatus
Standard Pitot tube and differential pressure gauge for flow rate calibration with standard Pitot—same as method 2.

Gas flow measuring device—permanently mounted type S Pitot tube or an orifice meter.

Barometer—same as in method 4.

Differential pressure gauge—water-filled U-tube manometer or equivalent, capable of measuring within 0.02 mmHg, for measuring the pressure of the pressure probes.

Procedure
LFG flow rate measurement. Determine the flow rate of LFG from the test wells continuously during testing with an orifice meter. Alternative methods to measure the LFG flow rate may be used with the approval of the EPA Administrator. Attach the wells to the blower-and-flare assembly. The individual wells may be ducted to a common header so that a single blower-and-flare assembly and flowmeter may be used. Use the proper procedures to calibrate the flowmeter.

Measure the LFG temperature and the static flow rate of each well once during static testing, using a flow measurement device, such as a type S Pitot tube, and measure the temperature of the landfill gas. The flow measurements should be made either just before or just after the measurements of the probe pressures and are used in determining the initial flow from the extraction well during the short-term testing. The temperature measurement is used in the check for infiltration.

Short-term testing. The purpose of short-term testing is to determine the maximum vacuum that can be applied to the wells without infiltration of air into the landfill. The short-term testing is done on one well at a time. During the short-term testing, burn LFG with a flare or incinerator.

At this maximum vacuum, measure $\overline{P}$ every 8 h for 24 h and record the LFG flow rate as Q_s and the probe gauge pressures for all the probes as P_f. Convert the gauge pressures of the deep probes to absolute pressures for each 8-h reading at Q_s.

For each probe, average the 8-h deep pressure probe readings and record as P_{fa}.

For each probe, compare the initial average pressure P_{ia} from section 3.6.1 to the final average pressure P_{fa}. Determine the farthest point from the wellhead along each radial arm where $P_{fa} \leq P_{ia}$. This distance is the maximum radius of influence (ROI), which is the distance from the well affected by the vacuum. Average these values to determine the average maximum radius of influence R_{ma}.

The average R_{ma} may also be determined by plotting on semilog paper the pressure differentials $P_{fa} - P_{ia}$ on the y axis (abscissa) versus the distances (3, 15, 30, and 45 m) from the wellhead on the x axis (ordinate). Use a linear regression analysis to determine the distance when the pressure differential is zero. Additional pressure probes may be used to obtain more points on the semilog plot of pressure differentials versus distances.

Calculate the depth D_{st} affected by the extraction well during the short-term test as follows. If the computed value of D_{st} exceeds the depth of the landfill, set D_{st} equal to the landfill depth.

Calculate the void volume for the extraction well V. Repeat for each well. Calculate the total void volume of the test wells V_v by summing the void volumes V of each well.

Long-term testing.   The purpose of long-term testing is to determine the methane generation rate constant k. Use the blower to extract LFG from the wells. If a single blower-and-flare assembly and common header system are used, open all control valves and set the blower vacuum equal to the highest stabilized blower vacuum demonstrated by any individual well. Every 8 h sample the LFG from the wellhead sample port; measure the gauge pressures of the shallow pressure probes, the blower vacuum, the LFG flow rate; and use the criteria for infiltration and method 3C to check for infiltration. If infiltration is detected, do not reduce the blower vacuum, but reduce the LFG flow rate from the well by adjusting the control valve on the wellhead. Adjust each affected well individually. Continue until the equivalent of two total void volumes V_v has been extracted, or until $V_t = 2V_v$.

Calculate V_t, the total volume of LFG extracted from the wells.

Record the final stabilized flow rate as Q_f. If, during the long-term testing, the flow rate does not stabilize, calculate Q_f by averaging the last 10 recorded flow rates.

For each deep probe, convert each gauge pressure to absolute pressure. Average these values and record as P_{sa}. For each probe, compare P_{ia} to P_{sa}. Determine the farthest point from the wellhead along each radial arm where $P_{sa} \leq P_{ia}$. This distance is the stabilized radius of influence. Average these values to determine the average stabilized radius of influence R_{sa}.

Determine the NMOC mass emission rate.

Method 3: Gas analysis for the determination of dry molecular weight

Applicability. This method is applicable for determining carbon dioxide and oxygen concentrations and dry molecular weight of a sample from a gas stream of a fossil-fuel combustion process. The method may also be applicable to other processes where it has been determined that compounds other than CO_2, O_2, CO, and N_2 are not present in concentrations sufficient to affect the results.

Other methods, as well as modifications to the procedure described here, are also applicable for some of or all the above determinations. Examples of specific methods and modifications include (1) a multipoint sampling method using an Orsat analyzer to analyze individual grab samples obtained at each point; (2) a method using CO_2 or O_2 and stoichiometric calculations to determine dry molecular weight; and (3) assignment of a value of 30.0 for dry molecular weight, in lieu of actual measurements, for processes burning natural gas, coal, or oil. These methods and modifications may be used, but are subject to the approval of the EPA Administrator.

Principle. A gas sample is extracted from a stack by one of the following methods: (1) single-point, grab sampling; (2) single-point, integrated sampling; or (3) multipoint, integrated sampling. The gas sample is analyzed for percent CO_2, percent O_2, and, if necessary, percent CO. For dry molecular weight determination, either an Orsat or a Fyrite analyzer may be used.

Apparatus. As an alternative to the sampling apparatus and systems described here, other sampling systems (e.g., liquid displacement) may be used provided they are capable of obtaining a representative sample and maintaining a constant sampling rate and are otherwise capable of yielding acceptable results. Use of such systems is subject to the approval of the Administrator.

Analysis. Use an Orsat or Fyrite type of combustion gas analyzer. For Orsat and Fyrite analyzer maintenance and operation procedures, follow the instructions recommended by the manufacturer, unless otherwise specified here.

Procedure. Place the probe in the stack, with the tip of the probe positioned at the sampling point; purge the sampling line long enough to allow at least five exchanges. Draw a sample into the analyzer, and immediately analyze it for percent CO_2 and percent O_2. Determine the percentage of the gas that is N_2 and CO by subtracting the sum of the percent CO_2 and percent O_2 from 100 percent. Calculate the dry molecular weight. Repeat the sampling, analysis, and calculation procedures until the dry molecular weights of any three grab samples differ from their mean by no more than 0.3 g/g mol (0.3 lb/lb mol). Average these three molecular weights, and report the results to the nearest 0.1 g/g mol (0.1 lb/lb mol).

Obtain one integrated flue gas sample during each pollutant emission rate determination. Within 8 h after the sample is taken, analyze it for percent CO_2 and percent O_2, using either an Orsat analyzer or a Fyrite-type combustion gas analyzer. If an Orsat analyzer is used, it is recommended that Orsat leak check be performed before this determination; however, the check is optional. Determine the percentage of the gas that is N_2 and CO by subtracting the sum of the percent CO_2 and percent O_2 from 100 percent. Calculate the dry molecular weight.

6.5.2   Inorganic parameters

Method 3A: Determination of oxygen and carbon dioxide concentrations in emissions from stationary sources (instrumental analyzer procedure)

Applicability.   This method is applicable to the determination of oxygen and carbon dioxide concentrations in emissions from stationary sources only when specified within the regulations.

Principle.   A sample is continuously extracted from the effluent stream; a portion of the sample stream is conveyed to the instrumental analyzer(s) for determination of O_2 and CO_2 concentration(s). Performance specifications and test procedures are provided to ensure reliable data.

Apparatus

Measurement system—any measurement system for O_2 or CO_2 that meets the specifications of this method. A schematic of an acceptable measurement system is shown in fig. 6C.1 of method 6C. The essential components of the measurement system are described below.

Sample probe—a leakfree probe of sufficient length to traverse the sample points.

Sample line—tubing to transport the sample gas from the probe to the moisture removal system. A heated sample line is not required for systems that measure the O_2 or CO_2 concentration on a dry basis or that transport dry gases.

Sample transport line, calibration valve assembly, moisture removal system, particulate filter, sample pump, sample flow rate control, sample gas manifold, and data recorder—same as in method 6C, except that the requirements to use stainless steel, Teflon, and nonreactive glass filters do not apply.

Gas analyzer—an analyzer to determine continuously the O_2 or CO_2 concentration in the sample gas stream. A means of controlling the analyzer flow rate and a device for determining proper sample flow rate (e.g., precision rotameter, pressure gauge downstream of all flow controls, etc.) shall be provided at the analyzer. The requirements for measuring and controlling the analyzer flow rate are not applicable if data are presented that demonstrate the analyzer is insensitive to flow variations over the range encountered during the test.

Method 3B: Gas analysis for the determination of emission rate correction factor or excess air

Applicability. This method is applicable for determining carbon dioxide, oxygen, and carbon monoxide concentrations of a sample from a gas stream of a fossil-fuel combustion process for excess air or emission rate correction factor calculations.

Other methods, as well as modifications to the procedure described here, are also applicable for all the above determinations. Examples of specific methods and modifications include (1) a multipoint sampling method using an Orsat analyzer to analyze individual grab samples obtained at each point and (2) a method using CO_2 or O_2 and stoichiometric calculations to determine excess air. These methods and modifications may be used, but are subject to the approval of the EPA Administrator.

Principle. A gas sample is extracted from a stack by one of the following methods: (1) single-point, grab sampling; (2) single-point, integrated sampling; or (3) multipoint, integrated sampling. The gas sample is analyzed for percent CO_2, percent O_2, and, if necessary, percent CO. An Orsat analyzer must be used for excess air or emission rate correction factor determinations.

Apparatus. The alternative sampling systems are the same as those mentioned in method 3.

Grab sampling and integrated sampling—same as in method 3.

Analysis—an Orsat analyzer only. For low CO_2 (less than 4.0%) or high O_2 (greater than 15.0%) concentrations, the measuring burette of the Orsat must have at least 0.1 percent subdivisions. For Orsat maintenance and operation procedures, follow the instructions recommended by the manufacturer, unless otherwise specified here.

Procedure

Single-point, grab sampling and analytical procedure.   Place the probe in the stack, with the tip of the probe positioned at the sampling point; purge the sampling line long enough to allow at least five exchanges. Draw a sample into the analyzer. For emission rate correction factor determinations, immediately analyze the sample for percent CO_2 or percent O_2. If excess air is desired, proceed as follows: (1) Immediately analyze the sample for percent CO_2, O_2, and CO; (2) determine the percentage of the gas that is N_2 by subtracting the sum of the percent CO_2, percent O_2, and percent CO from 100 percent; and (3) calculate percent excess air.

To ensure complete absorption of the CO_2, O_2, or, if applicable, CO, make repeated passes through each absorbing solution until two consecutive readings are the same. Several passes (three or four) should be made between readings. (If constant readings cannot be obtained after three consecutive readings, replace the absorbing solution.) *Note:* Since this single-point, grab sampling and analytical procedure is normally conducted in conjunction with a single-point, grab sampling and analytical procedure for a pollutant, only one analysis is ordinarily conducted. Therefore, great care must be taken to obtain a valid sample and analysis. Although in most cases only CO_2 or O_2 is required, it is recommended that both CO_2 and O_2 be measured.

Single-point, integrated sampling and analytical procedure. Obtain one integrated flue gas sample during each pollutant emission rate determination. For emission rate correction factor determination, analyze the sample within 4 h after it is taken for percent CO_2 or percent O_2. The Orsat analyzer must be leak-checked (see method 3) before the analysis. If excess air is desired, proceed as follows: (1) Within 4 h after the sample is taken, analyze it for percent CO_2, O_2, and CO; (2) determine the percentage of the gas that is N_2 by subtracting the sum of the percent CO_2, percent O_2, and percent CO from 100 percent; and (3) calculate percent excess air.

To ensure complete absorption of the CO_2, O_2, or, if applicable, CO, follow the procedure described in the method. *Note:* Although in most instances only CO_2 or O_2 is required, it is recommended that both CO_2 and O_2 be measured.

Multipoint, integrated sampling and analytical procedure.   Follow the procedures outlined above, except for the following: Traverse all sampling points, and sample at each point for an equal length of time.

Method 3C: Determination of carbon dioxide, methane, nitrogen, and oxygen from stationary sources

Applicability.   This method applies to the analysis of carbon dioxide, methane (CH_4), nitrogen, and oxygen in samples from municipal solid waste landfills and other sources when specified in an applicable subpart.

Principle. A portion of the sample is injected into a gas chromatograph (GC); and the CO_2, CH_4, N_2, and O_2 concentrations are determined by using a thermal conductivity detector (TCD) and integrator.

Apparatus

Gas chromatograph—one having at least the following components:

Separation column—appropriate column(s) to resolve CO_2, CH_4, N_2, O_2, and other gas components that may be present in the sample.

Sample loop—Teflon or stainless-steel tubing of the appropriate diameter. *Note:* Mention of trade names or specific products does not constitute endorsement or recommendation by the Environmental Protection Agency.

Conditioning system—to maintain the column and sample loop at constant temperature.

Thermal conductivity detector.

Recorder—recorder with linear strip chart. Electronic integrator (optional) is recommended.

Analysis. Purge the sample loop with sample, and allow to come to atmospheric pressure before each injection. Analyze each sample in duplicate, and calculate the average sample area *A*. The results are acceptable when the peak areas for two consecutive injections agree within 5 percent of their average. If they do not agree, run additional samples until consistent area data are obtained.

Method 4: Determination of moisture content in stack gases

Principle. A gas sample is extracted at a constant rate from the source; moisture is removed from the sample stream and determined either volumetrically or gravimetrically.

Applicability. This method is applicable for determining the moisture content of stack gas. Two procedures are given. The first is a reference method, for accurate determinations of moisture content (such as are needed to calculate emission data). The second (not described here) is an approximation method, which provides estimates of percent moisture to aid in setting isokinetic sampling rates prior to a pollutant emission measurement run. The approximation method described here is only a suggested approach; alternative means for approximating the moisture content, e.g., drying tubes, wet bulb–dry bulb techniques, condensation techniques, stoichiometric calculations, previous experience, etc., are also acceptable.

The reference method is often followed simultaneously with a pollutant emission measurement run; when it is, calculation of percent isokinetic, pollutant emission rate, etc., for the run shall be based upon the results of the reference method or its equivalent; these calculations shall not be based upon the results of the approximation method, unless the approximation method is shown, to the satisfaction of the EPA Administrator, to be capable of yielding results within 1 percent H_2O of the reference method.

Note: The reference method may yield questionable results when applied to saturated gas streams or to streams that contain water droplets. Therefore, when these conditions exist or are suspected, a second determination of the moisture content shall be made simultaneously with the reference method, as follows: Assume that the gas stream is saturated. Attach a temperature sensor [capable of measuring to within 1°C (2°F)] to the reference method probe. Measure the stack gas temperature at each traverse point during the reference method traverse; calculate the average stack gas temperature. Next, determine the moisture percentage, either by (1) using a psychrometric chart and making appropriate corrections if the stack pressure is different from that of the chart or (2) using saturation vapor pressure tables. In cases where the psychrometric chart or the saturation vapor pressure tables are not applicable (based on evaluation of the process), alternative methods, subject to the approval of the Administrator, shall be used.

Reference method

Apparatus. All components shall be maintained and calibrated according to the procedures in method 5.

Probe—stainless-steel or glass tubing, sufficiently heated to prevent water condensation and equipped with a filter, either in-stack (e.g., a plug of glass wool inserted into the end of the probe) or heated out-stack (e.g., as described in method 5), to remove particulate matter. When stack conditions permit, other metals or plastic tubing may be used for the probe, subject to the approval of the Administrator.

Condenser—See method 5 for a description of an acceptable type of condenser and for alternative measurement systems.

Cooling system—an ice bath container and crushed ice (or equivalent) to aid in condensing moisture.

Metering system—same as in method 5, except do not use sampling systems designed for flow rates higher than 0.0283 m³/min (1.0 ft³/min). Other metering systems, capable of maintaining a constant sampling rate to within 10 percent and determining sample gas volume to within 2 percent, may be used subject to the approval of the Administrator.

Barometer—mercury, aneroid, or other barometer capable of measuring atmospheric pressure to within 2.5 mm (0.1 in) Hg.

Graduated cylinder and/or balance—to measure condensed water and moisture caught in the silica gel to within 1 mL or 0.5 g. Graduated cylinders shall have subdivisions no greater than 2 mL. Most laboratory balances are capable of weighing to the nearest 0.5 g or less. These balances are suitable for use here.

Procedure. The following procedure is written for a condenser system (such as the impinger system described in method 5), incorporating volumetric analysis to measure the condensed moisture and silica gel and gravimetric analysis to measure the moisture leaving the condenser. Place known volumes of water in the first two impingers. Weigh and record the weight of the silica gel to the nearest 0.5 g, and transfer the silica gel to the fourth impinger; alternatively, the silica gel may first be transferred to the impinger, and the weight of the silica gel plus impinger recorded. Next, measure the volume of the moisture condensed to the nearest milliliter. Determine the increase in weight of the silica gel (or silica gel plus impinger) to the nearest 0.5 g. Record this information and calculate the moisture percentage, as described below.

Method 5: Determination of particulate emissions from stationary sources

Principle. Particulate matter (PM) is withdrawn isokinetically from the source and is collected on a glass fiber filter maintained at a temperature in the range of 120 ± 14°C (248 ± 25°F) or such other temperature as specified by an applicable subpart of the standards or approved by the EPA Administrator for a particular application. The PM mass, which includes any material that condenses at or above the filtration temperature, is determined gravimetrically after removal of uncombined water.

Applicability. This method is applicable for the determination of PM emissions from stationary sources.

Apparatus

Glass weighing dishes

Desiccator

Analytical balance—to measure to within 0.1 mg

Balance—to measure to within 0.5 g

Beakers—250 mL

Hygrometer—to measure the relative humidity of the laboratory environment

Temperature gauge—to measure the temperature of the laboratory environment

Procedure.   Record the data required on a sheet. Handle each sample container as follows:

Container 1.   Leave the contents in the shipping container, or transfer the filter and any loose PM from the sample container to a tared glass weighing dish. Desiccate for 24 h in a desiccator containing anhydrous calcium sulfate. Weigh to a constant weight, and report the results to the nearest 0.1 mg. For purposes of this section, 4.3, the term *constant weight* means a difference of no more than 0.5 mg or 1 percent of total weight less tare weight, whichever is greater, between two consecutive weighings, with no less than 6 h of desiccation time between weighings.

Alternatively, the sample may be oven-dried at 105°C (220°F) for 2 to 3 h, cooled in the desiccator, and weighed to a constant weight, unless otherwise specified by the Administrator. The tester may also opt to oven-dry the sample at 105°C (220°F) for 2 to 3 h, weigh the sample, and use this weight as a final weight.

Container 2.   Note the level of liquid in the container, and confirm on the analysis sheet whether leakage occurred during transport. If a noticeable amount of leakage has occurred, either void the sample or use methods, subject to the approval of the Administrator, to correct the final results. Measure the liquid in this container either volumetrically to ±1 mL or gravimetrically to ±0.5 g. Transfer the contents to a tared 250-mL beaker, and evaporate to dryness at ambient temperature and pressure. Desiccate for 24 h, and weigh to a constant weight. Report the results to the nearest 0.1 mg.

Container 3.   Weigh the spent silica gel (or silica gel plus impinger) to the nearest 0.5 g, using a balance. This step may be conducted in the field.

"Acetone blank" container.   Measure the acetone in this container either volumetrically or gravimetrically. Transfer the acetone to a tared 250-mL beaker, and evaporate to dryness at ambient temperature and pressure. Desiccate for 24 h, and weigh to a constant weight. Report the results to the nearest 0.1 mg.

Note.   At the option of the tester, the contents of container 2 as well as the acetone blank container may be evaporated at temperatures higher than ambient. If evaporation is done at an elevated temperature, the temperature must be below the boiling point of the solvent; also, to prevent "bumping," the evaporation process must be closely supervised, and the contents of the beaker must be swirled occasionally to maintain an even temperature. Use extreme care, as acetone is highly flammable and has a low flash point.

Method 5A: Determination of particulate matter emissions from the asphalt processing and asphalt roofing industry

Applicability. This method applies to the determination of particulate matter emissions from asphalt roofing industry process saturators, blowing stills, and other sources as specified in the regulations.

Principle. The PM is withdrawn isokinetically from the source and collected on a glass fiber filter maintained at a temperature of 42 ± 10°C (108 ± 18°F). The PM mass, which includes any material that condenses at or above the filtration temperature, is determined gravimetrically after removal of uncombined water.

Apparatus. For analysis, the following equipment is needed:

Glass weighing dishes, desiccator, analytical balance, balance, hygrometer, and temperature gauge—same as in method 5.

Beakers—glass, 250 and 500 mL.

Separatory funnel—100 mL or greater.

Analysis. Record the data required on a sheet. Handle each sample container as follows:

Container 1 (filter). Transfer the filter from the sample container to a tared glass weighing dish, and desiccate for 24 h in a desiccator containing anhydrous calcium sulfate. Rinse container 1 with a measured amount of TCE, and analyze this rinse with the contents of container 2. Weigh the filter to a constant weight. For the purpose of this section, the term *constant weight* means a difference of no more than 10 percent or 2 mg (whichever is greater) between two consecutive weighings made 24 h apart. Report the final weight to the nearest 0.1 mg as the average of these two values.

Container 2 (probe to filter holder). Before adding the rinse from container 1 to container 2, note the level of liquid in the container, and confirm on the analysis sheet whether leakage occurred during transport. If noticeable leakage occurred, either void the sample or take steps, subject to the approval of the Administrator, to correct the final results.

Measure the liquid in this container either volumetrically to ±1 mL or gravimetrically to ±0.5 g. Check to see whether there is any appreciable quantity of condensed water present in the TCE rinse (look for a boundary layer or phase separation). If the volume of condensed water appears larger than 5 mL, separate the oil-TCE fraction from the water fraction using a separatory funnel. Measure the volume of the water phase to the nearest milliliter; adjust the stack gas moisture content, if necessary (see sections 6.4 and 6.5). Next, extract the water phase with several 25-mL portions of TCE until, by visual observation, the TCE does not remove any additional organic material. Evaporate the remaining water fraction to dryness at 93°C (200°F), desiccate for 24 h, and weigh to the nearest 0.1 mg.

Treat the total TCE fraction (including TCE from the filter container rinse and water-phase extractions) as follows: Transfer the TCE and oil to a tared beaker, and evaporate at ambient temperature and pressure. The evaporation of TCE from the solution may take several days. Do not desiccate the sample until the solution reaches an apparent constant volume or until the odor of TCE is not detected. When it appears that the TCE has evaporated, desiccate the sample, and weigh it at 24-h intervals to obtain a "constant weight" (as defined for container 1 above). The *total weight* for container 2 is the sum of the evaporated PM weight of the TCE-oil and water-phase fractions. Report the results to the nearest 0.1 mg.

Container 3 (silica gel). This step may be conducted in the field. Weigh the spent silica gel (or silica gel plus impinger) to the nearest 0.5 g using a balance.

"TCE blank" container. Measure TCE in this container either volumetrically or gravimetrically. Transfer the TCE to a tared 250-mL beaker, and evaporate to dryness at ambient temperature and pressure. Desiccate for 24 h, and weigh to a constant weight. Report the results to the nearest 0.1 mg. *Note:* In order to facilitate the evaporation of TCE liquid samples, these samples may be dried in a controlled temperature oven at temperatures up to 38°C (100°F) until the liquid is evaporated.

Method 5B: Determination of nonsulfuric acid particulate matter from stationary sources

Applicability. This method is to be used for determining nonsulfuric acid particulate matter from stationary sources. Use of this method must be specified by an applicable subpart, or approved by the EPA Administrator for a particular application.

Principle. The PM is withdrawn isokinetically from the source using the method 5 train at 160°C (320°F). The collected sample is then heated in the oven at 160°C (320°F) for 6 h to volatilize any condensed sulfuric acid that may have been collected, and the nonsulfuric acid PM mass is determined gravimetrically.

Procedure. The procedure is identical to EPA method 5 except for the following:

Dry the probe sample at ambient temperature. Then oven-dry the probe and filter samples at a temperature of 160 ± 5°C (320 ± 10°F) for 6 h. Cool in a desiccator for 2 h, and weigh to constant weight. Use the applicable specifications and techniques of section 4.3 of method 5 for this determination.

Method 5D: Determination of particulate matter emissions from positive-pressure fabric filters

Applicability. This method applies to the determination of particulate matter emissions from positive-pressure fabric filters. Emissions are determined in terms of concentration (mg/m^3) and emission rate (kg/h).

The General Provisions of 40 CFR Part 60, §60.8(c), require that the owner or operator of an affected facility provide performance testing facilities. Such performance testing facilities include sampling ports, safe sampling platforms, safe access to sampling sites, and utilities for testing. It is intended that affected facilities also provide sampling locations that meet the specification for adequate stack length and minimal flow disturbances as described in method 1. Provisions for testing are often overlooked factors in designing fabric filters or are extremely costly. The purpose of this procedure is to identify appropriate alternative locations and procedures for sampling the emissions from positive-pressure fabric filters. The requirements that the affected facility owner or operator provide adequate access to performance testing facilities remain in effect.

Principle. PM is withdrawn isokinetically from the source and collected on a glass-fiber filter maintained at a temperature at or above the exhaust gas temperature up to a nominal 120°C (120 ± 14°C or 248 ± 25°F). The PM mass, which includes any material that condenses at or above the filtration temperature, is determined gravimetrically after removal of uncombined water.

Apparatus. The equipment requirements for the sampling train, sample recovery, and analysis are the same as specified in sections 2.1, 2.2, and 2.3, respectively, of method 5.

Procedure
Velocity determination. The velocities of exhaust gases from positive-pressure baghouses are often too low to measure accurately with the type S Pitot tube specified in method 2 [i.e., velocity head ≤ 1.3 mmH_2O (0.05 inH_2O)]. For these conditions, measure the gas flow rate at the fabric filter inlet following the procedures in method 2. Calculate the average gas velocity at the measurement site.

Use the average velocity calculated for the measurement site in determining and maintaining isokinetic sampling rates. *Note:* All sources of gas leakage, into or out of the fabric filter housing between the inlet measurement site and the outlet measurement site, must be blocked and made leaktight.

Velocity determinations at measurement sites with gas velocities within the range measurable with the type S Pitot [i.e., velocity head greater than 1.3 mmH$_2$O (0.05 inH$_2$O)] shall be conducted according to the procedures in method 2.

Sample analysis.   Follow the procedures specified in section 4.3 of method 5.

Method 5E: Determination of particulate emissions from wool fiberglass insulation manufacturing industry

Applicability.   This method is applicable for the determination of particulate matter (PM) emissions from wool fiberglass insulation manufacturing sources.

Principle.   The PM is withdrawn isokinetically from the source and collected on a glass-fiber filter maintained at a temperature in the range of 120 ± 14°C (248 ± 25°F) and in solutions of 0.1 N NaOH. The filtered particulate mass, which includes any material that condenses at or above the filtration temperature, is determined gravimetrically after removal of uncombined water. The condensed PM collected in the impinger solutions is determined as total organic carbon (TOC) by using a nondispersive infrared type of analyzer. The sum of the filtered PM mass and the condensed PM is reported as the total PM mass.

Apparatus.   The equipment list for analysis is the same as section 2.3 of method 5 with the additional equipment for TOC analysis as described below:

Sample blender or homogenizer—Waring type or ultrasonic.

Magnetic stirrer.

Hypodermic syringe—0- to 100-μL capacity.

Total organic carbon analyzer—Beckman Model 915 with 215 B infrared analyzer or equivalent and a recorder.

Beaker—30-mL.

Water bath—temperature controlled.

Volumetric flasks—1000-mL and 500-mL.

Procedure.   The procedures for analysis are the same as in section 4.3 of method 5 with exceptions noted as follows:

Container 1.   Determination of weight gain on the filter is the same as described for container 1 in section 4.3 of method 5 except that the filters must be dried at $20 \pm 6°C$ $(68 \pm 10°F)$ and ambient pressure.

Containers 2 and 3.   Analyze the contents of the containers 2 and 3 as described for container 2 in section 4.3 of method 5 except that evaporation of the samples must be at $20 \pm 6°C$ $(68 \pm 10°F)$ and ambient pressure.

Container 4.   Weigh the spent silica gel as described for container 3 in section 4.3 of method 5.

"Water and acetone blank" containers.   Determine the water and acetone blank values following the procedures for acetone blank container in section 4.3 of method 5. Evaporate the samples at ambient temperature $[20 \pm 6°C$ $(68 \pm 10°F)]$ and pressure.

Container 5.   For the determination of total organic carbon, perform two analyses on successive identical samples, i.e., total carbon and inorganic carbon. The desired quantity is the difference between the two values obtained. Both analyses are based on conversion of sample carbon to carbon dioxide for measurement by a nondispersive infrared analyzer. Results of analyses register as peaks on a strip-chart recorder.

The principal differences between operating parameters for the two channels involve the combustion tube packing material and temperature. In the total-carbon channel, a high-temperature $[950°C$ $(1740°F)]$ furnance heats a Hastelloy combustion tube packed with cobalt oxide–impregnated asbestos fiber. The oxygen in the carrier gas, the elevated temperature, and catalytic effect of the packing result in oxidation of both organic and inorganic carbonaceous material to CO_2 and steam. In the inorganic carbon channel, a low-temperature $[150°C$ $(300°F)]$ furnace heats a glass tub containing quartz chips wetted with 85% phosphoric acid. The acid liberates CO_2 and steam from inorganic carbonates. The operating temperature is below that required to oxidize organic matter. Follow the manufacturer's instructions for assembly, testing, calibration, and operation of the analyzer.

As samples collected in $0.1 N$ NaOH often contain a high measure of inorganic carbon that inhibits repeatable determinations of TOC, sample pretreatment is necessary. Measure and record the liquid volume of each sample. If the sample contains solids or immiscible liquid matter, homogenize the sample with a blender or ultrasonics until satisfactory repeatability is obtained. Transfer a representative portion of 10 to 15 mL to a 30-mL beaker, and acidify with about 2 drops of concentrated HCl to a pH of 2 or less. Warm the acidified sample at $50°C$ $(120°F)$ in a water bath for 15 min. While stirring the sample with a magnetic stirrer, withdraw a 20- to 50-μL sample from the beaker, and inject it into the total-carbon port of the analyzer. Measure the peak height. Repeat the injections until three consecutive peaks are obtained within 10 percent of the average.

Repeat the analyses for all the samples and the 0.1 N NaOH blank. Prepare standard curves for total carbon and for inorganic carbon of 10, 20, 30, 40, 50, 60, 80, and 100 mg/L by diluting with CO_2-free water 10, 20, 30, 40, and 50 mL of the two stock solutions to 1000 mL and 30, 40, and 50 mL of the two stock solutions to 500 mL. Inject samples of these solutions into the analyzer, and record the peak heights as described above. The acidification and warming steps are not necessary for preparation of the standard curve.

Ascertain the sample concentrations for the samples from the corrected peak heights for the samples by reference to the appropriate standard curve. Calculate the corrected peak height for the standards and the samples by deducting the blank correction as follows:

$$\text{Corrected peak height} = A - B$$

where A = peak height of standard or sample, mm or other appropriate unit, and B = peak height of blank, mm or other appropriate unit.

Note: If samples must be diluted for analysis, apply an appropriate dilution factor.

Method 5F: Determination of nonsulfate particulate matter from stationary sources

Applicability. This method is to be used for determining nonsulfate particulate matter from stationary sources. Use of this method must be specified by an applicable subpart of the standards or approved by the EPA Administrator for a particular application.

Principle. The PM is withdrawn isokinetically from the source using the method 5 train at 160°C (320°F). The collected sample is then extracted with water. A portion of the extract is analyzed for sulfate content. The remainder is neutralized with ammonium hydroxide before it is dried and weighed.

Apparatus. The apparatus is the same as that for method 5 with the following additions:

Erlenmeyer flasks—125-mL, with ground-glass joints.

Air condenser—with ground-glass joint compatible with the Erlenmeyer flasks.

Beakers—600-mL.

Volumetric flasks—1-L, 500-mL (one for each sample), 200-mL, and 50-mL (one for each sample and standard).

Pipette—5-mL (one for each sample and standard).

Ion chromatograph—one having at least the following components.

- *Columns*—An anion separation or other column capable of resolving the sulfate ion from other species present and a standard anion suppressor column. Suppressor columns are produced as proprietary items; however, one can be produced in the laboratory by using the resin available from BioRad Company, 32d and Griffin Streets, Richmond, California. Other systems which do not use suppressor columns may also be used.
- *Conductivity detector.*

Procedure

Sulfate (SO_4) analysis.   Allow the sample to settle until all solid material is at the bottom of the volumetric flask. If necessary, centrifuge a portion of the sample. Pipette 5 mL of the sample into a 50-mL volumetric flask, and dilute to 50 mL with water. Prepare a standard calibration curve according to section 5.1. Analyze the set of standards followed by the set of samples, using the same injection volume for both standards and samples. Repeat this analysis sequence followed by a final analysis of the standard set. Average the results. The two sample values must agree within 5 percent of their mean for the analysis to be valid. Perform this duplicate analysis sequence on the same day. Dilute any sample and the blank with equal volumes of water if the concentration exceeds that of the highest standard.

Document each sample chromatogram by listing the following analytical parameters: injection point, injection volume, sulfate retention time, flow rate, detector sensitivity setting, and recorder chart speed.

Sample residue.   Transfer the remaining contents of the volumetric flask to a tared 600-mL beaker. Rinse the volumetric flask, and add the rinsings to the tared beaker. Make certain that all particulate matter is transferred to the beaker. Evaporate the water in an oven, heated to 105°C until only about 100 mL of water remains. Remove the beakers from the oven, and allow them to cool.

After the beakers have cooled, add 5 drops of phenolphthalein indicator, and then add concentrated ammonium hydroxide until the solution turns pink. Return the samples to the oven at 105°C, and evaporate the samples to dryness. Cool the samples in a desiccator, and weigh the samples to constant weight.

Alternative procedures.   The following procedure may be used as an alternative to the procedure above.

Sample residue.   Place at least one clean glass-fiber filter for each sample in a Buchner funnel, and rinse the filters with water. Remove the filters from the funnel, and dry them in an oven at 105 ± 5°C; then cool in a desiccator. Weigh each filter to constant weight according to the procedure in method 5, section 4.3. Record the weight of each filter to the nearest 0.1 mg.

Assemble the vacuum filter apparatus, and place one of the clean, tared glass-fiber filters in the Buchner funnel. Decant the liquid portion of the extracted sample through the tared glass-fiber filter into a clean, dry, 500-mL filter flask. Rinse all the particulate matter remaining in the volumetric flask onto the glass-fiber filter with water. Rinse the particulate matter with additional water. Transfer the filtrate to a 500-mL volumetric flask, and dilute to 500 mL with water. Dry the filter overnight at 105 ± 5°C, cool in a desiccator, and weigh to the nearest 0.1 mg.

Dry a 350-mL beaker at 75 ± 5°C, and cool in a desiccator; then weigh to constant weight to the nearest 0.1 mg. Pipette 200 mL of the filtrate that was saved into a tared 250-mL beaker; add 5 drops of phenolphthalein indicator and sufficient concentrated ammonium hydroxide to turn the solution pink. Carefully evaporate the contents of the beaker to dryness at 75 ± 5°C. Check for dryness every 30 min. Do not continue to bake the sample once it has dried. Cool the sample in a desiccator, and weigh to constant weight to the nearest 0.1 mg.

Sulfate analysis. Adjust the flow rate through the ion-exchange column to 3 mL/min. Pipette a 20-mL aliquot of the filtrate onto the Stop of the ion-exchange column, and collect the eluate in a 500-mL volumetric flask. Rinse the column with two 15-mL portions of water. Stop collection of the eluate when the volume in the flask reaches 500 mL. Pipette a 20-mL aliquot of the eluate into a 250-mL Erlenmeyer flask, add 80 mL of 100% isopropanol and 2 to 4 drops of thorin indicator, and titrate to a pink endpoint, using 0.0100 N barium perchlorate. Repeat and average the titration volumes. Run a blank with each series of samples. Replicate titrations must agree within 1 percent or 0.2 mL, whichever is larger. Perform the ion-exchange and titration procedures on duplicate portions of the filtrate. Results should agree within 5 percent. Regenerate or replace the ion-exchange resin after 20 sample aliquots have been analyzed or if the endpoint of the titration becomes unclear.

Note: Protect the 0.0100 N barium perchlorate solution from evaporation at all times.

Method 5G: Determination of particulate emissions from wood heaters from a dilution tunnel sampling location

Applicability. This method is applicable for the determination of particulate matter emissions from wood heaters.

Principle. Particulate matter is withdrawn proportionally at a single point from a total collection hood and sampling tunnel that combines the wood heater exhaust with ambient dilution air. The particulate matter is collected on two glass-fiber filters in series. The filters are maintained at a temperature of no greater than 32°C (90°F). The particulate mass is determined gravimetrically after removal of uncombined water.

There are three sampling train approaches described in this method: (1) one dual-filter dry sampling train operated at about 0.015 m³/min, (2) one dual-filter plus impingers sampling train operated at about 0.015 m³/min, and (3) two dual-filter dry sampling trains operated simultaneously at any flow rate. Options 2 and 3 are referenced in section 7 of this method. The dual-filter sampling train equipment and operation, option 1, are described in detail in this method.

Apparatus. Glass weighing dishes, desiccator, analytical balance, beakers (250 mL or smaller), hygrometer, and temperature gauge as described in method 5, sections 2.3.1 through 2.3.3 and 2.3.5 through 2.3.7 are needed.

Procedure. Record the data required on a sheet. Use the same analytical balance for determining tare weight and final sample weights. Handle each sample container as follows:

Containers 1 and 2. Leave the contents in the sample containers, or transfer the filters and loose particulate to tared glass weighing dishes. Desiccate for no more than 36 h before the initial weighing, weigh to a constant weight, and report the results to the nearest 0.1 mg. For the purposes of this section, the term *constant* weight means a difference of no more than 0.5 mg or 1 percent of total sample weight (less tare weight), whichever is greater, between two consecutive weighings, with no less than 2 h between weighings.

Container 3. Note the level of liquid in the container, and confirm on the analysis sheet whether leakage occurred during transport. If a noticeable amount of leakage has occurred, either void the sample or use methods, subject to the approval of the EPA Administrator, to correct the final results. Determination of sample leakage is not applicable if sample recovery and analysis occur in the same room. Measure the liquid in this container either volumetrically to within 1 mL or gravimetrically to within 0.5 g. Transfer the contents to a tared 250-mL or smaller beaker, and evaporate to dryness at ambient temperature and pressure. Desiccate and weigh to a constant weight. Report the results to the nearest 0.1 mg.

"Acetone blank" container. Measure acetone in this container either volumetrically or gravimetrically. Transfer the acetone to a tared 250-mL or smaller beaker, and evaporate to dryness at ambient temperature and pressure. Desiccate and weigh to a constant weight. Report the results to the nearest 0.1 mg.

Alternative sampling and analysis procedure

Recovery and analysis of sample. Recover and analyze the samples from the two sampling trains separately.

For this alternative procedure, the probe and filter holder assembly may be weighed without sample recovery (use no solvents) described above, in order to determine the sample weight gains. For this approach, weigh the clean, dry probe and filter holder assembly upstream of the front filter (without filters) to the nearest 0.1 mg to establish the tare weights. The filter holder section between the front and second filter need not be weighed. At the end of the test run, carefully clean the outside of the probe, cap the ends, and identify the sample (label). Remove the filters from the filter holder assemblies as described for containers 1 and 2 above. Reassemble the filter holder assembly, cap the ends, identify the sample (label), and transfer all the samples to the laboratory weighing area for final weighing. Descriptions of capping and transport of samples are not applicable if sample recovery and analysis occur in the same room.

For this alternative procedure, filters may be weighed directly without a petri dish. If the probe and filter holder assembly are to be weighed to determine the sample weight, rinse the probe with acetone to remove moisture before desiccating prior to the test run. Following the test run, transport the probe and filter holder to the desiccator, and uncap the openings of the probe and the filter holder assembly. Desiccate no more than 36 h and weigh to a constant weight. Report the results to the nearest 0.1 mg.

Method 5H: Determination of particulate emissions from wood heaters from a stack location

Applicability. This method is applicable for the determination of particulate matter and condensible emissions from wood heaters.

Principle. Particulate matter is withdrawn proportionally from the wood heater exhaust and is collected on two glass-fiber filters separated by impingers immersed in an ice bath. The first filter is maintained at a temperature of no greater than 120°C (248°F). The second filter and the impinger system are cooled such that the exiting temperature of the gas is no greater than 20°C (68°F). The particulate mass collected in the probe, on the filters, and in the impingers is determined gravimetrically after removal of uncombined water.

Apparatus

Stack flow rate measurement system. It consists of the following components:

Sample probe—a glass or stainless-steel sampling probe.

Gas conditioning system—a high-density filter to remove particulate matter and a condenser capable of lowering the dewpoint of the gas to less than 5°C (40°F). Desiccant, such as Drierite, may be used to dry the sample gas. Do not use silica gel.

Pump—an inert (i.e., Teflon or stainless-steel heads) sampling pump capable of delivering more than the total amount of sample required in the manufacturer's instructions for the individual instruments. A means of controlling the analyzer flow rate and a device for determining proper sample flow rate (e.g., precision rotameter, pressure gauge downstream of all flow controls) shall be provided at the analyzer. The requirements for measuring and controlling the analyzer flow rate are not applicable if data are presented that demonstrate the analyzer is insensitive to flow variations over the range encountered during the test.

CO analyzer—any analyzer capable of providing a measure of CO in the range of 0 to 10 percent by volume at least once every 10 min.

CO_2 analyzer—any analyzer capable of providing a measure of CO_2 in the range of 0 to 25 percent by volume at least once every 10 min. *Note:* Analyzers with ranges less than those specified above may be used, provided actual concentrations do not exceed the range of the analyzer.

Manifold—a sampling tube capable of delivering the sample gas to two analyzers and handling an excess of the total amount used by the analyzers. The excess gas is exhausted through a separate port.

Recorders (optional)—to provide a permanent record of the analyzer outputs.

Proportional gas flow rate system.   To monitor stack flow rate changes and provide a measurement that can be used to adjust and maintain particulate sampling flow rates proportional to the stack flow rate. The proportional flow rate system consists of the following components:

Tracer gas injection system—to inject a known concentration of SO_2 into the flue. The tracer gas injection system consists of a cylinder of SO_2, a gas cylinder regulator, a stainless-steel needle valve or flow controller, a nonreactive (stainless-steel and glass) rotameter, and an injection loop to disperse the SO_2 evenly in the flue.

Sample probe—a glass or stainless-steel sampling probe.

Gas conditioning system—a combustor as described in method 16A, sections 2.1.5 and 2.1.6, followed by a high-density filter to remove particulate matter, and a condenser capable of lowering the dew-point of the gas to less than 5°C (40°F). Desiccant, such as Drierite, may be used to dry the sample gas. Do not use silica gel.

Pump.

SO_2 analyzer—any analyzer capable of providing a measure of the SO_2 concentration in the range of 0 to 1000 ppm by volume (or other range necessary to measure the SO_2 concentration) at least once every 10 min.

Recorder (optional)—to provide a permanent record of the analyzer outputs.

Note: Other tracer gas systems, including helium gas systems, are allowed for determining instantaneous proportional sampling rates.

Analysis. Weighing dishes, desiccator, analytical balance, beakers (250 mL or less), hygrometer or psychrometer, and temperature gauge as described in method 5, sections 2.3.1 through 2.3.7, are needed. In addition, a separatory funnel, glass or Teflon, 500 mL or greater, is needed.

Procedure. Handle each sample container as follows:

Containers 1 and 2. Leave the contents in the shipping container, or transfer both filters and any loose particulate from the sample container to a tared glass weighing dish. Desiccate for no more than 36 h. Weigh to a constant weight and report the results to the nearest 0.1 mg. For purposes of this section, 5.6, the term *constant weight* means a difference of no more than 0.5 mg or 1 percent of total weight less tare weight, whichever is greater, between two consecutive weighings, with no less than 2 h between weighings.

Container 3. Note the level of liquid in the container, and confirm on the analysis sheet whether leakage occurred during transport. If a noticeable amount of leakage has occurred, either void the sample or use methods, subject to the approval of the Administrator, to correct the final results. Determination of sample leakage is not applicable if sample recovery and analysis occur in the same room. Measure the liquid in this container either volumetrically to within 1 mL or gravimetrically to within 0.5 g. Transfer the contents to a tared 250-mL or smaller beaker, and evaporate to dryness at ambient temperature and pressure. Desiccate and weigh to a constant weight. Report the results to the nearest 0.1 mg.

Container 4. Note the level of liquid in the container, and confirm on the analysis sheet whether leakage occurred during transport. If a noticeable amount of leakage has occurred, either void the sample or use methods, subject to the approval of the Administrator, to correct the final results. Determination of sample leakage is not applicable if sample recovery and analysis occur in the same room. Measure the liquid in this container either volumetrically to within 1 mL or gravimetrically to within 0.5 g. Transfer the contents to a 500-mL or larger separatory funnel. Rinse the container with water, and add to the separatory funnel. Add 25 mL of dichloromethane to the separatory funnel, stopper, and vigorously shake 1 min. Let separate and transfer the dichloromethane (lower layer) into a tared beaker or evaporating dish. Repeat twice more. It is necessary to rinse container 4 with dichloromethane. This rinse is added to the impinger extract container. Transfer the remaining water from the separatory funnel to a tared beaker or evaporating dish, and evaporate to dryness at 220°F (105°C). Desiccate and weigh to a constant weight. Evaporate the combined impinger water extracts at ambient temperature and pressure. Desiccate and weigh to a constant weight. Report both results to the nearest 0.1 mg.

Container 5.   Weigh the spent silica gel (or silica gel plus impinger) to the nearest 0.5 g, using a balance.

"Acetone blank" container.   Measure acetone in this container either volumetrically or gravimetrically. Transfer the acetone to a tared 250-mL or smaller beaker, and evaporate to dryness at ambient temperature and pressure. Desiccate and weigh to a constant weight. Report the results to the nearest 0.1 mg.

"Dichloromethane" container.   Measure 75 mL of dichloromethane in this container, and treat it the same as the acetone blank.

"Water blank" container.   Measure 200 mL water into this container either volumetrically or gravimetrically. Transfer the water to a tared 250-mL beaker, and evaporate to dryness at 105°C (221°F). Desiccate and weigh to a constant weight.

Method 17: Determination of particulate emissions from stationary sources (in-stack filtration method).   Particulate matter is not an absolute quantity; rather, it is a function of temperature and pressure. Therefore, to prevent variability in particulate matter emission regulations and/or associated test methods, the temperature and pressure at which particulate matter is to be measured must be carefully defined. Of the two variables (i.e., temperature and pressure), temperature has the greater effect upon the amount of particulate matter in an effluent gas stream; in most stationary-source categories, the effect of pressure appears to be negligible.

In method 5, 250°F is established as a nominal reference temperature. Thus, where method 5 is specified in an applicable subpart of the standards, particulate matter is defined with respect to temperature. In order to maintain a collection temperature of 250°F, method 5 employs a heated glass sample probe and a heated filter holder. This equipment is somewhat cumbersome and requires care in its operation. Therefore, where particulate matter concentrations (over the normal range of temperature associated with a specified source category) are known to be independent of temperature, it is desirable to eliminate the glass probe and heating systems and to sample at stack temperature.

This method describes an in-stack sampling system and sampling procedures for use in such cases. It is intended to be used only when specified by an applicable subpart of the standards, and only within the applicable temperature limits (if specified), or when otherwise approved by the Administrator.

Principle. Particulate matter is withdrawn isokinetically from the source and collected on a glass-fiber filter maintained at stack temperature. The particulate mass is determined gravimetrically after removal of uncombined water.

Applicability. This method applies to the determination of particulate emissions from stationary sources for determining compliance with new-source performance standards only when specifically provided for in an applicable subpart of the standards. This method is not applicable to stacks that contain liquid droplets or are saturated with water vapor. In addition, this method shall not be used as written if the projected cross-sectional area of the probe extension-filter holder assembly covers more than 5 percent of the stack cross-sectional area.

Apparatus

Glass weighing dishes

Desiccator

Analytical balance—to measure to within 0.1 mg

Balance—to measure to within 0.5 mg

Beakers—250 mL

Hygrometer—to measure the relative humidity of the laboratory environment

Temperature gauge—to measure the temperature of the laboratory environment

Procedure. It is recommended that the impinger system described in method 5 be used to determine the moisture content of the stack gas. Alternatively, any system that allows measurement of both the water condensed and the moisture leaving the condenser, each to within 1 mL or 1 g, may be used. The moisture leaving the condenser can be measured either by: (1) monitoring the temperature and pressure at the exit of the condenser and using Dalton's law of partial pressures; or (2) passing the sample gas stream through a silica gel trap with exit gases kept below 20°C (68°F) and determining the weight gain.

Analysis.   Handle each sample container as follows:

Container 1.   Leave the contents in the shipping container, or transfer the filter and any loose particulate from the sample container to a tared glass weighing dish. Desiccate for 24 h in a desiccator containing anhydrous calcium sulfate. Weigh to a constant weight and report the results to the nearest 0.1 mg. For purposes of this section, 4.3, the term *constant weight* means a difference of no more than 0.5 mg or 1 percent of total weight less tare weight, whichever is greater, between two consecutive weighings, with no less than 6 h of desiccation time between weighings.

Alternatively, the sample may be oven-dried at the average stack temperature or 105°C (220°F), whichever is less, for 2 to 3 h, cooled in the desiccator, and weighed to a constant weight, unless otherwise specified by the Administrator. The tester may also opt to oven-dry the sample at the average stack temperature or 105°C (220°F), whichever is less, for 2 to 3 h, weigh the sample, and use this weight as a final weight.

Container 2.   Note the level of liquid in the container, and confirm on the analysis sheet whether leakage occurred during transport. If a noticeable amount of leakage has occurred, either void the sample or use methods, subject to the approval of the Administrator, to correct the final results. Measure the liquid in this container either volumetrically to ±1 mL or gravimetrically to ±0.5 g. Transfer the contents to a tared 250-mL beaker and evaporate to dryness at ambient temperature and pressure. Desiccate for 24 h and weigh to a constant weight. Report the results to the nearest 0.1 mg.

Container 3.   This step may be conducted in the field. Weigh the spent silica gel (or silica gel plus impinger) to the nearest 0.5 g, using a balance.

"Acetone blank" container.   Measure acetone in this container either volumetrically or gravimetrically. Transfer the acetone to a tared 250-mL beaker, and evaporate to dryness at ambient temperature and pressure. Desiccate for 24 h and weigh to a constant weight. Report the results to the nearest 0.1 mg.

Note: At the option of the tester, the contents of container 2 as well as the acetone blank container may be evaporated at temperatures higher than ambient. If evaporation is done at an elevated temperature, the temperature must be below the boiling point of the solvent; also, to prevent "bumping," the evaporation process must be closely supervised, and the contents of the beaker must be swirled occasionally to maintain an even temperature. Use extreme care, as acetone is highly flammable and has a low flash point.

Method 201: Determination of PM_{10} emissions (exhaust gas recycle procedure)

Applicability. This method applies to the in-stack measurement of particulate matter emissions equal to or less than an aerodynamic diameter of nominally 10 μm (PM_{10}) from stationary sources. The EPA recognizes that condensible emissions not collected by an in-stack method are also PM_{10}, and that emissions that contribute to ambient PM_{10} levels are the sum of condensible emissions and emissions measured by an in-stack PM_{10} method, such as this method or method 201A. Therefore, for establishing source contributions to ambient levels of PM_{10}, such as for emission inventory purposes, EPA suggests that source PM_{10} measurement include both in-stack PM_{10} and condensible emissions. Condensible emissions may be measured by an impinger analysis in combination with this method.

Principle. A gas sample is isokinetically extracted from the source. An in-stack cyclone is used to separate PM greater than PM_{10}, and an in-stack glass-fiber filter is used to collect the PM_{10}. To maintain isokinetic flow rate conditions at the tip of the probe and a constant flow rate through the cyclone, a clean, dried portion of the sample gas at stack temperature is recycled into the nozzle. The particulate mass is determined gravimetrically after removal of uncombined water.

Apparatus. *Note:* Method 5 as cited in this method refers to the method in 40 CFR Part 60, Appendix A.
Analysis. It is the same as in method 5, section 2.3.

Procedure
Analysis. It is the same as that in method 5, section 4.3, except handle EGR containers 1 and 2 as container 1 in method 5; EGR containers 3, 4, and 5 as container 3 in method 5; and EGR container 6 as container 3 in method 5.

Method 201A: Determination of PM_{10} emissions—constant-sampling-rate procedure

Applicability. This method applies to the in-stack measurement of particulate matter emissions equal to or less than an aerodynamic diameter of nominally 10 mm (PM_{10}) from stationary sources. The EPA recognizes that condensible emissions not collected by an in-stack method are also PM_{10}, and that emissions that contribute to ambient PM_{10} levels are the sum of condensible emissions and emissions measured by an in-stack PM_{10} method, such as this method or method 201. Therefore, for establishing source contributions to ambient levels of PM_{10}, such as for emission inventory purposes, EPA suggests that source PM_{10} measurement include both in-stack PM_{10} and condensible emissions. Condensible emissions may be measured by an impinger analysis in combination with this method.

Principle. A gas sample is extracted at a constant flow rate through an in-stack sizing device, which separates PM greater than PM_{10}. Variations from isokinetic sampling conditions are maintained within well-defined limits. The particulate mass is determined gravimetrically after removal of uncombined water.

Apparatus. *Note:* Methods cited in this method are part of 40 CFR Part 60, Appendix A.

Analysis. It is the same as that in method 5, section 2.3.

Procedure
Analysis. It is the same as that in method 5, section 4.3, except handle method 201A container 1 as container 1, method 201A containers 2 and 3 as container 2, and method 201A container 4 as container 3.

Method 202: Determination of condensible particulate emissions from stationary sources

Applicability. This method applies to the determination of condensible particulate matter (CPM) emissions from stationary sources. It is intended to represent condensible matter as material that condenses after passing through a filter and as measured by this method. (*Note:* The filter catch can be analyzed according to the appropriate method.)

This method may be used in conjunction with method 201 or 201A if the probes are glass-lined. In using method 202 in conjunction with method 201 or 201A, only the impinger train configuration and analysis are addressed by this method. The sample train operation and front-end recovery and analysis shall be conducted according to method 201 or 201A.

This method may also be modified to measure material that condenses at other temperatures by specifying the filter and probe temperature. A heated method 5 out-of-stack filter may be used instead of the in-stack filter to determine condensible emissions at wet sources.

Principle. The CPM is collected in the impinger portion of a method 17 (Appendix A, 40 CFR Part 60) type of sampling train. The impinger contents are immediately purged after the run with nitrogen (N_2) to remove dissolved sulfur dioxide gases from the impinger contents. The impinger solution is then extracted with methylene chloride ($MeCl_2$). The organic and aqueous fractions are taken to dryness and the residues weighed. The total of both fractions represents the CPM.

The potential for low collection efficiency exists at oil-fired boilers. To improve the collection efficiency at these types of sources, an additional filter placed between the second and third impinger is recommended.

Apparatus

Analysis.   The following equipment is necessary in addition to that listed in method 17, section 2.3:

Separatory funnel—glass, 1-L

Weighing tins—350-mL

Drying equipment—hot plate and oven with temperature control

Pipettes—5-mL

Ion chromatograph—same as in method 5F, section 2.1.6

Procedure

Analysis.   Handle each sample container as follows:

Containers 1, 2, and 3.   If filter catch is analyzed, follow method 17, section 4.3.

Containers 4 and 5.   Note the level of liquid in the containers, and confirm on the analytical data sheet whether leakage occurred during transport. If a noticeable amount of leakage has occurred, either void the sample or use methods, subject to the approval of the Administrator, to correct the final results. Measure the liquid in container 4 either volumetrically to $\pm$ 1 mL or gravimetrically to $\pm$ 0.5 g. Remove a 5-mL aliquot and set aside for later ion chromatograph (IC) analysis of sulfates. (*Note:* Do not use this aliquot to determine chlorides since the HCl will be evaporated during the first drying step; section 8.2 details a procedure for this analysis.)

Extraction.   Separate the organic fraction of the sample by adding the contents of container 5 ($MeCl_2$) to the contents of container 4 in a 1000-mL separatory funnel. After mixing, allow the aqueous and organic phases to fully separate, and drain off most of the organic/$MeCl_2$ phase. Then add 75 mL of $MeCl_2$ to the funnel, mix well, and drain off the lower organic phase. Repeat with another 75 mL of $MeCl_2$. This extraction should yield about 250 mL of organic extract. Each time, leave a small amount of the organic/$MeCl_2$ phase in the separatory funnel, ensuring that no water is collected in the organic phase. Place the organic extract in a tared 350-mL weighing tin.

Organic fraction weight determination (organic phase from containers 4 and 5). Evaporate the organic extract at room temperature and pressure in a laboratory hood. Following evaporation, desiccate the organic fraction for 24 h in a desiccator containing anhydrous calcium sulfate. Weigh to a constant weight, and report the results to the nearest 0.1 mg.

Inorganic fraction weight determination.   [*Note:* If NH_4Cl is to be counted as CPM, the inorganic fraction should be taken to near dryness (less than 1-mL liquid) in the oven and then allowed to air-dry at ambient temperature. If multiple acid emissions are suspected, the ammonia titration procedure in section 8.1 may be preferred.] Using a hot plate, or equivalent, evaporate the aqueous phase to approximately 50 mL; then evaporate to dryness in a 105°C oven. Redissolve the residue in 100 mL of water. Add 5 drops of phenolphthalein to this solution; then add concentrated (14.8 *M*) NH_4OH until the sample turns pink. Any excess NH_4OH will be evaporated during the drying step. Evaporate the sample to dryness in a 105°C oven, desiccate the sample for 24 h, weigh to a constant weight, and record the results to the nearest 0.1 mg. (*Note:* The addition of NH_4OH is recommended, but is optional when little or no SO_2 is present in the gas stream; i.e., when the pH of the impinger solution is greater than 4.5, the addition of NH_4OH is not necessary.)

Analysis of sulfate by IC to determine ammonium ion (NH_4^+) *retained in the sample.*   (*Note:* If NH_4OH is not added, omit this step.) Determine the amount of sulfate in the aliquot taken from container 4 earlier, as described in method 5F (Appendix A, 40 CFR Part 60). Based on the IC SO_4^{2-} analysis of the aliquot, calculate the correction factor to subtract the NH_4^+ retained in the sample and to add the combined water removed by the acid-base reaction (see section 7.2).

Analysis of water and $MeCl_2$ *blanks* (*containers 6 and 7*).   Analyze these sample blanks as described in sections 5.3.2.3 and 5.3.2.2, respectively.

Analysis of acetone blank (*container 8*).   It is the same as that in method 17, section 4.3.

Alternative procedures

Determination of NH_4^+ *retained in sample by titration.*   An alternative procedure to determine the amount of NH_4^+ added to the inorganic fraction by titration may be used. After dissolving the inorganic residue in 100 mL of water, titrate the solution with 0.1 *N* NH_4OH to a pH of 7.0, as indicated by a pH meter. The 0.1 *N* NH_4OH is made as follows: Add 7 mL of concentrated (14.8 *M*) NH_4OH to 1 L of water. Standardize against standardized 0.1 *N* H_2SO_4 and calculate the exact normality, using a procedure parallel to that described in section 5.5 of method 6 (Appendix A, 40 CFR Part 60). Alternatively, purchase 0.1 *N* NH_4OH that has been standardized against a National Institute of Standards and Technology (NIST) reference material.

Analysis of chlorides by IC.   At the conclusion of the final weighing as described in the method, redissolve the inorganic fraction in 100 mL of water. Analyze an aliquot of the redissolved sample for chlorides by IC, using techniques similar to those described in method 5F for sulfates. Previous drying of the sample should have removed all HCl. Therefore, the remaining chlorides measured by IC can be assumed to be NH_4Cl, and this weight can be subtracted from the weight determined for CPM.

Air purge to remove SO_2 from impinger contents. As an alternative to the posttest N_2 purge described in the method, the tester may opt to conduct the posttest purge with air at 20 L/min. *Note:* The use of an air purge is not as effective as an N_2 purge.

Chloroform-ether extraction. As an alternative to the methylene chloride extraction described in the method, the tester may conduct a chloroform-ether extraction. *Note:* Chloroform-ether was not as effective as $MeCl_2$ in removing the organics, but it was found to be an acceptable organic extractant. Chloroform and diethylether of ACS grade, with low blank values (0.001 percent), shall be used. Analysis of the chloroform and diethylether blanks shall be conducted according to section 5.3.3 for $MeCl_2$.

Add the contents of container 4 to a 1000-mL separatory funnel. Then add 75 mL of chloroform to the funnel, mix well, and drain off the lower organic phase. Repeat two more times with 75 mL of chloroform. Then perform three extractions with 75 mL of diethylether. This extraction should yield approximately 450 mL of organic extraction. Each time, leave a small amount of the organic/$MeCl_2$ phase in the separatory funnel, ensuring that no water is collected in the organic phase.

Add the contents of container 5 to the organic extraction. Place approximately 300 mL of the organic extract in a tared 350-mL weighing tin while storing the remaining organic extract in a sample container. As the organic extract evaporates, add the remaining extract to the weighing tin.

Determine the weight of the organic phase as described in the method.

Method 6: Determination of sulfur dioxide emissions from stationary sources

Principle. A gas sample is extracted from the sampling point in the stack. The sulfuric acid mist (including sulfur trioxide) and the sulfur dioxide (SO_2) are separated. The SO_2 fraction is measured by the barium-thorin titration method.

Applicability. This method is applicable for the determination of SO_2 emissions from stationary sources. The minimum detectable limit of the method has been determined to be 3.4 mg of SO_2/m^3 (2.12×10^{-7} lb/ft^3). Although no upper limit has been established, tests have shown that concentrations as high as 80,000 mg/m^3 of SO_2 can be collected efficiently in two midget impingers, each containing 15 mL of 3% hydrogen peroxide, at a rate of 1.0 L/min for 20 min. Based on theoretical calculations, the upper concentration limit in a 20-L sample is about 93,300 mg/m^3.

Possible interferents are free ammonia, water-soluble cations, and fluorides. The cations and fluorides are removed by glass-wool filters and an isopropanol bubbler, and hence do not affect the SO_2 analysis. When samples are being taken from a gas stream with high concentrations of very fine metallic fumes (such as found in inlets to control devices), a high-efficiency glass-fiber filter must be used in place of the glass-wool plug (i.e., the one in the probe) to remove the cation interferents.

Free ammonia interferes by reacting with SO_2 to form particulate sulfite and by reacting with the indicator. If free ammonia is present (this can be determined by knowledge of the process and noticing white particulate matter in the probe and isopropanol bubbler), alternative methods, subject to the approval of the EPA Administrator, are required.

Apparatus
Analysis

Pipettes—volumetric type, 5-mL, 20-mL (one per sample), and 25-mL

Volumetric flasks—100-mL (one per sample) and 1000-mL

Burettes—5- and 50-mL

Erlenmeyer flasks—250-mL (one for each sample, blank, and standard)

Dropping bottle—125-mL size, to add indicator

Graduated cylinder—100-mL

Spectrophotometer—to measure absorbance at 352 nm

Procedure
Sample analysis.   Note the level of liquid in the container, and confirm whether any sample was lost during shipment; note this on the analytical data sheet. If a noticeable amount of leakage has occurred, either void the sample or use methods, subject to the approval of the Administrator, to correct the final results.

Transfer the contents of the storage container to a 100-mL volumetric flask, and dilute to exactly 100 mL with water. Pipette a 20-mL aliquot of this solution into a 250-mL Erlenmeyer flask; add 80 mL of 100% isopropanol and 2 to 4 drops of thorin indicator; and titrate to a pink endpoint using 0.0100 N barium standard solution. Repeat, and average the titration volumes. Run a blank with each series of samples. Replicate titrations must agree within 1 percent or 0.2 mL, whichever is larger.

Protect the 0.0100 N barium standard solution from evaporation at all times.

Method 6A: Determination of sulfur dioxide, moisture, and carbon dioxide emissions from fossil-fuel combustion sources

Applicability. This method applies to the determination of sulfur dioxide emissions from fossil-fuel combustion sources in terms of concentration (mg/dscm) and emission rate (ng/J) and to the determination of carbon dioxide concentration (percent). Moisture, if desired, may also be determined by this method.

The minimum detectable limit, the upper limit, and the interferents of the method for the measurement of SO_2 are the same as those for method 6. For a 20-L sample, the method has a precision of 0.5 percent CO_2 for concentrations between 2.5% and 25% CO_2 and 1.0 percent moisture for moisture concentrations greater than 5%.

Principle. The principle of sample collection is the same as that for method 6 except that moisture and CO_2 are collected in addition to SO_2 in the same sampling train. Moisture and CO_2 fractions are determined gravimetrically.

Apparatus

Sample recovery and analysis. The equipment needed for sample recovery and analysis is the same as that required for method 6. In addition, a balance to measure within 0.05 g is needed for analysis.

Procedure

Sample analysis. The sample analysis procedure for SO_2 is the same as that specified in method 6, section 4.3.

Quality assurance (QA) audit samples. Obtain an audit sample set as directed in section 3.3.6 of method 6 only when this method is used for compliance determinations. Analyze the audit samples, and report the results as directed in section 4.4 of method 6. Acceptance criteria for the audit results are the same as those in method 6.

Emission rate procedure. If the only emission measurement desired is in terms of emission rate of SO_2 (ng/J), an abbreviated procedure may be used. The differences between the above procedure and the abbreviated procedure are described below.

Sample analysis. Analysis of the peroxide solution is the same as that described in section 4.3. Conduct an audit of the SO_2 analysis procedure as described in section 4.4 only when making compliance determinations.

Method 6B: Determination of sulfur dioxide and carbon dioxide daily average emissions from fossil-fuel combustion sources

Applicability. This method applies to the determination of sulfur dioxide emissions from combustion sources in terms of concentration (ng/dscm) and emission rate (ng/J), and for the determination of carbon dioxide concentration (percent) on a daily (24-h) basis.

The minimum detectable limits, upper limit, and the interferents for SO_2 measurements are the same as those for method 6. EPA-sponsored collaborative studies were undertaken to determine the magnitude of repeatability and reproducibility achievable by qualified testers following the procedures in this method. The results of the studies evolve from 145 field tests including comparisons with methods 3 and 6. For measurements of emission rates from wet flue gas desulfurization units in (ng/J), the repeatability (within laboratory precision) is 8.0 percent and the reproducibility (between laboratory precision) is 11.1 percent.

Principle. A gas sample is extracted from the sampling point in the stack intermittently over a 24-h or other specified time period. Sampling may also be conducted continuously if the apparatus and procedures are appropriately modified (see Note in section 4.1.1.4). The SO_2 and CO_2 are separated and collected in the sampling train. The SO_2 fraction is measured by the barium-thorin titration method, and CO_2 is determined gravimetrically.

Apparatus. The equipment required for this method is the same as that specified for method 6A, section 2, except the isopropanol bubbler is not used.

Procedure

Sample analysis. Analysis of the peroxide impinger solutions is the same as that in method 6, section 4.3.

Quality assurance audit samples. Only when this method is used for compliance determinations, obtain an audit sample set as directed in section 3.3.6 of method 6. Analyze the audit samples at least once for every 30 days of sample collection, and report the results as directed in section 4.4 of method 6. The analyst performing the sample analyses shall perform the audit analyses. If more than one analyst performs the sample analyses during the 30-day sampling period, each analyst shall perform the audit analyses and all audit results shall be reported. Acceptance criteria for the audit results are the same as those in method 6.

Emission rate procedure. The emission rate procedure is the same as that described in method 6A, section 7, except that the timer is needed and is operated as described. Perform the QA audit analyses only when this method is used for compliance determinations.

Method 6C: Determination of sulfur dioxide emissions from stationary sources (instrumental analyzer procedure)

Applicability. This method is applicable to the determination of sulfur dioxide concentrations in controlled and uncontrolled emissions from stationary sources only when specified within the regulations.

Principle. A gas sample is continuously extracted from a stack, and a portion of the sample is conveyed to an instrumental analyzer for determination of SO_2 gas concentration using an ultraviolet, nondispersive infrared (NDIR), or fluorescence analyzer. Performance specifications and test procedures are provided to ensure reliable data.

Apparatus

Gas analyzer. A uv or an NDIR absorption or fluorescence analyzer is used to determine continuously the SO_2 concentration in the sample gas stream. The analyzer shall meet the applicable performance specifications of section 4. A means of controlling the analyzer flow rate and a device for determining proper sample flow rate (e.g., precision rotameter, pressure gauge downstream of all flow controls, etc.) shall be provided at the analyzer. [*Note:* Housing the analyzer(s) in a clean, thermally stable, vibrationfree environment will minimize drift in the analyzer calibration.]

Data recorder. A strip-chart recorder, analog computer, or digital recorder is needed for recording measurement data. The data recorder resolution (i.e., readability) shall be 0.5 percent of span. Alternatively, a digital or analog meter having a resolution of 0.5 percent of span may be used to obtain the analyzer responses, and the readings may be recorded manually. If this alternative is used, the readings shall be obtained at equally spaced intervals over the duration of the sampling run. For sampling run durations of less than 1 h, measurements at 1-min intervals or a minimum of 30 measurements, whichever is less restrictive, shall be obtained. For sampling run durations greater than 1 h, measurements at 2-min intervals or a minimum of 96 measurements, whichever is less restrictive, shall be obtained.

Method 7: Determination of nitrogen oxide emissions from stationary sources

Applicability. This method is applicable to the measurement of nitrogen oxides emitted from stationary sources. The range of the method has been determined to be 2 to 400 mg NO_x (as NO_2) per dry standard cubic meter (dscm), without having to dilute the sample.

Principle. A grab sample is collected in an evacuated flask containing a dilute sulfuric acid–hydrogen peroxide absorbing solution, and the nitrogen oxides, except nitrous oxide, are measured colorimetrically using the phenoldisulfonic acid (PDS) procedure.

Apparatus

Analysis. For the analysis, the following equipment is needed:

Volumetric pipettes—two 1-mL, two 2-mL, one 3-mL, one 4-mL, two 10-mL, and one 25-mL for each sample and standard.

Porcelain evaporating dishes—175- to 250-mL capacity with lip for pouring, one for each sample and each standard. The Coors No. 45006 (shallow-form, 195-mL) has been found to be satisfactory. Alternatively, polymethyl pentene beakers (Nalge No. 1203, 150-mL) or glass beakers (150-mL) may be used. When glass beakers are used, etching of the beakers may cause solid matter to be present in the analytical step; the solids should be removed by filtration (see section 4.3).

Steam bath—low-temperature ovens or thermostatically controlled hot plates kept below 70°C (160°F) are acceptable alternatives.

Dropping pipette or dropper—three required.

Polyethylene policeman—one for each sample and each standard.

Graduated cylinder—100-mL with 1-mL divisions.

Volumetric flasks—50-mL (one for each sample and each standard), 100-mL (one for each sample and each standard, and one for the working standard KNO_3 solution), and 1000-mL (one).

Spectrophotometer—to measure at 410 nm.

Graduated pipette—10-mL with 0.1-mL divisions.

Test paper for indicating pH—to cover the pH range of 7 to 14.

Analytical balance—to measure to within 0.1 mg.

Procedure

Analysis. Note the level of the liquid in the container, and confirm whether any sample was lost during shipment; note this on the analytical data sheet. If a noticeable amount of leakage has occurred, either void the sample or use methods, subject to the approval of the Administrator, to correct the final results.

Immediately prior to analysis, transfer the contents of the shipping container to a 50-mL volumetric flask, and rinse the container twice with 5-mL portions of water. Add the rinse water to the flask, and dilute to mark with water; mix thoroughly. Pipette a 25-mL aliquot into the porcelain evaporating dish. Return any unused portion of the sample to the polyethylene storage bottle. Evaporate the 25-mL aliquot to dryness on a steam bath, and allow to cool. Add 2 mL phenoldisulfonic acid solution to the dried residue, and triturate thoroughly with a polyethylene policeman. Make sure the solution contacts all the residue. Add 1 mL water and 4 drops of concentrated sulfuric acid. Heat the solution on a steam bath for 3 min with occasional stirring. Allow the solution to cool, add 20 mL water, mix well by stirring, and add concentrated ammonium hydroxide, drop by drop, with constant stirring, until the pH is 10 (as determined by pH paper). If the sample contains solids, these must be removed by filtration (centrifugation is an acceptable alternative, subject to the approval of the Administrator), as follows: Filter through Whatman No. 41 filter paper into a 100-mL volumetric flask; rinse the evaporating dish with three 5-mL portions of water; filter these three rinses. Wash the filter with at least three 15-mL portions of water. Add the filter washings to the contents of the volumetric flask, and dilute to the mark with water. If solids are absent, the solution can be transferred directly to the 100-mL volumetric flask and diluted to the mark with water.

Mix the contents of the flask thoroughly, and measure the absorbance at the optimum wavelength used for the standards, using the blank solution as a zero reference. Dilute the sample and the blank with equal volumes of water if the absorbance exceeds A_4, the absorbance of the 400-μg NO_2 standard.

Determination of nitrogen oxide emissions from stationary sources (ion chromatographic method)

Applicability. This method applies to the measurement of nitrogen oxides emitted from stationary sources; it may be used as an alternative to method 7 [as defined in 40 CFR Part 60.8(b)] to determine compliance if the stack concentration is within the analytical range. The analytical range of the method is from 125 to 1250 mg NO_x/m^3 as NO_2 (65 to 655 ppm), and higher concentrations may be analyzed by diluting the sample. The lower detection limit is approximately 19 mg/m^3 (10 ppm), but may vary among instruments.

Principle. A grab sample is collected in an evacuated flask containing a dilute sulfuric acid–hydrogen peroxide absorbing solution. The nitrogen oxides, except nitrous oxide, are oxidized to nitrate and measured by ion chromatography.

Apparatus

Analysis.   For the analysis, the following equipment is needed. Alternative instrumentation and procedures will be allowed provided the calibration precision in section 5.2 and acceptable audit accuracy can be met.

Volumetric pipettes—class A; 1-, 2-, 4-, 5-mL (two for the set of standards and one per sample), 6-, 10-, and graduated 5-mL sizes.

Volumetric flasks—50-mL (two per sample and one per standard), 200-mL, and 1-L sizes.

Analytical balance—to measure to within 0.1 mg.

Ion chromatograph—The ion chromatograph should have at least the following components:

- *Columns*—an anion separation or other column capable of resolving the nitrate ion from sulfate and other species present and a standard anion suppressor column (optional). Suppressor columns are produced as proprietary items; however, one can be produced in the laboratory using the resin available from BioRad Company, 32d and Griffin Streets, Richmond, Calif. Peak resolution can be optimized by varying the eluent strength or column flow rate, or by experimenting with alternative columns that may offer more efficient separation. When guard columns are used with the stronger reagent to protect the separation column, the analyst should allow rest periods between injection intervals to purge possible sulfate buildup in the guard column.
- *Pump*—capable of maintaining a steady flow as required by the system.
- *Flow gauges*—capable of measuring the specified system flow rate.
- *Conductivity detector.*
- *Recorder*—compatible with the output voltage range of the detector.

Procedures

Analysis.   Prepare a standard calibration curve according to section 5.2. Analyze the set of standards followed by the set of samples, using the same injection volume for both standards and samples. Repeat this analysis sequence followed by a final analysis of the standard set. Average the results. The two sample values must agree within 5 percent of their mean for the analysis to be valid. Perform this duplicate analysis sequence on the same day. Dilute any sample and the blank with equal volumes of water if the concentration exceeds that of the highest standard.

Document each sample chromatogram by listing the following analytical parameters: injection point, injection volume, nitrate and sulfate retention times, flow rate, detector sensitivity setting, and recorder chart speed.

Method 7B: Determination of nitrogen oxide emissions from stationary sources (ultraviolet spectrophotometry)

Applicability. This method is applicable to the measurement of nitrogen oxides emitted from nitric acid plants. The range of the method as outlined has been determined to be 57 to 1500 mg NO_x (as NO_2) per dry standard cubic meter, or 30 to 786 ppm NO_x (as NO_2), assuming corresponding standards are prepared.

Principle. A grab sample is collected in an evacuated flask containing a dilute sulfuric acid–hydrogen peroxide absorbing solution; and the nitrogen oxides, except nitrous oxide, are measured by ultraviolet absorption.

Apparatus
Analysis. The following equipment is needed for analysis:

Volumetric pipettes—5-, 10-, 15-, and 20-mL to make standards and sample dilutions

Volumetric flasks—1000- and 100-mL for preparing standards and dilution of samples

Spectrophotometer—to measure ultraviolet absorbance at 210 nm

Analytical balance—to measure to within 0.1 mg

Procedures
Analysis. Pipette a 20-mL aliquot of sample into a 100-mL volumetric flask. Dilute to 100 mL with water. The sample is now ready to be read by ultraviolet spectrophotometry. Using the blank as zero reference, read the absorbance of the sample at 210 nm.

Method 7C: Determination of nitrogen oxide emissions from stationary sources—alkaline-permanganate/colorimetric method

Applicability. The method is applicable to the determination of NO_x emissions from fossil-fuel-fired steam generators, electric utility plants, nitric acid plants, or other sources as specified in the regulations. The lower detectable limit is 13 mg NO_x/m^3, as NO_2 (7 ppm NO_x) when sampling at 500 cm^3/min for 1 h. No upper limit has been established; however, when using the recommended sampling conditions, the method has been found to collect NO_x emissions quantitatively up to 1782 mg NO_x/m^3, as NO_2 (932 ppm N_x).

Principle. An integrated gas sample is extracted from the stack and collected in alkaline-potassium permanganate solution; NO_x ($NO + NO_2$) emissions are oxidized to NO_2^- and NO_3^-. Then NO_3^- is reduced to NO_2^- with cadmium, and the NO_2^- is analyzed colorimetrically.

Apparatus

Sample preparation and analysis

Hot plate—stirring type with 50- by 10-mm Teflon-coated stirring bars.

Beakers—400-, 600-, and 1000-mL capacities.

Filtering flask—500-mL capacity with side arm.

Buchner funnel—75-mm ID, with spout equipped with a 13-mm ID by 90-mm-long piece of Teflon tubing to minimize possibility of aspirating sample solution during filtration.

Filter paper—Whatman GF/C, 7.0-cm diameter.

Stirring rods.

Volumetric flasks—100-, 200- or 250-, 500-, and 1000-mL capacity.

Watch glasses—to cover 600- and 1000-mL beakers.

Graduated cylinders—50- and 250-mL capacities.

Pipettes—class A.

pH meter—to measure pH from 0.5 to 12.0.

Burette—50-mL with a micrometer-type stopcock. (The stopcock is catalogue no. 8225-t-05, Ace Glass, Inc., P.O. Box 996, Louisville, Ky. 50201.) Place a glass-wool plug in bottom of burette. Cut off burette at a height of 43 cm from the top of plug, and have a blower attach a glass funnel to top of burette such that the diameter of the burette remains essentially unchanged. Other means of attaching the funnel are acceptable.

Glass funnel—75-mm ID at the top.

Spectrophotometer—capable of measuring absorbance at 540 nm; 1-cm cells are adequate.

Metal thermometers—bimetallic thermometers, range 0 to 150°C.

Culture tubes—20- by 150-mm, Kimax no. 45048.

Parafilm "M"—obtained from American Can Company, Greenwich, Conn. 06830.

CO_2 measurement equipment—same as in method 3.

Procedure

Sample analysis. Pipette 10 mL of sample into a culture tube. (*Note:* Some test tubes give a high blank NO_2^- value, but culture tubes do not.) Pipette in 10 mL of sulfanilamide solution and 1.4 mL of NEDA solution. Cover the culture tube with parafilm, and mix the solution. Prepare a blank in the same manner, using the sample from treatment of the unexposed $KMnO_4/NaOH$ solution (section 3.1.2). Also, prepare a calibration standard to check the slope of the calibration curve. After a 10-min color development interval, measure the absorbance at 540 nm against water. Read μg NO_2^-/mL from the calibration curve. If the absorbance is greater than that of the highest calibration standard, pipette less than 10 mL and repeat the analysis. Determine the NO_2^- concentration using the calibration curve.

Method 7D: Determination of nitrogen oxide emissions from stationary sources—alkaline-permanganate/ion chromatographic method

Applicability. The method is applicable to the determination of NO_x emissions from fossil-fuel-fired steam generators, electric utility plants, nitric acid plants, or other sources as specified in the regulations. The lower detectable limit is similar to that for method 7C. No upper limit has been established; however, under the recommended sampling conditions, the method has been found to collect NO_x emissions quantitatively up to 1782 mg NO_x/m^3, as NO_2 (932 ppm NO_x).

Principle. An integrated gas sample is extracted from the stack and collected in alkaline-potassium permanganate solution; NO_x (NO + NO_2) emissions are oxidized to NO_3^-. Then NO_3^- is analyzed by ion chromatography.

Apparatus

Sample preparation and analysis

Magnetic stirrer—with 25- by 10-mm Teflon-coated stirring bars.

Filtering flask—500-mL capacity with sidearm.

Buchner funnel—75-mm ID, with spout equipped with a 13-mm ID by 90-mm-long piece of Teflon tubing to minimize possibility of aspirating sample solution during filtration.

Filter paper—Whatman GF/C, 7.0-cm diameter.

Stirring rods.

Volumetric flask—250-mL.

Pipettes—class A.

Erlenmeyer flasks—250-mL.

Ion chromatograph—equipped with an anion separator column to separate NO_3^-, H^+ suppressor, and necessary auxiliary equipment. Nonsuppressed and other forms of ion chromatography may also be used provided that adequate resolution of NO_3^- is obtained. The system must also be able to resolve and detect NO_2.

Procedure

Sample analysis The following chromatographic conditions are recommended: 0.003 M $NaHCO_3$/0.0024 Na_2CO_3 eluent solution (section 3.2.5), full-scale range, 3 μMHO; sample loop, 0.5 mL; flow rate, 2.5 mL/min. These conditions should give an NO_3^- retention time of approximately 15 min (figure 7D.1).

Establish a stable baseline. Inject a sample of water, and determine whether any NO_3^- appears in the chromatogram. If NO_3^- is present, repeat the water load/injection procedure approximately five times; then reinject a water sample, and observe the chromatogram. When no NO_3^- is present, the instrument is ready for use. Inject calibration standards. Then inject samples and a blank. Repeat the injection of the calibration standards (to compensate for any drift in response of the instrument). Measure the NO_3^- peak height or peak area, and determine the sample concentration from the calibration curve.

Method 7E: Determination of nitrogen oxide emissions from stationary sources (instrumental analyzer procedure)

Applicability. This method is applicable to the determination of nitrogen oxides (NO_x) concentrations in emissions from stationary sources only when specified within the regulations.

Principle. A sample is continuously extracted from the effluent stream; a portion of the sample stream is conveyed to an instrumental chemiluminescent analyzer for determination of NO_x concentration. Performance specifications and test procedures are provided to ensure reliable data.

Apparatus

Measurement system. Use any measurement system for NO_x that meets the specifications of this method. A schematic of an acceptable measurement system is shown in method 6C. The essential components of the measurement system are described below:

Sample probe, sample line, calibration valve assembly, moisture removal system, particulate filter, sample pump, sample flow rate control, sample gas manifold, and data recorder—same as in method 6C, sections 5.1.1 through 5.1.9, and 5.1.11.

NO$_2$-to-NO converter—That portion of the system that converts NO$_2$ in the sample gas to NO. An NO$_2$-to-NO converter is not necessary if the NO$_2$ portion of the exhaust gas is less than 5 percent of the total NO$_x$ concentration.

NO$_x$ analyzer—An analyzer based on the principles of chemiluminescence to determine continuously the NO$_x$ concentration in the sample gas stream. The analyzer must meet the applicable performance specifications of section 4. A means of controlling the analyzer flow rate and a device for determining proper sample flow rate (e.g., precision rotameter, pressure gauge downstream of all flow controls, etc.) must be provided at the analyzer.

Method 20: Determination of nitrogen oxides, sulfur dioxide, and oxygen emissions from stationary gas turbines

Applicability. This method is applicable for the determination of nitrogen oxides (NO$_x$), sulfur dioxide, and oxygen emissions from stationary gas turbines. For the NO$_x$ and O$_2$ determinations, this method includes (1) measurement system design criteria, (2) analyzer performance specifications and performance test procedures, and (3) procedures for emission testing.

Principle. A gas sample is continuously extracted from the exhaust stream of a stationary gas turbine; a portion of the sample stream is conveyed to instrumental analyzers for determination of NO$_x$ and O$_2$ content. During each NO$_x$ and O$_2$ determination, a separate measurement of SO$_2$ emissions is made, using method 6 or its equivalent. The O$_2$ determination is used to adjust the NO$_x$ and SO$_2$ concentrations to a reference condition.

Apparatus
Measurement system. Use any measurement system for NO$_x$ and O$_2$ that is expected to meet the specifications in this method. A schematic of an acceptable measurement system is shown in fig. 20.1. The essential components of the measurement system are described below:

Sample probe—heated stainless-steel, or equivalent, open-ended, straight tube of sufficient length to traverse the sample points.

Sample line—heated (>95°C) stainless-steel or Teflon tubing to transport the sample gas to the sample conditioners and analyzers.

Calibration valve assembly—a three-way valve assembly to direct the zero and calibration gases to the sample conditioners and to the analyzers. The calibration valve assembly shall be capable of blocking the sample gas flow and introducing calibration gases to the measurement system when in the calibration mode.

NO₂-to-NO converter—that portion of the system which converts NO_2 in the sample gas to NO. Some analyzers are designed to measure NO_x as NO_2 on a wet basis and can be used without an NO_2-to-NO converter or a moisture removal trap, provided the sample line to the analyzer is heated (>95°C) to the inlet of the analyzer. In addition, an NO_2-to-NO converter is not necessary if the NO_2 portion of the exhaust gas is less than 5 percent of the total NO_x concentration. As a guideline, an NO_2-to-NO converter is not necessary if the gas turbine is operated at 90 percent or more of peak load capacity. A converter is necessary under lower load conditions.

Moisture removal trap—a refrigerator-type condenser or other type of device designed to remove continuously condensate from the sample gas while maintaining minimal contact between any condensate and the sample gas. The moisture removal trap is not necessary for analyzers that can measure NO_x concentrations on a wet basis; for these analyzers, (1) heat the sample line up to the inlet of the analyzers, (2) determine the moisture content, using methods subject to the approval of the Administrator, and (3) correct the NO_x and O_2 concentrations to a dry basis.

Particulate filter—an in-stack or out-of-stack glass-fiber filter, of the type specified in method 5. However, an out-of-stack filter is recommended when the stack gas temperature exceeds 250 to 300°C.

Sample pump—a nonreactive leakfree sample pump to pull the sample gas through the system at a flow rate sufficient to minimize transport delay. The pump shall be made from stainless steel or coated with Teflon, or equivalent.

Sample gas manifold—a sample gas manifold to divert portions of the sample gas stream to the analyzers. The manifold may be constructed of glass, Teflon, stainless steel, or equivalent.

Diluent gas—an analyzer to determine the percent O_2 or CO_2 concentration of the sample gas stream.

Nitrogen oxides analyzer—an analyzer to determine the ppm NO_x concentration in the sample gas stream.

Data output—a strip-chart recorder, analog computer, or digital recorder for recording measurement data.

SO₂ analysis—method 6 apparatus and reagents.

Emission measurement test procedure
NO_x and diluent measurement. This test is to be conducted at each of the specified load conditions. Three test runs at each load condition constitute a complete test.

At the beginning of each NO_x test run and as applicable during the run, record turbine data. Also, record the location and number of the traverse points on a diagram.

Position the probe at the first point determined in the preceding section, and begin sampling. The minimum sampling time at each point shall be at least 1 min plus the average system response time. Determine the average steady-state concentration of diluent and NO_x at each point, and record the data.

After sampling the last point, conclude the test run by recording the final turbine operating parameters and by determining the zero and calibration drift, as follows: Immediately following the test run at each load condition, or if adjustments are necessary for the measurement system during the tests, reintroduce the zero and mid-level calibration gases as described in sections 4.3 and 4.4, one at a time, to the measurement system at the calibration valve assembly. (Make no adjustments to the measurement system until after the drift checks are made.) Record the analyzers' responses. If the drift values exceed the specified limits, the test run preceding the check is considered invalid and will be repeated following corrections to the measurement system. Alternatively, recalibrate the measurement system and recalculate the measurement data. Report the test results based on both the initial calibration and the recalibration data.

SO$_2$ measurement.   This test is conducted only at the 100 percent peak load condition. Determine SO_2 using method 6, or equivalent, during the test. Select a minimum of six total points from those required for the NO_x measurements; use two points for each sample run. The sample time at each point shall be at least 10 min. Average the O_2 readings taken during the NO_x test runs at sample points corresponding to the SO_2 traverse points, and use this average diluent concentration to correct the integrated SO_2 concentration obtained by method 6 to 15% O_2.

If the applicable regulation allows fuel sampling and analysis for fuel sulfur content to demonstrate compliance with the sulfur emission unit, emission sampling with method 6 is not required, provided the fuel sulfur content meets the limits of the regulation.

Method 8: Determination of sulfuric acid mist and sulfur dioxide emissions from stationary sources

Principle.   A gas sample is extracted isokinetically from the stack. The sulfuric acid mist (including sulfur trioxide) and the sulfur dioxide are separated, and both fractions are measured separately by the barium-thorin titration method.

Applicability.   This method is applicable for the determination of sulfuric acid mist [including sulfur trioxide (SO_3) in the absence of other particulate matter] and sulfur dioxide emissions from stationary sources. Collaborative tests have shown that the minimum detectable limits of the method are 0.05 mg/m^3 (0.03×10^{-7} lb/ft^3) for SO$_3$ and 1.2 mg/m^3 (0.74×10^{-7} lb/ft^3) for SO$_2$. No upper limits have been established. Based on theoretical calculations for 200 mL of 3% hydrogen peroxide solution, the upper concentration limit for SO$_2$ in a 1.0-m^3 (35.3-ft^3) gas sample is about 12,500 mg/m^3 (7.72×10^{-4} lb/ft^3). The upper limit can be extended by increasing the quantity of peroxide solution in the impingers.

Apparatus
Analysis

Pipettes—volumetric 25-mL, 100-mL

Burette—50-mL

Erlenmeyer flask—250-mL (one for each sample, blank, and standard)

Graduated cylinder—100-mL

Trip balance—500-g capacity, to measure to $\pm$ 0.5 g

Dropping bottle—to add indicator solution, 125-mL size

Procedure
Analysis.   Note the level of the liquid in containers 1 and 2, and confirm whether any sample was lost during shipment; note this on the analytical data sheet. If a noticeable amount of leakage has occurred, either void the sample or use methods, subject to the approval of the Administrator, to correct the final results.

Container 1.   Shake the container holding the isopropanol solution and the filter. If the filter breaks up, allow the fragments to settle for a few minutes before removing a sample. Pipette a 100-mL aliquot of this solution into a 250-mL Erlenmeyer flask, add 2 to 4 drops of thorin indicator, and titrate to a pink endpoint using 0.0100 N barium perchlorate. Repeat the titration with a second aliquot of sample, and average the titration values. Replicate titrations must agree within 1 percent or 0.2 mL, whichever is greater.

Container 2.   Thoroughly mix the solution in the container holding the contents of the second and third impingers. Pipette a 10-mL aliquot of sample into a 250-mL Erlenmeyer flask. Add 40 mL of isopropanol and 2 to 4 drops of thorin indicator, and titrate to a pink endpoint, using 0.0100 N barium perchlorate. Repeat the titration with a second aliquot of sample, and average the titration values. Replicate titrations must agree within 1 percent or 0.2 mL, whichever is greater.

Blanks. Prepare blanks by adding 2 to 4 drops of thorin indicator to 100 mL of 80% isopropanol. Titrate the blanks in the same manner as for the samples.

Method 10: Determination of carbon monoxide emissions from stationary sources

Principle. An integrated or continuous gas sample is extracted from a sampling point and analyzed for carbon monoxide content, using a Luft-type nondispersive infrared (NDIR) analyzer or equivalent.

Applicability. This method is applicable for the determination of carbon monoxide emissions from stationary sources only when specified by the test procedures for determining compliance with new-source performance standards. The test procedure will indicate whether a continuous or an integrated sample is to be used.

Apparatus

Analysis

Carbon monoxide analyzer—nondispersive infrared spectrometer, or equivalent. This instrument should be demonstrated, preferably by the manufacturer, to meet or exceed manufacturer's specifications and those described in this method.

Drying tube—to contain approximately 200 g of silica gel.

Calibration gas.

Filter—as recommended by NDIR manufacturer.

CO_2 *removal tube*—to contain approximately 500 g of ascarite.

Ice water bath—for ascarite and silica gel tubes.

Valve—needle valve, or equivalent, to adjust flow rate.

Rate meter—rotameter, or equivalent, to measure gas flow rate of 0 to 1.0 L/min (0 to 0.035 ft³/min) through NDIR analyzer.

Recorder (optional)—to provide permanent record of NDIR analyzer readings.

Procedure

CO analysis. Assemble the apparatus, calibrate the instrument, and perform other required operations as described in the method. Purge analyzer with N_2 prior to introduction of each sample. Direct the sample stream through the instrument for the test period, recording the readings. Check the zero and the span again after the test to ensure that any drift or malfunction is detected. Record the sample data.

Alternative procedure—Interference trap. The sample conditioning system described in method 101A, sections 2.1.2 and 4.2, may be used as an alternative to the silica gel and ascarite traps.

Method 10A: Determination of carbon monoxide emissions in certifying continuous emission monitoring systems at petroleum refineries

Applicability. This method applies to the measurement of carbon monoxide at petroleum refineries. This method serves as the reference method in the relative accuracy test for nondispersive infrared CO continuous emission monitoring systems (CEMSs) that are required to be installed in petroleum refineries on fluid catalytic cracking unit catalyst regenerators [40 CFR Part 60.105(a)(2)].

Principle. An integrated gas sample is extracted from the stack, passed through an alkaline-permanganate solution to remove sulfur and nitrogen oxides, and collected in a Tedlar bag. The CO concentration in the sample is measured spectrophotometrically using the reaction of CO with p-sulfaminobenzoic acid.

Apparatus
Analysis
Spectrophotometer—single- or double-beam to measure absorbance at 425 and 600 nm. Slit width should not exceed 20 nm.

Spectrophotometer cells—1-cm path length.

Vacuum gauge—U-tube mercury manometer, 1-m (39-in), with 1-mm divisions, or other gauge capable of measuring pressure to within 1 mmHg.

Pump—capable of evacuating the gas reaction bulb to a pressure equal to or less than 40 mmHg absolute, equipped with coarse and fine flow control valves.

Barometer—mercury, aneroid, or other barometer capable of measuring atmospheric pressure to within 1 mmHg.

Reaction bulbs—Pyrex glass, 100-mL with Teflon stopcock (fig. 10A.2), leakfree at 40 mmHg, designed so that 10 mL of the colorimetric reagent can be added and removed easily and accurately. Commercially available gas sample bulbs such as Supelco catalog no. 2-2161 may also be used.

Manifold—stainless-steel, with connections for three reaction bulbs and the appropriate connections for the manometer and sampling bag.

Pipettes—class A, 10-mL size.

Shaker table—reciprocating-stroke type such as Eberbach Corporation, model 6015. A rocking arm or rotary-motion type of shaker may also be used. The shaker must be large enough to accommodate at least six gas sample bulbs simultaneously. It may be necessary to construct a tabletop extension for most commercial shakers to provide sufficient space for the needed bulbs.

Valve—stainless-steel shutoff valve.

Analytical balance—capable of weighing to 0.1 mg.

Procedure

Analysis. Assemble the system, and record the information required as it is obtained. Pipette 10.0 mL of the colorimetric reagent into each gas reaction bulb, and attach the bulbs to the system. Open the stopcocks to the reaction bulbs, but leave the valve to the Tedlar bag closed. Turn on the pump, fully open the coarse-adjust flow valve, fine-adjust valve until the pressure is reduced to at least 40 mmHg. Now close the coarse adjust valve, and observe the manometer to be certain that the system is leakfree. Wait a minimum of 2 min. If the pressure has increased less than 1 mm, proceed as described below. If a leak is present, find and correct it before proceeding.

Record the vacuum pressure P_v to the nearest 1 mmHg, and close the reaction bulb stopcocks. Open the Tedlar bag valve, and allow the system to come to atmospheric pressure. Close the bag valve, open the pump coarse adjust valve, and evacuate the system again. Repeat this fill-and-evacuation procedure at least twice to flush the manifold completely. Close the pump coarse adjust valve, open the Tedlar bag valve, and let the system fill to atmospheric pressure. Open the stopcocks to the reaction bulbs, and let the entire system come to atmospheric pressure. Close the bulb stopcocks, remove the bulbs, record the room temperature and barometric pressure (P_{bar}, to nearest mmHg), and place the bulbs on the shaker table with their main axis either parallel or perpendicular to the plane of the tabletop. Purge the bulb-filling system with ambient air for several minutes between samples. Shake the samples for exactly 2 h.

Immediately after shaking, measure the absorbance A of each bulb sample at 425 nm if the concentration is less than or equal to 400 ppm CO or at 600 nm if the concentration is above 400 ppm. Use a small portion of the sample to rinse a spectrophotometer cell several times before taking an aliquot for analysis. If one cell is used to analyze multiple samples, rinse the cell several times between samples with water. Prepare and analyze standards and a reagent blank as described in section 5.3. Use water as the reference. Reject the analysis if the blank absorbance is greater than 0.1. All conditions should be the same for analysis of samples and standards. Measure the absorbances as soon after shaking is completed as possible. Determine the CO concentration of each bag sample, using the calibration curve for the appropriate concentration range as discussed in the method.

Method 10B: Determination of carbon monoxide emissions from stationary sources

Applicability. This method applies to the measurement of carbon monoxide emissions at petroleum refineries and from other sources when specified in an applicable subpart of the regulations.

Principle. An integrated gas sample is extracted from the sampling point and analyzed for CO. The sample is passed through a conditioning system to remove interferents and collected in a Tedlar bag. The CO is separated from the sample by gas chromatography (GC) and is catalytically reduced to methane (CH_4) prior to analysis by flame ionization detection (FID). The analytical portion of this method is identical to applicable sections in method 25 detailing CO measurement. The oxidation catalyst required in method 25 is not needed for sample analysis. Complete method 25 analytical systems are acceptable alternatives when calibrated for CO and operated by the method 25 analytical procedures.

Apparatus
Analysis
GC analyzer—A semicontinuous GC/FID analyzer capable of quantifying CO in the sample (see Fig. 6.1) and containing at least the following major components.

- *Separation column*—a column that separates CO from CO_2 and organic compounds that may be present. A ⅛-in-OD stainless-steel column packed with 5.5 ft of 60/80 mesh Carbosieve S-II (available from Supelco) has been used successfully for this purpose. The column listed in addendum 1 of method 25 is also acceptable.
- *Reduction catalyst*—same as in method 25, section 2.3.2.
- *Sample injection system*—same as in method 25, section 2.3.4, equipped to accept a sample line from the Tedlar bag.
- *Flame ionization detector*—linearity meeting the specifications in section 2.3.5.1 of method 25 where the linearity check is carried out using standard gases containing 20, 200, and 1000 ppm CO. The minimal instrument range shall span 10 to 1000 ppm CO.
- *Data recording system*—same as in method 25, section 2.3.6.

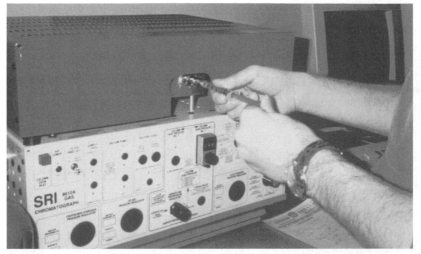

Figure 6.1 Gas sample analysis performed with a gas chromatograph flame ionization detector (GC-FID).

Procedure

Sample analysis. Purge the sample loop with sample, and then inject the sample. Analyze each sample in triplicate, and calculate the average sample area A. Determine the bag CO concentration.

Method 11: Determination of hydrogen sulfide content of fuel gas streams in petroleum refineries

Principle. Hydrogen sulfide (H_2S) is collected from a source in a series of midget impingers and absorbed in pH 3.0 cadmium sulfate ($CdSO_4$) solution to form cadmium sulfide (CdS). The latter compound is then measured iodometrically. An impinger containing hydrogen peroxide (H_2O_2) is included to remove SO_2 as an interfering species. This method is a revision of the H_2S method originally published in Federal Register (vol. 39, no. 47, March 8, 1974).

Applicability. This method is applicable for the determination of the H_2S content of fuel gas streams at petroleum refineries.

Apparatus
Analysis

Flask—glass-stoppered iodine flask, 500-mL

Burette—50-mL

Erlenmeyer flask—125-mL

Volumetric pipettes—one 25-mL; two each 50- and 100-mL

Volumetric flasks—one 1,000-mL; two 500-mL

Graduated cylinders—one each, 10- and 100-mL

Procedure

Analysis.   Note: Titration analyses should be conducted at the sample cleanup area in order to prevent loss of I_2 from the sample. Titration should never be made in direct sunlight.

Using 0.01 N $Na_2S_2O_3$ solution (or 0.01 N C_6H_5AsO, if applicable), rapidly titrate each sample in an iodine flask, using gentle mixing, until solution is light yellow. Add 4 mL of starch indicator solution, and continue titrating slowly until the blue color just disappears. Record V_{TT}, the volume of $Na_2S_2O_3$ solution used, or V_{AT}, the volume of C_6H_5AsO solution used, in milliliters.

Titrate the blanks in the same manner as the samples. Run blanks each day until replicate values agree within 0.05 mL. Average the replicate titration values which agree within 0.05 mL.

6.5.3   Inorganic chemicals

Method 12: Determination of inorganic lead emissions from stationary sources

Applicability.   This method applies to the determination of inorganic lead (Pb) emissions from specified stationary sources only.

Principle.   Particulate and gaseous Pb emissions are withdrawn isokinetically from the source and collected on a filter and in dilute nitric acid. The collected samples are digested in acid solution and analyzed by atomic absorption spectrometry using an air acetylene flame.

Apparatus

Analysis.   The following equipment is needed:

Atomic absorption spectrophotometer—with lead hollow cathode lamp and burner for air/acetylene flame

Hot plate

Erlenmeyer flasks—125-mL

Membrane filters—millipore SCWPO 4700, or equivalent

Filtration apparatus—millipore vacuum filtration unit, or equivalent, for use with the above membrane filter

Volumetric flasks—100-mL, 250-mL, and 1000-mL

Procedure

Analysis

Lead determination. Calibrate the spectrophotometer as described in the method, and determine the absorbance for each source sample, the filter blank, and 0.1 N HNO_3 blank. Analyze each sample three times in this manner. Make appropriate dilutions, as required, to bring all sample Pb concentrations into the linear absorbance range of the spectrophotometer.

If the Pb concentration of a sample is at the low end of the calibration curve and high accuracy is required, the sample can be taken to dryness on a hot plate and the residue dissolved in the appropriate volume of water to bring it into the optimum range of the calibration curve.

Check for matrix effects on the lead results. Since the analysis for Pb by atomic absorption is sensitive to the chemical composition and to the physical properties (viscosity, pH) of the sample (matrix effects), the analyst shall check at least one sample from each source, using the method of additions as follows:

Add or spike an equal volume of standard solution to an aliquot of the sample solution; then measure the absorbance of the resulting solution and the absorbance of an aliquot of unspiked sample.

Volume corrections will not be required if the solutions as analyzed have been made to the same final volume. Therefore, C_s and C_a represent Pb concentration before dilutions.

Method-of-additions procedures described on pages 9-4 and 9-5 of the section entitled "General Information" of the Perkin Elmer Corporation Atomic Absorption Spectrophotometry Manual, Number 303-0152 (see Bibliography), may also be used. In any event, if the results of the method-of-additions procedure used on the single source sample do not agree to within 5 percent of the value obtained by the routine atomic absorption analysis, then reanalyze all samples from the source using the method-of-additions procedure.

Container 3 (silica gel). The tester may conduct this step in the field. Weigh the spent silica gel (or silica gel plus impinger) to the nearest 0.5 g; record this weight.

Method 13A: Determination of total fluoride emissions from stationary sources (SPADNS zirconium lake method)

Applicability. This method applies to the determination of fluoride (F) emissions from sources as specified in the regulations. It does not measure fluorocarbons, such as Freons.

Principle. Gaseous and particulate F are withdrawn isokinetically from the source and collected in water and on a filter. The total F is then determined by the SPADNS Zirconium Lake colorimetric method.

Apparatus

Analysis.   The following equipment is needed:

Distillation apparatus—glass distillation apparatus

Bunsen burner

Electric muffle furnace—capable of heating to 600°C

Crucibles—nickel, 75- to 100-mL

Beakers—500-mL and 1500-mL

Volumetric flasks—50-mL

Erlenmeyer flasks or plastic bottles—500-mL

Constant-temperature bath—capable of maintaining a constant temperature of ±1.0°C at roomtemperature conditions

Balance—300-g capacity, to measure to ±0.5 g

Spectrophotometer—instrument that measures absorbance at 570 nm and provides at least a 1-cm light path

Spectrophotometer cells—1-cm path length

Procedure

Analysis

Containers 1 and 2.   After distilling suitable aliquots from containers 1 and 2 according to the method, dilute the distillate with the volumetric flasks to exactly 250 mL with water, and mix thoroughly. Pipette a suitable aliquot of each sample distillate (containing 10 to 40 μg fluoride/mL) into a beaker, and dilute to 50 mL with water. Use the same aliquot size for the blank. Add 10 mL of SPADNS mixed reagent, and mix thoroughly.

After mixing, place the sample in a constant-temperature bath containing the standard solutions for 30 min before reading the absorbance on the spectrophotometer.

Set the spectrophotometer to zero absorbance at 570 nm with the reference solution, and check the spectrophotometer calibration with the standard solution. Determine the absorbance of the samples, and determine the concentration from the calibration curve. If the concentration does not fall within the range of the calibration curve, repeat the procedure, using a different size aliquot.

Container 3 (silica gel).   Weigh the spent silica gel (or silica gel plus impinger) to the nearest 0.5 g, using a balance. The tester may conduct this step in the field.

Method 13B: Determination of total fluoride emissions from stationary sources (specific ion electrode method)

Applicability. This method applies to the determination of fluoride (F) emissions from stationary sources as specified in the regulations. It does not measure fluorocarbons, such as Freons.

Principle. Gaseous and particulate F are withdrawn isokinetically from the source and collected in water and on a filter. The total F is then determined by the specific ion electrode method.

Apparatus

Analysis. The following items are needed:

Distillation apparatus, bunsen burner, electric muffle furnace, crucibles, beakers, volumetric flasks, Erlenmeyer flasks or plastic bottles, constant-temperature bath, and balance—same as in method 13A, sections 5.3.1 to 5.3.9, respectively, except include also 100-mL polyethylene beakers.

Fluoride ion activity sensing electrode

Reference electrode—single-junction, sleeve type

Electrometer—a pH meter with millivolt scale capable of ±0.1-mV resolution, or a specific ion meter made especially for specific ion use

Magnetic stirrer and TFE fluorocarbon-coated stirring bars

Procedure

Analysis

Containers 1 and 2. Distill suitable aliquots from containers 1 and 2. Dilute the distillate in the volumetric flasks to exactly 250 mL with water, and mix thoroughly. Pipette a 25-mL aliquot from each of the distillate and separate beakers. Add an equal volume of TISAB, and mix. The sample should be at the same temperature as the calibration standards when measurements are made. If ambient laboratory temperature fluctuates more than ±2°C from the temperature at which the calibration standards were measured, condition samples and standards in a constant-temperature bath before measurement. Stir the sample with a magnetic stirrer during measurement to minimize electrode response time. If the stirrer generates enough heat to change the solution temperature, place a piece of temperature-insulating material, such as cork, between the stirrer and the beaker. Hold dilute samples (below 10^{-4} M fluoride ion content) in polyethylene beakers during measurement.

Insert the fluoride and reference electrodes into the solution. When a steady millivolt reading is obtained, record it. This may take several minutes. Determine concentration from the calibration curve. Between electrode measurements, rinse the electrode with water.

Container 3 (silica gel).   This is the same as in method 13A, section 7.4.2.

Method 14: Determination of fluoride emissions from roof monitors for primary aluminum plants

Applicability.   This method is applicable for the determination of fluoride emissions from stationary sources only when specified by the test procedures for determining compliance with new-source performance standards.

Principle.   Gaseous and particulate fluoride roof monitor emissions are drawn into a permanent sampling manifold through several large nozzles. The sample is transported from the sampling manifold to ground level through a duct. The gas in the duct is sampled using method 13A or 13B, determination of total fluoride emissions from stationary sources. Effluent velocity and volumetric flow rate are determined with anemometers located in the roof monitor.

Apparatus
Fluoride sampling train—same as that in method 13A or 13B.

Procedure
Analysis.   Use the analysis procedures described in method 13A or 13B.

Method 15: Determination of hydrogen sulfide, carbonyl sulfide, and carbon disulfide emissions from stationary sources

The method described below uses the principle of gas chromatographic separation and flame photometric detection (FPD). Since there are many systems or sets of operating conditions that represent usable methods of determining sulfur emissions, all systems which employ this principle but differ only in details of equipment and operation may be used as alternative methods, provided that the calibration precision and sample-line loss criteria are met.

Principle.   A gas sample is extracted from the emission source and diluted with clean, dry air. An aliquot of the diluted sample is then analyzed for hydrogen sulfide (H_2S), carbonyl sulfide (COS), and carbon disulfide (CS_2) by gas chromatographic separation and flame photometric detection.

Applicability.   This method is applicable for determination of the above sulfur compounds from tail gas control units of sulfur recovery plants.

Apparatus

Gas chromatograph.   The gas chromatograph must have at least the following components:

Oven—capable of maintaining the separation column at the proper operating temperature $\pm 1°C$.

Temperature gauge—to monitor column oven, detector, and exhaust temperature $\pm 1°C$.

Flow system.   Gas metering system to measure sample, fuel, combustion gas, and carrier gas flows.

Flame photometric detector.

Electrometer—capable of full-scale amplification of linear ranges of 10^{-9} to 10^{-4} A full-scale.

Power supply—capable of delivering up to 750 V.

Recorder—compatible with the output voltage range of the electrometer.

Rotary gas valves—multiport Teflon-lined valves equipped with sample loop. Sample loop volumes shall be chosen to provide the needed analytical range. Teflon tubing and fittings shall be used throughout to present an inert surface for sample gas. The gas chromatograph shall be calibrated with the sample loop used for sample analysis.

GC columns—the column system must be demonstrated to be capable of resolving three major reduced sulfur compounds: H_2S, COS, and CS_2.

To demonstrate that adequate resolution has been achieved, the tester must submit a chromatogram of a calibration gas containing all three reduced sulfur compounds in the concentration range of the applicable standard. Adequate resolution will be defined as baseline separation of adjacent peaks when the amplifier attenuation is set so that the smaller peak is at least 50 percent of full-scale. *Baseline separation* is defined as a return to zero ± 5 percent in the interval between peaks. Systems not meeting this criteria may be considered alternate methods subject to the approval of the Administrator.

Procedure

Analysis.   Aliquots of diluted sample are injected into the GC/FPD analyzer for analysis.

Sample run.   A sample run is composed of 16 individual analyses (injects) performed over a period of not less than 3 h or more than 6 h.

Observation for clogging of probe or filter.   If reductions in sample concentrations are observed during a sample run that cannot be explained by process conditions, the sampling must be interrupted to determine if the probe or filter is clogged with particulate matter. If either is found to be clogged, the test must be stopped and the results up to that point discarded. Testing may resume after cleaning or replacing of the probe and filter. After each run, the probe and filter shall be inspected and, if necessary, replaced.

Method 15A: Determination of total reduced sulfur emissions from sulfur recovery plants in petroleum refineries

Applicability. This method is applicable to the determination of total reduced sulfur (TRS) emissions from sulfur recovery plants where the emissions are in a reducing atmosphere, such as in Stretford units. The lower detectable limit is 0.1 ppm of sulfur dioxide when sampling at 2 L/min for 3 h or 0.3 ppm when sampling at 2 L/min for 1 h. The upper concentration limit of the method exceeds TRS levels generally encountered in sulfur recovery plants.

Principle. An integrated gas sample is extracted from the stack, and combustion air is added to the oxygen-deficient gas at a known rate. The TRS compounds (hydrogen sulfide, carbonyl sulfide, and carbon disulfide) are thermally oxidized to sulfur dioxide, collected in hydrogen peroxide as sulfate ion, and then analyzed according to the method 6 barium-thorin titration procedure.

Apparatus
Sample recovery and analysis. These are the same as in method 6, sections 2.2 and 2.3, except a 10-mL burette with 0.05-mL gradations is required for titrant volumes of less than 10.0 mL, and the spectrophotometer is not needed.

Procedure
Sample analysis. This is the same as in method 6, section 4.3. For compliance tests only, an EPA SO_2 field audit sample shall be analyzed with each set of samples. Such audit samples are available from the Quality Assurance Division, Environmental Monitoring Systems Laboratory, U.S. Environmental Protection Agency, Research Triangle Park, N.C. 27711.

Method 16: Semicontinuous determination of sulfur emissions from stationary sources. The method described below uses the principle of gas chromatographic separation and flame photometric detection. Since there are many systems or sets of operating conditions that represent usable methods of determining sulfur emissions, all systems which employ this principle but differ only in details of equipment and operation may be used as alternative methods, provided that the calibration precision and sample line loss criteria are met.

Principle. A gas sample is extracted from the emission source, and an aliquot is analyzed for hydrogen sulfide (H_2S), methyl mercaptan (MeSH), dimethyl sulfide (DMS), and dimethyl disulfide (DMDS) by gas chromatograph separation and flame photometric detection. These four compounds are know collectively as total reduced sulfur.

Applicability. This method is applicable for determination of TRS compounds from recovery furnaces, lime kilns, and smelt-dissolving tanks at kraft pulp mills.

Apparatus

Gas chromatograph. The gas chromatograph must have at least the following components:

Oven—capable of maintaining the separation column at the proper operating temperature $\pm 1°C$.

Temperature gauge—to monitor column oven, detector, and exhaust temperature $\pm 1°C$.

Flow system—gas metering system to measure sample, fuel, combustion gas, and carrier gas flows.

Flame photometric detector.

Electrometer—capable of full-scale amplification of linear ranges of 10^{-9} to 10^{-4} A full-scale.

Power supply—capable of delivering up to 750 V.

Recorder—compatible with the output voltage range of the electrometer.

Rotary gas valves—multiport Teflon-lined valves equipped with sample loop. Sample loop volumes shall be chosen to provide the needed analytical range. Teflon tubing and fittings shall be used throughout to present an inert surface for sample gas. The gas chromatograph shall be calibrated with the sample loop used for sample analysis.

Gas chromatogram columns—the column system must be demonstrated to be capable of resolving the four major reduced sulfur compounds: H_2S, MeSH, DMS, and DMDS. It must also demonstrate freedom from known interferences.

To demonstrate that adequate resolution has been achieved, the tester must submit a chromatogram of a calibration gas containing all four of the TRS compounds in the concentration range of the applicable standard. Adequate resolution will be defined as baseline separation of adjacent peaks when the amplifier attenuation is set so that the smaller peak is at least 50 percent of full-scale. Baseline separation is defined as a return to zero ± 5 percent in the interval between peaks. Systems not meeting these criteria may be considered alternate methods subject to the approval of the Administrator.

Procedure

Analysis.   Aliquots of sample are injected into the GC/FPD analyzer for analysis.

Sample run.   A sample run is composed of 16 individual analyses (injects) performed over a period of not less than 3 h or more than 6 h.

Observation for clogging of probe or filter.   If reductions in sample concentrations are observed during a sample run that cannot be explained by process conditions, the sampling must be interrupted to determine if the probe or filter is clogged with particulate matter. If either is found to be clogged, the test must be stopped and the results up to that point discarded. Testing may resume after cleaning or replacing of the probe and filter. After each run, the probe and filter shall be inspected and, if necessary, replaced.

Method 16A: Determination of total reduced sulfur emissions from stationary sources (impinger technique)

Applicability.   This method is applicable to the determination of total reduced sulfur emissions from recovery boilers, lime kilns, and smelt-dissolving tanks at kraft pulp mills, and from other sources when specified in an applicable subpart of the regulations. The TRS compounds include hydrogen sulfide, methyl mercaptan, dimethyl sulfide, and dimethyl disulfide.

The flue gas must contain at least 1% oxygen for complete oxidation of all TRS to sulfur dioxide. The lower detectable limit is 0.1 ppm SO_2 when sampling at 2 L/min for 3 h or 0.3 ppm when sampling at 2 L/min for 1 h. The upper concentration limit of the method exceeds TRS levels generally encountered at kraft pulp mills.

Principle.   An integrated gas sample is extracted from the stack. SO_2 is removed selectively from the sample using a citrate buffer solution. TRS compounds are then thermally oxidized to SO_2, collected in hydrogen peroxide as sulfate, and analyzed by the method 6 barium-thorin titration procedure.

Apparatus

Analysis.   It is the same as in method 6, section 2.3, except a 10-mL burette with 0.05-mL gradations is required and the spectrophotometer is not needed.

Procedure

Sample analysis. It is the same as in method 6, section 4.3, except for 1-h sampling. Take a 40-mL aliquot, add 160 mL of 100% isopropanol and 4 drops of thorin. Analyze an EPA SO_2 field audit sample with each set of samples. Such audit samples are available from the Source Branch, Quality Assurance Division, Environmental Monitoring Systems Laboratory, U.S. Environmental Protection Agency, Research Triangle Park, N.C. 27711.

Alternative procedures

Sample recovery and analysis apparatus
Erlenmeyer flasks—125- and 250-mL sizes

Pipettes—2-, 10-, 20-, and 100-mL volumetric

Burette—50-mL size

Volumetric flask—1-L size

Graduated cylinder—50-mL size

Wash bottle

Stirring plate and bars

Sample analysis. Sample treatment is similar to the blank treatment. Before detaching the stems from the bottoms of the impingers, add 20.0 mL of 0.01 N iodine through the stems of the impingers holding the zinc acetate solution, dividing it between the two (add about 15 mL to the first impinger and the rest to the second). Add 2 mL HCl solution through the stems, dividing it as with the iodine. Disconnect the sampling line, and store the impingers for 30 min. At the end of 30 min, rinse the impinger to a flask because this may result in a loss of iodine and cause a positive bias.

Method 16B: Determination of total reduced sulfur emissions from stationary sources

Applicability. This method is applicable to the determination of total reduced sulfur emissions from recovery furnaces, lime kilns, and smelt-dissolving tanks at kraft pulp mills, and from other sources when specified in an applicable subpart of the regulations. The TRS compounds include hydrogen sulfide (H_2S), methyl mercaptan, dimethyl sulfide, and dimethyl disulfide. The flue gas must contain at least 1% oxygen for complete oxidation of all TRS to sulfur dioxide.

Principle. An integrated gas sample is extracted from the stack. The SO_2 is removed selectively from the sample using a citrate buffer solution. The TRS compounds are then thermally oxidized to SO_2 and analyzed as SO_2 by gas chromatography using flame photometric detection.

Apparatus
Analysis
Dilution system (optional), gas chromatograph, oven, temperature gauges, flow system, flame photometric detector, electrometer, power supply, recorder, calibration system, tube chamber, and constant-temperature bath—same as in method 16, sections 5.2, 5.4, and 5.5.

Gas chromatograph columns—same as in method 16, section 12.1.4.1.1. Other columns with demonstrated ability to resolve SO_2 and be free from known interferences are acceptable alternatives.

Procedure
Analysis.   Pass aliquots of diluted sample through the SO_2 scrubber and oxidation furnace, and then inject into the GC/FPD analyzer for analysis. The rest of the analysis is the same as in method 16, sections 9.2.1 and 9.2.2.

Method 19: Determination of sulfur dioxide removal efficiency and particulate matter, sulfur dioxide, and nitrogen oxide emission rates

Applicability.   This method is applicable for (1) determining particulate matter, sulfur dioxide, and nitrogen oxides (NO_x) emission rates; (2) determining sulfur removal efficiencies of fuel pretreatment and SO_2 control devices; (3) determining overall reduction of potential SO_2 emissions from steam-generating units or other sources as specified in applicable regulations; and (4) determining SO_2 rates based on fuel sampling and analysis procedures.

Principle.   Pollutant emission rates are determined from concentrations of PM, SO_2, or NO_x, and oxygen or carbon dioxide along with F factors (ratios of combustion gas volumes to heat inputs).

An overall SO_2 emission reduction efficiency is computed from the efficiency of fuel pretreatment systems (optional) and the efficiency of SO_2 control devices.

The sulfur removal efficiency of a fuel pretreatment system is determined by fuel sampling and analysis of the sulfur and heat contents of the fuel before and after the pretreatment system.

The SO_2 removal efficiency of a control device is determined by measuring the SO_2 rates before and after the control device.

The inlet rates to SO_2 control systems, or when SO_2 control systems are not used, SO_2 emission rates to the atmosphere, may be determined by fuel sampling and analysis (optional).

Method 29: Determination of metals emissions from stationary sources

Applicability. This method is applicable to the determination of antimony (Sb), arsenic (As), barium (Ba), beryllium (Be), cadmium (Cd), chromium (Cr), cobalt (Co), copper (Cu), lead (Pb), manganese (Mn), mercury (Hg), nickel (Ni), phosphorus (P), selenium (Se), silver (Ag), thallium (Tl), and zinc (Zn) emissions from stationary sources. This method may be used to determine particulate emissions in addition to the metals emissions if the prescribed procedures and precautions are followed.

Mercury emissions can be measured, alternatively, using EPA method 101A of Appendix B, 40 CFR Part 61. Method 101A measures only Hg, but it can be of special interest to sources which need to measure both Hg and Mn emissions.

Principle. A stack sample is withdrawn isokinetically from the source, particulate emissions are collected in the probe and on a heated filter, and gaseous emissions are then collected in an aqueous acidic solution of hydrogen peroxide (analyzed for all metals including Hg) and an aqueous acidic solution of potassium permanganate (analyzed only for Hg). The recovered samples are digested, and appropriate fractions are analyzed for Hg by cold vapor atomic absorption spectroscopy (CVAAS) and for Sb, As, Ba, Be, Cd, Cr, Co, Cu, Pb, Mn, Ni, P, Se, Ag, Tl, and Zn by inductively coupled argon plasma (ICAP) emission spectroscopy, or atomic absorption spectroscopy (AAS). Graphite furnace atomic absorption spectroscopy (GFAAS) is used for analysis of Sb, As, Cd, Co, Pb, Se, and Tl if these elements require greater analytical sensitivity than can be obtained by ICAP emission spectroscopy. If one so chooses, AAS may be used for analysis of all listed metals if the resulting in-stack method detection limits meet the goal of the testing program. Similarly, inductively coupled plasma-mass spectroscopy (ICP-MS) may be used for analysis of Sb, As, Ba, Be, Cd, Cr, Co, Cu, Pb, Mn, Ni, As, Tl, and Zn.

Apparatus

Analysis

Volumetric flasks—100-mL, 250-mL, and 1000-mL. For preparation of standards and sample dilutions.

Graduated cylinders—for preparation of reagents.

Parr bombs or microwave pressure relief vessels with capping station (CEM Corporation model or equivalent)—for sample digestion.

Beakers and watch glasses—250-mL beakers, with watch glass covers, for sample digestion.

Ring stands and clamps—for securing equipment such as filtration apparatus.

Filter funnels—for holding filter paper.

Disposable Pasteur pipettes and bulbs.

Volumetric pipettes.

Analytical balance—accurate to within 0.1 mg.

Microwave or conventional oven—for heating samples at fixed power levels or temperatures, respectively.

Hot plates.

Atomic absorption spectrometer—equipped with a background corrector.

Graphite furnace attachment—with Sb, As, Cd, Co, Pb, Se, and Tl hollow cathode lamps (HCLs) or electrodeless discharge lamps (EDLs).

Cold vapor mercury attachment—with a mercury HCL or EDL, an air recirculation pump, a quartz cell, an aerator apparatus, and a heat lamp or desiccator tube. The heat lamp shall be capable of raising the temperature at the quartz cell by 10°C above ambient, so that no condensation forms on the wall of the quartz cell. This is the same as in method 7470. See Note 2, section 5.4.3, of the method for other acceptable approaches for analysis of Hg in which analytical detection limits of 0.002 ng/mL were obtained.

Inductively coupled argon plasma spectrometer—with either a direct or sequential reader and an alumina torch; same as EPA method 6010.

Inductively coupled plasma-mass spectrometer—same as EPA method 6020.

Procedure

Analysis.   For each sampling train sample run, seven individual analytical samples are generated—two for all desired metals except Hg, and five for Hg. The first two analytical samples, labeled analytical fractions 1A and 1B, consist of the digested samples from the front half of the train. Analytical fraction 1A is for ICAP, ICP-MS, or AAS analysis as described in the method. Analytical fraction 1B is for front-half Hg analysis as described in the method. The contents of the back half of the train are used to prepare the third through seventh analytical samples. The third and fourth analytical samples, labeled analytical fractions 2A and 2B, contain the samples from the moisture removal impinger 1, if used, and HNO_3/H_2O_2 impingers 2 and 3. Analytical fraction 2A is for ICAP, ICP-MS, or AAS analysis for target metals, except Hg. Analytical fraction 2B is for analysis for Hg. The fifth through seventh analytical samples, labeled analytical fractions

3A, 3B, and 3C, consist of the impinger contents and rinses from the empty impinger 4 and the $H_2SO_4/KMnO_4$ impingers 5 and 6. These analytical samples are for analysis for Hg as described in the method. The total back-half Hg catch is determined from the sum of analytical fractions 2B, 3A, 3B, and 3C. Analytical fractions 1A and 2A can be combined proportionally prior to analysis.

ICAP and ICP-MS analysis. Analyze analytical fractions 1A and 2A by ICAP using method 6010 or method 200.7 (40 CFR 136, Appendix C). Calibrate the ICAP, and set up an analysis program as described in method 6010 or method 200.7. Follow the quality control procedures described in the method.

AAS by direct aspiration and/or GFAAS. If analysis of metals in analytical fractions 1A and 2A by using GFAAS or direct aspiration AAS is needed, use the method to determine which techniques and procedures to apply for each target metal. Use the method, if necessary, to determine techniques for minimization of interferences. Calibrate the instrument according to the method, and follow the quality control procedures specified.

CVAAS Hg analysis. Analyze analytical fractions 1B, 2B, 3A, 3B, and 3C separately for Hg using CVAAS following the method outlined in EPA *SW-846* method 7470 or in *Standard Methods for Water and Wastewater Analysis,* 15th ed., method 303F, or optionally using Note 2 at the end of the method. Set up the calibration curve (0 to 1000 ng) as described in *SW-846* method 7470 or similar to method 303F using 300-mL BOD bottles instead of Erlenmeyer flasks. Perform the following for each Hg analysis: From each original sample, select and record an aliquot in the size range from 1 to 10 mL. If no prior knowledge of the expected amount of Hg in the sample exists, a 5-mL aliquot is suggested for the first dilution to 100 mL (see Note 1 at end of the method). The total amount of Hg in the aliquot shall be less than 1 μg and within the range (0 to 1000 ng) of the calibration curve. Place the sample aliquot into a separate 300-mL BOD bottle, and add enough water to make a total volume of 100 mL. Next add to it sequentially the sample digestion solutions, and perform the sample preparation described in the procedures of *SW-846* method 7470 or method 303F. (See Note 2 at the end of the method.) If the maximum readings are off-scale (because Hg in the aliquot exceeded the calibration range, including the situation where only a 1-mL aliquot of the original sample was digested), then dilute the original sample (or a portion of it) with 0.15% HNO_3 (1.5 mL concentrated HNO_3 per liter aqueous solution) so that when a 1- to 10-mL aliquot of the "0.15 HNO_3 percent dilution of the original sample" is digested and analyzed by the procedures described above, it will yield an analysis within the range of the calibration curve.

Note 1. When Hg levels in the sample fractions are below the in-stack detection limit, select a 10-mL aliquot for digestion and analysis as described.

Note 2. Optionally, Hg can be analyzed by using the CVAAS analytical procedures given by some instrument manufacturers' directions. These include calibration and quality control procedures for the Leeman model PS200, the Perkin Elmer FIAS systems, and similar models, if available, of other instrument manufacturers. For digestion and analyses by these instruments, perform the following two steps: (1) Digest the sample aliquot through the addition of the aqueous hydroxylamine hydrochloride/sodium chloride solution as described in the method: (The Leeman, Perkin Elmer, and similar instruments described in this note add automatically the necessary stannous chloride solution during the automated analysis of Hg.) (2) Upon completion of the digestion described in (1), analyze the sample according to the instrument manufacturer's directions. This approach allows multiple (including duplicate) automated analyses of a digested sample aliquot.

Method 101: Determination of particulate and gaseous mercury emissions from chlor-alkali plants (airstreams)

Applicability. This method applies to the determination of particulate and gaseous mercury (Hg) emissions from chlor-alkali plants and other sources (as specified in the regulations), where the carrier-gas stream in the duct or stack is principally air.

Principle. Particulate and gaseous Hg emissions are withdrawn isokinetically from the source and collected in acidic iodine monochloride (ICl) solution. The Hg collected (in the mercuric form) is reduced to elemental Hg, which is then aerated from the solution into an optical cell and measured by atomic absorption spectrophotometry.

Apparatus
Analysis. The following equipment is needed:
Atomic absorption spectrophotometer—Perkin-Elmer 303, or equivalent, containing a hollow-cathode mercury lamp and the optical cell described.

Optical cell—cylindrical shape with quartz end windows. Wind the cell with approximately 2 m of 24-gauge nichrome heating wire, and wrap with fiberglass insulation tape, or equivalent; do not let the wires touch each other.

Aeration cell—constructed according to the specifications in the method. Do not use a glass frit as a substitute for the blown-glass bubbler tip.

Recorder—matched to output of the spectrophotometer described in the method.

Variable transformer—to vary the voltage on the optical cell from 0 to 40 V.

Hood—for venting optical cell exhaust.

Flowmetering valve.

Flowmeter—rotameter, or equivalent, capable of measuring a gas flow of 1.5 L/min.

Aeration gas cylinder—nitrogen or dry, Hg-free air, equipped with a single-stage regulator.

Tubing—for making connections. Use glass tubing (ungreased ball-and-socket connections are recommended) for all tubing connections between the solution cell and the optical cell; do not use Tygon tubing, other types of flexible tubing, or metal tubing as substitutes. The tester may use Teflon, steel, or copper tubing between the nitrogen tank and flowmetering valve (section 5.3.7), and Tygon, gum, or rubber tubing between the flowmetering valve and the aeration cell.

Flow rate calibration equipment—bubble flowmeter or wet test meter for measuring a gas flow rate of 1.5 ± 0.1 L/min.

Volumetric flasks—class A with penny head standard taper stoppers; 100-, 250-, 500-, and 1000-mL.

Volumetric pipettes—Class A; 1-, 2-, 3-, 4-, and 5-mL.

Graduated cylinder—50-mL.

Magnetic stirrer—general-purpose laboratory type.

Magnetic stirring bar—Teflon-coated.

Balance—capable of weighing to ±0.5 g.

Alternative analytical apparatus—alternative systems are allowable as long as they meet the following criteria:

- A linear calibration curve is generated, and two consecutive samples of the same aliquot size and concentration agree within 3 percent of their average.
- A minimum of 95 percent of the spike is recovered when an aliquot of a source sample is spiked with a known concentration of Hg(II) compound.
- The reducing agent should be added after the aeration cell is closed.
- The aeration bottle bubbler should not contain a frit.
- Any Tygon tubing used should be as short as possible and conditioned prior to use until blanks and standards yield linear and reproducible results.

- If manual stirring is done before aeration, it should be done with the aeration cell closed.
- A drying tube should not be used unless it is conditioned as the Tygon tubing above.

Procedure

Analysis.   Calibrate the spectrophotometer and recorder, and prepare the calibration curve.

Mercury samples.   Repeat the procedure used to establish the calibration curve with an appropriately sized aliquot (1 to 5 mL) of the diluted sample until two consecutive peak heights agree within 3 percent of their average value. The peak maximum of an aliquot (except the 5-mL aliquot) must be greater than 10 percent of the recorder full-scale. If the peak maximum of a 1.0-mL aliquot is off scale on the recorder, further dilute the original source sample to bring the Hg concentration into the calibration range of the spectrophotometer.

Run a blank and standard at least after every five samples to check the spectrophotometer calibration; recalibrate as necessary.

It is recommended that at least one sample from each stack test be checked by the method of standard additions to confirm that matrix effects have not interfered in the analysis.

Container 2 (silica gel).   Weigh the spent silica gel (or silica gel plus impinger) to the nearest 0.5 g, using a balance. (This step may be conducted in the field.)

Method 101A: Determination of particulate and gaseous mercury emissions from sewage sludge incinerators.   This method is similar to method 101, except acidic potassium permanganate solution is used instead of acidic iodine monochloride for sample collection.

Applicability.   This method applies to the determination of particulate and gaseous mercury (Hg) emissions from sewage sludge incinerators and other sources as specified in the regulations.

Principle.   Particulate and gaseous Hg emissions are withdrawn isokinetically from the source and collected in acidic potassium permanganate ($KMnO_4$) solution. The Hg collected (in the mercuric form) is reduced to elemental Hg, which is then aerated from the solution into an optical cell and measured by atomic absorption spectrophotometry.

Apparatus

Analysis.   It is the same as in method 101, sections 5.3 and 5.4, except as follows:

Volumetric pipettes—class A; 1-, 2-, 3-, 4-, 5-, 10-, and 20-mL.

Graduated cylinder—25-mL.

Steam bath.

Atomic absorption spectrophotometer or equivalent—Any atomic absorption unit with an open sample presentation area in which to mount the optical cell is suitable. Use those instrument settings recommended by the particular manufacturer. Instruments designed specifically for the measurement of mercury using the cold-vapor technique are commercially available and may be substituted for the atomic absorption spectrophotometer.

Optical cell—Alternatively, a heat lamp mounted above the cell or a moisture trap installed upstream of the cell may be used.

Aeration cell—Alternatively, aeration cells available with commercial cold-vapor instrumentation may be used.

Aeration gas cylinder—nitrogen, argon, or dry, Hg-free air, equipped with a single-stage regulator. Alternatively, aeration may be provided by a peristaltic metering pump. If a commercial cold-vapor instrument is used, follow the manufacturer's recommendations.

Procedure
Analysis.   Calibrate the spectrophotometer and recorder, and prepare the calibration curve. Then repeat the procedure used to establish the calibration curve with appropriately sized aliquots (1 to 10 mL) of the samples until two consecutive peak heights agree within 3 percent of their average value. If the 10-mL sample is below the detectable limit, use a larger aliquot (up to 20 mL), but decrease the volume of water added to the aeration cell accordingly to prevent the solution volume from exceeding the capacity of the aeration bottle. If the peak maximum of a 1.0-mL aliquot is off scale, further dilute the original sample to bring the Hg concentration into the calibration range of the spectrophotometer. If the Hg content of the absorbing solution and filter blank is below the working range of the analytical method, use zero for the blank.

Run a blank and standard at least after every five samples to check the spectrophotometer calibration; recalibrate as necessary. It is also recommended that at least one sample from each stack test be checked by the method of standard additions to confirm that matrix effects have not interfered in the analysis.

Method 102: Determination of particulate and gaseous mercury emissions from chlor-alkali plants (hydrogen streams).   Although similar to method 101, method 102 requires changes to accommodate the sample's being extracted from a hydrogen stream. Conduct the test according to method 101, except as noted in the method.

Method 103: Beryllium screening method

Applicability. This procedure details guidelines and requirements for methods acceptable for use in determining beryllium (Be) emissions in ducts or stacks at stationary sources.

Principle. Beryllium emissions are isokinetically sampled from three points in a duct or stack. The collected sample is analyzed for Be using an appropriate technique.

Apparatus
Analysis. Use equipment necessary to perform an atomic absorption, spectrographic, fluorometric, chromatographic, or equivalent analysis.

Procedure. Guidelines for source testing are detailed in the following sections. These guidelines are generally applicable; however, most sample sites differ to some degree, and temporary alterations such as stack extensions or expansions often are required to ensure the best possible sample site. Further, since Be is hazardous, care should be taken to minimize exposure. Finally, since the total quantity of Be to be collected is quite small, the test must be carefully conducted to prevent contamination or loss of sample.

Analysis. Make the necessary preparation of samples and analyze for Be. Any currently acceptable method such as atomic absorption, spectrographic, fluorometric, chromatographic, or equivalent may be used.

Method 104: Determination of beryllium emissions from stationary sources

Applicability. This method is applicable for the determination of beryllium emissions in ducts or stacks at stationary sources. Unless otherwise specified, this method is not intended to apply to gas streams other than those emitted directly to the atmosphere without further processing.

Principle. Beryllium emissions are isokinetically sampled from the source, and the collected sample is digested in an acid solution and analyzed by atomic absorption spectrophotometry.

Apparatus
Analysis. The following equipment is needed:
Atomic absorption spectrophotometer—Perkin-Elmer 303, or equivalent, with nitrous oxide/acetylene burner

Hot plate

Perchloric acid fume hood

Procedure
Analysis—beryllium determination. Analyze the samples prepared in section 4.3.2 at 234.8 nm using a nitrous oxide/acetylene flame.

Aluminum, silicon, and other elements can interfere with this method if present in large quantities. Standard methods are available, however, that may be used to effectively eliminate these interferences.

Method 105: Determination of mercury in wastewater treatment plant sewage sludge

Applicability.   This method applies to the determination of total organic and inorganic mercury (Hg) content in sewage sludges. The range of this method is 0.2 to 5 $\mu g/g$; it may be extended by increasing or decreasing sample size.

Principle.   Time-composite sludge samples are withdrawn from the conveyor belt after dewatering and before incineration or drying. A weighed portion of the sludge is digested in aqua regia and oxidized by potassium permanganate ($KMnO_4$). Then Hg in the digested sample is measured by the conventional spectropho to metric cold-vapor technique.

Apparatus
Analysis.   It is the same as in method 101, sections 5.3 and 5.4, except for the following:
Balance—The balance of method 101, section 5.3.17, is not needed.

Filter paper—S and S no. 588 (or equivalent).

Procedure
Analysis for mercury.   It is the same as in method 101A, sections 7.4 and 8, except for the following variation:
Spectrophotometer and recorder calibration.   The mercury response may be measured by either peak height or peak area. *Note:* The temperature of the solution affects the rate at which elemental Hg is released from solution, and consequently, it affects the shape of the absorption curve (area) and the point of maximum absorbance (peak height). Therefore, to obtain reproducible results, bring all solutions to room temperature before use. Set the spectrophotometer wavelength to 253.7 nm. Make certain the optical cell is at the minimum temperature that will prevent water condensation from occurring. Then set the recorder scale as follows: Using a 25-mL graduated cylinder, add 25 mL of water to the aeration-cell bottle. Add 3 drops of Anitifoam B to the bottle, and then pipette 5.0 mL of the working Hg standard solution into the aeration cell.
Note: Always add the Hg-containing solution to the aeration cell after the 25 mL of water.
Place a Teflon-coated stirring bar in the bottle. Add 5 mL of 15% HNO_3 and 5 mL of 5% $KMnO_4$ to the aeration bottle, and mix well.

Next, attach the bottle section to the bubbler section of the aeration cell, and make certain that (1) the exit arm stopcock of the aeration cell (fig. 105.1) is closed (so that Hg will not prematurely enter the optical cell when the reducing agent is being added) and (2) there is no flow through the bubbler. Add 5 mL of sodium chloride–hydroxylamine solution to the aeration bottle through the side arm, and mix. If the solution does not become colorless, add additional sodium chloride–hydroxylamine solution in 1-mL increments until the solution is colorless. Now add 5 mL of tin(II) solution to the aeration bottle through the side arm. Stir the solution for 15 s, turn on the recorder, open the aeration cell exit arm stopcock, and then immediately initiate aeration with continued stirring. Determine the maximum absorbance of the standard, and set this value to read 90% of the recorder full-scale.

Method 108: Determination of particulate and gaseous arsenic emissions

Applicability. This method applies to the determination of inorganic arsenic (As) emissions from stationary sources as specified in the applicable subpart.

Principle. Particulate and gaseous arsenic emissions are withdrawn isokinetically from the source and collected on a glass matte filter and in water. The collected arsenic is then analyzed by means of atomic absorption spectrophotometry.

Apparatus
Analysis. The following equipment is needed:
Spectrophotometer—equipped with an electrodeless discharge lamp and a background corrector to measure absorbance at 193.7 nm. For measuring samples having less than 10 mg As/mL, use a vapor generator accessory or a graphite furnace.

Recorder—to match the output of the spectrophotometer.

Beakers—150-mL.

Volumetric flasks—glass, 50-, 100-, 200-, 500-, and 1000-mL; and polypropylene, 50-mL.

Balance—to measure within 0.5 g.

Volumetric pipettes—1-, 2-, 3-, 5-, 8-, and 10-mL.

Oven.

Hot plate.

Procedure
Analysis
Arsenic determination. Prepare standard solutions as directed under section 5.1, and measure their absorbances against 0.8 N

HNO_3. Then determine the absorbances of the filter blank and each sample using 0.8 N HNO_3 as a reference. If the sample concentration falls outside the range of the calibration curve, make an appropriate dilution with 0.8 N HNO_3 so that the final concentration falls within the range of the curve. Determine the arsenic concentration in the filter blank (i.e., the average of the two blank values from each lot). Next, using the appropriate standard curve, determine the arsenic concentration in each sample fraction.

Arsenic determination at low concentration. The lower limit of flame atomic absorption spectrophotometry is 10 mg As/mL. If the arsenic concentration of any sample is at a lower level, use the graphite furnace or vapor generator which is available as an accessory component. The analyst also has the option of using either of these accessories for samples whose concentrations are between 10 and 30 mg/mL. Follow the manufacturer's instructions in the use of such equipment.

Vapor generator procedure. Place a sample containing between 0 and 5 mg of arsenic in the reaction tube, and dilute to 15 mL with water. Since there is some trial and error involved in this procedure, it may be necessary to screen the samples by conventional atomic absorption until an approximate concentration is determined. After determining the approximate concentration, adjust the volume of the sample accordingly. Pipette 15 mL of concentrated HCl into each tube. Add 1 mL of 30% KI solution. Place the reaction tube into a 50°C water bath for 5 min. Cool to room temperature. Connect the reaction tube to the vapor generator assembly. When the instrument response has returned to baseline, inject 5.0 mL of 5% $NaBH_4$, and integrate the resulting spectrophotometer signal over a 30-s time period.

Graphite furnace procedure. Dilute the digested sample so that a 5-mL aliquot contains less than 1.5 mg of arsenic. Pipette 5 mL of this digested solution into a 10-mL volumetric flask. Add 1 mL of the 1% nickel nitrate solution, 0.5 mL of 50% HNO_3, and 1 mL of the 3% hydrogen peroxide; and dilute to 10 mL with water. The sample is now ready to inject in the furnace for analysis. Because instruments from different manufacturers are different, no detailed operating instructions are given here. Instead, the analyst should follow the instructions provided with the particular instrument.

Container 3 (silica gel). The tester may conduct this step in the field. Weigh the spent silica gel (or silica gel plus impinger) to the nearest 0.5 g; record this weight.

Method 108A: Determination of arsenic content in ore samples from nonferrous smelters

Applicability. This method applies to the determination of inorganic arsenic (As) content of process ore and reverberatory matte samples from nonferrous smelters and other sources as specified in the regulations.

Principle. Arsenic bound in ore samples is liberated by acid digestion and analyzed by atomic absorption spectrophotometry.

Apparatus
Analysis
Spectrophotometer and recorder—equipped with an electrodeless discharge lamp and a background corrector to measure absorbance at 193.7 nm. A graphite furnace may be used in place of the vapor generator accessory when measuring samples with low As levels. The recorder shall match the output of the spectrophotometer.

Volumetric flasks—class A, 50-mL (one needed per sample and blank).

Volumetric pipettes—class A; 1-, 5-, 10-, and 25-mL sizes.

Procedure
Analysis—arsenic determination. Determine the absorbance of each sample, using the blank as a reference. If the sample concentration falls outside the range of the calibration curve, make an appropriate dilution with 0.5 N HNO_3 so that the final concentration falls within the range of the curve. From the curve, determine the As concentration in each sample.

Method 108B: Determination of arsenic content in ore samples from nonferrous smelters

Applicability. This method applies to the determination of inorganic arsenic (As) content of process ore and reverberatory matte samples from nonferrous smelters and other sources as specified in the regulations. Samples resulting in an analytical concentration greater than 10 μg As/mL may be analyzed by this method.

Principle. Arsenic bound in ore samples is liberated by acid digestion and analyzed by flame atomic absorption spectrophotometry.

Apparatus
Analysis
Spectrophotometer—equipped with an electrodeless discharge lamp and a background corrector to measure absorbance at 193.7 nm

Beaker and watch glass—400-mL

Volumetric flask—1-L

Volumetric pipettes—1-, 5-, 10-, and 25-mL

Procedure

Analysis—arsenic determination.   Determine the absorbance of each sample, using the blank as a reference. If the sample concentration falls outside the range of the calibration curve, make an appropriate dilution with 2% $HClO_4$/10 percent HCl (prepared by diluting 2 mL concentrated $HClO_4$ and 10 mL concentrated HCl to 100 mL with water) so that the final concentration falls within the range of the curve. From the curve, determine the As concentration in each sample.

Method 108C: Determination of arsenic content in ore samples from nonferrous smelters

Applicability.   This method applies to the determination of inorganic arsenic content of process ore and reverberatory matte samples from nonferrous smelters and other sources as specified in the regulations. This method is applicable to samples having an analytical concentration less than 10 μg As/mL.

Principle.   Arsenic bound in ore samples is liberated by acid digestion and analyzed by the molybdenum blue photometric procedure.

Apparatus
Analysis
Photometer—capable of measuring at 660 nm

Volumetric flasks—50- and 100-mL

Procedure

Analysis.   Add 1 mL of $KBrO_3$ solution to the flask, and heat on a low-temperature hot plate to about 50°C to oxidize the arsenic and methyl orange. Add 5.0 mL of ammonium molybdate solution to the warm solution and mix. Add 2.0 mL of hydrazine sulfate solution, dilute until the solution comes within the neck of the flask, and mix. In a 400-mL beaker, place 80% full of boiling water, for 10 min. Enough heat must be supplied to prevent the water bath from cooling much below the boiling point upon inserting the volumetric flask. Remove the flask, cool to room temperature, dilute to the mark, and mix.

Transfer a suitable portion of the reference solution to an absorption cell, and adjust the photometer to the initial setting, using a light band centered at 660 nm. While maintaining this photometer adjustment, take the photometric readings of the calibration solutions followed by the samples.

Method 306: Determination of chromium emissions from decorative and hard chromium electroplating and anodizing operations

Applicability. This method applies to the determination of chromium (Cr) in emissions from decorative and hard chrome electroplating facilities and anodizing operations.

Principle. A sample is extracted isokinetically from the source, using an unheated method 5 sampling train (40 CFR Part 60, Appendix A), with a glass nozzle and probe liner, but with the filter omitted. The Cr emissions are collected in an alkaline solution: 0.1 N sodium hydroxide (NaOH) or 0.1 N sodium bicarbonate ($NaHCO_3$). The collected samples remain in the alkaline solution until analysis. Samples with high Cr concentrations may be analyzed by using inductively coupled plasma (ICP) emission spectrometry at 267.72 nm. Alternatively, if improved detection limits are required, a portion of the alkaline impinger solution is digested with nitric acid and analyzed by graphite furnace atomic absorption spectroscopy (GFAAS) at 357.9 nm.

If it is desirable to determine hexavalent chromium (Cr^{6+}) emissions, the samples may be analyzed by using an ion chromatograph equipped with a postcolumn reactor (IC/PCR) and a visible wavelength detector. To increase sensitivity for trace levels of Cr^{6+}, a preconcentration system can be used in conjunction with the IC/PCR.

Apparatus
Analysis.   For analysis, the following equipment is needed:
Analysis by GFAAS:
Chromium hollow-cathode lamp or electrodeless discharge lamp

Graphite furnace atomic absorption spectrophotometer

Analysis by ICP:

ICP spectrometer—computer-controlled emission spectrometer with background correction and radio-frequency generator

Argon gas supply—welding grade or better

Analysis by IC/PCR:

IC / PCR system—high-performance liquid chromatograph pump, sample injection valve, postcolumn reagent delivery and mixing system, and a visible detector, capable of operating at 520 nm, all with a nonmetallic (or inert) flow path. An electronic peak area mode is recommended, but other recording devices and integration techniques are acceptable provided the repeatability criteria and the linearity criteria for the calibration curve described in section

6.4.1 can be satisfied. A sample loading system will be required if preconcentration is employed.

Analytical column—a high-performance ion chromatograph (HPIC) nonmetallic column with anion separation characteristics and a high loading capacity designed for separation of metal chelating compounds to prevent metal interference. The resolution described in the method must be obtained. A nonmetallic guard column with the same ion-exchange material is recommended.

Preconcentration column—an HPIC nonmetallic column with acceptable anion retention characteristics and sample loading rates as described in the method.

0.45-μm filter cartridge—for the removal of insoluble material; to be used just prior to sample injection/analysis.

Procedure

Sample analysis by GFAAS.   The 357.9-nm wavelength line shall be used. Follow the manufacturer's operating instructions for all other spectrophotometer parameters.

Furnace parameters suggested by the manufacturer should be employed as guidelines. Since temperature-sensing mechanisms and temperature controllers can vary between instruments and/or with time, the validity of the furnace parameters must be periodically confirmed by systematically altering the furnace parameters while analyzing a standard. In this manner, losses of analyte due to higher-than-necessary temperature settings or losses in sensitivity due to less-than-optimum settings can be minimized. Similar verification of furnace parameters may be required for complex sample matrices.

Inject a measured aliquot of digested sample into the furnace and atomize. If the concentration found exceeds the calibration range, the sample should be diluted with the calibration blank solution (1.0% HNO_3) and reanalyzed. Consult the operator's manual for suggested injection volumes. The use of multiple injections can improve accuracy and help detect furnace pipetting errors.

Analyze a minimum of one matrix-matched reagent blank per sample batch to determine if contamination or any memory effects are occurring. Analyze a calibration blank and a midpoint calibration check standard after approximately every 10 sample injections.

Calculate the Cr concentrations (1) by the method of standard additions (see operator's manual), (2) from the calibration curve, or (3) directly from the instrument's concentration readout. All dilution or concentration factors must be taken into account. All results should be reported in μg Cr/mL with up to three significant figures.

Sample analysis by ICP.   The ICP measurement is performed directly on the alkaline impinger solution; acid digestion is not necessary provided the samples and standards are matrix-matched. However, ICP should only be used when the solution analyzed has a Cr concentration greater than 35 µg/L.

Two types of blanks are required for the analysis. The calibration blank is used in establishing the analytical curve, and the reagent blank is used to assess possible contamination resulting from sample processing. Use either 0.1 N NaOH or 0.1 N NaHCO$_3$, whichever was used for the impinger absorbing solution, for the calibration blank. The calibration blank can be prepared fresh in the laboratory; it does not have to be from the same batch of solution that was used in the field. Prepare a sufficient quantity to flush the system between standards and samples. The reagent blank is a sample of the impinger solution used for sample collection that is collected in the field during the testing program.

Set up the instrument with proper operating parameters including wavelength, background correction settings (if necessary), and interfering element correction settings (if necessary). The instrument must be allowed to become thermally stable before beginning performance of measurements (usually requiring at least 30 min of operation prior to calibration). During this warm-up period, the optical calibration and torch position optimization may be performed (consult the operator's manual).

Calibrate the instrument according to the instrument manufacturer's recommended procedures. Before analyzing the samples, reanalyze the highest calibration standard as if it were a sample. Concentration values obtained should not deviate from the actual values by more than 5 percent or the established control limits, whichever is lower. If they do, follow the recommendations of the instrument manufacturer to correct for this condition.

Flush the system with the calibration blank solution for at least 1 min before the analysis of each sample or standard. Analyze the midpoint calibration standard and the calibration blank after each 10 samples. Use the average intensity of multiple exposures for both standardization and sample analysis to reduce random error.

Dilute and reanalyze samples that are more concentrated than the linear calibration limit, or use an alternate, less sensitive Cr wavelength for which quality control data are already established.

If dilutions are performed, the appropriate factors must be applied to sample values. All results should be reported in µg Cr/mL with up to three significant figures.

Sample analyses by IC/PCR. The Cr^{6+} content of the sample filtrate is determined by IC/PCR. To increase sensitivity for trace levels of chromium, a preconcentration system is also used in conjunction with the IC/PCR.

Prior to preconcentration and/or analysis, filter all field samples through a 0.45-μm filter. This filtration should be conducted just prior to sample injection and analysis.

The preconcentration is accomplished by selectively retaining the analyte on a solid absorbent, followed by removal of the analyte from the absorbent. Inject the sample into a sample loop of the desired size (use repeated loadings or a larger loop for greater sensitivity). The Cr^{6+} is collected on the resin bed of the column. Switch the injection valve so that the eluent displaces the concentrated Cr^{6+} sample, moving it off the preconcentration column and onto the IC anion separation column. After separation from other sample components, the Cr^{6+} forms a specific complex in the postcolumn reactor with the diphenylcarbazide (DPC) reaction solution, and the complex is detected by visible absorbance at a wavelength of 520 nm. The amount of absorbance measured is proportional to the concentration of the Cr^{6+} complex formed. Compare the IC retention time and the absorbance of the Cr^{6+} complex with known Cr^{6+} standards analyzed under identical conditions to provide both qualitative and quantitative analyses.

Two types of blanks are required for the analysis. The calibration blank is used in establishing the analytical curve, and the reagent blank is used to assess possible contamination resulting from sample processing. Use either 0.1 N NaOH or 0.1 N $NaHCO_3$, whichever was used for the impinger solution, for the calibration blank. The calibration blank can be prepared fresh in the laboratory; it does not have to be from the same batch of solution that was used in the field. The reagent blank is a sample of the impinger solution used for sample collection in the field during the testing program.

Prior to sample analysis, establish a stable baseline with the detector set at the required attenuation by setting the eluent flow rate at approximately 1 mL/min and the postcolumn reagent flow rate at approximately 0.5 mL/min. *Note:* As long as the ratio of eluent flow rate to PCR flow rate remains constant, the standard curve should remain linear. Inject a sample of reagent water to ensure that no Cr^{6+} appears in the water blank.

First, inject the calibration standards prepared, to cover the appropriate concentration range, starting with the lowest standard first. Next, inject, in duplicate, the calibration reference standard, followed by the reagent blank and the field samples. Finally, repeat the injection of the calibration standards to assess instrument drift. Measure areas or heights of the Cr^{6+}/DPC complex chromatogram peaks. The response for replicate, consecutive injections of samples must be within 5 percent of the average response, or the injection should be repeated until the 5 percent criterion can be met. Use the average response (peak areas or heights) from the duplicate injections of calibration standards to generate a linear calibration curve. From the calibration curve, determine the concentrations of the field samples employing the average response from the duplicate injections.

Method 306A: Determination of chromium emissions from decorative and hard chromium electroplating and anodizing operations

Applicability. This method applies to the determination of chromium in emissions from decorative and hard chromium electroplating facilities and anodizing operations. The method is less expensive and less complex to conduct than method 306. Correctly applied, the precision and bias of the sample results will be comparable to those obtained with the isokinetic method 306. This method is applicable under ambient moisture, air, and temperature conditions.

Principle. A sample is extracted from the source at a constant sampling rate determined by a critical orifice and is collected in a probe and impingers. The sampling time at the sampling traverse points is varied according to the stack gas velocity at each point to obtain a proportional sample. The concentration is determined by the same analytical procedures used in method 306: inductively coupled plasma (ICP) emission spectrometry, graphite furnace atomic absorption spectrometry (GFAAS), or ion chromatography with a postcolumn reactor (IC/PCR).

Apparatus
Analysis.   It is the same as in method 306, section 3.3.

Procedure
Analysis.   Sample preparation and analysis procedures are identical to those in method 306, section 5.3.

6.5.4   Organic chemicals

Method 18: Measurement of gaseous organic compound emissions by gas chromatography. This method should not be attempted by persons unfamiliar with the performance characteristics of gas chromatography or by persons unfamiliar with source sampling. Particular care should be exercised in the area of safety concerning choice of equipment and operation in potentially explosive atmospheres.

Applicability. This method applies to the analysis of approximately 90 percent of the total gaseous organics emitted from an industrial source. It does not include techniques to identify and measure trace amounts of organic compounds, such as those found in building air and fugitive emission sources.

This method will not determine compounds that (1) are polymeric (high molecular weight), (2) can polymerize before analysis, or (3) have very low vapor pressures at stack or instrument conditions.

Principle. The major organic components of a gas mixture are separated by gas chromatography and individually quantified by flame ionization, photoionization, electron capture, or other appropriate detection principles.

The retention time of each separated component is compared with those of known compounds under identical conditions. Therefore, the analyst confirms the identity and approximate concentrations of the organic emission components beforehand. With this information, the analyst then prepares or purchases commercially available standard mixtures to calibrate the GC under conditions identical to those of the samples. The analyst also determines the need for sample dilution to avoid detector saturation, gas stream filtration to eliminate particulate matter, and prevention of moisture condensation.

Analysis development. Selection of GC parameters.

Column choice. Based on the initial contact with plant personnel concerning the plant process and the anticipated emissions, choose a column that provides good resolution and rapid analysis time. The choice of an appropriate column can be aided by a literature search, contact with manufacturers of GC columns, and discussion with personnel at the emission source.

Most column manufacturers keep excellent records on their products. Their technical service departments may be able to recommend appropriate columns and detector type for separating the anticipated compounds, and they may be able to provide information on interferences, optimum operating conditions, and column limitations.

Plants with analytical laboratories may be able to provide information on their analytical procedures.

Presurvey sample analysis. Before analysis, heat the presurvey sample to the duct temperature to vaporize any condensed material. Analyze the samples by the GC procedure, and compare the retention times to those of the calibration samples that contain the components expected to be in the stream. If any compounds cannot be identified with certainty by this procedure, identify them by other means such as GC/mass spectroscopy (GC/MS) or GC/infrared techniques. A GC/MS system is recommended. (See Fig. 6.2.)

Use the GC conditions determined for the first injection. Vary the GC parameters during subsequent injections to determine the optimum settings. Once the optimum settings have been determined, perform repeated injections of the sample to determine the retention time of each compound. To inject a sample, draw sample through the

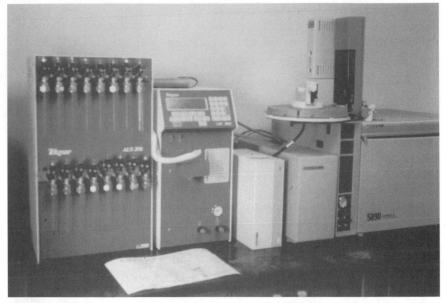

Figure 6.2   Gas chromatograph/mass spectrometer (GC/MS) shown with purge-and-trap sample extraction apparatus.

loop at a constant rate (100 mL/min for 30 s). Be careful not to pressurize the gas in the loop. Turn off the pump, and allow the gas in the sample loop to come to ambient pressure. Activate the sample valve, and record the injection time, loop temperature, column temperature, carrier flow rate, chart speed, and attenuator setting. Calculate the retention time of each peak, using the distance from injection to the peak maximum divided by the chart speed. Retention times should be repeatable within 0.5 s.

If the concentrations are too high for appropriate detector response, a smaller sample loop or dilutions may be used for gas samples; for liquid samples, dilution with solvent is appropriate. Use the standard curves to obtain an estimate of the concentrations.

Identify all peaks by comparing the known retention times of compounds expected to be in the retention times of peaks in the sample. Identify any remaining unidentified peaks which have areas larger than 5 percent of the total, using a GC/MS, or estimation of possible compounds by their retention times compared to known compounds, with confirmation by further GC analysis. (See Figs. 6.3 and 6.4.)

Final sampling and analysis procedure.   Considering safety (flame hazards) and the source conditions, select an appropriate sampling and

```
File:
Operator:        roballey
Date Acquired:   26 Feb 92     1:26 pm
Method File:     8240.M
Sample Name:
Misc Info:                          1 g soil, 10 ul spike
ALS vial:        1
```

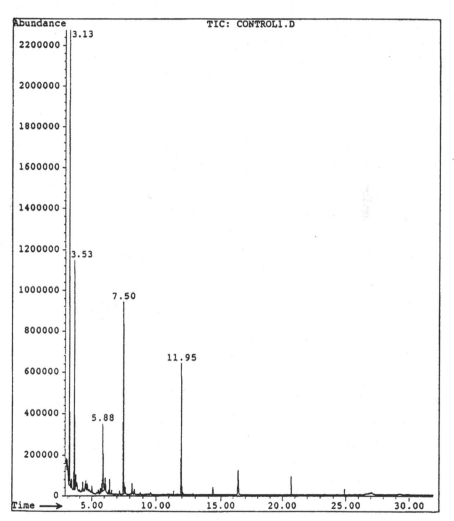

Figure 6.3 Analysis of a sample using GC/MS methods. Each *peak* is a distinct compound, which is detected at a particular, and unique, retention time.

```
Operator:        roballey
Date Acquired: 26 Feb 92    1:26 pm
Method File:    8240.M
Sample Name:
Misc Info:                   1 g soil, 10 ul spiKe
ALS vial:        1
```

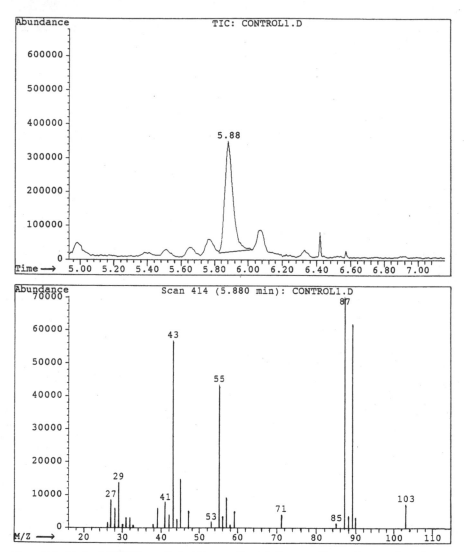

Figure 6.4 Typical GC/MS result, in this case for 1,4-dioxane. Upper diagram is a close-up of the peak from Fig. 6.3 at the 5.88-min retention time. The lower diagram is the unique mass scan of the peak which is used in the identification of the compound.

analysis procedure. In situations where a hydrogen flame is a hazard and no intrinsically safe GC is suitable, use the flexible bag collection technique or an adsorption technique. If the source temperature is below 100°C and the organic concentrations are suitable for the detector to be used, use the direct interface method. If the source gases require dilution, use a dilution interface and either the bag sample or adsorption tubes. The choice between these two techniques will depend on the physical layout of the site, the source temperature, and the storage stability of the compounds if collected in the bag. Sample polar compounds by direct interfacing or dilution interfacing to prevent sample loss by adsorption on the bag.

Integrated bag sample analysis—apparatus

Probe—stainless-steel, Pyrex glass, or Teflon tubing probe, according to the duct temperature, with 6.4-mm-OD Teflon tubing of sufficient length to connect to the sample bag. Use stainless-steel or Teflon unions to connect probe and sample line.

Quick connects—male (2) and female (2) of stainless-steel construction.

Needle valve—to control gas flow.

Pump—leakless Teflon-coated diaphragm-type pump or equivalent; to deliver at least 1 L/min.

Charcoal adsorption tube—tube filled with activated charcoal, with glass-wool plugs at each end, to adsorb organic vapors.

Flowmeter—0- to 500-mL flow range; with manufacturer's calibration curve.

Procedure

Analysis of bag samples. Establish proper GC operating conditions, and record all data. Prepare the GC so that gas can be drawn through the sample valve. Flush the sample loop with gas from one of the three Tedlar bags containing a calibration mixture, and activate the valve. Obtain at least two chromatograms for the mixture. The results are acceptable when the peak areas from two consecutive injections agree to within 5 percent of their average. If they do not agree, run additional samples or correct the analytical techniques until this requirement is met. Then analyze the other two calibration mixtures in the same manner. Prepare a calibration curve as described in the same manner. If the results are acceptable, analyze the other two calibration gas mixtures in the same manner. Prepare the calibration curve by using the least-squares method.

Analyze the two field audit samples by connecting each Tedlar bag containing an audit gas mixture to the sampling valve. Calculate the results; record and report the data to the audit supervisor. If the results are acceptable, proceed with the analysis of the source samples.

Analyze the source gas samples by connecting each bag to the sampling valve with a piece of Teflon tubing identified with that bag. Follow the restrictions on replicate samples specified for the calibration gases. Record the data. If certain items do not apply, use the notation *N.A.* If the bag has been maintained at an elevated temperature, determine the stack gas water content by method 4. After all samples have been analyzed, repeat the analysis of the calibration gas mixtures, and generate a second calibration curve. Use an average of the two curves to determine the sample second calibration curve gas concentrations. If the two calibration curves differ by more than 5 percent from their mean value, then report the final results by comparison to both calibration curves.

Direct interface sampling and analysis procedure. The direct interface procedure can be used provided that the moisture content of the gas does not interfere with the analysis procedure, the physical requirements of the equipment can be met at the site, and the source gas concentration is low enough that detector saturation is not a problem. Adhere to all safety requirements with this method.

Apparatus

Probe—constructed of stainless-steel, Pyrex glass, or Teflon tubing as required by duct temperature, 6.4-mm OD enlarged at duct end to contain glass-wool plug. If necessary, heat the probe with heating tape or a special heating unit capable of maintaining duct temperature.

Sample lines—6.4-mm-OD Teflon lines, heat-traced to prevent condensation of material.

Quick connects—to connect sample line to gas sampling valve on GC instrument and to pump unit used to withdraw source gas. Use a quick connect or equivalent on the cylinder or bag containing calibration gas to allow connection of the calibration gas to the gas sampling valve.

Thermocouple readout device—potentiometer or digital thermometer, to measure source temperature and probe temperature.

Heated gas sampling valve—of two-position, six-port design, to allow sample loop to be purged with source gas or to direct source gas into the GC instrument.

Needle valve—to control gas sampling rate from the source.

Pump—leakless Teflon-coated diaphragm-type pump or equivalent, capable of at least 1 L/min sampling rate.

Flowmeter—of suitable range to measure sampling rate.

Charcoal adsorber—to adsorb organic vapor collected from the source to prevent exposure of personnel to source gas.

Gas cylinders—carrier gas (helium or nitrogen), and oxygen and hydrogen for a flame ionization detector (FID) if one is used.

Gas chromatograph—capable of being moved into the field, with detector, heated gas sampling valve, column required to complete separation of desired components, and option for temperature programming.

Recorder/integrator—to record results.

Procedure

Analysis. After thorough flushing, analyze the sample, using the same conditions as for the calibration gas mixture. Repeat the analysis on an additional sample. Measure the peak areas for the two samples; and if they do not agree to within 5 percent of their mean value, analyze additional samples until two consecutive analyses meet this criterion. Record the data. After consistent results are obtained, remove the probe from the source and analyze a second calibration gas mixture. Record these calibration data and the other required data, deleting the dilution gas information.

Dilution interface sampling and analysis procedure. Source samples that contain a high concentration of organic materials may require dilution prior to analysis to prevent saturating the GC detector. The apparatus required for this direct interface procedure is basically the same as that described above, except a dilution system is added between the heated sample line and the gas sampling valve. The apparatus is arranged so that either a 10:1 or 100:1 dilution of the source gas can be directed to the chromatograph. A pump of larger capacity is also required, and this pump must be heated and placed in the system between the sample line and the dilution apparatus.

Apparatus. The equipment required in addition to that specified for the direct interface system is as follows:

Sample pump—leakless Teflon-coated diaphragm-type that can withstand being heated to 120°C and deliver 1.5 L/min.

Dilution pumps—two model A-150 Komhyr Teflon positive-displacement type delivering 150 cm³/min, or equivalent. As an option, cali brated flowmeters can be used in conjunction with Teflon-coated diaphragm pumps.

Valves—two Teflon three-way valves, suitable for connecting to 6.4-mm-OD Teflon tubing.

Flowmeters—two, for measurement of diluent gas, expected delivery flow rate to be 1350 cm^3/min.

Diluent gas with cylinders and regulators—gas can be nitrogen or clean dry air, depending on the nature of the source gases.

Heated box—suitable for being heated to 120°C, to contain the three pumps, three-way valves, and associated connections. The box should be equipped with quick-connect fittings to facilitate connection of (1) the heated sample line from the probe, (2) the gas sampling valve, (3) the calibration gas mixtures, and (4) diluent gas lines. (*Note:* Care must be taken to leak-check the system prior to the dilutions so as not to create a potentially explosive atmosphere.)

The heated box is designed to receive a heated line from the probe. An optional design is to build a probe unit that attaches directly to the heated box. In this way, the heated box contains the controls for the probe heaters; or if the box is placed against the duct being sampled, it may be possible to eliminate the probe heaters. In either case, a heated Teflon line is used to connect the heated box to the gas sampling valve on the chromatograph.

Procedure. Once the dilution system and GC operations are satisfactory, proceed with the analysis of source gas, maintaining the same dilution settings as used for the standards. Repeat the analyses until two consecutive values that do not vary by more than 5 percent from their mean value are obtained.

Repeat the analysis of the calibration gas mixtures to verify equipment operation. Analyze the two field audit samples, using the dilution system, or directly connect to the gas sampling valve as required. Record all data, and report the results to the audit supervisor.

Method 21: Determination of volatile organic compound leaks

Applicability. This method applies to the determination of volatile organic compound (VOC) leaks from process equipment. These sources include, but are not limited to, valves, flanges and other connections, pumps and compressors, pressure relief devices, process drains, open-ended valves, pump and compressor seal system degassing vents, accumulator vessel vents, agitator seals, and access door seals.

Principle. A portable instrument is used to detect VOC leaks from individual sources. The instrument detector type is not specified, but it must meet the specifications and performance criteria specified in the method. A leak definition concentration based on a reference compound is specified in each applicable regulation. This procedure is intended to locate and classify leaks only and is not to be used as a direct measure of mass emission rate from individual sources.

Apparatus
Monitoring instrument—specifications

- The VOC instrument detector shall respond to the compounds being processed. Detector types which may meet this requirement include, but are not limited to, catalytic oxidation, flame ionization, infrared absorption, and photoionization.

- The instrument shall be capable of measuring the leak definition concentration specified in the regulation.

- The scale of the instrument meter shall be readable to ±5 percent of the specified leak definition concentration.

- The instrument shall be equipped with a pump so that a continuous sample is provided to the detector. The nominal sample flow rate shall be 0.1 to 3.0 L/min.

- The instrument shall be intrinsically safe for operation in explosive atmospheres as defined by the applicable U.S. standards (e.g., National Electrical Code by the National Fire Protection Association).

- The instrument shall be equipped with a probe or probe extension for sampling not to exceed ¼-in OD, with a single end opening for admission of sample.

Method 23: Determination of polychlorinated dibenzo-*p*-dioxins and polychlorinated dibenzofurans from stationary sources

Applicability. This method is applicable to the determination of polychlorinated dibenzo-*p*-dioxins (PCDDs) and polychlorinated dibenzofurans (PCDFs) from stationary sources.

Principle. A sample is withdrawn from the gas stream isokinetically and collected in the sample probe, on a glass-fiber filter, and on a packed column of adsorbent material. The sample cannot be separated into a particle vapor fraction. The PCDDs and PCDFs are extracted from the sample, separated by high-resolution gas chromatography, and measured by high-resolution mass spectrometry.

Apparatus
Analysis
Sample container—125- and 250-mL flint glass bottles with Teflon-lined caps.

Test tube—glass.

Soxhlet extraction apparatus—capable of holding 43- × 123-mm extraction thimbles.

Extraction thimble—glass, precleaned cellulosic, or glass fiber.

Pasteur pipettes—for preparing liquid chromatographic columns.

Reacti-vials—amber glass, 2-mL, silanized prior to use.

Rotary evaporator—Buchi/Brinkman RF-121 or equivalent.

Nitrogen evaporative concentrator—N-Evap Analytical Evaporator model III or equivalent.

Separatory funnels—glass, 2-L.

Gas chromatograph—consisting of the following components:

- *Capillary columns*—a fused silica column, 60- × 25-mm inside diameter (ID), coated with DB-5 and a fused silica column, 30-m × 25-mm ID coated with DB-225.
- *Mass spectrometer*—capable of routine operation at a resolution of 1:10,000 with a stability of ±5 ppm. Other column systems may be used provided that the user is able to demonstrate using calibration and performance checks that the column system is able to meet the specifications of the method.

Analysis. All glassware shall be cleaned as described in section 3A of the "Manual of Analytical Methods for the Analysis of Pesticides in Human and Environmental Samples." All samples must be analyzed within 60 days of collection.

Analyze the sample with a gas chromatograph coupled to a mass spectrometer (GC/MS). Immediately prior to analysis, add a 20-μL aliquot of the recovery standard solution to each sample. A 2-μL aliquot of the extract is injected into the GC. Sample extracts are first analyzed using the DB-5 capillary column to determine the concentration of each isomer of PCDDs and PCDFs (tetra- through octa-). If tetra-chlorinated dibenzofurans are detected in this analysis, then analyze another aliquot of the sample in a separate run, using the DB-225 column to measure the 2,3,7,8 tetra-chloro dibenzofuran isomer. Other column systems may be used, provided that the user is able to demonstrate using calibration and performance checks that the column system is able to meet the specifications of the method.

Quantification.   The peak areas for the two ions monitored for each analyte are summed to yield the total response for each analyte. Each internal standard is used to quantify the indigenous PCDDs or PCDFs in its homologous series. For example, the $^{13}C_{12}$-2,3,7,8-tetra-chlorinated dibenzodioxin (TCDD) is used to calculate the concentrations of all other tetra-chlorinated isomers. Recoveries of the tetra-

and penta- internal standards are calculated using the $^{13}C_{12}$-1,2,3,4-TCDD. Recoveries of the hexa- through octa- internal standards are calculated using $^{13}C_{12}$-1,2,3,7,8,9-HxCDD. Recoveries of the surrogate standards are calculated using the corresponding homolog from the internal standard.

Method 24: Determination of volatile matter content, water content, density, volume solids, and weight solids of surface coatings

Applicability. This method applies to the determination of volatile matter content, water content, density, volume solids, and weight solids of paint, varnish, lacquer, or related surface coatings.

Principle. Standard methods are used to determine the volatile matter content, water content, density, volume solids, and weight solids of the paint, varnish, lacquer, or related surface coatings.

Procedure

Multicomponent coatings. Multicomponent coatings are coatings that are packaged in two or more parts which are combined before application. Upon combination a coreactant from one part of the coating chemically reacts, at ambient conditions, with a coreactant from another part of the coating. To determine the total volatile content, water content, and density of multicomponent coatings, follow the procedures in section 3.8 of the method.

Non-thin-film ultraviolet radiation-cured coating. To determine the volatile content of non-thin-film ultraviolet radiation-cured (uv radiation-cured) coatings, follow the procedures in section 3.9 of the method. Determine water content, density, and solids content of the uv-cured coatings. The uv-cured coatings are coatings which contain unreacted monomers that are polymerized by exposure to ultraviolet light.

Note: As noted in section 1.4 of ASTM D 5403-93, this method may not be applicable to radiation-curable materials wherein the volatile material is water. For all other coatings not covered by these first two sections, analyze as follows:

Volatile matter content. Use the procedure in ASTM D 2369-81 to determine the volatile matter content (may include water) of the coating.

Water content. For waterborne (water-reducible) coatings only, determine the weight fraction of water W_w, using either ASTM D 3792-79 or ASTM D 401781. A waterborne coating is any coating which contains more than 5 percent water by weight in its volatile fraction.

Coating density. Determine the density D_c kg/L of the surface coating, using the procedure in ASTM D 1475-60.

Solids content. Determine the volume fraction of solids V_s of the coating by calculation, using the manufacturer's formulation.

Exempt solvent content.   Determine the weight fraction of exempt solvents W_E by using ASTM method D 4457-85 (incorporated by reference—see §60.17).

To determine the total volatile content, water content, and density of multicomponent coatings, use the following procedures:

Prepare about 100 mL of sample by mixing the components in a storage container, such as a glass jar with a screw top or a metal can with a cap. The storage container should be just large enough to hold the mixture. Combine the components (by weight or volume) in the ratio recommended by the manufacturer. Tightly close the container between additions and during mixing to prevent loss of volatile materials. However, most manufacturers' mixing instructions are by volume. Because of possible error caused by expansion of the liquid in measuring the volume, it is recommended that the components be combined by weight. When weight is used to combine the components and the manufacturer's recommended ratio is by volume, the density must be determined.

Immediately after mixing, take aliquots from this 100-mL sample for determination of the total volatile content, water content, and density. To determine water content, follow section 3.4 of the method. To determine density, follow section 3.5 of the method. To determine total volatile content, use the apparatus and reagents described in ASTM D 2369-81, sections 3 and 4, respectively (incorporated by reference, and see §60.17).

Weigh and record the weight of an aluminum foil weighing dish. Add 3 ± 1 mL of suitable solvent as specified in ASTM D 2369-81 to the weighing dish. Using a syringe as specified in ASTM D 2369-81, weigh to 1 mg, by difference, a sample of coating into the weighing dish. For coatings believed to have a volatile content less than 40 weight percent, a suitable size is 0.3 ± 0.10 g; but for coatings believed to have a volatile content greater than 40 weight percent, a suitable size is 0.5 ± 0.1 g.

Note: If the volatile content determined pursuant to this method is not in the range corresponding to the sample size chosen, repeat the test with the appropriate sample size. Add the specimen drop by drop, shaking (swirling) the dish to disperse the specimen completely in the solvent. If the material forms a lump that cannot be dispersed, discard the specimen and prepare a new one. Similarly, prepare a duplicate. The sample shall stand for a minimum of 1 h, but no more than 24 h prior to being oven fired at $110 \pm 5°C$, for 1 h.

Heat the aluminum foil dishes containing the dispersed specimens in the forced-draft oven for 60 min at $110 \pm 5°C$. *Caution:* Provide adequate ventilation, consistent with accepted laboratory practice, to prevent solvent vapors from accumulating to a dangerous level.

Remove the dishes from the oven, place immediately in a desiccator, cool to ambient temperature, and weigh to within 1 mg.

UV-cured coating's volatile matter content. Use the procedure in ASTM D 5403-93 (incorporated by reference—see §60.17) to determine the volatile matter content of the coating, except the curing test described in Note 2 of ASTM D 5403-93 is required.

Method 24A: Determination of volatile matter content and density of printing inks and related coatings

Applicability. This method applies to the determination of the VOC content and density of solvent-borne (solvent-reducible) printing inks or related coatings.

Principle. Separate procedures are used to determine the VOC weight fraction and density of the coating and the density of the solvent in the coating. The VOC weight fraction is determined by measuring the weight loss of a known sample quantity which has been heated for a specified length of time at a specified temperature. The densities of both the coating and the solvent are measured by a standard procedure. From this information, the VOC volume fraction is calculated.

Weight fraction VOC
Apparatus
Weighing dishes—aluminum foil, 58 mm in diameter by 10 mm high, with a flat bottom. There must be at least three weighing dishes per sample.

Disposable syringe—5 mL.

Analytical balance—to measure to within 0.1 mg.

Oven—vacuum oven capable of maintaining a temperature of 120 ± 20°C and an absolute pressure of 510 ± 51 mmHg for 4 h. Alternatively, use a forced-draft oven capable of maintaining a temperature of 120 ± 2°C for 24 h.

Weight fraction VOC
Analysis. Shake or mix the sample thoroughly to ensure that all the solids are completely suspended. Label and weigh, to the nearest 0.1 mg, a weighing dish and record this weight (M_{x1}).

Using a 5-mL syringe without a needle, remove a sample of the coating. Weigh the syringe and sample to the nearest 0.1 mg, and record this weight M_{cy1}. Transfer 1 to 3 g of the sample to the tared weighing dish. Reweigh the syringe and sample to the nearest 0.1 mg, and record this weight (M_{cy2}). Heat the weighing dish and sample in a vacuum oven at an absolute temperature of 510 ± 2°C for 4 h. Alternatively, heat the weighing dish and sample in a forced-draft oven at a temperature of 120

± 2°C for 24 h. After the weighing dish has cooled, reweigh it to the nearest 0.1 mg and record the weight (M_{x2}). Repeat this procedure for a total of three determinations for each sample.

Coating density.   Determine the density of the ink or related coating according to the procedure outlined in ASTM D 1475-60 (reapproved 1980) (incorporated by reference—see §60.17).

Solvent density.   Determine the density of the solvent according to the procedure outlined in ASTM D 1475-60 (reapproved 1980). Make a total of three determinations for each coating. Report the density D_0 as the arithmetic average of the three determinations.

Method 25: Determination of total gaseous nonmethane organic emissions as carbon

Applicability.   This method applies to the measurement of volatile organic compounds as total gaseous nonmethane organics (TGNMOs) as carbon in source emissions. Organic particulate matter will interfere with the analysis, and therefore a particulate filter is required.

When carbon dioxide and water vapor are present together in the stack, they can produce a positive bias in the sample. The magnitude of the bias depends on the concentrations of CO_2 and water vapor. As a guideline, multiply the CO_2 concentration, expressed as volume percent, by the water vapor concentration. If this product does not exceed 100, the bias can be considered insignificant. For example, the bias is not significant for a source having 10% CO_2 and 10% water vapor, but it might be significant for a source having 10% CO_2 and 20% water vapor.

This method is not the only method that applies to the measurement of TGNMOs. Costs, logistics, and other practicalities of source testing may make other test methods more desirable for measuring VOC contents of certain effluent streams. Proper judgment is required in determining the most applicable VOC test method. For example, depending upon the molecular weight of the organics in the effluent stream, a totally automated semicontinuous nonmethane organics (NMOs) analyzer interfaced directly to the source may yield accurate results. This approach has the advantage of providing emission data semicontinuously over an extended period.

Direct measurement of an effluent with a flame ionization detector analyzer may be appropriate with prior characterization of the gas stream and knowledge that the detector responds predictably to the organic compounds in the stream. If present, methane (CH_4) will, of course, also be measured. The FID can be applied to the determination of the mass concentration of the total molecular structure of the organic emissions under any of the following limited conditions:

(1) where only one compound is known to exist; (2) when the organic compounds consist of only hydrogen and carbon; (3) where the relative percentages of the compounds are known or can be determined, and the FID responses to the compounds are known; (4) where a consistent mixture of the compounds exists before and after emission control and *only* the relative concentrations are to be assessed; or (5) where the FID can be calibrated against mass standards of the compounds emitted (solvent emissions, for example).

Another example of the use of a direct FID is as a screening method. If there is enough information available to provide a rough estimate of the analyzer accuracy, the FID analyzer can be used to determine the VOC content of an uncharacterized gas stream. With a sufficient buffer to account for possible inaccuracies, the direct FID can be a useful tool to obtain the desired results without costly exact determination.

In situations where a qualitative/quantitative analysis of an effluent stream is desired or required, a gas chromatographic FID system may apply. However, for sources emitting numerous organics, the time and expense of this approach will be formidable.

Principle. An emission sample is withdrawn from the stack at a constant rate through a heated filter and a chilled condensate trap by means of an evacuated sample tank. After sampling is completed, the TGNMOs are determined by independently analyzing the condensate trap and sample tank fractions and combining the analytical results. The organic content of the condensate trap fraction is determined by oxidizing the NMO to CO_2 and quantitatively collecting in the effluent in an evacuated vessel; then a portion of the CO_2 is reduced to CH_4 and measured by an FID. The organic content of the sample tank fraction is measured by injecting a portion of the sample into a GC column to separate the NMO from carbon monoxide, CO_2, and CH_4; the NMOs are oxidized to CO_2, reduced to CH_4, and measured by an FID. In this manner, the variable response of the FID associated with different types of organics is eliminated.

Apparatus

NMO analyzer. The NMO analyzer is a gas chromatograph with backflush capability for NMO analysis and is equipped with an oxidation catalyst, reduction catalyst, and FID. This semicontinuous GC/FID analyzer shall be capable of (1) separating CO, CO_2, and CH_4 from NMO; (2) reducing the CO_2 to CH_4 and quantifying as CH_4; and (3) oxidizing the NMO to CO_2, reducing the CO_2 to CH_4, and quantifying as CH_4. The analyzer consists of the following major components:

- *Oxidation catalyst*—a suitable length of 9.5-mm (⅜-in) OD Inconel 600 tubing packed with 5.1 cm (2 in) of 19% chromia on 3.2-mm (⅛-in) alumina pellets. The catalyst material is packed in the

center of the tube supported on either side by quartz wool. The catalyst tube must be mounted vertically in a 650°C furnace.
- *Reduction catalyst*—a 7.6-cm (3-in) length of 6.4-mm ($\frac{1}{4}$-in) OD Inconel tubing fully packed with 100-mesh pure nickel powder. The catalyst tube must be mounted vertically in a 400°C furnace.
- *Separation column(s)*—a 30-cm (1-ft) length of 3.2-mm ($\frac{1}{8}$-in) OD stainless-steel tubing packed with 60/80 mesh Unibeads 1S followed by a 61-cm (2-ft) length of 3.2-mm (1/8-in) OD stainless-steel tubing packed with 60/80 mesh Carbosieve G. The Carbosieve and Unibeads columns must be baked separately at 200°C with carrier gas flowing through them for 24 h before initial use.
- *Sample injection system*—a 10-port GC sample injection valve fitted with a sample loop properly sized to interface with the NMO analyzer (1-cm^3 loop recommended).
- *FID*—an FID meeting the specifications of the method.
- *Data recording system*—analog strip chart recorder or digital integration system compatible with the FID for permanently recording the analytical results.

Other analysis apparatus

Barometer—mercury, aneroid, or other barometer capable of measuring atmospheric pressure to within 1 mmHg

Thermometer—capable of measuring the laboratory temperature within 1°C

Vacuum pump—capable of evacuating to an absolute pressure of 10 mmHg

Syringes—10-μL and 50-μL liquid injection syringes

Liquid sample injection unit—316 SS U-tube fitted with an injection septum

Procedure

Analysis.   Before putting the NMO analyzer into routine operation, conduct an initial performance test. Start the analyzer, and perform all the necessary functions in order to put the analyzer into proper working order; then conduct the performance test. Once the performance test has been successfully completed and the CO_2 and NMO calibration response factors have been determined, proceed with sample analysis as follows:

Analysis of recovered condensate sample.   Purge the sample loop with sample, and then inject the sample. Under the specified operating conditions, the CO_2 in the sample will elute in approximately 100 s. As soon as the detector response returns to baseline following the CO_2 peak, switch the carrier gas flow to backflush, and raise the column oven temperature to 195°C as rapidly as possible. A rate of 30°C/min has been shown to be adequate. Record the value obtained for the condensible organic material C_{cm} measured as CO_2 and any measured NMO. Return the column oven temperature to 85°C in preparation for the next analysis. Analyze each sample in triplicate, and report the average C_{cm}.

Analysis of sample tank. Perform the analysis as described in section 4.4.3 of the method, but record only the value measured for NMO (C_{tm}).

Method 25A: Determination of total gaseous organic concentration using a flame ionization analyzer

Applicability. This method applies to the measurement of total gaseous organic concentration of vapors consisting primarily of alkanes, alkenes, and/or arenes (aromatic hydrocarbons). The concentration is expressed in terms of propane (or other appropriate organic calibration gas) or in terms of carbon.

Principle. A gas sample is extracted from the source through a heated sample line, if necessary, and glass-fiber filter to a flame ionization analyzer (FIA). Results are reported as volume concentration equivalents of the calibration gas or as carbon equivalents.

Apparatus. The essential components of the measurement system are described below:

Organic concentration analyzer—a flame ionization analyzer capable of meeting or exceeding the specifications in this method.

Sample probe—stainless-steel, or equivalent, three-hole rake type. Sample holes shall be 4 mm in diameter or smaller and located at 16.7, 50, and 83.3 percent of the equivalent stack diameter. Alternatively, a single-opening probe may be used so that a gas sample is collected from the centrally located 10 percent area of the stack cross section.

Sample line—stainless-steel or Teflon tubing to transport the sample gas to the analyzer. The sample line should be heated, if necessary, to prevent condensation in the line.

Calibration valve assembly—a three-way valve assembly to direct the zero and calibration gases to the analyzers is recommended. Other methods, such as quick-connect lines, to route calibration gas to the analyzers are applicable.

Particulate filter—an in-stack or an out-of-stack glass-fiber filter is recommended if exhaust gas particulate loading is significant. An out-of-stack filter should be heated to prevent any condensation.

Recorder—a strip-chart recorder, analog computer, or digital recorder for recording measurement data. The minimum data recording requirement is one measurement value per minute. *Note:* This method is often applied in highly explosive areas. Caution and care should be exercised in choice of equipment and installation.

Emission measurement test procedure—organic measurement. Begin sampling at the start of the test period, recording time and any required process information as appropriate. In particular, note on the recording chart periods of process interruption or cyclic operation.

Method 25B: Determination of total gaseous organic concentration using a nondispersive infrared analyzer

Applicability. This method applies to the measurement of total gaseous organic concentration of vapors consisting primarily of alkanes. (Other organic materials may be measured using the general procedure in this method, the appropriate calibration gas, and an analyzer set to the appropriate absorption band.) The concentration is expressed in terms of propane (or other appropriate organic calibration gas) or in terms of carbon.

Principle. A gas sample is extracted from the source through a heated sample line, if necessary, and glass-fiber filter to a nondispersive infrared analyzer. Results are reported as volume concentration equivalents of the calibration gas or as carbon equivalents.

Apparatus. The apparatus is the same as that for method 25A with the exception of the following:

Organic concentration analyzer—a nondispersive infrared analyzer designed to measure alkaline organics and capable of meeting the specifications in this method

Emission measurement test procedure—organic measurement. This is the same as in method 25A, section 7.1.

Method 25C: Determination of nonmethane organic compounds (NMOC) in municipal solid waste (MSW) landfill gases

Applicability. This method is applicable to the sampling and measurement of nonmethane organic compounds as carbon in MSW landfill gases.

Principle. A sample probe that has been perforated at one end is driven or augered to a depth of 1.0 m below the bottom of the landfill cover. A sample of the landfill gas is extracted with an evacuated cylinder. The NMOC content of the gas is determined by injecting a portion of the gas into a GC column to separate the NMOC from carbon monoxide, carbon dioxide, and methane; the NMOCs are oxidized to CO_2, reduced to CH_4, and measured by a flame ionization detector. In this manner, the variable response of the FID associated with different types of organics is eliminated.

Apparatus. This includes NMOC analyzer, barometer, thermometer, and syringes, the same as in sections 2.3, 2.4.1, 2.4.2, 2.4.4, respectively, of method 25.

Procedure

Analysis. The oxidation, reduction, and measurement of NMOCs are similar to those of method 25. Before putting the NMOC analyzer into routine operation, conduct an initial performance test. Start the analyzer, and perform all the necessary functions to put the analyzer into proper working order. Once the performance test has been successfully completed and the NMOC calibration response factor has been determined, proceed with sample analysis as follows:

Analysis of sample tank. Purge the sample loop with sample, and then inject the sample. Under the specified operating conditions, the CO_2 in the sample will elute in approximately 100 s. As soon as the detector response returns to baseline following the CO_2 peak, switch the carrier gas flow to backflush and raise the column oven temperature to 195°C as rapidly as possible. A rate of 30°C/min has been shown to be adequate. Record the value obtained for any measured NMOC. Return the column oven temperature to 85°C in preparation for the next analysis. Analyze each sample in triplicate, and report the average as C_{tm}.

Method 25D: Determination of the volatile organic concentration of waste samples. Performance of this method should not be attempted by persons unfamiliar with the operation of a flame ionization detector or an electrolytic conductivity detector (ECD) because knowledge beyond the scope of this presentation is required.

Applicability. This method is applicable for determining the volatile organic (VO) concentration of a waste sample.

Principle. A sample of waste is obtained at a point which is most representative of the unexposed waste (where the waste has had minimum opportunity to volatilize to the atmosphere). The sample is suspended in an organic/aqueous matrix, then heated and purged with nitrogen for 30 min in order to separate certain organic compounds. Part of the sample is analyzed for carbon concentration, as methane, with an FID, and part of the sample is analyzed for chlorine concentration, as chloride, with an ECD. The VO concentration is the sum of the carbon and chlorine contents of the sample.

Apparatus

Analysis. The following equipment is required.

Purging apparatus—for separating the VO from the waste sample. The purging apparatus consists of the following major components.

- *Purging flask*—a glass container to hold the sample while it is heated and purged with dry nitrogen. The cap of the purging flask is equipped with three fittings: one for a purging lance (fitting with the #7 Ace-thread), one for the Teflon exit tubing (side fitting, also a #7 Ace-thread), and a third (a 50-mm Ace-thread) to attach the base of the purging flask. The base of the purging flask is a 50-mm-ID cylindrical glass tube. One end of the tube is open while the other end is sealed.
- *Coalescing filter*—porous fritted disk incorporated into a container with the same dimensions as the purging flask.
- *Constant-temperature chamber*—a forced-draft oven capable of maintaining a uniform temperature around the purging flask and coalescing filter of 75 ± 2°C.
- *Three-way valve*—manually operated, stainless-steel; to introduce calibration gas into system.
- *Flow controllers*—two, adjustable. One is capable of maintaining a purge gas flow rate of 6 ± 0.06 L/min; the other is capable of maintaining a calibration gas flow rate of 1 to 100 mL/min.
- *Rotameter*—for monitoring the airflow through the purging system (0 to 10 L/min).
- *Sample splitters*—two heated flow restrictors (placed inside oven or heated to 120 ± 10°C). At a purge rate of 6 L/min, one will supply a constant flow to the first detector (the rest of the flow will be directed to the second sample splitter). The second splitter will split the analytical flow between the second detector and the flow restrictor. The approximate flow to the FID will be 40 mL/min and to the ECD will be 15 mL/min, but the exact flow must be adjusted to be compatible with the individual detector and to meet its linearity requirement. The two sample splitters will be connected to each other by $\frac{1}{8}$-in-OD stainless-steel tubing.
- *Flow restrictor*—stainless-steel tubing, $\frac{1}{8}$-in-OD, connecting the second sample splitter to the ice bath. Length is determined by the resulting pressure in the purging flask (as measured by the pressure gauge). The resulting pressure from the use of the flow restrictor shall be 6 to 7 lb/in^2 gauge.
- *Filter flask*—with one-hole stopper; used to hold ice bath. Excess purge gas is vented through the flask to prevent condensation in the flowmeter and to trap volatile organic compounds.

- *Four-way valve*—manually operated, stainless-steel; placed inside oven, used to bypass purging flask.
- *On/off valves*—two, stainless-steel. One is heat-resistant up to 130°C and is placed between oven and ECD. The other is a toggle valve used to control purge gas flow.
- *Pressure gauge*—range of 0 to 40 lb/in^2; to monitor pressure in purging flask and coalescing filter.
- *Sample lines*—Teflon, $\frac{1}{4}$-in OD, used inside the oven to carry purge gas to and from purging chamber as well as to and from coalescing filter to four-way valve. Also used to carry sample from four-way valve to first sample splitter.
- *Detector tubing*—stainless-steel, $\frac{1}{8}$-in OD, heated to 120 ± 10°C; used to carry sample gas from each sample splitter to a detector. Each piece of tubing must be wrapped with heat tape and insulating tape to ensure that no cold spots exist. The tubing leading to the ECD will also contain a heat-resistant on/off valve which shall also be wrapped with heat tape and insulation.
- *Volatile organic measurement system*—consisting of an FID to measure the carbon concentration of the sample and an ECD to measure the chlorine concentration.

Procedure

Sample analysis. Turn on the constant-temperature chamber and allow the temperature to equilibrate at 75 ± 2°C. Turn the four-way valve so that the purge gas bypasses the purging flask, the purge gas flowing through the coalescing filter and to the detectors (standby mode). Turn on the purge gas. Allow both the FID and the ECD to warm up until a stable baseline is achieved on each detector. Pack the filter flask with ice. Replace ice after each run, and dispose of the wastewater properly. When the temperature of the oven reaches 75 ± 2°C, start both integrators and record baseline. After 1 min, turn the four-way valve so that the purge gas flows through the purging flask, to the coalescing filter, and to the sample splitters (purge mode). Continue recording the response of the FID and the ECD. Monitor the readings of the pressure gauge and the rotameter. If the readings fall below established set points, stop the purging, determine the source of the leak, and solve the problem before resuming. Leaks detected during a sampling period invalidate that sample.

As the purging continues, monitor the output of the detectors to make certain that the analysis is proceeding correctly and that the results are being properly recorded. Every 10 min read and record the purge flow rate, the pressure, and the chamber temperature. Continue the purging for 30 min.

For each detector output, integrate over the entire area of the peak, starting at 1 min and continuing until the end of the run. Subtract the established baseline area from the peak area. Record the corrected area of the peak.

Method 25E: Determination of vapor-phase organic concentration in waste samples. Performance of this method should not be attempted by persons unfamiliar with the operation of a flame ionization detector or by those unfamiliar with source sampling because knowledge beyond the scope of this presentation is required.

Applicability. This method is applicable for determining the vapor pressure of samples which represent waste that is or will be managed in tanks.

Principle. The headspace vapor of the sample is analyzed for carbon content by a headspace analyzer, which uses an FID.

Apparatus

Analysis.   The following equipment is required.

Balanced pressure headspace sampler—Perkin-Elmer HS-6, HS-100, or equivalent, equipped with a glass bead column instead of a chromatographic column.

FID

Data recording system—analog strip-chart recorder or digital integration system compatible with the FID for permanently recording the output of the detector.

Thermometer—capable of reading temperatures in the range of 30 to 60°C with an accuracy of ±0.1°C.

Procedure

Analysis.   Allow 1 h for the headspace vials to equilibrate at the temperature specified in the regulation. Allow the FID to warm up until a stable baseline is achieved on the detector.

Check the calibration of the FID daily. Follow the manufacturer's recommended procedures for the normal operation of the headspace sampler and FID. Calculate the vapor-phase organic vapor pressure in the samples. Monitor the output of the detector to make certain that the results are being properly recorded.

Method 26: Determination of hydrogen chloride emissions from stationary sources

Applicability. This method is applicable for determining emissions of hydrogen halides (HX) [hydrogen chloride (HCl), hydrogen bromide (HBr), and hydrogen fluoride (HF)] and halogens (X_2) [chlorine (Cl_2)

and bromine (Br_2)] from stationary sources. Sources, such as those controlled by wet scrubbers, that emit acid particulate matter must be sampled using method 26A.

Principle. An integrated sample is extracted from the source and passed through a prepurged heated probe and filter into dilute sulfuric acid and dilute sodium hydroxide solutions, which collect the gaseous hydrogen halides and halogens, respectively. The filter collects other particulate matter including halide salts. The hydrogen halides are solubilized in the acidic solution and form chloride (Cl^-), bromide (Br^-), and fluoride (F^-) ions. The halogens have a very low solubility in the acidic solution and pass through to the alkaline solution where they are hydrolyzed to form a proton (H^+), the halide ion, and the hypohalous acid (HClO or HBrO). Sodium thiosulfate is added in excess to the alkaline solution to ensure reaction with the hypohalous acid to form a second halide ion such that two halide ions are formed for each molecule of halogen gas. The halide ions in the separate solutions are measured by ion chromatography (IC).

Apparatus
Analysis. The materials required for volumetric dilution and chromatographic analysis of samples are described below.

Volumetric flasks—class A, 100-mL size.

Volumetric pipettes—class A, assortment; to dilute samples into the calibration range of the instrument.

Ion chromatograph—suppressed or nonsuppressed, with a conductivity detector and electronic integrator operating in the peak area mode. Other detectors, strip-chart recorders, and peak height measurements may be used.

Procedure
Sample analysis. The IC conditions will depend upon analytical column type and whether suppressed or nonsuppressed IC is used.

Before sample analysis, establish a stable baseline. Next, inject a sample of water, and determine if any Cl^-, Br^-, or F^- appears in the chromatogram. If any of these ions are present, repeat the load-and-injection procedure until they are no longer present. Analysis of the acid and alkaline absorbing solution samples requires separate standard calibration curves. Ensure adequate baseline separation of the analyses.

Between injections of the appropriate series of calibration standards, inject in duplicate the reagent blanks, quality control sample, and field samples. Measure the areas or heights of the Cl^-, Br^-, and F^- peaks. Use the mean response of the duplicate injections to determine the concentrations of the field samples and reagent blanks, using the linear calibration curve. The values from duplicate injections should agree within 5 percent of their mean for the analysis to be valid. Dilute any sample and the blank with equal volumes of water if the concentration exceeds that of the highest standard.

Method 26A: Determination of hydrogen halide and halogen emissions from stationary sources—isokinetic method

Applicability. This method is applicable for determining emissions of hydrogen halides (HX) [hydrogen chloride (HCl), hydrogen bromide (HBr), and hydrogen fluoride (HF)] and halogens (X_2) [chlorine (Cl_2) and bromine (Br_2)] from stationary sources. This method collects the emission sample isokinetically and is therefore particularly suited for sampling at sources, such as those controlled by wet scrubbers, emitting acid particulate matter (e.g., hydrogen halides dissolved in water droplets).

Principle. Gaseous and particulate pollutants are withdrawn isokinetically from the source and collected in an optional cyclone, on a filter, and in absorbing solutions. The cyclone collects any liquid droplets and is not necessary if the source emissions do not contain them; however, it is preferable to include the cyclone in the sampling train to protect the filter from any moisture present. The filter collects other particulate matter including halide salts. Acidic and alkaline absorbing solutions collect the gaseous hydrogen halides and halogens, respectively. Following sampling of emissions containing liquid droplets, any halides/halogens dissolved in the liquid in the cyclone and on the filter are vaporized to gas and are collected in the impingers by pulling conditioned ambient air through the sampling train. The hydrogen halides are solubilized in the acidic solution and form chloride (Cl^-), bromide (Br^-), and fluoride (F^-) ions. The halogens have a very low solubility in the acidic solution and pass through to the alkaline solution where they are hydrolyzed to form a proton (H^+), the halide ion, and the hypohalous acid (HClO or HBrO). Sodium thiosulfate is added to the alkaline solution to ensure reaction with the hypohalous acid to form a second halide ion such that two halide ions are formed for each molecule of halogen gas. The halide ions in the separate solutions are measured by ion chromatography. If desired, the particulate matter recovered from the filter and the probe is analyzed following the procedures in method 5. (*Note:* If the tester intends to use this sampling arrangement to sample concurrently for particulate matter, the alternative Teflon probe liner, cyclone, and filter holder should not be used. The Teflon filter support must be used. The tester must also meet the probe and filter temperature requirements of both sampling trains.)

Apparatus

Analysis.   For analysis, the following equipment is needed:

Volumetric flasks—class A, various sizes.

Volumetric pipettes—class A, assortment, to dilute samples to calibration range of the ion chromatograph (IC).

Ion chromatograph—suppressed or nonsuppressed, with a conductivity detector and electronic integrator operating in the peak area
mode. Other detectors, a strip-chart recorder, and peak heights may
be used.

Procedure

Analysis.   Note the liquid levels in the sample containers and confirm on the analysis sheet whether leakage occurred during transport. If a noticeable leakage has occurred, either void the sample or
use methods, subject to the approval of the Administrator, to correct
the final results.

*Containers 1 and 2 and acetone blank (optional; particulate deter-
mination).*   This is the same as in method 5, section 4.3.

Container 5.   This is the same as in method 5, section 4.3, for silica
gel.

Containers 3 and 4 and absorbing solution and water blanks.
Quantitatively transfer each sample to a volumetric flask or graduated cylinder, and dilute with water to a final volume within 50 mL of
the largest sample.

The IC conditions will depend upon analytical column type and
whether suppressed or nonsuppressed IC is used. Prior to calibration
and sample analysis, establish a stable baseline. Next, inject a sample of water, and determine if any Cl^-, Br^-, or F^- appears in the chromatogram. If any of these ions are present, repeat the load-and-injection procedure until they are no longer present. Analysis of the acid
and alkaline absorbing solution samples requires separate standard
calibration curves. Ensure adequate baseline separation of the analyses.

Between injections of the appropriate series of calibration standards, inject in duplicate the reagent blanks and the field samples.
Measure the areas or heights of the Cl^-, Br^-, and F^- peaks. Use the
average response to determine the concentrations of the field samples
and reagent blanks, using the linear calibration curve. If the values
from duplicate injections are not within 5 percent of their mean, the
duplicate injection shall be repeated and all four values used to determine the average response. Dilute any sample and the blank with
equal volumes of water if the concentration exceeds that of the highest standard.

Method 106: Determination of vinyl chloride from stationary sources. Performance of this method should not be attempted by persons unfamiliar with the operation of a gas chromatograph or by those unfamiliar with source sampling, because knowledge beyond the scope of this presentation is required. Care must be exercised to prevent exposure of sampling personnel to vinyl chloride, a carcinogen.

Applicability. The method is applicable to the measurement of vinyl chloride in stack gases from ethylene dichloride, vinyl chloride, and polyvinyl chloride manufacturing processes. The method does not measure vinyl chloride contained in particulate matter.

Principle. An integrated bag sample of stack gas containing vinyl chloride (chloroethene) is subjected to GC analysis using a flame ionization detector.

Apparatus

Analysis. The following equipment is required:

Gas chromatograph—with FID, potentiometric strip-chart recorder and 1.0- to 5.0-mL heated sampling loop in automatic sample valve. The chromatographic system shall be capable of producing a response to 0.1 ppm vinyl chloride that is at least as great as the average noise level. (Response is measured from the average value of the baseline to the maximum of the waveform, while standard operating conditions are in use.)

Chromatographic columns—columns as listed below. The analyst may use other columns provided that the precision and accuracy of the analysis of vinyl chloride standards are not impaired and review information confirms that there is adequate resolution of the vinyl chloride peak. (*Adequate resolution* is defined as an area overlap of not more than 10 percent of the vinyl chloride peak by an interferent peak. Calculation of area overlap is explained in Appendix C, procedure 1, "Determination of adequate chromatographic peak resolution.")

- Column A—stainless-steel, 2.0-m by 3.2-mm, containing 80/100-mesh Chromasorb 102.
- Column B—stainless-steel, 2.0-m by 3.2-mm, containing 20 percent GE SF-96 on 60/80-mesh Chromasorb P AW; or stainless-steel, 1.0-m by 3.2-mm, containing 80/100-mesh Porapak T. Column B is required as a secondary column if acetaldehyde is present. If used, column B is placed after column A. The combined columns should be operated at 120°C.

Flowmeters—two, rotameter type, 100-mL/min capacity, with flow control valves.

Gas regulators—for required gas cylinders.

Thermometer—accurate to 1°C, to measure temperature of heated sample loop at time of sample injection.

Barometer—accurate to 5 mmHg; to measure atmospheric pressure around GC during sample analysis.

Pump—leakfree, with minimum of 100-mL/min capacity.

Recorder—strip-chart type, optionally equipped with either disk or electronic integrator.

Planimeter—optional, in place of disk or electronic integrator on recorder, to measure chromatograph peak areas.

Procedure

Analysis.   Set the column temperature to 100°C and the detector temperature to 150°C. When optimum hydrogen and oxygen flow rates have been determined, verify and maintain these flow rates during all chromatographic operations. Using zero helium or nitrogen as the carrier gas, establish a flow rate in the range consistent with the manufacturer's requirements for satisfactory detector operation. A flow rate of approximately 40 mL/min should produce adequate separations. Observe the baseline periodically, and determine that the noise level has stabilized and that baseline drift has ceased. Purge the sample loop for 30 s at the rate of 100 mL/min, shut off flow, allow the sample loop pressure to reach atmospheric pressure as indicated by the H_2O manometer, then activate the sample valve. Record the injection time (the position of the pen on the chart at the time of sample injection), sample number, sample loop temperature, column temperature, carrier gas flow rate, chart speed, and attenuator setting. Record the barometric pressure. From the chart, note the peak having the retention time corresponding to vinyl chloride. Measure the vinyl chloride peak area A_m by use of a disk integrator, electronic integrator, or a planimeter. Measure and record the peak heights H_m. Record A_m and retention time. Repeat the injection at least 2 times or until two consecutive values for the total area of the vinyl chloride peak do not vary more than 5 percent. Use the average value for these two total areas to compute the bag concentration.

Compare the ratio of H_m to A_m for the vinyl chloride sample with the same ratio for the standard peak that is closest in height. If these ratios differ by more than 10 percent, the vinyl chloride peak may not be pure (possibly acetaldehyde is present) and the secondary column should be employed.

Method 307: Determination of emissions from halogenated solvent vapor cleaning machines using a liquid-level procedure

Applicability. This method is applicable to the determination of halogenated solvent emissions from solvent vapor cleaners in the idling mode.

Principle. The solvent level in the solvent cleaning machine is measured using inclined liquid-level indicators. The change in liquid level corresponds directly to the amount of solvent lost from the solvent cleaning machine.

Method 311: Analysis of hazardous air pollutant compounds in paints and coatings by direct injection into a gas chromatograph

Applicability. This method is applicable for determination of most compounds designated by the EPA as volatile hazardous air pollutants (HAPs) (see Reference 1) that are contained in paints and coatings. Styrene, ethyl acrylate, and methyl methacrylate can be measured by ASTM D 4827-93 or ASTM D 4747-87. Formaldehyde can be measured by ASTM PS 9-94 or ASTM D 1979-91. Toluene diisocyanate can be measured in urethane prepolymers by ASTM D 3432-89. Method 311 applies only to those volatile HAPs which are added to the coating when it is manufactured, not to those which may form as the coating cures (reaction products or cure volatiles). A separate or modified test procedure must be used to measure these reaction products or cure volatiles in order to determine the total volatile HAP emissions from a coating. Cure volatiles are a significant component of the total HAP content of some coatings. The term *coating* used in this method shall be understood to mean paints and coatings.

Principle. The method uses the principle of gas chromatographic separation and quantification using a detector that responds to concentration differences. Because there are many potential analytical systems or sets of operating conditions that may represent usable methods for determining the concentrations of the compounds in the applicable matrices, all systems that employ this principle, but differ only in details of equipment and operation, may be used as alternative methods, provided that the prescribed quality control, calibration, and method performance requirements are met. Certified product data sheets (CPDSs) may also contain information relevant to the analysis of the coating sample including, but not limited to, separation column, oven temperature, carrier gas, injection port temperature, extraction solvent, and internal standard.

Summary of method. Whole coating is added to dimethylformamide, and a suitable internal standard compound is added. An aliquot of the sample mixture is injected onto a chromatographic column containing a stationary phase that separates the analytes from one another and from other volatile compounds contained in the sample. The concentrations of the analytes are determined by comparing the detector responses for the sample to the responses obtained using known concentrations of the analytes.

Apparatus

Analysis

Gas chromatograph—any instrument equipped with a flame ionization detector and capable of being temperature-programmed may be used. Optionally, other types of detectors (e.g., a mass spectrometer), and any necessary interfaces, may be used provided that the detector system yields an appropriate and reproducible response to the analytes in the injected sample. Autosampler injection may be used, if available.

Recorder—if available, an electronic data station or integrator may be used to record the gas chromatogram and associated data. If a strip-chart recorder is used, it must meet the following criteria: A 1- to 10-mV linear response with a full-scale response time of 2 s or less and a maximum noise level of ±0.03 percent of full-scale. Other types of recorders may be used as appropriate to the specific detector installed provided that the recorder has a full-scale response time of 2 s or less and a maximum noise level of ±0.03 percent of full-scale.

Column—The column must be constructed of materials that do not react with components of the sample (e.g., fused silica, stainless steel, glass). The column should be of appropriate physical dimensions (e.g., length, internal diameter) and contain sufficient suitable stationary phase to allow separation of the analytes. DB-5, DB-Wax, and FFAP columns are commonly used for paint analysis; however, it is the responsibility of each analyst to select appropriate columns and stationary phases.

Tube and tube fittings—supplies to connect the GC and gas cylinders.

Pressure regulators—devices used to regulate the pressure between gas cylinders and the GC.

Flowmeter—a device used to determine the carrier gas flow rate through the GC. Either a digital flowmeter or a soap-film bubble meter may be used to measure gas flow rates.

Septa—seals on the GC injection port through which liquid or gas samples can be injected using a syringe.

Liquid charging devices—devices used to inject samples into the GC such as clean and graduated 1-, 5-, and 10-μL capacity syringes.

Vials—containers that can be sealed with a septum in which samples may be prepared or stored. The recommended size is 25-mL capacity. Mininert valves have been found satisfactory and are available from Pierce Chemical Company, Rockford, Ill.

Balance—device used to determine the weights of standards and samples. An analytical balance capable of accurately weighing to 0.0001 g is required.

Procedure

Qualitative analysis.   An analyte (e.g., those cited in section 1.1) is considered tentatively identified if two criteria are satisfied: (1) elution of the sample analyte within ±0.05 min of the average GC retention time (RT) of the same analyte in the calibration standard and (2) either (a) confirmation of the identity of the compound by spectral matching on a gas chromatograph equipped with a mass selective detector or (b) elution of the sample analyte within ±0.05 min of the average GC retention time of the same analyte in the calibration standard analyzed on a dissimilar GC column.

The RT of the sample analyte must meet the criteria specified in section 9.3.3 of the method.

When doubt exists as to the identification of a peak or the resolution of two or more components possibly comprising one peak, additional confirmatory techniques must be used.

Quantitative analysis.   When an analyte has been identified, the quantification of that compound will be based on the internal standard technique.

6.5.5 Miscellaneous parameters— Radiological

Method 111: Determination of polonium 210 emissions from stationary sources.   Performance of this method should not be attempted by persons unfamiliar with the use of equipment for measuring radioactive disintegration rates.

Applicability.   This method is applicable to the determination of polonium 210 emissions in particulate samples collected in stack gases. Samples should be analyzed within 30 days of collection to minimize error due to growth of polonium 210 from any lead 210 present in the sample.

Principle. A particulate sample is collected from stack gases as described in method 5 of Appendix A to 40 CFR Part 60. The polonium 210 in the sample is put in solution, deposited on a metal disk, and the radioactive disintegration rate measured. Polonium in acid solution spontaneously deposits on surfaces of metals that are more electropositive than polonium. This principle is routinely used in the radiochemical analysis of polonium 210.

Apparatus

Alpha spectrometry system—consisting of a multichannel analyzer, biasing electronics, silicon surface barrier detector, vacuum pump, and chamber

Bath—constant-temperature bath at 85°C

Polished silver disks—3.8-cm diameter, 0.4-mm thick with a small hole near the edge

Glass beakers—400-mL, 150-mL

Hot plate—electric

Fume hood

Teflon beakers—150 mL

Magnetic stirrer

Stirring bar

Plastic or glass hooks—to suspend plating disks

Internal proportional counter—for measuring alpha particles

Nucleopore filter membranes—25-mm diameter, 0.2-μm pore size or equivalent

Planchets—stainless-steel, 32-mm diameter with 1.5-mm lip

Transparent plastic tape—2.5 cm wide with adhesive on both sides

Epoxy spray enamel

Suction filter apparatus—for 25-mm-diameter filter

Wash bottles—250-mL capacity

Plastic graduated cylinder—25-mL capacity

Procedure
Sample analysis

- Add the aliquot of solution to be analyzed to a suitable 200-mL container to be placed in a constant-temperature bath. Note, aliquot volume may require a larger container.

- If necessary, bring the volume to 100 mL with 1 M HCl. If the aliquot volume exceeds 100 mL, use total aliquot.

- Add 200 mg of ascorbic acid, and heat solution to 85°C in a constant-temperature bath.

- Stirring of the solution must be maintained while the solution is in the constant-temperature bath for plating.

- Suspend a silver disk in the heated solution, using a glass or plastic rod with a hook inserted through the hole in the disk. The disk should be totally immersed in the solution at all times.

- Maintain the disk in solution for 3 h while stirring.

- Remove the silver disk, rinse with distilled water, and allow to air-dry at room temperature.

Measurement of polonium 210.   Place the silver disk, with deposition side (unpainted side) up, on a planchet and secure with double-sided plastic tape. Place the planchet with disk in alpha spectrometry system and count for 1000 min.

Determination of procedure background.   Background counts used in all equations are determined by performing the specific analysis required using the analytical reagents only. This should be repeated every 10 analyses.

Determination of instrument background.   Instrument backgrounds of the internal proportional counter and alpha spectrometry system should be determined on a weekly basis. Instrument background should not exceed procedure background. If this occurs, it may be due to a malfunction or contamination.

Method 114: Test methods for measuring radionuclide emission from stationary sources.   This method provides the requirements for (1) stack monitoring and sample collection methods appropriate for radionuclides; (2) radiochemical methods which are used in determining the amounts of radionuclides collected by the stack sampling, and (3) quality assurance methods which are conducted in conjunction with these measurements. These methods are appropriate for emissions for stationary sources. A list of references is provided.

Many different types of facilities release radionuclides into air. These radionuclides differ in the chemical and physical forms, half-lives, and type of radiation emitted. The appropriate combination of sample extraction, collection, and analysis for an individual radionuclide is dependent upon many interrelated factors including the mixture of other radionuclides present. Because of this wide range of conditions, no single method for monitoring or sample collection and analysis of a radionuclide is applicable to all types of facilities. Therefore, a series of methods based on "principles of measurement" are described for monitoring and sample collection and analysis which are applicable to the measurement of radionuclides found in effluent streams at stationary sources. This approach provides the user with the flexibility to choose the most appropriate combination of monitoring and sample collection and analysis methods which are applicable to the effluent stream to be measured.

Radionuclide analysis methods. A series of methods based on principles of measurement are described which are applicable to the analysis of radionuclides collected from airborne effluent streams at stationary sources. These methods are applicable only under the conditions stated and within the limitations described. Some methods specify that only a single radionuclide be present in the sample or the chemically separated sample. This condition should be interpreted to mean that no other radionuclides are present in quantities which would interfere with the measurement.

Also identified (Table 6.3) are methods for a selected list of radionuclides. The listed radionuclides are most commonly used and have the greatest potential for causing harmful doses to members of the public. Use of methods based on principles of measurement other than those described in this section must be approved in advance of use by the Administrator. For radionuclides not listed in Table 6.3, any of the described methods may be used provided the user can demonstrate that the applicability conditions of the method have been met.

The type of method applicable to the analysis of a radionuclide is dependent upon the type of radiation emitted, i.e., alpha, beta, or gamma. Therefore, the methods described below are grouped according to principles of measurements for the analysis of alpha-, beta-, and gamma-emitting radionuclides.

Method A-1: Radiochemistry—alpha spectrometry

Principle. The element of interest is separated from other elements and from the sample matrix by using radiochemical techniques. The procedure may involve precipitation, ion exchange, or solvent extraction. Carriers (elements chemically similar to the element of interest) may be used. The element is deposited on a planchet in a very thin film by electrodeposition or by coprecipitation on a very small amount of carrier, such as lanthanum fluoride. The deposited element is then counted with an alpha spectrometer. The activity of the nuclide of interest is measured by the number of alpha counts in the appropriate energy region. A correction for chemical yield and counting efficiency is made using a standardized radioactive nuclide (tracer) of the same element. If a radioactive tracer is not available for the element of interest, a predetermined chemical yield factor may be used.

TABLE 6.3 List of Approved Methods for Specific Radionuclides

Radionuclide	Approved methods of analysis
Am 241	A-1, A-2, A-3, A-4
Ar 41	B-1, B-2, G-1, G-2, G-3, G-4
Ba 140	G-1, G-2, G-3, G-4
Br 82	G-1, G-2, G-3, G-4
C 11	B-1, B-2, G-1, G-2, G-3, G-4
C 14	B-5
Ca 45	B-3, B-4, B-5
Ce 144	G-1, G-2, G-3, G-4
Cm 244	A-1, A-2, A-3, A-4
Co 60	G-1, G-2, G-3, G-4
Cr 51	G-1, G-2, G-3, G-4
Cs 134	G-1, G-2, G-3, G-4
Cs 137	G-1, G-2, G-3, G-4
Fe 55	B-5, G-1
Fe 59	G-1, G-2, G-3, G-4
Ga 67	G-1, G-2, G-3, G-4
H 3 (H_2O)	B-5
H 3 (gas)	B-1
I 123	G-1, G-2, G-3, G-4
I 125	G-1
I 131	G-1, G-2, G-3, G-4
In 113m	G-1, G-2, G-3, G-4
Ir 192	G-1, G-2, G-3, G-4
Kr 85	B-1, B-2, B-5, G-1, G-2, G-3, G-4
Kr 87	B-1, B-2, G-1, G-2, G-3, G-4
Kr 88	B-1, B-2, G-1, G-2, G-3, G-4
Mn 54	G-1, G-2, G-3, G-4
Mo 99	G-1, G-2, G-3, G-4
N 13	B-1, B-2, G-1, G-2, G-3, G-4
O 15	B-1, B-2, G-1, G-2, G-3, G-4
P 32	B-3, B-4, B-5
Pm 147	B-3, B-4, B-5
Po 210	A-1, A-2, A-3, A-4
Pu 238	A-1, A-2, A-3, A-4
Pu 239	A-1, A-2, A-3, A-4
Pu 240	A-1, A-2, A-3, A-4
S 35	B-5
Se 75	G-1, G-2, G-3, G-4
Sr 90	B-3, B-4, B-5
Tc 99	B-3, B-4, B-5
Te 201	G-1, G-2, G-3, G-4
Uranium (total alpha)	A-1, A-2, A-3, A-4
Uranium (isotopic)	A-1, A-3
Uranium (natural)	A-5
Xe 133	G-1
Yb 169	G-1, G-2, G-3, G-4
Zn 65	G-1, G-2, G-3, G-4

Applicability. This method is applicable for determining the activity of any alpha-emitting radionuclide, regardless of what other radionuclides are present in the sample, provided the chemical separation step produces a very thin sample and removes all other radionuclides which could interfere in the spectral region of interest [APHA-605(2), ASTM-D-3972(13)].

Method A-2: Radiochemistry—alpha counting

Principle. The element of interest is separated from other elements and from the sample matrix by using radiochemistry. The procedure may involve precipitation, ion exchange, or solvent extraction. Carriers (elements chemically similar to the element of interest) may be used. The element is deposited on a planchet in a thin film and counted with an alpha counter. A correction for chemical yield (if necessary) is made. The alpha count rate measures the total activity of all emitting radionuclides of the separated element.

Applicability. This method is applicable for the measurement of any alpha-emitting radionuclide, provided no other alpha-emitting radionuclide is present in the separated sample. It may also be applicable for determining compliance when other radionuclides of the separated element are present, provided that the calculated emission rate is assigned to the radionuclide which could be present in the sample that has the highest dose conversion factor [IDO-1 2096(18)].

Method A-3: Direct alpha spectrometry

Principle. The sample, collected on a suitable filter, is counted directly on an alpha spectrometer. The sample must be thin enough and collected on the surface of the filter so that any absorption of alpha particle energy in the sample or the filter, which would degrade the spectrum, is minimal.

Applicability. This method is applicable to simple mixtures of alpha-emitting radionuclides and only when the amount of particulates collected on the filter paper is relatively small and the alpha spectra are adequately resolved. Resolutions should be 50-keV or better [ASTM-D-3084(16)].

Method A-4: Direct alpha counting (gross alpha determination)

Principle. The sample, collected on a suitable filter, is counted with an alpha counter. The sample must be thin enough that self-absorption is not significant, and the filter must be of such a nature that the particles are retained on the surface.

Applicability. Gross alpha determinations may be used to measure emissions of specific radionuclides only (1) when it is known that the sample contains only a single radionuclide, or the identity and isotopic ratio of the radionuclides in the sample are well known and (2) measurements using method A-1, A-2, or A-5 have shown that this method provides a reasonably accurate measurement of the emission rate. Gross alpha measurements are applicable to unidentified mixtures of radionuclides only for the purposes and under the conditions described in section 3.7 [APHA-601(3), ASTM-D-1943(10)].

Method A-5: Chemical determination of uranium

Principle. Uranium may be measured chemically by either colorimetry or fluorometry. In both procedures, the sample is dissolved and the uranium is oxidized to the hexavalent form and extracted into a suitable solvent. Impurities are removed from the solvent layer. For colorimetry, dibenzoylmethane is added, and the uranium is measured by the absorbance in a colorimeter. For fluorometry, a portion of the solution is fused with a sodium fluoride–lithium fluoride flux, and the uranium is determined by the ultraviolet-activated fluorescence of the fused disk in a fluorometer.

Applicability. This method is applicable to the measurements of emission rates of uranium when the isotopic ratio of the uranium radionuclides is well known [ASTM-E-318(15), ASTM-D-2907(14)].

Method A-6: Radon 222—continuous gas monitor

Principle. Radon 222 is measured directly in a continuously extracted sample stream by passing the airstream through a calibrated scintillation cell. Prior to the scintillation cell, the airstream is treated to remove particulates and excess moisture. The alpha particles from radon 222 and its decay products strike a zinc sulfide coating on the inside of the scintillation cell, producing light pulses. The light pulses are detected by a photomultiplier tube which generates electric pulses. These pulses are processed by the system electronics, and the readout is in pCi/L of radon 222.

Applicability. This method is applicable to the measurement of radon 222 in effluent streams which do not contain significant quantities of radon 220. Users of this method should calibrate the monitor in a radon calibration chamber at least twice per year. The background of the monitor should also be checked periodically by operating the instrument in a low-radon environment [EPA 520/1-89-009(24)].

Method A-7: Radon 222—alpha track detectors

Principle. Radon 222 is measured directly in the effluent stream using alpha track detectors (ATDs). The alpha particles emitted by radon 222 and its decay products strike a small plastic strip and produce submicrometer damage tracks. The plastic strip is placed in a caustic solution that accentuates the damage tracks which are counted using a microscope or automatic counting system. The number of tracks per unit area is correlated to the radon concentration in air, using a conversion factor derived from data generated in a radon calibration facility.

Applicability. Prior approval from the EPA is required for use of this method. This method is only applicable to effluent streams which do not contain significant quantities of radon 220, unless special detectors are used to discriminate against radon 220. This method may be used only when ATDs have been demonstrated to produce data comparable to data obtained with method A-6. Such data should be submitted to EPA when requesting approval for the use of this method. [EPA 520/1-89-009(24)].

Method B-1: Direct counting in flow-through ionization chambers

Principle. An ionization chamber containing a specific volume of gas which flows at a given flow rate through the chamber is used. The sample (effluent stream sample) acts as the counting gas for the chamber. The activity of the radionuclide is determined from the current measured in the ionization chamber.

Applicability. This method is applicable for measuring the activity of a gaseous beta-emitting radionuclide in an effluent stream that is suitable as a counting gas, when no other beta-emitting nuclides are present [DOE/EP-0096(17), NCRP-58(23)].

Method B-2: Direct counting with in-line or off-line beta detectors

Principle. The beta detector is placed directly in the effluent stream (in-line), or an extracted sample of the effluent stream is passed through a chamber containing a beta detector (off-line). The activities of the radionuclides present in the effluent stream are determined from the beta count rate and a knowledge of the radionuclides present and the relationship of the gross beta count rate and the specific radionuclide concentration.

Applicability. This method is applicable only to radionuclides with maximum beta particle energies greater than 0.2 MeV. This method may be used to measure emissions of specific radionuclides only when it is known that the sample contains only a single radionuclide or when the identity and isotopic ratio of the radionuclides in the effluent stream are well known. Specific radionuclide analysis of periodic grab samples may be used to identify the types and quantities of radionuclides present and to establish the relationship between specific radionuclide analyses and gross beta count rates.

This method is applicable to unidentified mixtures of gaseous radionuclides only for the purposes and under the conditions described below.

Method B-3: Radiochemistry—beta counting

Principle. The element of interest is separated from other elements and from the sample matrix by radiochemistry. This may involve precipitation, distillation, ion exchange, or solvent extraction. Carriers (elements chemically similar to the element of interest) may be used. The element is deposited on a planchet and counted with a beta counter. Corrections for chemical yield and decay (if necessary) are made. The beta count rate determines the total activity of all radionuclides of the separated element.

This method may also involve the radiochemical separation and counting of a daughter element, after a suitable period of ingrowth, in which case it is specific for the parent nuclide.

Applicability. This method is applicable for measuring the activity of any beta-emitting radionuclide with a maximum energy greater than 0.2 MeV, provided no other radionuclide is present in the separated sample [APHA-608(5)].

Method B-4: Direct beta counting (gross beta determination)

Principle. The sample, collected on a suitable filter, is counted with a beta counter. The sample must be thin enough that self-absorption corrections can be made.

Applicability. Gross beta measurements are applicable only to radionuclides with maximum beta particle energies greater than 0.2 MeV. Gross beta measurements may be used to measure emissions of specific radionuclides only (1) when it is known that the sample contains only a single radionuclide and (2) when measurements made using method B-3 show reasonable agreement with the gross beta measurement. Gross beta measurements are applicable to mixtures of radionuclides only for the purposes and under the conditions described in section 3.7 [APHA-602(4), ASTM-D-1890(11)].

Method B-5: Liquid scintillation spectrometry

Principle. An aliquot of a collected sample or the result of some other chemical separation or processing technique is added to a liquid scintillation "cocktail" which is viewed by photomultiplier tubes in a liquid scintillation spectrometer. The spectrometer is adjusted to establish a channel or "window" for the pulse energy appropriate to the nuclide of interest. The activity of the nuclide of interest is measured by the counting rate in the appropriate energy channel. Corrections are made for chemical yield where separations are made.

Applicability. This method is applicable to any beta-emitting nuclide when no other radionuclide is present in the sample or the separated sample provided that it can be incorporated in the scintillation cocktail. This method is also applicable for samples which contain more than one radionuclide, but only when the energies of the beta particles are sufficiently separated that they can be resolved by the spectrometer. This method is most applicable to the measurement of low-energy beta emitters such as tritium and carbon 14 [APHA-609(6), EML-LV-539-17(19)].

Method G-1: High-resolution gamma spectrometry

Principle. The sample is counted with a high-resolution gamma detector, usually either a Ge(Li) or a high-purity Ge detector, connected to a multichannel analyzer or computer. The gamma-emitting radionuclides in the sample are measured from the gamma count rates in the energy regions characteristic of the individual radionuclide. Corrections are made for counts contributed by other radionuclides to the spectral regions of the radionuclides of interest. Radiochemical separations may be made prior to counting but are usually not necessary.

Applicability. This method is applicable to the measurement of any gamma-emitting radionuclide with gamma energies greater than 20 keV. It can be applied to complex mixtures of radionuclides. The samples counted may be in the form of particulate filters, absorbers, liquids, or gases. The method may also be applied to the analysis of gaseous gamma-emitting radionuclides directly in an effluent stream by passing the stream through a chamber or cell containing the detector [ASTM-3649(9), IDO-12096(18)].

Method G-2: Low-resolution gamma spectrometry

Principle. The sample is counted with a low-resolution gamma detector, a thallium-activated sodium iodide crystal. The detector is coupled to a photomultiplier tube and connected to a multichannel analyzer. The gamma-emitting radionuclides in the sample are measured from the gamma count rates in the energy regions characteristic of the individual radionuclides. Corrections are made for counts contributed by other radionuclides to the spectral regions of the radionuclides of interest. Radiochemical separation may be used prior to counting to obtain less complex gamma spectra if needed.

Applicability. This method is applicable to the measurement of gamma-emitting radionuclides with energies greater than 100 keV. It can be applied only to relatively simple mixtures of gamma-emitting radionuclides. The samples counted may be in the form of particulate filters, absorbers, liquids, or gases. The method can be applied to the analysis of gaseous radionuclides directly in an effluent stream by passing the gas stream through a chamber or cell containing the detector [ASTM-D-2459(12), EMSL-LV-0539-17(19)].

Method G-3: Single-channel gamma spectrometry

Principle. The sample is counted with a thallium-activated sodium iodide crystal. The detector is coupled to a photomultiplier tube connected to a single-channel analyzer. The activity of a gamma-emitting radionuclide is determined from the gamma counts in the energy range for which the counter is set.

Applicability. This method is applicable to the measurement of a single gamma-emitting radionuclide. It is not applicable to mixtures of radionuclides. The samples counted may be in the form of particulate filters, absorbers, liquids, or gases. The method can be applied to the analysis of gaseous radionuclides directly in an effluent stream by passing the gas stream through a chamber or cell containing the detector.

Method G-4: Gross gamma counting

Principle. The sample is counted with a gamma detector, usually a thallium-activated sodium iodide crystal. The detector is coupled to a photomultiplier tube, and gamma rays above a specific threshold energy level are counted.

Applicability. Gross gamma measurements may be used to measure emissions of specific radionuclides only when it is known that the sample contains a single radionuclide or the identity and isotopic ratio of the radionuclides in the effluent stream are well known. When gross gamma measurements are used to determine emissions of specific radionuclides, periodic measurements using method G-1 or G-2 should be made to demonstrate that the gross gamma measurements provide reliable emission data. This method may be applied to analysis of gaseous radionuclides directly in an effluent stream by placing the detector directly in or adjacent to the effluent stream or passing an extracted sample of the effluent stream through a chamber or cell containing the detector.

Counting methods.   All the above methods with the exception of method A-5 involve counting the radiation emitted by the radionuclide. Counting methods applicable to the measurement of alpha, beta, and gamma radiations are listed below. The equipment needed and the counting principles involved are described in detail in ASTM-3648(8).

Alpha counting
Gas flow proportional counters—The alpha particles cause ionization in the counting gas, and the resulting electric pulses are counted. These counters may be windowless or have very thin windows.

Scintillation counters—The alpha particles transfer energy to a scintillator, resulting in a production of light photons which strike a photomultiplier tube, converting the light photons to electric pulses which are counted. The counters may involve the use of solid scintillation materials such as zinc sulfide or liquid scintillation solutions.

Solid-state counters—Semiconductor materials, such as silicon surface-barrier p-n junctions, act as solid ionization chambers. The alpha particles interact with the detector, producing electron-hole pairs. The charged pair is collected by an applied electric field, and the resulting electric pulses are counted.

Alpha spectrometers—Semiconductor detectors are used in conjunction with multichannel analyzers for energy discrimination.

Beta counting
Ionization chambers—These chambers contain the beta-emitting nuclide in gaseous form. The ionization current produced is measured.

Geiger-Muller (GM) counters or gas flow proportional counters—The beta particles cause ionization in the counting gas, and the resulting electric pulses are counted. Proportional gas flow counters which are heavily shielded by lead or other metal, and provided with an anticoincidence shield to reject cosmic rays, are called *low-background beta counters*.

Scintillation counters—The beta particles transfer energy to a scintillator, resulting in a production of light photons, which strike a photomultiplier tube, converting the light photon to electric pulses which are counted. This may involve the use of anthracene crystals, plastic scintillator, or liquid scintillation solutions with organic phosphors.

Liquid scintillation spectrometers—Liquid scintillation counters use two photomultiplier tubes in coincidence to reduce background counts. This counter may also electronically discriminate among pulses of a given range of energy.

Gamma counting

Low-resolution gamma spectrometers—The gamma rays interact with thallium-activated sodium iodide or cesium iodide crystal, resulting in the release of light photons which strike a photomultiplier tube, converting the light pulses to electric pulses proportional to the energy of the gamma ray. Multichannel analyzers are used to separate and store the pulses according to the energy absorbed in the crystal.

High-resolution gamma spectrometers—Gamma rays interact with a lithium-drifted [Ge(Li)] or high-purity germanium (HPGe) semiconductor detectors, resulting in a production of electron-hole pairs. The charged pair is collected by an applied electric field. A very stable low-noise preamplifier amplifies the pulses of electric charge resulting from the gamma photon interactions. Multichannel analyzers or computers are used to separate and store the pulses according to the energy absorbed in the crystal.

Single-channel analyzers—Thallium-activated sodium iodide crystals are used with a single window analyzer. Pulses from the photomultiplier tubes are separated in a single predetermined energy range.

Radiochemical methods for selected radionuclides. Methods for a selected list of radionuclides are listed in Table 6.3. The radionuclides listed are most commonly used and have the greatest potential for causing doses to members of the public. For radionuclides not listed in Table 6.3, methods based on any of the applicable principles of measurement described above may be used.

Applicability of gross alpha and beta measurements to unidentified mixtures of radionuclides. Gross alpha and beta measurements may be used as a screening measurement as a part of an emission measurement program to identify the need to do specific radionuclide analyses or to confirm or verify that unexpected radionuclides are not being released in significant quantities.

Gross alpha (method A-4) or gross beta (method B-2 or B-4) measurements may also be used for the purpose of comparing the measured concentrations in the effluent stream with the limiting concentration levels for environmental compliance in Table 2 of Appendix E. For unidentified mixtures, the measured concentration value shall be compared with the lowest environmental concentration limit for any radionuclide which is not known to be absent from the effluent stream.

Method 115: Monitoring for radon 222 emissions. This appendix describes the monitoring methods which must be used in determining the radon 222 emissions from underground uranium mines, uranium mill tailings piles, phosphogypsum stacks, and other piles of waste material emitting radon.

Radon 222 emissions from underground uranium mine vents. Each underground mine required to test its emissions, unless an equivalent or alternative method has been approved by the Administrator, shall use the following test methods:

- Test method 1 of Appendix A to Part 60 shall be used to determine velocity traverses. The sampling point in the duct shall be either the centroid of the cross section or the point of average velocity.

- Test method 2 of Appendix A to Part 60 shall be used to determine velocity and volumetric flow rates.

- Test method A-6 or A-7 of Appendix B and method 114 to Part 61 shall be used for the analysis of radon 222. Use of method A-7 requires prior approval of the EPA based on conditions described in Appendix B.

Radon 222 emissions from uranium mill tailings piles

Radon flux measurement. Measuring radon flux involves the adsorption of radon on activated charcoal in a large-area collector. The radon collector is placed on the surface of the pile area to be measured and is allowed to collect radon for 24 h. The radon collected on the charcoal is measured by gamma ray spectroscopy. The detailed measurement procedure provided in Appendix A of EPA 520/5-85-0029(1) shall be used to measure the radon flux on uranium mill tailings, except the surface of the tailings shall not be penetrated by the lip of the radon collector as directed in the procedure; rather, the collector shall be carefully positioned on a flat surface with soil or tailings used to seal the edge.

Radon 222 emissions from phosphogypsum stacks

Radon flux measurements. Measuring radon flux involves the adsorption of radon on activated charcoal in a large-area collector. The radon collector is placed on the surface of the stack area to be measured and is allowed to collect radon for 24 h. The radon collected on the charcoal is measured by gamma ray spectroscopy. The detailed measurement procedure provided in Appendix A of EPA 520/5-85-0029(1) shall be used to measure the radon flux on phosphogypsum stacks, except the surface of the phosphogypsum shall not be penetrated by the lip of the radon collector as directed in the procedure; rather, the collector shall be carefully positioned on a flat surface with soil or phosphogypsum used to seal the edge.

7

Fugitive Emissions

Taylor H. Wilkerson

7.1 Introduction

Fugitive emissions are defined as "those emissions which could not reasonably pass through a stack, chimney, vent, roof monitor, or other functionally equivalent opening." These emissions can include uncaptured process emissions, such as equipment leaks or dust from conveying systems; area sources, such as wastewater ponds or storage piles; and accidental releases, such as chemical spills. Fugitive emissions have been a regulatory concern since the passing of the Clean Air Act of 1970, but it was not until the early 1980s that specific regulations required facilities to estimate and control the fugitive emissions from various processes. In addition, unchecked fugitive emissions can lead to health and safety concerns, such as degradation of plant air quality, as well as material loss and increased maintenance problems.

7.2 History of Fugitive Emissions

The fact that certain emissions are fugitive has been known to both industry and regulatory agencies for a considerable time; yet, until recently, there was no means of estimating the emissions and, therefore, no means of regulating them.

In the 1970s and early 1980s, the EPA conducted a series of field tests to develop fugitive emissions estimates from various sources. These tests led to the development of the regulations which currently govern fugitive emissions and the current fugitive emissions measurement techniques.

TABLE 7.1   SOCMI Emission Factors

Source	Service	Average SOCMI emission factor
Valve	Gas/vapor	0.0123
	Light liquid	0.0156
	Heavy liquid	0.0005
Pumps	Light liquid	0.1087
	Heavy liquid	0.0471
Compressor seals	Gas/vapor	0.5016
Pressure relief valves	Gas/vapor	0.2288
Connections (flanges)	All	0.0018
Open-ended lines	All	0.0037

In 1986, Congress passed the Superfund Amendments and Reauthorization Act (SARA) which requires industries to quantify the levels of specific chemicals in their waste streams, including the loss of those chemicals to fugitive emissions. Most industries chose to estimate these emissions using either mass-balance equations or established emission factors, such as those developed by the Synthetic Organic Chemical Manufacturing Industry (SOCMI), shown in Table 7.1.

Also in the mid-1980s, the EPA began to develop a series of National Emissions Standards for Hazardous Air Pollutants (NESHAPs) which established emissions standards for industries most likely to emit chemicals listed as hazardous air pollutants (HAPs). Many NESHAPs include a leak detection and repair (LDAR) program to estimate and reduce fugitive emissions.

In 1990, Congress passed the Clean Air Act Amendments (CAAA) which are designed to further reduce air emissions under a nation-wide program. Both Title III, which addresses HAP emissions, and Title V, which addresses operating permits, of the CAAA regulate fugitive emissions.

7.3   Fugitive Emissions Regulations

Under Title V of the 1990 CAAA, fugitive emissions must be estimated at least every 6 months to determine a facility's compliance with operating permits. Facilities are free to use any acceptable means to estimate their fugitive emissions, including mass-balance calculations, emission factors, bagging, or any other method approved by the regulating agency. Title V does not require a facility to reduce fugitive emissions, only to estimate the quantity present.

Title III of the 1990 CAAA regulates the emissions of section 112(b) hazardous air pollutants and includes rules for specific source categories, some of which address fugitive emissions. These rules are known as *maximum achievable control technologies* (MACTs). The two MACTs written to date which most strongly address fugitive emissions are the Petroleum Refinery MACT and the Synthetic Organic Chemical Manufacturing Industry (SOCMI) MACT, also known as the Hazardous Organic NESHAP (HON). Both of these rules, along with many of the NESHAPs developed in the mid-1980s, require an LDAR program to measure and control fugitive emissions; however, each rule varies in the pollutant concentration which defines a leak, the components required to be inspected, and frequency of inspections. Most rules which include an LDAR program require that EPA Method 21 be used to inspect components for leaks.

The advantage of using an LDAR program is that the leaks are not merely estimated, but are actually located and repaired, which reduces the actual emissions. In addition, many of the regulations which address fugitive emissions offer incentives for a well-run LDAR program in the form of decreased monitoring frequency and penalties for poorly run programs in the form of increased monitoring frequency and fines.

It is important to be aware of and follow all regulations applicable to your facility, as new regulations are continually being proposed and promulgated and enforcement is becoming more stringent.

7.4   Sources of Fugitive Emissions

One of the most important factors in estimating and controlling fugitive emissions is an understanding of where in the facility the emissions originate. Sources of fugitive emissions can be generally grouped into process or area sources.

Process fugitive emissions are those that can be linked to a specific process and source. Process VOC fugitive emissions originate mostly from component and equipment leaks. A list of the common components which leak and where they are most likely to leak is included in Table 7.2. Process particulate emissions are mostly from material handling procedures such as the use of conveyors or dumping.

Area fugitive emission sources are those which cannot be linked to a specific process and usually originate from a large area, not a specific point. Area VOC sources include wastewater systems and open storage vessels. Storage tanks which are covered and have a vent are considered point sources, not fugitive sources. Wastewater system sources include open ponds, ditches, clarifiers, settling basins, and any other open vessels. Emissions from these sources come from

TABLE 7.2   Sources of SOCMI Fugitive Emissions

Component	Source of emissions
Valve	Packing glands and seals
Flange	Flange gasket seal
Connectors	Connection points
Pump	Pump seal
Compressor	Compressor seals
Pressure relief devices	Device packing or weakened spring allowance relief at a lower pressure
Drains	Valve bypass or improperly closed valves
Open-ended lines	Valve bypass or improperly closed valves
Accumulator vessel vents	Vent seals
Agitator seals	Around agitator seal
Access door seals	Around door seal

evaporation of chemicals in the liquid into the ambient air. Area particulate sources originate mostly from outdoor storage piles or vehicular traffic.

7.5   Measuring Fugitive Emissions

Several methods of estimating fugitive emissions from a source have been developed. These methods range from simple but overly conservative emission factors to complicated, in-depth process surveys which can be very accurate. Most of the more recent regulations governing fugitive emissions require that an LDAR program be used to measure and reduce fugitive emissions simultaneously.

7.5.1   VOC sources

The simplest method of calculating fugitive VOC emissions is to use emissions factors, such as those developed by SOCMI, which are based on several studies. These factors require only that a count of the components in a process (valves, flanges, pumps, etc.) be made along with the material each component carries. The component count is then multiplied by the applicable emission factor (see Table 7.1) and the hours of service to arrive at total fugitive emissions. This method of measuring fugitive emissions has been found to overestimate the emissions from a facility, and it also offers no means of locating the components which are leaking to implement repairs and reduce emissions. This method also does not account for the condition of a system; a system with no maintenance program at all will have the same amount of emissions as a system which is carefully monitored and maintained.

Figure 7.1 Leak/no leak detection program.

A more accurate method of measuring fugitive emissions from a source is to use leak/no leak emission factors (Fig. 7.1). This method requires that all components which can be reasonably tested be screened using EPA Method 21 for VOC leaks. Components which cannot be tested (because they are difficult to reach, they are insulated, or for some other reason the operator finds them unreasonable to test) are assigned a SOCMI emission factor, as discussed in the previous paragraph. Any component found to be leaking (a reading of >10,000 ppm in most instances) is multiplied by a leaking emission factor, and all other components are assigned a no-leak emission factor. This method is based on the premise that those components leaking at a rate of over 10,000 ppm are the major contributors to fugitive emissions. This method has several advantages over a simple emission factor method. Leaking components are located and, therefore, can be repaired to reduce the emissions. And the state of the individual process is taken into consideration; a process which has a better maintenance program or which is less corrosive will have a lower emission rate because fewer components will leak.

Similar to the leak/no leak emission factor method is the stratified emission factor method of calculating fugitive emissions. This method, like the leak/no leak method, requires all components which can reasonably be tested to be screened for VOC leaks by using

Method 21. The components are then grouped into three categories according to their leak rate: 0 to 1000 ppm, 1000 to 10,000 ppm, and over 10,000 ppm. The factors are based on the same studies which yielded the leak/no leak factors, but the stratified factors offer a slightly more accurate accounting of fugitive emissions than the leak/no leak method.

The next most accurate method is the EPA screening value correlation method. This method uses data obtained during the EPA field tests to create curves which relate the Method 21 screenings to a mass emission rate. As with the leak/no leak and stratified emission factor methods, all components which can reasonably be screened should be tested. The results of the screening are then applied to the correlation curves to obtain a mass emission rate for the process. This method, while more laborious than the stratified factor method, is much more accurate than the other methods discussed thus far, because emissions are linked to screening values on a continuous basis.

More accurate methods all require some degree of bagging. Bagging involves actually surrounding a component with a nonleaking material. The bags are equipped with two openings, one to let in clean air and the other to test the exhaust air. The bags are tested under either pressure or vacuum with a constant, known flow rate of air. The exhaust air is run through an analyzer to measure the VOC concentration, which is used with the flow rate to obtain a mass emission rate.

To obtain more accurate results than the EPA correlation factors, it is possible to create plant-specific correlation factors. Developing a process-specific correlation removes errors inherent in the EPA's correlation factors which do not allow for the individuality of every process. To develop factors, first all components must be screened using Method 21. A certain number of the components found leaking must then be bagged to obtain a mass emission rate. The mass emissions data is then analyzed against the screening readings for those components, and a curve relating Method 21 readings to mass emission rate is developed. If the values obtained are similar to the EPA correlation curves, then the EPA curves should be used because they are based on a larger data set; however, if the new values differ from the EPA curves, the new values should be used. This method, when used with future Method 21 screenings, should give a very accurate account of the emissions from a process.

Of course, the most accurate, yet costly, approach would be to bag every component to obtain a true mass emission rate. Although this method would yield an extremely accurate inventory of fugitive emissions, the cost required to achieve the increase of accuracy over that of other methods is generally considered prohibitive. In addition, this is the method that the EPA used to arrive at the various fugitive emission measurement methods, so bagging an entire process would only be repeating work needlessly.

7.5.2 Using Method 21

Method 21 was developed in the early 1980s by the EPA to be a standard for screening VOC components for leaks. Although it has become a standard for almost every method of measuring VOC fugitive emissions, Method 21 is far from perfect, and it is important to know how to use the method to best serve your needs. Each regulation requires that different components be screened and in slightly different ways, so it is critical to be familiar with the regulations to which you are subject and how they require you to screen your processes. Keep in mind that while you may have two processes subject to an LDAR program, they may require varying screening procedures.

Method 21 has a number of interferences which affect the screening values obtained. Among these are ambient temperature fluctuations, ambient winds, and distance from the component on which the screening is performed. In addition, EPA has audited several LDAR programs and found many to be deficient, usually as a result of the screening procedures. The most common violations found include missing plugs on open-ended lines, inaccurate component counts, data logging errors, and component tagging errors. For these reasons, it is important to screen a process in a conservative manner to minimize the risk of missing leaking components.

The most basic way of developing a conservative system is to internally define a leak at a lower screening value than the regulated one. The value which defines a leak internally should usually be about 90 percent of the regulated value. This will ensure that most of the leaks are found despite instrument error or other small interference.

Processes should also be screened on calm days or days with no winds, if possible, to minimize the effect of winds on the screening values. Studies have shown that a wind as low as 5 mi/h can greatly affect screening values by quickly removing the leaking material from the area tested. The instrument should be kept as close to the component being tested as possible, to minimize the amount of ambient air drawn into the instrument.

Also the instrument's pump should be set to draw air at the lower airflow limit set by the regulations, as this will give a higher reading. The regulations set a range for the instrument pump, but at a higher pumping rate, more ambient air can be drawn into the analyzer, resulting in lower readings.

If at all possible, processes should be screened on warm days so that the material being tested has a lower vapor pressure. Tests have shown that ambient temperature fluctuations can greatly affect screening values, yielding lower readings on cold days and higher readings on hot days.

Another important factor to consider is that the different instruments specified in Method 21 will read various materials differently. Therefore, a process should be screened with the instrument which will read the materials present at the highest values.

An LDAR program can also be improved with various administrative measures. The fugitive emissions coordinator should be someone with the authority and ability to ensure that the testing is performed properly and that all repairs are completed promptly in accordance with the regulations. Operators should be trained to recognize leaks as they walk through a system by using visual and olfactory techniques. All employees involved in any way with the LDAR program should understand the importance of implementing the program correctly. One of the easier ways to perform a quality assurance check on the internal program is to have it audited by an outside contractor. These measures will ensure a fugitive emissions measurement program which complies with applicable regulations and accurately measures the emissions from a facility.

Many companies, such as LeakTracker, have developed systems designed to manage an LDAR system. These systems usually include tags for each component which are either bar-coded or radio-frequency-encoded. Each tag is unique so that the individual component can be logged and recorded. A database of all the components is kept on a computer where the system can be managed. A self-contained field system consisting of an analyzer and a data logger is used to read the component identification and the leak rate for the component simultaneously. The data logger records all the information to be uploaded to the managing computer once the field work is done. This way, the field work is rather quick and easy to perform, and the time spent entering data is kept to a minimum. A system such as this can greatly reduce the time necessary to run and evaluate an LDAR program.

7.5.3   Area VOC sources

Area sources of VOC fugitive emissions are more difficult to characterize than process emission sources. While the emissions from area sources can be measured, the cost of conducting the tests is high and there are no very accurate methods at this time. However, there are acceptable means of estimating the emissions from area sources.

EPA has developed two software programs, CHEMDAT 7 and SIMS 2.0, which can be used to estimate the emissions from area sources.

CHEMDAT 7 can estimate the emissions from either wastewater or landfill sources. SIMS 2.0 can estimate emissions only from wastewater systems, but it can link a series of sources together in one data file to represent an entire wastewater treatment system. Both models use similar inputs and calculations to estimate the emissions, and the calculations are based on some idealistic assumptions but are still relatively accurate. In addition to these two programs, several universities and private companies have software which estimates area fugitive emissions.

There are several methods of obtaining actual emission data from area sources, but none can be applied to all sources or all conditions. The concentration profile method involves putting probes at varying heights on a mast over the source to measure VOC concentrations and then using those numbers to estimate the total emissions. The transit technique uses a line of analyzers perpendicular to the prevailing wind, set up downwind of the source, to measure concentrations across the emission plume. The isolation flux chamber method uses a small chamber which can be placed over a small area of the source. The chamber has two openings and is tested by creating a flow through it and analyzing the concentration of the outlet gas. The outlet concentration is multiplied by the flow to obtain a mass emission rate for that area, which can be prorated to the entire source. The broadband infrared and dual-beam laser methods both use light adsorption technology to measure the emissions. Both systems use a light transmitter on one side of the source and a receiver on the other. The loss of intensity of the transmitted light is used to measure the concentration distance of the VOC emissions in the air, and this can be used to measure the emissions from the source.

7.5.4 Controlling area sources

Emissions from wastewater systems can come from anyplace where the VOC contaminated water comes in contact with the ambient air. Since almost every wastewater treatment system differs in both design and the material being treated, there is not one good method of control which will work for every system. The most effective control technique is to reduce the level of VOCs in the waste stream. This can be done with a stripper or other common method of removing VOCs. The next best method is to reduce the turbulence and temperature of the wastewater wherever possible, to minimize the volatilization of the VOCs. Also, any area which can reasonably be covered should be covered, to avoid additional volatilization. Aside from these general rules, control of fugitive emissions from an area source must be looked at in a site-specific manner due to the large variation in treatment processes.

7.5.5 Particulate sources

Particulate fugitive emission sources, in general, are more difficult to measure than VOC emissions due to the large number of variables which affect particulate fugitive emissions, most importantly, local climate and the source of the fugitive emissions. Almost all methods of measuring fugitive emissions rely on emission factors and equations developed using several EPA studies. Examples of sources of particulate fugitive emissions and means of controlling them are shown in Tables 7.3 and 7.4. Emission factors and equations for estimating fugitive emissions are presented in Figs. 7.2 through 7.4 and Tables 7.5 through 7.7.

From these factors and equations, clearly the emissions are affected by such factors as wind speed, vehicle speed, particle size, and local rainfall, which vary from region to region. There are still no good means of estimating particulate fugitive emissions from either area or point sources, but some regulations still require that control techniques be implemented.

TABLE 7.3 Common Fugitive Emission Control Techniques

Source	Ventilation and collection	Process optimization	Wet suppression
Material handling			
Transferring or conveying	×	×	×
Loading and unloading	×	×	×
Storage	×	×	×
Bagging or packaging	×	...	×
Material crushing and screening	×	×	×
Mineral mining			
Drilling	×		
Blasting	...	×	
Extraction	...	×	×
Waste disposal (tailings)	...	...	×
Metallurgical operations			
Furnace charging	×	×	
Furnace tapping (product and slag)	×	×	
Casting	×	×	
Mold preparation	×	×	×
Casting shakeout	×	×	
Slag disposal	...	×	×
Sintering	×	×	
Coke oven charging	...	×	
Coking (leaks)	...	×	
Coke pushing	×	×	
Coke quenching	...	×	

TABLE 7.4 Common Vehicle Fugitive Emission Control Techniques

Source	Wet suppression	Stabilization	Speed reduction	Face cleaning/ transportation controls	Windbreaks	Good operating practices
Unpaved roads	X	X	X			
Construction activities	X			X		X
Dust from paved roads				X		
Off-road motor vehicles			X			
Overburden removal and storage	X					X
Reclamation efforts		X				X
Inactive tailings piles		X			X	
Disturbed soil surfaces		X			X	X
Agricultural tilling					X	

$$\text{EF} = P\left(E + \frac{0.20T}{4} + \frac{5.07T}{4}\right)$$

where EF = emission factor, g/VMT (grams/vehicle metric ton)
  P = fraction of particulate which will remain suspended (diameter less than 30 μm) from a paved road surface
    = 0.90
  E = particulate emission originating from vehicle exhaust (see Table 7.5)
  0.20 = tire wear in a four-wheeled vehicle, g/VMT
  5.07 = entrained dust in a four-wheeled vehicle, g/VMT
  T = number of tires per vehicle

Figure 7.2 Equation for calculating a specific emission factor for vehicles traveling on paved surfaces.

$$\text{EF} = P(0.81)(s)\left(\frac{S}{30}\right)\left(\frac{365-W}{365}\right)\left(\frac{T}{4}\right)$$

where EF = emission factor, lb/VMT
  P = fraction of particulate which will remain suspended (diameter less than 30 pm): 0.62 from a gravel road bed and 0.32 from a dirt road bed
  s = silt content of road bed material, percent; 12 percent approximate average value (values range between 5 and 15 percent)
  S = average vehicle speed, mi/h
  W = days with 0.01 in or more of precipitation
  T = average number of tires per vehicle

Figure 7.3 Equation for fugitive dust from unpaved surfaces using the EPA's published procedure.

$$\text{EF} = aIKCL^1 V^1$$
where EF = emission factor, ton/(acre · yr)
  a = portion of total wind erosion losses that would be measured as suspended particulate
  I = soil erodibility, ton/(acre · yr)
  K = surface roughness factor
  C = climatic factor
  L^1 = unsheltered field width factor
  V^1 = vegetative cover factor

In this equation, K, C, L, and V are all dimensionless. Some recent work has indicated that a and I are related to soil type. Values of a and I that might be applied to surface-mined areas during or following wind events are summarized in Table 7.7.

Figure 7.4 Wind erosion for soils.

TABLE 7.5 Emission Factors for Vehicles Traveling on Paved Surfaces (g/mi)

Vehicle type	Exhaust[a] E	Tire wear[b,c]	Reentrained dust[c]	Initial emission factor[d]	Final emission factor[e]
Average	0.53				
Light-duty gasoline (4-wheeled)	0.34	0.20	5.07	5.6	5.0
Heavy-duty gasoline (10-wheeled)	0.91	0.50	12.68	14.1	12.7
Heavy-duty diesel					
12-wheeled	1.30	0.60	15.21	17.1	15.4
18-wheeled	1.30	0.90	22.82	25.0	22.5

[a]Exhaust emissions are specific for fuel and vehicle type.

[b]The tire wear component is based upon 0.20 g/VMT (grams/vehicle metric ton) for a four-wheeled vehicle and can be adjusted upward for vehicles with large numbers of wheels.

[c]The reentrained dust component is estimated to be directly proportional to the number of tires. An additional multiplication factor of 2.5 should be applied to the tire wear and reentrained dust columns for large-wheeled equipment, i.e., mining haul trucks and wheeled tractors, loaders, or dozers.

[d]The initial emission factor is the sum of the exhaust, tire wear, and reentrained dust components.

[e]The final emission factor is the initial emission factor multiplied by 0.90. The factor of 0.90 accounts for that amount of particulate which will remain suspended.

TABLE 7.6 Fugitive Dust Emission Factors for Truck Loading Operations

Description of truck loading operation	Uncontrolled emission rate, lb/ton loaded	Reliability rating
Crushed rock (front-end loader)	0.05	E
Lignite coal* (shovel)	0.02	E
Coal† (shovel)	0.05	E
Coal‡ (shovel)	0.10	E
Coal§ (shovel)	0.04	E
Granite§ (unspecified)	Negligible	E

*Lignite coal from North Dakota mining sites.
†Coal from Colorado mines.
‡Unspecified location.
§Coal from Colorado mining operations.

TABLE 7.7   Summary of Variable Values for Wind Erosion
Equations

Surface soil type	a	I, ton/(acre · yr) (uncontrolled)
Rocky, gravelly	0.025	38
Sandy	0.010	134
Fine	0.041	52
Clay loam	0.025	47

The control of particulate fugitive emissions usually is achieved by either wetting surfaces, to reduce dust from wind and vehicle activity, or covering the fugitive emission source, to eliminate wind erosion. The wetting agents used are usually water, but in some instances, such as with dusts which react with water, a chemical wetting agent is used. Sources are usually covered to reduce emissions from piles which are not frequently used, transportation activities, such as trucking of materials, or conveying systems. In some cases, such as a small dumping operation, a vent hood can be put in place to capture any fugitive dust.

8

Air Quality Management Policy

E. Roberts Alley

8.1 Introduction

Ideally, air pollution should be managed regardless of air pollution regulations. As stated in the Introduction to this handbook, sustainable growth on this planet can be realized only if air pollution is controlled. Management is simply a tool for implementing control. Using the information presented in Part 1 of this handbook, each discharger of air pollutants should assess the quantity, quality, and variability of the pollutants to determine if air pollution management and subsequent control are necessary. Management includes, but should not be limited to, meeting regulations; therefore Part 2, Air Quality Management, mentions but is not organized around regulations.

Every organization which emits air pollution should have as a goal to control the impact of its pollution on the environment. To manage and demonstrate this control, many air pollution dischargers have developed an environmental policy, but few have implemented a system to ensure that their policy is appropriate, improvable, reviewable, and documentable. Most efforts toward setting an environmental control policy have begun with environmental reviews, assessments, or audits. But alone, these efforts have not normally resulted in sustained environmental compliance with either policies or regulations.

In its meeting of February 12, 1995, the International Standards Organization (ISO) adopted ISO/CD 14001 which makes recommendations for an environmental management system that describes the core elements for certification and registration of an organization's environmental management system. These standards can serve as a solid basis for an air quality management system that can be certified on an international basis to meet the guidelines suggested in this document.

This chapter recommends a system that has a policy approved by top management which is appropriate for the specific organization, is updatable, is periodically reviewed and reported to top management, and is documented so that results are provable. This system will work only if it is integrated with the overall management system of the organization.

8.2   Management Policy

Management is the act of implementing policy. A perfect policy is useless unless it is properly managed, but without policy there is nothing to manage. These two factors should enter into any successful program of air quality management. First, a policy should be set by top management. Historically, if top management was not involved intimately with setting and enforcing a policy, the policy was doomed to failure. Only top management can be responsible for balancing environmental compliance with production and ensuring that environmental goals are realistic, achievable, and internally enforceable.

The policy should be realistic in order to protect the environment, meet regulations, and conform to the goals and character of the organization. It must be practically achievable by the discharger, or it will falter because of a lack of compliance. It must be enforceable through documentation of compliance, periodic review of documentation, and response to and correction of noncompliance with policy.

In addition, a policy should be updatable. When process, regulation, or environmental changes occur, the policy may need to be altered. A discharger may decide to phase a policy so that goals can be achieved over a period of time in steps. If this approach is followed, the phases should have time deadlines in order to be internally enforceable.

The first step in the proper management of a policy is the assignment of an air quality manager. This manager should directly report to the top manager who initially set the policy. When conflicts occur between environmental policy compliance and process goals, top management must resolve the conflict, ideally within policy parameters. If this resolution cannot be made with these parameters, the policy should be made more realistic.

The policy should be publicized internally within the organization and externally to the air quality manager for tasks such as inventories, permit applications, monitoring, testing, and documentation.

The key to policy compliance enforcement is documentation. Documentation should span all areas of air quality management. The following are examples of occurrences which require documentation:

1. Unloading of raw materials which allows a release of air pollutants to the atmosphere such as leakage, spillage, or venting

2. Storage of raw materials which allows a release such as through vent pipes

3. Release of air pollutants from processes which are not managed through air pollution control devices

4. Operation of air pollution control devices including repairs, replacement, and failure

5. Monitoring of records from point sources and fugitive emissions

6. Testing records from monitoring points

7. Permit applications, reports, and replies

The policy should specify a method of documentation review. This method should be both routine in the form of reports to top management and periodic in the form of internal and external audits. The routine reports should include easily discernible noncompliance points so that top management has no need for detailed review.

The guidelines for establishing, maintaining, and implementing an air quality management policy are summarized in Table 8.1.

8.3 Assessment of Air Pollution

To effectively develop an air quality management policy, an assessment of existing air pollution quality and quantity should be made. An air pollution assessment should include the following elements:

1. The points of discharge of pollutants into the atmosphere, i.e., stacks, vents, etc.

2. The nonpoint or fugitive discharges of pollutants.

TABLE 8.1   Guidelines for an Air Quality Management Policy

- Set by top management
- Realistic
- Updatable
- Internally publicized
- Documented
- Reviewed

3. The characteristics of each point and nonpoint discharge in terms of strength and variability. As explained in Part 1 of this Handbook, these characteristics include airflow, temperature, moisture content, and concentration of various pollutants.

4. The characteristics of pollutants which are discharged, i.e., stack, velocity, stack height, etc.

5. Environmental, geographic, or atmospheric conditions which affect the distribution of the pollutants from the source of discharge into the air.

6. An analysis of the effect of the pollutants on the environment. This could be an effect on the stratosphere in the form of ozone depletion; an effect on the atmosphere in the form of ozone formation; or an effect on animal, plant, or human life due to the return of pollutants to the surface.

This complete air pollutant assessment is critical to the management process. If the assessment indicates that released air pollutants have a negative effect on the environment, an air quality control management system should be initiated regardless of whether environmental regulations are being met. This air quality control management system should begin with a policy adopted by top management that succinctly states its commitment to air quality control.

8.4   Air Quality Management Policy

The purpose of an air quality management policy for an organization is to set standards and goals for maintaining and improving its air quality management system. This policy should state the commitment to comply with all environmental regulations and continually improve compliance to such a point that air quality in the area affected by the organization is not diminished in any way. The policy should

1. Be easily understandable as to its scope or area of application

2. Be appropriate and practical for implementation

3. Be reviewable

4. Be documentable

5. Be improvable on a routine basis

6. Include a commitment to comply with all existing environmental regulations

7. Be distributed to all employees

8. Be available to the public

9. Be signed by top management

The air quality management policy should establish management programs and provide for program maintenance. Responsibility for each program should be delineated with direct reporting to top management, who must take legal responsibility for air quality management. The policy should also set objectives, targets, and time schedules for implementing the objectives. Management staff should be given appropriate authority which should be communicated to production, sales, and other arms of the organization. This authority should include the ability to shut down a process when goals are not met.

8.5 Training

All appropriate employees should receive general training on the air quality management policy similar to that received for other safety, health, and environmental programs. It should be emphasized that the policy is a management policy and not an enforcement policy, and therefore it must involve all employees associated in any way with possible releases of environmental pollutants into the atmosphere. Training should have the following objectives:

- Communicate the wording and the purpose of the policy

- Explain the commitment of top management to the policy

- Delineate the potential releases of pollutants associated with each employee's responsibilities

- List physical point sources of these releases

- Explain how these releases can be prevented, controlled, or minimized by the employee's direct or indirect actions

- Review the documentation procedure for which each employee is responsible

- Explain the annual internal and external audit procedures

- Invite comments and constructive criticism

- Document the employee's grasp and commitment of the training with a series of written questions

- Repeat the training annually for each appropriate employee and any visitors or contractors who could affect the policy

8.6   Communication

The policy is communicated to employees as explained above. It is recommended that there be further communication to demonstrate the lasting commitment including an annual summary of the documentation. This could be presented at the annual training session. As a minimum, this communication should include a general summary of the gap between the policy and the documentation for the entire organization and a more detailed summary for the employee's department or area of responsibility.

An organization may elect to communicate its policy to the public. If so, a series of periodic public meetings and/or press releases will go far toward informing neighbors, environmental groups, and the general public of the commitment to protection of public health, safety, and the environment.

Part 2

Air Quality Management

Air Management Programs

Taylor H. Wilkerson

9.1 Introduction

More and more, the emphasis of plantwide air pollution reduction programs is changing from emissions reduction after the source, such as pollutant removal systems, to before-source reductions and reduction through environmental management. These before-source reduction methods have the advantage of allowing greater control and understanding of parameters which contribute to emissions and often lead to increased process efficiency and product quality.

The key to reducing emissions through management procedures is to use a well-developed pollution management plan containing clearly defined goals and procedures. In addition, the plan must include a sufficient system of checks and balances to ensure proper implementation of the plan. Just as an after-source control system must be well designed and maintained for optimal performance, a pollution management plan must be well designed and maintained in order to work suitably and effectively.

This chapter describes steps involved in designing and implementing a pollution management plan. It is important to keep in mind that every plant and every process are unique; there is no one pollution management strategy which will be effective for every facility. The goal of this chapter is to present general steps involved in creating a pollution management plan. In addition, many sections of an environmental management plan will be based on existing operating procedures and environmental compliance procedures for most systems.

The basic steps covered in this chapter include

- Identification of pollution-causing activities
- Review of regulations which apply to the facility
- Audit of raw materials and their uses
- Characterization of waste products
- Quantification of emissions
- Audit of accident potential
- Procedures to control pollution and minimize waste
- Implementation and management of control procedures

9.2   Identification of Activities Contributing to Air Pollution

The first and most important step in a pollution management plan is to identify the sources of emissions. Once the sources are identified, the facility can concentrate on managing the emissions without getting lost in the other processes which do not contribute to emissions.

To establish which activities contribute to air emissions, an investigation of every process on site should be conducted. Several small sources, such as use of degreasing or cleaning solvent, can contribute emissions; so it is important to include all sources in the facility, no matter how small.

The investigation of each individual source should be conducted with personnel who have a solid understanding of the process operations in question along with personnel who are familiar with the federal, state, and local regulations applicable to the facility and its various processes.

9.2.1   Regulatory review

The investigation of each process should begin with a review of the current regulatory requirements of the equipment. The regulatory requirements can come in the form of a permit for the process or could be a blanket regulation which affects the source even though the source does not have a permit. This review will reveal many of the operational and record-keeping procedures required for the process and should also bring to attention many of the potential environmental impacts of the system.

When the regulatory review is undertaken, it is important to include all regulations which could apply to the system in question, such as

- General regulations which affect all sources
- Requirements specific to the system

  New-source performance standards (NSPSs)

  National Emissions Standards for Hazardous Air Pollutants (NESHAPs)

  Title III maximum achievable control technologies (MACTs)

  Other source-specific applicable rules

Each regulatory requirement should be investigated to determine why it is applicable to the source, whether it is due to the type of operation, the raw materials, the product, or some other factor.

It is important to keep in mind that permits are only a starting point. They will not magically reveal all regulations applicable to a source. The presence of a nonapplicable requirement or the omission of an applicable requirement on a permit is common. If the source is restricted by a requirement in a permit or by another restriction which does not appear to be applicable, measures should be taken to determine why the source is regulated by that requirement. In the event of a nonapplicable requirement, the facility should work with the permitting authority to get the permit updated.

Also, the fact that a source lacks a permit does not necessarily imply that the source does not require a permit or does not contribute to plantwide emissions. Several exemptions have been repealed since the passing of the 1990 Clean Air Act Amendments (CAAAs), and it is likely that a source could have mistakenly not been permitted for several reasons. Permits should not be taken as gospel. Although the permitting agency tries to include the correct applicable requirements on a permit, errors and omissions can and do occur.

Each requirement found should be investigated to determine whether it is being met. If the requirement is not being met, measures to bring the source into compliance should be considered and a plan to bring the source into compliance devised. These measures include

- Modifying operating procedures
- Repairing faulty equipment
- Adding a control system
- Altering the source's permits to account for the additional emissions

Once the source has been thoroughly reviewed for regulatory compliance, a plan for maintaining and, if necessary, improving compliance should be developed. This plan should include, among other things, record-keeping procedures and a review system to ensure that the process is maintaining compliance.

It is also important to evaluate the current internal environmental procedures for the system. Internal procedures are any operating procedures which are designed to reduce waste or emissions but are not required by any regulation. The procedures should be evaluated to determine (1) whether they are up to date and (2) why the system requires special procedures.

9.2.2   Raw material characterization

For each source, the raw materials used should be reviewed. It is important to understand how the use of a material contributes to the emissions. Once this is understood, it is much easier to investigate alternative raw materials and to determine their effect on the source's emissions. Alternative raw materials should also be evaluated to determine whether they will be able to reduce emissions without creating a serious economic impact or damaging product quality.

Raw materials should also be assessed to determine how they affect other environmental aspects of the source. If a material is not completely used by the source, the methods of disposing of the excess material should be reviewed. If the material is hazardous to store or handle, alternative materials might be investigated.

Equally important is to understand how the raw materials contribute to the source's waste stream. Many wastes can be reduced through better raw material management and utilization.

9.2.3   Waste characterization

The process wastes and by-products should be evaluated for how they are generated and how they are disposed of. For any waste that it disposed of off site, alternatives should be investigated, such as recycling or selling to another plant as a raw material. Methods of reducing the total waste generated by the system should be investigated.

Volatile wastes can contribute to air emissions, so these are very important to examine closely. Volatile wastes should be stored in sealed containers for both safety and environmental reasons.

9.2.4   Emissions characterization

Emissions should be reviewed to identify what materials are emitted to the air and how each is controlled. Emissions can be characterized by several means:

- Engineering equations
- Emission factors based on testing at similar sources
- Emissions tests performed on that source

If emission factors are used, they should be reviewed for accuracy and applicability. If a more accurate factor can be applied, it should be.

9.2.5 Potential accident characterization

Each process should be investigated to determine the potential for an environmental incident. An environmental incident includes any spills, gaseous release, or other accidental or emergency discharge. For each potential incident there should be a procedure implemented which takes into account the likelihood of occurrence and the effect of the incident (area of impact, potential health and environmental hazards).

Any past environmental incidents involving the system should be investigated. Each incident should be evaluated to determine how it occurred and what process improvements, whether by engineering or management, have been, are, or should be made to ensure that similar incidents do not happen again.

9.3 Procedures to Control Air Pollution—
Operation and Maintenance

No pollution management program is effective if there is no means of overseeing the implementation of the plan. Without a plan that has the full support of management, the procedures can be forgotten or so distorted that they are no longer effective. If there is no plan to explain why procedures must be carried out, they may be omitted in the future.

The first step to ensuring proper implementation is to establish a corporate policy regarding air pollution management. This policy should be set and enacted by the top management of the corporation and should be conveyed to all employees in a top-down fashion, with emphasis to employees on how the implementation of the policy affects them and their daily activities.

Most of all, the policy should allow for free communication between all levels of the corporate structure. The employee who works on a system every day may well have ideas for improving pollution management that someone familiar with only the theory of operation would not.

The policy should assign specific responsibilities to management representatives to help implement the plan. These responsibilities can include leading and designing training sessions, organizing data collection, and organizing implementation teams. Each plant will have different tasks to carry out its individual management plan, but it is crucial to have management involved for the plan to work smoothly.

In addition, management should make available all resources needed to enact the plan. Proper implementation of the plan can require the release of information on materials used and the processes, the purchasing of equipment for emergency release control, or the use of other resources. All these must be made available in order to have a functional plan.

9.4   Training

The policy should include provisions for training employees in their duties which are required by the plan. The training should include a clear explanation of why the procedures are necessary and should stress the importance of maintaining the policy.

The training should be extended to include any contract employees working on site. They should be trained in both general policies and any procedures specific to the areas in which they will work.

9.5   Responsibility

The responsibility for running a successful pollution management system falls to every employee in the corporation, from the CEO down. Each person's duties should be clearly defined in the environmental management plan, yet each person must be free to determine the best means to meet his or her responsibilities. That is, while a task may be important, the person who will know best how to carry it out is the one who performs the task. Therefore, that person should be given free reign to decide how to carry out the task as long as it is completed properly.

Delegation of responsibility for the plan should be handled carefully. People should be given responsibilities which are within their areas of expertise and which coincide as much as possible with their current job requirements. For instance, the maintenance manager should be responsible for keeping any required maintenance records and ensuring that the maintenance work is completed. Many systems and control devices require that a certain amount of preventive maintenance be performed periodically. Records of the maintenance performed on a system and when it was performed should be kept by the maintenance department.

Often many different records must be maintained at various places throughout a plant, and even for a small facility, the records can be more than one person alone can take care of. For that reason, it is beneficial to have the people for whom it is convenient record the information and then have an upper-management person ensure that the records are being kept properly and that all the information in the records is reasonable and not in violation of any requirements.

It is also beneficial for a company to have at least one person monitoring trade publications and vendor information for current technologies and materials which could help reduce emissions and possibly even costs.

9.6 Documentation

A written corporate environmental policy should be available to all employees. The core of the policy should describe the general corporate policy without addressing any specific activities. The core of the policy should also direct the reader to any specific information he/she desires. The more specific documentation should include information on all processes in the plant, organizational charts which specify who is responsible for the policy and their duties, internal standards and procedures, and all emergency plans.

A training and standard operating procedure (SOP) program should be developed for each system. These programs should be created by teams of personnel who understand both the environmental concerns and the operation of the system. The team should include employees who work on the system on a daily basis, including operators and maintenance personnel, so the plans will be smoothly integrated into daily activities.

Most record keeping is already being done in most cases. Record keeping should be delegated to someone who already keeps the records or to the person who has the closest contact with the information being recorded. For instance, if the record is to be kept of the material input to a system, then the person who loads material into the system should keep the applicable records.

9.7 Time Frames

One cannot expect a good environmental management plan to appear overnight. For that reason, it is important to set reasonable time-frame goals for the plan and for the people who share the responsibilities of the plan. The plan manager should keep in mind that the people contributing to the management process have other responsibilities to the company; therefore, to require frequent reporting can result in a work overload for these people.

The frequency of reports should be mandated by the frequency with which the data will be needed. If record keeping is required by the permit, then there is little choice as to when the data are needed. If the records are for internal use, it is best to have records submitted daily, weekly, or monthly, depending on the criticality of the data and the amount of data taken per day. The plan should also be updated periodically. This should be done at least yearly. The plan must be updated to include new processes and control techniques or to modify procedures that were found to be impractical.

A record-keeping manager should be assigned to ensure that all records are being maintained and submitted properly and in timely fashion. Depending on the size of the facility and the amount of records needed, this job may have to be divided among several people. It is very important to keep good records to monitor the progress of the plan and to prove that it is working properly.

If any sources are found to be out of compliance with a regulation or internal policy, a time line should be set up to bring them back into compliance. The time line should be developed based on what needs to be done to bring the source into compliance and how critical the noncompliance issue is.

9.8   Emergency Response

An environmental emergency can consist of any number of accidental or unplanned releases which are harmful either directly or indirectly to the environment or human health or life. It is extremely important for a facility to recognize the potential for an emergency to arise and to be able to either prevent or respond to all foreseeable emergencies in a manner which will minimize the risk to the environment and human health and life.

The first step in preventing an emergency is to identify potential emergencies around the facility. Emergencies can originate in many areas of a typical facility, including storage facilities, reactors, material handling procedures, or any other process which involves hazardous materials, including fuel combustion sources such as boilers and heaters. Each potential source should be evaluated by first determining what emergency episodes could occur and, from those, determining both the worst case and the most likely emergency scenarios.

For each source identified as a potential hazard, control methods should be evaluated and, where feasible or regulated, put in place. These methods first must prevent an accident from happening, and then if an accident does happen, these methods must prepare employees to respond to the incident such that hazard to human health or life and the environment is minimized. In addition, local police, fire, and emergency management agency (EMA) authorities should be informed of potential hazards at the facility so that they can respond to protect the surrounding community as efficiently as possible in the event of an emergency. It is also helpful to evaluate any past emergency incidents at the facility, if any exist. These incidents will shed light on what systems are potentially dangerous, how they are likely to fail, and what control procedures do or do not work at the facility.

When past incidents are evaluated, care should be taken to review how and why the incident occurred—whether it was due to human error, equipment failure, or design failure. After the probable cause is determined, measures to prevent the incident from recurring should be investigated. Most preventative measures consist of either procedural modifications or equipment modifications. Preventative measures can include containment and response procedures to minimize the impact, should the incident recur.

Containment and response measures can include such things as

- Dikes to prevent liquid spills from spreading
- Curbs to prevent liquid spills from spreading
- Double-walled pipes
- Tank enclosures
- Emergency response and cleanup procedures

In almost all cases, the response and cleanup activities must be carried out by people who are trained in OSHA 40-hour hazardous material (HAZMAT) handling and familiar with the process in question. These can be plant employees who are appropriately trained, local authorities who must be properly notified of the potential hazards so they can properly respond, or an outside contractor who is close enough to respond quickly and effectively.

If the emergency response is to be made in-house, whoever is in charge must make certain that the response team has on hand and readily accessible all the personal protection equipment (PPE) and other materials needed to respond to a worst-case scenario. In addition, it is extremely important that the people designated to respond to the emergency be aware of the potential hazards of the materials present, including flammability, reactivity, reactions with other substances, and toxicity.

Once adequate procedures for responding to and preventing emergencies are established, they should be written up in a manual. All employees should be made aware of their responsibilities in an emergency, even if that means to leave the area quickly. Also, local authorities need to have a copy of the plan and be made aware of their role, if any, in the procedures.

9.9 Updating Programs

One of the most important goals of a pollution management system is to keep the plan current. The plan should be reviewed at least once a year, because a pollution management plan is useless if it does not reflect the current configuration of the plant. In addition, the plan should be reviewed any time a process change is made, whether it is the installation of a new process, a raw material change, or a change in production schedule.

The yearly review should encompass everyone involved with the plan on one level or another. Each process should be analyzed to determine if any changes have occurred over the past review period. In addition, the pollution management plan sections which address that process should be reviewed for effectiveness and practicality. If there are any difficulties in implementing a section of the plan, everyone involved with that section should meet to discuss how to rectify the situation.

Regulatory changes can also have a great impact on the operation of a pollution management plan. If a regulatory change occurs which affects the facility, a full investigation of its impact should be made. Once the new regulation's impact is known, modifications to the plan should be made to accommodate the regulation.

Each review should also take into account any suggestions made to improve the implementation of the plan. Suggestions can make the plan more efficient and, therefore, cost-effective. Keeping the plan well maintained is as important as keeping any other piece of equipment well maintained. The plan needs to work smoothly to be effective.

9.10 Importance of a Management Plan

A pollution management plan can be a great advantage to a company. A well-designed, well-run plan can both help maintain compliance with regulatory agencies and reduce emissions and make the facility more efficient through greater understanding of the process operations and raw material use. Although these plans do require some effort to set up, they are usually well worth it.

Air Quality Audit

John Coulter

Industry management teams and environmental managers are finding it a very challenging and daunting task to ensure that their facilities are in compliance with today's environmental laws and regulations. This task is complex not only because of the enormous volume of applicable laws and regulations, but also because of the fact that these laws and regulations are growing at a exponential rate with no signs of decline on the horizon.

Industry is finding that the cost of compliance is becoming a significant expense and that the cost of noncompliance can be enormous and is an unacceptable alternative. Heavy fines for noncompliance and civil and criminal liabilities are on the increase. The industrial management teams must make every reasonable effort to ensure that their environmental programs are in place and effective.

10.1 Types of Audits

The audit is one of the primary tools to measure how well a facility is complying with current environmental regulations. Today the management team is seeing three basic types of audits: internal audits, regulatory audits, and customer audits (Fig. 10.1). Internal audits are con-

- Internal Audits
- Regulatory Audits
- Customer-Driven Audits

Figure 10.1 Types of environmental audits.

ducted by in-house personnel or by outside consultants to determine the compliance status of the facilities. The results of the internal audit are used to improve the environmental performance of the facility, to make improvements and modifications to the current environmental programs, to ensure continued compliance with all rules and regulations, to reduce the amount of risk imposed on the facility due to non-compliance, and finally as a tool to prepare for scheduled or unscheduled regulatory or other outside audits. The regulatory audit is conducted by a regulatory agency (state, EPA, or local agencies) to determine the compliance status of the facilities. These regulatory audits can be limited to a single area such as air emissions, water, or hazardous waste disposal; or they can be multimedia audits covering the entire environmental spectrum. Unsatisfactory results from the regulatory audit can be very unpleasant, ranging from a simple Notice of Violations to heavy fines and even civil and criminal litigation.

Customer audits are becoming more popular as the consumer population becomes aware of the environmental impact of their purchasing decisions. As a result, manufacturers are starting to audit their suppliers' environmental programs in order to demonstrate to the consumer that their products are as environmentally friendly as possible. Unsatisfactory results from the customer audit can be just as unpleasant as from the regulatory audit, resulting in lost sales and reduced revenue, an outcome that most management teams can ill afford.

The purpose of this chapter is to help the management team and environmental manager successfully prepare for regulatory and customer audits in order to prevent or limit any unpleasant results. As with most situations, good preparations are a prerequisite for success. The regulatory audit is not something to be left to chance. The efforts expended in preparation for the audit and the handling of the audit can make a very positive impact on the audit results. This chapter will also help the management team and environmental manager to develop an internal audit program. The internal audit program is the best tool available to ensure that the existing environmental programs are adequate and that the facilities are in compliance with existing rules and regulations.

10.2 Regulatory Audits

10.2.1 Preparing for the regulatory audit

Preparations are the key to any successful audit. These preparations must be an ongoing process that is integrated into the daily operations, and not a last-minute act of desperation once it is discovered that an audit has been scheduled. In most cases, the first inkling that the regulatory audit is to be conducted comes when the receptionist calls to say that someone from the state or EPA is in the front lobby and wants to see the responsible party. Even if one is fortunate enough to receive prior warning, most likely there will be only a few days in which to prepare. In today's fast-paced work environment, the luxury of being able to lay everything aside in order to concentrate on getting ready for an audit does not exist.

Environmental compliance is not one person's responsibility. It must be stressed that environmental compliance is everyone's responsibility, from the receptionist to the CEO. In preparing for the audit, everyone must understand their responsibilities and the impact of their jobs on the success of the organization's environmental programs. It takes a team effort to have a truly successful environmental program, and it takes a team effort to have a successful audit.

Preparations for a successful audit should start with the receptionists. A written operating procedure should be prepared which provides detailed instructions on how the auditors are to be received and treated. The receptionist should be instructed to be friendly and helpful. It should be clear who is to be contacted. If the primary contact is out of the office, there should be a second or even a third contact. Since the front lobby is generally the first thing that the auditor will see, it is extremely important that the receptionists give the auditor a very good first impression. The auditors should see an organization that is very professional and willing to help.

Documentation is probably the single most important issue of any audit. How the environmental records are kept and organized is often a direct reflection of the manner in which the environmental programs are maintained. If it is perceived that the records are kept in a lax manner, it may very well be assumed that the environmental programs are also kept in a lax manner. Maintaining the environmental records is an ongoing process that cannot be addressed on an occasional basis or when it is discovered that the auditor has finally arrived for the annual compliance audit. If responsible people have to hunt and search for various permits and records that are haphazardly filed in different locations or lost somewhere on a desk, this can give the auditor a very bad impression. Poor documentation and record archives can also result in possible Notice of Violations or worse if proof of compliance cannot be provided. As a result of the Clean Air Act of 1990, the burden of proving compliance is now placed more and more on industry. Permits are being issued that require certain documentation. If the company cannot readily provide this documentation, then it will be found to be out of compliance.

In preparing for audits, all pertinent records should be maintained in a professional manner and stored in an easily accessible area. They should be well documented and labeled and filed in a logical manner so that the documents can be easily retrieved. All the documents or copies of documents should be stored in a central location that is easily accessible during the audit. It is extremely important during the audit that any information requested by the regulator, such as permit information, background information used in the preparation of permit applications, current and past production records, and any pertinent correspondence concerning the environmental programs, be easily produced. Good documentation that is easily retrieved and comprehensive often makes a very positive and professional impression on the auditor. If the auditor can be shown that the environmental programs are well documented and professionally run, the actual inspection may not be as intensive as it could be. However, during the documentation phase, if there is a problem in finding the documentation requested by the regulator or the information cannot be located in a timely manner, or if the records are kept in a haphazard fashion, then the auditor's impression of the environmental programs may be more negative than it should be and may result in a more aggressive physical inspection.

The environmental files should be kept in a dedicated filing cabinet or in a set of well-marked three-ring looseleaf binders. The filing cabinet or binders should be kept in an area that is accessible to everyone who may need to review the files. Although access to these files is very important, security of these files should also be considered.

In addition to the environmental files, there should be an accurate drawing of the facility layout which shows the location of the various emissions sources and stacks. The purpose of this layout is to show accurate locations of various stacks during the audit and which sources they vent. It is also a very good idea to label the various stacks with a unique stack identification number so that there is no confusion during the audit. It can be very embarrassing during the audit when the stacks cannot be identified or the stack identities become confused. More importantly, the inability to properly identify the stack correctly may damage the credibility and the professional image presented to the auditor. If there is uncertainty about this aspect of the environmental program, in the eyes of the auditor, there may then be questions about the other areas which may require a closer, more intense investigation.

Although the environmental manager may know exactly the purpose of each stack and where to find them, an accurate, updated site plan with stack identification is still needed. The audit may take place when the environmental manager is not available, and a less experienced member of the team may be required to escort the auditor. The site map should allow the less experienced team member to escort the audit team and accurately find and describe the stack in a professional manner.

Another very important reason for maintaining an accurate and updated site plan with stack locations is that it provides the management team with an extremely helpful tool for managing the environmental programs. When there is an operational upset, the site map can help the team members quickly determine the cause of the emissions problem by properly identifying the stack. The emissions can then be brought into compliance either by correcting the problem or by shutting down the equipment until the problem can be identified and corrected.

Production records are vital to maintaining compliance with today's operating permits. Most operating permits today specify that a log of process inputs be kept and maintained for a specified number of years, in order to quantify the emissions of a source. The source is considered noncompliant if these documents cannot be located. Although it is not required that these records be kept in the same location as the environmental files, it is recommended that a hard copy be maintained alongside the environmental files. By maintaining a copy of the production records, you can verify on a regular basis that the necessary documentation is being kept. This will eliminate any surprises during the audit. Another benefit of keeping a copy of production records with the environmental records is that during the audit, the necessary records can be rapidly found and compliance proved. It can be extremely important during the audit not to have to expend a lot of time looking for people and searching for documents. Answer all questions as quickly and accurately as possible, and complete the audit as quickly as possible. The longer the auditor is at the site, the greater the opportunity to ask more questions and look at more areas. The audit should be as easy for the auditor as possible.

Maintenance records are also a vital component in today's operating permits. Environmental control equipment must be maintained in good operating condition if it is expected to efficiently control emissions and meet the permit specifications. Repairs and routine maintenance checks must be documented, and these records must be maintained for a specified number of years. As with the production records, these files are not required to be stored with the environmental files. However, for the same reasons given above with the production records, it is recommended that a hard copy be kept with the environmental files.

It cannot be overemphasized how important it is to maintain all the environmental files and supporting documentation in a neat, well-organized manner. All files should be clearly labeled so that the desired file can be easily located. All management team members should be familiar enough with the files that they can locate specific files during the absence of the environmental manager.

Plant cleanliness and safety are two issues that go hand in hand with the environmental program. As mentioned earlier, first impressions are very important for a successful audit. If a plant is slovenly and there are obvious safety problems, the auditor may find that the same attitude has carried over to the environmental programs. This is another area that requires complete teamwork and is an ongoing process. There will not be enough time or money to do a last-minute cleanup prior to the audit. The facilities must be cleaned daily. All spills must be cleaned up immediately and must not be allowed to accumulate. Waste management and hazardous waste management programs are not within the scope of this book. However, it is extremely important that none of the other environmental programs be overlooked, especially given the increase of multimedia inspections and audits.

It is in the best interest of the facility being audited to make the process as easy on the auditor as possible. On a regular basis, the environmental coordinator should conduct a self-audit. Start by reviewing all current permits. Check that all permits are current and do not require renewal. Review all permit conditions and determine how to prove compliance. Then assemble the necessary documentation; ensure that it is complete and that the required number of years is archived. If calculations are required, be prepared to reproduce them and be able to show the appropriate references. Be able to prove compliance in a very confident and professional manner.

Tour the facilities to check for any unpermitted sources or obvious violations of current permits. During the preliminary tour, determine the best routes that will allow the auditor to observe the various emission sources and at the same time avoid the areas that you prefer he or she not see, such as highly restricted areas, potentially unsafe areas, and other areas. This is not to say that you should deliberately hide anything from the auditor. However, you want to present the best possible picture of the facilities to the inspector while meeting her or his needs and requests.

Be prepared to help the inspector understand the process. Provide simple flow process diagrams to give the auditor a basic understanding of each emission source. Show how the raw and final process materials are stored and how they are transported and handled; show typical process rates, descriptions of emissions with typical emission levels, and the type and location of all emission control equipment. The idea is to provide enough information to facilitate the auditing process. Be careful not to give too much information that can lead to confusion or result in additional questions. Keep the process description as simple as possible, yet containing enough information that the auditor can grasp the basic understanding of the processes and emission sources. During the audit, answer any questions on the process quickly and accurately. If the questions cannot be answered immediately, tell the auditor that you don't know but will find out. Then ask another team member to find out the answer to the auditor's question as soon as possible, preferably prior to the completion of the audit.

The last step in making the audit as easy as possible for the auditor is to assemble all the necessary documentation discussed above. When the auditor asks for proof of compliance, it should be easy to locate the documents.

Again, answer all the auditor's questions as quickly as possible in an accurate and professional manner. The goal is to present an image of a very professional organization that is concerned with maintaining its environmental programs in an effective manner. Do not rush the auditor through the facilities. Answer the questions quickly, help the auditor to understand the process, and demonstrate compliance in a very confident and professional manner.

The regulatory audit is a fact of life. It will happen, and most likely at a time that is most inconvenient. The audit should be anticipated, and the company should be prepared for it at all times. As mentioned above, the audit cannot be prepared for once the receptionist announces that the auditor is in the lobby, waiting. See Fig. 10.2.

10.2.2 Handling the regulatory audit

When the auditor arrives at the facility, the receptionist contacts the appropriate team members in accordance with established written operating procedures for conducting external audits. The auditor should be

---

- Has the receptionist been properly trained in how to receive the regulatory auditor and who is to be contacted?
- Is all environmental documentation up to date?
  - a. Environmental records
  - b. Environmental permits
  - c. Maintenance records
  - d. Production records
  - e. Environmental correspondence file
- Are all files and documents neatly organized and easily accessible?
- Is there an up-to-date and accurate site map?
- Is the plant site clean? Is there any accumulation of trash, waste material, etc.?
- Are there any obvious safety issues that require immediate action?
- Has an internal audit of all environmental programs been conducted within a reasonable time?

Figure 10.2 Environmental audit checklist.

made comfortable until the responsible person is available. Do not allow the auditor to wait in the lobby area for an unreasonable time. The auditor has a limited amount of time and probably resents waiting and wasting time as much as anyone else. It is extremely important to meet with her or him as soon as possible. If the primary contact is unavailable, then the secondary contact should meet with the auditor.

The auditor should be escorted to a comfortable neutral location such as an empty office or conference room where you can meet undisturbed. If possible, do not use someone's office or workspace, the storage location of the environmental files, or a high-traffic area. The meeting should not be interrupted by telephone calls or other requests. Select a neutral location to prevent the auditor's seeing anything on a desk or in the files that may generate additional questions during the audit.

The auditor should be made as comfortable as possible, and the facility manager should then be notified if he or she has not already been informed. The first order of business is to determine the scope of the audit. If the auditor does not specify what she or he needs, then ask specifically what is needed and what the scope of the audit is. Once the scope of the audit has been determined, assemble all the documentation required. It must be stressed that this should be anticipated beforehand, and the documentation should be well organized and quickly obtained. The auditor should not be made to wait for an inordinate time while you search for documents.

Before the auditor is allowed to physically tour the facilities, conduct a safety review to ensure that all safety policies and procedures particular to the facilities are understood. If special personnel safety equipment is required, make sure that the auditor is properly fitted and understands how to use the equipment. Plant and personnel safety should not be placed at risk during any audit process.

During the plant tour, the auditor should be escorted at all times, both for safety reasons and to ensure that no plant processes are interrupted by the auditor's presence. The auditor's escort should be the primary source of answers to all the auditor's questions. In accordance with the previously established scope of the audit, ensure that the auditor receives all the information needed and requested. Do not volunteer additional information unless it reflects positively on the facility. If the designated escort does not know the answer to a question, the escort should get one as soon as possible. The auditor should not be allowed to disrupt production by asking questions of operators and distracting them from their duties.

Pictures should only be allowed in accordance with the facility's established policies. If the auditor requests pictures of a sensitive or secure area, volunteer to take the pictures of the area of concern.

Thus you can ensure that no proprietary processes or information is compromised by the pictures. When the film is processed, request double prints—one set for the facility files and one set for the auditor. Make sure that the prints are properly identified and labeled, including the date and time of the pictures. It is also a good idea to document all production information at the time the pictures are taken: which process lines were in operation, process inputs, production rates, other miscellaneous equipment in operation, and other pertinent information.

10.2.3 Preliminary findings

A postaudit interview to review the preliminary findings should be requested of the auditor at the completion of the inspection. During this brief meeting, the following should be accomplished:

- Determine the auditor's initial impression of the facility's compliance status.
- Determine whether there are any misconceptions concerning the process or operations.
- Find out whether the auditor requires any additional information to complete the audit.
- If additional data are required, determine the time frame for supplying the information.
- Find out whether any situations need immediate attention.
- Discover whether any major or minor discrepancies were uncovered during audit.
- Request a copy of the final audit report.

If possible, the facility's management team should be invited to attend the postaudit interview. This will allow the management team to participate in the audit process and to have a basic understanding of the preliminary findings, thereby eliminating any shocks due to the final audit report. Since the success of the environmental program is produced by a team effort, team leaders must understand their responsibilities, their impact on the outcome of the audit, and where they can help to make improvements.

10.2.4 Official notification of inspection results

Depending on the outcome of the regulatory audit, an official notification of the audit results may or may not be received. If no major or minor problems were found and a final audit report was not requested, no formal notification may be received, depending on the state or agency conducting the audit. However, if problems were found, then a Notice of Violation or other official report will most likely be received. Upon receipt of the audit report, it is imperative to communicate these findings to the appropriate people.

The findings should be reviewed as soon as possible to determine their validity. Was the plant actually out of compliance, or was there possibly a misinterpretation of the data or misconception of the process by the auditor? If the Notice of Violation or other findings are believed to be in error, then a letter should be drafted to the auditing agency describing in detail why it is believed that the facility was actually in compliance.

If the plant is actually in violation of regulations or its permits, then an *action plan* must be developed with a reasonable timetable to accomplish the work that will bring the facility back into compliance. The auditing and/or enforcement agency should be notified of the proposed action plan, and their approval of these plans should be requested. Some negotiations with the regulatory agency may be required in the development of this action plan and of the timetable for the completion of the work to ensure compliance with all regulations and permits. In addition to the Notice of Violations, the facility can be fined. Keep this in mind during the development of the action plan for possible use in negotiating a reduction in or elimination of the fine. Although the potential for a fine reduction may not be very large, it is to the plant's advantage to pursue this option.

The action plan must be implemented as soon as it has been accepted by the regulatory agency. The established timetable must be followed, and routine update reports on the progress of work to bring the facility into compliance must be made to the regulatory agency. Any problems in meeting the action plan should be dealt with immediately, and the regulatory agency should be informed of any changes or problems.

In addition to establishing the above action plan for the correction of existing problems, the existing management plans should be reviewed to determine why these problems were allowed to exist and what actions are needed to prevent similar problems from occurring in the future.

Updated management plans should be implemented, and routine updates provided to the management team on the status of these plans.

The regulatory audit, its findings, and resulting corrective actions should be well documented for future reference. These documents should include at a minimum

- Preliminary findings
- Immediate corrective actions
- Official inspection report with findings
- Action plans, if required
- Management plans, if required
- Capital expenditure summary
- Results of any postinspection negotiations and agreements
- Completion date of items in action plan

10.3   Customer Audits

Customer audits should be treated the same as regulatory audits. The preparations for the customer audit should be no different from those for the regulatory audit. The relationship with the customer audit team may possibly be a bit more relaxed than what is expected with the regulatory audit; however, the seriousness and importance of this audit cannot be overstressed.

As with the regulatory audit, the primary goal of the management team and environmental manager is to ensure that the customer audit team leaves the site with the perception that this is a very professional organization which is very serious about its environmental programs. Good preparations are mandatory to ensure a successful customer audit.

10.4   Internal Audits

As mentioned earlier, one of the best things that can be done to improve the effectiveness of the environmental programs is to implement an internal audit program. The internal audit does several things to assist in the management of these programs. First, it is an early-warning system that can point to potential problems before they blossom into major noncompliance issues. The internal audit can identify trends that may need to be corrected. The internal audit should focus on the importance of the environmental programs and how good teamwork is critical to its success. Internal audits can help

- Obtain management commitment.
- Develop audit plan.
- Develop audit timetable.
- Prepare preliminary audit questionnaire.
- Develop audit protocols.
- Conduct preaudit meeting.
- Conduct audit.
- Conduct postaudit meeting.
- Prepare and distribute final report.
- Develop action plan to address noncompliance issues.

Figure 10.3 Components of an internal environmental audit program.

in the preparation for an audit by a regulatory agency or others out-side the immediate organization and can provide time to correct any deficiencies. Lastly, an internal audit program can be an extremely effective training tool for management and work teams. See Fig. 10.3.

10.4.1 Management commitment

The first step in the development of the internal audit program is to obtain the commitment from management to provide the necessary resources for this program. For the program to be successful, manage-ment must be willing to supply the necessary workforce for the audit team and to pay the expenses that may be incurred during the audit process. It should be understood that this is a long-term commitment and not a one-time event. Management must have a thorough under-standing of why this program is needed and its expectations and goals. And finally, management must be prepared to act on any find-ings uncovered during the audit.

10.4.2 Internal audit plan

Once management has accepted that there is a need for the internal audit and have agreed to commit the necessary resources in work-force and expenses, the next step is to develop an internal audit plan. This plan should contain the following:

Internal audit plan

- Program objectives
- Scope of the audit
- Audit timetable
- Makeup of audit team

- Audit protocols or checklists
- Audit report format
- Procedures for report distribution

The audit plan should be very clear about the specific goals and objectives of the internal audit and what the expectations are of this program. This tool is to be used to help improve the performance of the organization's environmental programs and is *not* to be used to lay blame or in a punitive manner. One of the primary goals of this program is to create the perception in the workforce that the company is very serious about improving its environmental performance, that the workers' input and cooperation are both desired and appreciated, and that workers can provide information without fear of reprisals. If the function of the audit program is to find fault with individuals instead of to identify deficiencies, then the program will rapidly lose the willing cooperation of workers which is so critical to the success of this program.

The scope of the audit should be specified—what facilities will be included and which programs will be reviewed. This will become extremely important as the audit team makes preparations for the inspection.

10.4.3   Audit timetable

An audit timetable should be determined and promulgated. Once the timetable has been established, it should be adhered to as closely as possible. In establishing the timetable, the following items should be included as a minimum:

Audit timetable

- Establish audit date.
- Send out preliminary questionnaire to facility management with required return date.
- Make audit preparations based on above questionnaire.
- Set audit dates.
- Send preliminary audit report to facility management for review.
- Do final audit report.
- Set due date for action plan to address noncompliant areas.

The audit date should be established in accordance with the frequency of the internal audit program. Once the date has been agreed to, it should not be changed without good reason. The internal audit should be treated as if it were being conducted by an outside agency.

If the date is allowed to fluctuate without good reason, workers may perceive that the internal audit is of little value or importance.

10.4.4 Preliminary audit questionnaire

Once the date has been selected and promulgated, the audit team should develop a preliminary audit questionnaire which is then forwarded to the facility management team. The management team should be given a reasonable amount of time to respond to the questionnaire. However, it must be understood by all parties that the deadline must be met and that the questionnaire must be returned by the specified date. The questionnaire should request basic information about the facility so that the audit team can make the necessary preparations for the audit. The following items should be included in the questionnaire as a minimum:

Preliminary audit questionnaire items

1. Facility identification information—name, address, telephone, etc.
2. Facility description—area under roof, number of buildings, site plan if possible.
3. Description of area surrounding the facility, population characteristics, nearest residential area, schools, etc.
4. Management team—names, organizational structure, etc.
5. Technical contact
6. Brief description of process(es)
7. Brief descriptions of air emissions
    a. Number of permits
    b. Summary of emissions, type of controls, type and quantity of emissions
8. Brief description of solid waste stream
    a. Type of solid wastes
    b. Volume of wastes
    c. Method of disposal
    d. Name and quantity of hazardous wastes
    e. Method of disposal of all hazardous wastes
9. Wastewater discharge requirements
    a. NPDES permit?
    b. Description of all wastewater discharges
    c. Description of treatment system
    d. Description of test methods and required reports
10. Stormwater pollution prevention program (SWPPP)
    a. Is SWPPP plan updated?
    b. NPDES permit number and expiration date
11. Number of underground storage tanks and description

12. Potable water source
13. Is the Spill Prevention Control and Countermeasure plan up to date and certified?
14. Number and description of all spills
15. List of all hazardous materials stored on site
16. Has the facility received any regulatory notices violations or citations?
17. Are there any PCBs or PCB-contaminated equipment or materials on-site now, or were there at any time in the past?
18. Has an asbestos survey been conducted at the facility?
19. List of all documents to be made available for the audit: permits, permit applications, correspondence, air monitoring reports, water sampling reports, site plans, material safety data sheets, etc.

A brief letter should be attached to the preaudit questionnaire to explain how the audit will be conducted and what is expected of the facility being audited. The letter should define who will be conducting the audit, when it will take place, approximately how long it will take, what will happen at the conclusion of the audit, and who should be contacted if there are any questions. The letter should explain that the purpose of the preaudit questionnaire is to help prepare the auditors so that the audit can be conducted in such a manner as to minimize the time on-site.

10.4.5   Audit protocols and checklists

Based on the results of the preaudit questionnaire, the auditors should develop a set of protocols or checklists which will cover all appropriate federal, state, and local regulations that apply to the site. If desired, the protocols may be purchased from an outside source. Regardless of whether the protocols are developed in-house or purchased from an outside source, they must cover all areas of concern. The protocols should adequately cover the various requirements and should list the references so that the auditor can quickly look up the specific regulatory requirements.

Once the protocols have been developed, the audit team should assign the protocol responsibilities so that the individual audit team members can begin to prepare for the audit. The specific scope of the audit should be verified with the entire audit team.

In preparation, team members should study their assigned protocols and review all applicable regulations. Although the protocols provide a good background on the regulatory requirements, the actual regulations should be reviewed in their entirety prior to the audit. Based on the above research, auditors should determine the documentation required and develop a list of things that they will need to see, questions that they will need answered—basically how they will conduct the audit. Good auditors will have their homework completed prior to the day the audit is scheduled so that little time is wasted. Because of the resources in workforce committed to this audit process, it is extremely important that all preparations be completed well in advance of the scheduled date.

10.4.6   Conducting the audit

On the scheduled date of the audit, a preaudit meeting should be scheduled for all involved personnel. If the audit team is from another location, then the audit team leader should take time to introduce each member. On-site personnel who are responsible for the various areas should now be assigned to assist the audit team members during audit. The facility's management team should be present at the meeting and prepared to provide their support to the audit process and the audit team. At this time, the audit team leader should go over the proposed scope of the audit and should outline the various needs of the audit team. The audit schedule should be discussed, and any conflicts with the production schedule resolved to everyone's mutual satisfaction. Everyone should understand the proposed audit process and what will be expected; and a time and place should be established for conducting the postaudit meeting.

During the preaudit meeting, time should be set aside to discuss all pertinent safety issues and concerns that audit team members need to be aware of during their inspections. Personnel protection equipment should be provided as necessary. If there are any restricted access areas, these should be identified and discussed. It may be necessary to schedule an alternate date to review these areas during which a shutdown has been scheduled and access does not pose any dangers.

Time should also be taken to answer any questions from the audit team or those whose facilities are being audited. Everyone should have a clear picture of what is happening, what the audit will hopefully achieve, and what is expected of each person to ensure that the results of the audit are meaningful and worth the time and effort being expended.

At the completion of the preaudit meeting, the audit team should immediately start conducting the audit with the assistance of those assigned to escort them. First the auditor will conduct a thorough review of all pertinent documentation such as permits, permit applications, production records, training records as applicable, correspondence with regulatory agencies, and other such resources. The documentation should be well organized and complete without gaps in time or area. Should such gap be found, a complete and detailed explanation should be requested. The documentation will often point to areas of weakness or areas of concern that the auditor will want to consider in greater detail. Often the operating permits require that specific records be kept for a specified time. The auditor should verify that all required records are readily available and that they cover the required number of years. The records should be reviewed in detail to determine if there are periods that exceed the permit limitations. All such discrepancies should be recorded in detail.

Once the documentation has been thoroughly reviewed, the auditor will conduct an on-site inspection of all areas assigned. During this inspection, the auditor will pay close attention to the areas of concern found in the documentation review. The auditor should determine how the production records are maintained and what input the operators have in the production logs. Operators should be interviewed to determine their knowledge and understanding of all pertinent operating permit requirements in their assigned areas. Do they understand what is required and what is expected of them, and are they satisfactorily meeting these obligations? If not, the auditor should probe to determine reasons why the requirements or regulations are not being met.

As mentioned earlier, it is extremely important that everyone understand that the purpose of the internal audit is *not* to find fault with individuals, but to identify problems and shortcomings in the environmental programs. This must be emphasized time and time again if this program is to succeed.

Following the on-site inspection, the auditor should find a quiet space to review the notes and findings. Any findings which are suspected to be in violation of current regulations should be reviewed and confirmed with the applicable federal, state, or local regulations. The findings should be very detailed and the applicable source identified for future reference. It may be necessary at this point to go back and review some of the documentation or to go back into the facilities to confirm the details of the findings. When the audit team has completed their individual assignments, the team leader will then assemble all the auditors in order to review their work and to start putting together the rough draft.

It is important to complete the rough draft while the team is still on-site because the findings are still fresh in the auditor's mind. Once they leave the site or several days have passed, the level of detail that the auditor can recall becomes very limited. Also, after completing the rough draft at the conclusion of the audit, the audit teams can go back into the field to confirm any questionable details. But probably the most important reason to complete the rough draft is that it gets the work completed. Once the auditors go back to their regular duties, the demands of their jobs will naturally delay the completion of the audit.

Once the rough draft is complete, the audit team should prepare for the *exit interview*. The team leader should prepare a summary of all findings in the order of major findings followed by minor findings. The facility's management team and all other responsible personnel should be invited to the closing conference. The team leader should briefly reintroduce the audit team and specify which areas they covered. The goals, objectives, and scope of the audit should be reviewed to ensure that there is complete understanding throughout the organization. The team leader should then provide an overall summary of the audit, emphasizing all areas that were found to be commendable and any significant deficiencies. Then each individual area should be covered, and all findings should be reviewed. The audit team should ensure that all findings are based on the receipt of correct and complete information. Questions concerning the findings should be addressed and satisfactorily answered, if at all possible. All immediate health or safety problems that need to be corrected prior to the publication of the final report should be identified at this time. At the conclusion of the meeting, the facility's management team should be given a copy of the findings or a copy of the rough draft, if possible. This will allow the facility to start working on an action plan to address the various findings.

The final report should be completed as soon as possible after the audit and sent to the facility's manager for review and comment. A time limit for the comments should be established and adhered to as much as possible. Distribution of the final report should be limited to the facility manager or should follow corporate policies. Other distribution should be considered very carefully due to possible legal ramifications.

The facility's management team should now establish a comprehensive action plan with a reasonable and achievable timetable for addressing all discrepancies found during the audit. The action plan and status of all outstanding discrepancies should be reviewed on a regular basis with all appropriate personnel. All management plans should be reviewed and corrected to ensure that existing discrepancies are not recurring problems in the future. The final audit report, action plan, and associated documentation should be filed for future reference.

Part 3

Air Pollution Regulation

Air Quality

William L. Cleland

11.1 Introduction

When the 1990 Clean Air Act Amendments were enacted on November 15, 1990, the result was the most far-reaching authority for EPA regulations in the history of air pollution control. Rules that are in effect and some that will be forthcoming are to be made available with unprecedented access to the federal EPA, nearby states, and the public for the largest polluters. Rules already in place had resulted in setting standards that were thought to do all that was necessary to clean up the air. The strategies that had been used consisted of the following:

- Setting National Ambient Air Quality Standards (NAAQSs)
- Control by the states through state implementation plans (SIPs)
- Setting of new and modified source requirements
- Setting of new source performance standards (NSPSs)
- Setting of National Emission Standards for Hazardous Air Pollutants (NESHAPs)
- Setting of standards for the worst existing sources
- Control of new sources in nonattainment areas

11.2   National Ambient Air Quality Standards

As authorized in the Clean Air Act of 1963, a set of national ambient air quality standards was devised. These standards involved pollutants that were critical to producing harmful effects to public health and welfare. These *criteria pollutants* are

- Particulate matter
- Sulfur dioxide
- Ozone
- Nitrogen oxides
- Carbon monoxide
- Lead

This list is periodically evaluated and may be revised, as required by the act. A detailed summary of these pollutants and their effects can be found in Chap. 2, Basic Air Pollution Theory.

The NAAQSs were developed from the best scientific knowledge available regarding adverse effects on human health and welfare, and the current standards are shown in Table 11.1.

Ambient air monitoring at selected points across the country indicates whether control regions or smaller areas are within or above the NAAQSs. Those that remain below the standards are denoted *attainment areas* and those exceeding the standards are named *nonattainment areas* (NAAs) for each pollutant. Nonattainment areas are listed

TABLE 11.1   National Ambient Air Quality Standards

Pollutant	NAAQS		
	Averaging period	Primary	Secondary
Particulate matter (TSP)	Annual	75 μg/m^3	60 μg/m^3
	24 hr	266 μg/m^3	150 μg/m^3
Particulate matter (PM-10)	Annual	50 μg/m^3	50 μg/m^3
	24 hr	150 μg/m^3	120 μg/m^3
Sulfur dioxide (SO$_2$)	Annual	80 μg/m^3	—
	24 hr	365 μg/m^3	—
	3 hr	—	1300 μg/m^3
Carbon monoxide (CO)	8 hr	10,000 μg/m^3	—
	1 hr	—	40,000 μg/m^3
Nitrogen oxides (NO$_x$)	Annual	100 μg/m^3	100 μg/m^3
Ozone (O$_3$)	1 hr	235 μg/m^3 (0.12 ppm)	235 μg/m^3 (0.12 ppm)
Lead (Pb)	Calendar quarter	1.5 μg/m^3	1.5 μg/m^3

in 40 CFR 52 for each state and are updated periodically. Regulations for NAAs are discussed in detail in later sections.

11.3   State Implementation Plans

The methodology for maintaining attainment areas at or below the NAAQSs and bringing nonattainment areas into attainment was to require states to develop state implementation plans, to be submitted to the Environmental Protection Agency for approval, that outlined the procedures to accomplish this. Many components of SIPs are discussed below. As new laws are passed and, consequently, new rules promulgated, including changes in NAAQSs, the SIPs are required by the EPA to be revised to accommodate these changes.

These general requirements apply to all SIPs:

- Enforceable emissions limitations, control measures, economic incentives, schedules, and compliance timetables for sources that do not meet established standards.

- A permitting and enforcement program for new and modified sources. This should include at least permitting of major sources in attainment areas that ensures the prevention of significant deterioration (PSD) and in nonattainment areas that does not interfere with reasonable further progress (RFP).

- Ambient air quality monitoring and reporting systems.

- Provisions for funding for adequate personnel and statutory authority to implement the requirements of the Clean Air Act.

- Means to address the prevention of interstate and international air pollution.

- Provisions for changing the SIP to account for changes in NAAQSs, for new technologies, and when the EPA has determined that the SIP is deficient for implementing Clean Air Act requirements.

- Necessary monitoring of individual sources.

- Requirements to ensure proper nonattainment planning.

- Adequate ability to model source emissions to predict the effects on attainment.

- Provisions for collecting of permitting fees in order to cover the cost of reviewing, implementing, and enforcing permits. This was partly covered by federal grants prior to the installation of Title V programs, but the 1990 Clean Air Act Amendments (CAAAs) required these programs to be funded entirely by permitting fees.

- Provisions for working in conjunction with local officials.

- Control of emissions of stratospheric ozone-depleting chemicals. This is covered in Title VI of the Clean Air Act Amendments of 1990.

States must include in their SIPs everything necessary to comply with the Clean Air Act. Should the EPA decide, after approving a SIP, that there are deficiencies in this regard, it can issue a "SIP call." Since most SIPs in the past have been a collection of documents rarely organized or assembled in one location, EPA was required by procedures published in 1991 to publish by November 1995 a comprehensive document for each state, setting forth all requirements of the state's implementation plan. A SIP must also be guided by the state's statutory authority. By 1993, each state had revised its air quality statute to comply with federal law, and some are more stringent than federal requirements.

11.4   New and Modified Source Requirements

Control of major new sources was instituted as required by the 1977 Clean Air Act Amendments and was considered to aid attainment areas in preventing significant deterioration to the area so that the NAAQSs could be maintained and not endanger neighboring areas. Major source reviews (PSDs) are made involving the participation of the public, nearby states, federal land managers, and the Environmental Protection Agency. For regulated pollutants exceeding major thresholds, controls determined to be best available control technology (BACT) are to be imposed on a case-by-case basis, taking into account energy, environmental, and economic costs, including the cost of controls.

For sources on a list of 28 specific source categories (see Table 11.2), the major source threshold for each regulated pollutant is 100 tons/yr of potential emissions. For those sources not on this list of source categories, major source status is 250 tons/yr of potential emissions for each pollutant. *Potential to emit* (PTE) is defined as the potential to emit any pollutant when operating at maximum design capacity continuously, i.e., for 8760 h/yr or 365 days/yr, depending on the production or emission rate for which the source is permitted. If control equipment is required on a federally enforceable permit, then PTE can be calculated with the control device in operation. If the permit is not federally enforceable, then calculation of PTE must be made with the source uncontrolled.

Major source reviews are also required for modifications that exceed significant increases of potential emissions for sources that are

TABLE 11.2   Major Source Categories

I	Fossil-fuel-fired steam electric plants of more than 250 million BTU per heat input
II	Municipal incinerators (or combinations thereof) capable of charging more than 50 tons of refuse per day
III	Fossil-fuel boilers (or combinations thereof) totaling more than 250 million BTU per hour input
IV	Petroleum storage and transfer facilities with a total capacity exceeding 300,000 barrels
V	Coal cleaning plants (with thermal driers)
VI	Kraft pulp mills
VII	Portland cement plants
VIII	Primary zinc smelters
IX	Iron and steel mill plants
X	Primary aluminum ore reduction plants
XI	Primary copper smelters
XII	Hydrofluoric acid plants
XIII	Sulfuric acid plants
XIV	Nitric acid plants
XV	Petroleum refineries
XVI	Lime plants
XVII	Phosphate rock processing plants
XVIII	Coke oven batteries
XIX	Sulfur recovery plants
XX	Carbon black plants (furnace process)
XXI	Primary lead smelters
XXII	Fuel conversion plants
XXIII	Sintering plants
XXIV	Secondary metal production plants
XXV	Chemical process plants
XXVI	Taconite ore processing plants
XXVII	Glass fiber processing plants
XXVIII	Charcoal production plants

already major. Also, once a major source threshold for one pollutant is exceeded, all other pollutants exceeding a significant increase must be evaluated and BACT must be applied.

To demonstrate that a new major source will not exceed the NAAQSs or the established allowable increment for the subject pollutant, preconstruction computer modeling for the ambient impact of the pollutant must be done as part of the major source review. There are several EPA-approved computer models available for use as appropriate.

As noted above, notification and availability for comment are required for the public (normally by an advertisement in a local newspaper), nearby states, the National Park Service, and the National Forest Service if the area under their jurisdiction is located within 100 km of the source. EPA also participates in the process in an active oversight capacity of the permitting agency and reviews the agency's preliminary and final determinations, frequently requiring changes before issuance of the major source permit.

For major sources applying to locate in nonattainment areas, the major source review required is referred to as an NSR review, and the control level required is *lowest achievable emission rate* (LAER), which is defined as the most stringent emission limitation which is achieved in practice by a plant in the source category or the most stringent emission limitation which is contained in any SIP, whichever is more stringent. In addition to meeting LAER, new major sources or major modifications in nonattainment areas are required to obtain offsetting emission reductions such that the increases in emissions are offset by an equal or greater reduction in actual emissions from the same source or from other sources in the area. The amount of offset is dependent on the type of pollutant and the severity of the nonattainment area involved. The 1990 Clean Air Act Amendments rated nonattainment areas in severity for ozone, carbon monoxide, and PM_{10}. This is discussed at greater length below.

11.5   New Source Performance Standards

There are certain sources which, in the judgment of the EPA Administrator, are most reasonably expected to endanger public health or welfare. Beginning with the 1970 Clean Air Act Amendments, section 111 required that new sources in these categories be subject to specific standards, referred to as new source performance standards (NSPSs). These standards are promulgated in 40 CFR Part 60 and reflect "the degree of emission reduction achievable through application of the best system of emission reduction." Cost, nonair impacts, and energy requirements are to be taken into account.

The NSPSs are normally expressed as emission limitations, although equipment design, equipment operation, or work practice requirements may be specified. The standards are fixed as they apply to a source category. This distinguishes them from new source review permitting utilization of BACT or LAER, which are case-by-case determinations. NSPSs apply in both attainment and nonattainment areas. BACT and LAER must always be at least as stringent as NSPSs for a particular source category. Sometimes these determinations equal NSPSs.

A list of NSPS source categories and 40 CFR 60 subparts is shown in Table 11.3.

TABLE 11.3   Sources Subject to New Source Performance Standards*

Subpart	Type source
DDD	Fossil-fuel-fired steam generators for which construction commenced after August 17, 1971 (steam generators and lignite-fired steam generators)
Da	Electric utility steam-generating units for which construction commenced after September 18, 1978
Db	Industrial-commercial-institutional steam general generating units
Dc	Small industrial-commercial-industrial steam-generating units
E	Incinerators
Ea	Municipal waste combustion
F	Portland cement plants
G	Nitric acid plants
H	Sulfuric acid plants
I	Asphalt concrete plants
J	Petroleum refineries (all categories)
K	Storage vessels for petroleum liquids constructed after June 11, 1973 and prior to May 19, 1978
Ka	Storage vessels for petroleum liquids constructed after May 18, 1978 and prior to July 23, 1984
Kb	Volatile organic liquid storage vessels constructed after July 23, 1984
L	Secondary lead smelters
M	Secondary brass and bronze ingot production plants
N	Iron and steel plants
Na	Secondary emissions from basic oxygen process steelmaking facilities constructed after January 20, 1983
O	Sewage treatment plants
P	Primary copper smelters
Q	Primary zinc smelters
R	Primary lead smelters
S	Primary aluminum reduction plants
T	Phosphate fertilizer industry: wet process phosphoric acid plants
U	Phosphate fertilizer industry: phosphoric acid plants
V	Phosphate fertilizer industry: diammonium phosphate plants

TABLE 11.3   Sources Subject to New Source Performance Standards* (Continued)

Subpart	Type source
W	Phosphate fertilizer industry: triple superphosphate plants
X	Phosphate fertilizer industry: granular triple superphosphate
Y	Coal preparation plants
Z	Ferroalloy production facilities
AA	Steel plants: electric air furnaces
Aaa	Electric arc furnaces and argon-oxygen decarburization vessels in steel plants
BB	Kraft pulp mills
CC	Glass manufacturing plants
DD	Grain elevators
EE	Surface coating of metal furniture
GG	Stationary gas turbines
HH	Lime plants
KK	Lead acid battery manufacturing plants
LL	Metallic mineral processing plants
MM	Automobile and light-duty truck surface coating operations
NN	Phosphate rock plants
PP	Ammonium sulfate manufacturing plants
QQ	Graphic art industry publication rotogravure printing
RR	Pressure-sensitive tape and label surface coating operations
SS	Industrial surface coating: large appliances
TT	Metal coil surface coating
UU	Asphalt processing and asphalt roofing manufacturing
VV	Equipment leaks of volatile organic compounds in synthetic organic chemical manufacturing industry
WW	Beverage can surface-coating industry
XX	Bulk gasoline terminals
AAA	New residential wood heaters
BBB	Rubber tire manufacturing industry
DDD	VOC emissions from the polymer manufacturing industry
FFF	Flexible vinyl and urethane coating and printing
GGG	Equipment leaks of VOC in petroleum refineries
HHH	Synthetic fiber production facilities
III	VOC emissions from the synthetic organic chemical manufacturing industry (SOCMI) air oxidation unit processes
JJJ	Petroleum dry cleaners

TABLE 11.3   Sources Subject to New Source Performance Standards* (Continued)

Subpart	Type source
KKK	Equipment leaks of VOC from onshore natural gas processing plants
LLL	Onshore natural gas processing plants: SO_2 emissions
NNN	VOC emissions from SOCMI distillation oprerations
OOO	Nonmetallic mineral processing plants
PPP	Wool fiberglass insulation manufacturing plants
QQQ	VOC emissions from petroleum refinery wastewater systems
SSS	Magnetic tape coating facilities
TTT	Industrial surface coating: surface coating of plastic parts for business machines
VVV	Polymeric coating of supporting substrates facilities

*Sources are normally required to test to show compliance with the appropriate standards.

11.6   National Emission Standards for Hazardous Air Pollutants

Originally, section 112 of the Clean Air Act involved a health risk–based program of control for only seven substances: asbestos, beryllium, mercury, radionuclides, benzene, vinyl chloride, and inorganic arsenic. The resulting regulations are promulgated in 40 CFR Part 61. This program is viewed by many as being a failure. The 1990 CAAAs changed the approach to one that is technology-based and applies to many more hazardous air pollutants, originally a list of 189 substances. Both risk-based and technology-based standards are referred to as National Emission Standards for Hazardous Air Pollutants (NESHAPs). In the case of the new rules, the standards imposed are also called *maximum-achievable control technology* (MACT) for major sources and *generally available control technology* (GACT) for small or area sources. Regulations promulgated for the new rules appear at 40 CFR Part 63. The entire Chap. 13 covers the NESHAPs.

11.7   Existing Sources

While the main thrust for attaining and maintaining NAAQSs is the control of new sources, beginning with the 1977 amendments to the act, an existing major source in a nonattainment area is required to meet *reasonably available control technology* (RACT) if it is in a source category to which a standard applies. RACT is an evolving standard that is established by the state. EPA has defined RACT as the lowest emission limitation that a particular source is capable of meeting that is reasonable, considering the technological and economic feasibility. Therefore, given site-specific considerations, such as geographic constraints, RACT can differ for similar sources.

RACT is developed for a source from control technique guidelines (CTGs) that are published by the EPA for source categories. The practice has been to require states to have in their SIPs a requirement for setting RACT for sources in nonattainment areas to assist in bringing them into attainment, and only on source categories for which a CTG had been issued. Some states have taken RACT a little further and have required all new and existing major sources even in attainment areas to meet RACT for their source category, as a way to maintain attainment status.

Depending on the severity of a nonattainment area, the date on which the CTG is issued determines whether a source must meet RACT. The 1990 CAAAs required sources in marginal nonattainment to meet RACT only if the CTG for that source category was issued prior to 1990. Sources in more severe NAAs must meet the appropriate RACT no matter when the CTG is issued. Also, major sources of volatile organic compounds in any ozone NAA more severe than marginal or in the ozone transport region must meet RACT even if a CTG has not been issued for the source category.

States are required to submit information to the EPA about RACT that has been set. This information is found in the RACT/BACT/LAER Clearing House maintained by the EPA at Research Triangle Park, N.C.

11.8   Nonattainment Area Requirements

Each nonattainment program is expected to demonstrate progress toward attainment. Normally *reasonable further progress* (RFP) is defined as a percentage reduction in emissions which the SIP must achieve in order to reach attainment by a certain date. For some classifications, it is left to the state to define how progress is to be measured.

Obviously this begins by accomplishing a comprehensive, accurate, and current inventory of actual emissions of pollutants which contribute to the nonattainment condition. Pollutants which are precursors to the nonattainment pollutant must be included. Examples are volatile organic compounds and oxides of nitrogen, which are precursors for ground-level ozone. Growth of industry must also be taken into account when progress toward attainment is shown.

TABLE 11.4   Nonattainment Area Major Source Threshold

Ozone nonattainment area	Major source threshold, VOC tpy*	Major source threshold, NO_x tpy*	Minimum emissions offset ratio required
Extreme	10	10	1.5 to 1†
Severe	25	25	1.3 to 1
Serious	50	50	1.2 to 1
Moderate	100	100	1.15 to 1
Marginal	100	100	1.1 to 1

*tpy = tons per year.

†The minimum ratio is reduced to 1.2 to 1 if the SIP requires all VOC and NO_x major sources to use BACT.

The most important consideration in determining whether an area is progressing toward attainment is that new or modified sources in a nonattainment area must not be built until offsets in a greater than 1-to-1 ratio are obtained. A company wanting to build a new source or to modify an existing source must plan early in order to reduce emissions, by shutting down old sources or controlling the emissions from them. Sometimes it is possible to locate other sources within the company or outside the company from which offsets can be bought or otherwise obtained.

The more serious a nonattainment area is, i.e., the more it is above the NAAQSs, the higher the offset ratio required. Shown in Table 11.4 are ozone nonattainment classifications, major source thresholds, and offset ratios required by the 1990 CAAAs.

12

Mobile Sources

Charles Gallagher

12.1 Introduction

Title II of the Clean Air Act (CAA) of 1990 regulates emission standards for moving sources. Nine parts of the Title 40 Code of Federal Regulations (CFR) now regulate emission standards for moving sources.* These nine parts cover registration and regulation of fuels and fuel additives, emissions from motor vehicles, aircraft, clean-fuel vehicles, nonroad engines, and marine engines.

12.2 Emission Standards for New Motor Vehicles or New Motor Vehicle Engines

The EPA was charged by Congress to regulate standards applicable to the emission of any air pollutant from any class or classes of new motor vehicles or new motor vehicle engines, which cause or contribute to air pollution which may reasonably be anticipated to endanger public health or welfare. Such standards are applicable to vehicles and engines over the course of their useful life.

12.3 Prohibited Acts

The CAA prohibits any manufacturer or person from introducing any vehicle or engine for sale in the United States that has not been

---

*The following CFR Title 40 parts regulate emission standards for moving sources: Parts 79, 80, and 85 to 91.

issued a certificate of conformity as outlined by the compliance testing and certification section of the act (see Sec. 12.5). Manufacturers are required to keep records and paperwork on their vehicles and engines (see Sec. 12.7). No person or manufacturer may fail or refuse to permit the EPA access to such information upon request. No person may remove, tamper, or render inoperative any compliance device or élement of design installed on or in a motor vehicle or engine.

12.4   Civil Penalties

Any person who violates sections prohibited by the act may face penalties ranging from $2500 to $25,000 for each offense. In general, the $25,000 penalties apply to violations by manufacturers or dealers. The $2500 penalties apply to an individual tampering with a single vehicle as an isolated incident. Any civil action brought against an offender requires that the amount of the civil penalty be determined by the gravity of the violation, the economic benefit or savings (if any) resulting from the violation, the size of the violator's business, and the violator's history of compliance with the CAA.

12.5   Motor Vehicle and Motor Vehicle
Engine Compliance Testing and Certification

The EPA may test or require any new motor vehicle or engine to be tested for compliance with the regulations set forth by the act (see Sec. 12.2). If the vehicle or engine conforms to the regulations, the EPA will issue a certificate of conformity for the vehicle or engine. If the EPA later determines that a vehicle or engine being manufactured no longer conforms to the regulations under which the certificate was issued, the certificate can be revoked or suspended. Any manufacturer may request that the EPA grant the manufacturer a hearing as to whether the tests have been properly conducted or any sampling methods have been properly applied, and make a determination on the record with respect to any suspension or revocation. If the manufacturer is in disagreement with the EPA over any results, the manufacturer may seek judicial review by the U.S. Court of Appeals within 60 days. If a manufacturer finds that a vehicle or engine does not meet the requirements necessary to receive a certificate of conformity, the manufacturer may pay a nonconformance penalty in order to receive the certificate. No such penalty is allowed to be paid in order to receive the certificate of conformity if the nonconforming vehicle or engine is exceedingly out of compliance.

12.6 Compliance by Vehicles and Engines in Actual Use

Manufacturers are required to warrant their vehicles and engines to the ultimate purchaser and each subsequent purchaser that the vehicle or engine conforms to the regulations set forth by the act (see Sec. 12.2) and will continue to do so over the course of its useful life. Any replacement part, initially designed for emission controls as necessary to meet the regulations, which is scheduled for replacement over the useful life of the vehicle or engine must not exceed 2 percent of the retail price of the vehicle. If the 2 percent limit is exceeded, the manufacturer will bear the cost of replacement. If the EPA determines that a group or class of vehicles already in use does not conform to the requirements of the act, the manufacturer shall be required to submit a plan to remedy the problem. The manufacturer will be required to bear the cost of any necessary repairs. A graduated performance schedule for manufacturing standards applicable to emissions control of nonmethane hydrocarbons (NMHCs), carbon monoxide (CO), and NO_x was set forth by the CAA. This schedule will be in full effect by 1998 for light-duty vehicles (less than 6000 lb) and by 1999 for light-duty vehicles (more than 6000 lb).

12.7 Information Collection

Manufacturers of new motor vehicles or engines are required to maintain records, perform tests, make reports, and provide information to the EPA as necessary to make a determination of whether the vehicle or engine conforms to the requirements of the act. Any records or information provided to the EPA shall be made available to the public unless the manufacturer can show that release of such information would divulge trade secrets.

12.8 State Standards and Grants

In general, states are prohibited from adopting or attempting to enforce any standard relating to the control of emissions from new motor vehicles or engines which are subject to regulation by the CAA. However, upon approval by the EPA, a state may adopt standards for the control of emissions from new motor vehicles or engines if the standards are deemed to be at least as protective of health and welfare as the federal standards would otherwise be. In a similar manner, no state is allowed to adopt or attempt to enforce regulations for either nonroad engines or vehicles deemed as construction or farming equipment under 175 hp; nor may any standard be enforced for loco-motives. In the case of any other nonroad equipment, the state may enforce additional standards as approved by the EPA. One exception is California, which is allowed to operate under slightly different rules. The reader is referred to section 209 of the original act for fur-ther reading in regard to California. The EPA may make grants to appropriate state agencies in an amount up to two-thirds of the cost of developing and maintaining effective vehicle emission devices and systems inspection and emission testing and control programs.

12.9 Regulation of Fuels

The EPA was granted the power to regulate all fuel or fuel additives produced for sale or commerce. All manufacturers must register their fuels or fuel additives with the EPA. The manufacturer of any fuel is required to notify the EPA of the commercial identifying name and manufacturer of any additive contained in a fuel. The range of concen-tration of any additive in the fuel, the purpose-in-use of any such addi-tive, and the chemical composition of the additive must also be provid-ed. The EPA may request that the manufacturer of any fuel or additive conduct tests to determine potential public health effects of the fuel or additive. The manufacture and commerce of any fuel or fuel additive may be prohibited if the EPA finds that the fuel or additive causes, or contributes to, air pollution which may reasonably be anticipated to endanger the public health or welfare. Additionally, prohibitions on the manufacture and commerce of any fuel or fuel additive may occur if the emission products of the fuel or additive are found to impair to a significant degree the performance of any emission control device or system which is in general use. The act prohibits any person from using leaded fuel in a vehicle or engine intended by the manufacturer for unleaded fuel use. Furthermore, a maximum sulfur content has been established for diesel fuels. The act charged the EPA with pro-mulgating guidelines for *reformulated gasoline* to be used in nonat-tainment areas. Reformulated gasoline is intended to reduce emissions of ozone-forming volatile organic compounds and the emissions of toxic air pollutants. CFR Title 40, Part 80, subpart D specifically outlines the regulations for reformulated gasoline promulgated by the EPA.

12.10 Nonroad Engines and Vehicles

The EPA was charged by Congress to determine the extent to which nonroad vehicles and engines contribute to the emissions of carbon monoxide, oxides of nitrogen, and volatile organic compounds. If these contributions are found to be significant to the endangerment of public health and welfare, the EPA may impose regulations which require new nonroad vehicles and engines to meet stricter emission requirements. The act has also required the EPA to issue regulations which cover emission standards for locomotives. The reader is referred to CFR Title 40 Part 89 for complete reference to the control of emissions from new and in-use nonroad engines.

12.11   High-Altitude Performance Adjustments

The CAA requires that vehicle and engine manufacturers submit instructions for the products they sell which outline adjustments that should be made to vehicles operating in high-altitude areas. These adjustments are required to ensure that emission control performance is maintained at both high and low elevations. Furthermore, Congress instructed the EPA to establish a testing center in a high-altitude location which could determine if the regulations and intentions of the act were being achieved by vehicles and engines in high-altitude locations.

12.12   Study and Report on Fuel Consumption

Following each motor vehicle model year, the EPA is required to report to Congress respecting the motor vehicle fuel consumption associated with the standards applicable for the immediately preceding model year.

12.13   Motor Vehicle Compliance Program Fees and Prohibition on Production of Engines Requiring Leaded Gasoline

The CAA allows the EPA to collect fees necessary to recover all reasonable costs incurred during the certification of conformity, testing, and compliance monitoring of a new or in-use vehicle or engine. Such fees are applicable to both foreign and domestic manufacturers. The EPA was instructed to prohibit the manufacture, sale, or introduction into commerce of any engine that requires leaded gasoline. The prohibition has affected all vehicles produced beyond the 1992 model year.

12.14   Urban Bus Standards

Emission standards have been promulgated for urban buses manufactured for the 1994 model year and thereafter. A 50 percent reduction of particulate matter from the baseline set in 1990 was required by Congress. The EPA could decrease this reduction if it determined that such a reduction was not technologically feasible. If the EPA determines that urban buses do not meet the particulate matter standards over the course of their useful life, the EPA may require that any new vehicles purchased utilize low-polluting fuels. Furthermore, urban buses operating in areas with populations greater than 750,000 (1980 census) which replace or rebuild their engines must utilize retrofit technology to achieve emission standards specifically outlined by the EPA for retrofit and rebuilt vehicles. The reader is referred to CFR Title 40, Part 85, subpart O for in-depth coverage of retrofit and rebuild requirements.

12.15   Standards for Light-Duty Clean-Fuel Vehicles

In an attempt to reduce tailpipe emissions, Congress charged the EPA with promulgating standards for clean-fuel vehicles. Such vehicles would run on what Congress has termed a *clean alternative fuel,* being more specifically defined as any fuel (that includes methanol, ethanol, or other alcohols or any mixture thereof containing 85 percent or more by volume of such alcohol along with gasoline or other fuels, reformulated gasoline, diesel, natural gas, liquefied petroleum gas, and hydrogen) or power source (including electricity). Standards for light-duty clean-fuel trucks (up to 6000-lb gross vehicle weight) and light-duty clean-fuel vehicles are broken into two phases. The first phase was implemented in 1996. The second phase will take effect for the 2001 model year. Standards for light-duty clean-fuel trucks (greater than 6000-lb gross vehicle weight) will take effect for the 1998 model year. Flexible and dual-fuel vehicles which can operate on standard fuels as well as clean-fuel alternatives will be required to meet clean-fuel standards specific to dual-fuel vehicles set forth by the EPA for the 1996 model year as well as appropriate emission standards when operating on nonalternative fuel. For specific emission standards for light-duty clean-fuel vehicles, the reader should refer to CFR Title 40, Part 88, subpart A.

12.16   Standards for Heavy-Duty Clean-Fuel Vehicles

The CAA has required heavy-duty clean-fuel vehicles or engines manufactured for the model year 1998 and thereafter and having a gross vehicle weight rating (GVWR) greater than 8500 lb and up to 26,000 lb to follow standards for emissions of oxides of nitrogen (NO_x) and nonmethane hydrocarbons (NMHCs). There will be no standards for vehicles weighing more than 26,000 lb. For specific emission standards for heavy-duty clean-fuel vehicles, the reader should refer to CFR Title 40, Part 88, subpart A.

12.17 Centrally Fueled Fleets

Nonattainment areas for ozone or carbon monoxide are to implement fleet vehicle requirements as part of their state implementation plans. The plan would require fleet operators (including federal) to operate a portion of their fleet as clean-fuel vehicles. This requirement begins in model year 1998 at 30 percent and rises to 70 percent by the model year 2000 for light-duty trucks. The requirement for heavy-duty trucks is 50 percent beginning in model year 1998 and remains constant for later-model years. Fleet operators who purchase and operate larger percentages of clean-fuel vehicles may be issued credits. Credits may then be traded or sold for use by any other person to demonstrate compliance with the clean-fuel fleet requirements for nonattainment areas. Credits may also be held or banked for later use. Credits may be issued to fleet operators who use ultra-low-emission vehicles or zero-emission vehicles. Such vehicles operate at standards more stringent than would otherwise be typically applied to clean-fuel fleet vehicles. Credits may be issued on a "weighted" formula basis that accounts for the amount of extra emission reductions achieved by the vehicle. For specific emission standards for clean-fuel fleet vehicles, the reader should refer to CFR Title 40, Part 88, subpart C.

12.18 Vehicle Conversions

The requirements for clean-fuel fleets discussed above may be met by converting existing or new gasoline or diesel-powered vehicles to clean-fuel vehicles which comply with the applicable standards for clean-fuel vehicles. For specific emission standards for conversion requirements to clean-fuel vehicles, the reader should refer to CFR Title 40, Part 88, subpart C.

13

Hazardous Air Pollutants

William L. Cleland

13.1 Introduction

Title III of the 1990 Clean Air Act Amendments (CAAAs) changed the direction of the hazardous air pollution program. The existing program which began in 1970 had only addressed eight pollutants with regulations based on pollutant-by-pollutant risk to health. Section 112 of the CAAAs listed 189 hazardous air pollutants and defined a new process with regulations based on technology for source categories. A later phase of the new program was to bring in risk-based regulations after the technology-based program was in place and evaluated.

The list of hazardous air pollutants can be found in section 112(b) of the act. If a stationary source emits 10 tons/yr of any one pollutant or 25 tons/yr of any combination of these listed pollutants, it is considered a major source for Title III and is subject to the regulations. A list of 174 industry categories was initially issued on July 16, 1992. Standards called *maximum achievable control technology* (MACT) for each major source category were to be developed by EPA and promulgated over a 10-year period. Source categories were divided into four groups, based on seriousness of impact, with target years for MACT regulations of 1992, 1994, 1997, and 2000.

In addition to major sources, it was found that small, widely dispersed emissions of hazardous air pollutants also presented significant risks to public health. These are called *area sources* and are to be subject to less stringent standards, called generally available control technology (GACT), which are scheduled to be promulgated by November 2000.

Many of the regulations that were scheduled to be promulgated in 1992 and 1994 were late due to the heavy workload on EPA in a very short time. Consequently, many of these early regulations were mandated by the courts to meet slightly later deadlines. The EPA for the most part has been able to meet these mandated dates. At this writing, EPA reports that most of the MACT and GACT standards scheduled for 1997 and 2000 will be proposed by mid-1997.

There are provisions in the 1990 Title III for EPA to promulgate regulations to prevent accidental releases of hazardous air pollutants. Owners and operators producing, processing, handling, or storing at least 10,000 lb of pollutants listed in section 112(r) of the act will be required to develop a *risk management plan* (RMP). This plan is quite similar to the *process safety management* (PSM) plan already required by OSHA. It will include (1) identification of hazards resulting from a release, (2) the design and maintenance of a safe facility, and (3) minimization of consequences of accidental releases.

13.2   List of Hazardous Air Pollutants

A list of 189 hazardous air pollutants (HAPs) is found in the 1990 CAAAs in section 112(b). It contains organic chemicals, pesticides, metals, coke oven emissions, fine mineral fibers, and radionuclides. It does not contain criteria pollutants but does include precursors to criteria pollutants. This list is subject to review and can be changed from time to time by EPA. At this writing, the only change so far is the deletion of the compound Caprolactam. The original list of section 112(b) HAPs is provided in App. G.

13.3   List of Source Categories

Sources that emit one or more hazardous air pollutants were required to be listed by section 112(c) of the act. EPA published an initial list of 174 source categories on July 16, 1992. Of these, 166 were major sources and 8 were area sources. These lists are shown in App. H.

A *major* source, for the purposes of HAP regulation, is an individual or group of buildings, structures, or installations within a contiguous area under common control that emits or has the potential to emit 10 tons of any one or 25 tons of any combination of HAPs. An *area* source is anything else that emits HAPs, excluding mobile sources.

13.4 Maximum Achievable Control Technology Standards

The MACT standards apply to both new and existing sources of hazardous air pollutants. EPA is to consider health threshold levels that are established for particular compounds and elements. Standards may be based on any measures, processes, methods, systems or techniques, process changes, material substitutions, enclosures, collection and treatment of emissions, changes in work practices, operator training, or any combination of these. Cost of changes required is to be taken into consideration when setting control limits.

The MACT standard for a new source is to be as stringent as the best controlled similar source in its category has achieved in practice. A *new source* is defined as one that is constructed after the effective date of the rule promulgated for the source category. MACT standards for existing sources cannot be less stringent than the average of the best-performing 12 percent of existing sources in categories that contain 30 or more sources. If the category contains less than 30 existing sources, then the existing source MACT is set at the average limitation of the best 5 sources.

General provisions that apply to all MACT sources were promulgated on March 16, 1994, and contain the following common elements:

- General maintenance and operating procedures
- Performance testing and monitoring requirements
- Compliance report frequency
- Record-keeping requirements
- Allowance for extension of compliance for early reductions
- Notification requirements for construction of new or modified sources
- Description of the MACT case-by-case determination process
- General definitions

This regulation is shown in App. I.

13.5 Schedule for Promulgating MACT Standards

The original schedule for EPA to issue MACT standards was set forth in section 112(e) of the act. As noted earlier, many of these deadlines were not met. The general timetable was to have been as follows:

The 40 highest-priority source categories	11/15/92
Coke oven batteries	12/31/92
25 percent of listed categories	11/15/94
Another 25 percent of listed categories	11/15/97
Remaining categories	11/15/00

The factors that EPA was to consider in setting forth a source-specific schedule were (1) the adverse effects of HAPs on public health and the environment, (2) the quantity and locations of HAPs that the source category could emit, and (3) the efficiency of grouping categories or subcategories according to type of HAP or technology available.

A source-specific schedule for promulgating MACT standards was published on December 3, 1993, at 58 FR 63,941. This information is shown in App. J. Many of these dates have also been missed and replaced with court-mandated dates. Standards proposed and issued as of this writing are also shown in App. J. This is an ongoing process, but EPA appears to be catching up and should be on schedule by the end of 1997.

13.6   New and Modified Sources of Hazardous Air Pollutants

A new or modified source is prohibited from start-up after a state's Title V plan is approved unless it will meet MACT standards for its source category. If a standard has not been promulgated, the source must meet case-by-case MACT, required by section 112(g) of the act. A proposed rule for case-by-case MACT was published on April 1, 1994, at 59 FR 15,504. Since that time, there has been much controversy as to when this would go into effect. The current interpretation by EPA is that case-by-case MACT will not go into effect until the final rule is promulgated.

A new or modified source is required to meet a promulgated MACT standard on start-up. If a standard has been proposed but not promulgated, a new source would be required to meet the standard within 3 years.

An exception to case-by-case MACT was established in the April 1, 1994, guidance for de minimis level increases. The list of de minimis levels is shown in App. K.

13.7 Voluntary Early Reduction

An existing source that has applied BACT or LAER , prior to promulgation of the MACT to which it is subject, will not be required to meet MACT for a period of 5 years following the installation of BACT/LAER equipment. A 6-year extension could also have been obtained provided the source voluntarily reduced HAP emissions by 90 percent prior to proposal of its MACT. However, this 90 percent reduction must have been achieved prior to January 1, 1994.

13.8 Prevention of Accidental Releases

The purpose of section 112(r) of the act is to prevent the accidental release of substances known to cause death, injury, or serious adverse effects to human health or the environment. The most famous incident of this type was the release of methyl isocyanate in 1984 at Bhopal, India, which killed thousands of people.

The rule for accidental releases was promulgated on June 20,1996, at 61 FR 31668 and appears at 40 CFR Part 68. It imposes general rules on owners or operators producing, processing, handling, or storing listed substances. These rules consist of (1) identifying hazards which may result from accidental releases, (2) designing a safe facility and responding to accidental releases, and (3) minimizing the consequences of accidental releases that do occur. Any facility for which this rule applies is required to prepare a risk management plan (RMP) to be completed by June 20, 1999, or 3 years after the addition of any new substance to the list. This plan is similar in many respects to the process safety management plan already required by OSHA. The accidental releases rule is contained in App. L. The list of 77 toxic substances and 63 flammable gases and liquids was promulgated on January 31, 1994, at 59 FR 4478 and was amended on June 20, 1996, at 61 FR 31730. This final list is contained in App. L.

14

Acid Rain

Charles Gallagher

14.1 Introduction

The Clean Air Act of 1990 passed by Congress stated that the presence of acidic compounds and their precursors in the atmosphere and in deposition from the atmosphere represents a threat to natural resources, ecosystems, materials, visibility, and public health. This problem was recognized to have national and international significance. The principal sources of acidic compounds and their precursors in the atmosphere are emissions of sulfur and nitrogen oxides from the combustion of fossil fuels. With control technologies feasibly available at the time of passage of the 1990 Clean Air Act, Congress set forth to reduce the adverse effects of acid deposition through reductions in annual emissions of sulfur dioxide and nitrogen oxides. Many of these reductions were targeted to reduce precursor emissions from steam-electric generating units throughout the contiguous 48 states and the District of Columbia. The goal of the subchapter (Acid Deposition Control, subchapter IV-A of the 1990 Clean Air Act) was to reduce 1980 emissions levels of sulfur dioxide and nitrogen oxides by 10 million and 2 million tons, respectively.

A key aspect of the method of reaching that goal is embedded in emissions trading. Each source is to be allocated a certain emissions cap. If the source is over the cap, the emissions must be reduced by either switching to cleaner alternative energy sources, installing better pollution control equipment, or purchasing emissions allowances from another source that is not using its entire allowance. The trading mechanism in theory allows the combined group of all sources to achieve maximum emission reductions collectively at the lowest cost.

For instance, if company A has old and outdated pollution controls, yet is operating under its emissions limitation, it may wish to trade the remainder of its emission allowance. At the same time, company B is very large and has implemented excellent pollution control technologies to date. However, company B still finds itself over its emissions allowance cap. If company B implements additional control technologies to its facilities, it may achieve only minor reductions in overall emissions, since its current control strategy has already removed the vast majority of the pollution that company B generates. Furthermore, reduction targeted toward the small remaining fraction of a particular source's emissions is often expensive to achieve. The same amount of money could be used to upgrade company A's control equipment, which would result in a greater reduction in emissions from the perspective of company A combined with company B. In turn, company A would turn over all or a portion of the net savings in emission allowances to company B. These additional emissions allowances would be used by company B to meet its total emissions needs. The marketplace would dictate a price for emissions allowances, and companies would be free to choose how they would individually meet their respective emissions allowance caps.

14.2   Sulfur Dioxide Allowance Program for Existing and New Units

For the purposes of subchapter IV-A (Acid Deposition Control), existing units are units in operation prior to November 15, 1990. New units are units which commenced operation on or after November 15, 1990. The EPA allocated annual emissions allowances for existing units in accordance with phase I and phase II sulfur dioxide requirements (see Secs. 14.3 and 14.4). By January 1, 2000, the EPA will not be allowed to allocate total sulfur dioxide emissions in excess of 8.90 million tons. To achieve this goal, the basic phase II allowance allocation may be reduced, pro rata, by the EPA as necessary to attain the 8.90 million ton sulfur dioxide emissions limitation. A proposed list of phase II allocations has been published by the EPA, but currently is incomplete. (*Note:* An excellent source of up-to-date listings can be found on the Internet at the following address: http://www.epa.gov/ardpublc/acidrain.) The revised phase II allocation list is to be published by EPA no later than June 1, 1998. Allowances allocated by the EPA may be transferred among designated representatives of the owners or operators of affected sources as provided by the allowance system regulations (see 40 CFR) promulgated by the EPA. The regulations prohibit the use of any allowance prior to the calendar year for which the allowance was allocated. Unused allowances are to be identified by the EPA and carried forward and added to allowances allocated in subsequent years. Transfers of allowances are not effective until written certification of the transfer, signed by a responsible official of each party to the transfer, is received and recorded by the EPA. In the event that an existing unit is removed from commercial operation, the emission allowances granted to the unit are not to be forfeited. The owner or operator of the removed unit still retains the rights to the unit's emission allocations.

The EPA was also charged by Congress to evaluate the environmental and economic consequences of amending subchapter IV-A (Acid Deposition Control) to permit trading of sulfur dioxide allowances for nitrogen oxides allowances. After January 1, 2000, it shall be unlawful for a new utility unit to emit an annual tonnage of sulfur dioxide in excess of the number of allowances to emit held for the unit by the unit's owner or operator. New utility units may obtain their allowances either through trading mechanisms promulgated by the EPA or through an allowance based upon annual fuel consumption of the unit (see Sec. 14.9). The nature of any allowance granted to any unit is not to be construed as an absolute property right. The allowance may be terminated or limited by the authority of the United States.

14.3 Phase I Sulfur Dioxide Requirements

On January 1, 1995, phase I sulfur dioxide emission allowance requirements went into effect. The affected sources are listed in Table 14.1. Each unit (referenced by utility plant name and generator number) at a source has an individual sulfur dioxide emissions allowance. Each allowance unit is equivalent to 1 ton of sulfur dioxide emissions per calendar year. The unit may exceed the emission allowance as stated in Table 14.1 for a calendar year only if (1) the emissions reduction requirements applicable to the unit have been achieved (to be discussed later) or (2) the owner or operator of the unit holds additional allowances obtained through the trading system. The EPA was instructed by Congress to establish a reserve of sulfur dioxide emission allowance allocations not to exceed 3.50 million tons. The size of the reserve was based upon the total tonnage of reductions in emissions of sulfur dioxide from all utility units in the calendar year 1995 that occurred as a result of compliance with the emission limitations requirements from phase I. The reserve is to be allocated to phase I units until the reserve is depleted or December 31, 1999, whichever comes first. To receive a reserve allocation, the utility unit must be an "eligible phase I extension unit" (to be discussed later, this section). A special additional emission allowance allocation was granted to cer-

TABLE 14.1   Affected Sources and Units in Phase I and Their Sulfur Dioxide Allowances, tons

State	Plant name	Generator	Phase I allowances
Alabama	Colbert	1	13,570
		2	15,310
		3	15,400
		4	15,410
		5	37,180
	E. C. Gaston	1	18,100
		2	18,540
		3	18,310
		4	19,280
		5	59,840
Florida	Big Bend	1	28,410
		2	27,100
		3	26,740
	Crist	6	19,200
		7	31,680
Georgia	Bowen	1	56,320
		2	54,770
		3	71,750
		4	71,740
	Hammond	1	8,780
		2	9,220
		3	8,910
		4	37,640
	J. McDonough	1	19,910
		2	20,600
	Wansley	1	70,770
		2	65,430
	Yates	1	7,210
		2	7,040
		3	6,950
		4	8,910
		5	9,410
		6	24,760
		7	21,480
Illinois	Baldwin	1	42,010
		2	44,420
		3	42,550
	Coffees	1	11,790
		2	35,670
	Grand Tower	4	5,910
	Hennepin	2	18,410
	Joppa Steam	1	12,590
		2	10,770
		3	12,270
		4	11,860
		5	11,420
		6	10,620
	Kincaid	1	31,580
		2	33,810
	Meredosia	3	13,890
	Vermilion	2	8,880

TABLE 14.1   Affected Sources and Units in Phase I and Their Sulfur Dioxide
Allowances, tons (*Continued*)

State	Plant name	Generator	Phase I allowances
Indiana	Bailly	7	11,180
		8	15,630
	Breed	1	18,500
	Cayuga	1	33,370
		2	34,130
	Clifty Creek	1	20,150
		2	19,810
		3	20,410
		4	20,080
		5	19,360
		6	20,380
	E. W. Stout	5	3,880
		6	4,770
		7	28,610
	F. B. Culley	2	4,290
		3	16,970
	F. E. Ratts	1	8,330
		2	8,480
	Gibson	1	40,400
		2	41,010
		3	41,080
		4	40,320
	H. T. Pritchard	6	5,770
	Michigan City	12	23,310
	Petersburg	1	16,430
		2	32,380
	R. Gallagher	1	6,490
		2	7,280
		3	6,530
		4	7,650
	Tanners Creek	4	24,820
	Wabash River	1	4,000
		2	2,860
		3	3,750
		5	3,670
		6	12,280
	Warrick	4	26,980
Iowa	Burlington	1	10,710
	Des Moines	7	2,320
	George Neal	1	1,290
	M. L. Kapp	2	13,800
	Prairie Creek	4	8,180
	Riverside	5	3,990
Kansas	Quindaro	2	4,220
Kentucky	Coleman	1	11,250
		2	12,840
		3	12,340
	Cooper	1	7,450
		2	15,320

TABLE 14.1   Affected Sources and Units in Phase I and Their Sulfur Dioxide Allowances, tons (*Continued*)

State	Plant name	Generator	Phase I allowances
Kentucky	E. W. Brown	1	7,110
		2	10,910
		3	26,100
	Elmer Smith	1	6,520
		2	14,410
	Ghent	1	28,410
	Green River	4	7,820
	H. L. Spurlock	1	22,780
	Henderson II	1	13,340
		2	12,310
	Paradise	3	59,170
	Shawnee	10	10,170
Maryland	Chalk Point	1	21,910
		2	24,330
	C. P. Crane	1	10,330
		2	9,230
	Morgantown	1	35,260
		2	38,480
Michigan	J. H. Campbell	1	19,280
		2	23,060
Minnesota	High Bridge	6	4,270
Mississippi	Jack Watson	4	17,910
		5	36,700
Missouri	Asbury	1	16,190
	James River	5	4,850
	Labadie	1	40,110
		2	37,710
		3	40,310
		4	35,940
	Montrose	1	7,390
		2	8,200
		3	10,090
	New Madrid	1	28,240
		2	82,480
	Sibley	3	15,580
	Sioux	1	22,570
		2	23,690
	Thomas Hill	1	10,250
		2	19,390
New Hampshire	Merrimack	1	10,190
		2	22,000
New Jersey	B. L. England	1	9,060
		2	11,720
New York	Dunkirk	3	12,600
		4	14,060
	Greenridge	4	7,540
	Milliken	1	11,170
		2	12,410
	Northport	1	19,810
		2	24,110
		3	26,480

TABLE 14.1   Affected Sources and Units in Phase I and Their Sulfur Dioxide Allowances, tons (*Continued*)

State	Plant name	Generator	Phase I allowances
New York	Port Jefferson	3	10,470
		4	12,330
Ohio	Ashtabula	5	16,740
	Avon Lake	8	11,650
		9	30,480
	Cardinal	1	34,270
		2	38,320
	Conesville	1	4,210
		2	4,890
		3	5,500
		4	48,770
	Eastlake	1	7,800
		2	8,640
		3	10,020
		4	14,510
		5	34,070
	Edgewater	4	5,050
	Gen. J. M. Gavin	1	79,080
		2	80,560
	Kyger Creek	1	19,280
		2	18,560
		3	17,910
		4	18,710
		5	18,740
	Miami Fort	5	760
		6	11,380
		7	38,510
	Muskingum River	1	14,880
		2	14,170
		3	13,950
		4	11,780
		5	40,470
	Niles	1	6,940
		2	9,100
	Picway	5	4,930
	R. E. Burger	3	6,150
		4	10,780
		5	12,430
	W. H. Sammis	5	24,170
		6	39,930
		7	43,220
	W. C. Beckjord	5	8,950
		6	23,020
Pennsylvania	Armstrong	1	14,410
		2	15,430
	Brunner Island	1	27,760
		2	31,100
		3	53,820
	Cheswick	1	39,170
	Conemaugh	1	59,790
		2	66,450

TABLE 14.1   Affected Sources and Units in Phase I and Their Sulfur Dioxide Allowances, tons (*Continued*)

State	Plant name	Generator	Phase I allowances
Pennsylvania	Hatfield's Ferry	1	37,830
		2	37,320
		3	40,270
	Martins Creek	1	12,660
		2	12,820
	Portland	1	5,940
		2	10,230
	Shawville	1	10,320
		2	10,320
		3	14,220
		4	14,070
	Sunbury	3	8,760
		4	11,450
Tennessee	Allen	1	15,320
		2	16,770
		3	15,670
	Cumberland	1	86,700
		2	94,840
	Gallatin	1	17,870
		2	17,310
		3	20,020
		4	21,260
	Johnsville	1	7,790
		2	8,040
		3	8,410
		4	7,990
		5	8,240
		6	7,890
		7	8,980
		8	8,700
		9	7,080
		10	7,550
West Virginia	Albright	3	12,000
	Fort Martin	1	41,590
		2	41,200
	Harrison	1	48,620
		2	46,150
		3	41,500
	Kammer	1	18,740
		2	19,460
		3	17,390
	Mitchell	1	43,980
		2	45,510
	Mount Storm	1	43,720
		2	35,580
		3	42,430
Wisconsin	Edgewater	4	24,750
	La Crosse/Genoa	3	22,700

TABLE 14.1   Affected Sources and Units in Phase I and Their Sulfur Dioxide Allowances, tons (*Continued*)

State	Plant name	Generator	Phase I allowances
Wisconsin	Nelson Dewey	1	6,010
		2	6,680
	N. Oak Creek	1	5,220
		2	5,140
		3	5,370
		4	6,320
	Pulliam	8	7,510
	S. Oak Creek	5	9,670
		6	12,040
		7	16,180
		8	15,790

tain units in the states of Illinois, Indiana, and Ohio for each of the calendar years inclusive of 1995 through 1999.

To meet the sulfur dioxide emission limitations imposed, an owner or operator of an affected unit (see Table 14.1) may propose to the EPA to reassign, in whole or in part, the affected unit's sulfur dioxide reduction requirements to any other unit(s) under the control of the owner or operator. The "substitution" would be reviewed by the EPA and either granted or denied primarily on the basis of whether the substitution still ensures the emission reductions sought by subchapter IV-A. Specifically, the reassigned tonnage limits must, in total, achieve the same or greater emissions reduction as would have been achieved by the original affected unit and the substitute unit(s) without such substitution.

An owner or operator may petition for an affected unit to be designated as an "eligible phase I extension unit." If deemed eligible, the unit will receive a 2-year extension from the emission limitations imposed by phase I. To be eligible, the owner must hold additional emission allowances for the unit. The unit also must either employ a qualifying phase I technology or transfer its phase I emissions reduction obligation to a unit employing a qualifying phase I technology. *Qualifying phase I technology* is defined as a technological system of continuous emission reduction which achieves a 90 percent reduction in emissions of sulfur dioxide from the emissions that would have resulted from the use of fuels which were not subject to treatment prior to combustion. If the emissions reserve has not been depleted, additional allowances may be granted to eligible phase I extension units in the years 1997, 1998, and 1999. After January 1, 1997, in the event that a phase I extension unit exceeds its allowance for a given calendar year, the unit will receive a deduction in its annual allowance allocation for the following year equal to the net excess emissions of the previous year.

In an effort to encourage energy conservation and use of renewable energy resources, Congress provided 300,000 additional allowance units in the Conservation and Renewable Energy Reserve. For every ton of sulfur dioxide emissions avoided by an electric utility through the use of qualified energy conservation measures or qualified renewable energy, the EPA will grant the utility a single allowance on a first-come first-served basis. A *qualified energy conservation measure* is defined as a cost-effective measure, as identified by the EPA, that increases the efficiency of the use of electricity provided by the utility to its customers. *Qualified renewable energy* is defined as energy derived from biomass, solar, geothermal, or wind as identified by the EPA.

14.4   Phase II Sulfur Dioxide Requirements

On January 1, 2000, phase II sulfur dioxide emissions requirements will go into effect. The emission requirements are broken down into several categories:

1. Units equal to or above 75 MWe and 1.20 lb/(10^6 Btu)

2. Coal- or oil-fired units below 75 MWe and above 1.20 lb/(10^6 Btu)

3. Coal-fired units below 1.20 lb/(10^6 Btu)

4. Oil- and gas-fired units equal to or greater than 0.60 lb/(10^6 Btu) and less than 1.20 lb/(10^6 Btu)

5. Oil- and gas-fired units less than 0.60 lb/(10^6 Btu)

6. Units that commence operation between 1986 and December 31, 1995

7. Oil- and gas-fired units with less than 10 percent oil consumed

8. Units in high-growth states

Each unit will receive a basic phase II allowance allocation. In addition, in each year beginning in calendar year 2000 and ending in calendar year 2009 inclusive, the EPA shall allocate up to 530,000 phase II bonus allowances. Furthermore, in addition to basic phase II and phase II bonus allowance allocations, beginning January 1, 2000, the EPA shall allocate additional allowances for units listed in Table 14.1 from the states of Illinois, Indiana, Ohio, Georgia, Alabama, Missouri, Pennsylvania, West Virginia, Kentucky, and Tennessee. Three facilities are excluded from this third allowance: Kyger Creek, Clifty Creek, and Joppa Steam. Finally, another 45,000 additional allowances are available annually for units in high-growth states. These allowances are broken into two categories such that one group shall not exceed 40,000 allowances and the other group shall not exceed 5000 allowances.

14.5   Allowances for States with Emissions Rates at or below 0.80 lb/(10^6 Btu)

In lieu of phase II bonus allowances discussed previously, any state whose 1985 annual sulfur dioxide emissions rate is equal to or less than 0.80 lb/(10^6 Btu) (averaged over all fossil fuel–fired utility steam-generating units) can elect to receive an alternate bonus allowance. This allowance would apply to all affected units within the state and would equal 125,000 multiplied by the unit's pro rata share of electricity generated in calendar year 1985. For the alternate bonus allowance to be implemented, the governor of the state must notify the EPA of the state's decision to pursue the alternate allowance program. This alternate bonus allowance program will be in effect for the years 2000 through 2009, inclusive, after which there will be no bonus allowance program.

14.6   Nitrogen Oxides Emission Reduction Program

In general, on the date that a coal-fired utility unit becomes an affected unit pursuant to regulations restricting sulfur dioxide emissions, it is also classified as an affected unit pursuant to regulations restricting emissions of nitrogen oxides. Exceptions occur when a unit is granted a phase I sulfur dioxide emissions extension or is granted a repowered source extension (see Sec. 14.8).

Under the direction of Congress, EPA has set nitrogen oxide emission standards for several types of utility boilers. The maximum allowable emission rates are as follows:

Currently in effect:

- For tangentially fired boilers, 0.45 lb/(10^6 Btu)
- For dry-bottom wall-fired boilers (other than units applying cell burner technology), 0.50 lb/(10^6 Btu)

Proposed by EPA (currently awaiting final approval):

- For cell burner boilers, 0.68 lb/(10^6 Btu)
- For cyclone boilers, 0.94 lb/(10^6 Btu)
- For wet-bottom boilers, 0.86 lb/(10^6 Btu)

- For vertically fired boilers, 0.80 lb/(10^6 Btu)
- For fluidized-bed combustor boilers, 0.29 lb/(10^6 Btu)

New stationary-source performance standards for nitrogen oxide emissions from fossil fuel–fired steam-generating units have been promulgated by EPA.

The permitting authority (i.e., EPA or a state or local air pollution control agency, with an approved permitting program) may authorize less stringent nitrogen oxide emission limitations. To qualify, the owner must show that the unit is unable to meet the applicable limitation using low-NO_x burner technology or that the unit is unable to meet the applicable rate using the technology on which the EPA based the applicable emission limitation. The qualifying determination is to be based upon data from properly installed and operating equipment over a 15-month period. If the unit qualifies, the owner or operator shall specify an emission rate that can be met on an annual average basis. Units which receive an alternative emission limitation shall not be required to install any additional control technology beyond low-NO_x burners.

If an owner or operator has two or more units, a petition may be filed with the permitting authority requesting emissions averaging among units. Contemporaneous annual emission limitations for such units are to ensure that (1) the actual annual emission rate in pounds of nitrogen oxides per million Btu average over the units in question is a rate that is less than or equal to (2) the Btu-weighted average annual emission rate for the same units if they had been operated, during the same period of time in compliance with limitations set in accordance with the nitrogen oxide emission rates that would typically apply to the unit. Operating permits issued under the auspices of emissions averaging shall only remain in effect while both units continue operation under the conditions specified in their respective operating permits.

14.7   Permits and Compliance Plans

For specific guidance in permitting and compliance, the reader should refer to Chap. 15, Operating Permits, of this book. In general, the intent of the permit is to prohibit (1) annual emissions of sulfur dioxide in excess of the allowances held by the owner or operator for a specific unit; (2) excess emissions beyond applicable emission rates as outlined by the CAA of 1990 or promulgated by EPA; (3) the use of any allowance prior to the year for which it was allocated; and (4) contravention of any other provision of the operating permit. Typically, operating permits are to be issued for 5-year periods. They are required to contain a compliance plan specifically in regard to emission limitations associated with acid deposition control for the source. Permit applications for phase I (originally due November 1993) and phase II (originally due January 1996) sulfur dioxide and nitrogen oxide sources were required to have already been submitted. Permits for phase I sources should already have been issued by the permitting authority while phase II sources are scheduled to receive permits no later than December 31, 1997. New units are required to submit permit applications before the later of January 1, 1998, or the date the unit commences operation.

14.8 Repowered Sources

As defined by the CAA of 1990, the term *repowering* means replacement of an existing coal-fired boiler with one of the following clean coal technologies: atmospheric or pressurized fluidized-bed combustion, integrated gasification combined cycle, magnetohydrodynamics, direct and indirect coal-fired turbines, integrated gasification fuel cells, or, as determined by the EPA, a derivative of one or more of these technologies and any other technology capable of controlling multiple combustion emissions simultaneously with improved boiler or generation efficiency and with significantly greater waste reduction relative to the performance of technology in widespread commercial use as of November 15, 1990.

An owner or operator of an existing affected source above 1.2 lb/(10^6 Btu) wishing to repower (replace) the source shall notify the EPA of such an intent prior to December 31, 1997. The repowered source will be required to comply with the provisions outlined by the phase II sulfur dioxide emission limitation requirements. However, the repowered source will receive an extension from January 1, 2000, back to December 31, 2003, to achieve the limitation. Furthermore, to qualify for the extension, the owner must demonstrate no later than January 1, 2000, satisfactory documentation of a preliminary design and engineering effort for such repowering and an executed and binding contract for the majority of the equipment to repower the unit.

14.9 Election for Additional Sources

An owner or operator of a nonaffected unit which is a new, phase I, or phase II sulfur dioxide emissions source may elect to designate that unit as affected and become eligible to receive allowances for the unit. The owner or operator of a process source may also elect to receive designation for the source as an affected unit for the purpose of receiving allowances. However, allowances granted under these provisions carry certain restrictions involving transfer and forwarding of the allowance allocation. Small crude oil refineries (less than 18.25 million bbl throughput per year) may elect to receive allowances under this provision as well.

14.10   Excess-Emissions Penalty

The owner or operator of any unit or process source termed affected with respect to sulfur dioxide or nitrogen oxides is required to pay the EPA a penalty of $2000 (November 1990 CPI value), to be adjusted for inflation, per excessive ton of emissions per calendar year. Such a penalty is automatically due without demand from the EPA. Thus, the penalty is supposed to operate in a self-governing form. Furthermore, the excessive emissions are required to be offset by a subsequent reduction in emissions during the next calendar year equal to the net excess. A plan for such a reduction of emissions must be submitted to the EPA no later than 60 days into the new calendar year.

14.11   Monitoring, Reporting, and Record-Keeping Requirements

In general, affected sources for sulfur dioxide and nitrogen oxides are required to employ continuous emissions monitoring for each affected unit. If data from the monitoring device are unavailable for any unit, the unit may be deemed by the EPA as operating in an uncontrolled manner during the entire period for which the data were unavailable. The EPA will be charged with calculating the actual emissions for the period, and the owner shall be liable for any excess-emissions fees and offsets which result from such a calculation.

14.12   Clean Coal Technology Regulatory Incentives

As defined by the CAA of 1990, *clean coal technology* means any technology, including technologies applied at the precombustion, combustion, or postcombustion stage, at a new or existing facility which will achieve significant reductions in air emissions of sulfur dioxide or oxides of nitrogen associated with the utilization of coal in the generation of electricity, process steam, or industrial products, which was not in widespread use as of November 15, 1990.

Projects operated for 5 years or less and funded under the "Department of Energy—Clean Coal Technology" heading which complies with appropriate state implementation plans and National Ambient Air Quality Standards shall not be subject to the requirements of the standards of performance for new stationary sources.

Chapter 15

Operating Permits

William L. Cleland

15.1 Introduction

When the 1990 Clean Air Act Amendments (CAAA) were enacted, the primary program affecting industrial plants was referred to simply as *Title V.* It refers to a detailed and extensive requirement for regulating air pollution from industry in a federally enforceable operating permit system. Modeled after the National Pollutant Discharge Elimination System (NPDES), it imposes huge changes in Clean Air Act permitting. Previously unregulated sources are now required to be permitted, and all obligations for a facility are to be placed into one document.

The program is intended to be administered by states or local agencies with oversight provided by the federal Environmental Protection Agency (EPA).[1] New obligations of industry are the responsibility of applicability to applicable rules and compliance with those rules. Costs associated with the program are substantial even for small industries and businesses. The Title V program is in addition to the New Source Review programs (PSD, NSR, NSPS, NESHAPs) already in place prior to the 1990 CAAA. Title V permits are to include all requirements of these programs as well as new ones such as the new NESHAPs (MACT and GACT), acid rain, and ozone depletion rules.

The total cost for administering the program is borne by industry in the form of permitting fees paid to the permitting agency. The presumptive rate was set at $25 (in 1989–1990 dollars) per ton/year of

---

[1]All references to states in this document also apply to local authorities such as cities, counties, and sections that have obtained primacy from and operate under the oversight of a state and/or the EPA.

actual emissions and was to be adjusted according to the rise and fall of the consumer price index (CPI). Permitting agencies have some flexibility in this rate as necessary to run their programs. Some states use an annual work load analysis to set the actual/allowable rates.

Regulations for Title V are promulgated at 40 CFR Part 70 for state programs and Part 71 for the federal program. Part 71 comes into effect if a state's program is not approved by EPA. The original requirement was for states to have a program submitted to EPA by November 15, 1993, and approved by November 1994. There have been some delays, and these dates have changed for some agencies.

Several agencies have established bulletin boards and information centers for additional information. Some of these are the EPA's TTN BBS at Research Triangle, the Public Information Center, an Air Risk Information Center, a Control Technology Hotline, a Mobile Sources Hotline, and an Acid Rain Hotline. Appendix N describes how to access these sources.

15.2   State Operating Permit Programs

The Title V operating permit program is not intended to impose new air pollution control requirements. The assumption was that sources were generally not in compliance with existing rules since all nonattainment areas had not yet been brought into attainment with the NAAQSs. The Title V permit was to be an enforcement tool that contained all applicable requirements for a facility.

The final regulations for Title V for states were promulgated on July 21, 1992, at 57 FR 32,250; 40 CFR Part 70. Rules for the federal program, Part 71, have been proposed but have not been promulgated. Part 70 in its entirety is in App. E.

Many details of the permit system may vary from state to state due to differences in state law and in how a state administers and enforces its legal requirements. Consequently, a state program can be similar but not necessarily identical to Part 70. The required elements from Part 70 are as follows:

- Standard application forms. (There are no standard federal forms.)

- Requirements for monitoring and reporting.

- Annual fee requirements paid by source owners or operators to cover the cost of developing and administering the permit process, including enforcement and allowance for changes based on the CPI.

- Provisions for assessing penalties (up to 50 percent) for failure to pay and for allowing EPA to step in where the state has failed to do so.

- Authority for the permitting agency to issue permits for no more than 5 years for each term; to refuse permit issuance after an objection; and to terminate, revoke, or modify permits for cause.

- Authorization for imposing civil penalties of up to $10,000 per day for individual permit violations and for providing criminal penalties for "knowing violations."

- Provisions for reasonable and expeditious application review and permit issuance, allowing public comment. There must also be the opportunity for judicial review of final permit action.

- Provisions for judicial review in case the permitting agency does not act within the time period specified in Title V and Title VI of the act.

- Allowance for providing information to the public and nearby states.

- Authority for the permitting agency to require revised applications for updating permits to cover a new applicable standard or regulation whose effective date is prior to the expiration date of the permit.

- Provisions to allow a permit source to make changes in its facility which are not considered modifications as defined in Title I and do not exceed allowable permit limits, without requiring a permit revision.

All major sources are required to submit applications to the appropriate permitting authority (state or local) within 1 year after approval of the agency's plan by the EPA.

Each state was to have submitted its plan to EPA for approval by November 15, 1993, three years after enactment of the 1990 CAAA. Then EPA had a year to approve or disapprove the plan. There have been many delays by states in submitting their plans. Any state not submitting its plan by November 15, 1994, was subject to sanctions which include loss of highway funds and imposition of a 2-to-1 offset ratio for new or modified sources in nonattainment areas.

A state could be granted partial approval only if the plan submitted contained provisions to ensure compliance with acid rain requirements under Title IV, air toxic requirements under Title III, and SIP requirements under Title I. Interim approval could be granted even if all permit requirements in Title V were not complete. A state would have 2 years to submit a revised plan after partial or interim approval or be subject to federal sanctions and lose its Part 70 authority.

15.3   Sources Affected

Air pollution sources that are affected and required to obtain Title V federally enforceable operating permits are listed below:

1. All major sources. The general definition of *major source* is one that emits or has the potential to emit 100 tons/yr of any regulated pollutant. Lower thresholds for the major category apply for hazardous air pollutants and for sources located in nonattainment areas.
2. Any source regulated under the NSPS program.
3. Any source regulated under the NESHAPs program.
4. Sources subject to PSD.
5. Sources required to have a permit under Title I (nonattainment) or Title IV (acid rain).

Once a source is subject to permit requirement for one pollutant, all other pollutants emitted or having the potential to emit must be covered. Sources subject to NSPS or NESHAP rules that are below the major source threshold for Title III (described in Chap. 13) may be covered but are deferred until a later date.

15.4   Program Implementation Schedule

The Part 70 federal rules for the state program was promulgated on July 21,1992. States were required to submit their plan to EPA for approval originally by November 15, 1993. Due to delays for various reasons the deadline was extended to as late as November 15, 1994, which was the date sanctions could be imposed for failure to submit a plan. After submission of a state's plan, EPA was still allowed 1 year to give interim, partial, or final approval. Any of these approvals would start the clock for the permitting process to begin. Sources were required to submit a complete application within 1 year of approval. The schedule was to have one-third of permits issued during the first year, one-third during the second year, and one-third during the third year.

Interim or partial approval could be granted for up to 2 years. During that time EPA could give final approval or disapprove the plan. If the plan were disapproved, EPA would establish a federal plan for the state under Part 71 rules.

After the program is started, a new major source or a source modified into major source status is to submit an application within 1 year of commencing operation. The state then has 18 months to issue a Title V permit.

After an application is sent to the state, the state is allowed 60 days in which to determine whether the application is complete. When it has been determined to be complete, or after 60 days following receipt by the state, whichever occurs first, the state will issue a draft permit. Public notice will be given by newspaper advertisement, notice will be given to contiguous states by letter, and copies of the draft permit will be sent to EPA and to the applicant. After 30 days, provided there are no serious comments or objections, the state agency will submit a proposed permit to EPA. Then EPA has 45 days to approve or veto. If EPA vetos, the permit is sent back to the state for redrafting and the process starts over. If EPA approves, citizens (the public) are allowed the opportunity to petition for disapproval, but only for reasons that were brought up during the public comment period. There is also allowance for judicial action after permit approval.

Permits could be issued for a period up to a maximum of 5 years. On the first issuance cycle, because of the heavy load and compression of applications arriving very close together, many permits are being issued for less than 5 years to level out the workload, so that at renewal all could be given 5-year permits.

15.5   Preparing to Apply for a Title V Permit

The strategy that a source should use in preparing for a Title V permit involves, at the very least, familiarity with permitting regulations and determination of all applicable requirements. If the company is not large enough to have an individual who can devote full time to environmental matters, it is advisable to retain a consulting firm knowledgeable in regulations and permitting. It is also advisable to have access to an environmental legal counsel.

It will be extremely important to have or to develop an accurate emission inventory. This can be done by using material balance calculations, emission factors either developed at the source or from published data such as EPA's AP-42, detailed record keeping, meaningful engineering estimates, or some type of monitoring of emissions or monitoring of process parameters that can be related to emissions.

An evaluation of compliance with current rules is necessary. Be familiar with all requirements that are applicable to the source. These requirements will usually be covered by current permits. If not, it may be necessary to correct existing permits so that they include all appropriate limits. If a source is very old, documentation pertaining to dates of construction and/or modification should be found or estimated. This is normally necessary to determine regulatory limitations, since rules have been changing over the years and limits for an unmodified source could be less stringent than those for one built recently. How a source is monitored for emissions should be determined. If no monitoring or record-keeping is done, a way to prove or disprove emissions should be developed. Also, current control technology should be evaluated. If there is no control or attempt to control emissions, it may be important to analyze ways of eliminating or reducing emissions. Costs should be evaluated, and plans implementing either elimination or reduction of emissions should be drawn up.

Applicable requirements must be determined by someone very familiar with state and federal rules and regulations. State regulations established from the State Implementation Plan, New Source Performance Standards, National Emissions of Hazardous Air Pollutants, PSD, or NSR will normally cover most of these requirements. Depending on the status of state regulations, federal requirements may or may not have been adopted at any time and must be included.

There are other plans and decisions that must be made prior to preparing the Title V permit application. Whether worst-case conditions will always be used to establish limits or whether alternate scenarios is best for the source must be determined. The amount of fees to be paid is a factor in this determination as well as what limits are desired.

The preference for single or multiple permits should be determined. This normally is dependent on the size of the plant or could be determined by the complexity or lack of similarity between processes at the facility. If alternate scenarios or multiple permits are necessary, it should be kept in mind that either of these will require a complete application for each scenario or each permit, adding to the cost of preparing the permit application.

The designated official for Title V purposes must be determined. The definition of designated official is very specific in Part 70 regulations. He or she must be a vice president or division manager who is responsible for operations and has the authority at the facility to open or close down operations. For a manufacturing operation, the lowest in rank for the responsible official is a plant manager. It will be his or her responsibility to sign for completeness of an application and for compliance of the source to all applicable requirements.

It will be extremely important to meet the deadline for submitting a "timely and complete" application so that the application shield can be obtained. Since the state will have 60 days to determine completeness, a buffer of 90 days prior to the application deadline is a good target to shoot for. This will provide at least 30 days to respond with any additional information for completeness requested by the permitting agency.

During the initial cycle of Title V permits, it could be very beneficial for the source to prepare a draft permit to be included with the permit application. In this way it may be possible to receive conditions on the permit that are the most desirable. It should be kept in mind that it always is best to be as generic as possible in permit conditions to maximize flexibility.

An environmental data management procedures manual will be desirable or even absolutely necessary for showing continuous compliance with a Title V permit. The use of computers and spreadsheets makes the process much easier to handle.

15.6 Applying for a Title V Permit

The Title V permit plan for each state includes application forms that are unique to that state. There are no standard forms provided by the EPA. Certain information required in all applications is as follows:

- Facility identification
- Operations and flow diagrams
- Exhaust point (stack) locations and identification
- Air pollution control descriptions
- Compliance demonstrations
- Statement of completeness
- Statement of compliance

Typical application forms are shown in App. F. A list of permitting agencies is in App. C. The forms will normally be available in hard-copy form and in disk form. Application can be made by submitting hard copy or electronically. Instructions for completing applications will be provided by the agency. It is advisable to utilize permitting agency personnel as needed to ensure that the information entered on the forms is sufficient to satisfy the agency and the EPA with respect to both content and completeness.

Normally, to ensure that the application is complete, the check-off list (also in App. F) used by the agency to determine completeness will be made available to the applicant. A meeting with the permitting personnel prior to formal submission of the application would be advisable. This will help the applicant to include all the information needed for completeness. State agencies are usually helpful in this way, and most have been proactive all along in the permitting process.

A permit shield must be requested, and this can be normally done in a cover letter to the application. The permit application shield will be provided automatically if the application is submitted in a "timely and complete" manner. *Complete* means that the agency has been provided all information necessary to *begin* the permitting process, not necessarily all information that will be necessary to issue the permit. *Timely* means that all deadlines have been met and that all additional information is provided within the time reasonably requested by the agency. The important thing is to stay in communication with the permitting person assigned to the facility.

The permitting agency will determine completeness within 60 days of receipt of the application. If incompleteness has not been established within 60 days, the application must be deemed complete by regulation. Enough time should be allowed to respond to requests for additional information, normally 30 additional days. Thus, it would be advisable to allow at least 90 days prior to the deadline for submitting an application to ensure receipt of the application shield.

It should always be kept in mind that EPA or the permitting authority can open a Title V permit for cause.

15.7   Small Business Assistance

The 1990 CAAA and 40 CFR Part 70 contain provisions for assistance to small businesses in the Title V process. Each state is required to have established a division dedicated to this purpose. To be eligible for and to petition for small business assistance, a company must meet the following criteria:

- The facility employs 100 or fewer individuals.
- The company is a small business as defined in the Small Business Act.
- The facility does not emit 50 tons/yr or more of any regulated pollutant.
- The facility emits less than 75 tons/yr of all regulated pollutants.

A guide for small business is available from the EPA as publication 450-K-92-001, September 1992.

15.8   Permit Revisions

Any significant changes to source operations that would be inconsistent with permit conditions will require the permittee to obtain a Title V permit revision. This will be done through a process that is essentially the same as that for the original permit application, requiring opportunities for review by the public, the permitting authority, affected states, and EPA. Changes in monitoring, reporting, or record-keeping requirements could necessitate a permit revision as well.

Certain changes can be made to a source that do not require a permit revision. Minor modifications, off-permit exemptions, and administrative permit amendments are such changes. A considerable amount of controversy has arisen concerning minor permit modifications. Essentially, a minor permit modification is one that is not a *modification* as defined by Title I and an SIP.

Off-permit exemption is a change in operation of a source that is not addressed or prohibited by the Title V permit. Administrative amendments can be used to correct typographical errors, changes of name or address, increases in monitoring frequency, or change in ownership. These amendments can also be used to incorporate certain types of requirements from preconstruction review permits without opening the Title V permit.

16

Stratospheric Ozone Protection

Charles Gallagher

16.1 Introduction

After passage of the Clean Air Act of 1990, the EPA was instructed to list two classes of substances deemed detrimental to the stratospheric ozone layer which envelopes the earth's upper atmosphere. Class I substances (Table 16.1), broken down into five subgroups, included many chlorofluorocarbons, several halons, carbon tetrachloride, and methyl chloroform. Class II substances (Table 16.2) comprised a list of 33 hydrochlorofluorocarbons. The distinction between class I and class II substances was made in order to facilitate separate phaseout (production, importation, exportation, and usage) schedules of the two classes.

Additionally, the EPA was directed to add to the list any other substance found to cause or contribute significantly to the destruction of the stratospheric ozone layer. These additions are to be composed of all substances determined to have an ozone depletion potential of 0.2 or greater. The term *ozone depletion potential* means a factor established by the EPA to reflect the ozone depletion potential of a substance, on a mass per kilogram basis as compared to chlorofluorocarbon 11. The factor is based upon the substance's atmospheric lifetime, the molecular weights of bromine and chlorine, the substance's ability to be photolytically disassociated, and other factors determined to be an accurate measure of relative ozone depletion potential. The list of classified substances is to be updated no less often than every 3 years.

TABLE 16.1 Class I Controlled Substances

Controlled substances	Oxygen depletion potential
A. Group I	
$CFCl_3$—trichlorofluoromethane (CFC 11)	1.0
CF_2Cl_2—dichlorodifluoromethane (CFC 12)	1.0
$C_2F_3Cl_3$—trichlorotrifluoroethane (CFC 113)	0.8
$C_2F_4Cl_2$—dichlorotetrafluoroethane (CFC 114)	1.0
C_2F_5Cl—monochloropentafluoroethane (CFC 115)	0.6
All isomers of the above chemicals	
B. Group II	
CF_2ClBr—bromochlorodifluoromethane (Halon 1211)	3.0
CF_3Br—bromotrifluoromethane (Halon 1301)	10.0
$C_2F_4Br_2$—dibromotetrafluoroethane (Halon 2402)	6.0
All isomers of the above chemicals	
C. Group III	
CF_3Cl—chlorotrifluoromethane (CFC 13)	1.0
C_2FCl_5—CFC 111	1.0
$C_2F_2Cl_4$—CFC 112	1.0
C_3FCl_7—CFC 211	1.0
$C_3F_2Cl_6$—CFC 212	1.0
$C_3F_3Cl_5$—CFC 213	1.0
$C_3F_4Cl_4$—CFC 214	1.0
$C_3F_5Cl_3$—CFC 215	1.0
$C_3F_6Cl_2$—CFC 216	1.0
C_3F_7Cl—CFC 217	1.0
All isomers of the above chemicals	
D. Group IV: CCl_4—carbon tetrachloride	1.1
E. Group V	
$C_2H_3Cl_3$-1,1,1—trichloroethane (methyl chloroform)	0.1
All isomers of the above chemical except 1,1,2-trichloroethane	
F. Group VI: CH_3Br—bromomethane (methyl bromide)	0.7
G. Group VII	
$CHFBr_2$	1.00
CHF_2Br (HBFC 2201)	0.74
CH_2FBr	0.73
C_2HFBr_4	0.3–0.8
$C_2HF_2Br_3$	0.5–1.8
$C_2HF_3Br_2$	0.4–1.6
C_2HF_4Br	0.7–1.2
$C_2H_2FBr_3$	0.1–1.1
$C_2H_2F_2Br_2$	0.2–1.5
$C_2H_2F_3Br$	0.7–1.6
$C_2H_3FBr_2$	0.1–1.7
$C_2H_3F_2Br$	0.2–1.1
C_2H_4FBr	0.07–0.1
C_2HFBr_6	0.3–1.5
$C_3HF_2Br_5$	0.2–1.9
$C_3HF_3Br_4$	0.3–1.8
$C_3HF_4Br_3$	0.5–2.2
$C_3HF_5Br_2$	0.9–2.0
C_3HF_6Br	0.7–3.3
$C_3H_2FBr_5$	0.1–1.9
$C_3H_2F_2Br_4$	0.2–2.1
$C_3H_2F_3Br_3$	0.2–5.6

TABLE 16.1   Class I Controlled Substances (*Continued*)

Controlled substances	Oxygen depletion potential
G. Group VII	
$C_3H_2F_4Br_2$	0.3–7.5
$C_3H_2F_5Br$	0.9–14
$C_3H_3FBr_4$	0.08–1.9
$C_3H_3F_2Br_3$	0.1–3.1
$C_3H_3F_3Br_2$	0.1–2.5
$C_3H_3F_4Br$	0.3–4.4
$C_3H_4FBr_3$	0.03–0.3
$C_3H_4F_2Br_2$	0.1–1.0
$C_3H_4F_3Br$	0.07–0.8
$C_3H_5FBr_2$	0.04–0.4
$C_3H_5F_2Br$	0.07–0.8
C_3H_6FB	0.02–0.7

TABLE 16.2   Class II Controlled Substances

Controlled substances	ODP
$CHFCl_2$—dichlorofluoromethane (HCFC 21)	Reserved
CHF_2Cl—chlorodifluoromethane (HCFC 22)	0.05
CH_2FCl—chlorofluoromethane (HCFC 31)	Reserved
C_2HFCl_4 (HCFC 121)	Reserved
$C_2HF_2Cl_3$ (HCFC 122)	Reserved
$C_2HF_3Cl_2$ (HCFC 123)	0.02
C_2HF_4Cl (HCFC 124)	0.02
$C_2H_2FCl_3$ (HCFC 131)	Reserved
$C_2H_2F_2Cl_2$ (HCFC 132b)	Reserved
$C_2H_2F_3Cl$ (HCFC 133a)	Reserved
$C_2H_3FCl_2$ (HCFC 141b)	0.12
$C_2H_3F_2Cl$ (HCFC 142b)	0.06
C_3HCFCl_6 (HCFC 221)	Reserved
$C_3HF_2Cl_5$ (HCFC 222)	Reserved
$C_3HF_3Cl_4$ (HCFC 223)	Reserved
$C_3HF_4Cl_3$ (HCFC 224)	Reserved
$C_3HF_5Cl_2$ (HCFC 225ca)	Reserved
C_3HF_5Cl (HCFC 225cb)	Reserved
C_3HF_6Cl (HCFC 226)	Reserved
$C_3H_2FCl_5$ (HCFC 231)	Reserved
$C_3H_2F_2Cl_4$ (HCFC 232)	Reserved
$C_3H_2F_3Cl_3$ (HCFC 233)	Reserved
$C_3H_2F_4Cl_2$ (HCFC 234)	Reserved
$C_3H_2F_5Cl$ (HCFC 235)	Reserved
$C_3H_3FCl_4$ (HCFC 241)	Reserved
$C_3H_3F_2Cl_3$ (HCFC 242)	Reserved
$C_3H_3F_3Cl_2$ (HCFC 243)	Reserved
$C_3H_3F_4Cl$ (HCFC 244)	Reserved
$C_3H_4FCl_3$ (HCFC 251)	Reserved
$C_3H_4F_2Cl_2$ (IICFC 252)	Reserved
$C_3H_4F_3Cl$ (HCFC 253)	Reserved
$C_3H_5FCl_2$ (HCFC 261)	Reserved
$C_3H_5F_2Cl$ (HCFC 262)	Reserved
C_3H_6FCl (HCFC 271)	Reserved
All isomers of the above chemicals	

TABLE 16.3   Global Warming Potentials (Mass Basis) of Class I and Class II Substances*

Species (chemical)	Chemical formula	Global warming potential (time horizon), years		
		20	100	500
CFC 11	$CFCl_3$	5,000	4,000	1,400
CFC 12	CF_2Cl_2	7,900	8,500	4,200
CFC 13	$CClF_3$	8,100	11,700	13,600
CFC 113	$C_2F_3Cl_3$	5,000	5,000	2,300
CFC 114	$C_2F_4Cl_2$	6,900	9,300	8,300
CFC 115	C_2F_5Cl	6,200	9,300	13,000
H 1301	CF_3Br	6,200	5,600	2,200
Carbon tetrachloride	CCl_4	2,000	1,400	500
Methyl chloride	CH_3CCl_3	360	110	35
HCFC 22	CF_2HCl	4,300	1,700	520
HCFC 141b	$C_2FH_3Cl_2$	1,800	630	200
HCFC 142b	$C_2F_2H_3Cl$	4,200	2,000	630
HCFC 123	$C_2F_3HCl_2$	300	93	29
HCFC 124	C_2F_4HCl	1,500	480	150
HCFC 225ca	$C_3F_5HCl_2$	550	170	52
HCFC 225cb	$C_3F_5HCl_2$	1,700	530	170

*Referenced to the absolute GWP for the adopted carbon cycle model CO_2 decay response and future CO_2 atmospheric concentrations held constant at current levels. (Only direct effects are considered.)

Congress instructed that the EPA rank class I and class II substances in terms of their global warming potential (see Table 16.3). The global warming potential rankings are to be used for reference and are not allowed to be considered in policy making decisions by the EPA.

Any individual may petition the EPA to add a class I or class II substance. Such a petition would be reviewed and a response published within 180 days of the petition's submission. Any petition submitted is required to include data on the substance adequate to support the petition. No class I substances may be removed from the list. Class II substances may be removed only if they are simultaneously added to the class I list in lieu of the class II deletion.

16.2   Phaseout of Production and Consumption of Class I Substances

Production phaseout of class I substances began in January of 1991. The phaseout schedule of class I substances is listed in Table 16.4. Effective January 1, 2000, methyl chloroform will be the only remaining class I substance allowed to still be in production. By January 1, 2002, all class I substances, including methyl chloroform, will be unlawful for production. Consumption phaseout of class I substances is to reflect the same schedule of Table 16.4.

TABLE 16.4   Clean Air Act Amendments of 1990 Phaseout Schedule for Production of Ozone-Depleting Substances

Date	Carbon tetrachloride, %	Methyl chloroform, %	Other class substances, %
1994	70	85	65
1995	15	70	50
1996	15	50	40
1997	15	50	15
1998	15	50	15
1999	15	50	15
2000		20	
2001		20	

Exemptions to the production and consumption phaseout schedule of Table 16.4 are possible for several groups of consumers: medical usage, airline safety, developing countries, national security, and fire safety. Metal fatigue and corrosion testing of existing airline parts currently utilize several class I compounds. If an acceptable substitute is not found, limited production and consumption of these substances may be allowed specifically for such testing. Medical devices requiring class I substance production may also receive such an exemption. Any exemption is subject to the regulations of the Montreal Protocol, an international agreement which includes regulations on production and consumption of class I substances. Production of class I substances may also be allowed solely for export to developing nations which are attempting to import class I substances in accordance with article 5 of the Montreal Protocol. Such production shall not exceed quantities satisfying the basic domestic needs of the importing developing nation. The President may authorize production in the interest of national security. The EPA can issue the remaining exemption regarding production allowed for fire suppression and explosion prevention if alternative substitutes cannot be developed. However, such an exemption would exclude fire training or testing.

16.3 Phaseout of Production and Consumption of Class II Substances

Class II substances have less stringent time lines for phaseout requirements than class I substances. Effective January 1, 2015, it shall be unlawful for any person to introduce into interstate commerce or use any class II substance unless such substance (1) has been used, recovered, and recycled; (2) is used and entirely consumed (except for trace quantities) in the production of other chemicals; or

(3) is used as a refrigerant in appliances manufactured prior to January 1, 2020. Production phaseout of class II substances is to become effective on January 1, 2015. After this date it shall be unlawful for any person to produce any class II substance in an annual quantity greater than the amount produced during the baseline year. Class II production will cease to be effective January 1, 2030, after which time it shall be unlawful for any person to produce any class II substance. These dates are long-term goals set into law by Congress, and a more specific phaseout time line is required to be submitted by the EPA by December 31, 1999.

Exemptions to the above generalized time line set forth by Congress authorize the production and use of limited quantities of class II substances solely for purposes of use in medical devices as deemed necessary by the Commissioner of the Food and Drug Administration in consultation with EPA. Production of class II substances may also be authorized by EPA solely for export to developing countries participating in the Montreal Protocol. Such production shall not exceed quantities satisfying the basic domestic needs of the importing developing nation. Furthermore, any person producing class II substances for export to developing nations after January 1, 2015, and before January 1, 2030, shall not exceed 110 percent of the baseline production. After January 1, 2030, production for export to developing nations shall be limited to 15 percent of baseline production. By January 1, 2040, all production including exempted production will terminate. Any further exemptions would need to be reauthorized by Congress.

16.4   Accelerated Schedules

The EPA may accelerate production and consumption phaseout schedules of both class I and class II substances which become more stringent than those currently authorized by Congress provided that (1) based on an assessment of credible scientific information, a more stringent schedule may be necessary to protect human health and the environment from harmful effects on the stratospheric ozone layer associated with a class I or class II substance; (2) the EPA determines a more stringent schedule is practicable; and (3) the Montreal Protocol is modified to reflect similar accelerated time lines.

16.5   Exchange Authority

The EPA was required to issue rules allowing for the transfer and trading of production and consumption of class I and class II substances. The rules are to ensure that the transactions result in greater total reductions in production than would otherwise occur. The rules permit a production allowance for a substance for any year to be transferred for a production allowance for another substance for the same year on an ozone depletion–weighted basis. Allowances for substances in each group of class I substances may only be transferred for allowances for other substances within the same group. Groups set up by the EPA in class II are to function in a similar transfer manner. Interpollutant transfers occur within an individual producer's allowance; trades, however, occur between two or more persons. Consumption allowances are also set such that they may be transferred or traded in a similar manner.

16.6 National Recycling and Emission Reduction Program

The EPA was required to issue rules including requirements that reduce the use and emission of such substances to the lowest achievable level, and maximize the recapture and recycling of such substances. Such regulations may include requirements to use alternative substances in lieu of class I or class II substances. Furthermore, prior to disposal or recycling of any good, machine, or product containing class I or class II substances in bulk, the substance must be removed. Any new good, machine, or product containing class I or class II substances in bulk being manufactured, sold, or distributed must be equipped with a design feature to facilitate the recapture of such substances during service, repair, or disposal. Any other product or good containing class I or class II substances, not necessarily in bulk, shall be disposed of in a manner that reduces, to the maximum extent practicable, the release of such substance into the environment. Furthermore, it is unlawful for any person, in the course of maintaining, servicing, repairing, or disposing of an appliance or industrial process refrigeration, to knowingly vent or otherwise knowingly release or dispose of any class I, class II, or substitute for a class I or II substance used as a refrigerant in the said equipment in a manner which permits the substance to enter the environment.

16.7 Servicing of Motor Vehicle Air Conditioners

The 1990 Clean Air Act now requires that all persons servicing motor vehicle air conditioners acquire, and properly use, approved refrigerant recycling equipment and that each individual be properly trained and certified. Each certification requires the name and address of the person to be certified and the serial number of each unit of approved recycling equipment. Certifications may be made by submitting the required information to the EPA on a standard form provided by the manufacturer of the certified refrigerant recycling equipment. This section of the 1990 Clean Air Act also prohibits the offer, sale, or distribution of any class I or class II substance that is suitable for use as a refrigerant in a motor vehicle air conditioning system and that is in a container which contains less than 20 lb of such refrigerant.

16.8   Nonessential Products Containing Chlorofluorocarbons

Congress instructed the EPA to promulgate regulations that would ban nonessential products which contained class I (group I or III) or class II substances. EPA did so on January 15, 1993, and later revised the regulations in December of the same year. The ban is now listed as subpart C under Part 82 of Title 40. The ban prohibits the sale or distribution of a broad range of products which contain chlorofluorocarbons, such as noise horns, cleaning fluids for electronic and photographic equipment, packaging foams, and aerosol products. Exemptions from the ban are available for commercial users and medical needs. Any commercial user will be required to post information in a prominent area which states that resale to a noncommercial user is prohibited by law.

16.9   Labeling

Congress has required that all containers used to store or transport class I or class II substances contain labels stating that the contents of the container harm public health and the environment by destroying ozone in the upper atmosphere. A similar label is required for any product which contains a class I substance or requires a class I substance for production. These requirements went into effect on May 15, 1993. Beginning January 1, 2015, products which contain class II substances or were manufactured using class II substances will require labeling as well. More specific labeling specifications are outlined as subpart E under Part 82 of Title 40.

Enforcement and Administration

Charles Gallagher

17.1 Introduction

To enforce the regulations set forth by the Clean Air Act, Congress has given the EPA four enforcement powers. The situations to which these powers may be applicable vary with the intensity of the violation. These powers are as follows:

- The EPA may issue an order requiring compliance from an entity found in noncompliance.
- The EPA may issue an administrative penalty to an entity found in noncompliance.
- The EPA may bring a civil action against an entity found in non-compliance.
- The EPA may request the Attorney General to commence a criminal action against an entity found in noncompliance.

Possible infringement of compliance has been broken down into four categories:

- A person or entity could be failing to comply with a state implementation plan (SIP).
- A state could be failing to enforce a SIP or Title V permit program.
- A state could be violating new-source requirements relating to the construction of new sources or the modification of existing sources.

- A person or entity could be found in violation of any general prohibitions stated throughout the Titles of the CAA or federal regulations which have been promulgated under the authority of the CAA.

If the EPA finds that any person has violated any requirement or prohibition of an applicable implementation plan (SIP) or permit (Title V), the EPA will notify the person and the applicable state of the violation. Thirty days after the notification is issued, the EPA may implement an order requiring compliance, issue an administrative penalty, or bring civil action upon the violators.

If the EPA finds that violations of an applicable implementation plan (SIP) or an approved permit program (Title V) are so widespread that such violations appear to result from a failure of the state charged with enforcing them, the EPA will so notify the state. Upon such notification, the state will have 30 or 90 days to remedy problems regarding an implementation plan (SIP) or approved permit program (Title V), respectively. If the problem has not been remedied by the end of the applicable aforementioned period, the EPA will give public notice of such failure. Until such time as the EPA is satisfied that the state will enforce the implementation plan (SIP) or approved permit program (Title V), the EPA may enforce the plan or permit program on its own.

If the EPA finds that any general prohibition or requirement of the CAA not covered by the enforcement strategies outlined above for a SIP or approved permit program is being violated, the EPA may immediately issue an administrative penalty, issue an order requiring compliance, bring civil action upon the violators, and/or request that criminal actions be brought against the violators.

If the EPA finds that a state is failing to enforce requirements or prohibitions relating to the construction of new sources or the modification of existing sources, the EPA may issue an order prohibiting the construction or modification of any major stationary source in an applicable area, may issue an administrative penalty, and/or may bring civil action upon the violators.

17.2   Civil Actions

Civil actions brought against violators may be summarized as follows. The EPA may commence a civil action for a permanent or temporary injunction to halt further violations and/or assess and recover a civil penalty of not more that $25,000 per day for each violation. Any civil action is to be brought in the district court of the United States for the district in which the violation is alleged to have occurred, or in which the defendant resides, or where the defendant's principal place of business is located. Furthermore, the costs of attorneys' fees and expert witnesses may be recovered by the EPA against a defendant who loses a civil action.

17.3 Criminal Penalties

Criminal penalties vary in size and scope for violators. Convicted violators of a state implementation plan, requirements for new-source performance standards, operating permits, prohibitions relating to acid deposition control or stratospheric ozone depletion, and several other infractions can face fines and imprisonment not to exceed 5 years for a first offense. Any person who knowingly makes any false material statement, representation, certification; omits material information; alters or conceals documentation; or falsifies, renders inaccurate, or fails to install any monitoring device can face fines and imprisonment not to exceed 2 years. Any person who knowingly fails to pay any fee owed to the United States under the CAA can face fines and imprisonment not to exceed 1 year. Any person who negligently releases a hazardous air pollutant or hazardous substance into the air and at that time places a person in imminent danger of death or serious bodily injury may, upon conviction, receive fines and imprisonment not to exceed 1 year. Any person who knowingly commits the same offense and is convicted can face fines and imprisonment not to exceed 15 years. Repeat convictions of any infringement stated above may result in the maximum punishment's being doubled for both applicable fines and imprisonment.

Pollutant Control Systems

Ventilation

Barney Fullington

Ventilation is the collection and transportation of air contaminants from the work environment to the atmosphere for the purposes of reducing the exposure levels of the individual worker. An *air contaminant,* as described in this chapter, refers to any airborne substance that may be detrimental to the health of exposed workers. An air contaminant may be particulate matter, gases or vapors, excessive heat, or extreme cold.

Ventilation consists of supply systems (heating and air conditioning) and exhaust systems. This chapter focuses primarily on exhaust systems, yet several topics discussed here also pertain to climate control. An exhaust system includes a series of hoods (air collection devices), ductwork (containment devices), fans (transportation devices), and exhaust stacks. Exhaust systems are usually the most overlooked component of the total air pollution control system. Without proper capture efficiencies of contaminants, subsequent air pollution control devices have a lower overall effectiveness at controlling the air pollution from a given plant. Therefore ventilation plays an important role in a plant's air pollution management. The following chapter describes the devices used in exhaust systems, their design, and their application.

18.1 Fluid Mechanics

Any discussion of ventilation must include a basic understanding of fluid mechanics. *Fluid mechanics* in the context of ventilation refers to the characteristics of air as it is transported from the work environment through hoods, ductwork, fans, and exhaust stacks. This section gives a brief overview of those basic fundamentals.

There exist three main characteristics that describe any gas: volume, temperature, and pressure. The ideal gas law governs the relationship of these three characteristics. In practical applications, the ideal gas law is only an approximation of the actual conditions of a gas. For most ventilation applications, though, the approximation is very good. Equation (18.1) shows the general form of the ideal gas law:

$$PV = nRT \qquad (18.1)$$

where P = pressure of gas, psia (kPa)
   V = volume of gas, ft^3 (m^3)
   n = number of moles of gas, lb-mol (kg-mol)
   R = universal gas constant
   T = absolute temperature of the gas, °R (K)

The universal gas constant in both British and SI units is

$$10.73 \; \frac{\text{psia} \cdot \text{ft}^3}{\text{lb-mol} \cdot °\text{R}} \left(\frac{8.314 \; \text{kPa} \cdot \text{m}^3}{\text{kg-mol} \cdot \text{K}} \right)$$

The number of moles of gas can be determined from the following equation:

$$n = \frac{M}{MW} \qquad (18.2)$$

where M = weight of the gas, lbm (kg) and MW = molecular weight of the gas, lbm/(lb-mol) [kg/(kg-mol)].

The weight of the gas can be calculated from the following relationship:

$$M = \rho V \qquad (18.3)$$

where ρ = density of the gas, lbm/ft^3 (kg/m^3). Substituting Eq. (18.3) into Eq. (18.1) provides a more useful version of the ideal gas law:

$$P = \frac{\rho RT}{MW} \qquad (18.4)$$

The density of dry air is usually approximated to 29 lbm/(lb-mol) [29 kg/(kg-mol)]; however, the density varies with temperature and pressure. At standard atmospheric pressure, 14.7 psia (101.4 kPa), and standard temperature, 70°F (21°C), the density of dry air is 0.075 lbm/ft^3 (1.2 kg/m^3) (Perry et al., 1984). Table 18.1 gives the density at standard atmospheric pressure and various temperatures.

The flow of air is described by its volumetric flow rate in cubic feet per minute (ft^3/min, or cfm) or cubic meters per minute (m^3/min). The

TABLE 18.1   Density of Air at Various
Temperatures

Temperature, °F	Air density, lbm/ft^3
0	0.0863
10	0.0845
20	0.0827
30	0.0811
40	0.0794
50	0.0779
60	0.0763
70	0.0749
80	0.0734
90	0.0721
100	0.0708
110	0.0696
120	0.0684
130	0.0673
140	0.0662
150	0.0651
200	0.0601
250	0.0558
300	0.0521
350	0.0489
400	0.0460
450	0.0435
500	0.0412
600	0.0373
700	0.0341
800	0.0314
900	0.0295
1000	0.0275
1200	0.0238
1400	0.0212
1600	0.0192
1800	0.0175
2000	0.0161

volumetric flow rate is the velocity of the airstream multiplied by the area through which it is passing:

$$Q = vA \qquad (18.5)$$

where Q = volumetric flow rate, cfm (m^3/min)
$\quad v$ = air velocity, ft/min (m/min)
$\quad A$ = cross-sectional area, ft^2 (m^2)

As shown in Eq. (18.1), the volume of a gas is inversely proportional to its pressure and directly proportional to its temperature. Therefore, to compare volumetric flow rates at various temperatures and pressures, the flow rate at the actual condition [acfm (m^3/min)] is converted to a flow rate at standard condition [scfm (Nm³/min)] by Eq. (18.6):

$$Q_s = Q_a \times \frac{T_s}{T_a} \times \frac{P_a}{P_s}$$ (18.6)

where Q_s = volumetric flow rate at standard conditions, scfm (Nm³/min)
Q_a = volumetric flow rate at actual conditions, acfm (m³/min)
T_a = standard temperature = 70°F (21°C)
T_s = actual air temperature, °F (°C)
P_s = standard pressure = 14.7 psia (101.4 kPa)
P_a = actual air pressure, psia (kPa)

A very important design parameter for exhaust systems is the velocity pressure. This is the pressure required to accelerate air at rest (zero velocity) to a given velocity and is related to the kinetic energy of air at the given velocity. Therefore, the velocity pressure is present only in the direction of flow. The relationship between air velocity and velocity pressure is calculated by Eq. (18.7) (ACGIH, 1995):

$$v = F\sqrt{\frac{P_v}{\rho}}$$ (18.7)

where v = velocity of airstream, ft/min (m/min)
P_v = velocity pressure of airstream, inH$_2$O gauge (kPa gauge)
ρ = density of air, lbm/ft³ (kg/m³)
F = conversion factor = 1096 (2672 in SI)

At standard temperature and atmospheric pressure in which the density of air is 0.075 lbm/ft³ (1.2 kg/m³), Eq. (18.7) reduces to

$$V = F\sqrt{P_v}$$ (18.8)

where F = conversion factor = 4005 (2439 in SI).

Table 18.2 gives the velocity pressure at various air velocities at standard conditions.

Another form of pressure in the exhaust system is the static pressure. The *static pressure* of air in a hood, duct, fan, or stack is defined as the axial and radial pressure that is exerted by the air. The sum of the velocity pressure and the static pressure at a given point is the total pressure at that point.

18.2   Hoods

Hoods are devices that collect airborne contaminants in the exhaust system. Hoods use an induced draft to provide suction that captures

TABLE 18.2 Conversion of Velocity V (ft/min) to Velocity Pressure P_v (inH$_2$O gauge) at Standard Conditions

V	P_v	V	P_v	V	P_v	V	P_v	V	P_v
100	0.001	3,100	0.60	6,100	2.32	9,100	5.16	12,100	9.13
200	0.002	3,200	0.64	6,200	2.40	9,200	5.28	12,200	9.28
300	0.006	3,300	0.68	6,300	2.47	9,300	5.39	12,300	9.43
400	0.01	3,400	0.72	6,400	2.55	9,400	5.51	12,400	9.59
500	0.02	3,500	0.76	6,500	2.63	9,500	5.63	12,500	9.74
600	0.02	3,600	0.81	6,600	2.72	9,600	5.75	12,600	9.90
700	0.03	3,700	0.85	6,700	2.80	9,700	5.87	12,700	10.06
800	0.04	3,800	0.90	6,800	2.88	9,800	5.99	12,800	10.21
900	0.05	3,900	0.95	6,900	2.97	9,900	6.11	12,900	10.37
1,000	0.06	4,000	1.00	7,000	3.05	10,000	6.23	13,000	10.54
1,100	0.08	4,100	1.05	7,100	3.14	10,100	6.36	13,100	10.70
1,200	0.09	4,200	1.10	7,200	3.23	10,200	6.49	13,200	10.86
1,300	0.11	4,300	1.15	7,300	3.32	10,300	6.61	13,300	11.03
1,400	0.12	4,400	1.21	7,400	3.41	10,400	6.74	13,400	11.19
1,500	0.14	4,500	1.26	7,500	3.51	10,500	6.87	13,500	11.36
1,600	0.16	4,600	1.32	7,600	3.60	10,600	7.00	13,600	11.53
1,700	0.18	4,700	1.38	7,700	3.70	10,700	7.14	13,700	11.70
1,800	0.20	4,800	1.44	7,800	3.79	10,800	7.27	13,800	11.87
1,900	0.23	4,900	1.50	7,900	3.89	10,900	7.41	13,900	12.05
2,000	0.25	5,000	1.56	8,000	3.99	11,000	7.54	14,000	12.22
2,100	0.27	5,100	1.62	8,100	4.09	11,100	7.68	14,100	12.39
2,200	0.30	5,200	1.69	8,200	4.19	11,200	7.82	14,200	12.57
2,300	0.33	5,300	1.75	8,300	4.29	11,300	7.96	14,300	12.75
2,400	0.36	5,400	1.82	8,400	4.40	11,400	8.10	14,400	12.93
2,500	0.39	5,500	1.89	8,500	4.50	11,500	8.24	14,500	13.11
2,600	0.42	5,600	1.96	8,600	4.61	11,600	8.39	14,600	13.29
2,700	0.45	5,700	2.03	8,700	4.72	11,700	8.53	14,700	13.47
2,800	0.49	5,800	2.10	8,800	4.83	11,800	8.68	14,800	13.66
2,900	0.52	5,900	2.17	8,900	4.94	11,900	8.83	14,900	13.84
3,000	0.56	6,000	2.24	9,000	5.05	12,000	8.98	15,000	14.03

airborne contaminants. There exist many different types of hoods. The specific type of hood chosen is dependent on the process configuration, air contaminant discharge characteristics, and space availability. The most common types of hoods used for collection are canopy, capturing, and semienclosed hoods. Figure 18.1 shows the various types of hoods with design guidelines for each.

Canopy hoods are mounted a fixed distance above the emission source. Typically canopy hoods are used for processes that emit hot contaminants which are lighter than the ambient air. The density difference enables the hoods to readily collect the hotter, lighter contaminants with minimal airflow.

Capturing hoods are used when the source emissions have the same approximate density as the ambient air. Capturing hoods rely

HOOD TYPE	DESCRIPTION	ASPECT RATIO,W/L	AIR FLOW
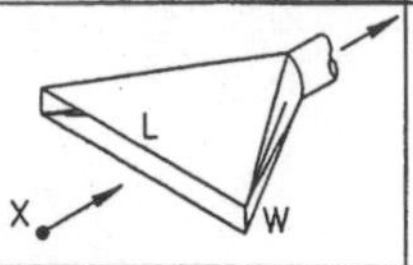	SLOT	0.2 OR LESS	$Q = 3.7\ LVX$
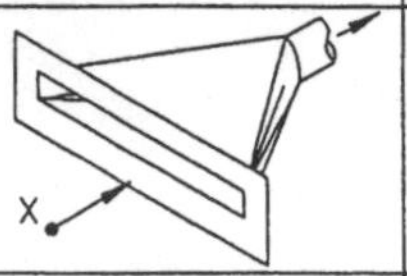	FLANGED SLOT	0.2 OR LESS	$Q = 2.6\ LVX$
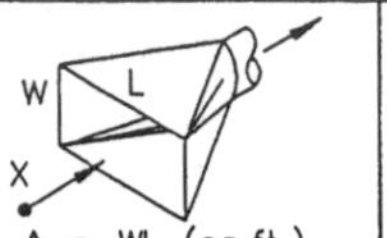 $A = WL$ (sq.ft.)	PLAIN OPENING	0.2 OR GREATER AND ROUND	$Q = V(10X^2 + A)$
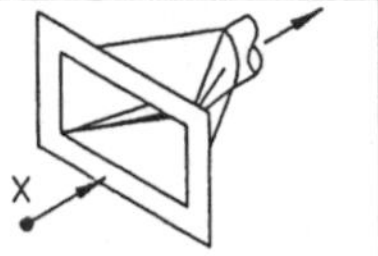	FLANGED OPENING	0.2 OR GREATER AND ROUND	$Q = 0.75V(10X^2 + A)$
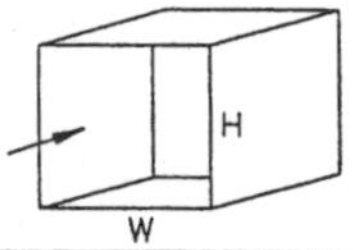	BOOTH	TO SUIT WORK	$Q = VA = VWH$
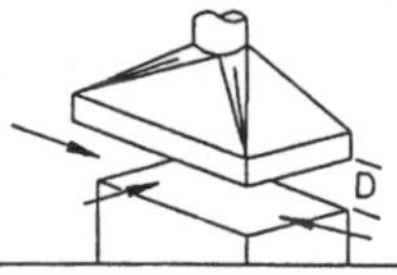	CANOPY	TO SUIT WORK	$Q = 1.4\ PVD$ SEE VS– 99-03 $P = $ PERIMETER $D = $ HEIGHT ABOVE WORK
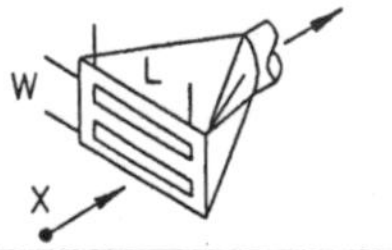	PLAIN MULTIPLE SLOT OPENING 2 OR MORE SLOTS	0.2 OR GREATER	$Q = V(10X^2 + A)$
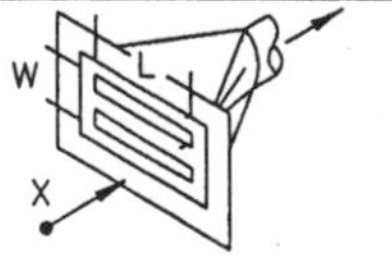	FLANGED MULTIPLE SLOT OPENING 2 OR MORE SLOTS	0.2 OR GREATER	$Q = 0.75V(10X^2 + A)$

Figure 18.1 Exhaust hood types. (*From American Conference of Governmental Industrial Hygienists, Industrial Ventilation, A Manual of Recommended Practice, 22d ed., Cincinnati, Ohio, 1995.*)

on larger capture velocities to collect the airborne contaminants from the emission source. A common type of capturing hood is the hood adjacent to a wood table saw. Another type of capturing hood is the slotted hood that uses one or more slots to increase the capture veloci-

TABLE 18.3   Recommended Capture Velocities for Adequate Ventilation

Conditions of generation, dispersion, or release of contaminants	Minimum capture velocity, ft/min	Example of process or operation
Released with no significant velocity into relatively quiet air	100	Evaporation from tanks, degreasing, plating
Released with low initial velocity into moderately quiet air	100–200	Spray paint booths, intermittent dumping of dry material into container, welding
Released with considerable initial velocity into zone of rapid air movement	200–500	Spray painting in shallow booth, active container filling, conveyor loading
Released with high velocity into zone of very rapid air movement	500–2000	Grinding, abrasive blasting, tumbling

SOURCE: Allen D. Brandt, *Heating and Ventilating,* vol. 42, no. 5, 1945.

ty by reducing the flow area. A typical use for slotted hoods is for collecting mist emissions from a plating tank in which the slots are located at the edge of the tank.

Semienclosed hoods are collection devices that surround the source of the airborne contaminant as completely as the process allows. Due to the variation in machinery and process demands, semienclosed hoods are designed specifically for a given process. An advantage of semienclosed hoods over canopy hoods is that higher capture efficiencies are possible since the enclosure provides much less space for the contaminants to disperse into the ambient plant air. An example of a semienclosed hood is a spray booth used for the application of paints and other coatings.

18.2.1   Design factors

The velocity at which air is pulled into hoods is called the *capture velocity*. Capture velocities sufficient for venting emission sources are dependent on several factors, such as the velocity of the contaminant emission from the source, conditions of the surrounding plant air, toxicity or nuisance of contaminant, and contaminant generation rate. Table 18.3 gives recommended capture velocities for various conditions. With the recommended capture velocities, hood size, and the equation for flow rate given in Fig. 18.1, the airflow rate needed for each hood is calculated.

The hood size is dependent on the emission source size and should be selected to (1) provide adequate coverage area for the source and (2) be small enough to reduce the airflow rates needed to achieve the required capture velocities. The following example demonstrates the procedure for preliminary hood design.

Example 18.1   Spray Booth Design for Staining Wood Furniture   Wood furniture travels on a conveyor through several open-sided spray booths in which stain, lacquer, sealer, and glaze are applied by manually operated spray guns. The conveyor has a speed of 9 linear ft/min (2.74 m/min) while the time needed to adequately coat a wooden chair with stain is 1.5 min (Simpson, 1996). Therefore, the space needed for the staining process is 13.5 linear ft (4.11 m). The spray booth must accommodate the height of the operators; a typical booth height of 7 ft (2.13 m) is recommended. Therefore the spray booth dimensions should be 13.5 ft (4.11 m) by 7 ft (2.13 m) tall. The air is exhausted horizontally through the rear of the spray booth which has an area of 94.5 ft^2 (8.8 m^2). The recommended capture velocity for this type of operation is 150 ft/min (45.7 m/min) from Table. 18.3. Therefore, the required airflow rate, using the flow rate equation for spray booths from Table 18.2, is

$$Q = 94.5 \text{ ft}^2 \times 150 \text{ ft/min } (8.8 \text{ m}^2 \times 45.7 \text{ m/min})$$

$$= 14{,}175 \text{ cfm } (401 \text{ m}^3/\text{min})$$

This required airflow rate is then used to size the remainder of the exhaust system which includes the connecting ductwork, exhaust fan, and stack.

18.2.2   Hood losses

The main design parameters used in sizing ventilation equipment are the flow rate, which can be calculated by using the method described in the previous section, and the total pressure of the system. The total pressure of the system is calculated as the pressure developed by the air velocity plus the pressure (head) losses of the air as it travels through the system. Two types of head losses are encountered in exhaust systems: friction losses and fitting losses. Simple hoods have plain, flanged, or rounded-duct openings; canopy hoods are hoods in which the airflow encounters only one significant energy loss. As the air enters the hood, the streamlines converge at some distant point downstream of the hood entrance (or orifice). The point at which the streamlines become parallel is called the *vena contracta*, as shown in Fig. 18.2.

After the vena contracta, the air velocity decreases as the air expands to the size of the duct. This uncontrolled expansion of the air causes a loss in energy. The smaller the vena contracta, the larger the pressure loss across the orifice. An analogy to the energy loss from a vena contracta would be to drive a car around a corner. If the corner

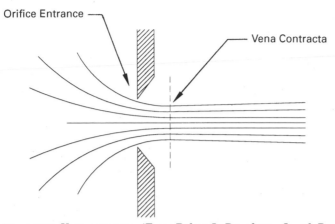

Figure 18.2   Vena contracta. (*From Robert L. Daugherty, Joseph B. Franzini, and E. John Finnemore, Fluid Mechanics with Engineering Application, McGraw-Hill, New York, 1985.*)

requires one to make a U-turn similar to Fig. 18.3*a*, the car must slow down considerably. If the corner is sharp as in Fig. 18.3*b,* the decrease in speed is less; and if the corner is well rounded (Fig. 18.3*c*), the car is able to round the corner with a significantly lower loss in speed. The head loss due to the vena contracta can be expressed in terms of a hood loss factor. Multiplying the hood loss factor by the velocity pressure of the airflow gives the hood entry loss in inches of water, as shown in the following equation:

$$h_e = F_h P_v \qquad (18.9)$$

where h_e = head loss associated with hood entry, inH_2O gauge (kPa gauge)
F_h = head loss factor, dimensionless
P_v = duct velocity pressure, inH_2O gauge (kPa gauge)

The hood static pressure is the pressure required to accelerate the air from rest to the recommended capture velocity and to overcome the hood entry head loss. The hood static pressure is calculated by the following equation:

$$P_s = P_v + h_e \qquad (18.10)$$

where P_s = hood static pressure, inH_2O gauge (kPa gauge).
This method of hood static pressure calculation is applicable only for simple hoods with capture velocities less than 1000 ft/min (305 m/min) (ACGIH, 1995). For those with capture velocities greater than 1000 ft/min (305 m/min) and hoods with more than one source of head loss

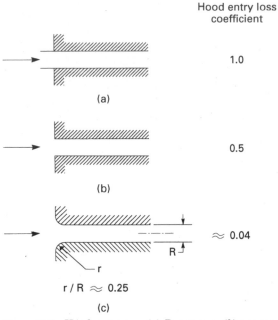

Figure 18.3   Hood entrance. (*a*) Reentrant, (*b*) square-edged, and (*c*) well rounded. (*From Fox, Robert W., and McDonald, Alan T., Introduction to Fluid Mechanics, Wiley, New York, 1978.*)

(compound hoods), each source of head loss is added to the duct velocity pressure to determine the hood static pressure (ACGIH, 1995). An example of a compound hood is shown in Fig. 18.1 in which a slot hood with a plenum is followed by a transition section. The losses occur at the slot orifice and at the transition section. Table 18.4 gives head loss fac-

TABLE 18.4   Loss Factors for Tapered Hood Entry

Degree of taper	Round	Rectangular
15	$0.15P_v$	$0.25P_v$
30	$0.08P_v$	$0.16P_v$
45	$0.06P_v$	$0.15P_v$
60	$0.08P_v$	$0.17P_v$
90	$0.15P_v$	$0.25P_v$
120	$0.26P_v$	$0.35P_v$
150	$0.40P_v$	$0.48P_v$
180	$0.50P_v$	$0.50P_v$

SOURCE: American Conference of Governmental Industrial Hygienists, *Industrial Ventilation: A Manual of Recommended Practice*, 22d ed., Cincinnati, Ohio, 1995.

tors for different tapered hoods. The calculation of the head loss associated with the slot and the tapered transition follows Eq. (18.9). The total hood static pressure from a compound hood follows

$$P_s = P_v + h_{es} + h_{et} \qquad (18.11)$$

where h_{es} = head loss due to the slot, inH_2O gauge (kPa gauge) and h_{et} = head loss due to the tapered transition, inH_2O gauge (kPa gauge). The hood static pressure is used along with the airflow rate calculated in Sec. 18.2.1 to correctly size the exhaust fan needed for the process.

18.2.3 Cost of hoods

The purchased equipment cost of hoods is a factor of the size of the hood, the materials of construction, and the fabrication labor. Purchased equipment cost estimation of hoods can be based on equations developed by Vatavuk (1990), adjusted for inflation to 1996 dollars. For circular canopy hoods, the equation for cost estimation is as follows:

$$\text{PEC} = Fd_h^{1.62} \qquad (18.12)$$

where PEC = purchased equipment cost of canopy hoods, 1996 dollars
d_h = diameter of circular hood, ft (m)
F = conversion factor = 58.9 (404 in SI)

This equation is based on the use of carbon steel with a thickness of $\frac{3}{16}$ in (0.47 mm) or less and a 35° tapered hood. For thicker and more exotic materials of construction, the price will obviously increase. Steeper slope angles of taper would also increase PEC. For rectangular canopy hoods with the same degree of taper and the same materials of construction, the cost estimation equation is

$$\text{PEC} = FL^{1.80} \qquad (18.13)$$

where L = length of the hood, ft (m), and F = conversion factor = 48.8 (414 in SI). For this equation it is assumed that the length-to-width ratio of the hood is 1. For higher length-to-width ratios at any given length, the purchased equipment price would decrease. These cost equations give only a rough estimate of the cost of canopy-type hoods. Vendors of ventilation systems should be consulted for more accurate numbers and for prices on semienclosed and other hood configurations.

18.3 Ductwork

Ductwork is the device through which the collected contaminants and air travel from the collection point to the point of exhaust from the plant. The selection of the type of ductwork needed is based on the physical and chemical characteristics of the exhaust gas stream. These characteristics include temperature, pressure, and corrosivity of the air-contaminant mixture. Two standard materials used for ducting are black iron and galvanized sheet metal. These materials can be used over a wide range of conditions. For extreme conditions, fabricated plastics or specially coated metal should be used. For specific applications, duct suppliers should be consulted.

The shape of the duct is dependent on the cost and space constraints. Typically circular ducts have higher capital costs than rectangular ducts but provide lower friction losses than rectangular ducts of the same material. Therefore, the increased capital cost may be offset by the decrease in fan capital and operating costs.

18.3.1   Design factors of ductwork

The primary design factors for ductwork are the airflow rate and minimum duct velocity. The flow rate will dictate the size of duct needed to maintain minimum necessary duct velocity according to Eq. (18.5). For exhaust systems that handle particulate-laden emissions, a minimum duct velocity is required to prevent settling and plugging of the duct by the particulate matter. Table 18.5 gives recommended minimum duct velocities for several types of contaminant-laden airstreams. The values in Table 18.5 are higher than the theoretical and experimental minimal duct velocity requirements to account for differences between actual operation and design conditions.

18.3.2   Duct losses

The head losses in a duct consist of friction and fitting losses. Friction losses are those that are caused by friction between the air and the duct material. The friction losses for circular ducts are calculated from

$$H_f = \frac{fLP_v}{d} \qquad (18.14)$$

where H_f = friction loss, inH_2O gauge (kPa gauge)
$\quad f$ = friction factor, dimensionless
$\quad L$ = length of duct, ft (m)
$\quad P_v$ = duct velocity pressure, inH_2O gauge (kPa gauge)
$\quad d$ = duct diameter, ft (m)

Friction factors for common duct materials are given in Tables 18.6 and 18.7. Friction factors for nonstandard duct material should be obtained from the material manufacturer. Calculating the friction

TABLE 18.5   Minimum Duct Velocities

Nature of pollutant	Duct velocity, ft/min	Example of pollutant
Vapors, gases, smoke	1000–2000	All vapors, gases, and smoke
Fumes	2000–2500	Welding
Very fine light dust	2500–3000	Cotton lint, wood flour, litho powder
Dry dust and powders	3000–4000	Fine rubber dust, jute lint, cotton dust, light shavings, soap dust, leather shavings
Average industrial dust	3500–4000	Grinding dust, buffing lint (dry), wool jute dust, coffee beans, shoe dust, granite dust, silica flour, general material handling, brick cutting, clay dust, general foundry dust, limestone dust
Heavy dust	4000–4500	Sawdust, metal turnings, foundry tumbling barrels and shake-out, sand blast dust, wood blocks, hog waste, cast iron boring dust lead dust
Very heavy or moist dust	4500 and up	Lead dust with small chips, moist cement dust, asbestos chunks, buffin lint (sticky), quicklime dust

SOURCE: American Conference of Governmental Industrial Hygienists, *Industrial Ventilation: A Manual of Recommended Practice*, 22d ed., Cincinnati, Ohio, 1995.

losses from noncircular ducts requires finding an equivalent diameter, using Eq. (18.15):

$$D_{equiv} = F\frac{(ab)^{0.625}}{(a+b)^{0.25}} \tag{18.15}$$

where D_{equiv} = equivalent round duct size of rectangular duct, in (m)
a = one side of rectangular duct, in (m)
b = adjacent side of rectangular duct, in (m)
F = conversion factor = 1.3 (51.1 in SI)

Once an equivalent diameter is obtained, Eq. (18.14) is used to calculate the losses due to friction.

Fitting losses are the losses due to turbulence and friction caused by disruptions in the airflow from duct fittings. There are two methods for calculating the head losses from fittings. The first method is called the *velocity-pressure method*. In this method, the losses are calculated using the following equation:

TABLE 18.6   Tabulated Friction Loss Factors for Galvanized Sheet Metal Ducts

Diameter, in	Friction loss, No. VP per foot*			Diameter, in	Friction loss, No. VP per foot*		
	2000 ft/min	4000 ft/min	6000 ft/min		2000 ft/min	4000 ft/min	6000 ft/min
0.5	0.9549	0.904	0.8755	33	0.0057	0.0054	0.0052
1	0.4088	0.387	0.3748	34	0.0055	0.0052	0.005
1.5	0.2489	0.2356	0.2282	35	0.0053	0.005	0.0048
2	0.175	0.1657	0.1604	36	0.0051	0.0048	0.0047
2.5	0.1332	0.1261	0.1221	37	0.0049	0.0047	0.0045
3	0.1065	0.1009	0.0977	38	0.0048	0.0045	0.0044
3.5	0.0882	0.0835	0.0809	39	0.0046	0.0044	0.0042
4	0.0749	0.0709	0.0687	40	0.0045	0.0042	0.0041
4.5	0.0649	0.0614	0.0595	41	0.0043	0.0041	0.004
5	0.057	0.054	0.0523	42	0.0042	0.004	0.0039
5.5	0.0507	0.048	0.0465	43	0.0041	0.0039	0.0038
6	0.0456	0.0432	0.0418	44	0.004	0.0038	0.0036
7	0.0378	0.0358	0.0346	45	0.0039	0.0037	0.0036
8	0.0321	0.0304	0.0294	46	0.0038	0.0036	0.0035
9	0.0278	0.0263	0.0255	47	0.0037	0.0035	0.0034
10	0.0244	0.0231	0.0224	48	0.0036	0.0034	0.0033
11	0.0217	0.0206	0.0199	49	0.0035	0.0033	0.0032
12	0.0195	0.0185	0.0179	50	0.0034	0.0032	0.0031
13	0.0177	0.0168	0.0162	52	0.0032	0.0031	0.003
14	0.0162	0.0153	0.0148	54	0.0031	0.0029	0.0028
15	0.0149	0.0141	0.0136	56	0.003	0.0028	0.0027
16	0.0137	0.013	0.0126	58	0.0028	0.0027	0.0026
17	0.0127	0.0121	0.0117	60	0.0027	0.0026	0.0025
18	0.0119	0.0113	0.0109	62	0.0026	0.0025	0.0024
19	0.0111	0.0105	0.0102	64	0.0025	0.0024	0.0023
20	0.0104	0.0099	0.0096	66	0.0024	0.0023	0.0022
21	0.0098	0.0093	0.009	68	0.0023	0.0022	0.0021
22	0.0093	0.0088	0.0085	70	0.0023	0.0021	0.0021
23	0.0088	0.0083	0.0081	72	0.0022	0.0021	0.002
24	0.0084	0.0079	0.0077	74	0.0021	0.002	0.0019
25	0.008	0.0075	0.0073	76	0.002	0.0019	0.0019
26	0.0076	0.0072	0.0069	78	0.002	0.0019	0.0018
27	0.0072	0.0069	0.0066	80	0.0019	0.0018	0.0018
28	0.0069	0.0066	0.0063	82	0.0019	0.0018	0.0017
29	0.0066	0.0063	0.0061	84	0.0018	0.0017	0.0017
30	0.0064	0.006	0.0058	86	0.0018	0.0017	0.0016
31	0.0061	0.0058	0.0056	88	0.0017	0.0016	0.0016
32	0.0059	0.0056	0.0054	90	0.0017	0.0016	0.0015

*VP = velocity pressure (foot water column, gauge).

SOURCE: American Conference of Governmental Industrial Hygienists, *Industrial Ventilation, A Manual of Recommended Practice,* 22d ed., Cincinnati, Ohio, 1995.

TABLE 18.7   Tabulated Friction Loss Factors for Black Iron, Aluminum, Stainless-
Steel, and PVC Ducts

Diameter, in	Friction loss, No. VP per foot*			Diameter, in	Friction loss, No. VP per foot*		
	2000 ft/min	4000 ft/min	6000 ft/min		2000 ft/min	4000 ft/min	6000 ft/min
0.5	0.7963	0.7242	0.6850	33	0.0051	0.0047	0.0044
1	0.3457	0.3143	0.2974	34	0.0050	0.0045	0.0043
1.5	0.2121	0.1929	0.1825	35	0.0048	0.0043	0.0041
2	0.1500	0.1364	0.1291	36	0.0046	0.0042	0.0040
2.5	0.1147	0.1043	0.0987	37	0.0045	0.0041	0.0038
3	0.0921	0.0837	0.0792	38	0.0043	0.0039	0.0037
3.5	0.0765	0.0696	0.0658	39	0.0042	0.0038	0.0036
4	0.0651	0.0592	0.0560	40	0.0041	0.0037	0.0035
4.5	0.0565	0.0514	0.0486	41	0.0040	0.0036	0.0034
5	0.0498	0.0453	0.0428	42	0.0038	0.0035	0.0033
5.5	0.0444	0.0404	0.0382	43	0.0037	0.0034	0.0032
6	0.0400	0.0363	0.0344	44	0.0036	0.0033	0.0031
7	0.0332	0.0302	0.0286	45	0.0035	0.0032	0.0030
8	0.0283	0.0257	0.0243	46	0.0034	0.0031	0.0030
9	0.0245	0.0223	0.0211	47	0.0034	0.0030	0.0029
10	0.0216	0.0197	0.0186	48	0.0033	0.0030	0.0028
11	0.0193	0.0175	0.0166	49	0.0032	0.0029	0.0027
12	0.0174	0.0158	0.0149	50	0.0031	0.0028	0.0027
13	0.0158	0.0143	0.0136	52	0.0030	0.0027	0.0026
14	0.0144	0.0131	0.0124	54	0.0028	0.0026	0.0024
15	0.0133	0.0121	0.0114	56	0.0027	0.0025	0.0023
16	0.0123	0.0112	0.0106	58	0.0026	0.0024	0.0022
17	0.0114	0.0104	0.0098	60	0.0025	0.0023	0.0021
18	0.0106	0.0097	0.0092	62	0.0024	0.0022	0.0021
19	0.0100	0.0091	0.0086	64	0.0023	0.0021	0.0020
20	0.0094	0.0085	0.0081	66	0.0022	0.0020	0.0019
21	0.0088	0.0080	0.0076	68	0.0021	0.0020	0.0018
22	0.0084	0.0076	0.0072	70	0.0021	0.0019	0.0018
23	0.0079	0.0072	0.0068	72	0.0020	0.0018	0.0017
24	0.0075	0.0068	0.0065	74	0.0019	0.0018	0.0017
25	0.0072	0.0065	0.0062	76	0.0019	0.0017	0.0016
26	0.0068	0.0062	0.0059	78	0.0018	0.0017	0.0016
27	0.0054	0.0059	0.0056	80	0.0018	0.0016	0.0015
28	0.0063	0.0057	0.0054	82	0.0017	0.0016	0.0015
29	0.0060	0.0055	0.0052	84	0.0017	0.0015	0.0014
30	0.0058	0.0052	0.0050	86	0.0016	0.0015	0.0014
31	0.0055	0.0050	0.0048	88	0.0016	0.0014	0.0014
32	0.0053	0.0048	0.0046	90	0.0015	0.0014	0.0013

*VP = velocity pressure (foot water column, gauge).
SOURCE: American Conference of Governmental Industrial Hygienists, *Industrial Ventilation: A Manual of Recommended Practice,* 22d ed., Cincinnati, Ohio, 1995.

$$H_l = fP_v \qquad\qquad (18.16)$$

where H_l = fitting losses, inH_2O gauge (kPa gauge)
f = fitting loss coefficient, dimensionless
P_v = duct velocity pressure, inH_2O gauge (kPa gauge)

Loss coefficients for a variety of fittings can be obtained from a ventilation vendor.

The second method used to calculate fitting losses is called the *equivalent-length method*. In this method, fittings are assigned an equivalent length of straight duct that would have the same head loss as the fitting. The equivalent lengths are based on duct size and velocity. Equivalent lengths for various fittings assuming a duct velocity of 4000 ft/min (1219 m/min) are presented in App. O (ACGIH, 1995). Once the equivalent length is determined, Eq. (18.14) is used to determine the head loss due to the fitting. The friction factor used in Eq. (18.14) should be chosen for ducts that are constructed of the same material as the fitting.

18.3.3 Types of ductwork

There are two main types of exhaust system ductwork: the conventional, tapered duct system and the plenum exhaust system. The conventional, tapered duct system uses branch ducts and main ducts. The main ducts are sized to handle the exhaust flow rate from several branch ducts. As more branch ducts are connected to the main duct, the size of the main duct progressively increases to handle the larger flows. Both the branch and main ducts are designed to maintain the design duct velocities from Table 18.6. There are two methods to maintain proper flow rates throughout a conventional, tapered duct system: balance by design and the blast gate method. Since air flows in the direction of lowest pressure, at each junction, air will distribute itself according to the pressure losses of each duct. To maintain proper design flow rates using the balance by design method, the static pressure in the ducts at each junction must be the same. The blast gate method utilizes sliding dampers that can be opened or closed with varying degrees of obstruction to the airflow. This method permits a greater degree of operational flexibility than the balance by design method but can lead to greater maintenance problems.

The second type of ventilation system is the plenum exhaust system. This type of system utilizes an oversized main duct, or plenum. The plenum (pressure-equalizing chamber) does not maintain minimum duct velocity requirements. The function of the plenum is to provide a path of minimal head loss from the branch ducts to the fan. Since the main duct is oversized, duct velocities are at most one-half of the design branch velocity and typically less than 1000 ft/min (305 m/min) (ACGIH, 1995). With the low velocities, the main duct in a plenum system has a particulate removal mechanism. The mechanism may be self-cleaning or may require manual cleaning. The self-cleaning mechanisms typically are drag chains or belt conveyors located at the bottom of the duct. Manually cleaned main ducts should be generously overdesigned to facilitate the cleaning process. The hoods and branch ducts of a plenum exhaust system are designed just as those used in a conventional, tapered duct system.

18.3.4   Cost of ductwork

The cost associated with ductwork is dependent on the size of the ducts, length of ductwork, material of construction, and complexity of ductwork system. Cost estimation equations were developed by Vatavuk (1990) for PVC, FRP, carbon-steel, and stainless-steel ducts. The cost equations from Vatavuk (1990) have been corrected to account for inflation. Equation (18.17) shows the correlation between duct diameter and capital cost for PVC ductwork:

$$C = aD^b \qquad (18.17)$$

where C = capital cost of duct (1996 \$/ft)
　　a, b = correlation factors, dimensionless
　　D = duct diameter, in (m)

The correlation parameters for PVC ducts between 6 and 12 in (0.15 and 0.30 m) are 1.10 (51.7 in SI) and 1.05 for a and b, respectively. The correlation parameters for PVC ducts between 14 and 24 in (0.35 and 0.60 m) are 0.093 (133.9 in SI) and 1.98 for a and b, respectively.

　　The cost estimation calculation for FRP duct is presented in Eq. (18.18):

$$C = FD \qquad (18.18)$$

where D = duct diameter, ft (m), and F = conversion factor = 30 (98 in SI). This equation applies only for duct diameters from 2 to 5 ft (0.6 to 1.5 m). For other sizes of FRP ducts, consult a supplier of ductwork.

　　Ducts constructed of carbon steel can be estimated by Fig. 18.4. Stainless-steel ducts can be estimated by applying multipliers to the cost of the carbon-steel ducts. These multipliers are shown in Table 18.8.

18.4   Fans

The exhaust fan is the device used to move the air-contaminant mixture through the exhaust system. The exhaust fan provides both the flow rate needed to develop adequate capture velocities in the hoods

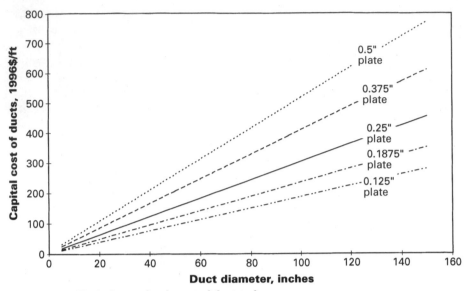

Figure 18.4 Capital cost of carbon-steel ductwork.

TABLE 18.8 Cost Multipliers for Stainless-Steel Ducts

Type of stainless steel	Multiplier
304	3.6
304L	3.7
316	4.3
316L	4.4

SOURCE: William M. Vatavuk, *Estimating Costs of Air Pollution Controls,* Lewis Publishers, Chelsea, Mich., 1990.

and the necessary pressure to overcome the pressure losses in the ductwork. There are two basic types of fans used for exhaust: axial fans and centrifugal fans. The main difference between the two is that the flow of air is axial to the fan motor shaft in an axial fan and radial to the fan motor shaft in a centrifugal fan.

Axial fans consist of three designs based on the blade and housing type: propeller fans, tube axial fans, and vane axial fans. Figure 18.5 shows the three types of axial fans. Propeller fans are used for low-pressure applications between 0 and -1 inH$_2$O gauge (0 and -0.25 kPa gauge) (ACGIH, 1995). A common use for propeller fans is in the wall vent where no duct is present. Tube axial fans are used for mod-

PROPELLER

These are designed to move air from one enclosed space to another or from outdoors to
indoors or vice versa in a wide range of volumes at low pressure (0 to 1 in. water,
static). Automatic shutter on discharge is
not part of fan. It protects fan from wind,
rain, snow, cold during shutdown. Belt or
direct drive

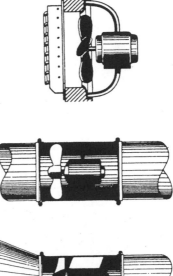

TUBEAXIAL

Axial-flow wheel in cylinder moves wide
range of air or gas volumes at medium pressures (¼ to 2½ in. water, static). Belt or
direct drive. Fan may be mounted in air or
gas duct. Blades may be disk or airfoil type

VANEAXIAL

This fan has a set of air guide vanes mounted
in cylinder before or behind airfoil-type wheel.
It moves air over wide range of volumes and
pressures. Usual pressure range is ½ to 6
in.; special designs can go up to 60 in. or
higher. Wheel often made of sheet metal but
cast metal is also used. Belt or direct drive

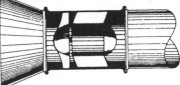

Figure 18.5 Axial fan types. (*From Clifford, George, Modern Heating, Ventilating, and Air Conditioning, Prentice-Hall, Englewood Cliffs, N.J., 1990.*)

erate static pressure applications less than -2 inH$_2$O gauge (-0.5 kPa gauge) (ACGIH, 1995). Tube axial fans are used when the airflow does not encounter any major pressure losses from the source to the fan. A tube axial fan uses a blade that has a hub which is less than 50 percent of the total blade diameter. Vane axial fans are similar to tube axial fans except that vane axial fans have air guide vanes mounted on the drive housing, and the hub of the fan blade is usually greater than 50 percent of the total diameter of the blade. Vane axial fans are used for high pressures up to -8 inH$_2$O gauge (-2 kPa gauge) (ACGIH, 1995). Special designs can operate at pressures up to -60 inH$_2$O gauge (-15 kPa gauge) (Clifford, 1994). These fans can be used when there are some moderate pressure losses from the source to the fan.

Centrifugal fans are designed to move air over a wide range of airflows at moderate to high pressures. Centrifugal fans differ from axial fans in that the airflow enters the blade housing axially and exits the housing tangentially to the fan blades (Fig. 18.6). Centrifugal fans are capable of operating at a higher pressure range than typical axial fans. Centrifugal

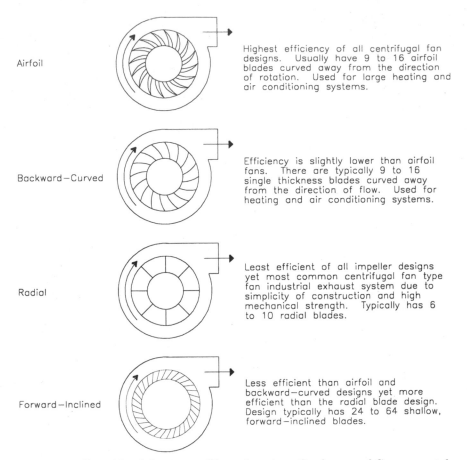

Airfoil

Highest efficiency of all centrifugal fan designs. Usually have 9 to 16 airfoil blades curved away from the direction of rotation. Used for large heating and air conditioning systems.

Backward—Curved

Efficiency is slightly lower than airfoil fans. There are typically 9 to 16 single thickness blades curved away from the direction of flow. Used for heating and air conditioning systems.

Radial

Least efficient of all impeller designs yet most common centrifugal fan type fan industrial exhaust system due to simplicity of construction and high mechanical strength. Typically has 6 to 10 radial blades.

Forward—Inclined

Less efficient than airfoil and backward—curved designs yet more efficient than the radial blade design. Design typically has 24 to 64 shallow, forward—inclined blades.

Figure 18.6 Centrifugal fan types. (*From American Conference of Governmental Industrial Hygienists, Industrial Ventilation: A Manual of Recommended Practice, 22d ed., Cincinnati, Ohio, 1995.*)

fans differ by the type of blade used: airfoil, backward-curved, forward-curved, or radial blade. The different blade types are used in a variety of applications. Airfoil blades are highly efficient and are used primarily in HVAC systems. Backward-curved blades are only slightly less efficient and are used in some industrial applications. Forward-curved blades are used primarily for low-pressure HVAC systems, and radial blades are the most common type used in industrial applications due to their structural integrity and ease of maintenance.

18.4.1   Design factors

Fan performance is based on the fan's flow rate and the fan static or total pressure. Using the fan inlet flow rate, an inlet velocity can be calculated from Eq. (18.5). From the inlet velocity, the fan velocity pressure can be determined by using Eq. (18.8). The fan static pressure is then calculated by Eq. (18.19):

$$\text{Fan } P_s = P_{s,o} - P_{s,i} - P_{v,i} \tag{18.19}$$

where fan P_s = fan static pressure, inH_2O gauge (kPa gauge)
$P_{s,o}$ = static pressure at outlet, inH_2O gauge (kPa gauge)
$P_{s,i}$ = static pressure at inlet, inH_2O gauge (kPa gauge)
$P_{v,i}$ = velocity pressure at inlet, inH_2O gauge (kPa gauge)

The fan's total pressure is calculated by Eq. (18.20):

$$\text{Fan } P_t = P_{s,o} + P_{v,o} - (P_{s,i} - P_{v,i}) \tag{18.20}$$

where fan P_t = fan total pressure and $P_{v,o}$ = velocity pressure at outlet.

From the required airflow rate and fan static or total pressure, the fan size, operating speed, and brake horsepower needed for sizing the motor can be obtained from fan rating tables. Table 18.9 shows a typical fan rating table. The type of pressure used depends on the format of the manufacturer's rating tables. The rating tables are calculated based on a standard air density of 0.075 lbm/ft^3 (1.2 kg/m^3). If the density of the air in a specific application varies from the standard air density by greater than 5 percent, the static pressure at the fan inlet should be corrected to actual conditions by Eq. (18.21) (ACGIH, 1995):

$$P_e = P_a \frac{\rho_s}{\rho_a} \tag{18.21}$$

where P_e = equivalent static pressure, inH_2O gauge (kPa gauge)
P_a = actual static pressure, inH_2O gauge (kPa gauge)
ρ_a = actual air density, lbm/ft^3 (kg/m^3)
ρ_s = air density at standard conditions = 0.075 lb/ft^3 (1.2 kg/m^3)

The equivalent pressure can be either static or total depending on the format of the manufacturer's rating tables. The power requirement must also be corrected for nonstandard density air, as shown in Eq. (18.22):

$$\text{PWR}_a = \text{PWR}_t \times \frac{\rho_a}{\rho_s} \tag{18.22}$$

where PWR_a = actual power requirement, hp, and PWR_t = uncorrected power requirement from rating table, hp. As a rule of thumb, the highest mechanical efficiency at a given fan pressure will occur in the

TABLE 18.9 Typical Fan Rating Table

454LS		Inlet diameter: 26-in OD; outlet area: 3.69 ft² inside; wheel diameter: 45⅛ in; wheel circumference: 11.81 ft								
		2″SP*		4″SP		6″SP		8″SP		10″SP	
ft³/min	OV*	rpm	bhp	rpm	bhp	rpm	bhp	rpm	bhp	rpm	bhp
3,690	1,000	421	2.02	590	4.42	720	7.18	832	10.3	930	13.6
4,428	1,200	427	2.35	592	4.97	721	7.93	832	11.2	930	14.8
5,166	1,400	433	2.72	596	5.56	724	8.74	833	12.2	930	16.0
5,904	1,600	441	3.13	600	6.20	727	9.60	835	13.3	931	17.2
6,642	1,800	450	3.59	606	6.89	731	10.5	837	14.4	933	18.6
7,380	2,000	460	4.09	613	7.65	736	11.5	841	15.6	936	20.0
8,118	2,200	471	4.63	621	8.45	741	12.5	847	16.9	940	21.5
8,856	2,400	482	5.24	629	9.32	749	13.7	852	18.2	944	23.0
9,594	2,600	495	5.90	638	10.2	756	14.8	858	19.6	949	24.7
11,070	3,000	521	7.41	659	12.3	774	17.5	873	22.7	962	28.2
12,546	3,400	549	9.20	682	14.7	793	20.4	890	26.2	977	32.1
14,022	3,800	580	11.4	706	17.3	814	23.6	908	29.9	994	36.4
15,498	4,200	612	13.9	733	20.4	837	27.2	928	34.0	1,012	41.1
16,974	4,600	646	16.8	761	23.9	861	31.2	951	38.6	1,033	46.2
18,450	5,000	681	20.2	791	27.8	888	35.6	975	43.6	1,055	51.8
19,926	5,400	717	24.1	822	32.2	915	40.6	999	49.1	1,077	57.7
21,402	5,800	754	28.6	854	37.2	944	46.0	1,025	55.0	1,102	64.3

*SP = static pressure (inches of water column, gauge), OV = outlet velocity (feet per minute).
SOURCE: New York Blower Company, 1997.

middle third of the flow rate column of the rating table (ACGIH, 1995). For a more exact mechanical efficiency, Eq. (18.23) is used:

$$\eta = \frac{Q \times \text{fan } P_t}{F \times \text{PWR}}$$ (18.23)

where η = fan mechanical efficiency, %
  Q = volumetric flow rate, acfm (m³/min)
  PWR = fan power requirement, hp (kW)
  F = conversion factor = 6356 (60.2 in SI)

18.4.2 Cost of fans

The costs associated with exhaust fans are dependent on the type of fan, the pressure at which the fan is capable of operating, and the flow rate of the fan. The costs increase significantly for high-pressure, high-flow-rate fans compared to low-pressure, low-flow-rate fans. The cost of an exhaust fan will include the cost of a motor capable of producing the required speed and brake horsepower (bhp) from the fan rating tables. For accurate cost estimation of the fans and fan motors,

12"SP		14"SP		16"SP		18"SP		20"SP		22"SP*	
rpm	bhp	rpm	bhp	rpm	bhp	rpm	bhp	rpm	bhp	rpm	bhp
1,019	17.2	1,101	21.0	1,177	24.9	1,248	29.1	1,316	33.4	1,381	38.0
1,017	18.5	1,099	22.5	1,176	26.7	1,247	31.1	1,315	35.6	1,379	40.4
1,018	20.0	1,099	24.2	1,174	28.5	1,245	33.1	1,313	37.8	1,377	42.7
1,019	21.4	1,099	25.8	1,174	30.4	1,245	35.1	1,312	40.1	1,377	45.3
1,020	22.9	1,100	27.5	1,175	32.4	1,245	37.2	1,313	42.5	1,376	47.7
1,022	24.6	1,101	29.3	1,176	34.4	1,245	39.5	1,313	44.9	1,374	50.2
1,025	26.3	1,103	31.2	1,177	36.4	1,247	41.8	1,313	47.2	1,376	53.0
1,029	28.0	1,107	33.2	1,179	38.6	1,249	44.1	1,314	49.9	1,377	55.7
1,032	29.9	1,109	35.2	1,182	40.8	1,250	46.5	1,317	52.6	1,378	58.6
1,043	33.8	1,120	39.7	1,190	45.7	1,259	51.9	1,322	58.2	1,384	64.8
1,056	38.2	1,131	44.5	1,202	51.0	1,268	57.7	1,331	64.4	1,391	71.3
1,072	43.1	1,145	49.8	1,214	56.7	1,279	63.7	1,341	70.9	1,402	78.4
1,090	48.3	1,160	55.4	1,229	62.9	1,293	70.4	1,353	78.0	1,413	85.8
1,108	53.9	1,179	61.7	1,244	69.5	1,308	77.5	1,369	85.7	1,426	93.9
1,128	59.9	1,197	68.2	1,262	76.5	1,325	85.1	1,383	93.7	1,440	102
1,149	66.4	1,219	75.5	1,283	84.4	1,343	93.4	1,400	102		
1,173	73.6	1,240	83.1	1,302	92.3	1,362	102	1,419	111		

contact a fan vendor. To expedite the estimation, be prepared to give to the vendor the flow rate and pressure information from which the vendor can determine cost estimates for both the fan and the fan motor. If the application has special requirements such as high particulate loading or high corrosive potential, specify the type of fan needed in addition to the operating flow rate and pressure.

18.5 Stacks

A *stack* is a vertical shaft or duct through which exhaust air is discharged to the atmosphere. Stacks can be divided into two categories: short and tall stacks. Short stacks are usually less than 100 ft (30.5 m) tall and are constructed of carbon steel or fabricated plastics (U.S. EPA, 1985). Exhaust stacks from industry typically fall into the short stack category with tall stacks being used for large furnaces and boilers. Tall stacks are generally over 100 ft (30.5 m) tall and consist of an outer shell made of reinforced concrete with an inner shell constructed of either carbon steel or brick (U.S. EPA, 1985). For structural stability of tall stacks, the base diameter of the stack is larger than the top diameter. In general, the taller the stack, the greater the difference between the base and top diameters.

18.5.1 Design factors for stacks

The design parameters used in sizing stacks are the stack height and stack diameter. The stack height is chosen based on good engineering practice (GEP), which is defined in 40 CFR 51 Section 123 as "the height necessary to insure that emissions from the stack do not result in excessive concentrations of the pollutant in the immediate vicinity of the source as a result of atmospheric downwash, eddies, or wakes which may be created by the source itself, nearby structure or nearby terrain obstacles." Excessive concentration is defined as at least 40 percent in excess of the maximum ground-level concentration experienced in the absence of atmospheric downwash, eddies, or wakes (U.S. EPA, 1985). Ground-level concentrations of pollutants emitted from an industry are calculated by using the dispersion modeling described in Chap. 3, Atmospheric Dispersion Models, or the ambient air monitoring as described in Chap. 4, Ambient Air Monitoring.

The practical application of GEP to stack height determination is described in 40 CFR 51 Section 123 as no taller than 2.5 times the height of nearby structures, measured from the base of the stack, unless a demonstration is performed that shows a justification for a taller stack. Nearby structures that should be taken into account are those which are at a distance from the stack that is 5 times greater than the tallest nearby structure. Nearby structures to be excluded from stack height determination are other stacks and radio or TV transmission towers.

Engineering judgment is used to determine the necessity of following the stack height guidelines described above. State regulatory agencies make the final decision on when the guidelines should be used. For many applications, the stack height guidelines are not necessary. In general, the guidelines should be followed for sources such as large boilers, furnaces, and other sources that emit large quantities of sulfur dioxide, oxides of nitrogen, and chemicals that are hazardous to human health.

The diameter of a stack is designed to provide adequate stack exit velocities that will help prevent plume downwash. A rule of thumb for calculating the necessary stack exit velocity is 1.5 times the maximum expected wind speed at the stack exit (U.S. EPA, 1985). Stack diameter can be adjusted to reduce the capital cost of the stack. At some point, the decrease in capital cost is offset by the increase in fan operating cost due to the higher pressure drop across the stack.

18.5.2 Cost of stacks

The costs associated with stacks are dependent on the material of construction, stack height, and stack diameter. Stack costs can be estimated by using techniques developed in Vatavuk (1990). Figure

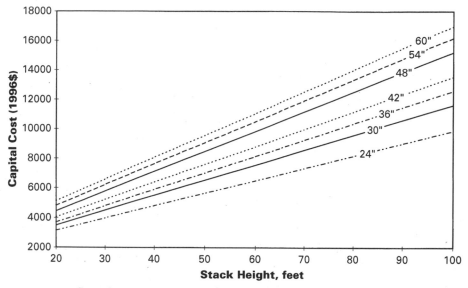

Figure 18.7   Capital cost estimate for short stacks.

18.7 shows the approximations of capital costs for short stacks made of ¼-in carbon steel. The major costs associated with tall stacks are material costs and installation labor. The installed costs of stacks over 200 ft (61 m) can exceed $1 million while the installed cost of stacks over 500 ft (153 m) can exceed several million dollars. For a construction project as large as building a tall stack, contractor quotes should be obtained for more specific cost estimation.

References

American Conference of Governmental Industrial Hygienists (ACGIH): *Industrial Ventilation, A Manual of Recommended Practice,* 22d ed., Cincinnati, Ohio, 1995.

Brandt, Allen D.: *Heating and Ventilating,* vol. 42, no. 5, 1945.

Clifford, George: *Modern Heating, Ventilating, and Air Conditioning,* Prentice-Hall, Englewood Cliffs, N.J., 1990.

Daugherty, Robert L., Franzini, Joseph B., and Finnemore, E. John: *Fluid Mechanics with Engineering Applications,* McGraw-Hill, New York, 1985.

Fox, Robert W., and McDonald, Alan T.: *Introduction to Fluid Mechanics,* Wiley, New York, 1978.

Perry, Robert H., Green, Don W., and Maloney, James O.: *Perry's Chemical Engineering Handbook,* McGraw-Hill, New York, 1984.

Simpson, Roger, Director of Health and Safety: Shelby Williams Industries, Inc., personal communications, 1996.

U.S. EPA, *Guideline for Determination of Good Engineering Practice Stack Height,* Office of Air Quality Planning and Standards, Research Triangle Park, N.C., 1985.

Vatavuk, William M.: *Estimating Costs of Air Pollution Control,* Lewis Publishers, Chelsea, Mich., 1990.

19

Control of Particulate Emissions

Lem B. Stevens, III

19.1 Definition of Particulate

Particulate pollution is defined as any solids or liquid droplets released to the ambient atmosphere from industrial activity. For the purposes of stack testing, particulate is anything caught on the filter at 248°F because EPA method 5 for testing particulates requires that the filter used to determine the concentration of particulate in a gas be maintained at 248°F. Therefore any droplets which will evaporate at 248°F will not be counted as particulate in a source test.

19.2 Particle Distribution

The first step in specifying a control device for particulate is to determine the characteristics of both the particulate and the gas stream. The most important characteristics of the particulate are the size and the weight distribution of the particles. Generally, the aerodynamic diameter of the particles is used as a measure of the size of the particulate. This dimension is usually measured in micrometers or 10^{-9} m. This unit of measure is also called a micron and is well suited for the description of particulate pollution because many of the particles that stay suspended in air and present a hazard have a mean diameter between 0.1 and 10 μm. Larger particles tend to settle out of the air quickly and are therefore less of a concern. In fact, the recent Title V air permit program requires industry to permit emissions of particu-

TABLE 19.1   Particle Distribution

Size range, 10^{-9} m	Average size, 10^{-9} m	Particle count	Size-range volume	Volume percent
30–50	40	0	0	0.6
10–30	20	2	8,378	17.6
5–10	7.5	124	27,391	57.5
3–5	4	232	7,774	16.3
1–3	2	589	2,467	5.2
0.5–1	0.75	5,030	1,111	2.3
0–0.5	0.25	65,400	535	1.1

late with a mean diameter of less than or equal to 10 μm (PM_{10}) and does not require permitting of total suspended particulate (TSP).

The weight distribution of particulates is a measure of the percentage of the total mass of particulate in an air sample in specific size ranges. A number of size ranges are chosen, and the total mass of particulate in each size range is determined. An example of a distribution is given in Table 19.1. In many cases, the density of the particles can be assumed to be constant. If this assumption is valid, then the volume percent of the size range is equal to the weight percent. An interesting fact is illustrated in this table: The weight percent and the particle count for a size range can be vastly different. The range of 0 to 0.5 μm has 65,400 particles but accounts for only 1.1 percent of the total weight.

19.3   Collection Efficiency

In particulate control, we are interested in controlling the mass of particulate emissions, not the number of particles. The collection efficiency of a device is based on the mass fraction of the particles that it controls. The equation for overall efficiency is

$$n = \frac{M_i - M_e}{M_i} \tag{19.1}$$

where M_i = total mass input to the device and M_e = total mass emitted from the device. When a particle size distribution is known, the overall efficiency of a control device can be calculated by performing a weighted average of the efficiency of the device for each size range.

$$n = \sum_{z=1}^{n} \frac{M_{iz}}{M_i} \left(\frac{M_{iz} - M_{ez}}{M_{iz}} \right) \tag{19.2}$$

where M_i = total mass input to device
  M_{iz} = mass input of size range z
  M_{ez} = mass emitted of size range z

Control devices typically have high collection efficiencies in specific size ranges and may have poor efficiencies in other size ranges. Therefore, knowledge of the distribution of particles is very important when a control device is chosen. A control device with a high efficiency for the size ranges that have the greatest weight of particles should be chosen for a specific particle distribution. For example, a particle weight distribution skewed toward larger sizes may be effectively controlled by a cyclone. But the same cyclone may be markedly inefficient on a particle weight distribution skewed toward the smaller sizes.

The consistency of the particles is also very important. Sticky, oily particles must be treated with a control device that is not negatively affected by their consistency. If oily particles are controlled by a baghouse, then the baghouse will quickly blind, resulting in lower system airflow. Low airflow results in poor ventilation of the process controlled by the baghouse and poor capture of particulate. Furthermore, blinded oily bags can represent a significant fire hazard.

19.4 Gas Stream Characteristics

The characteristics of the gas stream containing the particulates are also very important. The temperature, moisture content, and gas content must be known to avoid problems with control devices. For example, an aluminum smelter can produce a gas stream of particulates and gaseous hydrochloric acid with a high temperature. Typically, a baghouse is used to control particulate from a smelter. However, in colder months, the temperature of the baghouse may be lower than the dew point of the gas stream. Under this condition, liquid hydrochloric acid forms in the baghouse and attacks the baghouse structure, resulting in unacceptably short life. Corrective action must be taken to keep the baghouse above the dew point or to remove hydrochloric acid from the gas stream.

19.5 Control Devices

The remainder of this chapter considers examples of control devices for particulates. Each section presents the basic theory behind the control devices and gives an example of the control device in service.

19.5.1   Settling chambers

The simplest control device available for particulate control is a settling chamber. The settling chamber for control of particulate in gas
takes advantage of the gravitational force on a particle and the
resulting downward velocity of that particle, called the *settling* or *terminal velocity*.

Settling chambers are advantageous if the particle distribution contains large particles, because they have very low maintenance and low
pressure drops. A settling chamber is most often used as a precleaner
to a more efficient device. In some cases, a settling chamber can be
used to protect a secondary device. This is the case for many arc furnaces and smelters, which use a settling chamber to remove large, hot
particles that can burn holes in a fabric filter and then a baghouse to
polish the gas stream before it is discharged. In a settling chamber,
the velocity of the gas is reduced by enlarging the cross-sectional area
of the flow. The settling chamber is designed so that the gas remains
at the lower velocity long enough for a high percentage of the particles
to settle to the laminar flow region near the bottom of the settling
chamber. The velocity of the gas in the settling chamber must be low
enough to avoid reentrainment of the particles. From a practical
standpoint, settling chambers are used only for large particles, which
have significant settling velocities. For the present discussion, a large
particle is 50 μm or greater. Small particles with low settling velocities
require too long a settling period, and the settling chamber must be
very large to keep the gas velocity low for this period.

The settling velocity of a particle can be calculated with Stoke's law.
The form of this law used is

$$V_s = \frac{gpd^2}{18u} \tag{19.3}$$

where V_s = settling velocity
  g = acceleration of gravity, 9.806 m/s^2
  p = density of particle, kg/m^3
  d = diameter of particle, m
  u = viscosity of gas, kg/(m · s)

The distance that the particle will settle is calculated as the product
of the settling velocity and the residence time in the settling chamber.
The residence time T in seconds is calculated according to the continuity equation

$$T = \frac{WLH}{Q} \tag{19.4}$$

where W = width of settling chamber, m
L = length of settling chamber, m
H = height of settling chamber, m
Q = actual flow rate of gas, m³/s

The settling distance must be great enough for a particle to reach the laminar region of flow. Otherwise, the particle will remain entrained in the gas stream and will exit the settling chamber. In actual practice, not all the particles calculated to settle will be removed, because high gas flow rates exist at specific localities in all but theoretical settling chambers. A simple settling chamber is shown in Fig. 19.1. An improvement to the simple settling chamber can be made by adding flat plates across the width and length of the settling chamber, as shown in Fig. 19.2. In this type of settling chamber, the particles must settle only the distance between plates to be removed from the airstream. This device can be thought of as a stack of parallel settling subchambers where the height of each subchamber is the distance between the plates. The residence time in this type of settling chamber is not affected significantly by the plates, but is similar to that of a simple settling chamber.

19.5.2 Inertial separators

Inertial separators take advantage of the difference between the density of the gas and the particles in the gas to separate the particles from the gas. Inertial separators cause the gas stream to make high-velocity or abrupt turns that only the gas molecules and extremely

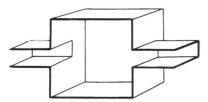

Figure 19.1 Settling chamber.

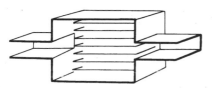

Figure 19.2 Improved settling chamber.

small particles can make. Larger particles have enough mass that their inertia prevents them from making the turn, and these particles are separated from the gas stream. Thus the efficiency of an inertial separator usually depends upon the velocity at which the gas stream spins. The inertial separator then has some method of retaining the particles and preventing them from being reentrained in the gas stream.

Cyclones. The most common inertial separator is the cyclone. A typical cyclone is pictured in Fig. 19.3. The gas stream enters the cyclone in such a way that it spirals around the inside of the cyclone and slings the particles with greater inertia to the outside wall, where they encounter the laminar boundary layer of gas and fall down the cone to the collection hopper below. There are two main methods of imparting spin to the gas. One is to introduce the gas to the cyclone in a tangential manner so that the gas must curve around the inside of the cyclone, as in Fig. 19.3. Another method is to install axial vanes on the inlet to the cyclone which cause the gas to spin as it flows past. An example of such a cyclone is pictured in Fig. 19.4.

An advantage of a vane-axial cyclone is that it can be produced in small sizes, which aids inertial separation because the gas must make tighter turns. Many small vane-axial cyclones can be used in parallel to treat a large flow, producing a high-efficiency control device called a *multiclone*. A multiclone generally has a higher removal efficiency than a single, large cyclone with a tangential inlet,

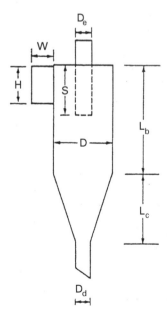

Figure 19.3   Standard cyclone dimensions.

Clean
gas

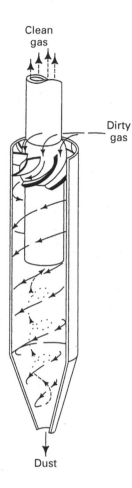

Dirty
gas

Dust

Figure 19.4 Vane-axial cyclone.

because a large cyclone has an inherently larger radius and therefore lower efficiency. Multiclones are often used as particulate control devices for coal-fired boilers because multiclones are capable of removing fly ash and are resistant to degradation by heat in the flue gas. One drawback of multiclones is that the pressure drop across them is relatively high compared to that in a single large cyclone. Therefore, a large cyclone is often specified when there are many large particles in a gas stream and high removal efficiency of small particles is not required. An example of this type of service is a wastepaper-handling system in magazine production (Fig. 19.5). As the magazines are trimmed by paper-cutting machines, the trimmings can be collected in an airstream. A large cyclone is a good device to separate the paper from the airstream because it has an acceptable removal efficiency on the paper trimmings, which are large particles, and has a relatively low pressure drop. Large cyclones

Figure 19.5   Large paper-handling cyclone.

also find use in woodworking operations to remove wood chips and sawdust for the same reasons.

Chevron blade demisters.   Another inertial separator is a chevron blade demister. This device has a series of plates with sine wave curves. The plates are set parallel so that the gas must flow between them and turn with curves in the plates. As the gas turns, the inertia of particles in the gas stream causes the particles to impact the plates, where the particles are separated from the gas. A drawing of chevron blades is shown in Fig. 19.6. To prevent reentrainment of the particles, the particles must move down the blades and out of the high-velocity gas stream. For this reason, chevron blade demisters are most often used to remove mist from a gas stream. The droplets impact the blades and form a liquid film, which forms in the laminar boundary layer at the surface of the blade and flows toward the bottom of the device, where the liquid can be drained away.

19.5.3   Impingement separators

Impingement separators are similar to inertial separators because they remove particles from the gas stream by physical interaction of

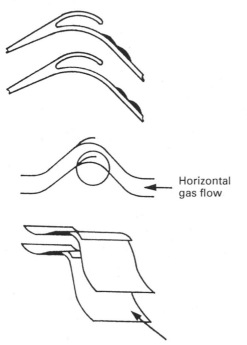

Horizontal
gas flow

Figure 19.6 Chevron blades.

the particles and the device. However, impingement separators do not use centrifugal forces to separate particles. Instead, the gas flows through a mesh around the individual filaments of the mesh. A stream which flows around rocks in its path is analogous to an impingement separator. Separation of the particles occurs when the particles impinge on the individual filaments in the mesh and are not reentrained in the gas stream. As the gas bends around the filament, the inertia of the particle causes it to flow straight into the filament. (See Fig. 19.7.)

The designer must take care to provide the correct gas flow for an impingement separator to function properly, because the velocity with which the gas flows past the filaments is a critical parameter. A low velocity will result in fewer particles contacting the filaments. A high gas velocity will reentrain particles. Both of these conditions degrade the efficiency of the device.

Impingement separators work best with mists because the liquid can form on the filaments and flow down the filaments to a sump, where it is drained away. It is difficult to prevent solid particles from being reentrained in the gas stream. This process must not be confused with filtration. An impingement separator will remove particles far smaller than the cross-sectional area of the smallest path through the mesh.

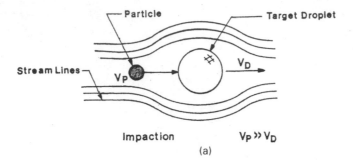

Impaction $V_P \gg V_D$

(a)

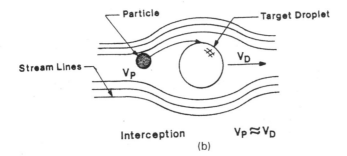

Interception $V_P \approx V_D$

(b)

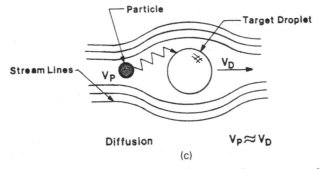

Diffusion $V_P \approx V_D$

(c)

Figure 19.7 Particle and filament. (*a*) Most particles are removed by direct impaction into a droplet. (*b*) Other particles come close to the droplet and are intercepted. (*c*) Diffusion.

Impingement separators called *composite mesh pad demisters* are used to control emissions of chromic acid mist from chrome electroplating operations. A composite mesh pad demister usually employs three mesh pads to accomplish separation. The first mesh pad acts as a filter to remove larger droplets of chromic acid mist. The mist that passes through the first pad is very fine and impinges on the second pad, which is denser and thicker than the first pad. In theory, some of the mist collected on the second pad agglomerates into large droplets that are carried by the gas to the third pad, which polishes the gas stream.

In practice, it has been shown that keeping the second pad damp can greatly improve the removal efficiency of the composite mesh pad demister. The author believes that the improvement derives from two areas. First, the damp filaments have a film of water around their circumference, which presents a larger cross section to the gas flow. The gas must turn harder around the filament, and smaller particles will impinge on the filament due to their inertia. Second, the film of water provides a better surface for impingement. A droplet of chromic acid mist that contacts the film of water is more likely to be collected than if the droplet contacted a solid filament, because chromic acid is soluble in water.

19.5.4 Fabric filters

Fabric filters are the most common collection device used to control solid particulate. The collection efficiency of a properly operating fabric filter is often well above 99.9 percent. In fact, many applications of fabric filters involve transfer of raw and finished materials in industry. For example, in the silica products industry, sand flour and silica products are conveyed with air from storage bins to the process and back in very high concentrations. Fabric filters are used to separate the materials from the conveying air at the bins and to drop the material into the appropriate bin (Fig. 19.8). Losses of the material from the fabric filters are so small as to be insignificant.

The operation of fabric filters appears uncomplicated at first. Many engineers assume that the method of separation is simple filtration, where the openings in the fabric determine what size particles are retained. However, tests show that particles with sizes much smaller than the openings in the fabric are removed with extremely high efficiencies. An attempt to explain the actual operation of a fabric filter is offered below.

At first, the fabric is clean and does not have any dust cake on its surface. This condition occurs after each cleaning cycle, and the efficiency of the fabric filter is lowest during this period. As particle-laden gas passes through the fabric, larger particles are filtered out of the gas by the fabric and begin to form a dust layer, or cake, on the upstream side of the fabric. As the thickness of this layer increases, the clear path through the dust layer and the fabric becomes more tortuous and is called a *pore*. Smaller particles begin to adhere to the cake as they travel through the pores by inertial separation. Electrostatic forces are also at work in the process, causing a phenomenon

Figure 19.8   Fabric filters.

called *bridging*. Bridging occurs when particles stick together through electrostatic attraction and build chains which bridge an opening through the cake. As the opening is bridged, a smaller opening is created. The process is repeated for the smaller opening until there is no path through the cake at that location. Electrostatic forces are affected by humidity. In practice, the humidity of the gas stream has a marked affect on the performance of a fabric filter, which underlines the importance of the bridging effect.

As some pores are closed, the pressure drop across the fabric increases and the velocity through the remaining pores increases. Some of these pores will also be closed as the cake thickness increases. However, the velocity through the open pores becomes so high that some pores will remain open indefinitely, and these are called *leakage sites*. Leakage sites occur in a random fashion and are many different sizes over millions of pores. The result of this phenomenon is a fairly even control efficiency for a wide range of particle sizes, which cannot be explained by simple filtration.

Theoretical relationships have been developed in an attempt to predict the performance of fabric filters (Fig. 19.9). Most of these rela-

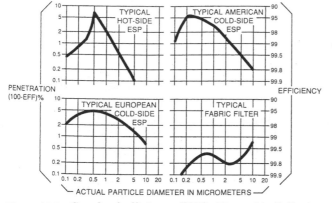

Figure 19.9 Graph of efficiency EPRI. (*From Air Pollution Engineering Manual, Air and Waste Management Association, Sewickley, Pa., 1992, p. 115.*)

tionships consist of two terms: one for the fabric itself and one for the cake. The term for the cake must be time-dependent because the cake thickness increases with time. Laboratory tests with such relationships have generated favorable results, but tight controls must be kept on parameters such as particle size, temperature, humidity, fabric characteristics, and airflow. The efficiency of a fabric filter is greatly affected by these parameters, so that in the real environments, with varying parameters, the theoretical relationships cannot be used to predict performance. The factors that affect performance are too complex and time dependent. Thankfully, although the performance of a fabric filter changes continually, removal efficiency is above 99 percent most of the time in a proper installation.

The most successful method of specifying a fabric filter for an operation calls for experience. Reputable manufacturers have large databases of installations and their success to rely on. While theoretical relationships may help the engineer understand why the device works, empirical data provide the best measure of actual effectiveness. The more information about the process that is compared to empirical data, the higher the probability of a successful installation. The following parameters are important considerations in determining what kind of fabric filter to use:

The *temperature of the gas stream* is an important consideration for the type of fabric used. Acceptable temperatures for fiberglass fabrics are as high as 500°F, while polypropylene should not be subjected to temperatures above 200°F. Many gas streams have temperature peaks which are significantly higher than the average gas temperature. These peaks must be characterized because some filter media can withstand high temperatures for short periods while others cannot.

Moisture in the gas can affect the performance of a fabric filter in two ways. First, moisture can cause particles to stick together in the cake and can prevent the cleaning mechanism from functioning properly. The resulting buildup of cake results in an objectionably high pressure drop across the fabric and a low flow rate. In extreme cases, the fabric may be breached. Second, moisture may affect the fabric itself and cause premature degradation.

An estimate of the amount of particulate in the gas stream is a very important parameter in the design of a fabric filter. Heavy concentrations of particulate require greater cloth area than do lower concentrations. If a gas stream containing a very high concentration of particulate is cleaned by a small fabric filter, then cake will build up very quickly and the cleaning cycle will be short. The fast cake buildup will cause changes in the flow rate through the device which may be unacceptable, and the frequent cleaning will shorten the life of the fabric.

Particle distribution is also important. A fabric filter is best suited to small particles ranging from 50 μm to submicron particles. If the particle distribution is heavily weighted toward larger particles, then a fabric filter alone may not be the best control device for the gas stream. Instead, a cyclone may be appropriate or a cyclone in series with a fabric filter. The cyclone can be designed to remove the large particulate with a relatively low pressure drop, and the fabric filter can be used for polishing.

Characteristics of the particles can affect a fabric filter. By characteristics of the particles, the author means the *behavior* of the particles in a cake. If the nature of the particles is to stick together, forming a solid mass, then a fabric filter is not a good control device. As an example, consider the exhaust of a tenter frame used in the textile industry. The exhaust from many tenter frames is lint from the fabric being processed mixed with oil that was used to facilitate weaving. With lint and oil in the gas stream, a fabric filter will quickly blind. The lint is actually very large particulate which will bridge the openings in the fabric, and the oil will quickly saturate this felt, creating a thick, sticky mass which is impossible to clean and may be a fire hazard. A wet collector would be a better choice for such a gas stream.

The *corrosiveness of gas constituents* must be considered. Many industrial exhausts contain constituents which are corrosive to a fabric filter and can cause accelerated degradation if they are not properly handled. For example, aluminum smelters produce a fume which can be adequately controlled with a fabric filter. However, hydrochloric acid is often found in the exhaust from aluminum smelters. If the temperature of the gas falls below the dew point of hydrochloric acid, then acid will form on the interior of the fabric filter and on the fabric itself. If the fabric and device structure cannot economically be fabricated from materials resistant to the acid, then the acid may create holes in the fabric and may quickly corrode the structure of the fabric filter. One method of preventing this problem is to heat the gas above the acid dew point to prevent the liquid acid from forming. Another method is to scrub the gas of the hydrochloric acid.

20

Absorption of Gaseous Emissions

Lem B. Stevens, III

Absorption is the process of removing a gas from an exhaust stream by preferentially absorbing the gas into a liquid solvent. Many exhaust streams are comprised mostly of air, which is relatively insoluble. Absorption takes advantage of the solubility of the target compound to remove it from the insoluble air. An example of an application for an absorber is the scrubbing of hydrogen chloride gas from an exhaust stream with a sodium hydroxide solvent. Absorbers are mainly used for control of gases, but may also be used in conjunction with a separation device to reclaim the pollutant. For example, methanol can be removed from an exhaust gas and dissolved into water with an absorber. Then the methanol-water mixture can be separated by distillation and reclaimed.

20.1 Mass Transfer

Absorbers operate by creating an advantageous environment for mass transfer of the target gas to the liquid or solvent. The theory for this mass transfer is fairly complex, and most designers of absorbers make use of empirical data and experience for the specific gas and solvent to be used when designing an absorber. Therefore, the actual mass-transfer theory is left to more specialized texts. Instead, the physical movement of the gas from the exhaust stream to the solvent will be traced and described. Refer to Fig. 20.1. The line in the middle of the figure is the gas-liquid interface. The curve to the left of the

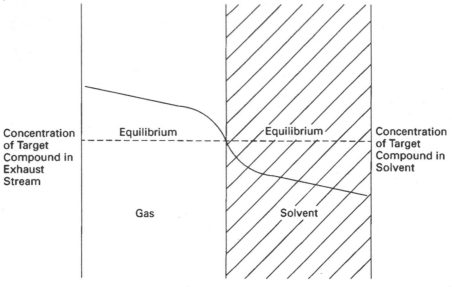

Figure 20.1   Mass transfer.

gas-liquid interface represents the concentration of the target compound in the exhaust stream. The curve to the right of the gas-liquid interface represents the concentration of the target compound in the solvent. Molecules of the target component move from the exhaust stream to the solvent in the following manner: When the exhaust gas and the solvent come into contact, molecules of the target gas diffuse across the gas-liquid interface because the solvent and gas are not in equilibrium for the target substance. Once in liquid phase, the target substance disperses through the liquid and away from the gas-liquid interface toward lower concentrations, in an attempt to achieve a consistent concentration throughout the liquid. The exhaust gas concentration of the target substance at the gas-liquid interface is now lower than the average concentration in the exhaust stream because of the molecules diffusing out of the gas. Molecules of the target compound diffuse toward the gas-liquid interface in an attempt to achieve a consistent concentration of target compound in the exhaust stream. This process continues as long as the equilibrium concentrations of the target compound in the exhaust stream and the solvent are not satisfied.

The goal of an absorber is to keep the concentration of target substance in the solvent as far from equilibrium concentration as possible, to achieve the highest rate of mass transfer possible. There are several design considerations which must be optimized to achieve good mass transfer. Each of the design considerations is discussed below.

20.2   Exhaust Gas Turbulence

As stated earlier, a local region of low target gas concentration occurs as the target gas diffuses across the gas-liquid boundary. This low concentration reduces the concentration gradient across the phase boundary and therefore reduces mass transfer. The highest possible concentration of target gas exists at the phase boundary when the exhaust gas is fully mixed and the concentration is the same at the boundary and elsewhere in the gas stream. The most effective force to mix the exhaust gas is turbulence. Turbulence causes eddy currents on a microscopic scale which effectively mix the gas and create consistent concentration of target gas. Therefore, turbulence is very important in absorbers.

20.3   Exhaust Gas–Solvent Contact

Mass transfer in an absorber occurs across a gas-liquid boundary. The collective phase boundary is best described geometrically as an area or surface. All else being equal, the greater the area of this phase boundary, the greater the mass transfer.

20.4   Contact Time

Mass transfer is not an instantaneous reaction to a concentration gradient. Instead, mass transfer takes time. The longer a concentration gradient has to drive mass transfer, the more mass transfer will occur. This statement is true when we discuss the relatively short time frame in an absorber. Of course, if a concentration gradient acts over a sufficiently long time, equilibrium will occur and net mass transfer will cease.

20.5   Solvent Turbulence

On the solvent side of the phase boundary, there is a locally high concentration of the target liquid as mass transfer occurs. This high concentration reduces the concentration gradient which drives the mass transfer. It is therefore advantageous to keep the local concentration as low as possible. This goal is best achieved by mixing the solvent in a turbulent fashion, causing the target liquid concentration throughout the solvent to approach an average, low value, much like the exhaust gas turbulence.

20.6   Solvent Recirculation Rate

In the design of an absorber, the engineer chooses how much solvent to use in relation to the exhaust gas flow rate. The more solvent used, the more mass transfer of the target compound can occur because the concentration of target liquid in the solvent will be lower, all other things being equal. Since the target gas is usually a small percentage of the exhaust gas, the liquid phase of the target component represents a very small liquid flow rate. Therefore, when clean solvent passes through an absorber, the exiting solvent is still very clean. In most cases, the exiting solvent is clean enough to be reused in the absorber, and it is recirculated through the absorber. To ensure that an unwanted high concentration of target liquid does not build up in the solvent, fresh makeup solvent is added to the recirculating solvent and an equal amount of "dirty" solvent is removed from the system. The removed solvent is often called *blowdown*.

The amount of solvent recirculated is a function of the amount of contact desired between the exhaust gas and the solvent. Usually, a solvent flow rate through the absorber is calculated based on theoretical relationships to obtain the desired removal of the target gas. Then that flow rate is increased by a factor of 1.5 to 3.0 to yield the actual solvent flow rate, and this flow rate is pumped through the absorber. Care must be taken to offer the solvent enough flow area to avoid flooding. If too much solvent flows through the absorber, then it can defeat the design of the absorber and flood the absorber, resulting in restricted gas flow and smaller liquid-solvent interface. Absorber design dictates how much solvent can be pumped through the absorber without flooding.

The required blowdown rate is a function of the mass transfer of the target compound to the solvent. The mass flow rate of the target compound in the blowdown should equal the mass transfer to the solvent. If the preceding condition is met, then the concentration of target compound in the solvent will remain constant. The blowdown for many absorbers can be automated in a relatively simple fashion. For example, consider an absorber which removes hydrochloric acid from an exhaust stream with a basic solvent. A pH meter can be installed in the solvent recirculation line and essentially reads how much acid has been absorbed in the solvent. If the pH falls below a low set point, then the makeup and blowdown valves are opened until the pH rises to a high set point. The acid concentration of the solvent is effectively controlled in this manner.

20.7   Absorber Configuration

From a theoretical standpoint, if the above conditions are met in a device, then successful absorption will occur. From a practical standpoint, most absorbers in use today can be classified as a packed tower absorber. Packed towers are so named because they employ packing to create a large surface area for the gas-liquid interface. A general schematic of a packed tower is presented in Fig. 20.2.

The solvent is fed into the top of the tower and is evenly distributed over the packing. The solvent flows by gravity down through the packing. In tall towers, the solvent may be collected on intermediate plates designed to redistribute the solvent and prevent channeling. Generally, redistribution plates are installed every 10 ft of drop through the packing. Channeling occurs when the solvent finds a preferential path through the packing and floods that channel, starving other parts of the packing and reducing the overall gas-liquid

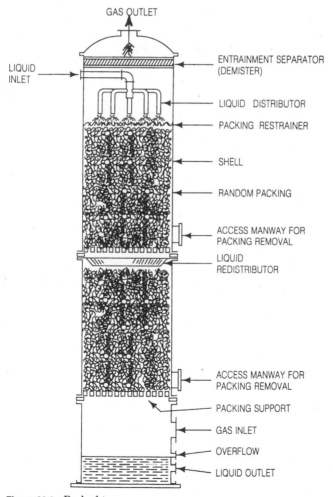

Figure 20.2 Packed tower.

interface for mass transfer. The solvent is then collected in a sump below the packing and is pumped from this sump to the top of the tower. The gas enters the tower below the packing and rises up through the packing, where the target compound is absorbed into the solvent. The clean gas exits the tower above the level where the solvent is injected.

20.8 Packing

The heart of the packed tower is the packing. Packing material is designed to cause liquid poured over it to spread out and form a film as it flows around the packing, thus presenting as much surface area as possible for mass transfer. There are several successful packing designs on the market, and new designs are presented as engineers discover better ways to make solvent form sheet flow around geometric shapes. Several types of packing are shown in Table 20.1. Packing

TABLE 20.1   Some Typical Packings and Applications

Packing	Application features
Raschig rings	Most popular type, usually cheaper per unit cost, but sometimes less efficient than others; available in widest variety of materials to fit service; very sound structurally; usually packed by dumping wet or dry, with larger 4- to 6-in sizes sometimes hand-stacked; wall thickness varies among manufacturers, also some dimensions; available surface changes with wall thickness; produces considerable side thrust on tower; usually has more internal liquid channeling and directs more liquid to walls of column.
Berl saddles	More efficient that Raschig rings in most applications, but more costly; packing nests together and creates "tight" spots in bed, which promotes channeling, but not as much as Raschig rings; does not produce much side thrust and has lower unit pressure drops with higher flooding point than Raschig rings; easier to break in bed than Raschig rings.
Intalox saddles	One of the most efficient packings, but more costly, very little tendency or ability to nest and block areas of bed; gives fairly uniform bed; higher flooding limits and lower pressure drop than Raschig rings or Berl saddles; easier to break in bed than Raschig rings.

TABLE 20.1   Some Typical Packings and Applications (*Continued*)

Packing	Application features
Pall rings	Lower pressure drop (less than half) than Raschig rings; higher flooding limit; good liquid distribution, high capacity; considerable side thrust on column wall; available in metal, plastic, and ceramic.
Spiral rings	Usually installed as stacked, taking advantage of internal whirl of gas-liquid and offering extra contact surface over Raschig rings. Lessing rings, or cross-partition rings, available in single, double, and triple internal spiral designs, higher pressure drop, wide variety of performance data not available.
Teller rosette (Tellerette) 	Available in plastic; lower pressure drops, higher flooding limits than Raschig rings or Berl saddles; very low unit weight, low side thrust; relatively expensive.
Cross-partition rings	Usually used stacked, and as first layers on support grids for smaller packing above; pressure drop relatively low, channeling reduced for comparative stacked packings; no sidewall thrust.
Lessing rings	Not many performance data available, but in general slightly better than Raschig rings; pressure drop slightly higher; high sidewall thrust.

TABLE 20.1 Some Typical Packings and Applications (*Continued*)

Packing	Application features
Ceramic balls	Tend to fluidize in certain operating ranges, self-cleaning, uniform bed structure, higher pressure drop, and better contact efficiency than Raschig rings; high side thrust; not many commercial data.
Goodloe packing and wire mesh packing	Available in metal only, used in large and small columns for distillation, absorption, scrubbing, liquid extraction; high efficiency, low pressure drop.

manufacturers have data on the effectiveness of the packing for different applications as well as the limit on the amount of solvent flow the packing can support without flooding. Most packing manufacturers also have data on the pressure drop through the packing for the exhaust gas. Data from the manufacturers should be used in the design of a packed tower.

21

Adsorption of Gaseous Compounds

E. Roberts Alley, Jr.

One of the traditional techniques for controlling the release of volatile organic compounds is adsorption. Adsorption is primarily used for the reduction of organic emissions, although metals removal has been demonstrated in some wastewater applications. *Adsorption* is a mass-transfer process that can generally be defined as the accumulation of material at the interface between two phases. For example, a contaminant in an airstream while passing through a bed of carbon is transferred from the gas phase to the surface of the carbon. More generally, chemicals in the gas phase preferentially accumulate on an unsaturated solid surface, causing the chemical to be removed from the gas phase. The material upon which the chemical is adsorbed (e.g., carbon) is known as the *adsorbent*. The material that is adsorbed (typically the contaminant) is known as the *adsorbate*.

Adsorption is a thermodynamic system in which the various components are striving for equilibrium. The process of adsorption occurs in both steady-state and unsteady-state conditions. An example of a steady-state condition is dry lime injection for the removal of acid gases in a gas stream. Unsteady-state conditions are more common, involving a fluid flowing continuously over or through a fixed adsorbent bed. The primary forces driving the interaction between the adsorbate and the adsorbent are the electrostatic attraction and repulsion between molecules of the adsorbate and the adsorbent. These driving forces can be either physical or chemical.

Physical adsorption is a result of intermolecular forces that interact between the adsorbate and the adsorbent. These physical electrostatic forces include the *van der Waals force,* consisting of weak attraction and repulsion through dipole-dipole interactions and dispersing interactions, and *hydrogen bonding.* Dipole-dipole interactions are the result of polar compounds orienting themselves so that their charges result in a lower combined free energy. Dispersing interactions are the result of attractive forces between electrons and nuclei of molecular systems. If the molecules come too close to one another, repulsive forces can push the molecules apart. Hydrogen bonding is a special case of dipole-dipole interaction in which the hydrogen atom in a molecule has a partial positive charge, attracting another atom or molecule with a partial negative charge. For gas-phase systems, the van der Waals force is the primary physical force driving adsorption. Physical adsorption is a readily reversible reaction and includes both mono- and multilayer coverage. Because physical adsorption does not involve the sharing of electrons, it generally has a low adsorption energy and is not site-specific. The heat of adsorption for the reaction is on the order of 40 Btu/(lb · mol of the adsorbate). When the intermolecular forces between a chemical molecule in a gas stream and a solid (the adsorbent) are greater than the forces between the molecules of the gas stream, the chemical is adsorbed onto the adsorbent surface.

Chemical adsorption (chemisorption), like physical adsorption, is also based upon electrostatic forces. The mechanisms of chemical adsorption are similar to those of physical adsorption, yet are often stronger (approaching the adsorption energies of chemical bonds). Chemical adsorption is produced by the transfer of electrons and the formation of chemical bonds between the adsorbate and the adsorbent. It may be an irreversible reaction and have high adsorption energies. The heat of adsorption is significantly greater than for physical adsorption, ranging from 80 to 400 Btu/(lb · mol). It is not unusual for the adsorbate to have chemically changed due to the reaction. Chemical adsorption involves only monolayer coverage and is a site-specific reaction occurring at specific functional group locations. Functional groups are distinctive arrangements of atoms in organic compounds that give that compound its specific chemical and physical properties. See Basic Air Pollution Theory (Chap. 2) for more information on functional groups.

Some reactions have adsorption energies that are higher than those of physical adsorption, but lower than those of chemical adsorption. Although not common, these interactions are referred to as *specific adsorption.* Specific adsorption involves an interaction with a specific functional group on the adsorbent surface, but it does not result in the formation of a true chemical bond. The primary indication of specific adsorption is that the adsorption energies are between chemical (strong) and physical (weak).

21.1 Materials

It can be argued that, to some degree, all solids adsorb gases, although the adsorption might not be measurable in significant quantities. Generally, adsorption is directly proportional to the surface area of a given material. Common materials used for adsorption are materials that exhibit high surface areas per unit mass, such as charcoal (activated carbon), molecular sieves, and silica gels.

21.1.1 Activated carbon

The most frequently used adsorbent is activated carbon. This is primarily because a wide range of organic chemicals can be adsorbed economically. It is also used for solvent recovery, odor control, and for gas stream purification. Table 21.1 lists some of the classes of organic chemicals that are readily treated with activated carbon.

Activated carbon is produced from porous, carbonaceous materials such as wood charcoal, coal, peat, lignite, recycled tires, petroleum coke, and coconut shells. In general, the production process involves first charring the material to remove hydrocarbons and then activat-

TABLE 21.1 Classes of Organic Compounds Amenable to Adsorption on Activated Carbon

Aromatic solvents
    Benzene, toluene, xylene
Polynuclear aromatics
    Naphthalene, biphenyls
Chlorinated aromatics
    Cholorobenzene, toxaphene, DDT
Phenolics
    Phenol, cresol, resorcinol
High-molecular-weight aliphatic amines and aromatic amines
    Aniline, toluene diamine
Surfactants
    Alkyl benzene sulfonates
Soluble organic dyes
    Methylene blue, textile dyes
Fuels
    Gasoline, kerosene, oil
Chlorinated solvents
    Carbon tetrachloride, percholoroethylene
Aliphatic and aromatic acids
    Tar acids, benzoic acids

SOURCE: Calgon Carbon Corporation.

ing the carbon. The most common method for activation in the United States is achieved through high-temperature (750 to 950°C) steam activation in an oxygen-depleted atmosphere. The reaction between the steam and the carbon is promoted by a dehydrating agent such as zinc chloride or phosphoric acid. In Europe, it is more common to use a chemical activation process. Chemical activation occurs at a lower temperature (400 to 600°C) using the dehydrating agent without steam. The actual production processes commercially used vary and are proprietary. Materials used for activated carbon have large surface areas, typically 800 to 1400 m^2/g, as shown in Table 21.2. However, petroleum coke–based activated carbon can have surface areas in excess of 3000 m^2/g.

A typical carbon particle consists of a highly developed, microscopic network of pores within a crystalline matrix. The pore diameters typically range from less than 2 Å to greater than 500 Å, with an average of 15 to 25 Å. The carbon particles are crushed, graded, acid-washed, and washed with water before distribution. Activated carbon comes in four general forms: extrudates (pellets), beads, granules, and powder. For air control applications, granular activated carbon is most commonly used. Granular activated carbon (GAC) has a particle diameter that ranges from 0.1 to 2 mm (typically 1.2 mm). Powdered activated carbon (PAC) has a particle diameter of less than 0.1 mm and is typically 0.05 to 0.075 mm.

TABLE 21.2   Typical Surface Areas of Activated Carbon

Carbon	Base material	Surface area, m^2/g
PCC SGL	Bituminous coal	1000–1200
PCC BPL	Bituminous coal	1000–1200
PCC RB	Bituminous coal	1200–1400
PCC GW	Bituminous coal	800–1000
Calgon Filtrasorb 300		950–1050
Calgon Filtrasorb 400		1000–1200
Columbia CXA/SXA	Coconut shell	1100–1300
Columbia AC	Coconut shell	1200–1400
Columbia G	Coconut shell	1100–1150
Darco S 51	Lignite	500–550
Darco G60	Lignite	750–800
Darco KB	Wood	950–1000
Hydro Darco	Lignite	550–600
Nuchar Aqua	Pulp mill residue	550–650
Nuchar C	Pulp mill residue	1050–1100
Nuchar (various)	Pulp mill residue	300–1400
Norit (various)	Wood	700–1400

SOURCE: Calgon Carbon Corporation, "Basic Concepts of Adsorption on Activated Carbon," p. 2.

One of the unique characteristics of the use of activated carbon is that after breakthrough is achieved, the adsorbent is not normally discarded. Activated carbon can be reactivated, restoring the carbon to approximately its original adsorptive capacity. As a rule of thumb, more than 90 percent of the activated carbon is recovered during reactivation. Losses are primarily due to spillage and overburning. Reactivation can occur in both on-line applications, where the bed remains in place, and off-line applications, where the bed is physically removed from service for reactivation. Off-line reactivation is commonly used for smaller (less than 2000-lb) carbon beds that are rented or leased at a site. Carbon is often reactivated off-line by passing through a high-temperature, multiple-hearth furnace. Temperatures in the furnace reach 1800°F in order to thermally destroy the organic contaminants. Other regeneration techniques include the use of steam or a vacuum (particularly with regenerative applications), solvent extraction (hexane), and bioregeneration.

A typical example of activated carbon used for gas stream treatment is Centaur HSV developed by Calgon Carbon Corporation. The Centaur product was specifically designed for odor control at sewage treatment plants. The product catalyzes H_2S to water-soluble sulfur compounds, in addition to adsorbing volatile organics in the gas stream. See Fig. 21.1.

21.1.2   Activated alumina

Activated alumina is primarily used to remove moisture from a gas stream, particularly under pressurized applications. However, activated alumina is sometimes used in industry for solvent recovery. Prepared by heating alumina trihydrate to 400°C, activated alumina has a typical surface area of 300 m²/g. It is available in both pellet and granule form. Average pore diameter is 18 to 48 Å. The bulk density ranges from 38 to 42 lb/ft³ for granules, and from 54 to 58 lb/ft³ for pellets.

21.1.3   Molecular sieves

While activated carbon, silica gel, and activated alumina are materials with amorphous structures, molecular sieves are crystalline structures in which the molecules are arranged in a definite pattern. Molecular sieves are manufactured from aluminosilicate gels which are dehydrated. The most common molecular sieve is based on anhydrous aluminosilicate. It is also common to refer to molecular sieves as zeolites. They are used for both contaminant and odor removal. Molecular sieves are often impregnated with potassium permanganate or other proprietary mixtures of compounds to improve removal. Molecular

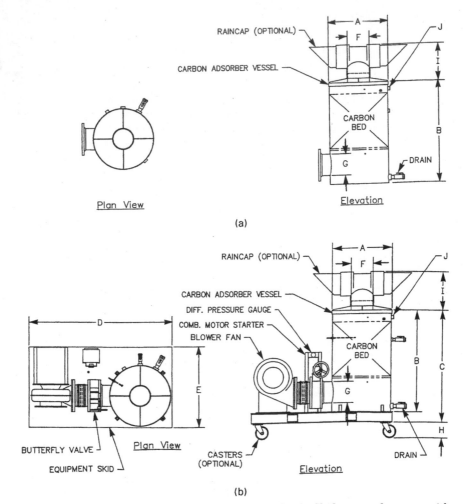

Figure 21.1 (*a*) Canister with standard features only. (*b*) Skid-mounted system with standard features only. (*Courtesy of Calgon Carbon Corporation, Pittsburgh, Pa.*)

sieves are most often found in granular form, having regularly spaced cavities with interconnecting pores of a known and definite size. Aluminosilicates are effective at removing sulfur compounds, mercaptans, alcohols, hydrogen sulfide, and formaldehyde. Typical surface areas are on the order of 1200 m^2/g. Bulk density for anhydrous aluminosilicate is approximately 38 lb/ft^3, with an effective pore diameter of 13 Å. Other materials used for molecular sieves include anhydrous sodium aluminosilicate and anhydrous calcium aluminosilicate. The bulk density for both these materials is 44 lb/ft^3. The average pore diameter of anhydrous sodium aluminosilicate is 4 Å and for anhydrous calcium aluminosilicate is 5 Å.

Molecular sieves can be regenerated with either a thermal swing process or a pressure swing process. The thermal swing process is used more often, passing a hot gas through the adsorbent bed in a countercurrent direction. Regeneration processes are more fully described in Sec. 21.4.2.

21.1.4 Silica gel

Silica gel is produced in a process that begins with the neutralization of sodium silicate with mineral acid. Subsequent steps include washing, drying, grading, and roasting. Silica gel is most often used in its granular form. The average surface area is 750 m^2/g, with an average pore diameter of 22 Å. The bulk density is 44 to 46 lb/ft^3. Silica gel is primarily used for gas drying, the removal of sulfur from gas streams, and gas purification.

21.1.5 Resins

Resin adsorbents are typically used for removal of organic contaminants from wastewater streams, rather than for air pollution control. However, a few systems have been developed for gas stream treatment. A typical example is the PADRE system developed by Purus, Inc.

Resins are made from many different monomeric compounds with varying degrees of polarity. Because of these polarity differences, each resin can be specifically designed to adsorb a unique compound or contaminant. For example, nonpolar organic compounds adsorb effectively onto hydrophobic resins via van der Waal's forces. And polar organics adsorb to acrylic resins with dipole-dipole interactions. Polystyrene resins are used to remove organics and recover antibiotics. Polyacrylic ester resins purify pulping wastewater, and phenolic resins are used to decolorize and deodorize waste streams.

One of the drawbacks to resins is that resins typically have a significantly smaller surface area than that of activated carbon (about 100 to 700 m²/g). In addition, the bonding forces of the contaminants to the resins are generally weaker than those with GAC. This tends to make the resin adsorption capacities smaller. In contrast to GAC, however, the resin adsorbents can be designed to be very selective of the compounds adsorbed. Additionally, resins have a low ash content and tend to be very resistant to bacterial growth.

The typical resin particle is spherical and approximately 0.5 mm in diameter. Resins are more expensive than GAC. Thus resin adsorbents are not generally used with waste streams involving multiple contaminants with no recovery value. If the adsorbed material is worth recovering, resin adsorbers may be an economically viable, alternative form of treatment.

The spent resin adsorbents can be regenerated with considerable ease. A solvent wash followed by distillation can be used to recover the adsorbed material without the dangers that thermal regeneration presents for GAC. The resins can be regenerated in situ using simple aqueous solutions and solvents.

An example of a resin adsorption system developed for gas stream treatment is the PADRE Vapor Treatment Process, developed by Purus, Inc., San Jose, California. The system is intended to work with soil vacuum extraction wells and wastewater air strippers as controls for volatile organic compound (VOC) emissions. The system has also proved viable for chlorinated solvent emissions for paint booth operations and remediation system discharges.

The PADRE technology has been purchased and enhanced by Thermatrix, Knoxville, Tennessee, by combining it with thermal oxidation technology. These enhancements allow for cost-effective removal of concentrated low- and high-flow emissions. One or more resin filter beds are used continuously in the process to treat VOCs. The VOCs adsorb to the resin in the first bed, while the resin in the second bed is regenerated and the contaminants recovered with a combination of temperature, pressure, and nitrogen application. More beds are added as required to handle higher flows. Regeneration can be handled sequentially or in combination, depending on loading and breakthrough. The treatment and regeneration steps are automatically alternated between the beds. The regenerated resin is rotated to the adsorption step, while the contaminants are condensed to a liquid for recycling or disposal.

This system solves the problem of costly off-site regeneration of spent carbon. In addition, while the activated-carbon adsorption efficiency is significantly hampered by humidity in the airstream, resin is not affected as much. A humid airstream can decrease the original carbon efficiency to 30 percent, whereas the resin adsorption efficiency holds at more than 75 percent.

Test beds of the PADRE process have been shown to withstand recycling more than 2000 times with no significant loss of efficiency or resin. See Figs. 21.2 and 21.3.

21.2 Theory

There are a number of factors that affect adsorption. However, as a generalization, the adsorptive capacity of a given adsorbent material is proportional to the surface area available. In selecting an adsorbent material, several factors (design variables) must be considered and balanced. Some of the more important design variables include capacity, selectivity, ability to be regenerated, compatibility, and cost. These design variables are described more specifically below.

21.2.1 Polarity

For activated carbon, nonpolar chemicals are preferentially adsorbed because activated carbon is itself nonpolar. Polarity is influenced by both physical and chemical forces such as dipole-dipole interactions, dispersing interactions, and hydrogen bonding. Polarity can be a significant factor with some of the adsorbent resins.

Figure 21.2 Thermatrix PADRE A3100 unit. (*Courtesy of Thermatrix, Knoxville, Tenn.*)

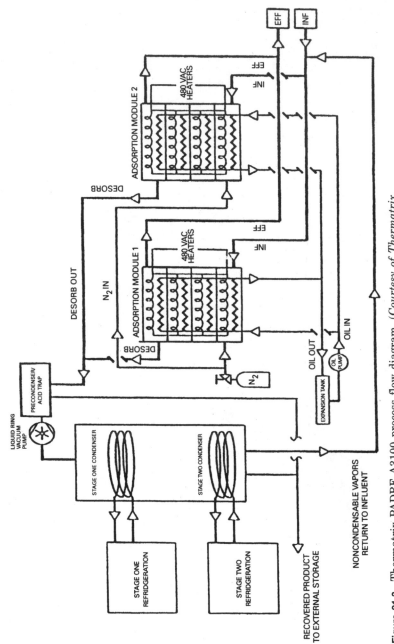

Figure 21.3 Thermatrix PADRE A3100 process flow diagram. (*Courtesy of Thermatrix, Knoxville, Tenn.*)

21.2.2 Charge

Adsorption of charged (ionized) chemicals is less significant than the adsorption of uncharged particles.

21.2.3 Molecular weight

In general, larger molecules are better adsorbed, unless the size of the molecule is greater than the diameter of micropores within the carbon particle.

21.2.4 Temperature

The temperature of a system is especially important in the adsorption of volatile organic chemicals. Adsorption capacity is inversely proportional to temperature, increasing as temperature decreases. This principle is based upon Gibbs' free energy. While a rigorous examination of Gibbs' free energy is beyond the scope of this text, a brief introduction is provided. Gibbs' free energy G is a measure of the spontaneous change in a system. The change in Gibbs' free energy of a system, at constant temperature and pressure, is defined by the following equation:

$$\Delta G = \Delta H - T \, \Delta S \qquad (21.1)$$

where ΔG = change in Gibbs' free energy
     ΔH = change in enthalpy
      T = temperature of system
     ΔS = change in entropy

When $\Delta G < 0$, a system reacts spontaneously and adsorption occurs. The adsorption of molecules from the gas phase to the adsorbent involves a change to a lower degree of disorder ($\Delta S < 0$). Decreasing T will produce a "less positive" $T \, \Delta S$ term, resulting in a more negative ΔG because ΔH is negative. A compound that does not "react," or adsorb, at a given temperature ($\Delta G > 0$) may absorb at a lower temperature (if $\Delta G < 0$). However, as the temperature decreases, the kinetic reaction rate also decreases.

Adsorption is an exothermic reaction. As the zone of adsorption moves though an adsorbent bed, the temperature of the bed increases and heat is released to the gas stream. Likewise, when the gas stream leaves the area of adsorption activity, the gas transfers the heat back to the bed. For certain chemicals and adsorbents, the heat transfer can be significant. In physical adsorption situations, the amount of heat released during the adsorption process is approximately equal to the latent heat of condensation of the adsorbate and the heat of wetting of the adsorbent by the adsorbate. During chemical adsorption, the heat released is approximately equal to the heat of chemical reaction. The temperature differential during the operation of the bed can be estimated as follows:

$$\Delta T = \frac{6.1}{(C_p/C_I) \times 10^5 + 0.51 C_A/q_e}$$ (21.2)

where ΔT = temperature rise, °F
C_p = heat capacity of air, Btu/(ft³ · °F)
C_I = influent concentration of adsorbate, ppm
C_A = heat capacity of adsorbent, Btu/(ft³ · °F)
q_e = equilibrium loading of adsorbent, lb/100 lb

See Table 21.3.

21.2.5   Surface area

Physical characteristics are very important in selecting adsorbent materials. One of the major considerations in selecting the adsorbent is the surface area of the material. Adsorbent materials used for waste treatment are highly porous. The majority of the surface area of an activated carbon particle is provided by the pore structure. Another surface-area property affecting adsorption is the distribution of pore size diameters. The larger adsorbate molecules can only adsorb in the larger-diameter pores. Two carbon particles with different pore size distributions will exhibit different adsorption performance. The primary method for determining the surface area of a unit volume of activated carbon is iodine adsorption by the Brunauer-Emmett-Teller (BET) method, as described in Sec. 21.3.2.

A related characteristic is the adsorbent particle diameter. The particle diameter influences the rate of adsorption. Smaller-diameter particles such as powdered activated carbon (PAC) have a shorter diffusion path, resulting in more rapid adsorption. The mass-transfer rate increases in inverse proportion to the particle diameter $d^{3/2}$. The internal adsorption rate increases in inverse proportion to d^2. The use

TABLE 21.3   Typical Adsorbent Heat Capacities

Adsorbent	Adsorbent heat capacities, Btu/(ft³ · °F)
Activated carbon	0.25
Alumina	0.21
Molecular sieve	0.25

SOURCE: A. J. Buonicore and W. T. Davis, eds., *Air Pollution Engineering Manual,* Van Nostrand Reinhold, New York, 1992.

of smaller-diameter adsorbent particles results in a higher pressure drop across the adsorbate bed than is exhibited with larger-diameter particles. As the Reynolds number (the measure of laminar or turbulent fluid flow) for the gas stream increases (becomes more turbulent), the pressure drop across the bed increases. An ideal system with minimal pressure drop occurs with adsorbent particles that are uniform in size and spherical.

21.2.6 Pore size distribution

Pore size distribution is a measurement of the percentage of the space of a particle occupied by micropores (pore diameter < 2 Å), mesopores (pore diameter > 20 Å and < 500 Å), and macropores (pore diameter > 500 Å). Figure 21.4 illustrates a portion of a typical particle cross

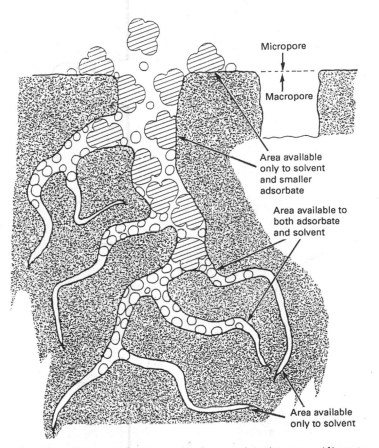

Figure 21.4 Concept of molecular screening in micropores (diameter range = 10 to 1000 Å). (*From Buonicore and Davis, 1992.*)

section. A molecule cannot penetrate into a pore smaller than a given minimum diameter (depending upon the size of the molecule). This process screens out larger molecules, and allows smaller molecules to penetrate further into the adsorbent particle or to adsorb in the smaller-diameter pores.

21.2.7   Other factors

In addition to the above-mentioned parameters which affect adsorption, other factors that may be significant in the design of an adsorption control system include resistance to airflow, as determined by the size and shape of the adsorbent particles; adsorbent bed depth; gas velocity; relative humidity of the gas stream; and desired removal efficiency.

21.3   Design

The general design procedure for gas-phase adsorption equipment involves two steps: (1) developing adsorption isotherms by collecting experimental data and applying various equations to the data and (2) applying the isotherm information to the anticipated adsorbate concentrations, flow rates, and other relevant properties to determine the optimal design.

21.3.1   Isotherm generation

Isotherms are a measure of the capacity of an adsorbent as a function of the concentration of the adsorbate in a feed stream. The key to determining the capacity and life of an adsorbent is the development of adsorption isotherms. Treatability studies provide the most accurate means of determining the specific adsorption properties of a waste stream for a given adsorbent. Experimental data can be determined on a volumetric basis, gravimetrically, or chromatographically. According to the volumetric method, an adsorbent is exposed to several adsorbate concentrations, and a mass balance is used to determine the adsorbed concentration. The volumetric method is most commonly used because of its low cost and straightforward procedures. A typical volumetric experimental procedure is described in Table 21.4. In performing the isotherm experiments, one should avoid activated-carbon dosages of less than 50 mg due to statistical errors in calculating q_{ei}. Measurements near the analytical detection limits may also contribute to errors in calculating q_{ei}. Also, make sure that the isotherm is developed for the concentration range of the particular application, as the isotherm parameters may exhibit a concentration dependence.

TABLE 21.4   Typical Volumetric Experimental Procedure

1. Determine preliminary isotherm parameters for design of the experiment. Include some experimental data which can be used for initial concentrations.
2. Using the following equation, select a range of 10 to 15 values of C_e per isotherm:

$$M_i = \frac{V(C_0 - C_e)}{KC_e^{1/n}} \qquad (21.3)$$

where M_i = mass of activated carbon
   V = volume of bottle
   C_0 = initial concentration
   C_e = final (equilibrium) concentration
   K = Freundlich constant
   n = Freundlich constant

3. Vary C_0 and V to determine the best range of values of M_i.
4. Set up 10 to 15 airtight bottles of a constant volume (V) with different masses of activated carbon (M_i) determined in step 3.
5. Add the same chemical concentration C_0 to each bottle.
6. Gently shake the bottles and wait for equilibrium to be established. This will typically take 4 to 8 h, although it could take longer.
7. Measure the equilibrium concentration in each bottle C_{ei}.
8. Calculate the activated-carbon loading q_{ei} in each bottle, using the following equation:

$$q_{ei} = \frac{V(C_0 - C_{ei})}{M_i}$$

9. Plot q_{ei} versus C_{ei} and determine the isotherm constants. Data for the Freundlich isotherm will need to be plotted on log-log scale (preferable) or the analysis made using nonlinear regression techniques.

For gas-phase adsorption, the gravimetric method is the most accurate means for determining adsorption isotherm data. In the gravimetric method, an adsorbate is diluted with an inert gas such as helium and is injected into a cell containing a few milligrams of the adsorbate. During the experiment, the cell is maintained at a constant temperature and adsorbate potential pressure while suspended from an electrobalance. The weight of the cell and the adsorbent is recorded until saturation is reached. The chromatographic method is a gas-phase screening method that involves crushing the adsorbent, placing it into a chromatographic column, and extracting it into a carrier gas for analysis (Cooper and Alley, 1986).

In its most general form, the equilibrium concentration of an adsorbate can be expressed as follows:

$$q^* = f(p) \qquad (21.4)$$

where q^* = the equilibrium concentration in moles per unit weight of adsorbent and p = partial pressure of the gaseous adsorbate. Experimental data derived from adsorption isotherm experiments will typically fit one of five different curves. The five types of curves are shown in Fig. 21.5, where n_m is defined as either the monolayer or

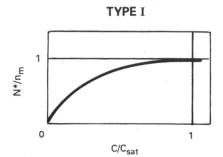

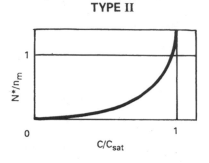

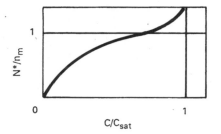

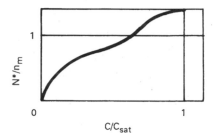

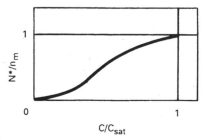

Figure 21.5 Adsorption isotherms can assume any of several shapes. In types II, III, and IV, n_m can stand for either monolayer or maximum loading (with the value at n^*/n_m = 1 in those types pertaining to the monolayer loading). In types IV and V, the light path is part of the isotherm that pertains to the regeneration portion of the adsorption cycle, whereas the dark path pertains to the adsorption portion. (*From Knaebel, 1995.*)

maximum loading, C is the adsorbate concentration, and C_{sat} is the saturation concentration of the adsorbent in the gas phase. Type I adsorption is convex upward throughout the curve and is considered favorable for adsorption. Type II adsorption is concave upward throughout the curve and is considered unfavorable for adsorption. Types II, IV, and V adsorption curves have multiple inflection points and therefore both concave and convex portions. Type IV and type V curves can exhibit hysteresis, a condition in which desorption occurs along a different isotherm from adsorption. Hysteresis only becomes an item of concern to the design when the system in question will have built-in regenerative capability. A sixth condition that may be observed is a linear adsorption curve, as modeled by the Langmuir equation, described below.

Based upon the plotted data, equations are fit to the data to determine the best predictive model. Once the isotherm that most accurately fits the data has been determined, a user may accurately predict, based upon the concentration of the adsorbate in the gas phase, the amount of adsorbent required to "capture" a given amount of adsorbate.

21.3.2 Single-component isotherms

Several isotherms have been developed for use in modeling the adsorption equilibrium. While numerous adsorption isotherms exist and are useful, the three most common equations used to describe the equilibrium between a surface (adsorbent) and a chemical in solution (adsorbate) are the Langmuir, BET, and Freundlich adsorption equations. As will be noted in the following equations, gas-phase reactions are commonly presented in terms of partial pressure p rather than concentration C, the common form in water applications. A number of other equations have been developed either as extensions of the three equations mentioned above or as new adsorption equations altogether. Among the other single-component equations encountered, although not presented here, are the BDDT (Brunauer, Deming, Deming, and Teller), Langmuir-Freundlich, Unilan, Toth, and three "Dubinin" equations. The more complex equations will typically more accurately model the adsorption data observed, although they necessitate the collection of more data, which can be an expensive and time-consuming process. These other methods all concentrate on various aspects of adsorption isotherms that exhibit different levels of accuracy over various ranges of data. In choosing an equation, a designer should lean toward the simplest equation that can account for the "nonideal" conditions observed in the data. In many cases, either the Langmuir or Freundlich equations are sufficient. For additional information on these other isotherm equations, see Knaebel (1995).

Langmuir equation. The Langmuir equation treats the interaction between the adsorbent and the adsorbate as a linear, reversible, monolayer chemical reaction. Developed by Langmuir in 1915, this equation is a relatively straightforward model that assumes that the adsorbent surface is completely homogeneous, each adsorbent "site" can bind a maximum of one adsorbate molecule, and there are no interactions between molecules of the adsorbate. The Langmuir adsorption model can be expressed in the following form:

$$q_e = \frac{Q_0 K_L p}{1 + K_L p} \tag{21.5}$$

where q_e = equilibrium loading on adsorbent
$\quad Q_0$ = ultimate adsorption capacity of adsorbent
$\quad K_L$ = relative energy of adsorption, also known as equilibrium constant—typically empirically determined
$\quad p$ = partial pressure of gaseous adsorbate

Advantages of the Langmuir equation include simplicity and applicability to a wide range of data. Limitations to the model include the monolayer assumption, the reversibility of bonding, and the constant uptake rates.

Brunauer-Emmett-Teller equation. The Brunauer-Emmett-Teller (BET) equation was first published in 1938 and is an extension of the Langmuir equation. The BET model extends the monolayer assumption of the Langmuir model to incorporate multiple adsorbate layers. Each adsorbate layer, regardless of thickness, is assumed to equilibrate with the layer immediately below. The BET adsorption model can be expressed in the following form:

$$q_e = \frac{q_0 K_L p}{(1 + K_L p + p/P)(1 - p/P)} \tag{21.6}$$

where q_e = equilibrium loading on adsorbent
$\quad q_0$ = ultimate adsorption capacity of adsorbent
$\quad K_L$ = relative energy of adsorption
$\quad p$ = partial pressure of gaseous adsorbate
$\quad P$ = vapor pressure

The BET model is primarily used in two cases: gas-solid systems that approach condensation and the estimation of adsorbent surface areas by iodine (or nitrogen) adsorption. The BET model, like the

Langmuir model, assumes that every adsorption site has an equivalent energy of adsorption. In reality, variations in an adsorbent surface will cause the energy of adsorption to vary.

Freundlich equation.   The Freundlich equation is an empirically derived logarithmic model that attempts to factor in the effects of various adsorption energy levels. The model assumes that the number of sites associated with a particular free energy of adsorption decreases exponentially as the free energy level increases. The Freundlich adsorption model can be expressed in the following form on a partial pressure basis:

$$q_e = K_F p^{1/n} \tag{21.7}$$

where q_e = equilibrium loading on adsorbent
$\quad K_F$ = adsorption capacity at unit concentration
$\quad 1/n$ = adsorption intensity—typically empirically determined and sometimes noted as ß
$\quad p$ = partial pressure of gaseous adsorbate

The equation fits a straight line when plotted on a log-log basis. The Freundlich equation can then be written in the following form:

$$\log q_e = \log K + \frac{1}{n} \log p \tag{21.8}$$

The Freundlich equation is commonly used by environmental engineers for empirical data and is helpful in quickly providing some general information about the tendency of a compound to be adsorbed. For irreversibly adsorbed chemicals, $1/n$ (the slope of the line) is zero. For favorably adsorbed chemicals, $1/n$ is between 0 and 1. For unfavorably adsorbed chemicals, $1/n$ is greater than 1. For chemicals that are not adsorbed, $1/n$ approaches infinity (or is very large). In many cases, intermediate conditions between the Freundlich and Langmuir isotherms are observed, requiring the designer to use a different isotherm equation to determine which of the two equations better fits the applicable data range.

Isotherm prediction.   Because isotherm data are not always readily available, several predictive models have been developed to allow estimation of adsorptive capacities without experimental data. The most common approaches utilize quantitative structure-activity relationship (QSAR) techniques, such as the Dubinin-Polyani adsorption potential theory and the Polyani-Reducskevich correlation theory. The QSAR techniques are based upon two general assumptions: (1) given the isotherm of one chemical on an adsorbent, the isotherm of any other (similar) chemical can be calculated; and (2) given the isotherm of a chemical at a given temperature, the isotherm at any other temperature can be calculated.

When isotherm data are available, several equations should be fit to empirical data and the optimum fit determined statistically. Most computer spreadsheet programs have analytical packages that include linear and nonlinear regression methods. Comparison of the sample correlation coefficients between different adsorption equations will give the designer a good indication of the model that most accurately predicts the adsorption isotherm. As mentioned above, other equations may account better for data observed over various ranges of concentrations. The designer should be sure that the isotherm equation used accurately models the adsorption over the range of conditions applicable.

21.3.3   Multicomponent isotherms

It is always important to consider that a gas stream with multiple components may exhibit the preferential adsorption of one compound over another. Individual chemicals do in fact compete with one another for adsorption sites. A common model for competitive adsorption is the *ideal adsorbed solution* (IAS) theory. The IAS model uses the isotherms of each of the single components to predict the competition among the chemicals. The general form of the IAS model is as follows:

$$p_i = \frac{q_i}{\displaystyle\sum_{j=1}^{N} n_j} \left(\frac{\displaystyle\sum_{j=1}^{N} n_j q_j}{n_i K_{Fi}} \right)^{n_i} \tag{21.9}$$

where p_i = partial pressure of ith chemical
$\quad\quad q_i$ = equilibrium loading on adsorbent for ith chemical
$\quad\quad K_{Fi}$ = Freundlich adsorption capacity constant for ith chemical
$\quad\quad n_i$ = reciprocal of Freundlich adsorption intensity for ith chemical
$\quad\quad N$ = number of components

For a multiple-component system composed of chemicals that individually follow Langmuir isotherms, the following equation can be used to predict the equilibrium relationship:

$$q_{ei} = \frac{q_{0i} K_{Li} p_i}{1 + \displaystyle\sum_{J}^{N-1} K_{Lj} p_j} \tag{21.10}$$

where q_{ei} = equilibrium loading on adsorbent
$\qquad q_{0i}$ = ultimate adsorption capacity of adsorbent for ith chemical
$\qquad K_{Li}$ = relative energy of adsorption of ith chemical, also known as equilibrium constant—typically empirically determined
$\qquad p_i$ = partial pressure of ith chemical

Multiple-component adsorption calculations are typically performed using computer models due to the complexity of the calculations.

21.3.4 Other sources of isotherm data

When a column treatability study is not practicable, or when preliminary information is all that is required, standard "pure" isotherms are available. Sources available, besides a review of current scientific literature, include Valenzuela's (1989) *Adsorption Equilibrium Data Handbook,* Dobbs and Cohen's (1980) *Carbon Adsorption Isotherms for Toxic Organics* (using the Freundlich isotherm), "Adsorption-Capacity Data for 283 Organic Compounds" (Yaws and Nijhawan, 1995), and manufacturers' literature. The designer should always remember that when literature isotherms are used, the isotherm is based upon ideal, noncompetitive conditions, and that actual field conditions may vary from the predicted results.

21.4 Operational Overview—Column Operation

21.4.1 Mass-transfer zones

During operation, an adsorbent bed consists of three zones of activity, as shown in Fig. 21.6. The majority of adsorption occurs in the primary adsorption zone, which is also referred to as the *mass-transfer zone* (MTZ). Behind the mass-transfer zone, the adsorbent is saturated with the adsorbate. Ahead of the MTZ, the bed is essentially free of the chemical. The thickness of the MTZ is a function of concentration, adsorbent, reaction kinetics, and contact time, and is proportional to the gas velocity. A larger MTZ is often due to higher gas flow rates and results in decreased adsorbent bed life. A large MTZ is typically indicative of poor bed utilization. Smaller-diameter adsorbent particles and "flatter isotherms" (small Freundlich $1/n$ values) are characteristic of small MTZ, and indicative of lower velocities and result in good adsorbent bed utilization. The mass-transfer zone can be determined using the following equation:

$$L_{\mathrm{MTZ}} = \frac{L}{t_{\mathrm{st}}/(t_{\mathrm{st}} - t_B) - X} \qquad (21.11)$$

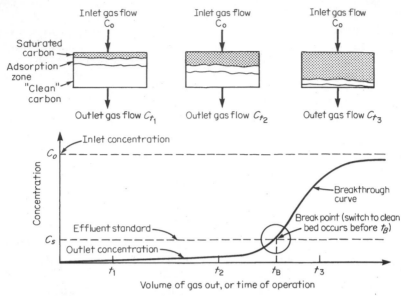

Figure 21.6 The adsorption wave and breakthrough curve. (*From Cooper and Alley, 1986.*)

where L_{MTZ} = length, or thickness, of mass-transfer zone
L = adsorbent bed length
t_{st} = time required for adsorbent saturation
t_B = time required for adsorbent break point
X = degree of saturation in mass-transfer zone

As the adsorbent is saturated, the MTZ moves through the bed. When a portion of the MTZ reaches the outlet of the bed (i.e., the effluent concentration is equal to the influent concentration), breakthrough is said to have occurred. Prior to breakthrough, however, the gas stream effluent concentration will equal the permitted effluent discharge standard. This point is called the *break point*. The time required for adsorbent break point and adsorbent saturation is most commonly empirically determined or estimated based upon experience.

Under actual operating conditions, bed adsorption capacity, known as the *operating* or *dynamic capacity*, will typically be some fraction of the theoretical adsorption capacity predicted according to the isotherms. These losses in efficiency are due to several factors, primarily the removal of the adsorption column from the treatment system at the break point, before the adsorbent has become fully saturated. Other losses in the theoretical adsorptive capacity of an adsorbent bed are due to the moisture in a gas stream, residual moisture from on-line regeneration, and the heat of adsorption due to the motion of the MTZ through the bed (Fig. 21.7, also see Sec. 21.2.4, Tempera-

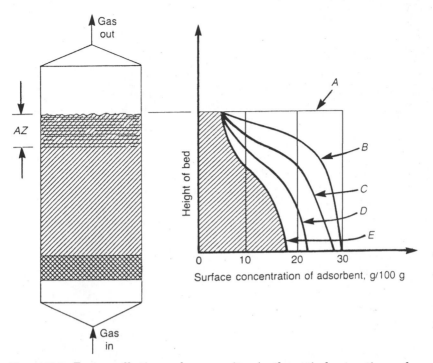

Figure 21.7 Factors affecting carbon capacity: *A*—theoretical saturation under equilibrium conditions; *B*—loss due to adsorption zone; *C*—loss due to heat wave; *D*—loss due to moisture; *E*—loss due to residual moisture or carbon. (*From Cooper and Alley, 1986.*)

ture). If empirical information is available, the following equation can be used to determine the operating capacity of the bed:

$$q_{\text{oper}} = q_e \left(\frac{L - L_{\text{MTZ}}}{L} \right) + 0.5 \left(\frac{L_{\text{MTZ}}}{L} \right) - \text{heel} \qquad (21.12)$$

where q_{oper} = the operating loading on the adsorbent, and heel = residual adsorbate present after regeneration, referred to as the "heel." Because the data required for the above equation are not often readily available, the operating capacity is assumed to be some fraction of the equilibrium loading. From a practical point of view, most systems operate at 25 to 50 percent of the adsorbent equilibrium capacity, with typical operation at 30 to 40 percent. In multicomponent applications, the following equation can be used to calculate the operating loading of the adsorbent:

$$q_{\text{oper}} = \frac{1}{\displaystyle\sum_{i=1}^{N} (w_i / q_{ei})} \qquad (21.13)$$

where q_{ei} = equilibrium loading on adsorbent of the ith chemical
$\quad\quad w_i$ = mass fraction of ith chemical in N components
$\quad\quad N$ = number of components in gas stream, exclusive of carrier
$\quad\quad\quad$ gas

If the designer is concerned that a more accurate number for column breakthrough is necessary for the application, pilot-scale tests under actual operating conditions can be used to determine the operating capacity of the adsorbate.

In a gas stream with multiple chemicals, several MTZs will be present. These zones may or may not overlap, and they are also subject to competitive adsorption. Under competitive conditions, displacement of less readily adsorbed chemicals may occur. One of the ways that this displacement can manifest itself is through a sudden increase in the exhaust concentrations (greater than the inlet concentration) of the less readily adsorbed compound. Additional off-gas controls or different operating strategies may be necessary if the displacement is a problem.

Factors that will influence the adsorption capacity of a carbon column include the temperature and the relative humidity. The adsorptive capacity decreases with increasing system temperature. Likewise, a relative humidity greater than 50 percent can reduce the adsorptive capacity of the column because the adsorption sites are occupied by water molecules. See Fig. 21.7.

21.4.2   Bed regeneration

Once the break point has been attained, the bed must be regenerated or replaced. When one is working with activated carbon, on-line bed regeneration is accomplished by applying low-pressure steam or a vacuum to the bed. As a rule of thumb, approximately 90 percent of the granular carbon remains after regeneration. Two common techniques for on-line regeneration are *thermal swing regeneration* and *pressure swing regeneration*. Thermal swing regeneration involves heating the bed directly with either a hot inert gas or low-pressure steam or indirectly with surface contact. The adsorbent is heated to between 300 and 600°F, and the gas is applied to purge the adsorbent. The thermal swing regeneration process requires a period in the operational cycle for cooling. Pressure swing regeneration involves applying either a low pressure or a vacuum to draw the adsorbate from the bed. Operated under essentially isothermal conditions, the pressure swing regeneration process allows for shorter cycle times and the recovery of a high-purity product. Low-pressure steam is the process most often used, except in solvent recovery applications where water would contaminate the recovered solvent. Applied at a rate of 1 to 4 lb of steam per pound of adsorbate, the saturated steam will rapidly heat the bed without polymerizing the adsorbed chemicals. During the regeneration process, condensation will occur in the adsorbent bed, leaving a portion of the adsorbate in the bed. This fraction is referred to as the *heel*. The most common off-line regeneration technique, particularly for activated carbon, involves passing the adsorbent through a high-temperature, multiple-hearth furnace.

Other factors to consider in designing the regeneration system include bed ignition and explosive solvent concentrations. Most safety and insurance regulations specify that the inlet vapor concentration of a solvent not exceed 25 percent of the *lower explosive limit* (LEL). Bed ignition, while not a common problem, is known to occur. Thermocouples mounted on the effluent side of the bed are used to detect temperature surges. When the temperature exceeds a predetermined level, the bed can immediately be flooded with an inert gas or water.

21.5 Equipment

Gas-phase carbon adsorption systems can be both regenerative and nonregenerative. Regenerative systems can be configured as fixed-, moving-, or fluidized-bed systems. Fixed-bed systems are the most common for gas-phase organic control. Fixed-bed systems can be operated on either a continuous or an intermittent flow basis, with flow rates ranging from 1000 to greater than 100,000 ft³/min. Intermittent operation allows a single carbon column to be regenerated during the off-line periods of operation. For continuous flow operation, multiple beds are used in parallel so that at least one carbon column is in use while another column is being regenerated or is otherwise off-line. A typical VOC removal system would also incorporate prefiltration to remove particulate, dehumidifiers or driers, cooling to prevent fires from the hydrocarbons and to maintain the optimum operating temperature, and recirculation systems. The *empty bed contact time* (EBCT) is a measure of the column detention time. The EBCT is the total volume of the column divided by the flow rate, and it is typically on the order of 1 to 10 min. Columns with larger EBCTs tend to have lower operational costs while having higher capital costs.

21.5.1 Design process

A primary concern is whether an adsorption unit should be designed and built by the user or purchased as a unit from a manufacturer. As a rule of thumb, adsorption systems with a capacity of less than

20,000 ft^3/min can be more cost-effectively purchased as a unit from a reputable manufacturer. It is usually more cost-effective to custom-design larger systems. The steps used in designing an adsorption system are as follows:

1. Establish the required exhaust concentration (typically available in regulations or existing facility permits). Based upon the influent gas stream concentration (measured), determine the mass per unit time period that will need to be removed (i.e., pounds per hour).

2. Determine chemical characteristics of the various components in the gas stream.

3. Select an adsorbent material, including size. This can be accomplished by consulting with vendors, literature review, or though the designer's prior experience.

4. Develop adsorption isotherm data and the best equation to fit the data.

5. Determine gas stream operating conditions, such as flow rate, temperature, gas stream pressure, and partial pressure of the various chemicals of concern.

6. Given the adsorption isotherm, determine the theoretical adsorption capacity, defined as the mass of adsorbate per mass of adsorbent (often lb adsorbate/100 lb adsorbent).

7. Determine the operating capacity of the carbon, as a percentage of the theoretical adsorption capacity. This can be done either empirically or by using a percentage of the theoretical adsorption capacity.

The remaining steps in the design process are somewhat iterative, and they may require several revisions to the calculations.

8. Select a regeneration cycle time for the system. Experience plays a key role in determining the appropriate cycle time. Set the adsorption time equal to the regeneration time, for one complete cycle. Most systems operate with a regeneration time of 90 min or less.

9. Select an operating velocity v for the system. The velocity is usually around 80 ft/min, although it can range as high as 100 ft/min for solvent recovery, where thick adsorbate beds are used, and as low as 40 ft/min for thin beds. The maximum velocity for an adsorbent is determined by the *crushing velocity* for a particular adsorbent material. Crushing velocity information is available from the manufacturer of the adsorbent.

10. Calculate the mass of adsorbent required for one-half of an operational cycle, the adsorption phase of the cycle

$$M = \frac{QC_i t_{ads}}{q_{oper}} \tag{21.14}$$

where M = mass of adsorbent required
 Q = gas flow rate
 C_i = adsorbate influent concentration
 t_{ads} = adsorption phase of cycle time

11. Calculate the volume of the adsorbent required by dividing the mass of adsorbent M by the adsorbent bulk density ρ_B.

12. Calculate the surface area of the adsorbent bed A, using the gas stream flow rate and the volume, and the bed length L, using the volume and area. As a rule of thumb, a length-to-diameter ratio on the order of 3-to-4 is an acceptable target. Keep in mind that longer beds usually result in greater removal.

13. Determine the steam loading rates for regeneration.

14. Calculate the empty bed contact time, typically on the order of 1 to 10 min.

15. If the EBCT is out of the acceptable range, repeat steps 8 through 14.

Other factors to consider during the design of the system are that the gas stream must be evenly distributed at the entrance to the adsorbent bed for maximum bed efficiency and that prefilters are required to remove particulate matter and excessive moisture which can foul an adsorbent bed. Due to size constraints, the system can typically be installed vertically if the flow rate is less than 2500 actual ft³/min.

21.5.2 Fixed-bed systems

There are three primary types of adsorption systems for air pollution control. The systems are described by the bed arrangement. The three systems include fixed, or stationary, beds, moving beds, and fluidized beds. A fixed-bed system consists of a square or cylindrical chamber containing the adsorbent. In many installations, fixed beds are operated in several stages, either in parallel or in series. A good design practice is to install a second unit in series to catch any of the contaminants if breakthrough occurs. Gas flow in fixed-bed systems is either downward, in vertical installations, or across, in horizontal installations. Smaller systems are usually fixed-bed types. The adsorbent is packed into drums or other containers, which are removed when breakthrough occurs. Adsorbent manufacturers can furnish complete packaged systems, and they will often offer to regenerate the spent adsorbent when necessary. Gas adsorption media are available in a wide range of packaging. Many adsorbents are available in 1-ft^3 (or smaller) corrugated cartons for use in interior applications. Typical sizes for off-gas systems range from 55-gal plastic drums to 12,000-lb and larger deep-bed filtration systems, which can be made up of multiple filter beds in series.

21.5.3   Moving-bed systems

A moving-bed system consists of coaxial rotating cylinders. Both the outer and inner cylinders are impervious to gas flow. The outer cylinder has slots near one end for the gas stream to enter. The adsorbent is located between the two cylinders. The inner cylinder has slots near the opposite end for the gas to exit after it has passed through the adsorbent bed. These slots in the inner cylinder also serve as inlet steam ports for regeneration. As the cylinder rotates, a portion of the cylinder is being regenerated, while the remainder of the cylinder is being used for adsorption, allowing a more efficient use of the adsorbent. Because the moving bed is continuously being regenerated, the regeneration time required for reactivation of the adsorbent is reduced, allowing a shorter adsorbent bed to be used. These changes result in more compact designs and a reduced pressure drop across the shorter adsorbent bed.

21.5.4   Fluidized-bed systems

Fluidized-bed systems are based upon the principle of continuously recirculating the adsorbent through the adsorption and regeneration cycles. The gas flow is upward through the adsorption chamber. The saturated adsorbent migrates downward and is transferred up to a surge bin, where it passes into the regeneration chamber. The regenerated adsorbent is then metered back into the top of the adsorption chamber. Carbon is the most common adsorbent used in a fluidized-bed system. The adsorbent flows countercurrent to the gas stream, which must have a velocity of approximately 240 ft/min for bed fluidization. Fluidized-bed systems exhibit more efficient use of the adsorbent—the adsorbent is saturated just before it is discharged from the adsorption chamber, and the majority of the adsorbent is fully utilized. The increased efficiency also results in a reduced mass of adsorbent required for an application, translating to reduced system size. A disadvantage of the fluidized-bed system is the attrition losses of the adsorbent due to the fluidization of the beds, causing "crushing" of the carbon particles.

References

Bioclimatic, Inc., "Gas Phase Contaminant Control—Adsorption," Technical Document 198, December 1993.

Buonicore, A. J., and Davis, W. T., eds., *Air Pollution Engineering Manual,* Air and Waste Management Association, Van Nostrand Reinhold, New York, 1992.

Calgon Carbon Corporation, "Basic Concepts of Adsorption on Activated Carbon," Pittsburgh, Pa., 1985

Cooper, C. D., and Alley, F. C., *Air Pollution Control: A Design Approach,* Waveland Press, Inc., Prospect Heights, Ill., 1986.

Dobbs, Richard A., and Cohen, Jesse M., *Carbon Adsorption Isotherms for Toxic Organics,* EPA-600/8-80-023, 1980.

Dow Chemical Company, *Dowex Monosphere Resins.*

Knaebel, Kent S., "For Your Next Separation Consider Adsorption," *Chemical Engineering,* November 1995, pp. 92–102.

Montgomery, James M., Consulting Engineers, Inc., *Water Treatment Principles and Design,* Wiley, New York, 1985.

Noyes, Robert, ed., *Unit Operations in Environmental Engineering,* Noyes Publications, Park Ridge, N.J., 1994.

Perry's Chemical Engineers Handbook, 6th ed., McGraw-Hill, New York, 1984.

Purus, Inc., *PADRE Vapor Treatment Process,* www.nttc.edu/env/site95/demo/ongoing/purus.html, February 20, 1997.

Speitel, Gerald E., Jr., University of Texas at Austin, notes, 1990-1.

Swindell-Dressler Company, *Process Design Manual for Carbon Adsorption,* for the EPA Technology Transfer, Program #17020 GNR, 1971.

Yaws, Bu, and Nijhawan, "Adsorption-Capacity Data for 283 Organic Compounds," *Environmental Engineering World,* May–June 1995.

Zanitsch, R. H., "Control of Volatile Organic Compounds Using Granular Activated Carbon," presented at Air Pollution Control Association, Southern section, 10th Annual Meeting, September 1979.

Chapter 22

Incineration of Gaseous Emissions

Barney Fullington

Incineration is used as an air pollution control device for the control of gaseous emissions from industrial sources. In the context of air pollution control, the term *incineration* refers to the oxidation of undesirable air pollutants through the process of combustion. Of the variety of gaseous pollutants emitted by industry, volatile organic compounds (VOCs) are the most common pollutant controlled by incineration. The chemicals known as VOCs are known precursors to the formation of photochemical smog and thus are regulated under the Clean Air Act. The two major categories of incineration for air pollution control are thermal and catalytic oxidation. The theory, design, and application of these two technologies are discussed in this chapter.

22.1 Thermal Oxidation

Thermal oxidation of waste gas streams is a widely used method for controlling VOC emissions from industrial sources. The proven efficiency of thermal oxidizers for destroying VOCs is the principal reason for their widespread use. Removal efficiencies in excess of 99.99 percent are possible with efficiencies greater than 95 percent commonly achieved (Frame, 1996; Klobucar, 1995; Straitz, 1996). The major drawbacks of thermal oxidation are the high capital and operating costs associated with this technology. The capital costs associated with a thermal oxidizer include costs for the burner, combustion chamber, refractory, and associated air pollution control devices. The major cost associated with operating a thermal oxidizer is due to the use of an auxiliary fuel. Other operating costs include electrical costs from the instrumentation system and fan. Today the common use of heat recovery units has helped minimize the operating cost associated with the use of an auxiliary fuel. The decision to use thermal incineration for VOC control is dependent on several factors of the waste gas stream. Figure 22.1 gives a flowchart for determining the suitability of thermal oxidation and the need for auxiliary fuel and oxygen requirements.

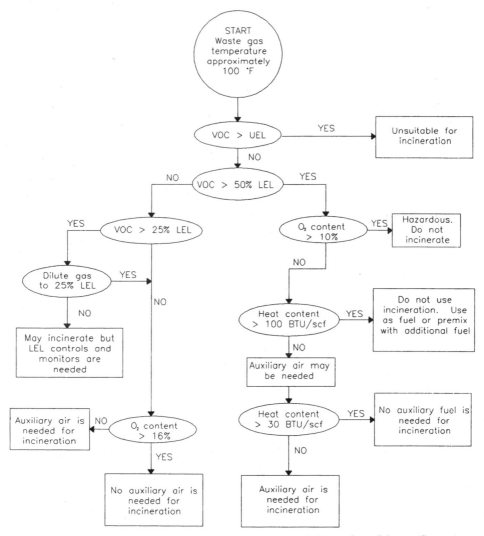

Figure 22.1 Flowchart for categorizing waste gas incinerability and need for auxiliary air. (*Vishnu S. Katari, William M. Vatavuk, and Albert Wehe, "Incineration Techniques for Control of Volatile Organic Compound Emissions," JAPCA, vol. 37, no. 1, p. 91, January 1987.*)

Combustion is a chemical process in which oxygen is rapidly combined with various chemical compounds at elevated temperatures. The rate of oxidation is directly dependent on the temperature at which the reaction occurs. Normal operating temperatures for combustion of organic pollutants in thermal oxidizers range from 1000 to 1500°F (540 to 820°C). For complete oxidation, the pollutants must be maintained at the reaction temperature for a specific period referred to as the *residence* (or *detention*) *time* of the oxidizer. Typical detention times for thermal oxidizers range from 0.5 to 2.0 s (Klobucar, 1995).

Proper combustion is also dependent on complete mixing of the auxiliary fuel with oxygen and the pollutants with the heat from the combustion of the fuel. If complete mixing is not achieved, short-circuiting of the combustion chamber may occur. This short-circuiting will result in incomplete combustion and a concomitant decrease in destruction efficiency and an increase in the formation of products of incomplete combustion (PICs).

The use of thermal incineration to control VOC emissions presents several problems involving safety. The *lower explosive limit* (LEL) of an organic vapor is defined as the VOC concentration of the gas stream at which insufficient organics are present to sustain combustion. The *upper explosive limit* (UEL) is defined as the VOC concentration above which there exists insufficient oxygen to sustain combustion. Waste gases that contain VOCs between the LEL and UEL are sufficiently mixed with oxygen to sustain combustion. VOCs in this range are very dangerous and cannot be combusted in a thermal incinerator due to the possibility of flashback. *Flashback* is the combustion of the gas stream back from the incinerator and into the connecting ductwork and processes. For safety reasons, the VOC concentration in the waste gas stream to be incinerated cannot be above the LEL and is typically limited to 25 percent or less of the LEL (Vatavuk, 1990; Brunner, 1984). A list of the upper and lower explosion limits for various compounds is shown in Table 22.1 (U.S. EPA, 1973).

22.1.1 Theory of incineration

Thermal oxidation is a process in which organic compounds are oxidized through combustion. The principal products of this reaction are carbon dioxide and water which can safely be discharged to the atmosphere. The general expression for combustion of hydrocarbons is

$$C_X H_Y + \left(\frac{4x + y}{4} \right) O_2 \xrightarrow{\text{Heat}} x CO_2 + \frac{y}{2} H_2O \qquad (22.1)$$

where x and y are the number of atoms of carbon and hydrogen, respectively, in the hydrocarbon. Of course, the pollutants to be com-

TABLE 22.1   Lower and Upper Explosion Limits for Pure Gases and Vapors

Chemical	Lower explosion limit, % by volume	Upper explosion limit, % by volume
Acetaldehyde	4.0	57.0
Acetone	2.5	12.8
Acetylene	2.5	80.0
Allyl alcohol	2.5	—
Ammonia	15.5	26.6
Amyl acetate	1.0	7.5
Amylene	1.6	7.7
Benzene	1.3	6.8
Benzyl chloride	1.1	—
Butene	1.8	8.4
Butyl acetate	1.4	15.0
Butyl alcohol	1.7	—
Carbon disulfide	1.2	50.0
Carbon monoxide	12.5	74.2
Chlorobenzene	1.3	7.1
Cresol, m- or p-	1.1	—
Crotonaldehyde	2.1	15.5
Cyclohexane	1.3	8.4
Cyclohexanone	1.1	—
Cyclopropane	2.4	10.5
Cymene	0.7	—
Dichlorobenzene	2.2	9.2
Dichloroethylene (1,2)	9.7	12.8
Diethyl selenide	2.5	—
Dimethyl formamide	2.2	—
Dioxane	2.0	22.2
Ethane	3.1	15.5
Ether (diethyl)	1.8	36.5
Ethyl acetate	2.2	11.5
Ethyl alcohol	3.3	19.0
Ethyl bromide	6.7	11.3
Ethyl cellosolve	2.6	15.7
Ethyl chloride	4.0	14.8
Ethyl ether	1.9	48.0
Ethyl lactate	1.5	—
Ethylene	2.7	28.6
Ethylene dichloride	6.2	15.9
Ethyl formate	2.7	16.5
Ethyl nitrite	3.0	50.0
Ethylene oxide	3.0	80.0
Furfural	2.1	—
Gasoline	1.4	7.6
Heptane	1.0	6.0
Hexane	1.2	6.9
Hydrogen cyanide	5.6	40.0
Hydrogen	4.0	74.2
Hydrogen sulfide	4.3	45.5
Illuminating gas	5.3	33.0
Isobutyl alcohol	1.7	—
Isopentane	1.3	—

TABLE 22.1   Lower and Upper Explosion Limits for Pure Gases and Vapors (*Continued*)

Chemical	Lower explosion limit, % by volume	Upper explosion limit, % by volume
Isopropyl acetate	1.8	7.8
Isopropyl alcohol	2.0	—
Kerosene	0.7	5.0
Methane	5.0	15.0
Methyl acetate	3.1	15.5
Methyl alcohol	6.7	36.5
Methyl bromide	13.5	14.5
Methyl butyl ketone	1.2	8.0
Methyl chloride	8.2	18.7
Methyl cyclohexane	1.1	—
Methyl ether	3.4	18.0
Methyl ethyl ether	2.0	10.1
Methyl ethyl ketone	1.8	9.5
Methyl formate	5.0	22.7
Methyl propyl ketone	1.5	8.2
Mineral spirits no. 10	0.8	—
Naphthalene	0.9	—
Nitrobenzene	1.8	—
Nitroethane	4.0	—
Nitromethane	7.3	—
Nonane	0.8	2.9
Octane	1.0	3.2
Paraldehyde	1.3	—
Pentane	1.4	7.8
Propane	2.1	10.1
Propyl acetate	1.8	8.0
Propyl alcohol	2.1	13.5
Propylene	2.0	11.1
Propylene dichloride	3.4	14.5
Propylene oxide	2.0	22.0
Pyridine	1.8	12.4
Toluene	1.3	7.0
Turpentine	0.8	—
Vinyl ether	1.7	27.0
Vinyl chloride	4.0	21.7
Xylene	1.0	6.0

Source: USEPA (1973).

busted may contain other components such as halogens, sulfur, phosphorus, and metals. The following primary reactions occur during the proper combustion of a waste stream:

- Carbon in waste stream is converted to carbon dioxide.

$$C + O_2 \rightarrow CO_2$$

- Halogens in waste stream are converted to halogenated acids.

$$H + Cl \rightarrow HCl$$
$$H + F \rightarrow HF$$

- Remaining hydrogen converts to water vapor.

$$4H + O_2 \rightarrow 2H_2O$$

- Nonalkali metals convert to oxides.

$$S + O_2 \rightarrow SO_2$$
$$4P + 5O_2 \rightarrow 2P_2O_5$$

Several secondary reactions may also take place during combustion. These reactions include the generation of carbon monoxide (CO), oxides of nitrogen (NO_x), polychlorinated dibenzo-p-dioxins (dioxins), and dibenzofurans (furans). These compounds are major pollutants, and their emission to the atmosphere should be minimized.

Carbon monoxide is formed during the combustion process when inadequate amounts of oxygen are present for all the carbon in the gas stream to be converted to carbon dioxide. This may be the result of either not feeding enough air to the oxidizer or insufficient mixing inside the combustion chamber. With insufficient oxygen the following reaction will occur:

$$C + O_2 \rightarrow CO \tag{22.2}$$

Oxides of nitrogen (NO_x) are formed during combustion by two mechanisms. The first mechanism is the thermal fixation of atmospheric N_2, which follows the general equilibrium equation

$$N_2 + O_2 \rightarrow 2NO + heat \tag{22.3}$$

Here the final concentration of NO is directly dependent on the combustion temperature and the amount of excess air in the combustion chamber. The second mechanism of NO_x formation during combustion is the oxidation of fuel-bound nitrogen which follows the general equilibrium equation

$$\text{Fuel-N} + 2O_2 \rightarrow 2NO + heat + 2(\text{fuel}) - O \tag{22.4}$$

The generation of NO in this reaction is primarily dependent on the amount of fuel-bound nitrogen and free O_2 in the combustion chamber. This reaction is less dependent on the combustion temperature than thermal fixation and can occur at modest combustion temperatures. The NO_x generation in combustion processes cannot be theoretically determined and is highly dependent on the operating conditions of the unit.

Polychlorinated dibenzo-p-dioxins and dibenzofurans may form under incomplete combustion conditions inside the combustion chamber or may form downstream of the combustion chamber. Although the formation reactions are not completely understood, it is known that the following precursors must be present in order for dioxins and furans to form (Williamson, 1994):

- Temperature window between 300 and 700°F (150 and 370°C)
- Sufficient residence time in the required temperature window
- Presence of a suitable catalyst (typically copper or lead)
- Presence of inert particulate matter in the gas stream
- Presence of chlorine in the waste gas stream

All these requirements must be met prior to dioxin or furan formation; therefore, to reduce their potential for formation, one or more of the above requirements must be controlled. The presence of the precursors can be controlled through prefiltration of the gas stream, rapid cooling of the gas downstream of the combustion chamber, and regulation of the types of compounds that enter the oxidizer.

There are several steps to control the generation of undesirable pollutants during thermal oxidation of VOC-laden gas streams. Good combustion practices will usually minimize the majority of the CO and NO_x emissions from the oxidizer. Dioxin and furan emissions are usually more difficult to control due to their complex formation reactions, which are not completely understood; but if the known precursors and formation conditions are monitored and controlled, dioxin and furan emissions from a thermal oxidizer should be minimal.

22.1.2 Design of thermal oxidation systems

Complete combustion is a result of the proper design and operation of the oxidizer. The following requirements must be met for proper combustion to take place:

- Sufficiently high combustion temperature
- Adequate detention time of gas stream at the combustion temperature
- Complete mixing of gas stream with heat released from auxiliary fuel combustion
- Abundant supply of oxygen for combustion

compounds undergo during oxidation. The thermal stability ranking for the 320 organics can be found in App. Q. This ranking does not give absolute destruction efficiencies of the compounds but ranks only their relative incinerability. The results from the first-order kinetic model and the thermal stability ranking can be compared, and approximate design parameters can be determined. In general, the determination of the temperature and residence time needed for proper combustion is achieved experimentally, although the approximate design parameters obtained from the method given above may be a satisfactory starting point for obtaining experimental data.

Turbulence is one of the most difficult aspects of the incineration process to control. The phenomenon of turbulence is easily visualized yet is difficult to quantify. Turbulence is necessary for proper combustion for two reasons. The two objectives of sufficient turbulence in the combustion chamber are to adequately mix the waste gas stream with the available oxygen and to satisfactorily mix the incoming waste gas stream with the heat released from combusting the auxiliary fuel.

For complete oxidation of the pollutants in the waste gas stream, a sufficient quantity of oxygen is needed. Without sufficient oxygen, incomplete combustion occurs. Depending on the waste gas stream, additional combustion air may not be needed if the waste gas stream contains enough oxygen in excess of the stoichiometric amount needed for complete oxidation. If the waste gas stream does not contain a high enough amount of oxygen, supplemental air is needed. This air is most often obtained from the atmosphere and is usually preheated to minimize fuel consumption.

Due to the elevated temperatures needed for thermal oxidation, some form of heat recovery is usually employed to minimize auxiliary fuel consumption. There are typically two types of heat recovery in use today, recuperative and regenerative. Recuperative heat recovery involves preheating the incoming gas stream with the combustion flue gas, using an air-to-air heat exchanger. Regenerative heat recovery involves exchanging heat from the combustion flue gas to the incoming waste gas by using beds of heat-resistant exchange media. These two heat recovery technologies are discussed below. Another type of heat recovery creates steam by passing the hot flue gas through a waste heat boiler. Figure 22.2 shows a typical thermal oxidizer with a waste heat boiler. The boiler is the long horizontal cylinder above the combustion chamber (here shown in the foreground). Depending on the application, thermal oxidizers can be used without heat recovery. Figure 22.3 shows a typical thermal oxidizer without heat recovery.

Figure 22.4 shows a schematic of a typical recuperative heat exchanger. In Fig. 22.4, the waste gas enters the oxidizer at the inlet. Once inside the oxidizer, the waste gas flows around the heat exchang-

Figure 22.2 Thermal oxidizer with waste heat boiler.

Figure 22.3 Thermal oxidizer without heat recovery.

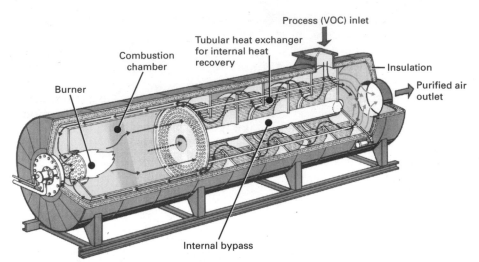

Figure 22.4   Recuperative thermal oxidizer (*Wahlco, Inc./LTG, product literature, 1996*).

er tubes where the waste gas is preheated. After the heat exchanger, the preheated waste gas enters the combustion chamber, where a natural gas-fired burner heats the gas to the combustion temperature. Once the gas has reached the combustion temperature, oxidation occurs, destroying the majority of the VOCs in the gas stream. From the combustion chamber the clean flue gas enters the tubes of the heat exchanger, where a portion of the heat of the flue gas is transferred to the incoming waste gas. The cooled flue gas then exits the oxidizer where it can be either exhausted to the atmosphere or sent to additional air pollution control devices (APCDs). If the incoming gas stream VOC content is significantly higher than the design concentration, the flue gas can reach temperatures that cannot be handled by the heat exchanger. In this case, the flue gas will bypass the heat exchanger to prevent thermal deterioration of the tube material. Figure 22.4 shows an installed recuperative thermal oxidizer.

The amount of heat recovered using a recuperative thermal oxidizer can vary greatly depending on the size of the heat exchanger, but is limited in practice to 70 percent (Klobucar, 1995; Straitz, 1996). Higher recoveries are theoretically possible but are not practical for two reasons. (1) The higher the heat recovery, the larger the heat exchanger needs to be. At some point, the cost savings associated with the decrease in fuel consumption is offset by the increase in capital costs associated with the larger heat exchanger. (2) Thermal recoveries greater than 70 percent are not used in recuperative thermal oxidizers because at higher temperatures, autoignition of the organics in the incoming waste gas stream is possible. Autoignition within the heat recovery equipment could damage the heat exchanger due to overheating and corrosion.

Figure 22.5 shows a typical regenerative thermal oxidizer. In the figure, the incoming waste gas stream enters the air distribution system, where it is routed to the middle canister. The waste gas then flows upward through a preheated regenerator where the gas temperature is raised to approximately 90 to 95 percent of the combustion temperature (Klobucar, 1995; Straitz, 1996; REECO, 1991). Once preheated, the gas stream passes through the burner area where more heat is added. From the burner area, the gas (now at the combustion temperature) enters the oxidation chamber, where the oxidation reaction is completed. From the oxidation chamber, the clean flue gas passes through cool exchange media where 90 to 95 percent of the gas stream's heat is transferred to the bed. Once cooled, the flue gas can be exhausted to the stack or sent to additional APCDs.

After approximately 1 to 4 s (Klobucar, 1995), the airflow is reversed and the incoming waste gas stream passes through the third canister and the hot exchange media, where it is preheated. Meanwhile the hot flue gas passes through the first canister that contains cooled bed media. In this configuration, the exchange medium in the middle canister is then purged of the captured VOCs to prevent a release of unreacted VOCs every time the airflow is reversed. The purged gas is next combined with the incoming waste gas upstream of the oxidizer. After

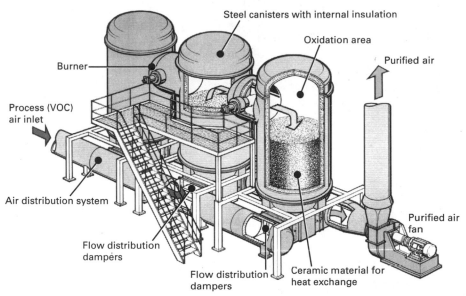

Figure 22.5 Regenerative thermal oxidizer (*Wahlco, Inc./LTG, product literature, 1996*).

another 1 to 4 s, the incoming gas is redirected to the first canister that contains hot exchange media while the flue gas transfers heat to the middle canister's bed. In this cycle, the exchange medium in the third canister is then purged of unreacted VOCs. This cycle is repeated indefinitely 15 to 60 times per minute depending on the design.

The basic design of the recuperative and regenerative thermal oxidizer may change depending on the needs of the industry. Small, compact designs can be used to oxidize low flows of gas. A compact regenerative thermal oxidizer is shown in Fig. 22.6. This type of design can treat a gas with a flow rate up to 20,000 actual ft³/min (566 m³/min) (Frame, 1996). Regenerative oxidizers can expand to accommodate very large flows using multiple canisters.

22.1.3 Application of thermal oxidation systems

A multitude of companies design and manufacture thermal oxidizers. These companies offer the capability of providing all the equipment

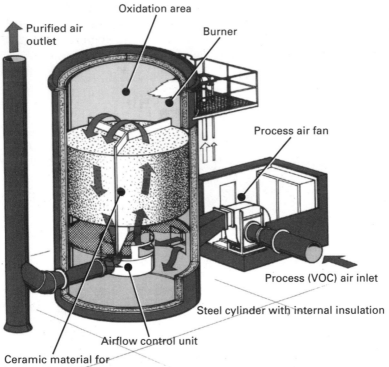

Figure 22.6 Compact regenerative thermal oxidizers (*Wahlco, Inc. / LTG, product literature, 1996*).

needed to build and operate a thermal oxidizer including the burner, combustion chamber, appropriate ductwork, instrumentation, and controls. For applications with small flow rates, typically less than 20,000 actual ft^3/min (566 m^3/min), most companies provide package units complete with all necessary equipment (Frame, 1996). For larger flow rates, thermal oxidizer manufacturers can custom-design units to fit the application.

The capital cost for thermal incinerator systems depends on the heat content of the gas flow rate, the heat content of the waste gas, the degree of heat recovery, the residence time, and the presence of halogens or other contaminants that require further treatment after incineration. The most important parameters in determining the costs of a thermal incinerator are the waste gas flow rate and the degree of heat recovery. Figure 22.7 shows typical capital costs of a recuperative thermal oxidizer for various waste gas flow rates (Vatavuk, 1990).

Regenerative thermal oxidizer capital costs depend upon the same factors as the recuperative thermal oxidizers. The more complex controls associated with regenerative units cause the capital cost to be higher than that of units with recuperative heat recovery. Figure 22.8 shows typical installed costs for regenerative thermal oxidizers over a wide range of waste gas flow rates.

The annual costs of operating and maintaining a thermal oxidation system are for operating labor, supervisory labor, maintenance labor

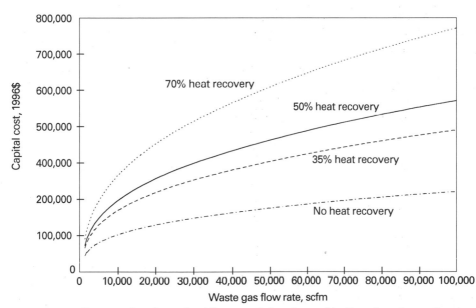

Figure 22.7 Recuperative thermal oxidizer: capital cost as a function of waste gas flow rate.

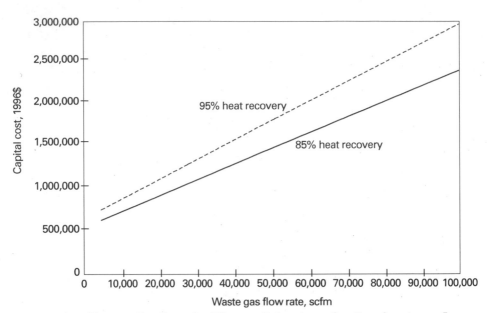

Figure 22.8   Regenerative thermal oxidizer: capital cost as a function of waste gas flow rate.

and materials, electricity, and fuel requirements. A thermal oxidizer usually requires operator attention for 0.5 to 1.0 h per shift with supervisory labor being 10 to 20 percent of the operation labor (Vatavuk, 1990). A thermal oxidizer also requires 0.5 to 1.0 h of maintenance per shift (Vatavuk, 1990). The major electrical cost of a thermal oxidizer is associated with operating the fan and is dependent on the gas flow rate and the pressure drop across the incinerator, heat recovery, subsequent control devices, and connecting ductwork. The amount of auxiliary fuel required for combustion is dependent on the degree of heat recovery. Figure 22.9 shows the annual operating cost as a function of the waste gas flow rate. As can be seen from the graph, the operating costs of thermal oxidizers with little or no heat recovery become very large.

Thermal oxidation can be ideally suited for many applications in which relatively dilute VOCs are present in process streams. The following is a list of advantages of thermal oxidation over other VOC control technologies:

- High removal efficiency of organic gases and vapors
- High removal efficiency of submicron, organic particulate
- Ability to control a widely varying waste gas stream
- Ease of construction

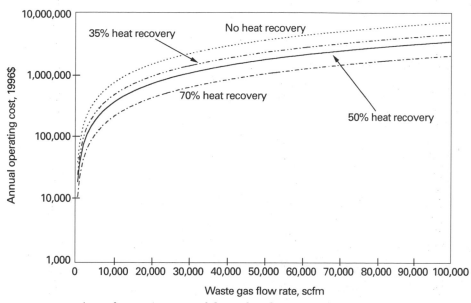

Figure 22.9 Annual operating costs of thermal oxidation as a function of waste gas flow rate.

- Small space requirements
- Low maintenance costs

Just as with any air pollution control technology, thermal incineration has some distinct disadvantages for certain applications:

- High operating costs
- Fire hazards
- Possibility of flashback
- Possible generation of undesirable pollutants

22.2 Catalytic Oxidation

Catalytic oxidation employs a catalyst that accelerates the rate of the oxidation reactions and allows the reactions to proceed at much lower temperatures than thermal oxidation. The lower temperature enables the use of lighter materials of construction and reduces the supplementary fuel requirements. The acceleration of the oxidation reaction also allows for a shorter residence time through the combustion chamber. These benefits may be partially offset by the added capital cost of the catalysts and typically higher maintenance costs.

22.2.1   Theory of catalytic oxidation

Catalytic oxidation is very similar to thermal oxidation in that at elevated temperatures organic compounds are oxidized to more desirable end products (CO_2 and H_2O). The difference between catalytic and thermal oxidation is that the reaction in a catalytic oxidizer takes place in the presence of a catalyst. The catalyst accelerates the rate of reaction without undergoing a chemical change itself. The reaction follows the general equation:

$$C_XH_Y + \left(\frac{4x + y}{4} \right) O_2 \xrightarrow{\text{Heat+catalyst}} xCO_2 + \frac{y}{2}\, H_2O \qquad (22.10)$$

Catalytic oxidation follows these steps:

1. Diffusion of the VOCs through a stagnant fluid layer surrounding the catalyst surface
2. Adsorption of the VOCs onto the catalyst surface
3. Oxidation of VOCs while adsorbed onto the catalyst [Eq. (22.10)]
4. Desorption of the reaction products from the catalyst surface
5. Diffusion of the reaction products through the stagnant fluid layer into the bulk gas

Although the general reactions for both are similar, catalytic oxidation and thermal oxidation have many differences. The major operational difference is that catalytic oxidation requires much lower temperatures and residence time for the oxidation reaction to take place. Typical residence times are an order of magnitude less than those for thermal oxidation. The design parameter for catalytic oxidizers is called the *space velocity* with units of inverse time (1/h). The space velocity is defined as the volumetric flow rate of the waste gas stream at standard conditions, divided by the catalyst volume. Therefore, larger space velocities denote greater catalyst reactivity while lower space velocities signify that more catalyst is needed for a given destruction efficiency.

The most common types of catalyst are precious-metal-based (platinum or palladium). The catalyst is applied to the surface of a substrate that is typically a monolithic honeycomb made of either stainless steel or ceramic. Stainless-steel substrates offer superior structural integrity compared to ceramic substrates whereas ceramic substrates are preferred where corrosion is a possibility.

A less common type of catalyst used for catalytic oxidation is base-metal catalysts. These catalysts are usually oxides of chromium or manganese. The base-metal catalysts are typically supplied in pelletized form. The drawbacks of using base-metal catalysts are that they have higher pressure drops and that pellet abrasion can generate particulate matter which could plug the bed. Also, in the case of chromium-based catalysts, pellet abrasion can cause the emission of hexavalent chromium (a hazardous air pollutant) to the atmosphere.

One of the major drawbacks to catalytic oxidation technologies is catalyst deactivation, which is a degradation of a catalyst's efficiency. Deactivation can be either reversible (particulate fouling) or irreversible (poisoning, thermal deactivation). Catalyst fouling occurs when particulate matter in the waste gas stream physically blocks the pores of the catalyst bed. Fouled catalyst may be regenerated thermally, physically, or chemically. Thermal cleaning is used when the fouled particulate is organic in nature and can be volatilized at higher temperatures. Physical cleaning involves dislodging the particulates by applying a vacuum or by blowing compressed air across the catalyst. Physical cleaning has limited effectiveness when it is used on particulates that have strongly adhered to the catalyst surface. Chemical cleaning is the most common method for cleaning catalyst from an oxidizer. Chemical cleaning involves immersing the catalyst in an acid or alkaline cleaning solution.

The primary reactions in a catalytic oxidation system are almost identical to those listed for thermal oxidation. The difference comes with the secondary reactions. Because catalytic oxidation occurs at a much lower temperature than thermal oxidation, the only generation of NO_x occurs when nitrogen is present in the auxiliary fuel. Due to the potential for fouling of the catalyst, typical waste gases treated by catalytic oxidation do not contain a large amount of particulate matter. Therefore, the generation of dioxins and furans would be minimized due to the lack of available formation sites. The formation of carbon monoxide is possible if the catalyst is deactivated, resulting in incomplete combustion.

22.2.2 Design of catalytic oxidation systems

Due to the presence of a catalyst, combustion temperatures in a catalytic oxidizer are significantly lower than the combustion temperatures needed for thermal oxidation. Combustion temperatures in a catalytic oxidizer may vary from 300 to 1200°F (150 to 650°C) (Wark and Warner, 1981; Tichenor and Palazzolo, 1987; Chu and Windawi, 1996) with typical temperatures between 700 and 900°F (370 and 480°C) (Katari et al., 1987).

Since the oxidation reaction takes place on the surface of the catalyst, the gas stream must be well mixed prior to diffusing through the stagnant fluid layer that surrounds the catalyst. Adequate mixing in a catalytic oxidizer is typically accomplished through two methods. The first method is typically a distribution grate which (1) distributes the waste gas stream evenly over the cross section of the catalyst and (2) promotes mixing of the VOCs and oxygen. The second method of mixing is the catalyst bed itself in which the tortuous path the waste gas stream takes through the catalyst allows further mixing of the VOCs and oxygen in the gas.

Residence times in a catalytic oxidizer are very small; therefore, the inverse of the residence time, or space velocity, is used as the design parameter. The space velocity in a catalytic oxidizer may vary from 15,000 to 80,000 h^{-1} (Tichenor and Palazzolo, 1987) with a typical range of 30,000 to 40,000 h^{-1} for precious-metal catalysts and 10,000 to 15,000 h^{-1} for base-metal catalysts (Katari et al., 1987). The difference is due to the fact that precious-metal catalysts are more reactive and thus require less oxidation time than base-metal catalysts.

Catalyst design is dependent on several characteristics of the waste gas stream such as type and content of VOCs, flow rate, temperature, and pressure drop requirements. Other factors that affect catalyst design are the destruction efficiency needed, durability requirements, and space constraints. In practice, the catalyst designs are typically performed by the catalytic oxidizer manufacturer. The manufacturers use small-scale testing and experience to provide the size and type of catalyst suitable for each application.

Due to the elevated temperatures at which catalytic oxidation occurs, recovery of heat in the exhaust gas can significantly decrease auxiliary fuel consumption. As with thermal oxidation, the two major types of heat recovery used in air pollution control applications are recuperative and regenerative heat recovery. Figure 22.10 shows a catalytic recuperative oxidizer. As shown in the schematic, the incoming waste gas stream enters the heat recovery unit where it is preheated by the hot combustion from the oxidizer. Once preheated, the temperature of the waste gas is increased to the required reaction temperature. Upon reaching the required reaction temperature, the waste gas enters the catalytic chamber, where the gas flow is evenly distributed over the entire cross section of the catalyst bed. The waste gas then enters the catalyst bed where the oxidation takes place. Having passed through the catalyst bed, the treated gas flows to the heat recovery unit, where the heat of the flue gas is transferred to the incoming waste gas stream. To prevent thermal deterioration of the heat exchanger, a bypass may be included in the design. The bypass would be opened when the incoming waste gas contained a spike of VOCs that would cause the combustion gas temperature to reach a critical maximum.

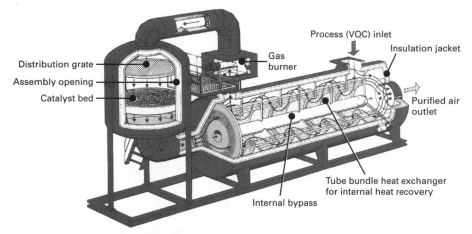

Figure 22.10   Catalytic recuperative oxidizer (*Wahlco, Inc./LTG, product literature, 1996*).

Catalytic oxidation systems can also employ regenerative heat recovery. The regenerative catalytic oxidizers (RCOs) operate in a similar fashion to the regenerative thermal oxidizers described in the preceding section. RCOs can be specially designed for low-flow applications, as shown in Fig. 22.11. The incoming waste gas stream

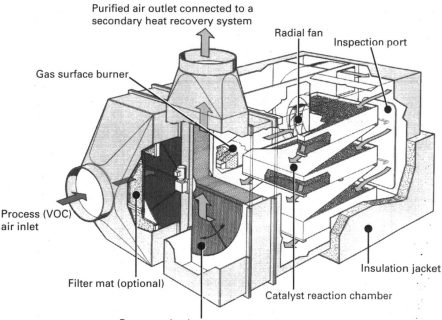

Figure 22.11   Compact regenerative catalytic oxidizer with rotary heat exchanger (*Wahlco, Inc./LTG, product literature, 1996*).

enters the unit and passes through a rotating regenerator that preheats the gas. Once the gas is through the regenerator, a burner heats the gas further prior to its entering the catalyst. In this design, the waste gas is divided and passes through three catalyst beds where the oxidation occurs. After being oxidized, the hot cleaned gas passes through the rotating regenerator, where a portion of the heat is transferred to the regenerator material. Once through the regenerator, the cooled gas stream exits the unit and is exhausted to the atmosphere, sent to subsequent air pollution controls, or sent to a secondary heat recovery unit.

22.2.3   Application of catalytic oxidation systems

The companies that manufacture thermal oxidizers most often also supply catalytic oxidation systems. The information required when specifying catalytic oxidizers is identical to the information needed to specify thermal oxidizers. Many of the catalytic and thermal oxidation companies provide prefabricated units for small flow rates and will custom-design catalytic oxidizers for larger applications. The capital costs for catalytic oxidation units are typically higher than those for similar-sized thermal oxidation systems. Since most catalytic oxidation units employ precious metal–based catalysts, the capital costs tend to be much higher than those of similarly sized thermal oxidation units. A major advantage of catalytic oxidation is that the operating costs can be extremely low due to the relatively low oxidation temperature. Depending on the application, supplementary fuel may not be required at all times for regenerative catalytic oxidizers due to the heat released from the oxidation reaction and the efficient heat recovery. In these designs, the electrical costs to operate the fan and the instrumentation and controls are the only significant operating costs. As the efficiency of the heat recovery decreases, the supplementary fuel requirements increase.

Economic tradeoffs exist between catalytic and thermal oxidation systems. Operating costs of both thermal and catalytic oxidation systems may be very high due to electrical and fuel requirements. Catalytic oxidization as an air pollution control technology benefits from relatively low operating costs but is hindered by large capital costs. On the other hand, thermal oxidation benefits from relatively low capital costs but may have very high operating costs. Figure 22.12 shows the relationship between waste gas flow rate and capital costs for recuperative catalytic oxidation systems. Catalytic oxidation units require more maintenance than thermal oxidizers yet require less fuel for operation. As fuel is the major component in the operating cost of the oxidation system, the annual operating costs for cat-

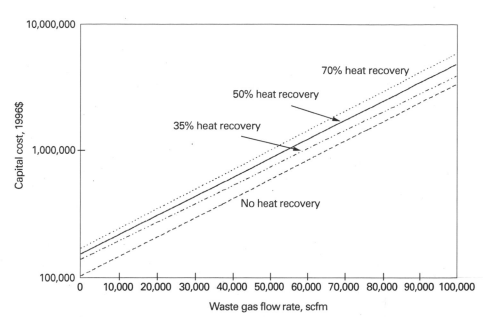

Figure 22.12 Recuperative catalytic oxidizer capital cost as a function of waste gas flow rate.

alytic oxidizers are lower than those of thermal oxidation systems. Figure 22.13 shows annual operating costs of catalytic oxidizers for various waste gas flow rates.

The following is a list of advantages of catalytic oxidation over thermal oxidation:

- Lower fuel requirements
- Lower operating temperatures
- Smaller space requirement
- Reduced risk of fire
- Reduced risk of flashback

The disadvantages of catalytic oxidation compared to thermal oxidation systems are as follows:

- Higher capital cost than thermal oxidizers
- Higher maintenance costs
- Catalyst deactivation
  1. Catalyst poisoning
  2. Fouling of catalyst with particulates
  3. Thermal deactivation

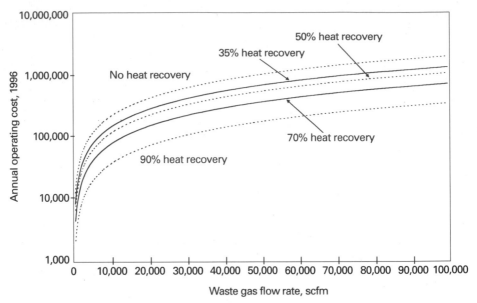

Figure 22.13 Annual operating costs of thermal oxidation as a function of waste gas flow rate.

22.3 Specifying Thermal and Catalytic Oxidation Systems

The first step in selecting an incineration system vendor is to solicit budget estimates from a wide variety of equipment vendors. To do this, a process specification sheet should be developed. This specification sheet will describe the waste gas that needs to be treated and will inform the vendor of any special requirements needed (such as space availability, materials of construction, and/or safety issues). The specification sheet should be as specific as possible when describing the waste gas characteristics and process conditions. One of the most important aspects of selecting an equipment vendor is to supply an accurate and explicit process specification. A process specification sheet should include the following:

- Waste gas flow rate, actual ft³/min (m³/min)
- Waste gas temperature, °F (°C)
- Waste gas density, lb/ft³ (kg/m³)
- Waste gas pressure, psig or inH$_2$O gauge (kPa gauge)
- Waste gas composition including weight percent of VOCs, water, and oxygen
- Particulate content of waste gas, lb/h (kg/h)

- Hours of operation

- Operating fluctuations (flow rate and VOC content)

- Auxiliary fuel available (including pressure)

- VOC destruction efficiency needed, percent

- Heat recovery needs

- Physical limitations (space availability, indoor/outdoor)

Once developed, the process specification sheet should be sent to several incineration system vendors. When the vendors return a budget quote based on the process specification sheet, the quotes and equipment descriptions should be examined to determine the best economical and practical fit to the application. From the budget quotes, two or three systems should be selected for further inspection. A meeting with the remaining vendors should be scheduled to discuss the application and any possible problems that could not be addressed on a specification sheet or budget quote. From the meeting, request that the remaining vendors submit a firm estimate of installation of the respective system. Once the firm estimates are received, choose a system which best suits the application in terms of the cost of the system, quality of construction, ease of operation, and reputation of the manufacturer.

References

Brunner, Calvin R., *Incineration Systems, Selection and Design,* Van Nostrand Reinhold, New York, 1984.

Chu, Wilson, and Windawi, Hassan, "Control VOCs via Catalytic Oxidation," *Chemical Engineering Progress,* vol. 92, no. 3, p. 37, March 1996.

Frame, Robert D., Vice President, Wahlco, Inc.–VOC Technologies, personal communication, 1996.

Katari, Vishnu S., Vatavuk, William M., and Wehe, Albert H., "Incineration Techniques for Control of Volatile Organic Compound Emissions," *JAPCA,* vol. 37, no. 1, p. 91, January 1987.

Klobucar, Joseph M., "Choose the Best Heat Recovery Method for Thermal Oxidizers," *Chemical Engineering Progress,* vol. 91, no. 4, p. 57, April 1995.

Klobucar, Joseph M., Process Engineering Manager, Durr Industries, Inc., personal communication, 1996.

Perry, Robert H., Green, Don W., and Maloney, James O., *Perry's Chemical Engineering Handbook,* McGraw-Hill, New York, 1984.

Research Cottrell (REECO), product literature, 1991.

Straitz, III, John F., "Use Incineration to Destroy Toxic Gases Safely," *Environmental Engineering World,* vol. 2, no. 6, November–December 1996.

Taylor, Philip H., Dellinger, Barry, and Lee, C. C., "Development of a Thermal Stability Based Ranking of Hazardous Organic Compound Incinerability," *Environmental Science and Technology,* vol. 24, no. 3, p. 316, 1990.

Tichenor, Bruce A., and Palazzolo, Michael A., "Destruction of Volatile Organic Compounds via Catalytic Incineration," *Environmental Progress,* vol. 6, no. 8, p. 172, August 1987.

USEPA, *Recommended Methods of Reduction, Neutralization, Recovery or Disposal of Hazardous Waste*, vol. 3, *Disposal Process Descriptions, Ultimate Disposal, Incineration and Pyrolysis Processes,* USEPA 670/2-73-0536, August 1973.

Vatavuk, William M., *Estimating Costs of Air Pollution Control*, Lewis Publishers, Chelsea, Mich., 1990.

Wahlco, Inc./LTG, product literature, 1996.

Wark, Kenneth, and Warner, Cecil F., *Air Pollution, Its Origin and Control*, 2d ed., HarperCollins Publishers, New York, 1981.

Williamson, Peter, "Production and Control of Polychlorinated Dibenzo-*p*-Dioxins and Dibenzofurans in Incineration Systems: A Review," 87th Annual Meeting and Exhibition of the Air and Waste Management Association, Cincinnati, Ohio, 1994.

Biofiltration of Gaseous Compounds

E. Roberts Alley, Jr.

Biofiltration is the use of microorganisms located in a filter-type medium to biologically degrade contaminants from a waste stream. Theoretically, a microorganism exists that can degrade any organic chemical. Microbes exist that survive, and even thrive, in practically every environment encountered, except perhaps the extremes of temperature and pressure. It is well known that various microbes can degrade or metabolize hydrocarbons, chlorinated organics, pesticides, explosives, and even metals. The task that has faced, and continues to face, engineers in the field of biological treatment turns on how to utilize microbes so that the removal of compounds of concern can be controlled and predicted. Therefore, although any degradation reaction may be theoretically possible, and the appropriate microorganisms may be present in nature, the challenge is to optimize the process to achieve efficient and cost-effective degradation.

A key aspect of harnessing microbial degradation for the treatment of a waste stream is the determination of the proper environment for a cost-effective treatment process. Factors that may need to be adjusted include electron acceptors and donors present, substrate available, temperature, and the presence of other microbes.

Biodegradation requires contact between the microbes and the waste stream to be degraded. This can occur when the microbes are suspended in a liquid solution (such as in wastewater treatment), attached to soil particles, or attached to other media. Specific requirements aside, the microbes are generally not too particular about the specifics of their environment as long as the conditions necessary for growth are present. Therefore, if a particular waste is biodegradable in one environment, such as in a wastewater treatment plant, that same waste is biodegradable in another environment, such as a biofilm reactor, as long as the conditions exist to allow microbial growth to occur. Establishing the proper environment for the microbes is the key to successful biodegradation.

A well-designed and efficient biofilm reactor uses microbes to completely degrade the chemicals of concern to end products of carbon dioxide, water, inorganic salts, and cellular matter. The degradation of compounds containing inorganic molecules such as chlorine or sulfur results in the additional production of various mineral salts or acids. The fact that the contaminants are actually destroyed is one of the primary advantages of a biofiltration system compared to traditional off-gas treatment systems such as incineration or adsorption. Therefore, there is no need for media regeneration or the landfilling of waste products. Other advantages over traditional granular activated carbon (GAC) or thermal/catalytic oxidation off-gas control include operation at ambient temperature and low operation and maintenance costs.

Biofilm reactors have been used to treat both contaminated water and air from pharmaceutical plants, print shops, foundries, and wastewater treatment plants. Generally, the process involves passing the contaminated stream along a biological film that degrades the contaminants into carbon dioxide and water. There are three general mechanisms involved in transporting the organic compound to the microorganisms for degradation, often used in combination. The organic compounds in the gas stream can be adsorbed to the media before they are degraded by the microorganisms, they can be adsorbed by the moisture attached to the media before they diffuse into the microorganism, or the organic compounds can diffuse directly into the microorganism from the gas stream. For example, the "media adsorption" process uses soil or other organic matter as the filter media. As the contaminated gas stream passes through the filter media, the organic compounds adsorb onto the available sites. These compounds are then degraded by the available microorganisms in the media. As the organics are degraded, the adsorption sites are freed up onto which new compounds in the gas stream can adsorb.

Biofiltration has been actively used in Europe and Asia since the 1960s. There are more than 500 active biofiltration systems in Europe and Asia, while there are only a few dozen in the United States. Long used for odor control, biofilters have recently been applied to the treatment of gas streams containing various volatile organic compounds such as styrene, benzene, and toluene. Biofiltration is particularly effective at treating large volumes of air with relatively low concentrations (5 to 5000 mg/L) of volatile organic compounds (VOCs).

A biofilter consists of a "film" of microorganisms attached to a medium that is contained within a reactor vessel. The film, referred to as the *biofilm,* is a gel-like material that consists of a mass of microorganisms, any extracellular enzymes produced, and degradation byproducts. The biofilm may completely encapsulate the medium (particularly with synthetic media) or may be present throughout the medium (as with natural media such as soil and peat). The contaminants in the airstream desorb from the gas phase onto the medium or into the biofilm itself. These sorbed contaminants are then degraded by the microorganisms.

For the purposes of this chapter, *biodegradation* is used to describe the metabolic process by which a microbe breaks down a compound, using specific enzymes for the process. *Bioremediation* is used to describe the engineered process of using microbes to biodegrade a waste stream to less toxic components.

23.1 Theory

To properly understand and apply biodegradation to the treatment of VOCs in a gas stream, a basic grasp of microbiology is necessary. This chapter does not attempt to make the reader fully aware of the complexities of the microbial world and its interactions or to be an exhaustive text on microbial systems; there are many excellent references available for an in-depth study of biodegradation, including *Bioremediation Engineering* by Cookson and the *Bioremediation* series published by Battelle Press. This text does, however, provide an overview of the concepts and principles that underlie biodegradation and are important to designing a remedial system. This chapter will give the reader a strong foundation in microbial systems if it becomes necessary to pursue further information for system design or research.

In general, biodegradation is the process by which microbes either directly or indirectly (by cometabolism) degrade compounds of interest. This has been occurring throughout the history of the world, and is constantly happening around us. Only recently—in roughly the last hundred years—humans have artificially generated favorable conditions so that microbes in a particular "environment" (such as a reactor) will rapidly break down environmental contaminants. Biological treatment was pioneered in the wastewater treatment industry, and has spread to the remediation industry (soil and groundwater) and most recently to the air pollution control industry.

In a biodegradation application, ideally the microbes present are used to degrade the contaminants to end products such as carbon dioxide, inorganic salts, cellular matter, and water. This is referred to as *ultimate* degradation. Because a contaminated waste stream does not often offer the optimal conditions for the growth of the microbes necessary for biodegradation, the addition of a substrate or various macro- and micronutrients may be required to achieve the optimum degradation rates. Likewise, one particular strain of microbes may not be able to ultimately degrade the contaminant(s) of concern in a waste stream. Successful biodegradation consists of a wide range of microbes working together in a complex network of interrelationships. This network of interrelated microbes is commonly referred to as a *microbial consortium*. While it is not usually necessary to precisely define the exact type of each microbe in a consortium, it is important to determine whether the proper microbes are present. For example, methanotrophic bacteria, able to degrade many chlorinated and brominated compounds, are not very useful if the contaminant of concern is a petroleum product.

Most people associate biodegradation with bacteria—single-celled organisms that are essentially omnipresent in the environment. And the majority of examples of bioremediation for waste treatment do use bacteria of one form or another. Recently, however, researchers have utilized multicellular organisms such as fungi and various plants for waste treatment. Of particular note are the white rot fungi being used for PCB degradation.

Field studies and laboratory investigations have shown a wide range of organic compounds to be biodegradable under the proper conditions. A list of some of these compounds is seen in Table 23.1. This list of biodegradable compounds is being increased every year, as scientists and engineers constantly seek to understand and better apply biodegradation to the waste treatment problems we face. Biofiltration systems in particular have been applied to the treatment of gas streams containing BTEX, glycol ethers, styrene, methanol, and reduced sulfides such as hydrogen sulfide, mercaptans, and dimethyl sulfide. The reader should also keep in mind that, as stated before, microorganisms generally are not particular about whether they are in a water system, soil, or fixed film, as long as the proper nutrients, substrate, and other environmental conditions necessary for growth are present.

23.1.1 Microbial classification

Early work could not satisfactorily classify microorganisms in either the animal or plant kingdom, the two traditional major classifications. In 1866, E. Haeckel proposed that a third kingdom, the *protists,* be established for bacteria, algae, fungi, and protozoa. Protists were

TABLE 23.1   Partial List of Degradable Organic Compounds

Aliphatics, nonhalogenated	4-Chlorophenol
Acrylonitrile*	1,2-Dichlorobenzene
Aliphatics, halogenated	2,3-Dichlorobenzene
Bromochloromethane	1,4-Dichlorobenzene
Bromodichloromethane	3,4-Dichlorobenzoate
Bromoform	3-5,Dichlorobenzoate
Chloroethane	Hexachlorobenzene
bis-(2-chloroisopropyl) ether	3-Methyl benzoate
Dibromochloromethane	Monochlorobenzoate
Dichloroethane	Monochlorophenol
1,1-Dichloroethylene	Pentachlorophenol
1,2-Dichloroethylene	Trichlorobenzene
1,3-Dichloropropylene	1,2,3-Trichlorobenzene
1,2-trans-Dichloroethylene	1,2,4-Trichlorobenzene
Methyl chloride	**Nitrosamines**
Methylene chloride	Dimethylnitrosamines
1,1,2,2,-Tetrachloroethane	**Pesticides**
Tetrachloroethylene	Acrolein
Tetrachloromethane	Aldrin
Trichloroethane	Chlordimeform
Trichloroethylene	DDT
Trichlorofluoromethane	Dieldrin
Trichloromethane	Diuron
Vinylidiene chloride	Endosulfan
Aromatics, nonhalogenated	Endrin
Benzene	Heptachlorobornane
di-n-Butylphthalate	Kepone
Creosol	Lindane
2,4-Dinitrotoluene	Methoxychlor
2,6-Dinitrotoluene	Parathion
Diphenylhydrazine	Pentachloronitrobenzene
Nitrobenzene	Phorate sulfoxide
p-Nitrophenol	Toxaphene
Phenol	**Polycyclic aromatics hydrocarbons,**
Toluene	**halogenated**
Aromatics, halogenated	PCBs (mono- and dichlorobiphenyls)
m-chlorobenzoate	4-Chlorobiphenyl
o-chlorobenzoate	4,4-Dichlorobiphenyl
p-chlorobenzoate	3,3'-Dichlorobiphenyl

SOURCE: Information from Kobayashi and Rittmann, 1982.

either unicellular or multicellular with little differentiation of cells and tissues, and they included algae, protozoa, fungi, and bacteria. In the 1950s, the advent of the electron microscope allowed a more in-depth examination of the structure of cells and led to the realization that organisms tend to fall into two types of cell structure: procaryotic (simple) and eucaryotic (complex). Members of the protist kingdom were either procaryotic or eucaryotic. This led to the current classification of organisms into three kingdoms: Eucaryotes, Eubacteria, and Archaebacteria. Eucaryotes can be multicellular, with extensive differentiation of cells and tissues (i.e., plants and animals), or unicellular, coencytic, or mycelial, with little differentiation of cells or tissues (i.e., protists, such as algae, fungi, and protozoa). Eubacteria are procaryotic organisms with a cellular chemistry similar to that of eucaryotes and include most bacteria. Archaebacteria are procaryotic organisms with a unique cell chemistry that is not fully understood. Examples of Archaebacteria include methanogens, thermophiles, and halophiles. Many of the distinguishing characteristics of the different kingdoms of organisms are based upon chemical and structural differences in the cell membranes, cytoplasm, cytoskeletal elements, and chromosomes.

Microbes can be classified according to a wide range of characteristics. These characteristics include morphology, oxygen requirements, carbon and energy source, and temperature. Other methods of classification include Gram stain results, spore formation, motility, or a number of other specific functions. The method of classification used in large part depends upon the purpose of the identification; i.e., for what will the information be used? If all that is of concern is whether the microbes are aerobic or anaerobic, classifying a microbe according to the energy source is not necessary.

Morphology.   The classification of organisms according to size and shape is known as *morphology*. Bacteria are typically less than 5 μm in length and are single-celled organisms, most often shaped as a rod, sphere, or "spiral rod." As mentioned above, bacteria are procaryotic organisms that are part of the Eubacteria kingdom. Other characteristics of Eubacteria include a lack of internal cellular compartmentalization and the presence of only one chromosome. Eucaryotes are more complex organisms, often multicellular with extensive differentiation of cells and tissue (i.e., plants and animals), although they can be unicellular, such as the protists. They are typically greater than 20 μm in length and contain internal membranes and many chromosomes. Examples of eucaryotes include fungi, protozoa, and other higher forms of life.

Oxygen requirement.   In general, microbes fall under two broad classifications of oxygen use: aerobic and anaerobic. Aerobic microbes require the presence of oxygen as a terminal oxidizing agent. Anaerobic microbes require the absence of oxygen. An example of the anaerobic process is fermentation. Within these two general classifications, microbes are grouped according to the following:

Strict. Can only survive under one set of conditions; also referred to as *obligate.*

Facultative. Can survive with or without oxygen.

A facultative anaerobe prefers anaerobic conditions, but can survive and grow in the presence of oxygen—just not as efficiently as in the absence of oxygen. An example of a facultative anaerobe is enteric bacteria, which can shift from aerobic respiration when oxygen is available to fermentation in the absence of oxygen. It is not unusual for compounds to be degradable under both aerobic and anaerobic conditions by the respective microbes, although at different rates. The presence or absence of oxygen is important to bioremediation for several reasons, among them the degradation rates and the degradation by-products. An example of this is the biodegradation of chlorinated compounds such as trichloroethylene (TCE). Methanotrophic bacteria are aerobic microbes that produce an enzyme which degrades methane as an energy source while simultaneously being able to degrade many single- and double-bonded chlorinated compounds. TCE is degraded through several steps to carbon dioxide, water, and cellular matter. However, in the absence of oxygen, degradation of TCE is accomplished by methanogens, and results in the production of vinyl chloride, which is more toxic than TCE. It can therefore be very important to know the oxygen conditions of a biodegradation reaction.

Cellular material and energy source.   All microorganisms differ in their preference for a particular substrate for carbon or the other nutrients necessary for growth.

A common form of microbial classification that is useful in bioremediation applications sorts microorganisms according to the sources of cellular material and energy, also known as *nutritional classification.* Carbon is the principal component of cellular material and is therefore essential to microbial growth. *Heterotrophic* microbes satisfy their carbon needs from the metabolism of organic matter. Because organic compounds are used as the carbon source at approximately the same oxidation state as that of organic cellular material, a net input of energy is not required. *Autotrophic* microbes derive their carbon from inorganic compounds such as carbon dioxide. This is a reductive process, requiring a net input of energy. Autotrophic microbes are the first step in the food chain, converting inorganic carbon to organic material. Many heterotrophic and autotrophic microorganisms are able to utilize a range of carbon sources that are similar to one another. The designer should keep in mind that a significant change in the substrate to a biological treatment system such as a biofilter may result in a lag in the degradation of the contaminants of concern. The lag is a result of the production of any new enzymes that are necessary to degrade the new substrate. This lag time can vary from hours to months.

It quickly became apparent to scientists that the two classifications, autotrophic and heterotrophic, did not accurately reflect the diversity of microorganisms and their cellular material and energy requirements. A more useful system of classification describes both the carbon and energy source. The energy source for microbes is either light (photoheterotrophic or photoautotrophic) or chemical reactions (chemoheterotrophic or chemoautotrophic). Chemoautotrophs derive their energy from the oxidation of reduced inorganic compounds. Chemoheterotrophs derive their energy, and often their carbon, from the metabolism of the same organic compound. Photosynthesis is the process by which pigments in the cell adsorb light, triggering a molecular reaction which allows the storage of energy in the cell. Chlorophyll is the most common pigment, and it gives plants the green color. The chemical reactions providing energy may be either aerobic or anaerobic and may involve either organic or inorganic chemicals. The nutritional classification can be further subdivided according to the physical state in which the nutrients enter the cell. As is the case with the oxygen requirements for microbes, heterotrophic and autotrophic microbes can be either obligate (restrictive) or facultative (versatile).

In regard to the usable sources of carbon, microorganisms can be highly versatile, such as the *Pseudomonas* group, which can use more than 90 organic compounds, or the microorganisms can be very specific, using only one or two sources of carbon. When a microorganism can use several organic substrates and more than one substrate is present, *sequential substrate uptake* will occur. Sequential substrate uptake, also referred to as *preferential substrate uptake,* describes the process by which the microorganism degrades the available substrates in a particular order of preference. Generally, the microorganism will degrade first the compound that yields the highest amount of energy. The two general mechanisms driving sequential uptake are described as follows:

Catabolite repression. Degradation products inhibit the production of the enzymes required to metabolize the primary substrate. This may be due to the presence of other, more readily metabolized substrates.

End-product repression. Intermediate or end products formed due to the degradation of a substrate inhibit the production of enzymes used for (typically the first step of) substrate degradation. This can affect either the same or other substrates. It is also referred to as *end-product inhibition.*

Overall, every naturally occurring organic compound can be used as a carbon or energy source by some microorganism. This provides for two additional terms used to classify microorganisms: *phototrophy* and *auxotrophy.* Phototrophy defines the use of a principal carbon source to meet all the carbon requirements of the microorganism. Auxotrophy requires the use of one or more growth factors in addition to the principal carbon source.

Temperature.   Different microorganisms survive and perform better in different temperature ranges. The minimum, optimum, and maximum temperatures for bacteria vary widely, from -10 to $250°C$ (14 to $482°F$). Many microorganisms have proved to be viable after they are thawed from storage at temperatures as low as $-194°C$ ($-317°F$). The typical temperature range for the majority of microorganisms is 0 to $80°C$ (32 to $176°F$), with the optimal metabolism occurring between 20 and $50°C$ (68 to $122°F$).

The *minimum* temperature is defined as the temperature at which the organism no longer grows due to the effect of reduced temperature on the cellular chemical reactions. The *maximum* temperature is the temperature at which microbial growth is eliminated due to the thermal inactivation of proteins or cell membranes. The *optimum* temperature is the temperature at which the microbial growth rate is maximized. These temperature-based classifications are particularly important when microbes are used in a surface reactor, such as a biofilm reactor, exposed to the temperature extremes often experienced in the environment. The most efficient microbe for the degradation of a particular compound may not function well or even at all at the system design temperatures. The three most common temperature classes of procaryotic organisms are psychrophiles, mesophiles, and thermophiles. Psychrophillic microbes have their optimum temperature range from 0 to $6°C$ (32 to $43°F$). Mesophyllic microbes, the most common, have their optimum temperature range from 25 to $30°C$ (77 to $86°F$). Thermophillic microbes have their optimum temperature range from 58 to $62°C$ (136 to $144°F$). As with the other methods of classifying microorganisms, the three temperature classes are not as clear-cut in nature. Microorganisms exist that operate over a temperature range that can cross between classifications or that may function over a very narrow range of temperature. During design, consideration should be given to the actual site temperatures and the impact this may have on microbial growth and biodegradation rates. The effect of temperature on biodegradation rates is discussed further in Sec. 23.1.3.

23.1.2   Microbial nutrition

All cells require building materials and an energy source to grow, manufacture enzymes, maintain cellular functions, and reproduce. The chemical composition of a cell generally indicates the requirements of a microorganism for growth, although the requirements are unique to the culture. The majority of cellular matter is water (approximately 80 to 90 percent by weight). Water is a key element necessary in the environment for a microorganism to grow and properly function. This environmental moisture is the primary vehicle for transporting nutrients into the interior of a microorganism, where they can be utilized. About 95 percent of the remaining (solid) cellular matter consists of hydrogen, oxygen (derived from water), carbon, nitrogen, phosphorus, and sulfur. Twenty-four different elements in various concentrations and ratios are required for life. These building materials are referred to as *macro-* and *micronutrients*. The general composition of the solid cellular matter of a microorganism is shown in Table 23.2.

TABLE 23.2   Approximate Dry-Weight Cellular Composition

Element	Percent
Carbon	50
Oxygen	20
Nitrogen	14
Hydrogen	8
Phosphorus	3
Sulfur	1
Potassium	1
Sodium	1
Calcium	0.5
Magnesium	0.5
Chlorine	0.5
Iron	0.2
Manganese	<0.1
Cobalt	<0.1
Copper	<0.1
Molybdenum	<0.1
Zinc	<0.1

SOURCE: Stanier et al., 1986; A.R. Bowers, personal correspondence, 1986.

Macronutrients. The materials present in the highest concentrations are called macronutrients and include carbon, nitrogen, and phosphorus. These three elements are typically present in a ratio of 17:5:1. Although a particular organism may only use one source for the supply of nutrients, a wide variety of sources exist in nature. As mentioned previously, carbon is supplied in a wide range of forms, from organic carbon, such as in glucose or phenols, to inorganic carbon, such as carbon dioxide gas. The source of nitrogen can be ammonia, nitrate, molecular nitrogen, organic nitrogen, or an amino group "clipped" from an amino acid. Phosphorus can be derived from orthophosphates or from mineralized inorganic phosphates.

Micronutrients. Less than 5 percent of all cellular material is made up of micronutrients. The micronutrients are essential as enzyme cofactors and as enzyme constituents. Other elements required for microbial growth are vitamins (specific organic compounds required for normal cellular activity) and steroids (controllers of biological functions). The nutrients and other elements necessary must be available to the microbes in solution for the cells to properly function. While insufficient levels of the nutrients or other elements can inhibit microbial degradation, many microbes have the ability to store excess concentrations of the materials. This allows for some microbial stability when conditions may vary somewhat.

Enzymes. A cell uses enzymes to catalyze (increase the rate of) a chemical reaction that, while thermodynamically favorable, is slow. Chemically, an enzyme is a long-chain protein molecule with one or more sites that are specific for particular molecules or reactions. Enzymes consist of *cofactors* and *coenzymes*. Cofactors are nonprotein compounds which combine with an inactive protein (coenzyme) to form an active enzyme complex. Enzymes are both natural and synthetic, although natural enzymes can be several orders of magnitude faster than the equivalent commercial enzymes. The three primary functions of enzymes are, in no particular order, the breakdown of toxins, the conversion of existing molecules to new molecules, and isomerization. The degradation of organic contaminants is typically mediated by bacterial enzymes. Enzyme activity can be influenced by pH, temperature, ionic strength, the presence of specific ions, or other chemical or physical factors. Most enzymes exist with the cellular cytoplasm and require the compounds to diffuse into the cell. Some enzymes, however, are *extracellular*, in that they diffuse out of the cell to attack the compound in the surrounding environment.

Growth factors. Organic compounds required by an organism as either a precursor or a constituent of its organic cellular material are known as *growth factors,* which cannot be synthesized from the organism's carbon source. The growth factors must be provided as a nutrient. Examples of growth factors include amino acids (constituents of proteins), vitamins (constituents of certain enzymes), and purines and pyrimidines (constituents of nucleic acids). Growth factors are required by a cell in small quantities.

A final factor to consider in dealing with microbial growth is the effect of *toxicity.* Every microorganism responds differently to chemical or physical stress. The presence of a contaminant at high concentrations may result in the inhibition of growth rates, the inhibition of certain metabolic functions, or toxicity. While general conclusions can be determined from current literature, the most accurate method for determining the effects of high levels of contaminants is through treatability studies designed to simulate the application and waste stream as specifically as possible.

Cometabolism.   In most cases, the enzymes produced by the microorganism degrade compounds that provide a direct benefit to the microorganism, such as the addition of cellular matter or energy. However, some microorganisms produce enzymes that, in addition to degrading beneficial compounds, degrade compounds with no apparent benefit to the microorganism. This is called *fortuitous degradation,* or *cometabolism.* Cometabolism is the degradation of a compound that occurs only when in the presence of another organic that serves as the primary energy source. Because the two compounds are competing for the enzyme (substrate competitive inhibition), it is important that the concentration of the primary substrate be kept at sufficiently low levels. Cometabolism is an important factor in the degradation of certain halogenated aromatics. For example, a group of bacteria known as methanotrophs (aerobic microorganisms that grow on methane) produce an enzyme that, in addition to attacking methane, attacks many chlorinated and brominated compounds. The proposed degradation pathway of TCE is presented in Fig. 23.1.

23.1.3 Microbial kinetics

To accurately design a biological treatment system, the designer must have a reasonable estimate of the microbial growth rates and the contaminant degradation rates. This information can be directly applied to the sizing of reactors for sufficient degradation. It is possible to design a system without knowing the degradation rates beforehand; however, this will involve significant trial and error, resulting in greater cost, and will leave the designer not truly understanding how the process is working and how it can be improved.

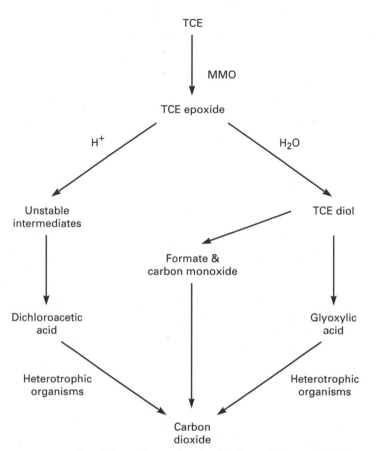

Figure 23.1 Possible pathway for the biodegradation of TCE by methanotrophic bacteria. (*Little et al., 1988.*)

Growth.   One measure of microbial activity is the growth of a microbial culture. In a medium to which a microbial culture has become acclimated, the microorganisms are said to be in a state of *balanced growth*. An increase in cellular biomass is accompanied by a similar increase in all other measurable cellular properties (number of cells, DNA, intracellular water, etc.). Balanced growth is a first-order, or exponential, reaction which is described as follows: The rate of the increase of bacteria at any given time is proportional to the number or mass of bacteria present at that time. In its differential form, the following equation describes the rate of growth of any cellular property:

$$\frac{dZ}{dt} = kZ \tag{23.1}$$

where Z = any measurable cellular property
    k = growth rate constant
    t = time

Typical cellular properties measured include mass of cells per milliliter and number of cells per millimeter. When integrated, the growth of a first-order microbial culture is described by the following equation, which is easier to use in practice:

$$\ln Z - \ln Z_0 = k(t - t_0) \tag{23.2}$$

By measuring Z and Z_0, the growth rate can be easily determined. There is a linear relationship predicted between the natural logarithm of cell numbers (or other measurable property) and time, with the slope of the line equal to $k/2.303$. The intercept of the line is Z_0. This linear or "balanced" growth is considered the "ideal" state of microbial growth.

Typical microbial growth occurs in four phases. The four phases are graphically shown in Fig. 23.2. The first phase is called the *lag phase*. Exposure of a microbial culture to either a new substrate or a different concentration of the same substrate causes the microorganisms to undergo a period of adjustment before they are capable of growth. This phase of adjustment may involve the synthesis of new enzymes or a change in the rate of enzyme production. The lag phase is also

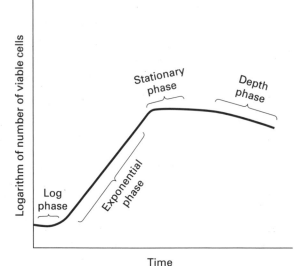

Figure 23.2 Generalized growth curve of a bacterial culture. (*Stanier et al., 1986.*)

referred to as *acclimation*. The *exponential phase* is, obviously, the most rapid phase of growth, and it continues at the high rate of growth until limited by the availability of the substrate or nutrients, or by the accumulation of toxic metabolic products. The exponential phase can mathematically be described by the following equation:

$$Z = Z_0 \times 10^{k(t-t_0)/2.303} \tag{23.3}$$

where Z = any measurable cellular property (such as number of cells)
$\quad$ k = growth rate constant
$\quad$ t = time

As the growth of the microbial population is limited by the environmental conditions, the rate of growth is reduced until the rate approaches zero. The microbial population is then said to be in the phase of *stationary growth*. The transition between the exponential and stationary growth phases is marked by unequal rates of synthesis of cellular components. After a period in the stationary phase, the cells will begin to die. This is known as the *death phase* and is an exponential function, similar to the exponential growth phase. The primary cause of the onset of the death phase is the depletion of either the substrate or cellular reserves of energy. The rate of the culture's death phase is dependent upon many environmental conditions as well as the characteristics of the organisms themselves.

Degradation.   The ultimate goal of any biological treatment system is the transformation of a harmful contaminant into a less harmful compound. As described previously, this is accomplished by the microorganism's use of enzymes to transform an organic compound, often deriving energy or cellular matter from the reaction. There are three general types of degradation: *primary, acceptable,* and *ultimate.* These three types are more fully described as follows:

Primary. A structural change in the parent compound that changes the molecular integrity of that compound. Also referred to as *degradation.*

Acceptable. Degradation of a compound to the extent that the toxicity is reduced. Also referred to as *detoxification.*

Ultimate. The total conversion of organic compounds into inorganic compounds and normal metabolic products (CO_2, NH_3, H_2O, etc.). Often referred to as *mineralization.*

Primary degradation is an element of both acceptable and ultimate degradation. Similarly, primary degradation and acceptable degradation are necessary elements of ultimate degradation. It is important for the designer to be aware of the degradation pathway for a particular compound to avoid intermediate products which may be harmful or which may limit the ultimate degradation of a compound. The pathway for the methanotrophic degradation of TCE is shown in Fig. 23.1.

To properly design or monitor a biological treatment system, the designer must be able to model the degradation of the contaminant. Biological reactions such as degradation are analogous to chemical kinetics. The model must incorporate concentration, time, the reaction (degradation) rate constant, and a factor called the *reaction order*. The degradation rate constant describes the overall rate (mass per unit time) at which the contaminant will be degraded by the microorganisms, and the reaction order describes the shape of the degradation curve (essentially, how much degradation occurs at what time). Both factors, the degradation rate constant and the reaction order, are important in the proper design of a biological treatment system. The degradation rate and reaction order, when used properly, will allow the designer to determine the contact time necessary to achieve the desired removal. The degradation reaction in its general form is mathematically described by

$$\frac{dC}{dt} = -kC^n \qquad (23.4)$$

where C = concentration of compound
$\quad\ k$ = degradation rate constant
$\quad\ n$ = reaction order, typically 0, 1, or 2
$\quad\ t$ = time

Unfortunately, when presented with concentration data from a treatability study or from a field application, this differential equation is not very practical for determining the degradation rate and reaction order. The reaction order and degradation rate can be determined numerically or graphically. The traditional method is to analyze the data graphically. The graphical method is less precise, but is often easier to use. Concentration versus time is plotted on an XY graph. The shape of the curve is determined by the reaction order. The slope of the line is the degradation rate. The y intercept is the appropriate form of the concentration (C_0, $\ln C_0$, $1/C_0$ for zero-, first-, and second-order linear plots). To make an accurate determination of the degradation order for a reaction, concentration versus time data must be collected over at least 90 percent removal of the compound in question.

A zero-order reaction is characterized by a linear plot of the concentration versus time data and is defined by Eq. (23.5). The first-order reaction is typified by an exponential decay curve [(Eq. (23.6)]. Typical zero-order, first-order, and second-order graphs are presented in Fig. 23.3. For demonstration purposes, the first-order curve is presented in two forms, natural logarithm of the concentration versus time (linear) and concentration versus time (exponential decay). The second-order curve is presented in a similar form, with a plot of the inverse concentration versus time (linear) and the plot of concentration versus time.

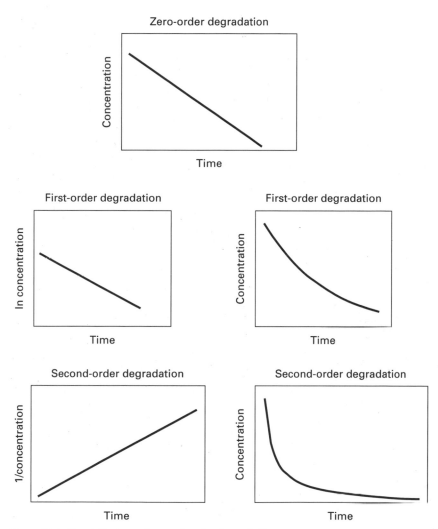

Figure 23.3 Zero-, first-, and second-order degradation curves.

Zero-order

$$C = C_0 - kt \qquad (23.5)$$

First-order

$$C = C_0 e^{-kt} \qquad \text{or} \qquad \ln C = \ln C_0 - kt \qquad (23.6)$$

Second-order

$$\frac{1}{C} = \frac{1}{C_0} + kt \qquad (23.7)$$

where C = concentration of the compound at time t
C_0 = concentration of the compound at time t_0, the initial concentration
k = degradation rate constant
t = time

The alternative to determining the degradation rate and the reaction order consists of numerical methods. Two general approaches can be used to determine the reaction rate and the reaction order numerically—correlation comparison and curve fitting. The use of computer programs allows either method to be used accurately and relatively painlessly. In the correlation comparison method, linear and nonlinear regressions with $n = 0$ or 1 are performed on the data, and values of the correlation coefficient r^2 are compared. The method with the higher correlation coefficient is considered the more accurate. This method assumes that degradation occurs in either zero- or first-order reactions. In many cases, this assumption is accurate, or accurate enough for modeling. Recent advances in statistical software packages are allowing researchers and engineers to accurately determine both the degradation rates and the reaction orders, which are not necessarily integers. In the majority of field applications, the assumption that the reaction is zero- or first-order is adequate. However, as more difficult compounds are degraded and more complex waste streams are treated biologically, designers are realizing that it is important to accurately know the true degradation constants as well as the reaction order.

Another factor to consider with degradation rate constants is temperature. The degradation rates determined for a system are, by definition, determined at a particular temperature, whether ambient or that of the laboratory. Degradation rates determined at one temperature in a treatability study may not be applicable to a pilot-scale or field biofiltration unit. The system temperature directly affects the rate of degradation. The logarithm of the reaction velocity is inversely proportional to the temperature (kelvins). An idealized reaction as a function of inverse temperature is linear with a negative slope. In actuality, the effect of temperature on microbial degradation rates is linear over only a portion of the curve. At low temperatures, the growth rate decreases sharply as the temperature approaches the freezing point of water. At high temperatures, the rate decreases sharply due to the thermal deactivation of proteins. A relatively straightforward formula can be used to estimate the effect of temperature on degradation rates. This equation is

$$k_2 = k_1 \Theta(T_2 - T_1) \qquad (23.8)$$

where k_1 = first-order degradation rate at T_1 (°C)
k_2 = first-order degradation rate at T_2 (°C)
Θ = temperature coefficient

The temperature coefficient must be experimentally determined. Temperature coefficients are available in the literature for many compounds.

In studying the biodegradation of a single compound, or particularly a group of compounds, the designer must be concerned with *refractory organics*. Refractory organics are compounds that resist biodegradation and therefore tend to persist in the environment. These compounds can be either natural or synthetic. A compound can be refractory for a number of reasons, including these:

- The microorganisms have not developed the appropriate enzymes (for synthetic compounds).
- The compound may not have sites for the enzyme to "attack."
- The compound may be too large to diffuse into the cell cytoplasm where the enzyme is located.
- The enzyme cannot migrate outside the microorganism.
- Oxygen, substrate, or nutrient levels are nonoptimal.
- Concentration of the compound is toxic.

Some of these factors can be influenced more easily than others. It is important for the designer to consider that while all organics are biodegradable, they are not necessarily readily degraded under the existing conditions.

23.2 Treatability Studies

There are two different types of treatability study: research and design. The steps and issues to be addressed are very similar, although the goals of the two studies are slightly different. Each type of study is appropriate at different stages of the design process. The research study is more abstract, is often used in the early stages of investigation, and is concerned with whether a compound can be biodegraded under a particular set of conditions. Or, looked at differently, under what conditions can a compound be biodegraded? A design study is perhaps more practical: Given that a compound can be biodegraded under these conditions, how might the degradation rates be optimized, and how can this be done practically? In the bottom line, is this cost-effective for our application? The design study may evolve out of a research study, or both may be conducted in the same study. As stated previously, both types of study are very important in the application of bioremediation.

Planning a treatability study can be defined by a six-step protocol. These six steps may not be defined for each study; they may be implicit (particularly when similar studies are performed), or they may be predetermined by the project. It is important, however, for the designer to consider each of these elements when designing the study, to ensure that valid conclusions can be drawn. An example of the protocol for a treatability study involving the degradation of TCE is presented below to illustrate the planning process:

1. *Define the objectives of the study.*

    Determine if methanotrophic bacteria are present in soils collected from the test site. If such microorganisms are present, can they become acclimated to a set concentration of TCE in a gas stream so that sufficient degradation can occur? The study will be conducted under ideal conditions and will be used to determine if this approach is technologically and economically feasible.

    To determine if methanotrophic degradation can occur, the system will be set up as a column of soil in a reactor, which will allow the contaminated gas stream to pass through the soil, coming into contact with the microorganisms.

    The primary questions to be addressed:

    Under what condition is biodegradation feasible?

    What is the degradation rate for each of the contaminants of concern?

    What is the required contact time to achieve the necessary degradation?

2. *Determine the steps of the study.*

2.1 Determine the operating parameters for the study. Establish the moisture content, nutrient levels, substrate levels, and contaminant concentration.

2.2 Collect soil samples from the site.

2.3 Prepare the contaminated gas stream (TCE).

2.4 Prepare the nutrient solution for the microorganisms.

2.5 Determine the method for the supply of the substrate (methane).

2.6 Determine the endpoint. What conditions (time, concentration) will determine when the study has been concluded?

2.7 Determine appropriate level of quality control.

2.8 Set up five reactors to operate under different conditions for the treatability study: control, nutrient level 1, nutrient level 2, substrate level 1, substrate level 2.

2.9 Collect samples from the reactors during the test at a predefined interval (every second day).

2.10 Analyze samples according to specific analytical methods.

2.11 Prepare report, including all data and records.

Some of the steps defined in this section may need to be resolved. What is the design of the reactor that will be used? (Continuous flow? Batch flow? See Fig. 23.4 for a typical gas treatment reactor.) Do appropriate analytical methods exist? (EPA SW-846 method 8260) How is the waste stream delivered to the microorganisms? (Recirculation from reservoir to the reactor with a pump)

3. *Perform the initial treatability setup.*

3.1 Prepare contaminant streams and nutrient solutions determined as part of step 2.

3.2 Construct reactors designed in step 2.

3.3 Prepare and calibrate analytical equipment, including method development if necessary.

3.4 Collect soil samples from the test site.

4. *Conduct the treatability study.*

4.1 Prepare the reactors by filling with the soil from the test site. Label as appropriate.

4.2 Add the nutrient solution to the soil.

4.3 Add the contaminated waste stream and substrate to the reactor.

4.4 Start the pumps, allowing the substrate (methane gas) and the contaminant stream to circulate through the reactor.

4.5 Collect the initial samples (at $t = 0$) for analysis.

4.6 Collect and analyze periodic samples as defined in step 2.

5. *Results and data reporting.*

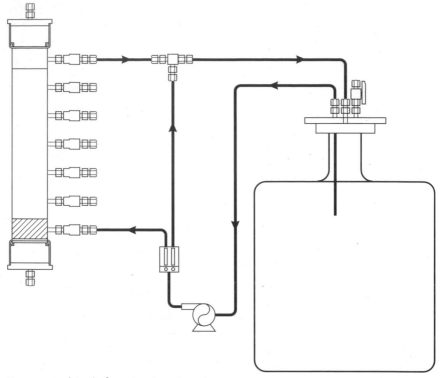

Figure 23.4 A typical gas treatment reactor.

What did the study determine? (That the indigenous microorganisms can degrade TCE at a particular level) Of primary concern are the degradation rates and required contact times for the contaminants. Present data, statistical analyses, and other pertinent information.

6. *Conclusions and recommendations.*
   This is perhaps the most important element of the treatability study, and often the most difficult. Based upon the results reported in step 5, what then? If the treatability study was successful, how can this be implemented in the field, and under what conditions? Or how should the study have been performed differently?

The above protocol can be used to develop the procedure for a treatability study investigating the biodegradation of any compound under any conditions. As stated in step 6, the conclusions and recommendations are arguably the most important step for the designer, because decisions will be made based upon this report. The designer, having conducted or overseen the study, is the best person to tell whether an application is feasible or appropriate. Although others (upper management, clients, regulators) may make the decision as to whether to implement the results of the treatability study as a biofilm reactor, they will look to the designer for guidance.

23.3 Design

Based upon the results of the treatability studies, the next step is to begin the design process. Depending upon the type of project, a pilot-scale system may be beneficial; or the designer may proceed to a full-scale system. Pilot-scale units are intended to allow the proposed treatment system to be tested under the actual field conditions, while typically operating at a lower flow rate. What works in the laboratory does not always work the same in the field. A pilot-scale system allows the operation of the treatment unit to be adjusted and optimized at reduced capital and operating costs. However, for some smaller-scale systems, a pilot-scale system may not be cost-effective. A "small" full-scale treatment system can be constructed and operated as cost-effectively as many pilot-scale units, and can be optimized just as easily. As with many engineering decisions, capital and operating costs should be compared to the end result (parameter optimization, operating information, etc.) of the proposed work (pilot-scale unit) to determine whether the work is necessary.

In a biofilm reactor, the degradation of the contaminants is ultimately limited by the rate of the mass transfer from the gas phase to the biofilm and by the rate of diffusion within the biomass. The contaminant, as it passes though the biofilm reactor in the gas stream, will adsorb onto either the filter medium or the biomass itself. The organics are then degraded. A schematic of the adsorption, diffusion, and transformation within a biofilm reactor is shown in Fig. 23.5.

In general, the degradation rate for various chemicals increases with chemical complexity. Alcohols degrade faster than ketones, followed by alkanes and aromatics. The lower the degradation rate, the more time that is required to degrade the compound in question. This translates to a longer retention time, also known as the *empty bed contact time* (EBCT). The EBCT is measured as the volume of the reactor divided by the flow rate.

A wide range of filter media are being used in biofiltration systems. The medium materials used include soil, compost, peat, wood chips, bark, and synthetic pelletized, ring, or channelized media. Each has advantages and disadvantages. For example, treatability studies conducted with synthetic pelletized medium exhibit a significantly lower EBCT than reactors with compost or soil as the medium. However, the biofilters with pelletized medium would exhibit a significant

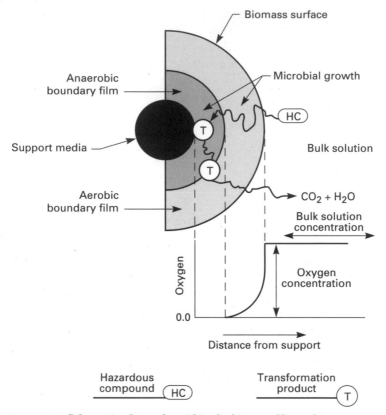

Figure 23.5 Schematic of transfer within the biomass film and oxygen concentration for aerobic systems. (*Cookson, 1995.*)

increase in head loss due to the accumulation of biomass on the surface of the pellets (more directly in the flow path of the gas stream), which was not seen with the compost medium. Other effects of the type of support medium on the efficiency of a biofilter, summarized by Bishop and Govind, include these:

1. The geometry and structure of the medium establish the surface area per unit volume of the filter medium.

2. The thickness of the biofilm growth is dependent upon the ability of the microbial growth to adhere to the support medium.

3. The use of an adsorptive medium provides an opportunity for an adsorbed substrate to diffuse from the adsorbent into the biofilm, creating a more constant contaminant level.

4. Porous media can adsorb moisture and nutrients to provide a more constant supply of nutrients to the microorganisms.

Figure 23.6 Installation of biofiltration tower, showing tray packing media. (*Courtesy of EG&G Biofiltration, Saugerties, N.Y.*)

Figure 23.7 Installation of biofiltration tower, showing completed tray stack. (*Courtesy of EG&G Biofiltration, Saugerties, N.Y.*)

Figure 23.8 Completed biofiltration tower, showing control panel. (*Courtesy of EG&G Biofiltration, Saugerties, N.Y.*)

Figure 23.9 Typical biofiltration installation. (*Courtesy of EG&G Biofiltration, Saugerties, N.Y.*)

Figure 23.10   Typical biofiltration installation. (*Courtesy of EG&G Biofiltration, Saugerties, N.Y.*)

Figure 23.11   Typical biofiltration installation. (*Courtesy of EG&G Biofiltration, Saugerties, N.Y.*)

Figure 23.12 Typical biofiltration installation. (*Courtesy of EG&G Biofiltration, Saugerties, N.Y.*)

Synthetic media that have been used include plastic saddles, pellets, and extruded supports; granular activated carbon; and ceramic extruded media. One of the advantages of synthetic media is the ability to apply an adsorbent material such as carbon or a resin to the support medium. The addition of this adsorbent material provides excess capacity for the filter to better accommodate variations in the influent gas concentration. Studies using adsorbent medium as the biofilm support have also exhibited shorter start-up times, greater system stability, and greater biofilm mass (and therefore greater removal efficiency).

Currently, the most commonly used material is an organic material such as compost, soil, or some combination thereof. Organic materials have several advantages including the interior pore structure of the material (see the discussion of adsorption, Sec. 23.2), resulting in a very large surface area; the ready availability of nutrients in the organic matter; and the presence of microorganisms. Evens et al. (1995) found compost medium to have higher microbial populations than either peat or activated carbon. The systems with compost medium also exhibited greater contaminant removal. The disadvantage of the compost medium—the head loss observed across the filter—was alleviated by using a mixture of compost and perlite. If the microorganisms naturally present in the medium have not been exposed to the organic contaminants, or if a significant fraction of the medium is synthetic, the biofilter may need to be seeded with an acclimated microbial culture to facilitate the start-up of the treatment system.

Biofilters with a single direction of flow have traditionally exhibited a tendency to stratify into a zone with significant microbial growth, often producing significant head loss, potentially resulting in filter plugging, followed by zones with no microbial growth. The zone nearest the inlet degrades most of the nutrients and contaminants, while the remaining areas are "starved." Several methods have been used to attempt to control the uneven or excessive microbial growth in the columns, including regular backwashing, nutrient limitations, and bidirectional (or countercurrent) flow. Directional switching involves reversing the inlet and outlet flows, allowing the excess microbial growth to be reduced due to the lack of nutrients and food. The end result of the bidirectional flow is reduced filter plugging, lower head loss across the column, and therefore lower system operating costs.

Gas flow rates range from 500 to 100,000 ft^3/min. The optimal internal temperature for the biofilter is around 100°F.

One common measure of reactor capacity is the EBCT. The EBCT is the total volume of the column divided by the flow rate, and for most biofilters it is on the order of 1 to 10 min. Larger EBCT values typically result in lower system operational costs, although they have higher capital costs.

It is important to note that the EBCT is not necessarily the same as the time required for the degradation of the contaminant (the contaminant residence time). As mentioned previously, the gas-phase contaminant either is adsorbed onto the filter medium or diffuses into the biofilm directly. The biofilm can grow and expand to areas of the medium that do not have an active microbial culture. Once the contaminant comes into contact with the microorganisms in the biofilm, it is degraded. As the contaminant is degraded by the microorganisms within the reactor's biofilm, additional "contact points" become available for either adsorption onto the medium or diffusion directly into the biofilm. So the actual removal process from a contaminated gas stream is a combination of adsorption and biodegradation.

A primary concern is whether a biofiltration unit should be designed and built by the user or purchased as a unit from a manufacturer. Many times this decision must be based upon the designer's experience. The steps used in designing an adsorption system are as follows:

1. Establish the allowable effluent concentration (typically available in regulations or existing facility permits). Based upon the influent gas stream concentration (measured), determine the mass per unit time period that it will need to be removed (i.e., pounds per hour).

2. Determine the chemical characteristics of the various components of the gas stream over time.

3. Select the filter medium.

4. Perform a treatability study to determine the degradation rates, or use appropriate and applicable rates generated previously.

5. Determine gas stream operating conditions, such as flow rate (scfm and acfm), relative humidity, particulate loading, and temperature.

6. Calculate the surface area of the adsorbent bed A, using the gas stream flow rate and the volume, and the bed length L, using the volume and area. As a rule of thumb, a length-to-diameter ratio on the order of 3-to-4 is an acceptable target. Keep in mind that longer beds usually result in greater removal.

7. Calculate the empty bed contact time, a measure of the column detention time. The EBCT should be long enough to ensure that sufficient time is allowed for degradation of the contaminants (based upon the degradation rates determined in treatability studies).

There are several other factors to consider during the design of the system. The system must be designed with enough flexibility to allow the operating parameters to be varied to optimize system performance on a full-scale basis. The gas stream must be evenly distributed at the entrance to the filter bed for maximum bed efficiency. Variations in the influent VOC concentration must be considered as to the impact on the microorganisms in the filter. The microorganisms will strive to adapt to the average concentration of the influent stream. As long as the VOC concentration fluctuation cycles several times each hour (or more rapidly), the microorganisms should not be adversely affected. If the cycle is less frequent, modifications to the system (such as combining several similar waste streams) must be made to increase the efficiency. The system must be tested for compliance by sampling the inlet and outlet using the EPA method 25a sampling procedure (Chap. 5). The frequency of the tests will be established by the regulatory agency having jurisdiction over the site. To properly test the outlet, the system must be enclosed and must have a discharge "stack." Leachate from the system must be captured and prevented from contaminating the underlying ground.

The capital costs of biofiltration systems are usually on the same order as those of other VOC treatment systems. The major cost savings are achieved in the operational costs.

References

Anthony, C., *The Biochemistry of Methylotrophs,* Academic Press, London, 1982.

Bishop, D. F., and Govind, R., "Development of Novel Biofilters for Treatment of Volatile Organic Compounds," *Biological Unit Processes for Hazardous Waste Treatment,* vol. 3, edited by Hinchee, R. E., Skeen, R. S., and Sayles, G. D., Battelle Press, Columbus, Ohio, 1995.

Cookson, J. T., Jr., *Bioremediation Engineering: Design and Application,* McGraw-Hill, New York, 1995.

Evens, P. J., Bourbonais, K. A., Peterson, L. E., Lee, J. H., and Laakso, G. L., "Vapor-Phase Biofiltration: Laboratory and Field Experience," *Biological Unit Processes for Hazardous Waste Treatment,* vol. 3, edited by Hinchee, R. E., Skeen, R. S., and Sayles, G. D., Battelle Press, Columbus, Ohio, 1995.

Farmer, R. W., Chen, J. S., Kopchynski, D. M., and Maier, W. J., "Reactor Switching: Proposed Biomass Control Strategy for the Biofiltration Process," *Biological Unit Processes for Hazardous Waste Treatment,* vol. 3, edited by Hinchee, R. E., Skeen, R. S., and Sayles, G. D., Battelle Press, Columbus, Ohio, 1995.

King, R. B., Long, G. M., and Sheldon, J. K., *Practical Environmental Bioremediation,* Lewis Publishers, Boca Raton, Fla., 1992.

Kinney, K. A. et al., "Performance of a Directionally-Switching Biofilter Treating Toluene Contaminated Air" (96-RP87C.05), Air and Waste Management Association, 89th Annual Meeting, Nashville, Tenn., 1996.

Kobayashi, H., and Rittmann, B. E., "Microbial Removal of Hazardous Organic Compounds," *Environmental Science and Technology,* vol. 16, no. 3, 1982, pp. 170A–183A.

Lackey, L., and Holt, T., "Not for the Birds," *Industrial Wastewater,* May/June, 1996, pp. 31–33.

Large, P. J., *Methylotrophy and Methanogenesis,* Van Nostrand Reinhold (UK) Co. Ltd., Hong Kong, published in United States by American Society for Microbiology, 1983.

Little, C. D., Palumbo, A. V., Herbes, S. E., Lindstrom, M. E., Tyndall, R. L., and Gilmer, P. J., "Trichloroethylene Biodegradation by a Methane-Oxidizing Bacterium," *Applied and Environmental Microbiology,* vol. 54, no. 4, 1988, pp. 951–956.

Singleton, B., Zeni, A., and Cha, S., "Biofiltration of Odors Caused by Reduced Sulfur Compounds at a Pulping Process, Analysis and Design" (96-RA87B.05), Air and Waste Management Association, 89th Annual Meeting, Nashville, Tenn., 1996.

Smith, F. L. et al., "Development of High-Rate Trickle Bed Biofilter," http://128.6.70.23/html_docs/rrel/brenner.html.

Stanfelder, S., "Evaluating Biofiltration," *Environmental Technology,* July/August, 1996, pp. 26–32.

Stanier, R. Y. et al., *The Microbial World,* 5th ed., Prentice-Hall, Englewood Cliffs, N.J., 1986.

Yudelson, J. M., and Tinari, P. D., "Economics of Biofiltration for Remediation Projects," *Biological Unit Processes for Hazardous Waste Treatment,* vol. 3, edited by Hinchee, R. E., Skeen, R. S., and Sayles, G. D., Battelle Press, Columbus, Ohio, 1995.

24

Condensation of Gaseous Emissions

Lem B. Stevens, III

24.1 Condensation

Condensation is a phase change of a substance from a vapor to a liquid. It is the opposite of boiling. The ideal gas model describes a gas as a collection of molecules of a substance with enough energy and space between the molecules that they move freely in an individual fashion. Condensation occurs when the individual molecules are forced close enough together that the attractive forces between the molecules overcome repulsive forces and cause the molecules to stick together, forming a liquid. This phenomenon can be accomplished by cooling the gas until the molecules have a free energy lower than the attractive forces, or by pressurizing the gas until the molecules are forced so close together that they become a liquid. Pressurization is usually accomplished with a compressor. Most gas flow rates from industrial processes are large enough that pressurization is not feasible. Therefore, this chapter focuses on condensation through cooling.

24.2 Application

Condensation is used in two main fashions in air pollution control. First, it is used to control gases by changing them to a liquid and draining them away from the airstream. An example is the condensation of vapors in the vent of a chemical reactor. Vapors of reactants form in the head space of a reactor and can escape through the reactor vent. A condenser can be installed on the vent to collect these vapors. Once condensed to liquid form, the reactants can be drained back into the reactor for a more efficient reaction with respect to the amount of reactants consumed.

Second, condensation is used as a precontrol device for more expensive control devices. As a vapor changes to a liquid, its volume is greatly reduced. A gas stream may be routed through a condenser prior to an incinerator to remove gases that are harmful to the incinerator and to reduce the volumetric flow of the gas stream so that a smaller incinerator may be used. The benefit is an incinerator with lower capital and operating costs that will last longer than an incinerator subjected to harmful gases.

24.3   Types of Condensers

There are two main categories of condensers: surface condensers (Fig. 24.1) and contact condensers (Fig. 24.2). The difference is most easily described by their operation. In a contact condenser, the gas being controlled is directly mixed with the coolant. The advantages of a contact condenser are very high heat transfer and no fouling. However, in many cases it is not practical to recycle the coolant, and the spent coolant represents a new waste stream. In a surface condenser, the condensed gas and the coolant are separated by a heat-transfer surface. The heat-transfer surface is usually a bundle of tubes through which the coolant flows. The gas condenses on the cool, outer surface of the tubes. In surface condensers, the coolant is almost always recycled, which is a major advantage compared to contact condensers. However, surface condensers are subject to heat-transfer resistance through the tube walls and to fouling of the tubes. Fouling can greatly reduce the heat-transfer efficiency of the device.

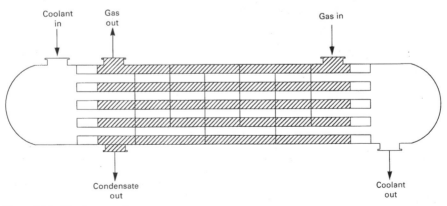

Figure 24.1 Surface condenser.

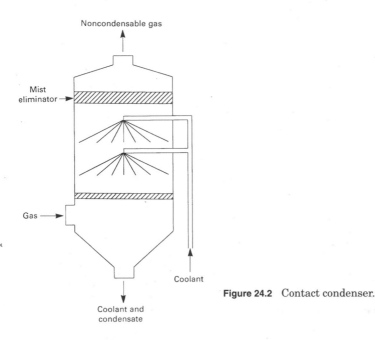

Figure 24.2   Contact condenser.

24.4   Determination of Gas Temperatures

The first step in specifying a condenser is to determine the inlet and exit temperatures of the vapor to be controlled in the gas stream. The inlet temperature of the gas stream is measured if the gas stream exists or predicted if the process has not been built. To specify the exit temperature, two steps are taken. The dew point temperature is either calculated or obtained from engineering or chemical data. The dew point can be calculated by a trial-and-error procedure, using the molecular fraction and vapor pressures of each component in its liquid and vapor states. It is important to note that the dew point of a gas depends upon its pressure, temperature, and components. It is beyond the scope of this text to calculate the dew point. The author suggests consultation with a reputable condenser manufacturer, who is experienced in such calculations and may have experience with the gas to be controlled. Then the amount of subcooling is chosen. Most condensers are designed to subcool the vapor to ensure good control efficiency. The extent of subcooling is left up to the design engineer, but 5°F of subcooling is a good rule of thumb.

24.5   Coolant

Next, the coolant for the condenser must be chosen. In most cases, the coolant is a liquid because heat transfer is much more efficient than

TABLE 24.1   Temperature Range of
Coolants

Coolant	Condensation temperature, °F
Water	80 to 100
Chilled water	45 to 60
Brine solution	−30 to 45
Refrigerants	−90 to −30

for a gas coolant. First, the coolant must have a freezing point below the temperature specified for the gas leaving the condenser, so that the coolant does not freeze and clog the condenser. Second, the coolant must be benign to the condenser and the components which complete the coolant system. In many cases water can be used. If the application requires a freezing point below that of water, then salts may be added to the water to create a brine. The addition of salts to water has the effect of lowering the freezing point. Table 24.1 lists the temperature range of some common coolants. Once the coolant has been chosen, the specific heat capacity of the coolant is determined from engineering data.

24.6   Coolant Flow Rate

The coolant flow rate is determined by equating the heat that must be removed from the gas stream to accomplish the specified temperature drop with the heat that the coolant must absorb. There are two terms in the equation which describe the change in heat energy of the gas stream. The first term accounts for the heat removed to decrease the temperature from the inlet temperature to the exit temperature. This is called the *sensible heat* and is of the form

$$E_s = m_v C_{pv}(T_i - T_o) \qquad (24.1)$$

where E_s = sensible energy
m_v = mass flow rate of vapor
C_{pv} = specific heat of vapor
T_i = temperature of vapor at inlet to condenser
T_o = temperature of vapor at outlet or exit from condenser

The second term accounts for the energy required for the vapor to change from a gas phase to a liquid phase. This is called the *latent heat of vaporization*. The adjective *latent* is used because little change in measured temperature is observed while the vapor is changing phases, even though heat is removed. Vaporization refers to the change in state from a liquid to a gas. The amount of energy required to vaporize a specific liquid into a gas equals the energy which must be removed from the gas to change it back to a liquid. Therefore, it is correct to use the term *heat of vaporization* in discussing condensation, which is the process of changing the phase of a substance from a gas to a liquid. The latent heat of vaporization is expressed in the form

$$E_v = m_v h_v \qquad (24.2)$$

where E_v = latent heat of vaporization
  m_v = mass flow rate of vapor
  h_v = specific latent heat of vaporization

Therefore, the total energy removed from the vapor is the sensible energy plus the latent heat of vaporization and is described by

$$E_t = E_s + E_v = m_v h_v + m_v C_{pv}(T_i - T_o) \qquad (24.3)$$

This amount of energy must be removed from the vapor to condense it. Therefore, this amount of energy must be transferred to the coolant in the condenser. Since the coolant usually does not change phase in the condenser, only the sensible energy term applies, and it is of the form

$$E_t = m_c C_{pc}(t_o - t_i) \qquad (24.4)$$

where m_c = mass flow rate of coolant
  C_{pc} = specific heat of coolant
  t_o = temperature of coolant as it exits condenser
  t_i = temperature of coolant as it enters condenser

Now the expressions for the energy lost by the vapor and gained by the coolant are equated and the terms rearranged to yield

$$m_c = \frac{m_v h_v + m_v C_{pv}(T_i - T_o)}{C_{pc}(t_o - t_i)} \qquad (24.5)$$

From this expression, the mass flow rate of the condensate can be determined. However, not all the terms in the expression are known. The temperature t_i is usually known from the source of the coolant. The temperature t_o cannot be known until the design of the condenser is determined. Notice that if the condenser is a contact condenser, then $t_o = T_o$ because the temperature of the coolant must equal the temperature of the condensate at the exit of the condenser. In fact, the relationships discussed up to this point are all that is required to design a contact condenser. However, in a surface condenser, this condition is not necessarily true. The design of the surface condenser

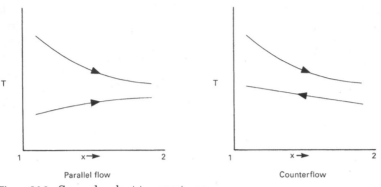

Parallel flow   Counterflow

Figure 24.3   Gas and coolant temperatures.

may dictate that the outlet temperature of the coolant be colder or hotter than the outlet temperature of the condensate. The configuration of the condenser is the major factor which determines the outlet temperature of the coolant. Figure 24.3 shows general temperature charts for counterflow and parallel-flow condensers.

24.7   Log Mean Temperature Difference

Based on the configuration of condenser being used, t_o can be assumed. Later in the design, t_o is iteratively corrected until the equations governing the condenser are satisfied. At any point in the condenser, the driving force for heat transfer is the difference in temperature between the coolant and the gas stream. Since this temperature changes for each location in the condenser, the solution can be very tedious. To avoid this problem, engineers use an overall heat-transfer relationship for a condenser and an average temperature difference between the coolant and the gas stream. This average temperature difference is called the *log mean temperature difference* and is calculated according to the following relationship:

$$dT_{lm} = \frac{(T_i - t_i) - (T_o - t_o)}{\ln[(T_i - t_i)/(T_o - t_o)]} \qquad (24.6)$$

Cross-flow and multipass condensers have correction factors to dT_{lm} which are beyond the scope of this chapter.

24.8   Overall Heat-Transfer Coefficient

The next step in the design of the condenser is to determine the overall heat-transfer coefficient U, where U accounts for the resistances to heat transfer in the condenser. A resistance is associated with the convection of heat through the coolant inside the tubes, the material of the tube wall, and convection of heat through the condensate on the outside of the tubes. A general expression for U is

$$U = \frac{1}{1/h_i + L/k + 1/h_o} \qquad (24.7)$$

where h_i = coolant convection coefficient
      L = length of tube in gas stream
      h_o = condensate convection coefficient

This expression assumes clean surfaces in the condenser. With time, clean surfaces become an unrealistic assumption, and buildup of fouling on the tube walls occurs. Fouling can have a marked effect on heat transfer. In practice, heat exchanger manufacturers have obtained data for the overall heat-transfer coefficient for many applications and materials, which account for some fouling. The author suggests the use of these data instead of the calculation of an overall heat-transfer coefficient when it is available.

24.9 Condenser Surface Area

Once dT_{lm} and U are known, the area of heat-transfer surface in the condenser may be calculated according to

$$Q = UA\, dT_{lm} \qquad (24.8)$$

where Q = heat transferred = E_t
      U = overall heat-transfer coefficient
      A = area of heat-transfer surface
   dT_{lm} = log mean temperature difference

The area A determines the size of the condenser required. It is from this area that the designer picks which condensers will perform satisfactorily in the application.

24.10 General Considerations

As a word of caution, great care should be taken to design the coolant system correctly. The temperature and flow rate of the coolant must meet design specifications for the condenser to function properly. The coolant pump must be matched to the resistance of the coolant system, including pressure losses associated with the condenser. The exit gas temperature should be monitored daily. Where condenser operation is critical, the author suggests installation of a temperature gauge for coolant at the exit of the condenser. Monitoring the coolant temperature is suggested because it is the first item to check if the condenser performance degrades. Coolant exit temperature is also a secondary check on coolant flow because low flow will result in a high coolant exit temperature. Fouling is a major concern in the operation of a surface condenser. Buildup of scale, corrosion, or material on the heat-transfer surface can greatly degrade the performance of the condenser. The telltale sign of fouling is normal coolant temperatures with a high exit gas temperature. Fouling generally occurs gradually over time. Engineers sometimes specify a slightly oversized condenser or high coolant flow so that the clean condenser will operate better than is required to accomplish condensation and will maintain acceptable operation as fouling occurs for a period before cleaning is required. All condensers should be designed to allow cleaning.

The condensate in a condenser must be separated from the gas stream in order to achieve maximum removal. Generally, some measure is taken to drain the condensate away from the airstream by gravity. Contact condensers use some type of mist eliminator to separate the condensate from the gas. In the case of a condenser on a chemical process reactor vent, the reaction efficiency may be increased by draining the condensate back into the reactor. For other applications, the condensate is a waste stream which must be treated appropriately.

References

1. F. P. Incropera and D. P. DeWitt, *Fundamentals of Heat Transfer,* John Wiley & Sons, New York, 1981.
2. A. J. Buonicore and W. T. Davis, *Air Pollution Engineering Manual,* Air and Waste Management Association, Sewickely, Pa., 1992.
3. K. B. Schnelle, professor of chemical and environmental engineering, Vanderbilt University, Nashville, Tenn., personal communication, 1994.

25

Control of Nitrogen Oxide Emissions

Mike Ayers

According to the EPA, in 1987, mobile sources of NO_x were responsible for approximately 43 percent of the nation's atmospheric NO_x, and stationary sources were responsible for 57 percent of the NO_x. Of the stationary sources, 70 percent were the result of existing coal-fired utility boilers.

When nitrogen is oxidized, it can form NO, NO_2, NO_3, N_2O, N_2O_3, N_2O_4, and N_2O_5. The term NO_x is used to refer to all the oxides of nitrogen. However, NO and NO_2 are the most important to air pollution control because of the volumes in which they are emitted. There are two sources of NO_x emissions—mobile and stationary sources. While mobile sources (automobiles) do represent a significant portion (approximately 43 percent) of the NO_x problem, the majority (approximately 57 percent) of the NO_x emitted is from stationary sources. Of the stationary sources of NO_x emissions, fuel combustion (mostly electric utility generators) sources represent the single largest source of NO_x.

NO_x causes decreased visibility, crop damage, eye irritation, bad odor, and rubber deterioration. Photochemical reactions of nitrogen oxides with hydrocarbons form ozone, which is designated a priority pollutant by the Clean Air Act (CAA). The past 25 years have shown great developments in the understanding of nitrogen oxides formation and control. The complex chemical reactions involved in its formation make NO_x one of the most expensive and hardest to control of the major air pollutants.

25.1   Major Sources

The major stationary sources of NO_x production are as follows:

- High-temperature combustion
  Fixation of atmospheric nitrogen
  Fuel nitrogen
- Nitric acid manufacture and concentration
- Organic nitration reactions
- Internal combustion and diesel engine exhaust

25.2 NO_x Chemistry of Formation

NO_x is formed as thermal NO_x or fuel NO_x. When fuel containing nitrogen is combusted, that nitrogen forms fuel NO_x. But when NO_x is formed from nitrogen in the air reacting at high temperatures with the combustion products, this is called *thermal* NO_x.

25.2.1   Thermal NO_x

As a general rule, nitrogen reacts with oxygen according to the following reaction:

$$O_2 + N_2 \rightleftharpoons 2NO \tag{25.1}$$

The temperature at which this reaction takes place can have a major effect on the equilibrium. The more the reaction temperature is increased, the more the equilibrium of the reaction shifts toward the product, and a greater percentage of NO is produced. Equation (25.1) illustrates the production of thermal NO_x. For lower temperatures, the balance of the reaction is shifted toward the reactants, and O_2 and N_2 are produced more rapidly. But this increased reactant production hits optimum at about 870°C (1600°F), and further cooling will not increase the production of N_2 and O_2.

In most reactors, though, the gas stream spends so little time in the reaction chamber that the gases never have enough time to come to equilibrium. In this case all three species end up in the flue gas. With as little as 5 to 10 percent excess air, coal or oil burners often produce NO concentrations of 1000 to 2000 ppmv.

The chemical model for the postcombustion reactions of nitrogen and oxygen was developed by Zeldovich (1946):

$$N_2 + O \rightleftharpoons NO + N \tag{25.2}$$

$$N + O_2 \rightleftharpoons NO + O \tag{25.3}$$

also

$$N + OH \rightleftharpoons NO + H \tag{25.4}$$

The basic equations for the thermodynamics of NO_x formation are as follows:

$$N_2 + O_2 \rightleftharpoons 2NO \tag{25.5}$$

$$NO + \tfrac{1}{2}O_2 \rightleftharpoons NO_2 \tag{25.6}$$

Equation (25.5) has a very high activation energy and is usually the rate-limiting reaction. This equation is highly dependent on temperature in addition to the concentration of N_2 and O_2. It is believed that hydrocarbon radicals can facilitate the NO production process with a possible intermediate being HCN.

The equilibrium constants for reactions (25.5) and (25.6) are the following:

$$K_{p_1} = \frac{(\overline{P}_{NO})^2}{\overline{P}_{N_2}\overline{P}_{O_2}} = \frac{y_{NO}^{\;2}}{y_{N_2}y_{O_2}} \tag{25.7}$$

and

$$K_{p_2} = \frac{\overline{P}_{NO_2}}{\overline{P}_{NO_2}\overline{P}_{O_2}} \frac{(P_T)^{-1/2}}{y_{NO}(y_{O_2})^{1/2}} y_{NO_2} \tag{25.8}$$

where K_p = equilibrium constant
$\overline{P}_i$ = partial pressure of component i, atm
y_i = mole fraction of component i
P_T = total pressure, atm

Based on equilibrium constant data, at flame-zone temperatures, those in the range of 3000 to 3600°F, concentrations of approximately 6000 to 10,000 ppm NO_x are produced and significantly more NO is produced than NO_2. Furthermore, at flue gas temperatures, which are much lower (300 to 600°F), NO_x concentrations are less than 1 ppm and more NO_2 is produced than NO.

However, in industrial settings, the conditions are somewhat different. Flue gas concentrations are typically much higher, as is the NO-to-NO_2 ratio. Therefore, there are other factors besides equilibrium that influence the production of NO_x.

MacKinnon (1974) observed that NO formation increased dramatically with temperature up to 3615°F ± 46°F. At this point the rate of NO formation began to decrease. He developed an expression that relates NO concentration to temperature and nitrogen and oxygen concentrations at 1 atm:

$$C_{\text{NO}} = 5.2 \times 10^{17}\, e^{-72,300/T}\, y_{\text{N}_2}\, y_{\text{O}_2}{}^{1/2}\, t \qquad (25.9)$$

where C_{NO} = NO concentration, ppm
$\qquad y_i$ = mole fraction of component i
$\qquad T$ = absolute temperature, K
$\qquad t$ = time, s

This equation is a good approximation for NO formation in the flame zone. The reaction occurs quickly (less than 0.5 s). Other compounds such as of hydrogen, carbon, sulfur, and hydroxide can significantly alter the system, however. Once past the flame zone, temperatures drop and reaction rates fall by orders of magnitude, virtually halting the reaction.

The formation of NO_2 is not as affected by temperature as the formation of NO. In order to reduce NO_x emission, especially NO formation, the peak temperature, the oxygen concentration in the high-temperature zone, and the gas residence time should all be reduced.

25.2.2 Fuel NO_x

In addition to burning the fuel with regular air, which contains nitrogen, some fuels contain nitrogen themselves. When this fuel-bound nitrogen is combusted, fuel NO_x is formed. NO_x production can be greatly increased during a combustion process when fuel-bound nitrogen is present.

Nitrogen found in the fuel, usually bonded to a carbon atom (H-C), can make up as much as 50 percent of the total NO_x (Cooper, 1990). The kinetic mechanisms are not entirely known. The H-C bonds are weaker than N-N bonds and thus form NO_x more easily.

Boilers and furnaces can be described by firing mode or ash characteristics. Firing mode boilers include single or opposed wall, cyclonic, and tangentially fired. Ash removal is either dry-bottom (solid ash removal) or wet-bottom (molten ash removal). Typically cyclone-fired boilers burn coal. Others also burn oil and gas.

Different kinds of coal contain different amounts of nitrogen. And different boilers produce different amounts of NO_x.

Nitrogen rate depends on the air-to-fuel ratio as well as quantity of nitrogen in fuel, generally 0.5 to 2 percent for U.S. coal (Cooper, 1990). Greater mixing also converts more of the bound nitrogen to NO. The efficiencies of conversion are thus 10 to 60 percent. Unlike thermal NO_x, fuel NO_x is not so affected by temperature changes.

25.3   Stationary Sources

NO_x emissions can be reduced by modifying the combustion process that generated the NO_x or by treating the flue gas that contains the NO_x. Most EPA regulations could potentially be met through combustion modifications only.

The following methods are effective at reducing NO_x formation during combustion:

1. Low excess air

   Reducing excess oxygen to 5 to 10 percent will greatly limit NO formation, but must be accompanied by good combustion condition design.
2. Two-stage combustion

   The first stage is a high-temperature, fuel-rich stage that increases NO_x reduction. The second stage is operated at lower temperatures (1000 to 1100°C) that contribute to oxidation.
3. Flue gas recirculation

   a. Combustion temperature is lowered.

   b. The O_2 concentration is diluted, thus oxidizing less nitrogen.

   c. NO is added to the reaction mixture (from the previous combustion) and shifts the equilibrium in the reverse direction [Eq. (25.1)].
4. Combustion modification

   a. Burner placement

   b. Rate of heat removal from flame

A combination of two or more of the above methods will further reduce NO_x formation.

Retrofit operations for tangentially fired and wall-fired boilers typically include low-NO_x burners (LNBs) and overfired air (OFA). Cyclone-fired furnaces (boilers) are usually retrofitted with flue-gas reburning.

Retrofitting experiments have shown that a 30 percent reduction of NO_x can be accomplished with overfired air when using 10 to 20 percent of the total combustion air as overfired air. However, if overfired air is combined with low-NO_x burners, up to 50 percent NO_x reduction can be achieved.

Excess-air reduction is also effective for further NO_x reduction. More finely ground pulverized coal burns much more efficiently. This permits reduction of excess air further, resulting in unburned carbon in the effluent.

The EPA has created the AP-42 Compilation of Air Pollution Control Emission Factors to assist industry in reducing NO_x emissions from coal-fired (Table 25.1) and oil-fired (Table 25.2) boilers as well as postcombustion (Table 25.3).

TABLE 25.1 Combustion Modification NO$_x$ Controls for Stoker Coal-Fired Industrial Boilers

Control technique	Description of technique	Effectiveness of control (% NO$_x$ reduction)	Range of application	Commercial availability/R&D status	Comments
Low excess air (LEA)	Reduction of airflow under stoker bed	5–25	Excess oxygen limited to 5 to 6% minimum	Available now but need R&D on lower limit of excess air	Danger of overheating grate, clinker formation, corrosion, and high CO emissions
Staged combustion [LEA + overfired air (OFA)]	Reduction of under-grate airflow and increase of overfired airflow	5–25	Excess oxygen limited to 5% minimum	Most stokers have OFA ports as smoke control devices but may need better air-flow control devices	Need research to determine optimum location and orientation of OFA ports for NO$_x$ emission control. Overheating grate, corrosion, and high CO emission can occur if under-grate airflow is reduced below acceptable level as in LEA
Load reduction (LR)	Reduction of coal and air feed to stoker	Varies from 49% decrease to 25% increase in NO$_x$ (average 15% decrease)	Has been used down to 25% load	Available	Only stokers that can reduce load without increasing excess air. Not a desirable technique because of loss in boiler efficiency.
Reduced-air pre-heat (RAP)	Reduction of combustion air temperature	8	Combustion air temperature reduced from 473 to 453 K	Available now if boiler has combustion air heater	Not a desirable technique because of loss in boiler efficiency
Ammonia injection	Injection of NH$_3$ in convective section of boiler	40–40 (from gas- and oil-fired boiler experience)	Limited by furnace geometry. Feasible NH$_3$ injection rate limited to 1.5 NH$_3$/NO	Commercially offered but not yet demonstrated	Elaborate NH$_3$ injection, monitoring, and control system required. Possible load restrictions on boiler and air preheater fouling by ammonium bisulfate

TABLE 25.2 Combustion Modification NO$_x$ Controls for Oil-Fired Boilers

Control technique	Description of technique	Effectiveness of control (Percent NO$_x$ reduction)		Range of application	Commercial availability/R&D status	Comments
		Residual oil	Distillate oil			
Low excess air (LEA)	Reduction of combustion air	0–28	0–24	Generally excess O$_2$ can be reduced to 2.5%, representing a 3% drop from baseline	Available	Added benefits included increase in boiler efficiency. Limited by increase in CO, HC, and smoke emissions
Staged combustion (SC)	Fuel-rich firing burners with secondary combustion air ports	20–50	17–44	70 to 90% burner stoichiometries can be used with proper installation of secondary-air ports	Technique is applicable on package and field-erected units. However, not commercially available for all design types	Best implemented on new units. Retrofit is probably not feasible for most units, especially packaged ones
Burners out of service (BOOS)	One or more burners on air only. Remainder firing fuel rich	10–30	NA	Applicable only for boilers with minimum of four burners. Best suited for square burner pattern with top burner or burners out of service. Only for retrofit application	Available. Retrofit requires careful selection of BOOS pattern and control of airflow	Retrofit often requires boiler derating unless fuel delivery system is modified
Flue gas recirculation (FGR)	Recirculation of portion of flue gas to burners	15–30	58–73	Up to 25 to 30% of flue gas recycled. Can be implemented on all design types	Available. Requires extensive modifications to the burner and wind box.	Best suited for new units. Costly to retrofit. Possible flame instability at high FGR rates.

TABLE 25.2 Combustion Modification NO$_x$ Controls for Oil-Fired Boilers (Continued)

Control technique	Description of technique	Effectiveness of control (Percent NO$_x$ reduction) Residual oil	Effectiveness of control (Percent NO$_x$ reduction) Distillate oil	Range of application	Commercial availability/R&D status	Comments
Flue gas recirculation plus staged combustion	Combined techniques of FGR and staged combustion	25–53	73–77	Max. FGR rates set at 25% for distillate oil and 20% for residual oil	Combined techniques are still at experimental stage.	Retrofit may not be feasible. Best implemented on new units.
Load reduction (LR)	Reduction of air, and fuel flow to all burners in service	33% decrease to 25% increase in NO$_x$	31% decrease to 17% increase in NO$_x$	Applicable to all boiler types and sizes. Load can be reduced to 25% of maximum	Available now as a retrofit application. Better implemented with improved firebox design	Technique not effective when it necessitates an increase in excess O$_2$ levels. LR possibly implemented in new designs as reduced combustion intensity (enlarged furnace plan area)
Low-NO$_x$ burners (LNBs)	New burner designs with controlled air/fuel mixing and increased heat dissipation	20–50	20–50	New burners described generally applicable to all boilers. More specific information needed	Commercially offered but not demonstrated	Specific emissions data from industrial boilers equipped with LNB are lacking
Ammonia injection	Injection of NH$_3$ as a reducing agent in flue gas	40–70	40–70	Applicable for large package and field-erected water-tube boilers. May not be feasible for fire-tube boilers	Commercially offered but not demonstrated	Elaborate NH$_3$ injection, monitoring, and control system required. Possible load restrictions on boiler and air preheater fouling when burning high-sulfur oil
Reduced-air preheat (RAP)	Bypass of combustion air preheater	5–16	NA	Combustion air temperature can be reduced to ambient conditions (340 K)	Available. Not implemented because of significant loss in thermal efficiency	Application of this technique on new boilers requires installation of alternate heat recovery system (e.g., an economizer)

*NA 5 not applicable.

TABLE 25.3 Postcombustion NO$_x$ Reduction Technologies

Technique	Description	Advantages	Disadvantages
Urea injection	Injection of urea into furnace to react with NO$_x$ to form N$_2$ and H$_2$O	Low capital cost Relatively simple system Moderate NO$_x$ removal (30 to 60%) Nontoxic chemical Typically, low-energy injection sufficient	Temperature-dependent Design must consider boiler operating conditions and design Reduction may be decreased at lower loads
Ammonia injection (Thermal DeNOx)	Injection of ammonia into furnace to react with NO$_x$ to form N$_2$ and H$_2$O	Low operating cost Moderate NO$_x$ removal (30 to 60%)	Moderately high capital cost Ammonia handling, storage, vaporization, and injection systems required (ammonia is a toxic chemical)
Air heater (AH) SCR	Air heater baskets replaced with catalyst-coated baskets. Catalyst promotes reaction of ammonia with NO$_x$	Moderate NO$_x$ removal (40 to 65%) Moderate capital cost No additional ductwork or reactor required Low pressure drop Can use urea as ammonia feedstock Rotating air heater assists mixing, contact with catalyst	Design must address pressure drop, maintain heat transfer Due to rotation of air heater, only 50% of catalyst is active at any time
Duct SCR	A smaller version of conventional SCR is placed in existing ductwork	Moderate capital cost Moderate NO$_x$ removal (30%) No additional ductwork required	Duct location unit specific temperature, access-dependent Some pressure drop must be accommodated

TABLE 25.3 Postcombustion NO$_x$ Reduction Technologies (*Continued*)

Technique	Description	Advantages	Disadvantage
Activated-carbon SCR	Activated-carbon catalyst, installed downstream of air heater, promotes reaction of ammonia with NO$_x$ at low temperature	Active at low temperature High surface area reduces reactor size Low cost of catalyst Can use urea as ammonia feedstock Activated carbon is nonhazardous material SO$_x$ removal as well as NO$_x$ removal	High pressure drop Not a fully commercial technology
Conventional SCR	Catalyst located in flue gas stream (usually upstream of air heater) promotes reaction of ammonia with NO$_x$	High NO$_x$ removal (90%)	Very high capital cost High operating cost Extensive ductwork to/from reactor Large-volume reactor must be sited Increased pressure drop may require ID fan or larger FD fan Reduced efficiency Ammonia sulfate removal equipment for air heater Water treatment of air heater wash

25.3.1 Reburning

During a reburn operation, fuel is added to the exiting flue gas and is fed to another burner in such a way that existing NO_x will be dissociated back into N_2 and O_2. The oxygen released from the reaction is quickly consumed by the burning process.

Reburning is most effective with fine pulverized coal, but may also be used with fuel oil or natural gas. Combining reburning with low-NO_x burners and overfired air can yield 75 percent NO_x reduction in retrofitting.

25.3.2 Combustion modifications

As described above, it is desirable to reduce the peak temperature in the flame zone, gas residence time in the flame zone, and oxygen concentrations in the flame zone. To accomplish these tasks, modifications can be made to existing equipment, or new low-NO_x burners can be used. See Table 25.4.

TABLE 25.4   How to Treat Each Step

To reduce the peak temperatures:
- Use a fuel-rich primary flame zone.
- Increase the rate of flame cooling.
- Decrease the adiabatic flame temperature by dilution.

To reduce the gas residence time:
- Change the shape of the flame zone.
- Use a fuel-rich primary flame zone.
- Increase the rate of flame cooling.
- Decrease the adiabatic flame temperature by dilution.

To reduce the oxygen concentration:
- Decrease the excess air.
- Maintain controlled fuel/air mixing.
- Use a fuel-rich primary flame zone.

The following procedures can be used to modify existing equipment to eliminate the costs associated with new equipment:
1. Low-excess-air (LEA) firing
2. Off-stoichiometric combustion (OSC) including overfired air
3. Flue gas recirculation (FGR)
4. Reduced-air preheat
5. Reduced firing rates
6. Water injection

25.3.3 Low-excess-air (LEA) firing

This is very simple and also effective. In the past, it was common to fire furnaces at 50 to 100 percent excess air, that is, 1.5 to 2 times the amount of air that is stoichiometrically required to completely burn the fuel. The reason for this is that perfect mixing can never really be achieved, and excess air is a way to ensure that complete combustion will occur. But today controls are available that enable us to use less than 15 to 30 percent excess air and still achieve the same combustion efficiency. Simply reducing the excess air a few percentage points can significantly reduce NO_x emissions.

25.3.4 Off-stoichiometric combustion (OSC)

OSC, or staged combustion, is a control technology by which the combustion process is broken down into two or more steps. The first step is known as the *fuel-rich* step, and each following step is a *fuel-lean* step. Modifications to existing systems can be made in one of three ways. First, use some of the burners for the fuel-rich step and others for the fuel-lean steps. Second, some of the burners might be taken out of service and only be allowed to admit air to the furnace. Finally, fire all the burners as fuel-rich, and then add overfired air over the flame zone. This method requires careful monitoring to keep the CO and smoke down, but yields the best results—up to 35 percent reduction in NO_x.

25.3.5 Flue gas recirculation (FGR)

This technique reduces the furnace temperature by recirculating flue gas back to the furnace. Because of the potentially large cost of new ductwork, fans, dampers, and controls, this process is more appropriate for new designs than for retrofitting.

25.3.6 Reduced-air preheat and reduced firing rates

This method reduces thermal NO_x by reducing the peak flame zone temperature. Reduced firing rates can also be applied. Both have the problem of less energy production and usually require more excess air to reduce smoke. Furthermore, the operation's flexibility is diminished.

25.3.7 Water injection

Water or steam injection reduces flame temperatures and hence thermal NO_x. This is very effective for gas turbines (an 80 percent NO_x

reduction is possible). The energy loss is only about 1 percent, but can be as high as 10 percent for utility boilers.

Low-NO_x burners (LNBs) are very popular for both new and existing plants. LNBs work by carefully controlling the mixing of fuel and air. They have been proved to reduce NO_x by 40 to 60 percent. New plants usually utilize this type of equipment because it is cost-effective. However, older plants often cannot justify the cost of upgrading to these systems. About 85 percent of today's plants were built prior to 1971, and many still contain outdated technology that does not effectively reduce NO_x emissions. Because today's furnaces are larger and placed a greater distance from one another, temperature is reduced and residence time increased, and these contribute to more complete combustion and NO_x reduction.

25.3.8 Flue gas treatment (FGT)

FGT is used if further NO_x reduction is required or if combustion controls do not apply (for example, HNO_3 plants). These are common in Japan where most of the research is done.

Flue gas treatment for NO_x control can be classified as dry or wet. Dry control systems for NO_x control are catalytic reduction, noncatalytic reduction, adsorption, and irradiation. Wet control systems include various flue gas scrubbing techniques.

25.3.9 Adsorption

Dry adsorption can generally be used to control both SO_x and NO_x in the same process. For example, activated carbon and ammonia injection can be used to convert NO_x to N_2 by reduction and SO_2 to H_2SO_4 by oxidation. The adsorption system typically operates around 220 to 230°C, and the carbon must be regenerated to remove the H_2SO_4. This generally creates a concentrated SO_2 stream which needs to be treated.

Adsorption via copper oxide catalyst is also effective. The catalyst converts SO_2 to copper sulfate, and then both the copper oxide and the copper sulfate selectively catalyze NO_x to NH_3. An SO_2 stream is produced from the regeneration of the catalyst beds with hydrogen. This process can achieve approximately 90 percent SO_x removal coupled with 70 percent NO_x removal. Elemental sulfur or H_2SO_4 is produced from the SO_2 streams.

25.3.10 Irradiation

Electron beam irradiation can be used to reduce NO_x and SO_x to nitrate and sulfate compounds. If ammonia is added to the flue gas, the result is NH_4NO_3 and $(NH_4)_2SO_4$, which are fertilizers ("Electron Guns Clean Up SO_2 and NO_x," 1983). The process involves several steps including particulate removal, followed by flue gas humidification. Next the stream is cooled to below 100°C, at which point the ammonia is injected. Then the electron beam hits the stream and creates the nitrate compounds and then the sulfate compounds. The solid fertilizer particles are recovered via a baghouse.

25.3.11   Wet absorption

Also called *wet scrubbing,* this process is a dual control process for SO_x and NO_x. Because NO has a very low solubility in water, this process is effective only if the NO stream in the flue gas is first converted to NO_2.

Counce and Perona (1983) modeled the NO_x, HNO_x, H_2O system from 0.01- to 0.10-atm partial pressure.

Gas phase

$$NO + \tfrac{1}{2}O_2 \rightarrow NO_2 \tag{25.10}$$

$$2NO_2 \rightleftharpoons N_2O_4 \tag{25.11}$$

$$NO + NO_2 \rightleftharpoons N_2O_3 \tag{25.12}$$

$$NO + NO_2 + H_2O \rightleftharpoons 2HNO_2 \tag{25.13}$$

Liquid phase

$$N_2O_3 + H_2O \rightarrow 2HNO_2 \tag{25.14}$$

$$N_2O_4 + H_2O \rightarrow HNO_2 + HNO_3 \tag{25.15}$$

$$2NO_2 + H_2O \rightarrow HNO_2 + HNO_3 \tag{25.16}$$

$$2HNO_2 \rightleftharpoons HNO_3 + H_2O + 2NO \tag{25.17}$$

These studies were conducted in NO_x-rich environments. In order to achieve the efficiency that they did, NO must be oxidized to NO_2 and then absorbed in a caustic scrubbing solution. Or this can be reversed, such that the NO is first absorbed in the caustic solution and then converted to NO_2 in the liquid phase.

Uchida, Kobayashi, and Kageyama (1983) modeled absorption of NO in $KMnO_4/NaOH$ and $Na_2SO_3/FeSO_4$. The formulas for absorption of NO with $KMnO_4/NaOH$ are as follows.

At high pH:

$$NO + MnO_4^- + 2OH^- \rightarrow NO_2^- + MnO_4^- + H_2O \tag{25.18}$$

At low or neutral pH:

$$NO + MnO_4^- \rightarrow NO_3^- + MnO_2 \qquad (25.19)$$

Solid MnO_2 is formed that reduces gas transfer rates by floating on top. Absorption with $Na_2SO_3/FeSO_4$ was found to use the following pathway:

$$FeSO_4 + NO \rightleftharpoons Fe(NO)SO_4 \qquad (25.20)$$

$$Fe(NO)SO_4 + 2Na_2SO_3 + 2H_2O \rightarrow Fe(OH)_3 + Na_2SO_4 \\ + NH(SO_3Na)_2 \qquad (25.21)$$

Typical flue gas NO concentrations were used (399, 900, and 1790 ppm). NO absorption was found to be about 1 to 10 mol/(s · cm²) for the $KMnO_4/NaOH$ solution and 1 to 4 mol/(s · cm²) for the $Na_2SO_3/FeSO_4$ system. Coupled with SO_x scrubbing this could be useful.

Low-NO_x burners, overfired air, and other current (1992) NO_x technologies are not likely to reduce emissions to the limits spelled out in CAAA. What is required now is postcombustion techniques such as selective catalytic reduction (SCR) and selective noncatalytic reduction (SNCR). SCR is very expensive, but when either of these processes is combined with other reduction operations, it can be made cost-effective.

25.4 Selective Catalytic Reduction

Selective catalytic reduction reduces NO_x to N_2. Nonselective catalytic reduction (NSCR) is also effective, but large amounts of oxygen are reduced by the process. The reducing gas is usually NH_3 or urea, and a titanium or vanadium oxide is used for the catalyst. For gas-fired furnaces, the catalyst is made in pellet form, whereas for coal- and oil-fired units, it is made of a honeycomb structure. The stoichiometric equations for SCR are (EPA, 1983)

$$4NO + 4NH_3 + O_2 \rightarrow 4N_2 + 6H_2O \qquad (25.22)$$

$$3NO_2 + 4NH_3 + O_2 \rightarrow 3N_2 + 6H_2O \qquad (25.23)$$

SCR systems operate optimally in a temperature range of 300 to 400°C (600 to 800°F). Vaporized ammonia is injected from the economizer which is the boiler feedwater preheater. Approximately 80 percent reduction can be achieved with SCR technology.

Of postcombustion NO_x reduction techniques, SCR is the most widely used. Japan has applied this technology to gas- and oil-fired boilers which account for more than 10,000 MW. Updating the coal-fired boilers (approximately 60 percent of the boilers) with SCR created difficulties including catalyst poisoning and shortened catalyst life. But the installations are now successful. East Germany and Europe have also updated their coal-fired boilers with SCR technology. More than 30,000 MW of power is now produced from coal-fired boilers utilizing SCR technology. The strong interest in the technology in Europe is due in part to regulations that have set NO_x emission limits as low as 50 ppmv.

A platinum-based catalyst is typically used in SCR technology and usually is carried on an alumina support. This alumina support comes in two major forms. These include a plate containing rectangular passages or a honeycomb shape. The cost of retrofitting is the main consideration in deciding on a particular catalyst. The properties of the catalyst must be weighed against the properties of and the impurities in the coal. Because it is unlikely that the plant supervision will change the type and grade of coal to fit the catalyst type, the catalyst must be fit to the coal type. If space is a consideration, the honeycomb catalyst contains more contact surface area per volume of catalyst. However, if the pressure drop of the flue gas is a concern, the plate type has lower pressure drop and may be preferred.

25.5 Selective Noncatalytic Reduction

Because of the requirements and demands of the catalyst, many processes have been developed which do not require a catalyst for activating the reaction. Instead, the reaction is run at higher temperatures. These processes are called SNCR (selective noncatalytic reduction) technology.

SNCR is the process by which ammonia or urea is injected into the flue gas stream at a temperature between 900 and 1000°C. Because of the high temperature, this process does not use a catalyst to initiate the reactions. Ammonia is injected at a rate of about 1:1 to 2:1 mol of NH_3 to mol of NO_x reduced. This will only yield approximately 40 to 60 percent reduction in NO_x emissions. In addition to the low-NO_x reduction rate, the process is hard to control thermally. Too hot and the NH_3 is oxidized to NO. If it gets too cold, then the ammonia does not all react and NH_3 is emitted with the NO_x.

Prior to the catalyst, NO_3 must be added to the flue gas in a quantity that is proportional to the gas's NO content. Gas distribution is very important because poor distribution of the NO_3 can result in improper NO_x reduction in the presence of a localized low of NH_3. On the other hand, an ammonia slip is produced if there is an excess of NH_3. In this case, not all the NH_3 present in the gas was reacted, and some was emitted to the atmosphere. Most applications have an ammonia slip of approximately 1 to 2 ppmv in the off-gas. A slip of approximately 10 ppmv is considered to be the maximum that the system can take.

Most successful applications, especially in Europe, use coal containing less than 1 percent sulfur and producing an alkaline ash. Some coal contaminants poison the catalyst and decrease its efficiency. These include, but are not limited to, arsenic, potassium, and a number of trace metals. It is also generally harmful to let the catalyst absorb water vapor during plant downtime. In this case, it is worthwhile to utilize an air dryer to dehumidify the air in the system.

Although the effect of sulfur on the catalyst has not been accurately ascertained, it is likely that coals containing moderate to high sulfur contents will not work effectively with the catalyst. To account for these grades of coal, it is necessary to use an additional amount of catalyst or at least set aside some space for adding more catalyst in the future.

Catalyst prices are fairly high and can range from $300 to $900 per cubic foot. Thus, a power plant that produces 500 MW can expect to pay $2 million to $3 million just for a catalyst that has an expected useful life of only 3 to 5 years. As cheaper, more effective catalysts continue to be developed, this technology will be come more affordable.

To minimize catalysts costs, it is often necessary to reduce as much of the NO_x in the gas stream as possible prior to the catalyst through such operations as low-NO_x burners, low excess air, and reburning. This requires less catalyst, puts less of a demand on it, and uses less ammonia.

Most of the SCR installations in Europe and Japan have been hot-side, high-dust catalyst systems. In these types of installations, the SCR reactor is installed between the exit of the economizer and the inlet to the air preheater. This places it ahead of any particulate collection equipment. SCR technology is being explored along with FGD (flue gas desulfurization), such as wet limestone slurry scrubbing. This is similar to the European systems, in a hot and dusty atmosphere, except the catalyst would be subjected to SO_2.

However, if the SCR unit were installed after the FGD, the catalyst would be protected from both sulfur and dust. Unfortunately, after the scrubbing process, the gas temperature is too low for the SCR catalyst to be active and effective. Thus, gas reheating is required after scrubbing. And although this is more costly, the catalyst charge is smaller, making the catalyst life much longer.

The problems associated with SCR and wet scrubbing have provided the U.S. power industry with greater incentive to develop "dry scrubbing" techniques including alkaline absorbents. This type of operation would enable the use of efficient SCR technology in conjunction with desulfurization and filter collection of particulates because the gas stream temperature would not have to be reduced greatly.

SNCR was invented and pioneered in the 1970s. The process reacts the NO_x with a nitrogen-based reagent (usually ammonia, NH_3, or urea) at high temperatures to produce nitrogen and water with no solids. The design parameters are

1. Flue gas temperature profile

2. NO_x baseline

3. Percentage of NO_x reduction target

4. Residence time

5. By-product emissions limit

Early SCNR processes introduced anhydrous ammonia to the gas stream at a ratio of 1 mol NH_3 to 1 mol NO_x at 1700 to 1900°F. On the plus side of this process, anhydrous ammonia is inexpensive. On the negative side, NH_3 is listed under section 301 of the CAAA as an "extremely hazardous" chemical requiring permits for storage in a system that requires proper temperature, pressure, and vapor controls as well as safety precautions such as showers, alarms, and escape plans.

According to SARA Title III, fugitive emissions of NH_3 are limited to 1 lb before a hazardous response team must be notified. Because of this rule, ammonia-based SCNR processes are prohibited in some states and local areas.

Under ideal conditions, the SCNR system can reduce NO_x emissions by 60 percent. Those conditions are as follows:

1. Flue gas temperature between 1700 and 1900°F

2. NO_x baseline greater than 200 ppm

3. Residence time greater than 250 ms

Unfortunately, this produces a large ammonia slip—the process by which ammonia escapes to the atmosphere prior to reacting with the NO_x. Both inefficient NH_3 distribution and reacting the ammonia below the above temperature range can increase ammonia slip. As much as 3 percent of the total flue gas is carrier gas needed for good distribution of the ammonia. But this extra gas promotes excess feed and inefficiency.

25.5.1   Thermal DeNOx

Among the oldest of the SCNR processes is Thermal DeNOx. Developed by Exxon, the Thermal DeNOx process injects ammonia into the gas stream at a fairly high temperature. To efficiently reduce the NO_x, the ammonia must be added in a very specific temperature window. If the process is run between 870°C (1600°F) and 1000°C (1800°F), then NO_x can be reduced by as much as 70 to 80 percent. Higher temperatures may be used, but NO_x removal is less complete. The ammonia does not completely react below 870°C unless H_2 is also injected or is already present in the gas stream.

Again, as discussed above with SCR processes, distribution of the gases is very important. The ammonia must be evenly distributed and proportioned throughout the stream to effectively treat the NO_x and prevent ammonia slip.

The petroleum industry is the major user of the Thermal DeNOx process. It has been used primarily on gas-fired boilers and on still heaters. Other applications include oil- and coal-fired boilers, glass furnaces, and some municipal solid waste incinerators that typically yield 40 to 60 percent NO_x removal.

25.5.2   Urea injection

The urea injection process was invented and patented by EPRI and marketed by Fuel Tech, Inc. under the name NO_xOUT. By injecting only urea into the gas stream, significant NO_x reduction is obtained between 925°C (1700°F) and 1040°C (1900°F). The temperature window can be enlarged with exhancers to 815°C (1500°F) to 1150°C (2100°F).

Prior to injection, the urea must be dissolved in water to create a 50 percent solution. The solution is then injected into the gas stream inside the appropriate temperature window. The quantity of urea must be proportioned accurately to the NO_x present in the gas stream. Proper distribution of the gas is very important. By varying the urea and enhancer concentrations, an optimal NO_x reduction can be obtained. Once the optimal concentrations are determined, the stream characteristics should change as little as possible. For this reason, urea injection is not practical for processes with large load variations. The process is also not recommended for gas turbines. This process can typically reduce NO_x from 35 to 70 percent in commercial installations.

25.5.3   Urea-methanol injection

Aurea-methanol injection process was patented by Emcotek. This two-stage process injects a urea solution into the gas stream followed by a methanol solution injection. The methanol injection reduces the ammonia slip and helps to prevent deposits from forming in the air preheater. At a temperature range of 1500 to 1900°F, this system can reduce NO_x by as much as 65 to 80 percent with an ammonia slip below 5 ppmv.

25.5.4 Other SNCR research

Injecting such chemicals as ammonium sulfate and cyanuric acid into the gas stream is currently being studied as a method of reducing NO_x.

25.6 Nitric Acid Manufacture and Organic Nitrations

The process by which nitric acid is manufactured, ammonia oxidation, releases a significant amount of NO. A catalytic NO reduction process has been developed to deal with this emission source. In the manufacturing process, ammonia gas and air are passed over a platinum-rhodium gauze catalyst at high temperature and pressure and are oxidized to NO.

The process releases some NO, which must be reoxidized. As the NO is reoxidized, the NO released again into the gas stream becomes progressively more dilute. A condition is eventually reached where it is not economical to reoxidize and absorb any more of the NO. This stream containing NO was initially simply released to the atmosphere. Now the stream must be treated.

The process developed to treat this gas stream is a catalytic process that reduces NO and NO_2 to nitrogen. Either hydrogen or methane is added to the stream prior to passing it over a nonselective reducing catalyst. The concentration of the NO_x stream is typically reduced to 300 to 400 ppmv after a single-stage catalyst pass. After another pass, the NO_x concentration can be reduced to as low as 100 to 200 ppmv.

However, NO cannot be reduced by the catalyst in the presence of O_2. Therefore, excess O_2 must be consumed by the fuel prior to NO_x reduction. This can often require significant amounts of fuel. In the past, to save fuel, manufacturers carried out the reaction only halfway and released a significant portion of the NO to the atmosphere. Because NO is colorless, no visible plume was produced. But the NO released will again be reoxidized to NO_2, and so this process has accomplished nothing. Additional problems can be encountered if considerable O_2 is present. Burning off the excess oxygen from the gas stream generates a significant amount of heat and raises the temperature of the off-gas. The temperature at this point may damage the catalyst.

25.7 Organic Nitrations

Many organic nitrate compounds are produced by the reaction of organic compounds with nitric acid. This process releases quantities of nitrogen dioxide. Recovery of the NO_2 by absorption in water or weak nitric acid in packed columns can be done, but the yield is often quite low because the NO_2 concentrations are very dilute and there is very little of a pressure driving force. Thus it is usually futile to attempt to absorb all the released NO_x. For this type of source, a catalytic reduction method is able to control the NO_2 emission with efficiencies found in fuel combustion sources.

25.8 Costs

Low-NO_x equipment is usually cost-effective and can reduce NO_x emissions by approximately 50 percent from those of coal-fired boilers with about $5 per kilowatt capital cost. Retrofit can vary quite a bit ($1 to $20 per kilowatt).

The capital cost for an SCR process is higher than that of low-NO_x equipment—$100 per kilowatt for new systems, and retrofits are more. Wet scrubbing to remove both NO_x and SO_x can cost $200 to $400 per kilowatt.

References

Benitez, J.: *Process Engineering and Design for Air Pollution Control,* PTR Prentice-Hall, Englewood Cliffs, N.J., 1993.

Breen, B. P., Bell, A. W., De Volo, N. B., Bagwell, F. A., and Rosenthal, K.: "Combustion Control for Elimination of Nitric Oxide Emissions from Fossil Fuel Power Plants," *13th Symposium (International) on Combustion,* The Combustion Institute, Pittsburgh, Pa., 1971, p. 391.

Cooper, C. D., Alley, F. C.: *Air Pollution Control—A Design Approach,* Waveland Press, Inc., Prospect Heights, Ill., 1990.

Counce, R. M., and Perona, J. J.: "Scrubbing of Gaseous Nitrogen Oxides in Packed Towers," *Journal of American Institute of Chemical Engineers,* 29(1):26–32, January 1983.

"Electron Guns Clean Up SO_2 and NO_x," *Chemical Week,* 133(21), Nov. 23, 1983.

Environmental Protection Agency: *Control Techniques for Nitrogen Oxide Emissions from Stationary Sources.* EPA-450/3-83-002. U.S. Environmental Protection Agency, Research Triangle Park, N.C., 1983.

Lewis, W. H.: *Nitrogen Oxides Removal,* Noyes Data Corporation, Park Ridge, N.J., 1975.

MacKinnon, D. J.: "Nitric Oxide Formation at High Temperatures," *Journal of the Air Pollution Control Association,* 24(3), March 1974.

Pickens, R.: "All SNCR Processes Are Not Created Equal," *Nalco Duel Tech,* Naperville, Ill.

Uchida, S., Kobayashi, T., and Kageyama, S.: "Absorption of Nitrogen Monoxide into Aqueous $KMnO_4$/NaOH and Na_2SO_3/$FeSO_4$ Solutions," *Industrial and Engineering Chemistry: Process Design and Development,* 22(2), 1983.

Yaverbaum, L. H.: *Nitrogen Oxides Control and Removal,* Noyes Data Corporation, Park Ridge, N.J., 1979.

Zeldovich, J.: "The Oxidation of Nitrogen in Combustions and Explosions," *Acta. Physiochim.,* 21(4), 1946.

Chapter 26

Control of SO$_2$ Emissions

Taylor H. Wilkerson

26.1   Introduction

Sulfur dioxide (SO$_2$) has long been an air pollution concern due to its role in acid rain development. Sulfur dioxide in the atmosphere reacts with the water in clouds to form sulfuric acid, which can harm wildlife and vegetation by lowering the pH of soils and upsetting ecosystems. Controlling SO$_2$ emissions is an important step toward eliminating the damage caused by acid rain, and, as an added benefit, some systems can create by-products which can be sold to offset a portion of the cost of the control system. (See Table 26.1.)

26.2   Regulations

As with most other emissions reductions, sulfur dioxide emissions reduction is primarily driven by government regulations. State and federal agencies are continually regulating sulfur dioxide more stringently. As a result, industry is required to develop and implement better control and reduction techniques to maintain compliance with the regulations.

Regulation of sulfur dioxide emissions began when sulfur dioxide was established as one of the five "criteria" pollutants—those which are considered to contribute most to air quality degradation. At the same time, a National Ambient Air Quality Standard (NAAQS) was developed for SO$_2$. The NAAQSs establish an ambient air quality level which defines acceptable ambient air for an area. If an area does not meet this level, sources in that area must take measures to

TABLE 26.1   Postcombustion SO_2 Controls for Coal Combustion Sources

Control technology	Process	Typical control efficiencies, %	Remarks
Wet scrubber	Lime/limestone	80–95+	Applicable to high-sulfur fuels, wet sludge product
	Sodium carbonate	80–98	1 to 125 MW (5 to 430 million Btu/h) typical application range, high reagent costs
	Magnesium oxide/hydroxide	80–95+	Can be regenerated
	Dual alkali	90–96	Uses lime to regenerate sodium-based scrubbing liquor
Spray drying	Calcium hydroxide	70–90	Applicable to low and slurry, vaporizes in medium-sulfur fuels, spray vessel produces dry product
Furnace injection	Dry calcium carbonate/hydrate injection in upper furnace cavity	25–50	Commercialized in Europe, several U.S. demonstration projects under way
Duct injection	Dry sorbent injection into duct, sometimes combined with water spray	25–50+	Several R&D and demonstration projects under way, not yet commercially avail-

reduce current and future emissions to bring the area below the defined level.

Original control methods for meeting NAAQSs usually involved a source's increasing its stack height to disperse the SO_2 higher in the atmosphere; however, raising the stack merely transported the problem out of the source's NAAQS defined area, but it did not actually reduce emissions. EPA now prefers flue gas desulfurization as a more reasonable means of reducing sulfur emissions. This has prompted most states to put specific SO_2 emissions limits on SO_2 sources.

The Clean Air Act Amendments of 1990 require coal-fired plants to have emissions below 1.2 lb per million Btu for phase 2 of the Clean Air Act. In addition, many new sources must comply with New Source Performance Standards (NSPS) regulations. These regulations are specific to various source categories and require either specific control equipment or minimum levels of reduction from the source. These regulations are based on the reasonable available control technology (RACT) for the source category.

26.2.1 History of sulfur dioxide control

Large coal-fired electric generator boilers have traditionally been the largest producers of SO$_2$; therefore, the majority of SO$_2$ control technology has been built around the power industry. The British did some of the earliest work on flue gas desulfurization in the 1930s to reduce fuel burning emissions.

In fact, the first recorded SO$_2$ removal system was installed on the Battersea A power plant in England. This system used water from the Thames River, which is slightly caustic, as a medium for scrubbing SO$_2$ from the exhaust gas. The wastewater from this system was returned to the Thames with no pretreatment. The system was taken off-line after a short time because the wastewater degenerated the quality of the Thames. Another disadvantage of the Battersea A plant was that the flue gas came out of the scrubber saturated with water and cold. This caused the gas to lack buoyancy and to coat the surrounding area in an acidic film rather than rise to the upper atmosphere and disperse.

Despite its shortcomings, the Battersea A SO$_2$ control system was effective at reducing sulfur dioxide emissions from the combustion of coal. As a result, the use of an alkali solution to remove sulfur dioxide from exhaust gas streams was proved effective and has been the primary focus of most sulfur reduction technologies.

Around the 1930s, interest also grew in reducing SO$_2$ from smeltering operations and in using flue gas desulfurization systems which produce a marketable product. Most early experimentation in flue gas sulfur removal was performed by British and U.S. scientists.

26.3 Sources of SO$_2$ Emissions

Sulfur dioxide emissions originate from several sources including combustion of sulfur-bearing fuels, sulfuric acid plants, oleum manufacturing plants, sludge acid purification processes, sulfur recovery plants, nonferrous smelters, and pulp and paper manufacturing. Each one of these processes is unique in the volume and concentration of sulfur dioxide emitted; therefore, no one method of SO$_2$ reduction will work for all sources of emissions. (See Table 26.2.)

The two main types of SO$_2$ sources are fuel combustion and various other sources. Fuel-burning sources emit SO$_2$ in relatively low concentrations but at high volume, while other sources emit higher concentrations but generally lower exhaust volumes than those found in fuel burning. Typical uncontrolled coal boiler flue gas consists of

- N$_2$
- CO$_2$, 10 to 15 percent

TABLE 26.2   Sources of SO_2 in the United States (1978)

Source	SO_2 emissions, 10^6 tons/yr	
Electricity generation		20.48
Coal	18.32	
Oil	2.07	
Other	0.09	
Industrial combustion		4.00
Coal	2.50	
Oil	1.48	
Other	0.02	
Other combustion		0.96
Industrial processes		6.23
Primary metals	3.82	
Petroleum	0.97	
Chemical manufacturing	0.91	
Other	0.53	
Transportation		0.71
Other		0.44
Grand total		32.82

- O_2, 3 to 10 percent
- H_2O, 5 to 10 percent
- SO_2, 500 to 5000 ppm

Process SO_2 sources usually have higher concentrations (up to 2 percent) than fuel-burning operations. Since SO_2 generated from fuel burning is present in low concentrations, it is generally considered more difficult to obtain high removal efficiencies. Because of this and the fact that fuel burning creates roughly two-thirds of the SO_2 emissions in the United States, most SO_2 reduction technologies have evolved around the fuel-burning industries, especially coal-fired power generation boilers.

For SO_2 created during fuel burning, emissions can be reduced in several ways. When combusted, the sulfur in the fuel is oxidized to SO_2 by the following equation:

$$S + O_2 \rightarrow SO_2$$

For this reason, one of the most obvious methods of reducing SO_2 emissions is to burn fuels with lower sulfur contents.

Switching to low-sulfur fuels is not always as easy as it sounds, however. Most high-sulfur coals are mined east of the Mississippi. Switching to lower-sulfur coals is not always easy. Boilers on the east coast are designed for the bituminous coals mined in those areas;

western subbituminous coals have lower heating values, lower ash fusion temperatures, higher moisture, and higher ash content. All these can combine to change a boiler's heat absorption profile, can increase slag formation, and can foul a boiler's transfer surfaces. Switching to a different type of coal can reduce emissions, but care should be taken to ensure that other aspects of the operation will not be adversely affected.

As an alternative to switching to a lower-sulfur coal, sulfur can be removed from coal and other fossil fuels. Desulfurization of a fuel often involves a hydrogen treatment which removes sulfur by forming hydrogen sulfide

$$H_2 + XS \rightarrow H_2S + X$$

where X is the chemical formula for the material being desulfurized.

In coals, which are usually high in sulfur, roughly 50 percent of the sulfur is in an inorganic or pyretic state. This sulfur can be removed by a crushing and washing procedure. The organically occurring sulfur is much more difficult to remove and often cannot be removed economically. However, coal can be converted to natural gas, leaving the sulfur behind as a by-product.

Tests have been performed with advanced horizontal cyclone combustors which replace the existing combustion chambers in a boiler. These cyclonic combustors remove more than 90 percent of the sulfur dioxide from the combustion gases and 70 to 75 percent of the fly ash produced in the combustor, and they reduce NO$_x$ emissions to below 200 ppm at 3 percent O$_2$. The capital and operating costs of systems testing this method are similar to those of oil and natural gas units but use lower-cost coal and other solid waste fuels.

The cyclonic combustor is constructed with an air-cooled ceramic lining. Typically, oil and gas are used to preheat the boiler during start-up. Dry pulverized coal, air, and powdered limestone are injected tangentially toward the wall of the combustion chamber through several tubes in the closed-end region of the chamber, to create the cyclonic action which produces a swirling cyclonic flame and helps to retain total solid particulates (TSP) in the combustor. Secondary air is fed into the combustor after passing through the ceramic liner, to adjust the stochiometry and to cool the ceramic liner. The slag is kept liquid so that it can flow freely out of the slag tap at the bottom of the combustion chamber.

Sulfur dioxide emissions were reduced by more than 80 percent by injecting calcium sorbent into the stack. At most, 33 percent of the coal sulfur was retained in the dry ash removed from the combustor, and 11 percent was retained in the slag. The metals and other elements in the slag were all in low enough concentrations to allow disposal in a regular landfill. Other tests of similar systems on different boilers achieved sulfur dioxide reductions up to 95 percent. The last option is to remove the sulfur dioxide from the flue gas after the combustion source.

26.4 SO_2 Control Techniques

The majority of modern SO_2 control systems are based on flue gas scrubbing due to the historical success of this method. Scrubbing is also the most widely commercialized control method, making it the most proven and obtainable.

Sulfur dioxide can be removed with a simple scrubber using water as a medium; however, this method is not practical, as the volume of water required to get the desired removal efficiency is tremendous. To increase the efficiency of the scrubbing system, an alkali agent is usually added to the water to provide increased SO_2 solubility and thus more efficient removal.

In scrubbing systems, many different scrubber types are employed, including spray towers, venturi scrubbers, turbulent beds of hollow or solid spheres, modified sieve trays, and open egg crate packing. The type of scrubber which works best is dependent on several factors such as airflow, sulfur concentration, and waste characteristics desired.

Most modern sulfur reduction technologies can be broken down into throwaway or regenerative systems, and then further broken down into wet or dry processes. The throwaway systems consist of an absorption medium which is both unmarketable and nonrecyclable after it is spent, and, therefore, must be disposed of. The wastes from these systems are usually either pumped into a pond or landfilled. Intentional air oxidation can also be used to produce gypsum for sale or to obtain better disposal properties. See Fig. 26.1.

A great deal of interest has been shown lately in developing regenerative systems. Regenerative systems use the spent scrubber liquor to produce sulfur, sulfuric acid, or other marketable products. These systems are becoming attractive because of the reduced waste and the potential for the end product to offset a portion of the operating costs.

The most common method of removal in wet systems is the absorption of the SO_2 using an alkali solution or slurry in an absorption column. The spent scrubber liquor from these systems either is regenerated or produces $CaSO_3$ or gypsum, which must be disposed of in a landfill or pond. The waste is also usually considered hazardous due to the toxic impurities in the fly ash. Dry systems have been experimented with which use a powdered lime or limestone either mixed

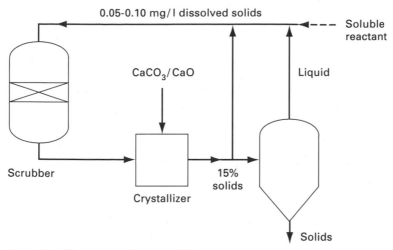

Figure 26.1 Throwaway slurry scrubbing.

with the coal during burning or in an absorption column followed by a particulate control device. The biggest hindrance to dry systems is the capital cost of handling fixed or moving beds of solids. Some of the dry systems which are currently being experimented with include carbon absorption, catalytic oxidation, and CuO absorption with hydrogen regeneration. Injection of limestone into the boiler was tested but rejected because of recurring plugging problems in the boiler.

Many traditional throwaway systems which produce CaSO$_3$ waste have started to use forced aeration to upgrade the waste to gypsum (CaSO$_4$). This conversion is performed for several reasons. Most importantly, gypsum is easier to dewater than CaSO$_3$, and therefore gypsum is landfill quality, making disposal much simpler and economical. Gypsum can also be used as a fertilizer which returns sulfur and calcium to the soil. Attempts have been made to use the gypsum from an SO$_2$ scrubber to produce wallboard and other building products; however, gypsum of high enough quality for this purpose is difficult to obtain due to impurities in the scrubber waste. Gypsum can also be used as an ingredient in concrete.

Despite all the innovations in sulfur removal technology, slurry scrubbing is still the dominant technology of choice. This is because it has been found to be the most effective and the most reliable. Also, most early SO$_2$ reduction technologies were scrubber-based, and therefore slurry scrubbing is more widely researched and the most established. Most scrubber systems employ a crystallizer to precipitate out the CaSO$_3$ and CaSO$_4$ or other waste products from the scrubber bottom's stream. The crystallizer is usually a simple stirred tank or several tanks in series. Clarifiers, filters, centrifuges, and settling ponds are also used for liquid-solid separation.

Most commercial systems have problems with solids forming on scrubber internals and within the mist eliminator. Steps can be taken to avoid scaling in the collection system, many of which involve simple process monitoring. The combination of crystallizer residence time and $CaSO_4$ solids concentration must be sufficient to reduce the gypsum concentration in the scrubber feed to below the saturation level. The liquid circulation rate must be high enough to avoid high $CaSO_4$ concentrations, especially to keep them below saturation, at scrubber exit. $CaSO_3$ scaling is avoided by limiting the amount of unreacted $CaCO_3$ of CaO solid that is returned to the scrubber.

A scrubbing medium can also be used which creates a chemical reaction with SO_2 to further increase the solubility. When the sulfur undergoes a reaction to convert it to a new substance, more "room" is created for sulfur dioxide in the scrubbing solution before it reaches saturation level. Thus, the SO_2 is absorbed more quickly, as the reaction does not get slowed by its approaching the saturation level. This type of operation can be seen in the magnesium oxide and Bechtel's seawater scrubbing systems.

26.4.1 Throwaway systems

By far, the most common SO_2 removal technologies are the throwaway processes, which yield a scrubber waste which cannot be reused and must be disposed of. The most widely used of these processes are the limestone and lime processes. These both offer the advantages of lower capital costs and a greater proliferation of the technology, making them more dependable and reliable than other available technologies. The obvious disadvantage of these technologies is the cost of disposing of the wet sludge, which must be either transported off site or neutralized before release to the environment, usually in a landfill or pond.

One of the most abundant scrubbing media to use is seawater, which is fairly efficient. However, the effluent from a seawater scrubbing system typically has low pH and must be aerated to pH 6 or 7 before disposal. The aeration process can cause more SO_2 to be released (Fig. 26.4). In addition, seawater scrubbing requires great deals of water to achieve the desired reductions.

Solid waste disposal.   A typical 1000-MW power plant burning high-sulfur coal will produce around 760,000 tons of wet $CaSO_3$/$CaSO_4$ sludge per year plus an additional 340,000 tons/yr of fly ash waste. This material is generally 50 percent water and thixotropic, so it cannot be landfilled. Therefore, the waste is usually disposed of in a pond. This amount of sludge requires a pond which is approximately 20 ft deep and covers 42 acres. The pond must be clay-lined to prevent hazardous elements of the waste from entering the groundwater. Fixation of the sludge with pozzolanic reactions with lime and fly ash can be used to produce landfill-suitable waste. The CaSO$_3$ can also be oxidized to form gypsum which is more easily dewatered for disposal.

Limestone/lime scrubbing. One of the most common and the most proven SO$_2$ removal technologies is limestone scrubbing. In this process, a limestone slurry contacts the flue gas in a spray tower, producing calcium sulfite (CaSO$_3$) and calcium sulfate (CaSO$_4$) according to the following equations:

$$CaCO_{3\,(s)} + H_2O + 2SO_2 \rightarrow Ca^{2+} + 2HSO_3^- + CO_{2\,(g)} \qquad (26.1)$$

$$CaCO_3 + 2HSO_3^- + Ca^{2+} \rightarrow 2CaSO_3 + CO_2 + H_2O \qquad (26.2)$$

$$CaSO_3 + \tfrac{1}{2}O_2 \rightarrow CaSO_4 \qquad (26.3)$$

This system is widely used because the absorbent is abundant and inexpensive. See Fig. 26.2.

Lime scrubbing systems are very similar to limestone systems in many respects. The main advantage of using lime (CaO) is that lime is more reactive than limestone, as shown by these equations:

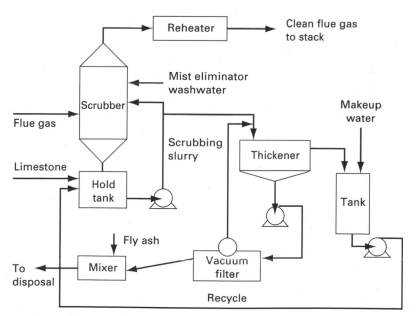

Figure 26.2 Limestone-based system.

$$CaO + H_2O \rightarrow Ca(OH)_2 \qquad (26.4)$$

$$SO_2 + H_2O \rightarrow H_2SO_3 \qquad (26.5)$$

$$H_2SO_3 + Ca(OH)_2 \rightarrow CaSO_3 \cdot 2H_2O \qquad (26.6)$$

$$CaSO_3 \cdot 2H_2O + \tfrac{1}{2}O_2 \rightarrow CaSO_4 \cdot 2H_2O \qquad (26.7)$$

The higher reactivity of lime makes it more efficient at removing SO_2; however, lime is also more expensive than limestone. Lime contains less impurities than limestone and dissolves in a slurry better.

The main operating parameter of lime/limestone systems is the scrubber liquor pH. For limestone systems, the optimal operating pH ranges from 5.8 and 6.2. In lime systems the optimal pH is approximately 8.0.

The drawbacks of a limestone system are that scaling due to buildup of $CaSO_4$ in the slurry can occur inside the scrubber and demister. Equipment plugging from excessive $CaSO_3$, which can precipitate as a soft, leafy solid, can occur as well. Scaling of $CaSO_3$ hemihydrate is rare because it is pH-dependent. In the absence of limestone, $CaSO_3$ will dissolve in the scrubber:

$$CaSO_3 + SO_2 + \tfrac{1}{2}H_2O \rightarrow Ca^{2+} + 2HSO_3^-$$

and precipitate when $CaCO_3$ or CaO is added to the crystallizer:

$$Ca^{2+} + 2HSO_3^- + CaCO_3 \rightarrow 2CaSO_3 + CO_2 + H_2O$$

With an ideal amount of unreacted limestone, $CaSO_3$ neither dissolves nor crystallizes in the scrubber, but $CaCO_3$ dissolves with the stochiometry:

$$CaCO_3 + 2SO_2 + H_2O \rightarrow Ca^{2+} + 2HSO_3^- + CO_2$$

With excessive $CaCO_3$ in the scrubber, $CaSO_3$ precipitates in the scrubber:

$$CaCO_3 + SO_2 \rightarrow CaSO_3 + CO_2$$

So, even though increased limestone reactivity enhances SO_2 removal, it must be limited to prevent $CaSO_3$ scaling in the scrubber and mist eliminator. Usually limestone usage should be greater than 75 percent. Also, the caustic atmosphere inside the system can accelerate corrosion with various components causing maintenance costs to rise.

One of the biggest operating interferences with lime and limestone systems is the formation of gypsum ($CaSO_4$), which forms a hard, stubborn scale inside the scrubbing systems and can lead to excessive downtime for cleaning and maintenance. The solubility of $CaSO_3$ tends to increase as pH decreases, meaning that the transformation of $CaSO_3$ to $CaSO_4$ increases as pH decreases. Limestone operations operating at a pH of 6.0 can usually avoid excessive gypsum formation. However, at a pH above 8.0, lime systems start to experience "soft plugging," in which formation of large, leafy masses of $CaSO_3$ inside the scrubber can plug openings and interrupt system operation.

Several methods can be used to prevent scaling in lime/limestone systems.

1. *Coprecipitation.* Removal of calcium sulfate as part of the calcium sulfite/sulfate solution. When the system is operated with maximum oxidation in slurry circuit of about 16 percent, liquor is subsaturated with calcium sulfate and hard scaling does not occur. More oxidation causes higher calcium sulfate, and it can be supersaturated and scaling can occur.

2. *Desupersaturation.* Removal of calcium sulfate with calcium sulfate seed crystals. Seed crystals control calcium sulfate in a closed-loop supersaturated system. Sulfate is removed as gypsum.

3. *Magnesium additives.* An increase in magnesium ions causes an increase in the liquid-phase alkalinity of scrubbing slurry and increases the amounts of sulfate and sulfite that the liquid can hold before saturation. This causes subsaturated scrubbing liquor with higher SO_2 removal.

4. *Other additives.* Organic acids are used to aid the mass transfer by buffering the pH at the liquid interface and act as a base because they are weaker than the sulfurous acid. Acids used are carboxylic, benzoic, and adiptic.

Several other methods to improve efficiency include the use of dampers to adjust the flow for optimal performance, cooling gas to its adiabatic saturation temperature before the scrubber, and using presaturators and quenchers which have the added advantage of reducing corrosion by taking out particulate before the scrubber.

Scrubber, hold tank, and crystallizer sizes must be adequate to provide system reliability. Small scrubbers have entrainment problems. Small hold tanks can result in $CaSO_4$ and/or $CaSO_3$ deposition and scaling in the scrubber.

The mist eliminator in a limestone or lime scrubbing system is usually a chevron blade rather than mesh pad type, to reduce the buildup of solids on the internals. The mist eliminator is also almost continuously washed down with water to avoid solids buildup. This wash-down can be a significant fraction of the total water used in the system.

Other considerations which must be addressed during the design of a lime or limestone system include piping sizes and construction materials. Pipes must be sized such that the flow through the pipe is rapid enough to avoid settling of solids from the slurry inside the piping. The scrubber internals and most other sections of the system must be constructed with organic or corrosion-resistant liners to prevent excessive corrosion from the caustic slurry.

Also, the flue gas leaves the scrubber at a low temperature and saturated with water vapor. Because of this, the exhaust gas must be reheated to preserve buoyancy and to prevent acidic condensation from building up in the stack; otherwise, the surrounding area can be coated in a caustic film, as with the Battersea A plant, and more rapid corrosion of the stack can occur. This reheating, of course, reduces the thermal efficiency of the system.

Dual alkali process.   The dual alkali processes (Fig. 26.3) eliminates many of the scaling problems associated with limestone and lime

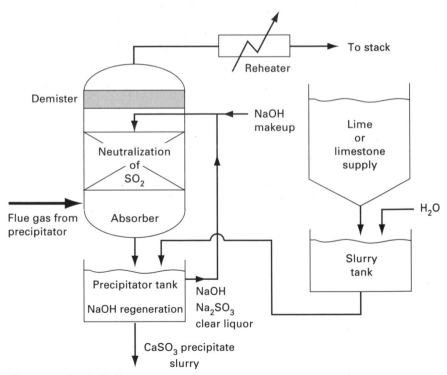

Figure 26.3   Dual alkali process.

scrubbing systems by eliminating calcium in the scrubber. The actual scrubber liquor is an alkali agent, usually sodium sulfite, which absorbs SO_2. This slurry is then regenerated by mixing it with a lime slurry, which joins with the SO_2 to form a solid that settles out in a crystallizer.

If less than 15 to 20 percent of the SO_2 is oxidized in the scrubber, $CaSO_4$ is precipitated with the $CaSO_3$. Otherwise, sulfate must be removed as $NaSO_4$ solids. The $CaSO_x$ solids are separated by clarification along with filtration or centrifugation. The filter cake is washed with makeup water to reclaim some of the sodium to reduce the need for Na_2CO_3 makeup.

The dual alkali system is more efficient than slurry and more reliable with less downtime for periodic maintenance. However, the liquid-solid separation is a more complex procedure, and more expensive lime, not limestone, must be used to achieve acceptable results.

Bechtel's seawater scrubbing process. Bechtel's seawater process alleviates some of the more common difficulties associated with conventional seawater scrubbing (Fig. 26.4). In this system, less than 2 percent of the cooling seawater from the plant condensers flows through the scrubber. The rest is used to dissolve gypsum created in the scrubbing system. An alkali is added to the seawater as a lime and limestone mix. Naturally occurring magnesium in the seawater ($MgCl_2$ and $MgSO_4$) reacts with the alkali mixture to produce magnesium hydroxide, which absorbs SO_2 from the flue gas according to the following equations:

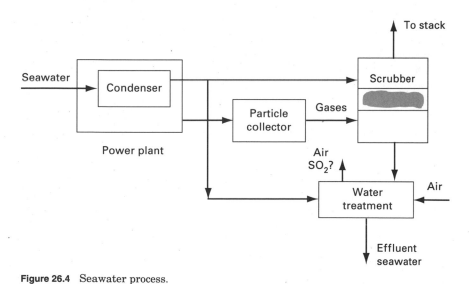

Figure 26.4 Seawater process.

$$SO_2 + H_2O \rightarrow H_2SO_3 \qquad (26.8)$$

$$H_2SO_3 + Mg(OH)_2 \rightarrow MgSO_3 + 2H_2O \qquad (26.9)$$

$$MgSO_3 + H_2SO_3 \rightarrow Mg(HSO_3)_2 \qquad (26.10)$$

$$Mg(HSO_3)_2 + Mg(OH)_2 \rightarrow 2MgSO_3 + 2H_2O \qquad (26.11)$$

$$MgSO_3 + \tfrac{1}{2}O_2 \rightarrow MgSO_4 \qquad (26.12)$$

In the regeneration process, $MgSO_4$ is reacted with calcium hydroxide to form gypsum and magnesium hydroxide:

$$MgSO_4 + Ca(OH)_2 \rightarrow Mg(OH)_2 + CaSO_4 \cdot 2H_2O \qquad (26.13)$$

$$MgCl_2 + Ca(OH)_2 \rightarrow Mg(OH)_2 + CaCl_2 \qquad (26.14)$$

This system has many advantages. Magnesium hydroxide and magnesium sulfite react quickly, reducing the recirculation rate of the slurry by a factor of approximately 4. Also, the liquor is well buffered, reducing the change in pH as the system operates. Because there is no calcium hydroxide formed, scaling is mostly eliminated.

The liquid effluent from this system is mostly seawater with low concentrations of dissolved gypsum, fly ash, and trace metals. This can be safely released back to the source without affecting the ecology.

Dowa process.   The Dowa process (Fig. 26.5) is a dual alkali system which uses basic aluminum sulfate, $Al_2(SO_4)_3$, to absorb sulfur dioxide and then uses a limestone slurry to regenerate the raw materials. SO_2 is absorbed in aqueous solution buffered with aluminum sulfate to pH 3 to 4. After passing through the scrubber, the solution is oxidized by air to convert dissolved SO_2 to sulfate. A slipstream from the scrubber loop is neutralized by lime or limestone to crystallize gypsum, which is then centrifuged.

The advantages of this system include the use of a clear medium instead of a slurry, a reduced limestone requirement compared to traditional limestone scrubbing, and production of gypsum instead of $CaSO_3$.

The disadvantages are (1) it is a more complex system; (2) the pH of the scrubbing solution is much lower, around 3 compared to 5 to 6 for limestone systems; and (3) the equipment needs to have higher resistivity to acidic materials.

Lime spray drying.   Lime spray drying (Fig. 26.6) is a semidry process, meaning that the scrubbing medium is wet, but it evaporates and forms a solid when it enters the scrubber and reacts with the SO_2. The flue gas and lime mix in a spray dryer, where the lime

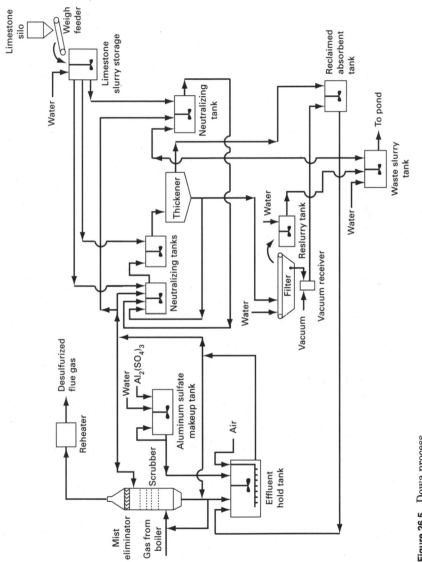

Figure 26.5 Dowa process.

26.15

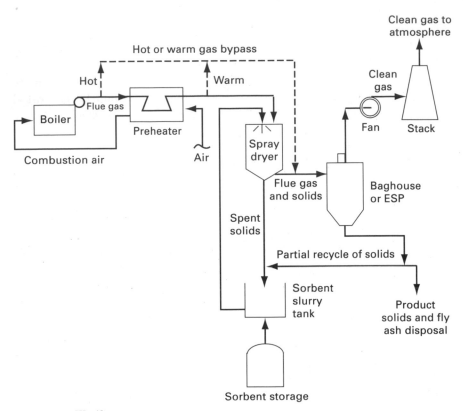

Figure 26.6 Wet/dry process.

reacts with SO_2 in the exhaust gas much the same way as it reacts with a lime slurry in wet collection systems to form a solid, which is collected along with the fly ash and other particulate contaminates in a fabric filter. The capital, maintenance, energy, and operating costs for this type of system are generally lower than those for conventional limestone scrubbing. However, if the flue gas approaches saturation temperature, the particulates can form a pastelike substance which will collect on the fabric filters and blind them. Also, scaling can occur inside the spray dryer. Slurrying the lime with fly ash can enhance the SO_2 reaction and has been shown to do so in tests.

Dry scrubbing. Dry scrubbing systems use a dry sorbent injected into the gas stream to remove SO_2. The sorbent remains solid as it collects SO_2 and then is collected in a particulate control device along with all other particulates.

In the most common dry scrubbing system, a $Ca(OH)_2$ slurry or an Na_2CO_3 solution is sprayed into the hot flue gas in a dry scrubbing chamber. The resulting dry solids along with the fly ash are removed by either a fabric filter or an electrostatic precipitator. One advantage of this type of system is that SO$_2$ is absorbed during the scrubbing stage and is further removed when the air passes through the sorbent collected in the particulate control device. Based on systems run to date, good SO$_2$ removal requires the flue gas to be nearly saturated with H$_2$O. Typically 20 to 100 percent excess Na$_2$CO$_3$ is used.

In a dry scrubber, the gas/liquid contactor is usually a large version of a conventional spray dryer. The liquid scrubbing solution is atomized by high-speed rotary atomizers or two fluid nozzles. Good liquid-gas distribution prevents the solution from reaching the vessel walls.

This type of system has many advantages over conventional wet scrubbing. Dry systems do not require the water load that wet systems do, and the waste does not need to be dewatered before disposal. The control system equipment can be constructed of mild steel instead of corrosion-resistant materials because no wet caustic comes in contact with the system. Waste solids are powder instead of sludge, allowing for much easier disposal and lowered capital costs because not as much equipment is needed for water handling.

However, dry scrubbing does not come without some disadvantages. Dry scrubbing systems must use lime instead of limestone, which is more expensive and can only provide 70 to 90 percent removal. In addition, the technology is immature and has not been proved in very many full-scale facilities; therefore, the true costs and effectiveness of this type of control cannot be known until it is actually installed.

Because much of the SO$_2$ reduction does take place as exhaust gas passes through the particulate control device, these systems are most attractive when used with fabric filtration control because this allows greater contact between the exhaust and the already captured particulates.

Nahcolite or Trona injection process.   The Nahcolite or Trona injection process is a true dry scrubbing operation using either Nahcolite or Trona as a sorption medium. Nahcolite is naturally occurring in NaHCO$_3$, and Trona is naturally occurring in Na$_2$CO$_3$.

The operation of these systems is very similar to the dry scrubbing process. The pulverized reagent is injected into the flue gas where dry sorption of SO$_2$ occurs. Further sorption occurs as the exhaust gas flows through the reagent-laden fabric filters.

Like the dry scrubbing process, this process has the advantage of lower capital and maintenance costs; however, the reagent costs are currently quite high due to the low demand for the reagents, which also creates a lack of interest in mining and refining the materials.

There are also possible disposal problems associated with the leaching of the soluble sodium salts from the system's wastes. There is limited full-scale use of this process, but it does show some good potential for future use.

Chiyoda Thoroughbred-121.   Southern Company in Atlanta, Georgia, has developed the Chiyoda Thoroughbred-121 (CT-121) scrubbing system to control SO_2 emissions. Tests of this system have found it to remove up to 98 percent of SO_2 and more than 99 percent of ash particulate. As an added advantage, the system produces gypsum as a marketable product.

In many ways, the basic principle of the CT-121 system is similar to that of conventional wet scrubbers, with the main exception of the oxidation of the scrubber wastes in the scrubber itself, reducing a great deal of extraneous refining equipment. The exhaust gas is first sent through an electrostatic precipitator to remove fly ash before being piped to a fiber-reinforced plastic (FRP) vessel, known as a jet bubbling reactor, which can hold up to 140,000 gal. FRP was chosen as a construction material because it is less expensive than rubber-lined steel or specialty-alloy vessels. The jet bubbling reactor is filled with a slurry of ground limestone and water. The gas is piped into the solution through 1000 spargers or pipes to form calcium sulfite. Air is also pumped into the solution from the bottom of the reactor, forming bubbles which oxidate the slurry and form gypsum. An agitator aids the reaction by creating turbulence and allowing greater residence time in the reactor. The slurry is piped to a dyked area where the gypsum solids settle, allowing clear water to flow to the retention pond. The water from the retention pond is recycled to the system. Virtually all the water from this system is reused, eliminating a wastewater stream. See Fig. 26.7.

The CT-121 scrubber can be adapted to almost any coal-fired boiler, as it is not greatly affected by the type of coal being burned or its sulfur content or the grade of limestone used in the slurry.

Advanced flue gas desulfurization.   Air Products and Chemicals in Allentown, Pennsylvania, designed the advanced flue gas desulfurization (AFGD) system. This system was designed to offer a low-cost alternative to conventional scrubbing.

The system is similar to the CR-121 jet bubbler because it combines a prequencher, absorber, and sludge oxidizer into a single vessel, but with some key differences, such as vessel size, sparging equipment, and gypsum processing. The vessel is a 1,000,000-gal vessel made with carbon-steel lining. The system uses an air rotary sparger (ARS) to oxidize and agitate the gypsum. The ARS is a pinwheel-shaped sparger with holes at the bottom. When the ARS is rotated, air is

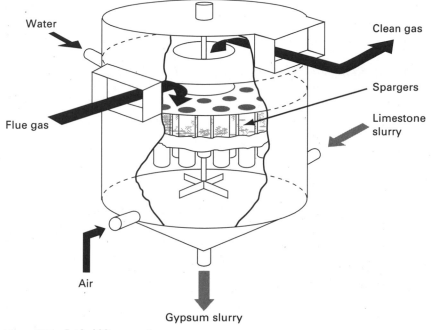

Figure 26.7   Jet bubbling reactor.

introduced into the sparging nozzles simultaneously. Very small bubbles are formed by the vacuum created from the arms moving through the slurry (cavitation).

Gypsum settles in the bottom of the reactor vessel and is pumped to a centrifuge to remove water. The water is recycled to the reactor vessel, and the remaining water is sent to an on-site wastewater treatment plant.

The wastewater can also be evaporated by injecting it into the flue gas upstream of the existing electrostatic precipitator. The electrostatic precipitator collects water impurities and eliminates the need for a wastewater treatment facility. During a 3-year test, SO_2 removal averaged 94.71 percent with the maximum removal efficiency exceeding 98 percent. The test facility, a 528-MW power plant, produces an average of 7 tons/h of gypsum with an average purity of 97.56 percent.

26.4.2   Regenerative systems

Regenerative systems have piqued great interest recently because they have the potential not only to reduce sulfur emissions, but also to create a by-product which can be sold to offset a portion of the system operating costs and to reduce waste. Most of these systems are still in the experimental stages, but they show great promise. These systems are generally designed to produce elemental sulfur, sulfuric acid, or liquid SO_2. In most cases, sulfur is the preferred product due to its easy storage and transportation parameters as well as the fact that, should there be no current market, it is easily disposed of. At this time, liquid-SO_2 markets are fairly limited, making it an unattractive product.

Currently, the energy required and the complexity of the processing usually make regenerative systems more expensive than throwaway systems. Thus, there are not many regenerable processes in commercial operation at this time; but as the technology advances and the demand for systems with less waste increases, systems should be developed which offer an economically attractive alternative to throwaway systems without sacrificing removal efficiency.

Several processes have also been developed which can yield elemental sulfur as an end product. These include citric acid, phosphate, and carbonate processes.

Wellman-Lord process. Of the regenerative processes, only the Wellman-Lord process is of note for practical purposes, as it is the only regenerative process currently in use in the United States. Therefore, it is the only system with good data on its operating parameters and removal efficiencies as well as the only one which has been field-proven. See Fig. 26.8 and Table 26.3.

The Wellman-Lord process consists of five basic steps. First the flue gas goes through a flue gas pretreatment system, which usually is a venturi prescrubber with a water medium. This stage removes residual particulate, hydrochloric acid, and SO_3 which can interfere. The first stage also cools and humidifies the exhaust stream which improves scrubbing efficiency.

The second stage is the primary SO_2 scrubber. This stage usually consists of a spray tower arrangement using sodium sulfite as a scrubbing medium and forming sodium bisulfite:

$$Na_2SO_3 + SO_2 + H_2O \rightarrow 2NaHSO_3$$

Some of the sulfite is oxidized inside the scrubber to sulfate, which reacts with SO_3 in the flue gas to form aqueous sulfate according to the following equations:

$$Na_2SO_3 + \tfrac{1}{2}O_2 \rightarrow Na_2SO_4 \qquad (26.15)$$

$$2Na_2SO_3 + SO_3 + H_2O \rightarrow Na_2SO_4 + 2NaHSO_3 \qquad (26.16)$$

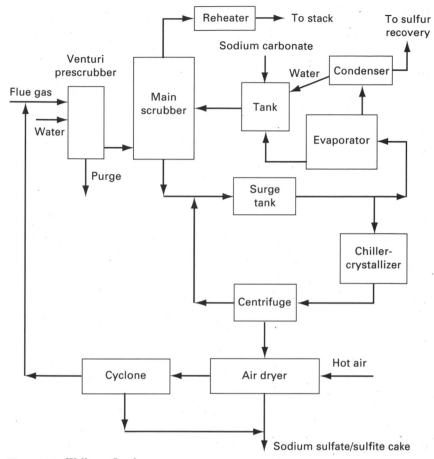

Figure 26.8 Wellman-Lord process.

The sodium sulfate is removed because it hinders the further reduction of SO$_2$ in the exhaust stream.

The bottom's stream from the Wellman-Lord process is rich in bisulfite. To remove this, the purge from the scrubber is sent to a chiller-crystallizer where less soluble sodium sulfite crystals are formed. The crystallizer should be operated below 100°C to avoid an excessive disproportion of bisulfite to sulfate and thiosulfate. The remaining slurry is then centrifuged, and the resulting solids are air-dried and then discarded. The centrifugate is returned to the scrubber.

The remainder of the absorber bottom's stream is sent to a heated evaporator-crystallizer where SO$_2$ is removed and sodium sulfite crystals are formed according to the reaction

TABLE 26.3   Regenerable Scrubbing Processes

Generic classification	Process description
Thermal regeneration	
Wellman-Lord	Na_2SO_3 solution
	Evaporative crystallization
Steam stripping	Acid/base buffer or nonvolatile aldehyde solvent
MgO	$MgO/MgSO_3$ slurry
	Centrifuge, dry, and calcine
ZnO	ZnO/Na_2SO_3 slurry
	Centrifuge, dry, and calcine
Acid decomposition of bisulfate	
Stone and Webster/Ionics	NaOH solution
	Electrolytic production of $NaOH/H_2SO_4$
Ammonium bisulfate	$(NH_4)_2SO_3$ solution
	Thermal decomposition of $(NH_4)_2SO_4$ to NH_3 and NH_4HSO_4
H_2S regeneration to S	
Citrate	Na citrate buffer
Aquaclaus	Na phosphate buffer
NH_3-IFP	$(NH_4)_2SO_3$ solution
	Evaporative decomposition
	Aqueous reaction of H_2S with SO_2
CO regeneration	
Consol	K_2SO_3 solution
Sulfoxel	Reaction with CO
	Na_2SO_3 solution
	Reaction with CO
High-temperature reduction	
Kel-S	$CaCO_3$ slurry
	Reaction with coal to CaS
	Carbonation
Aqueous carbonate	Na_2CO_3 solution
	Reduction of molten Na_2SO_4 with coal to make Na_2S
	Carbonation

$$2NaHSO_3 \rightarrow Na_2SO_3 + SO_2 + H_2O$$

The water vapor is condensed and returned to the process. The regenerated sodium sulfite is returned to the process. The remaining component is concentrated SO_2 which can be further reduced to elemental sulfur or oxidized to produce sulfuric acid. Because a portion of the sodium is removed in the sodium sulfate purge, soda ash (Na_2CO_3) is added to the process to provide makeup sodium:

$$Na_2CO_3 + SO_2 \rightarrow Na_2SO_3 + CO_2$$

A typical makeup material rate for this process is 1 mol of sodium for every 42 mol of SO$_2$ removed.

Magnesium oxide process.   The magnesium oxide process (Fig. 26.9) is very similar to Bechtel's seawater process in the absorption phase. The magnesium oxide process, however, uses a scrubbing medium of Mg(OH)$_2$ which produces MgSO$_3$ and MgSO$_4$ solids. These solids are centrifuged to separate the crystals.

The centrifuge cake is sent to a dryer where it is calcinated in the presence of coke or some other reducing agent to generate SO$_2$ and MgO. Anhydrous MgSO$_3$ is shipped to an H$_2$SO$_4$ plant where it is converted to MgO and SO$_2$ by direct flame heating in a rotary or fluidized-bed calcinator at 1500 to 2000°F. The MgO solids are shipped back to the scrubber where they are regenerated.

$$MgSO_3 \rightarrow MgO + SO_2 \qquad (26.17)$$

$$MgSO_4 + \tfrac{1}{2}C \rightarrow MgO + SO_2 + \tfrac{1}{2}CO_2 \qquad (26.18)$$

$$MgO + H_2O \rightarrow Mg(OH)_2 \qquad (26.19)$$

The main disadvantage found in this type of system is the difficulty of coordinating the efforts of the control system with the operation of

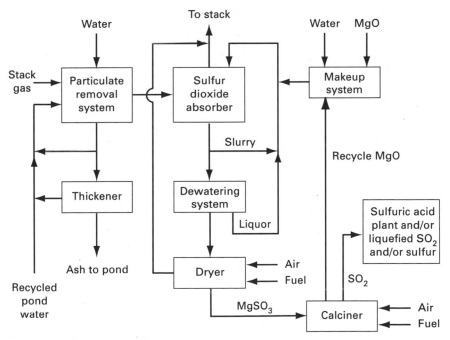

Figure 26.9 Magnesium oxide process.

the regeneration facility. Other than that, the process is fundamentally sound and economically attractive. Most units currently use a spray dryer as an absorber.

Shell flue gas treating system.   The Shell flue gas treating system (Fig. 26.10) offers the advantage of simultaneous removal of SO_2 and NO_x. The system consists of two or more reactors in parallel. The reactors contain copper oxide (CuO) supported on alumina (Al_2O_3). The system operates one reactor at a time. The SO_2 reacts with CuO to form copper sulfate ($CuSO_4$):

$$CuO + \tfrac{1}{2}O_2 + SO_2 \rightarrow CuSO_4$$

and $CuSO_4$ acts as a catalyst with ammonia to reduce NO:

$$4NO + 4NH_3 + O_2 \rightarrow 4N_2 + 6H_2O$$

When the reactor becomes saturated with $CuSO_4$, the exhaust gas is sent to a new reactor while the first is regenerated using hydrogen to reduce $CuSO_4$ to copper:

$$CuSO_4 + 2H_2 \rightarrow Cu + SO_2 + 2H_2O$$

The SO_2 is sufficiently concentrated to easily convert it to elemental sulfur or sulfuric acid. The copper is oxidized to CuO and recycled to the system.

Sulfuric acid plants.   Sulfuric acid plants fall into two main categories: conventional contact sulfuric acid plants and oleum (SO_3 in a sulfuric acid solution) plants. These processes generally have SO_2 emissions on the order of 1800 ppmv and can go up to 3500 to 4000 ppmv.

The traditional control for these plants has been a four-pass catalyst plant to convert the SO_2 to SO_3 and recycle it back to the process. However, this process is equilibrium-based, and a buildup of SO_3 in the exhaust gas can actually cause the reaction to start reversing and result in decreased control efficiency despite the addition of more control equipment. The solution to this is interpass absorption where SO_3 is removed between the third and fourth passes by a scrubber to allow the additional passes to be much more efficient. Through this method, exit gas can be reduced to around 400 ppmv, a figure which does rise to 450 to 500 ppmv during start-up, shutdown, and emergencies. To further reduce the SO_2 concentrations, sodium, calcium, or ammonium alkali scrubbing can be employed after the catalytic system. Ordinary SO_2 removal systems are typically not used on H_2SO_4 plants because the SO_3 from the catalytic system is used as a raw material

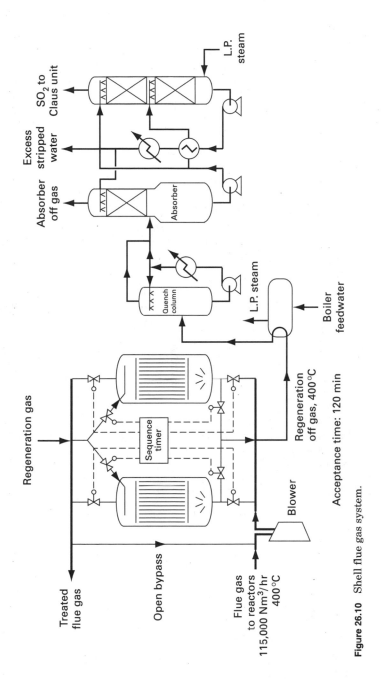

Regeneration gas

Treated flue gas

Open bypass

Flue gas to reactors 115,000 Nm³/hr 400°C

Blower

Sequence timer

Regeneration off gas, 400°C

Acceptance time: 120 min

Boiler feedwater

L.P. steam

Quench column

Absorber off gas

Excess stripped water

Absorber

SO$_2$ to Claus unit

L.P. steam

Figure 26.10 Shell flue gas system.

26.25

for the plant where traditional scrubbing would result in the sulfur being thrown away in costly disposal activities.

Claus recovery plants.   Claus recovery plants are used in many petroleum refineries to convert H_2S formed during the refining process to elemental sulfur which can be sold off-site. These plants have traditionally achieved 94 percent conversion with a two-catalyst-pass system. A third catalyst pass can bring the conversion up to approximately 97.5 percent, which approaches the chemical equilibrium for the process.

Until recently, the unconverted H_2S was simply oxidized to form sulfur dioxide and water; however, most plants are now required to achieve 99.5 percent conversion efficiency, which is beyond the capabilities of a Claus system on its own. The two most common ways to achieve the additional conversion are to employ tail gas scrubbing to remove the excess sulfur or to further process the exhaust gas.

Scrubbing the SO_2 out of the tail gas has a lower capital cost than additional processing, but the operational costs and waste disposal costs can be rather large. Further processing generally consists of either the shell process or the Beavon process.

In the shell process, all sulfur compounds in the exhaust stream are converted to hydrogen sulfide by a hydrogen reaction. Also H_2S is separated from CO_2 through absorption in an amine. The newly formed H_2S is then stripped from the amine and returned to the Claus unit for processing into sulfur. The Beavon process uses a similar hydrogen reaction to the one used in the shell process to convert SO_2 to H_2S, but H_2S is converted to sulfur using a Stretford process.

The capital cost for the additional 2 to 4 percent control achieved by these add-on systems is usually roughly equivalent to the cost of the Claus system itself.

26.4.3   Nonferrous smelting operations

Another significant source of sulfur dioxide emissions is the smelting of nonferrous metals. For these processes, control has traditionally consisted of feeding the effluent gas to a sulfuric acid plant where it was processed and sold off-site. This method, however, proved difficult for rural plants due to the costs involved with transporting or neutralization and disposal of the acid. As an alternative, the gas can be treated with most of the traditional SO_2 control processes used for fuel-burning sources. The SO_2 can also be removed with natural gas and a Claus process in the case of fluid-bed roasting of nickel ores.

SO_2 emissions from nonferrous smelting operations are not difficult to control, as they are easy to remove using the same control systems used for other sources. The difficulty facing these plants with SO$_2$ control is mainly due to the cost of adding control to the system and the disposal of wastes from the control systems once the SO$_2$ is removed.

26.4.4 Pulp and paper industry

The pulp and paper industry has traditionally been a large source of sulfur emissions. Most pulp and paper plant sulfur emissions stem from the loss of SO$_2$, the release of sulfur-bearing compounds, and the release of H$_2$S, which gives these plants their familiar foul odor. Several methods can be used to control the SO$_2$ emissions from pulp and paper plants. The most common of these includes process modifications which limit sulfur losses to the atmosphere, more careful control of the process, leak prevention control to reduce fugitive emissions, and conversion from calcium-based pulping to magnesium-based, which has lower emissions.

26.5 The Future of SO$_2$ Control

Sulfur dioxide emissions control has come a long way from the water scrubbing systems first created to current systems being experimented with which yield a marketable end product. Although many of the systems discussed in this chapter are still experimental and may never see commercial popularity, they all represent significant steps toward more efficient reduction of sulfur emissions while reducing the cost of control. As these systems are investigated further, better techniques will be developed, resulting in increased air quality and greater efficiency for many processes.

A

Acronyms

AA	Atomic absorption
ACGIH	American Conference of Governmental Industrial Hygienists
AIRS	Aerometric information retrieval system
ALAPCO	Associations of Local Air Pollution Control Offices
APTI	Air Pollution Training Institute
A/Q	Air quality
AQCR	Air quality control region
*AQDHS-II	(AQ-II)—Air Quality Data Handling System (version 2)
*AQDM	Air quality display model
AQMA	Air quality maintenance area
AQMP	Air quality maintenance plan
ASTM	American Society for Testing and Materials
AWMA	Air and Waste Management Association
BACT	Best-available control technology
BAT	Best-available treatment
BDT	Best-demonstrated technology (air)
BID	Background information document
Btu	British thermal unit
CAA	Clean Air Act
CAAA	Clean Air Act Amendments
CAFE	Corporate average fuel economy
CAS	Chemical Abstract Service
*CDS	Compliance data system

*These are acronyms used in reference to computer work.

CEM	Continuous emission monitoring
CEQ	Council on Environmental Quality
CFC	Chlorofluorocarbon
CFM	Cubic feet per minute
CFR	Code of Federal Regulations
CNG	Compressed natural gas
CO	Carbon monoxide
CO_2	Carbon dioxide
COG	Council of Governments
CTC	Control technology center
CTG	Control techniques guidelines
DE	Destruction efficiency
DOE	Department of Energy
DOT	Department of Transportation
DRE	Destruction and removal efficiency
dscf	Dry standard cubic feet
EDS	Electronic data system
EGR	Exhaust gas recirculation
EPA	Environmental Protection Agency
ERC	Emission reduction credit
*EIS	Emissions inventory system
*EMS	Emission management system
ESP	Electrostatic precipitator
ETP	Emissions trading policy
FEA	Federal Energy Administration
FEMA	Federal Emergency Management Agency
FGD	Flue gas desulfurization
FHWA	Federal Highway Administration
FID	Flame ionization detector
FIP	Federal implementation plan
FLM	Federal Land Manager
FMVCP	Federal Motor Vehicle Control Program
FPC	Federal Power Commission
FR	Federal Register
FY	Fiscal year
GACT	Generally available control technology
GAO	General Accounting Office
GC	Gas chromatograph

GEMS	Global environmental monitoring system
GEP	Good engineering practice
gr	Grain
HAP	Hazardous air pollutant
HC	Hydrocarbon
HCFC	Hydrochlorofluorocarbon
HEPA	High-efficiency particle air
HHS	(Department of) Health and Human Services (formerly HEW)
HHV	Higher heating value
Hi-VOL	High-volume sampler
HON	Hazardous organic NESHAP
HUD	Housing and Urban Development
I&M	Inspection and maintenance (automobiles)
ID	Infrared detector
IDL	Instrument detection limit
IR	Infrared
IRIS	Integrated risk information system
LAER	Lowest-achievable emission rate
LEL	Lower explosive limit
LPG	Liquefied petroleum gas
MACT	Maximum-achievable control technology
MSA	Metropolitan statistical area
MSDS	Material safety data sheet
MTBE	Methyl tertiary butyl ether
NAA	Nonattainment area
NAAQS	National Ambient Air Quality Standard
NAPCTAC	National Air Pollution Control Techniques Advisory Committee
NATICH	National Air Toxic Information Clearinghouse
NCC	National Computer Center
NEDS	National emission data system
NESHAP	National Emission Standards for Hazardous Air Pollutant
NIOSH	National Institute of Occupational Safety and Health
NMHC	Nonmethane hydrocarbon
NOV	Notice of violation
NO_x	Nitrogen oxide
NPDES	National Pollutant Discharge Elimination System

NPRM	Notice of Proposed Rulemaking
NPS	National Park Service
NRC	National Research Council; Nuclear Regulatory Commission
NRDC	Natural Resources Defense Council
NSD	Nonsignificant deterioration
NSPS	New-Source Performance Standard
NSR	New-Source Review
NTIS	National Technical Information Service
O_3	Ozone
OMB	Office of Management and Budget
OSHA	Occupational Safety and Health Act
O_x	Oxidant
PAN	Peroxyacetyl nitrate
PM	Particulate matter
PM_{10}	Particulate matter—nominally 10-μm diameter or less
ppb	Parts per billion
ppm	Parts per million
PSD	Prevention of significant deterioration
PSI	Pollutant standards index
PTE	Potential to emit
PTPLU	Point-source Gaussian diffusion model
QA	Quality assurance
RACT	Reasonable available control technology
RAM	Urban air quality model for point and area sources in EPA UNAMAP series
RFP	Reasonable further progress
RTP	Research Triangle Park, North Carolina
S&A	Surveillance and analysis (EPA)
*SAROAD	Storage and retrieval of aerometric data
SCC	Source classification code
SCR	Selective catalytic reduction
SCS	Supplementary control system
SIC	Standard industrial classification
SIP	State implementation plan
SOCMI	Synthetic Organic Chemicals Manufacturing Industry
SO_x	Sulfur oxide
*SPSS	Statistical Package for Social Sciences

STALAPCO	State and local air pollution control official
STAPPA	State and territorial air pollution control administrator
STEL	Short-term exposure limit
TAMS	Toxic air monitoring system
TCP	Transportation control plan
TLV	Threshold limit value
TPY	Tons per year
TSM	Transportation systems management
TSP	Total suspended particulate
TWA	Time-weighted average
UEL	Upper explosive limit
UMTA	Urban Mass Transportation Administration
UNAMAP	User's network for applied modeling of air pollution
*UTM	Universal transverse mercator
VEE	Visible emissions evaluation
VMT	Vehicle miles traveled
VOC	Volatile organic compound
VOST	Volatile organic sampling train

B

Definitions

accuracy   The degree of agreement between a measured value and the true value, usually expressed as ± percent of scale.

Act   The Clean Air Act (42 U.S.C. 1857-28572), as amended.

activation energy   The amount of energy above the average energy level that the reactants must have in order for combustion to proceed.

actual emissions   The actual rate of emissions of an air containment from an emissions unit, as determined in accordance with items 1 through 3 below.

1. In general, actual emissions as of a particular date shall equal the average rate, in tons per year, at which the emissions unit actually emitted the air contaminant during a 2-year period which precedes the particular date and which is representative of normal source operation. The permitting agency may allow the use of a different time period upon a determination that is more representative of normal source operation. Actual emissions shall be calculated using the unit's actual operating hours, production rates, and types of materials processed, stored, or combusted during the selected time period.

2. In the absence of reliable data, the permitting agency may presume that permitted-specific allowable emissions for the emissions unit are equivalent to the actual emissions of the emissions unit.

3. For any emissions unit which has not begun normal operations on the particular data, actual emissions shall equal the potential to emit of the unit on that date.

Administrator   The Administrator of the Environmental Protection Agency.

affected facility   With reference to a stationary source, any apparatus to which a standard is applicable.

Agency   Environmental Protection Agency.

air contaminant   Particulate matter, dust, fumes, gas, mist, smoke, or vapor, or any combinations thereof, total suspended particulates, PM_{10}, sulfur dioxide, carbon dioxide, ozone, nitrogen, lead, and gaseous fluorides expressed as HF.

allowable emissions   The emissions rate of stationary source calculated using the maximum rated capacity of the source (unless the source is subject to legally enforceable limits which restrict the operating rate, or hours of operations, or both) and the most stringent of the applicable standards set forth in:

1. The New Source Performance Standards (NSPS), or
2. The National Emission Standard for Hazardous Air Pollutants (NESHAP), or
3. Limits established pursuant to any CAA applicable standards, or
4. In the State Implementation Plan, emissions rates, specified as legally enforceable permit conditions including those with a future compliance date.

allowance   An authorization, allocated by the Administrator under the Acid Rain program, to emit up to 1 ton of sulfur dioxide during or after a specified calendar year.

ambient air   That portion of the atmosphere, external to buildings, to which the general public has access.

autoignition   The combustion of a gas or vapor without the use of an outside fuel source.

automated method or analyzer   A method for measuring concentrations of an ambient air pollutant in which sample collection, analysis, and measurement are performed automatically.

auxiliary fuel   The additional source of fuel for combustion. Used when the waste gas stream does not have sufficient heating value to sustain combustion. Typically natural gas or fuel oil.

begin actual construction   In general, initiation of physical on-site construction activities on an emissions unit which are of a permanent nature. Such activities include, but are not limited to, installation of building supports and foundations, laying of underground pipe work, and construction of permanent storage structures. With respect to a change in method of operation, this term refers to those on-site activities, other than preparatory activities, which mark the initiation of the change.

building, structure, facility, or installation   All the air contaminant-emitting activities which belong to the same industrial grouping, are located on one or more contiguous or adjacent properties, and are under the control of the same person (or persons under common control). Air contaminant-emitting activities shall be considered part of the same industrial grouping if they belong to the same **major group** (i.e., have the same two-digit code) which is specified in the Standard Industrial Classification manual, 1972, as amended by the 1977 Supplement (U.S. Government Printing Office stock numbers 4101-0065 and 003-005-00176-0, respectively).

Candidate method   A method of sampling and analyzing the ambient air for an air pollutant for which an application for a reference method determination or an equivalent method determination is submitted.

capacity factor   The ratio of the average load on a machine or equipment for the period of time considered to be the capacity rating of the machine or equipment.

capital expenditure   An expenditure for a physical or operational change to an existing facility which exceeds the product of the applicable "annual asset guideline repair allowance percentage" specified in the latest edition of

Internal Revenue Service (IRS) Publication 534 and the existing facility's basis, as defined by section 1012 of the Internal Revenue Code.

capture efficiency   The percent of airborne contaminants emitted from a process which are actually collected by the exhaust system.

capture velocity   The velocity at which airborne contaminants are transported into an exhaust hood from the work environment.

carbon monoxide (CO)   A priority pollutant generated from incomplete combustion of fuel.

catalyst deactivation   The degradation of a catalyst's efficiency due to particulate fouling, poisoning by certain chemicals, and application of excess temperatures.

catalyst oxidation   The destruction of gaseous pollutants through rapid oxidation at elevated temperatures in the presence of a catalyst.

catalyst substrate   The structure on which a catalyst is coated. Typically stainless steel or ceramic, and monolith in shape.

combustion   The rapid oxidation of a gas or vapors at elevated temperatures.

commence construction   As applied to a major stationary source or major modification, the owner operator has all necessary construction permits and either has begun, or caused to begin, a continuous program of actual on-site construction of the stationary source, to be completed within reasonable time, or entered into binding agreements or contractual obligations, which cannot be canceled or modified without substantial loss to the owner or operator, to undertake a program of actual construction of the stationary source to be completed within a reasonable time.

compliance schedule   A chronology of actions to be taken by a noncomplying source to bring it into full compliance. Generally speaking, compliance schedule increments will be divided into (1) engineering evaluation of problem solution, (2) procurement of the equipment and/or services necessary to solve the problem, (3) on-site delivery of the equipment, (4) completion of the equipment's installation, including start-up of said equipment, and (5) source testing to establish the air contaminant emission levels of the completion installation.

compliance use date   The first calendar year in which an allowance may be used for purposes of meeting a unit's sulfur dioxide emissions limitation requirements.

Consumer Price Index (CPI)   The United States government's primary indicator of the monetary inflation rate, published monthly by the U.S. Department of Labor, Bureau of Labor Statistics, Consumer Price Indices Branch, in the *CPI Detailed Report* and in the *Monthly Labor Review*.

continuous monitoring system   The total equipment, used to sample and condition (if applicable), to analyze, and to provide a permanent record of emissions or process parameters.

control strategy   A combination of measures designated to achieve the aggregate reduction of emissions necessary for attainment and maintenance of national standards.

destruction efficiency   The percentage of a pollutant destroyed in an incinerator.

dioxins   Polychlorinated dibenzo-*p*-dioxins, air pollutant which may be formed during combustion.

dispersion technique Any technique which attempts to affect the concentration of a pollutant in the ambient air.

distributor Any person who transports or stores or causes the transportation or storage of gasoline or diesel fuel at any point between any gasoline or diesel fuel refinery or importer's facility and any retail outlet or wholesale purchaser-consumer's facility.

duct velocity The velocity at which airborne contaminants are transported in a duct.

ductwork The device through which collected airborne contaminants are transported in an exhaust system.

effective date The date of promulgation in the Federal Register of an applicable standard or other regulation.

emission limitation and emission standard A requirement established by a state, a local government, or the Administrator which limits the quantity, rate, or concentration of emissions of air pollutants on a continuous basis, including any requirements which limit the level of opacity, prescribe equipment, set fuel specifications, or prescribe operation or maintenance procedures for a source to ensure a continuous emission reduction.

emission unit Any part of a stationary source which emits, or would have the potential to emit, any air contaminant subject to regulation.

equivalent method A method of sampling and analyzing the ambient air for an air pollutant that has been designated as producing results comparable to those of the reference method.

excess air An amount of air used for combustion that is in excess of the stoichiometric quantity of air required to complete the reaction.

excess emissions Emissions of an air pollutant in excess of an emission standard.

exchange medium A bed of inorganic material used to transfer heat in a regenerative heat recovery system.

existing source Any stationary source which is not a new source.

fall time (90 percent) The interval between initial response and time to 90 percent response after a step decrease in concentration.

fan The device used to move airborne contaminants through an exhaust system.

flashback The ignition of a VOC-laden gas upstream of the combustion chamber in the connecting ductwork and processes. Originates from the combustion chamber.

fly ash Inorganic particulate in a waste gas stream which cannot be combusted.

fossil-fuel-fired steam generator A furnace or boiler used in the process of burning fossil fuel for the primary purpose of producing steam by heat transfer.

fuel gas The gas stream generated during combustion.

fugitive emissions Those emissions which could not reasonably pass through a stack, chimney, vent, or other functionally equivalent opening.

full-scale The maximum measuring limit for a given operational range.

furans Polychlorinated dibenzofurans, an air pollutant which may be formed during combustion.

gasoline   Any fuel sold in any state for use in motor vehicles and motor vehicle engines, and commonly or commercially known or sold as gasoline.

Good engineering practice   As applied to stack height, the greater of:

1. 65 m, measured from the ground-level elevation at the base of the stack, or
2. The height of the following stack:
   a. For a stack in existence on January 12, 1979, and for which the owner or operator had obtained all applicable permits or approvals required under 40 CFR 51 and 52 (July 1, 1993 CFR)

$$H_g = 2.5H$$

   provided the owner or operator produces evidence that this equation was actually relied on in establishing an emission limitation.
   b. For all other stacks,

$$H_g = H + 1.5L$$

   where H_g = good engineering practice stack height, measured from ground-level elevation at base of stack. This is the height at which structural downwash no longer influences computer-modeled ambient impacts

   H = height of nearby structure(s) measured from ground-level elevation at base of stack

   L = lesser dimension, height or projected width, of nearby structure(s)

   provided that the permitting agency may require the use of a field study of fluid model to verify GEP stack height for the source; or
   c. The height demonstrated by a fluid model or a field study approved by the permitting agency, which ensures that the emissions from a stack do not result in excessive concentrations of any air pollutant as a result of atmospheric downwash, wakes, or eddy effects created by the source itself, nearby structures, or nearby terrain features.

heat input   The total gross calorific value.

hood   The device used to capture airborne contaminants from the work environment.

hydrocarbons   Chemical compounds consisting of hydrogen and carbon atoms.

incinerability   The relative ease or difficulty of destroying a chemical compound via combustion.

incineration   The combustion of undesirable materials for pollution control.

interference   An undesired positive or negative output caused by a substance other than the one being measured.

isokinetic sampling   Sampling in which the linear velocity of the gas entering the sampling nozzle is equal to that of the undisturbed gas stream at the sample point.

lag time   The time interval from a step change in the input concentration of the instrument inlet to a reading of 90 percent of the ultimate recorded concentration.

lead additive   Any substance containing lead or lead compounds.

leaded gasoline   Gasoline which is produced with the use of any lead additive or which contains more than 0.05 g/gal of lead or more than 0.005 g/gal of phosphorus.

legally enforceable   All limitations and conditions which are enforceable by the permitting agency and the EPA Administrator.

local agency   Any local government agency other than the state agency which is charged with responsibility for carrying out a portion of the plan.

lower explosive limit (LEL)   Concentration of volatile organic compounds in a gas stream below which combustion cannot be sustained due to an insufficient quantity of O_2 present.

lowest-achievable emission rate (LAER)   For any major stationary source or major modifications, the more stringent rate of emissions based on the following:

1. The most stringent emissions limitation which is contained in the applicable standards under any state implementation plan for such class or category of stationary source, unless the owner or operator of the proposed source demonstrates that such limitations are not achievable; or

2. The most stringent emissions limitation which is achieved in practice by such class or category of stationary source. This limitation, when applied to a modification, means the lowest-achievable emissions rate for the new or modified emissions units within the stationary source. In no event shall the application of this term permit a proposed new or modified stationary source to emit any air contaminant in excess of the amount allowable under applicable New Source Standards of Performance.

major modification

1. Any physical change in, or change in the method of operation of, a major stationary source that would result in a significant net emissions increase for any pollutant subject to regulations under the Clean Air Act Amendments of 1990.

2. Any net emissions increase that is considered significant for volatile organic compounds or nitrogen oxides shall be considered significant for ozone.

3. A physical change or change in the method of operation shall not include
   a. Routine maintenance, repair, and replacement.
   b. Use of an alternative fuel or raw material by reason of any order under section 2(a) and (b) of the Energy Supply and Environmental Coordination Act of 1974 (or any superseding legislation) or by reason of a natural gas curtailment plan pursuant to the federal power act.
   c. Use of an alternative fuel by reason of an order or rule under section 125 of the Clean Air Act Amendments, August 7, 1977.
   d. Use of an alternative fuel at a steam generating unit (burning equipment of 250 MBtu/h or larger) to the extent that the fuel is generated from municipal solid waste.
   e. Use of an alternative fuel or raw material by a stationary source which the source was capable of accommodating before December 12, 1976, unless such change would be prohibited under a legally enforceable per-

mit condition which was established after December 12, 1976, pursuant to 40 CFR 52.21 (July 1, 1993 CFR), or under regulations approved pursuant to 40 CFR subpart I or 51.166 (July 1, 1993 CFR).

f. An increase in the hours of operation or in the production rate, unless such change would be prohibited under a legally enforceable permit condition which was established after December 21, 1976, pursuant to 40 CFR 52.21 (July 1, 1993 CFR), or regulations approved pursuant to 40 CFR Part 51 subpart I or 40 CFR 51.166 (July 1, 1993 CFR).

g. Any change in ownership at a stationary source.

Major stationary source

1. Any stationary source of air contaminants which emits, or has the potential to emit, 100 tons/yr or more of any air contaminants regulated.
2. Any physical change that would occur at a stationary source not qualifying under item (1) as a major stationary source, if the change constituted a major stationary source by itself.
3. A major stationary source that is major for volatile organic compounds or nitrogen oxides shall be considered major for ozone.
4. The fugitive emissions of a stationary source shall not be included in determining for any of the purposes of this item, whether it is a major stationary source, unless the source belongs to one of the following categories of stationary sources:

   a Coal cleaning plants (with thermal dryers)
   b. Kraft pulp mills
   c. Portland cement plants
   d. Primary zinc smelters
   e. Iron and steel mills
   f. Primary aluminum ore reduction plants
   g. Primary copper smelters
   h. Municipal incinerators (or combination thereof) capable of charging more than 50 tons/day of refuse
   i. Hydrofluoric, sulfuric, or nitric acid plants
   j. Petroleum refineries
   k. Lime plants
   l. Phosphate rock processing plants
   m. Coke oven batteries
   n. Sulfur recovery plants
   o. Carbon-black plants (furnace process)
   p. Primary lead smelters
   q. Fuel conversion plants
   r. Sintering plants
   s. Secondary metal production plants
   t. Chemical process plants
   u. Fossil-fuel boilers (or combinations thereof) totaling more than 250 MBtu/h heat input
   v. Petroleum storage and transfer units with a total storage capacity exceeding 300,000 bbl
   w. Taconite ore processing plants

$x.$ Glass fiber processing plants
$y.$ Charcoal production plants
$z.$ Fossil-fuel-fired steam electric plant of more than 250 MBtu/h heat input
     $aa.$ Any other stationary source category which, as of August 7, 1980, is being regulated under section 111 or 112 of the Act

malfunction   Any sudden and unavoidable failure of air pollution control equipment or process equipment or of a process to operate in a normal or usual manner.

manual method   A method for measuring concentrations of an ambient air pollutant in which sample collection, analysis, or measurement, or some combination thereof, is performed manually.

minor modification

1. Any modification which is not a major modification, or
2. Any modification which is a physical change in, or a change in the method of operation of, a minor stationary source provided the change would not constitute a major stationary source by itself.

minor stationary source   Any source which is not a major stationary source.

modification or modified source   Any physical change in, or change in the method of operation of, a stationary source which increases the emission rate of any pollutant for which a national standard has been promulgated or which results in the emission of any such pollutant not previously emitted, except that

1. Routine maintenance, repair, and replacement shall not be considered a physical change.
2. The following shall not be considered a change in the method of operation:
    a An increase in the production rate, if such increase does not exceed the operating design capacity of the source
    $b.$ An increase in the hours of operation
    $c.$ Use of an alternative fuel or raw material if the source is designed to accommodate such alternative use

monitoring device   The total equipment used to measure and record (if applicable) process parameters.

National air monitoring stations (NAMS)   A subset of the SLAMS air quality network.

National Ambient Air Quality Standard (NAAQS)   The ambient air quality standard set by the EPA in order to safeguard human health (primary standards) and welfare (secondary standards).

national standard   Either a primary or secondary standard.

net air quality benefit   An improvement in air quality within the boundaries of a nonattainment area, as demonstrated by confirmed monitored air quality data or predictions of an air quality modeling analysis.

net emissions increase   The amount by which the sum of the following exceeds zero:

1. Any increase in actual emissions from a particular physical change or change in the method of operation at a stationary source

2. Any other increases and decreases in actual emissions at the stationary source that are contemporaneous with the particular change and are otherwise creditable

new source Any stationary source, the construction or modification of which is commenced after the publication in the Federal Register of proposed national emission standards.

nitrogen oxides (NO$_x$) All oxides of nitrogen except nitrous oxide (N$_2$O).

noise Spontaneous deviations from a mean output not caused by input concentration changes.

nonattainment area Any area that does not meet (or that contributes to ambient air quality in a nearby area that does not meet) any ambient air quality standard for the pollutant. The demonstration required under section 165(a)(3) of the 1990 Clean Air Act shall not apply to maximum allowable increases for class II areas in the case of an expansion or modification of a major emitting facility which was in existence on the date of enactment of the Clean Air Act Amendments of 1977, and whose allowable emissions of air pollutants are established as required in subsection 165(a)(4) of the 1990 Clean Air Act.

one-hour period Any 60-min period commencing on the hour.

opacity The degree to which emissions reduce the transmission of light and obscure the view of an object in the background.

owner or operator Any person who owns, leases, operates, controls, or supervises a facility, building, structure, or installation which directly or indirectly results in or may result in emissions of any air pollutant for which a national standard is in effect.

oxidation A chemical reaction involving the combination of oxygen with other elements.

ozone precursor Generally volatile organic compound and/or nitrogen oxide. A proposed new source or a net emissions increase at an existing source in an ozone transport region (or an ozone nonattainment area) can be classified as major based on either VOC or NO$_x$ emissions or both (but not in combination). That is, the determination of major must be made individually for each pollutant, since VOC and NO$_x$ emissions cannot be added to meet the minimum level required for such a demonstration.

particulate matter Any airborne, finely divided solid or liquid material with an aerodynamic diameter smaller than 100 μm.

particulate matter emissions All finely divided solid or liquid material, other than uncombined water, emitted to the ambient air.

photochemical smog Smog resulting from an atmospheric reaction of nitrogen oxides and volatile organic compounds in the presence of sunlight.

plan An implementation plan approved or promulgated.

plenum The main duct in a plenum exhaust system which acts as a pressure-equalizing chamber.

PM$_{10}$ Particulate matter with an aerodynamic diameter less than or equal to a nominal 10 μm.

PM$_{10}$ emissions Finely divided solid or liquid material, with an aerodynamic diameter less than or equal to a nominal 10 μm, emitted to the ambient air.

potential to emit The maximum capacity of a stationary source to emit an air contaminant under its physical and operational design. Any physical or operational limitation on the capacity of the source to emit an air contaminant, including air contaminant control equipment and restriction on hours of operation or on the type or amount of material combusted, stored, or processed, shall be treated as part of its design only if the limitation or the effect it would have on emissions is "legally enforceable." Secondary emissions do not count in determining the potential to emit of a stationary source.

precision   The degree of agreement between repeated measurements of the same concentration, expressed as the average deviation of the single results from the mean.

prevention of significant deterioration (PSD)   By Part D of the Clean Air Act Amendments of 1977, certain new major stationary sources and major modifications are subject to a preconstruction review which includes an ambient air quality analysis.

primary standard   A national primary ambient air quality standard.

products of incomplete combustion (PICs)   Chemical compounds which are generated from improper combustion of fuels.

PSD station   An ambient air monitoring station operated for the purpose of establishing the effect on air quality of the emissions from a proposed source for the prevention of significant deterioration of the ambient air quality prior to increased emissions.

reasonable further progress (RFP)   Such annual incremental reductions in emissions of the relevant air pollutant as are required by this part or may reasonably be required by the permitting agency for the purpose of ensuring attainment of the applicable ambient air quality standard by the applicable date.

recuperative heat recovery   The recovery of heat from incinerator's flue gas using an air-to-air heat exchanger to preheat the incoming waste gas.

reference method   A method of sampling and analyzing the ambient air for an air pollutant which is stated to be the standard against which alternative methods must be compared for equivalent performance.

refinery   A plant at which gasoline or diesel fuel is produced.

regenerative heat recovery   The recovery of heat from incinerator flue gas using regenerating exchange media to preheat the incoming waste gas.

region   An area designated as an air quality control region.

Regional Administrator   The Administrator of one of the 10 EPA regional offices or his or her authorized representative.

regional office   One of the 10 EPA regional offices.

residence time   The amount of time which a waste gas stream is subject to combustion conditions in an incinerator. Synonym: **detention time.**

retail outlet   Any establishment at which gasoline or diesel fuel is sold or offered for sale for use in motor vehicles.

rise time (90 percent)   The interval between initial response and time to 90 percent response after a step increase in the inlet concentration.

secondary emissions   Emissions which would occur as a result of the construction or operation of a major stationary source or major modification, but do not come from the major stationary source or major modification itself.

short-circuiting   The tendency of a portion of a fluid to exhibit pseudo-plug flow in a mixed reactor.

shutdown   The cessation of operation of an affected facility for any purpose.

significant   In reference to a net emissions increase or the potential of a source to emit any of the following air contaminants, a rate of emissions that would equal or exceed any of the following rates:

Air contaminant	Emissions rate
Carbon monoxide	100 tons/yr
Nitrogen oxides	40 tons/yr
Sulfur dioxide	40 tons/yr
Ozone	40 tons/yr of an ozone precursor
Lead	0.6 ton/yr
PM_{10}	15 tons/yr

significant impact   The contribution by a new stationary source or modification to the air quality in a nonattainment area in concentrations equal to or greater than the amount as follows:

	Averaging time and applicable concentration				
Pollutant	Annual	24-h	3-h	8-h	1-h
Sulfur dioxide	$1\ \mu g/m^3$	$5\ \mu g/m^3$	$25\ \mu g/m^3$		
PM_{10}	$1\ \mu g/m^3$	$5\ \mu g/m^3$			
Carbon monoxide				$500\ \mu g/m^3$	$2000\ \mu g/m^3$
Nitrogen oxide	$1\ \mu g/m^3$				

six-minute period   Any one of the 10 equal parts of a 1-h period.

SLAMS   State or local air monitoring station.

space velocity   The inverse of the residence time of a waste gas stream in a catalytic oxidizer.

span drift   The change in instrument output over a stated time period, usually 24 h, of unadjusted continuous operation when the input concentration is a stated upscale value. Span drift is usually measured at 80 percent of scale and is usually expressed as percent of full scale.

stack   Any point in a source designed to emit solids, liquids, or gases into the air, including a pipe or duct but not including flares.

standard   A standard of performance proposed or promulgated.

standard conditions   A temperature of 293 K (68°F) and a pressure of 101.3 kPa (29.92 inHg).

start-up   The setting in operation of a source for any purpose.

state agency   The air pollution control agency primarily responsible for development and implementation of a plan under the Act.

State implementation plan (SIP)   The mechanism by which states set emission limits and allocate pollution control responsibility among sources to meet the limits on ambient concentrations in the NAAQSs. Among others, the requirement to maintain systems to monitor and report on ambient air quality.

state or local air monitoring stations (SLAMS)   Monitoring station which measures ambient concentrations of those pollutants for which standards have been established. The SLAMSs make up the air quality monitoring network which is required to be provided for in the state's implementation plan.

static pressure   The axial and radial components of pressure in a duct which tend to burst or collapse the duct.

stationary source   Any building, structure, facility, or installation which emits or may emit an air pollutant for which a national standard is in effect.

storage and retrieval of aerometric data (SAROAD) system   A computerized system which stores and reports information relating to ambient air quality.

thermal fixation   A chemical reaction involving the oxidation of atmospheric nitrogen at elevated temperatures.

thermal oxidation   The destruction of a gaseous pollutant through rapid oxidation at elevated temperatures using direct heat.

time period   Any period of time designated by hour, month, season, calendar year, averaging time, or other suitable characteristics, for which ambient air quality is estimated.

total rated capacity   The sum of the rated capacities of all fuel-burning equipment connected to a common stack. The rated capacity shall be the maximum guaranteed by the equipment manufacturer or the maximum normally achieved during use, whichever is greater.

total suspended particulate   Particulate matter.

traceable   A local standard has been compared and certified, either directly or via not more than one intermediate standard, to a primary standard such as a National Bureau of Standards standard reference material (NBS SRM) or a USEPA/NBS-approved certified reference material (CRM).

turbulence   The rapid mixing of one or more fluids.

unleaded gasoline   Gasoline which is produced without the use of any lead additive and which contains not more than 0.05 g/gal of lead and not more than 0.005 g/gal of phosphorus.

upper explosive limit (UEL)   The concentration of volatile organic compounds in a gas stream above which combustion cannot be sustained due to an insufficient quantity of O_2 present.

variance   The temporary deferral of a final compliance date for an individual source subject to an approved regulation, or a temporary change to an approved regulation as it applies to an individual source.

velocity pressure   The pressure created in the ventilation system due to the velocity of the exhausted air. The velocity pressure acts parallel to the direction of fluid flow.

vena contracta   An area downstream of an orifice where the flow lines of a fluid are closest together.

ventilation   The collection and transport of airborne contaminants from the work environment to the atmosphere for purposes of reducing exposure to individual workers.

volatile organic compound (VOC)   Any compound of carbon, excluding carbon monoxide, carbon dioxide, carbonic acid, metallic carbides or carbonates, and ammonium carbonate, which participates in atmospheric photochemical reactions.

I. This includes any such organic compound other than the following, which have been determined to have negligible photochemical reactivity: Methane; ethane; methylene chloride (dichloromethane); 1,1,1-trichloroethane (methyl chloroform); 1,1,1-trichloro-2,2,2-trifluoroethane (CFC 113); trichlorofluoramethane (CFC 11); dichlorodifluormethane (CFC 12); chlorodifluoromethane (CFC 22); trifluoromethane (FC 23); 1,2-dichloro 1,1,2,2-tetrafluoroethane (CFC 114); chloropentafluoroethane (CFC 115); 1,1,1-trifluoro 2,2-dichloroethane (HCFC 123); 1,1,1,2-tetrafluoroethane (HFC-134a); 1,1-dichloro 1-fluoroethane (HCFC 141b); 1-chloro

1,1, difluoroethane (HCFC 142b); 2-chloro-1,1,1,2-tetrafluoroethane (HCFC 124); pentafluoroethane (HFC 125); 1,1,2,2-tetrafluoroethane (HFC 134); 1,1,1-trifluoroethane (HFC 143a); 1,1-difluoroethane (HFC 125a); and perfluorocarbon compounds which fall into these classes:

A. Cyclic, branched, or linear, completely fluorinated alkanes
B. Cyclic, branched, or linear, completely fluorinated ethers with no unsaturations
C. Cyclic, branched, or linear, completely fluorinated tertiary amines with no unsaturations
D. Sulfur-containing perfluorocarbons with no unsaturations and with sulfur bonds only to carbon and fluorine

waste gas   The gas stream which is generated by one or more specific industrial processes.

zero drift   The change in instrument output over a stated time period, usually 24 h, of unadjusted continuous operation when the input concentration is zero. Zero drift is usually expressed as percent of full scale.

C

State and Territorial Air Pollution Control Agencies

This is a list of state and territorial air pollution control agencies. States may designate another state agency to administer the Part 70 operating program, but in most instances the program will be delegated to the agencies listed below.

Alabama Dept. of Environmental
Management
Air Division
1751 Cong. Dickenson Drive
Montgomery, AL 36109-2608
(334) 271-7861

Alaska Dept. of Environmental
Conservation
Division of Air and Water Quality
410 Willoughby Avenue, Suite 105
Juneau, AK 99801-1795
(907) 465-5100

American Samoa
Environmental Quality Commission
Governor's Office
Pago Pago, Am. Samoa 96799
011-(684) 633-4116

Arizona Dept. of Environmental
Quality
Office of Air Quality
3033 N. Central
Phoenix, AZ 85012
(602) 207-4518

Arkansas Dept. of Pollution Control
and Ecology
Air Division
8001 National Drive, P.O. Box 8913
Little Rock, AR 72219-8913
(501) 682-0730

California Air Resources Board
Executive Office
2020 L Street
Sacramento, CA 95814
(916) 445-4383

Colorado Dept. of Public Health and
Environment
Air Pollution Control Division
4300 Cherry Creek Drive South
Denver, CO 80222
(303) 692-3100

Connecticut Dept. of Environmental
Protection
Bureau of Air Management
79 Elm Street
Hartford, CT 06106-5127
(860) 424-3026

Delaware Dept. of Nat. Resources
and Environmental. Control
Division of Air and Waste
Management
89 Kings Highway, P.O. Box 1401
Dover, DE 19903
(302) 739-4764

Dist. of Columbia Dept. Consumer
and Regulatory Affairs
Air Quality Control and Monitoring
Branch
2100 Martin Luther King Ave, SE
Washington, DC 20020
(202) 645-6093

Florida Dept. of Environmental
Regulation
Division of Air Resources
Management
2600 Blair Stone Road
Tallahassee, FL 32399-2400
(850) 488-0114

Georgia Dept. of Natural Resources
Air Protection Branch
4244 International Parkway, Suite 120
Atlanta, GA 30354
(404) 363-7016

Guam Environmental Protection
Agency
Complex Unit D-107
130 Rojas Street
Harmon, Guam 96911
011-(671) 646-8863

Hawaii State Dept. of Health
Clean Air Branch
919 Ala Moana Boulevard
Honolulu, HI 96814
(808) 586-4200

Idaho Division of Environmental
Quality
Air Quality Bureau
1410 North Hilton
Boise, ID 83720
(208) 334-0502

Illinois Environmental Protection
Agency
Division of Air Pollution Control
2200 Churchill Road, P.O. Box 19276
Springfield, IL 62794-9276
(217) 782-7326

Indiana Dept. of Environmental
Management
Office of Air Management
100 No. Senate
P.O. Box 6015
Indianapolis, IN 46206-6015
(317) 233-0178

Iowa Dept. of Natural Resources
Air Quality Bureau
7900 Hickman Avenue
Des Moines, IA 50319
(515) 281-6061

Kansas Dept. of Health and
Environment
Bureau of Air and Radiation
Forbes Field, Building 283
Topeka, KS 66620-0001
(785) 296-1593

Kentucky Dept. for Environmental
Protection
Division for Air Quality
803 Schenkel Lane
Frankfort, KY 40601-1403
(502) 573-3382

Louisiana Dept. of Environmental
Quality
Office of Air Quality and Radiation
Protection
Air Quality Division, P.O. Box 82135
Baton Rouge, LA 70884-2135
(504) 765-0110

Maine Dept. of Environmental
Protection
Bureau of Air Quality Control
17 State House Station
Augusta, ME 04333-0017
(207) 287-2437

Maryland Dept. of the Environment
Air and Radiation Management
Administration
2500 Broening Highway
Baltimore, MD 21224
(410) 631-3255

Massachusetts Dept. of
Environmental Protection
Division of Air Quality Control
One Winter Street, 8th floor
Boston, MA 02108
(617) 292-5915

Michigan Dept. of Environmental
Quality
Air Quality Division
P.O. Box 30260, 106 W. Allegan
Lansing, MI 48909-7760
(517) 373-7023

Minnesota Pollution Control Agency
Air Quality Division
520 Lafayette Road
Saint Paul, MN 55155-4194
(612) 296-7331

Mississippi Dept. of Environmental
Quality
Air Division, Office of Pollution Control
P.O. Box 10385
Jackson, MS 39289
(601) 961-5171

Missouri Dept. of Natural Resources
Air Pollution Control Program
P.O. Box 176
Jefferson City, MO 65102
(573) 751-4817

Montana Dept. of Environmental
Quality
Air and Waste Management Bureau
1520 E. Sixth Avenue
Helena, MT 59620
(406) 444-3490

Nebraska Dept. of Environmental
Control
Air Quality Division
P.O. Box 98922
Lincoln, NE 68509-8922
(402) 471-2189

Nevada Division of Environmental
Protection
Bureau of Air Quality
123 West Nye Lane
Carson City, NV 89710
(702) 687-5065

New Hampshire Dept. Env. Services
Air Resources Division
64 N. Main Street, Box 2033
Concord, NH 03302-2033
(603) 271-1370

New Jersey Dept. of Env. Prot. &
Energy
Air Pollution Control Program
401 East State Street
Trenton, NJ 08625
(609) 292-6710

New Mexico Environmental
Department
Environmental Protection Division
Air Quality Bureau
P.O. Box 26110
Santa Fe, NM 87502
(505) 827-0031

New York State Dept. of
Environmental Conservation
Division of Air Resources
50 Wolf Road
Albany, NY 12223-3250
(518) 457-7230

North Carolina Department of
Environment and Natural Resources
Division of Air Quality
P.O. Box 29580
Raleigh, NC 27626-0580
(919) 733-3340

North Dakota State Dept. of Health
Division of Environmental
Engineering
1200 Missouri Avenue
Bismarck, ND 58506-5520
(701) 328-5188

Ohio Environmental Protection
Agency
Division of Air Pollution Control
1600 WaterMark Drive
Columbus, OH 43215-1034
(614) 644-2270

Oklahoma Dept. of Envrionmental
Quality
Air Quality Division
4545 N. Lincoln, Suite 250
Oklahoma City, OK 73105-3483
(405) 290-8247

Oregon Dept. of Environmental
Quality
Air Quality Division
811 SW 6th Avenue
Portland, OR 97204
(503) 229-5359

Pennsylvania Dept. of
Environmental Protection
Bureau of Air Quality
Rachel Carson State Office Bldg.,
16th Floor
P.O. Box 2063
Harrisburg, PA 17105-2063
(717) 787-9702

Puerto Rico Environmental Quality
Board
Air Program
P.O. Box 11488
Santurce, PR 00910
(809) 767-8071

Rhode Island Dept. of Envrnmntl.
Mgmt.
Office of Air Resources
235 Promenade St.
Providence, RI 02908-5767
(401) 222-2808

South Carolina Dept. of Health and
Env. Control
Bureau of Air Quality
2600 Bull Street
Columbia, SC 29201
(803) 734-4750

South Dakota Dept. of Environment
and Nat. Resources, Air Quality
523 East Capitol Avenue
Pierre, SD 57501
(605) 773-3351

Tennessee Dept. of Environment and
Conservation
Division of Air Pollution Control
Life & Casualty Tower Annex, 9th floor
401 Church Street
Nashville, TN 37243-1531
(615) 532-0554

Texas Natural Resource
Conservation Commission
Office of Air Quality
12100 Park 35 Circle
Austin TX 78753
(512) 239-5440

Utah Dept. of Environmental Quality
Division of Air Quality
1950 West North Temple
Salt Lake City, UT 84114-4820
(801) 536-4015

Vermont Agency of Natural
Resources
Air Pollution Control Division
103 S. Main Street, building 3 south
Waterbury, VT 05671-0402
(802) 241-3840

Virgin Islands Dept. Planning/Nat.
Resources
Div. of Environmental Protection
Watergut Homes 1118 Christiansted
St. Croix, VI 00820-5065
(809) 773-0565

Commonwealth of Virginia
Department of Envrionmental
Quality
P.O. Box 10089
Richmond, VA 23240
(804) 698-4311

Washington State Department of
Ecology
Air Quality Program
P.O. Box 47600
Olympia, WA 98504-7600
(360) 407-6873

West Virginia Division of
Environmental Protection
1558 Washington St. East
Charleston, WV 25311
(304) 558-4022

Wisconsin Dept. of Natural
Resources
Bureau of Air Management
Box 7921
Madison, WI 53707
(608) 266-7718

Wyoming Dept. of Environmental
Quality
Air Quality Division
122 W. 25th Street
Cheyenne, WY 82002
(307) 777-7391

D

Regional Air Division Directors

1. Region I, U.S. Environmental Protection Agency, room 2203, John F. Kennedy Federal Building, Boston, MA 02203. (Connecticut, Maine, Massachusetts, New Hampshire, Rhode Island, and Vermont)

2. Region II, U.S. Environmental Protection Agency, room 900, 26 Federal Plaza, New York, NY 10278. (New Jersey, New York, Puerto Rico, and the Virgin Islands)

3. Region III, U.S. Environmental Protection Agency, 841 Chestnut Street, Philadelphia, PA 19107. (Delaware, Maryland, Pennsylvania, Virginia, West Virginia, and the District of Columbia)

4. Region IV, U.S. Environmental Protection Agency, 345 Courtland Street NE, Atlanta, GA 30365. (Alabama, Florida, Georgia, Kentucky, Mississippi, North Carolina, South Carolina, and Tennessee)

5. Region V, U.S. Environmental Protection Agency, 230 South Dearborn Street, Chicago, IL 60604. (Illinois, Indiana, Michigan, Minnesota, Ohio, and Wisconsin)

6. Region VI, U.S. Environmental Protection Agency, 1201 Elm Street, Dallas, TX 75270. (Arkansas, Louisiana, New Mexico, Oklahoma, and Texas)

7. Region VII, U.S. Environmental Protection Agency, 726 Minnesota Avenue, Kansas City, KS 66101. (Iowa, Kansas, Missouri, and Nebraska)

8. Region VIII, U.S. Environmental Protection Agency, 999 18th Street, One Denver Place, Denver, CO 80202. (Colorado, Montana, North Dakota, South Dakota, Utah, and Wyoming)

9. Region IX, U.S. Environmental Protection Agency, 215 Freemont Street, San Francisco, CA 94105. (Arizona, California, Hawaii, Nevada, American Samoa, Trust Territories of the Pacific Islands, Guam, Wake Islands, and the Northern Marianas)

10. Region X, U.S. Environmental Protection Agency, 1200 Sixth Avenue, Seattle, WA 98101. (Alaska, Idaho, Oregon, and Washington)

Appendix

E

State Operating Permit Programs CFR Title 40, Part 70

PART 70—STATE OPERATING PERMIT PROGRAMS

AUTHORITY: 42 U.S.C. 7401, *et seq.*

SOURCE: 57 FR 32295, July 21, 1992, unless otherwise noted.

§70.1   Program overview.

(a) The regulations in this part provide for the establishment of comprehensive State air quality permitting systems consistent with the requirements of title V of the Clean Air Act (Act) (42 U.S.C. 7401, *et seq.*). These regulations define the minimum elements required by the Act for State operating permit programs and the corresponding standards and procedures by which the Administrator will approve, oversee, and withdraw approval of State operating permit programs.

(b) All sources subject to these regulations shall have a permit to operate that assures compliance by the source with all applicable requirements. While title V does not impose substantive new requirements, it does require that fees be imposed on sources and that certain procedural measures be adopted especially with respect to compliance.

(c) Nothing in this part shall prevent a State, or interstate permitting authority, from establishing additional or more stringent requirements not inconsistent with this Act. The EPA will approve State program submittals to the extent that they are not inconsistent with the Act and these regulations. No permit, however, can be less stringent than necessary to meet all applicable requirements. In the case of Federal intervention in the permit process, the Administrator reserves the right to implement the State operating permit program, in whole or in part, or the Federal program contained in regulations promulgated under title V of the Act.

(d) The requirements of part 70, including provisions regarding schedules for submission and approval or disapproval of permit applications, shall apply to the permitting of affected sources under the acid rain program, except as provided herein

or modified in regulations promulgated under title IV of the Act (acid rain program).

(e) Issuance of State permits under this part may be coordinated with issuance of permits under the Resource Conservation and Recovery Act and under the Clean Water Act, whether issued by the State, the U.S. Environmental Protection Agency (EPA), or the U.S. Army Corps of Engineers.

§70.2   Definitions.

The following definitions apply to part 70. Except as specifically provided in this section, terms used in this part retain the meaning accorded them under the applicable requirements of the Act.

Act means the Clean Air Act, as amended, 42 U.S.C. 7401, *et seq.*

Affected source shall have the meaning given to it in the regulations promulgated under title IV of the Act.

Affected States are all States:

(1) Whose air quality may be affected and that are contiguous to the State in which a part 70 permit, permit modification or permit renewal is being proposed; or

(2) That are within 50 miles of the permitted source.

Affected unit shall have the meaning given to it in the regulations promulgated under title IV of the Act.

Applicable requirement means all of the following as they apply to emissions units in a part 70 source (including requirements that have been promulgated or approved by EPA through rulemaking at the time of issuance but have future-effective compliance dates):

(1) Any standard or other requirement provided for in the applicable implementation plan approved or promulgated by EPA through rulemaking under title I of the Act that implements the relevant requirements of the Act, including any revisions to that plan promulgated in part 52 of this chapter;

(2) Any term or condition of any preconstruction permits issued pursuant to regulations approved or promulgated through rulemaking under title I, including parts C or D, of the Act;

(3) Any standard or other requirement under section 111 of the Act, including section 111(d);

(4) Any standard or other requirement under section 112 of the Act, including any requirement concerning accident prevention under section 112(r)(7) of the Act;

(5) Any standard or other requirement of the acid rain program under title IV of the Act or the regulations promulgated thereunder;

(6) Any requirements established pursuant to section 504(b) or section 114(a)(3) of the Act;

(7) Any standard or other requirement governing solid waste incineration, under section 129 of the Act;

§ 70.2

(8) Any standard or other requirement for consumer and commercial products, under section 183(e) of the Act;

(9) Any standard or other requirement for tank vessels under section 183(f) of the Act;

(10) Any standard or other requirement of the program to control air pollution from outer continental shelf sources, under section 328 of the Act;

(11) Any standard or other requirement of the regulations promulgated to protect stratospheric ozone under title VI of the Act, unless the Administrator has determined that such requirements need not be contained in a title V permit; and

(12) Any national ambient air quality standard or increment or visibility requirement under part C of title I of the Act, but only as it would apply to temporary sources permitted pursuant to section 504(e) of the Act.

Designated representative shall have the meaning given to it in section 402(26) of the Act and the regulations promulgated thereunder.

Draft permit means the version of a permit for which the permitting authority offers public participation under § 70.7(h) or affected State review under § 70.8 of this part.

Emissions allowable under the permit means a federally enforceable permit term or condition determined at issuance to be required by an applicable requirement that establishes an emissions limit (including a work practice standard) or a federally enforceable emissions cap that the source has assumed to avoid an applicable requirement to which the source would otherwise be subject.

Emissions unit means any part or activity of a stationary source that emits or has the potential to emit any regulated air pollutant or any pollutant listed under section 112(b) of the Act. This term is not meant to alter or affect the definition of the term "unit" for purposes of title IV of the Act.

The EPA or the Administrator means the Administrator of the EPA or his designee.

Final permit means the version of a part 70 permit issued by the permitting authority that has completed all review procedures required by §§ 70.7 and 70.8 of this part.

Fugitive emissions are those emissions which could not reasonably pass through a stack, chimney, vent, or other functionally-equivalent opening.

General permit means a part 70 permit that meets the requirements of § 70.6(d).

Major source means any stationary source (or any group of stationary sources that are located on one or more contiguous or adjacent properties, and are under common control of the same person (or persons under common control)) belonging to a single major industrial grouping and that are described in paragraph (1), (2), or (3) of this definition. For the purposes of defining "major source,"

a stationary source or group of stationary sources shall be considered part of a single industrial grouping if all of the pollutant emitting activities at such source or group of sources on contiguous or adjacent properties belong to the same Major Group (i.e., all have the same two-digit code) as described in the Standard Industrial Classification Manual, 1987.

(1) A major source under section 112 of the Act, which is defined as:

(i) For pollutants other than radionuclides, any stationary source or group of stationary sources located within a contiguous area and under common control that emits or has the potential to emit, in the aggregate, 10 tons per year (tpy) or more of any hazardous air pollutant which has been listed pursuant to section 112(b) of the Act, 25 tpy or more of any combination of such hazardous air pollutants, or such lesser quantity as the Administrator may establish by rule. Notwithstanding the preceding sentence, emissions from any oil or gas exploration or production well (with its associated equipment) and emissions from any pipeline compressor or pump station shall not be aggregated with emissions from other similar units, whether or not such units are in a contiguous area or under common control, to determine whether such units or stations are major sources; or

(ii) For radionuclides, "major source" shall have the meaning specified by the Administrator by rule.

(2) A major stationary source of air pollutants, as defined in section 302 of the Act, that directly emits or has the potential to emit, 100 tpy or more of any air pollutant (including any major source of fugitive emissions of any such pollutant, as determined by rule by the Administrator). The fugitive emissions of a stationary source shall not be considered in determining whether it is a major stationary source for the purposes of section 302(j) of the Act, unless the source belongs to one of the following categories of stationary source:

(i) Coal cleaning plants (with thermal dryers);

(ii) Kraft pulp mills;

(iii) Portland cement plants;

(iv) Primary zinc smelters;

(v) Iron and steel mills;

(vi) Primary aluminum ore reduction plants;

(vii) Primary copper smelters;

(viii) Municipal incinerators capable of charging more than 250 tons of refuse per day;

(ix) Hydrofluoric, sulfuric, or nitric acid plants;

(x) Petroleum refineries;

(xi) Lime plants;

(xii) Phosphate rock processing plants;

(xiii) Coke oven batteries;

(xiv) Sulfur recovery plants;

(xv) Carbon black plants (furnace process);

(xvi) Primary lead smelters;

(xvii) Fuel conversion plants;

(xviii) Sintering plants;

(xix) Secondary metal production plants;

(xx) Chemical process plants;

(xxi) Fossil-fuel boilers (or combination thereof) totaling more than 250 million British thermal units per hour heat input;

(xxii) Petroleum storage and transfer units with a total storage capacity exceeding 300,000 barrels;

(xxiii) Taconite ore processing plants;

(xxiv) Glass fiber processing plants;

(xxv) Charcoal production plants;

(xxvi) Fossil-fuel-fired steam electric plants of more than 250 million British thermal units per hour heat input; or

(xxvii) All other stationary source categories regulated by a standard promulgated under section 111 or 112 of the Act, but only with respect to those air pollutants that have been regulated for that category;

(3) A major stationary source as defined in part D of title I of the Act, including:

(i) For ozone nonattainment areas, sources with the potential to emit 100 tpy or more of volatile organic compounds or oxides of nitrogen in areas classified as "marginal" or "moderate," 50 tpy or more in areas classified as "serious," 25 tpy or more in areas classified as "severe," and 10 tpy or more in areas classified as "extreme"; except that the references in this paragraph to 100, 50, 25 and 10 tpy of nitrogen oxides shall not apply with respect to any source for which the Administrator has made a finding, under section 182(f) (1) or (2) of the Act, that requirements under section 182(f) of the Act do not apply;

(ii) For ozone transport regions established pursuant to section 184 of the Act, sources with the potential to emit 50 tpy or more of volatile organic compounds;

(iii) For carbon monoxide nonattainment areas:

(A) That are classified as "serious," and

(B) in which stationary sources contribute significantly to carbon monoxide levels as determined under rules issued by the Administrator, sources with the potential to emit 50 tpy or more of carbon monoxide; and

(iv) For particulate matter (PM–10) nonattainment areas classified as "serious," sources with the potential to emit 70 tpy or more of PM–10.

Part 70 permit or *permit* (unless the context suggests otherwise) means any permit or group of permits covering a part 70 source that is issued, renewed, amended, or revised pursuant to this part.

Part 70 program or *State program* means a program approved by the Administrator under this part.

Part 70 source means any source subject to the permitting requirements of this part, as provided in §§ 70.3(a) and 70.3(b) of this part.

Permit modification means a revision to a part 70 permit that meets the requirements of § 70.7(e) of this part.

Permit program costs means all reasonable (direct and indirect) costs required to develop and administer a permit program, as set forth in § 70.9(b) of this part (whether such costs are incurred by the permitting authority or other State or local agencies that do not issue permits directly, but that support permit issuance or administration).

Permit revision means any permit modification or administrative permit amendment.

Permitting authority means either of the following:

(1) The Administrator, in the case of EPA-implemented programs; or

(2) The State air pollution control agency, local agency, other State agency, or other agency authorized by the Administrator to carry out a permit program under this part.

Potential to emit means the maximum capacity of a stationary source to emit any air pollutant under its physical and operational design. Any physical or operational limitation on the capacity of a source to emit an air pollutant, including air pollution control equipment and restrictions on hours of operation or on the type or amount of material combusted, stored, or processed, shall be treated as part of its design if the limitation is enforceable by the Administrator. This term does not alter or affect the use of this term for any other purposes under the Act, or the term "capacity factor" as used in title IV of the Act or the regulations promulgated thereunder.

Proposed permit means the version of a permit that the permitting authority proposes to issue and forwards to the Administrator for review in compliance with § 70.8.

Regulated air pollutant means the following:

(1) Nitrogen oxides or any volatile organic compounds;

(2) Any pollutant for which a national ambient air quality standard has been promulgated;

(3) Any pollutant that is subject to any standard promulgated under section 111 of the Act;

(4) Any Class I or II substance subject to a standard promulgated under or established by title VI of the Act; or

(5) Any pollutant subject to a standard promulgated under section 112 or other requirements established under section 112 of the Act, including sections 112(g), (j), and (r) of the Act, including the following:

(i) Any pollutant subject to requirements under section 112(j) of the Act. If the Administrator fails to promulgate a standard by the date established pursuant to section 112(e) of the Act, any pollutant for which a subject source would be major shall be considered to be regulated on the date 18

§ 70.3

months after the applicable date established pursuant to section 112(e) of the Act; and

(ii) Any pollutant for which the requirements of section 112(g)(2) of the Act have been met, but only with respect to the individual source subject to section 112(g)(2) requirement.

Regulated pollutant (for presumptive fee calculation), which is used only for purposes of § 70.9(b)(2), means any "regulated air pollutant" except the following:

(1) Carbon monoxide;

(2) Any pollutant that is a regulated air pollutant solely because it is a Class I or II substance to a standard promulgated under or established by title VI of the Act; or

(3) Any pollutant that is a regulated air pollutant solely because it is subject to a standard or regulation under section 112(r) of the Act.

Renewal means the process by which a permit is reissued at the end of its term.

Responsible official means one of the following:

(1) For a corporation: a president, secretary, treasurer, or vice-president of the corporation in charge of a principal business function, or any other person who performs similar policy or decision-making functions for the corporation, or a duly authorized representative of such person if the representative is responsible for the overall operation of one or more manufacturing, production, or operating facilities applying for or subject to a permit and either:

(i) The facilities employ more than 250 persons or have gross annual sales or expenditures exceeding $25 million (in second quarter 1980 dollars); or

(ii) The delegation of authority to such representatives is approved in advance by the permitting authority;

(2) For a partnership or sole proprietorship: a general partner or the proprietor, respectively;

(3) For a municipality, State, Federal, or other public agency: Either a principal executive officer or ranking elected official. For the purposes of this part, a principal executive officer of a Federal agency includes the chief executive officer having responsibility for the overall operations of a principal geographic unit of the agency (e.g., a Regional Administrator of EPA); or

(4) For affected sources:

(i) The designated representative in so far as actions, standards, requirements, or prohibitions under title IV of the Act or the regulations promulgated thereunder are concerned; and

(ii) The designated representative for any other purposes under part 70.

Section 502(b)(10) changes are changes that contravene an express permit term. Such changes do not include changes that would violate applicable requirements or contravene federally enforceable permit terms and conditions that are monitoring (including test methods), recordkeeping, reporting, or compliance certification requirements.

State means any non-Federal permitting authority, including any local agency, interstate association, or statewide program. The term "State" also includes the District of Columbia, the Commonwealth of Puerto Rico, the Virgin Islands, Guam, American Samoa, and the Commonwealth of the Northern Mariana Islands. Where such meaning is clear from the context, "State" shall have its conventional meaning. For purposes of the acid rain program, the term "State" shall be limited to authorities within the 48 contiguous States and the District of Columbia as provided in section 402(14) of the Act.

Stationary source means any building, structure, facility, or installation that emits or may emit any regulated air pollutant or any pollutant listed under section 112(b) of the Act.

Whole program means a part 70 permit program, or any combination of partial programs, that meet all the requirements of these regulations and cover all the part 70 sources in the entire State. For the purposes of this definition, the term "State" does not include local permitting authorities, but refers only to the entire State, Commonwealth, or Territory.

§ 70.3 Applicability.

(a) *Part 70 sources.* A State program with whole or partial approval under this part must provide for permitting of at least the following sources:

(1) Any major source;

(2) Any source, including an area source, subject to a standard, limitation, or other requirement under section 111 of the Act;

(3) Any source, including an area source, subject to a standard or other requirement under section 112 of the Act, except that a source is not required to obtain a permit solely because it is subject to regulations or requirements under section 112(r) of this Act;

(4) Any affected source; and

(5) Any source in a source category designated by the Administrator pursuant to this section.

(b) *Source category exemptions.* (1) All sources listed in paragraph (a) of this section that are not major sources, affected sources, or solid waste incineration units required to obtain a permit pursuant to section 129(e) of the Act, may be exempted by the State from the obligation to obtain a part 70 permit until such time as the Administrator completes a rulemaking to determine how the program should be structured for nonmajor sources and the appropriateness of any permanent exemptions in addition to those provided for in paragraph (b)(4) of this section.

(2) In the case of nonmajor sources subject to a standard or other requirement under either section 111 or section 112 of the Act after July 21, 1992 publication, the Administrator will determine whether to exempt any or all such applicable sources from the requirement to obtain a part 70 permit at the time that the new standard is promulgated.

(3) Any source listed in paragraph (a) of this section exempt from the requirement to obtain a permit under this section may opt to apply for a permit under a part 70 program.

(4) Unless otherwise required by the State to obtain a part 70 permit, the following source categories are exempted from the obligation to obtain a part 70 permit:

(i) All sources and source categories that would be required to obtain a permit solely because they are subject to part 60, subpart AAA—Standards of Performance for New Residential Wood Heaters; and

(ii) All sources and source categories that would be required to obtain a permit solely because they are subject to part 61, subpart M—National Emission Standard for Hazardous Air Pollutants for Asbestos, § 61.145, Standard for Demolition and Renovation.

(c) *Emissions units and part 70 sources.* (1) For major sources, the permitting authority shall include in the permit all applicable requirements for all relevant emissions units in the major source.

(2) For any nonmajor source subject to the part 70 program under paragraph (a) or (b) of this section, the permitting authority shall include in the permit all applicable requirements applicable to emissions units that cause the source to be subject to the part 70 program.

(d) *Fugitive emissions.* Fugitive emissions from a part 70 source shall be included in the permit application and the part 70 permit in the same manner as stack emissions, regardless of whether the source category in question is included in the list of sources contained in the definition of major source.

§ 70.4 State program submittals and transition.

(a) *Date for submittal.* Not later than November 15, 1993, the Governor of each State shall submit to the Administrator for approval a proposed part 70 program, under State law or under an interstate compact, meeting the requirements of this part. If part 70 is subsequently revised such that the Administrator determines that it is necessary to require a change to an approved State program, the required revisions to the program shall be submitted within 12 months of the final changes to part 70 or within such other period as authorized by the Administrator.

(b) *Elements of the initial program submission.* Any State that seeks to administer a program under this part shall submit to the Administrator a letter of submittal from the Governor or his designee requesting EPA approval of the program and at least three copies of a program submission. The submission shall contain the following:

(1) A complete program description describing how the State intends to carry out its responsibilities under this part.

(2) The regulations that comprise the permitting program, reasonably available evidence of their procedurally correct adoption, (including any notice of public comment and any significant comments received on the proposed part 70 program as requested by the Administrator), and copies of all applicable State or local statutes and regulations including those governing State administrative procedures that either authorize the part 70 program or restrict its implementation. The State shall include with the regulations any criteria used to determine insignificant activities or emission levels for purposes of determining complete applications consistent with § 70.5(c) of this part.

(3) A legal opinion from the Attorney General for the State, or the attorney for those State, local, or interstate air pollution control agencies that have independent legal counsel, stating that the laws of the State, locality, or interstate compact provide adequate authority to carry out all aspects of the program. This statement shall include citations to the specific states, administrative regulations, and, where appropriate, judicial decisions that demonstrate adequate authority. State statutes and regulations cited by the State Attorney General or independent legal counsel shall be in the form of lawfully adopted State states and regulations at the time the statement is signed and shall be fully effective by the time the program is approved. To qualify as "independent legal counsel," the attorney signing the statement required by this section shall have full authority to independently represent the State agency in court on all matters pertaining to the State program. The legal opinion shall also include a demonstration of adequate legal authority to carry out the requirements of this part, including authority to carry out each of the following:

(i) Issue permits and assure compliance with each applicable requirement and requirement of this part by all part 70 sources.

(ii) Incorporate monitoring, recordkeeping, reporting, and compliance certification requirements into part 70 permits consistent with § 70.6.

(iii) Issue permits for a fixed term of 5 years in the case of permits with acid rain provisions and issue all other permits for a period not to exceed 5 years, except for permits issued for solid waste incineration units combusting municipal waste

§ 70.4

subject to standards under section 129(e) of the Act.

(iv) Issue permits for solid waste incineration units combusting municipal waste subject to standards under section 129(e) of the Act for a period not to exceed 12 years and review such permits at least every 5 years. No permit for a solid waste incineration unit may be issued by an agency, instrumentality or person that is also responsible, in whole or in part, for the design and construction or operation of the unit.

(v) Incorporate into permits all applicable requirements and requirements of this part.

(vi) Terminate, modify, or revoke and reissue permits for cause.

(vii) Enforce permits, permit fee requirements, and the requirement to obtain a permit, as specified in § 70.11.

(viii) Make available to the public any permit application, compliance plan, permit, and monitoring and compliance, certification report pursuant to section 503(e) of the Act, except for information entitled to confidential treatment pursuant to section 114(c) of the Act. The contents of a part 70 permit shall not be entitled to protection under section 115(c) of the Act.

(ix) Not issue a permit if the Administrator timely objects to its issuance pursuant to § 70.8(c) of this part or, if the permit has not already been issued, to § 70.8(d) of this part.

(x) Provide an opportunity for judicial review in State court of the final permit action by the applicant, any person who participated in the public participation process provided pursuant to § 70.7(h) of this part, and any other person who could obtain judicial review of such actions under State laws.

(xi) Provide that, solely for the purposes of obtaining judicial review in State court for failure to take final action, final permit action shall include the failure of the permitting authority to take final action on an application for a permit, permit renewal, or permit revision within the time specified in the State program. If the State program allows sources to make changes subject to post hoc review [as set forth in §§ 70.7(e)(2) and (3) of this part], the permitting authority's failure to take final action within 90 days of receipt of an application requesting minor permit modification procedures (or 180 days for modifications subject to group processing requirements) must be subject to judicial review in State court.

(xii) Provide that the opportunity for judicial review described in paragraph (b)(3)(x) of this section shall be the exclusive means for obtaining judicial review of the terms and conditions of permits, and require that such petitions for judicial review must be filed no later than 90 days after the final permit action, or such shorter time as the

State shall designate. Notwithstanding the preceding requirement, petitions for judicial review of final permit actions can be filed after the deadline designated by the State, only if they are based solely on grounds arising after the deadline for judicial review. Such petitions shall be filed no later than 90 days after the new grounds for review arise or such shorter time as the State shall designate. If the final permit action being challenged is the permitting authority's failure to take final action, a petition for judicial review may be filed any time before the permitting authority denies the permit or issues the final permit.

(xiii) Ensure that the authority of the State/local permitting Agency is not used to modify the acid rain program requirements.

(4) Relevant permitting program documentation not contained in the State regulations, including the following:

(i) Copies of the permit form(s), application form(s), and reporting form(s) the State intends to employ in its program; and

(ii) Relevant guidance issued by the State to assist in the implementation of its permitting program, including criteria for monitoring source compliance (e.g., inspection strategies).

(5) A complete description of the State's compliance tracking and enforcement program or reference to any agreement the State has with EPA that provides this information.

(6) A showing of adequate authority and procedures to determine within 60 days of receipt whether applications (including renewal applications) are complete, to request such other information as needed to process the application, and to take final action on complete applications within 18 months of the date of their submittal, except for initial permit applications, for which the permitting authority may take up to 3 years from the effective date of the program to take final action on the application, as provided for in the transition plan.

(7) A demonstration, consistent with § 70.9, that the permit fees required by the State program are sufficient to cover permit program costs.

(8) A statement that adequate personnel and funding have been made available to develop, administer, and enforce the program. This statement shall include the following:

(i) A description in narrative form of the scope, structure, coverage, and processes of the State program.

(ii) A description of the organization and structure of the agency or agencies that will have responsibility for administering the program, including the information specified in this paragraph. If more than one agency is responsible for administration of a program, the responsibilities of each agency must be delineated, their procedures for coordination must be set forth, and an agency shall be designated as a "lead agency" to facilitate communications between EPA and the other agencies having program responsibility.

(iii) A description of the agency staff who will carry out the State program, including the number, occupation, and general duties of the employees. The State need not submit complete job descriptions for every employee carrying out the State program.

(iv) A description of applicable State procedures, including permitting procedures and any State administrative or judicial review procedures.

(v) An estimate of the permit program costs for the first 4 years after approval, and a description of how the State plans to cover those costs.

(9) A commitment from the State to submit, at least annually to the Administrator, information regarding the State's enforcement activities including, but not limited to, the number of criminal and civil, judicial and administrative enforcement actions either commenced or concluded; the penalties, fines, and sentences obtained in those actions; and the number of administrative orders issued.

(10) A requirement under State law that, if a timely and complete application for a permit renewal is submitted, consistent with § 70.5(a)(2), but the State has failed to issue or deny the renewal permit before the end of the term of the previous permit, then:

(i) The permit shall not expire until the renewal permit has been issued or denied and any permit shield that may be granted pursuant to § 70.6(f) may extend beyond the original permit term, until renewal; or

(ii) All the terms and conditions of the permit including any permit shield that may be granted pursuant to § 70.6(f) shall remain in effect until the renewal permit has been issued or denied.

(11) A transition plan providing a schedule for submittal and final action on initial permit applications for all part 70 sources. This plan shall provide that:

(i) Submittal of permit applications by all part 70 sources (including any sources subject to a partial or interim program) shall occur within 1 year after the effective date of the permit program;

(ii) Final action shall be taken on at least one-third of such applications annually over a period not to exceed 3 years after such effective date;

(iii) Any complete permit application containing an early reduction demonstration under section 112(i)(5) of the Act shall be acted on within 9 months of receipt of the complete application; and

(iv) Submittal of permit applications and the permitting of affected sources shall occur in accordance with the deadlines in title IV of the Act and the regulations promulgated thereunder.

(12) Provisions consistent with paragraphs (b)(12)(i) through (iii) of this section to allow changes within a permitted facility without requiring a permit revision, if the changes are not modifications under any provision of title I of the Act and the changes do not exceed the emissions allowable under the permit (whether expressed therein as a rate of emissions or in the terms of total emissions): *Provided*, That the facility provides the Administrator and the permitting authority with written notification as required below in advance of the proposed changes, which shall be a minimum of 7 days, unless the permitting authority provides in its regulations a different time frame for emergencies. The source, permitting authority, and EPA shall attach each such notice to their copy of the relevant permit. The following provisions implement this requirement of an approvable part 70 permit program:

(i) The program shall allow permitted sources to make section 502(b)(10) changes without requiring a permit revision, if the changes are not modifications under any provision of title I of the Act and the changes do not exceed the emissions allowable under the permit (whether expressed therein as a rate of emissions or in terms of total emissions).

(A) For each such change, the written notification required above shall include a brief description of the change within the permitted facility, the date on which the change will occur, any change in emissions, and any permit term or condition that is no longer applicable as a result of the change.

(B) The permit shield described in § 70.6(f) of this part shall not apply to any change made pursuant to this paragraph (b)(12)(i) of this section.

(ii) The program may provide for permitted sources to trade increases and decreases in emissions in the permitted facility, where the applicable implementation plan provides for such emissions trades without requiring a permit revision and based on the 7-day notice prescribed in this paragraph (b)(12)(ii) of this section. This provision is available in those cases where the permit does not already provide for such emissions trading.

(A) Under this paragraph (b)(12)(ii) of this section, the written notification required above shall include such information as may be required by the provision in the applicable implementation plan authorizing the emissions trade, including at a minimum, when the proposed change will occur, a description of each such change, any change in emissions, the permit requirements with which the source will comply using the emissions trading provisions of the applicable implementation plan, and the pollutants emitted subject to the emissions trade. The notice shall also refer to the provisions with which the source will comply in the applica-

§ 70.4

ble implementation plan and that provide for the emissions trade.

(B) The permit shield described in § 70.6(f) of this part shall not extend to any change made under this paragraph (b)(12)(ii) of this section. Compliance with the permit requirements that the source will meet using the emissions trade shall be determined according to requirements of the applicable implementation plan authorizing the emissions trade.

(iii) The program shall require the permitting authority, if a permit applicant requests it, to issue permits that contain terms and conditions, including all terms required under § 70.6 (a) and (c) of this part to determine compliance, allowing for the trading of emissions increases and decreases in the permitted facility solely for the purpose of complying with a federally-enforceable emissions cap that is established in the permit independent of otherwise applicable requirements. The permit applicant shall include in its application proposed replicable procedures and permit terms that ensure the emissions trades are quantifiable and enforceable. The permitting authority shall not be required to include in the emissions trading provisions any emissions units for which emissions are not quantifiable or for which there are no replicable procedures to enforce the emissions trades. The permit shall also require compliance with all applicable requirements.

(A) Under this paragraph (b)(12)(iii) of this section, the written notification required above shall state when the change will occur and shall describe the changes in emissions that will result and how these increases and decreases in emissions will comply with the terms and conditions of the permit.

(B) The permit shield described in § 70.6(f) of this part may extend to terms and conditions that allow such increases and decreases in emissions.

(13) Provisions for adequate, streamlined, and reasonable procedures for expeditious review of permit revisions or modifications. The program may meet this requirement by using procedures that meet the requirements of § 70.7(e) or that are substantially equivalent to those provided in § 70.7(e) of this part.

(14) If a State allows changes that are not addressed or prohibited by the permit, other than those described in paragraph (b)(15) of this section, to be made without a permit revision, provisions meeting the requirements of paragraphs (b)(14) (i) through (iii) of this section. Although a State may, as a matter of State law, prohibit sources from making such changes without a permit revision, any such prohibition shall not be enforceable by the Administrator or by citizens under the Act unless the prohibition is required by an applicable requirement. Any State procedures

implementing such a State law prohibition must include the requirements of paragraphs (b)(14) (i) through (iii) of this section.

(i) Each such change shall meet all applicable requirements and shall not violate any existing permit term or condition.

(ii) Sources must provide contemporaneous written notice to the permitting authority and EPA of each such change, except for changes that qualify as insignificant under the provisions adopted pursuant to § 70.5(c) of this part. Such written notice shall describe each such change, including the date, any change in emissions, pollutants emitted, and any applicable requirement that would apply as a result of the change.

(iii) The change shall not qualify for the shield under § 70.6(f) of this part.

(iv) The permittee shall keep a record describing changes made at the source that result in emissions of a regulated air pollutant subject to an applicable requirement, but not otherwise regulated under the permit, and the emissions resulting from those changes.

(15) Provisions prohibiting sources from making, without a permit revision, changes that are not addressed or prohibited by the part 70 permit, if such changes are subject to any requirements under title IV of the Act or are modifications under any provision of title I of the Act.

(16) Provisions requiring the permitting authority to implement the requirements of §§ 70.6 and 70.7 of this part.

(c) *Partial programs.* (1) The EPA may approve a partial program that applies to all part 70 sources within a limited geographic area (e.g., a local agency program covering all sources within the agency's jurisdiction). To be approvable, any partial program must, at a minimum, ensure compliance with all of the following applicable requirements, as they apply to the sources covered by the partial program:

(i) All requirements of title V of the Act and of part 70;

(ii) All applicable requirements of title IV of the Act and regulations promulgated thereunder which apply to affected sources; and

(iii) All applicable requirements of title I of the Act, including those established under sections 111 and 112 of the Act.

(2) Any partial permitting program, such as that of a local air pollution control agency, providing for the issuance of permits by a permitting authority other than the State, shall be consistent with all the elements required in paragraphs (b) (1) through (16) of this section.

(3) Approval of any partial program does not relieve the State from its obligation to submit a whole program or from application of any sanc-

tions for failure to submit a fully-approvable whole program.

(4) Any partial program may obtain interim approval under paragraph (d) of this section if it substantially meets the requirements of this paragraph (c) of this section.

(d) *Interim approval.* (1) If a program (including a partial permit program) submitted under this part substantially meets the requirements of this part, but is not fully approvable, the Administrator may be rule grant the program interim approval.

(2) Interim approval shall expire on a date set by the Administrator (but not later than 2 years after such approval), and may not be renewed. Sources shall become subject to the program according to the schedule approved in the State program. Permits granted under an interim approval shall expire at the end of their fixed term, unless renewed under a part 70 program.

(3) The EPA may grant interim approval to any program if it meets each of the following minimum requirements and otherwise substantially meets the requirements of this part:

(i) *Adequate fees.* The program must provide for collecting permit fees adequate for it to meet the requirements of § 70.9 of this part.

(ii) *Applicable requirements.* (A) The program must provide for adequate authority to issue permits that assure compliance with the requirements of paragraph (c)(1) of this section for those major sources covered by the program.

(B) Notwithstanding paragraph (d)(3)(ii)(A) of this section, where a State or local permitting authority lacks adequate authority to issue or revise permits that assure compliance with applicable requirements established exclusively through an EPA-approved minor NSR program, EPA may grant interim approval to the program upon a showing by the permitting authority of compelling reasons which support the interim approval.

(C) Any part 70 permit issued during an interim approval granted under paragraph (d)(3)(ii)(B) of this section that does not incorporate minor NSR requirements shall:

(*1*) Note this fact in the permit;

(*2*) Indicate how citizens may obtain access to excluded minor NSR permits;

(*3*) Provide a cross reference, such as a listing of the permit number, for each minor NSR permit containing an excluded minor NSR term; and

(*4*) State that the minor NSR requirements which are excluded are not eligible for the permit shield under § 70.6(f).

(D) A program receiving interim approval for the reason specified in (d)(3)(ii)(B) of this section must, upon or before granting of full approval, institute proceedings to reopen part 70 permits to incorporate excluded minor NSR permits as terms of the part 70 permits, as required by § 70.7(f)(1)(iv).

Such reopening need not follow full permit issuance procedures nor the notice requirement of § 70.7(f)(3), but may instead follow the permit revision procedure in effect under the State's approved part 70 program for incorporation of minor NSR permits.

(iii) *Fixed term.* The program must provide for fixed permit terms, consistent with paragraphs (b)(3) (iii) and (iv) of this section.

(iv) *Public participation.* The program must provide for adequate public notice of and an opportunity for public comment and a hearing on draft permits and revisions, except for modifications qualifying for minor permit modification procedures under § 70.7(e) of this part.

(v) *EPA and affected State review.* The program must allow EPA an opportunity to review each proposed permit, including permit revisions, and to object to its issuance consistent with § 70.8(c) of this part. The program must provide for affected State review consistent with § 70.8(b) of this part.

(vii) *Permit issuance.* The program must provide that the proposed permit will not be issued if EPA objects to its issuance.

(vii) *Enforcement.* The program must contain authority to enforce permits, including the authority to assess penalties against sources that do not comply with their permits or with the requirement to obtain a permit.

(viii) *Operational flexibility.* The program must allow changes within a permitted facility without requiring a permit revision, if the changes are not modifications under any provision of title I of the act and the changes do not exceed the emissions allowable under the permit, consistent with paragraph (b)(12) of this section.

(ix) *Streamlined procedures.* The program must provide for streamlined procedures for issuing and revising permits and determining expeditiously after receipt of a permit application or application for a permit revision whether such application is complete.

(x) *Permit application.* The program submittal must include copies of the permit application and reporting form(s) that the State will use in implementing the interim program.

(xi) *Alternative scenarios.* The program submittal must include provisions to insure that alternate scenarios requested by the source are included in the part 70 permit pursuant to § 70.6(a)(9) of this part.

(e) *EPA review of permit program submittals.* Within 1 year after receiving a program submittal, the Administrator shall approve or disapprove the program, in whole or in part, by publishing a notice in the FEDERAL REGISTER. Prior to such notice, the Administrator shall provide an opportunity for public comment on such approval or disapproval. Any EPA action disapproving a pro-

§ 70.4

gram, in whole or in part, shall include a statement of the revisions or modifications necessary to obtain full approval. The Administrator shall approve State programs that conform to the requirements of this part.

(1) Within 60 days of receipt by EPA of a State program submission, EPA will notify the State whether its submission is complete enough to warrant review by EPA for either full, partial, or interim approval. If EPA finds that a State's submission is complete, the 1-year review period (i.e., the period of time allotted for formal EPA review of a proposed State program) shall be deemed to have begun on the date of receipt of the State's submission. If EPA finds that a State's submission is incomplete, the 1-year review period shall not begin until all the necessary information is received by EPA.

(2) If the State's submission is materially changed during the 1-year review period, the Administrator may extend the review period for no more than 1 year following receipt of the revised submission.

(3) In any notice granting interim or partial approval, the Administrator shall specify the changes or additions that must be made before the program can receive full approval and the conditions for implementation of the program until that time.

(f) *State response to EPA review of program—*
(1) *Disapproval.* The State shall submit to EPA program revisions or modifications required by the Administrator's action disapproving the program, or any part thereof, within 180 days of receiving notification of the disapproval.

(2) *Interim approval.* The State shall submit to EPA changes to the program addressing the deficiencies specified in the interim approval no later than 6 months prior to the expiration of the interim approval.

(g) *Effective date.* The effective date of a part 70 program, including any partial or interim program approved under this part, shall be the effective date of approval by the Administrator.

(h) *Individual permit transition.* Upon approval of a State program, the Administrator shall suspend the issuance of Federal permits for those activities subject to the approved State program, except that the Administrator will continue to issue phase I acid rain permits. After program approval, EPA shall retain jurisdiction over any permit (including any general permit) that it has issued unless arrangements have been made with the State to assume responsibility for these permits. Where EPA retains jurisdiction, it will continue to process permit appeals and modification requests, to conduct inspections, and to receive and review monitoring reports. If any permit appeal or modification request is not finally resolved when the federally-issued permit expires, EPA may, with the

consent of the State, retain jurisdiction until the matter is resolved. Upon request by a State, the Administrator may delegate authority to implement all or part of a permit issued by EPA, if a part 70 program has been approved for the State. The delegation may include authorization for the State to collect appropriate fees, consistent with § 70.9 of this part.

(i) *Program revisions.* Either EPA or a State with an approved program may initiate a program revision. Program revision may be necessary when the relevant Federal or State statutes or regulations are modified or supplemented. The State shall keep EPA apprised of any proposed modifications to its basic statutory or regulatory authority or procedures.

(1) If the Administrator determines pursuant to § 70.10 of this part that a State is not adequately administering the requirements of this part, or that the State's permit program is inadequate in any other way, the State shall revise the program or its means of implementation to correct the inadequacy. The program shall be revised within 180 days, or such other period as the Administrator may specify, following notification by the Administrator, or within 2 years if the State demonstrates that additional legal authority is necessary to make the program revision.

(2) Revision of a State program shall be accomplished as follows:

(i) The State shall submit a modified program description, Attorney General's statement, or such other documents as EPA determines to be necessary.

(ii) After EPA receives a proposed program revision, it will publish in the FEDERAL REGISTER a public notice summarizing the proposed change and provide a public comment period of at least 30 days.

(iii) The Administrator shall approve or disapprove program revisions based on the requirements of this part and of the Act.

(iv) A program revision shall become effective upon the approval of the Administrator. Notice of approval of any substantial revision shall be published in the FEDERAL REGISTER. Notice of approval of nonsubstantial program revisions may be given by a letter from the Administrator to the Governor or a designee.

(v) The Governor of any State with an approved part 70 program shall notify EPA whenever the Governor proposes to transfer all or part of the program to any other agency, and shall identify any new division of responsibilities among the agencies involved. The new agency is not authorized to administer the program until the revision has been approved by the Administrator under this paragraph.

(3) Whenever the Administrator has reason to believe that circumstances have changed with respect to a State program, he may request, and the State shall provide, a supplemental Attorney General's statement, program description, or such other documents or information as he determines are necessary.

(j) *Sharing of information.* (1) Any information obtained or used in the administration of a State program shall be available to EPA upon request without restriction and in a form specified by the Administrator, including computer-readable files to the extent practicable. If the information has been submitted to the State under a claim of confidentiality, the State may require the source to submit this information to the Administrator directly. Where the State submits information to the Administrator under a claim of confidentiality, the State shall submit that claim to EPA when providing information to EPA under this section. Any information obtained from a State or part 70 source accompanied by a claim of confidentiality will be treated in accordance with the regulations in part 2 of this chapter.

(2) The EPA will furnish to States with approved programs the information in its files that the State needs to implement its approved program. Any such information submitted to EPA under a claim of confidentiality will be subject to the regulations in part 2 of this chapter.

(k) *Administration and enforcement.* Any State that fails to adopt a complete, approvable part 70 program, or that EPA determines is not adequately administering or enforcing such program shall be subject to certain Federal sanctions as set forth in § 70.10 of this part.

[57 FR 32295, July 21, 1992, as amended at 61 FR 31448, June 20, 1996]

EFFECTIVE DATE NOTE: At 61 FR 31448, June 20, 1996, in § 70.4, paragraphs (d)(3) introductory text and (d)(3)(ii) were revised, effective July 22, 1996. For the convenience of the user, the superseded text is set forth as follows.

§ 70.4   State program submittals and transition.

*   *   *   *   *

(d) * * *

(3) The EPA will grant interim approval to any program if it meets each of the following minimum requirements:

*   *   *   *   *

(ii) *Applicable requirements.* The program must provide for adequate authority to issue permits that assure compli-

ance with the requirements of paragraph (c)(1) of this section for those major sources covered by the program.

*   *   *   *   *

§ 70.5   Permit applications.

(a) *Duty to apply.* For each part 70 source, the owner or operator shall submit a timely and complete permit application in accordance with this section.

(1) *Timely application.* (i) A timely application for a source applying for a part 70 permit for the first time is one that is submitted within 12 months after the source becomes subject to the permit program or on or before such earlier date as the permitting authority may establish.

(ii) Part 70 sources required to meet the requirements under section 112(g) of the Act, or to have a permit under the preconstruction review program approved into the applicable implementation plan under part C or D of title I of the Act, shall file a complete application to obtain the part 70 permit or permit revision within 12 months after commencing operation or on or before such earlier date as the permitting authority may establish. Where an existing part 70 permit would prohibit such construction or change in operation, the source must obtain a permit revision before commencing operation.

(iii) For purposes of permit renewal, a timely application is one that is submitted at least 6 months prior to the date of permit expiration, or such other longer time as may be approved by the Administrator that ensures that the term of the permit will not expire before the permit is renewed. In no event shall this time be greater than 18 months.

(iv) Applications for initial phase II acid rain permits shall be submitted to the permitting authority by January 1, 1996 for sulfur dioxide, and by January 1, 1998 for nitrogen oxides.

(2) *Complete application.* The program shall provide criteria and procedures for determining in a timely fashion when applications are complete. To be deemed complete, an application must provide all information required pursuant to paragraph (c) of this section, except that applications for permit revision need supply such information only if it is related to the proposed change. Information required under paragraph (c) of this section must be sufficient to evaluate the subject source and its application and to determine all applicable requirements. The program shall require that a responsible official certify the submitted information consistent with paragraph (d) of this section. Unless the permitting authority determines that an application is not complete within 60 days of receipt of the application, such application shall be deemed to be complete, except as otherwise pro-

§ 70.5

vided in § 70.7(a)(4) of this part. If, while processing an application that has been determined or deemed to be complete, the permitting authority determines that additional information is necessary to evaluate or take final action on that application, it may request such information in writing and set a reasonable deadline for a response. The source's ability to operate without a permit, as set forth in § 70.7(b) of this part, shall be in effect from the date the application is determined or deemed to be complete until the final permit is issued, provided that the applicant submits any requested additional information by the deadline specified by the permitting authority.

(3) *Confidential information.* In the case where a source has submitted information to the State under a claim of confidentiality, the permitting authority may also require the source to submit a copy of such information directly to the Administrator.

(b) *Duty to supplement or correct application.* Any applicant who fails to submit any relevant facts or who has submitted incorrect information in a permit application shall, upon becoming aware of such failure or incorrect submittal, promptly submit such supplementary facts or corrected information. In addition, an applicant shall provide additional information as necessary to address any requirements that become applicable to the source after the date it filed a complete application but prior to release of a draft permit.

(c) *Standard application form and required information.* The State program under this part shall provide for a standard application form or forms. Information as described below for each emissions unit at a part 70 source shall be included in the application. The Administrator may approve as part of a State program a list of insignificant activities and emissions levels which need not be included in permit applications. However, for insignificant activities which are exempted because of size or production rate, a list of such insignificant activities must be included in the application. An application may not omit information needed to determine the applicability of, or to impose, any applicable requirement, or to evaluate the fee amount required under the schedule approved pursuant to § 70.9 of this part. The permitting authority may use discretion in developing application forms that best meet program needs and administrative efficiency. The forms and attachments chosen, however, shall include the elements specified below:

(1) Identifying information, including company name and address (or plant name and address if different from the company name), owner's name and agent, and telephone number and names of plant site manager/contact.

(2) A description of the source's processes and products (by Standard Industrial Classification Code) including any associated with alternate scenario identified by the source.

(3) The following emission-related information:

(i) All emissions of pollutants for which the source is major, and all emissions of regulated air pollutants. A permit application shall describe all emissions of regulated air pollutants emitted from any emissions unit, except where such units are exempted under this paragraph (c) of this section. The permitting authority shall require additional information related to the emissions of air pollutants sufficient to verify which requirements are applicable to the source, and other information necessary to collect any permit fees owed under the fee schedule approved pursuant to § 70.9(b) of this part.

(ii) Identification and description of all points of emissions described in paragraph (c)(3)(i) of this section in sufficient detail to establish the basis for fees and applicability of requirements of the Act.

(iii) Emissions rate in tpy and in such terms as are necessary to establish compliance consistent with the applicable standard reference test method.

(iv) The following information to the extent it is needed to determine or regulate emissions: Fuels, fuel use, raw materials, production rates, and operating schedules.

(v) Identification and description of air pollution control equipment and compliance monitoring devices or activities.

(vi) Limitations on source operation affecting emissions or any work practice standards, where applicable, for all regulated pollutants at the part 70 source.

(vii) Other information required by any applicable requirement (including information related to stack height limitations developed pursuant to section 123 of the Act).

(viii) Calculations on which the information in paragraphs (c)(3 (i) through (vii) of this section is based.

(4) The following air pollution control requirements:

(i) Citation and description of all applicable requirements, and

(ii) Description of or reference to any applicable test method for determining compliance with each applicable requirement.

(5) Other specific information that may be necessary to implement and enforce other applicable requirements of the Act or of this part or to determine the applicability of such requirements.

(6) An explanation of any proposed exemptions from otherwise applicable requirements.

(7) Additional information as determined to be necessary by the permitting authority to define alternative operating scenarios identified by the

source pursuant to § 70.6(a)(9) of this part or to define permit terms and conditions implementing § 70.4 (b) (12) or § 70.6 (a) (10) of this part.

(8) A compliance plan for all part 70 sources that contains all the following:

(i) A description of the compliance status of the source with respect to all applicable requirements.

(ii) A description as follows:

(A) For applicable requirements with which the source is in compliance, a statement that the source will continue to comply with such requirements.

(B) For applicable requirements that will become effective during the permit term, a statement that the source will meet such requirements on a timely basis.

(C) For requirements for which the source is not in compliance at the time or permit issuance, a narrative description of how the source will achieve compliance with such requirements.

(iii) A compliance schedule as follows:

(A) For applicable requirements with which the source is in compliance, a statement that the source will continue to comply with such requirements.

(B) For applicable requirements that will become effective during the permit term, a statement that the source will meet such requirements on a timely basis. A statement that the source will meet in a timely manner applicable requirements that become effective during the permit term shall satisfy this provision, unless a more detailed schedule is expressly required by the applicable requirement.

(C) A schedule of compliance for sources that are not in compliance with all applicable requirements at the time of permit issuance. Such a schedule shall include a schedule of remedial measures, including an enforceable sequence of actions with milestones, leading to compliance with any applicable requirements for which the source will be in noncompliance at the time of permit issuance. This compliance schedule shall resemble and be at least as stringent as that contained in any judicial consent decree or administrative order to which the source is subject. Any such schedule of compliance shall be supplemental to, and shall not sanction noncompliance with, the applicable requirements on which it is based.

(iv) A schedule for submission of certified progress reports no less frequently than every 6 months for sources required to have a schedule of compliance to remedy a violation.

(v) The compliance plan content requirements specified in this paragraph shall apply and be included in the acid rain portion of a compliance plan for an affected source, except as specifically superseded by regulations promulgated under title IV of the Act with regard to the schedule and

method(s) the source will use to achieve compliance with the acid rain emissions limitations.

(9) Requirements for compliance certification, including the following:

(i) A certification of compliance with all applicable requirements by a responsible official consistent with paragraph (d) of this section and section 114(a)(3) of the Act;

(ii) A statement of methods used for determining compliance, including a description of monitoring, recordkeeping, and reporting requirements and test methods;

(iii) A schedule for submission of compliance certifications during the permit term, to be submitted no less frequently than annually, or more frequently if specified by the underlying applicable requirement or by the permitting authority; and

(iv) A statement indicating the source's compliance status with any applicable enhanced monitoring and compliance certification requirements of the Act.

(10) The use of nationally-standardized forms for acid rain portions of permit applications and compliance plans, as required by regulations promulgated under title IV of the Act.

(d) Any application form, report, or compliance certification submitted pursuant to these regulations shall contain certification by a responsible official of truth, accuracy, and completeness. This certification and any other certification required under this part shall state that, based on information and belief formed after reasonable inquiry, the statements and information in the document are true, accurate, and complete.

§ 70.6 Permit content.

(a) *Standard permit requirements.* Each permit issued under this part shall include the following elements:

(1) Emission limitations and standards, including those operational requirements and limitations that assure compliance with all applicable requirements at the time of permit issuance.

(i) The permit shall specify and reference the origin of and authority for each term or condition, and identify any difference in form as compared to the applicable requirement upon which the term or condition is based.

(ii) The permit shall state that, where an applicable requirement of the Act is more stringent than an applicable requirement of regulations promulgated under title IV of the Act, both provisions shall be incorporated into the permit and shall be enforceable by the Administrator.

(iii) If an applicable implementation plan allows a determination of an alternative emission limit at a part 70 source, equivalent to that contained in the plan, to be made in the permit issuance, renewal, or significant modification process, and the

§ 70.6

State elects to use such process, any permit containing such equivalency determination shall contain provisions to ensure that any resulting emissions limit has been demonstrated to be quantifiable, accountable, enforceable, and based on replicable procedures.

(2) *Permit duration.* The permitting authority shall issue permits for a fixed term of 5 years in the case of affected sources, and for a term not to exceed 5 years in the case of all other sources. Notwithstanding this requirement, the permitting authority shall issue permits for solid waste incineration units combusting municipal waste subject to standards under section 129(e) of the Act for a period not to exceed 12 years and shall review such permits at least every 5 years.

(3) *Monitoring and related recordkeeping and reporting requirements.* (i) Each permit shall contain the following requirements with respect to monitoring:

(A) All emissions monitoring and analysis procedures or test methods required under the applicable requirements, including any procedures and methods promulgated pursuant to sections 114(a)(3) or 504(b) of the Act;

(B) Where the applicable requirement does not require periodic testing or instrumental or noninstrumental monitoring (which may consist of recordkeeping designed to serve as monitoring), periodic monitoring sufficient to yield reliable data from the relevant time period that are representative of the source's compliance with the permit, as reported pursuant to paragraph (a)(3)(iii) of this section. Such monitoring requirements shall assure use of terms, test methods, units, averaging periods, and other statistical conventions consistent with the applicable requirement. Recordkeeping provisions may be sufficient to meet the requirements of this paragraph (a)(3)(i)(B) of this section; and

(C) As necessary, requirements concerning the use, maintenance, and, where appropriate, installation of monitoring equipment or methods.

(ii) With respect to recordkeeping, the permit shall incorporate all applicable recordkeeping requirements and require, where applicable, the following:

(A) Records of required monitoring information that include the following:

(*1*) The date, place as defined in the permit, and time of sampling or measurements;

(*2*) The date(s) analyses were performed;

(*3*) The company or entity that performed the analyses;

(*4*) The analytical techniques or methods used;

(*5*) The results of such analyses; and

(*6*) The operating conditions as existing at the time of sampling or measurement;

(B) Retention of records of all required monitoring data and support information for a period of at least 5 years from the date of the monitoring sample, measurement, report, or application. Support information includes all calibration and maintenance records and all original strip-chart recordings for continuous monitoring instrumentation, and copies of all reports required by the permit.

(iii) With respect to reporting, the permit shall incorporate all applicable reporting requirements and require the following:

(A) Submittal of reports of any required monitoring at least every 6 months. All instances of deviations from permit requirements must be clearly identified in such reports. All required reports must be certified by a responsible official consistent with § 70.5(d) of this part.

(B) Prompt reporting of deviations from permit requirements, including those attributable to upset conditions as defined in the permit, the probable cause of such deviations, and any corrective actions or preventive measures taken. The permitting authority shall define ''prompt'' in relation to the degree and type of deviation likely to occur and the applicable requirements.

(4) A permit condition prohibiting emissions exceeding any allowances that the source lawfully holds under title IV of the Act or the regulations promulgated thereunder.

(i) No permit revision shall be required for increases in emissions that are authorized by allowances acquired pursuant to the acid rain program, provided that such increases do not require a permit revision under any other applicable requirement.

(ii) No limit shall be placed on the number of allowances held by the source. The source may not, however, use allowances as a defense to noncompliance with any other applicable requirement.

(iii) Any such allowance shall be accounted for according to the procedures established in regulations promulgated under title IV of the Act.

(5) A severability clause to ensure the continued validity of the various permit requirements in the event of a challenge to any portions of the permit.

(6) Provisions stating the following:

(i) The permittee must comply with all conditions of the part 70 permit. Any permit noncompliance constitutes a violation of the Act and is grounds for enforcement action; for permit termination, revocation and reissuance, or modification; or for denial of a permit renewal application.

(ii) Need to halt or reduce activity not a defense. It shall not be a defense for a permittee in an enforcement action that it would have been necessary to halt or reduce the permitted activity in order to maintain compliance with the conditions of this permit.

(iii) The permit may be modified, revoked, reopened, and reissued, or terminated for cause. The filing of a request by the permittee for a permit modification, revocation and reissuance, or termination, or of a notification of planned changes or anticipated noncompliance does not stay any permit condition.

(iv) The permit does not convey any property rights of any sort, or any exclusive privilege.

(v) The permittee shall furnish to the permitting authority, within a reasonable time, any information that the permitting authority may request in writing to determine whether cause exists for modifying, revoking and reissuing, or terminating the permit or to determine compliance with the permit. Upon request, the permittee shall also furnish to the permitting authority copies of records required to be kept by the permit or, for information claimed to be confidential, the permittee may furnish such records directly to the Administrator along with a claim of confidentiality.

(7) A provision to ensure that a part 70 source pays fees to the permitting authority consistent with the fee schedule approved pursuant to § 70.9 of this part.

(8) *Emissions trading.* A provision stating that no permit revision shall be required, under any approved economic incentives, marketable permits, emissions trading and other similar programs or processes for changes that are provided for in the permit.

(9) Terms and conditions for reasonably anticipated operating scenarios identified by the source in its application as approved by the permitting authority. Such terms and conditions:

(i) Shall require the source, contemporaneously with making a change from one operating scenario to another, to record in a log at the permitted facility a record of the scenario under which it is operating;

(ii) May extend the permit shield described in paragraph (f) of this section to all terms and conditions under each such operating scenario; and

(iii) Must ensure that the terms and conditions of each such alternative scenario meet all applicable requirements and the requirements of this part.

(10) Terms and conditions, if the permit applicant requests them, for the trading of emissions increases and decreases in the permitted facility, to the extent that the applicable requirements provide for trading such increases and decreases without a case-by-case approval of each emissions trade. Such terms and conditions:

(i) Shall include all terms required under paragraphs (a) and (c) of this section to determine compliance;

(ii) May extend the permit shield described in paragraph (f) of this section to all terms and con-

ditions that allow such increases and decreases in emissions; and

(iii) Must meet all applicable requirements and requirements of this part.

(b) *Federally-enforceable requirements.* (1) All terms and conditions in a part 70 permit, including any provisions designed to limit a source's potential to emit, are enforceable by the Administrator and citizens under the Act.

(2) Notwithstanding paragraph (b)(1) of this section, the permitting authority shall specifically designate as not being federally enforceable under the Act any terms and conditions included in the permit that are not required under the Act or under any of its applicable requirements. Terms and conditions so designated are not subject to the requirements of §§ 70.7, 70.8, or of this part, other than those contained in this paragraph (b) of this section.

(c) *Compliance requirements.* All part 70 permits shall contain the following elements with respect to compliance:

(1) Consistent with paragraph (a)(3) of this section, compliance certification, testing, monitoring, reporting, and recordkeeping requirements sufficient to assure compliance with the terms and conditions of the permit. Any document (including reports) required by a part 70 permit shall contain a certification by a responsible official that meets the requirements of § 70.5(d) for this part.

(2) Inspection and entry requirements that require that, upon presentation of credentials and other documents as may be required by law, the permittee shall allow the permitting authority or an authorized representative to perform the following:

(i) Enter upon the permittee's premises where a part 70 source is located or emissions-related activity is conducted, or where records must be kept under the conditions of the permit;

(ii) Have access to and copy, at reasonable times, any records that must be kept under the conditions of the permit;

(iii) Inspect at reasonable times any facilities, equipment (including monitoring and air pollution control equipment), practices, or operations regulated or required under the permit; and

(iv) As authorized by the Act, sample or monitor at reasonable times substances or parameters for the purpose of assuring compliance with the permit or applicable requirements.

(3) A schedule of compliance consistent with § 70.5(c)(8) of this part.

(4) Progress reports consistent with an applicable schedule of compliance and § 70.5(c)(8) of this part to be submitted at least semiannually, or at a more frequent period if specified in the applicable requirement or by the permitting authority. Such progress reports shall contain the following:

§ 70.6

(i) Dates for achieving the activities, milestones, or compliance required in the schedule of compliance, and dates when such activities, milestones or compliance were achieved; and

(ii) An explanation of why any dates in the schedule of compliance were not or will not be met, and any preventive or corrective measures adopted.

(5) Requirements for compliance certification with terms and conditions contained in the permit, including emission limitations, standards, or work practices. Permits shall include each of the following:

(i) The frequency (not less than annually or such more frequent periods as specified in the applicable requirement or by the permitting authority) of submissions of compliance certifications;

(ii) In accordance with § 70.6(a)(3) of this part, a means for monitoring the compliance of the source with its emissions limitations, standards, and work practices;

(iii) A requirement that the compliance certification include the following:

(A) The identification of each term or condition of the permit that is the basis of the certification;

(B) The compliance status;

(C) Whether compliance was continuous or intermittent;

(D) The method(s) used for determining the compliance status of the source, currently and over the reporting period consistent with paragraph (a)(3) of this section; and

(E) Such other facts as the permitting authority may require to determine the compliance status of the source;

(iv) A requirement that all compliance certifications be submitted to the Administrator as well as to the permitting authority; and

(v) Such additional requirements as may be specified pursuant to sections 114(a)(3) and 504(b) of the Act.

(6) Such other provisions as the permitting authority may require.

(d) *General permits.* (1) The permitting authority may, after notice and opportunity for public participation provided under § 70.7(h) of this part, issue a general permit covering numerous similar sources. Any general permit shall comply with all requirements applicable to other part 70 permits and shall identify criteria by which sources may qualify for the general permit. To sources that qualify, the permitting authority shall grant the conditions and terms of the general permit. Notwithstanding the shield provisions of paragraph (f) of this section, the source shall be subject to enforcement action for operation without a part 70 permit if the source is later determined not to qualify for the conditions and terms of the general permit. General permits shall not be authorized for

affected sources under the acid rain program unless otherwise provided in regulations promulgated under title IV of the Act.

(2) Part 70 sources that would qualify for a general permit must apply to the permitting authority for coverage under the terms of the general permit or must apply for a part 70 permit consistent with § 70.5 of this part. The permitting authority may, in the general permit, provide for applications which deviate from the requirements of § 70.5 of this part, provided that such applications meet the requirements of title V of the Act, and include all information necessary to determine qualification for, and to assure compliance with, the general permit. Without repeating the public participation procedures required under § 70.7(h) of this part, the permitting authority may grant a source's request for authorization to operate under a general permit, but such a grant shall not be a final permit action for purposes of judicial review.

(e) *Temporary sources.* The permitting authority may issue a single permit authorizing emissions from similar operations by the same source owner or operator at multiple temporary locations. The operation must be temporary and involve at least one change of location during the term of the permit. No affected source shall be permitted as a temporary source. Permits for temporary sources shall include the following:

(1) Conditions that will assure compliance with all applicable requirements at all authorized locations;

(2) Requirements that the owner or operator notify the permitting authority at least 10 days in advance of each change in location; and

(3) Conditions that assure compliance with all other provisions of this section.

(f) *Permit shield.* (1) Except as provided in this part, the permitting authority may expressly include in a part 70 permit a provision stating that compliance with the conditions of the permit shall be deemed compliance with any applicable requirements as of the date of permit issuance, provided that:

(i) Such applicable requirements are included and are specifically identified in the permit; or

(ii) The permitting authority, in acting on the permit application or revision, determines in writing that other requirements specifically identified are not applicable to the source, and the permit includes the determination or a concise summary thereof.

(2) A part 70 permit that does not expressly state that a permit shield exists shall be presumed not to provide such a shield.

(3) Nothing in this paragraph or in any part 70 permit shall alter or affect the following:

(i) The provisions of section 303 of the Act (emergency orders), including the authority of the Administrator under that section;

(ii) The liability of an owner or operator of a source for any violation of applicable requirements prior to or at the time of permit issuance;

(iii) The applicable requirements of the acid rain program, consistent with section 408(a) of the Act; or

(iv) The ability of EPA to obtain information from a source pursuant to section 114 of the Act.

(g) *Emergency provision*—(1) *Definition.* An "emergency" means any situation arising from sudden and reasonably unforeseeable events beyond the control of the source, including acts of God, which situation requires immediate corrective action to restore normal operation, and that causes the source to exceed a technology-based emission limitation under the permit, due to unavoidable increases in emissions attributable to the emergency. An emergency shall not include noncompliance to the extent caused by improperly designed equipment, lack of preventative maintenance, careless or improper operation, or operator error.

(2) *Effect of an emergency.* An emergency constitutes an affirmative defense to an action brought for noncompliance with such technology-based emission limitations if the conditions of paragraph (g)(3) of this section are met.

(3) The affirmative defense of emergency shall be demonstrated through properly signed, contemporaneous operating logs, or other relevant evidence that:

(i) An emergency occurred and that the permittee can identify the cause(s) of the emergency;

(ii) The permitted facility was at the time being properly operated;

(iii) During the period of the emergency the permittee took all reasonable steps to minimize levels of emissions that exceeded the emission standards, or other requirements in the permit; and

(iv) The permittee submitted notice of the emergency to the permitting authority within 2 working days of the time when emission limitations were exceeded due to the emergency. This notice fulfills the requirement of paragraph (a)(3)(iii)(B) of this section. This notice must contain a description of the emergency, any steps taken to mitigate emissions, and corrective actions taken.

(4) In any enforcement proceeding, the permittee seeking to establish the occurrence of an emergency has the burden of proof.

(5) This provision is in addition to any emergency or upset provision contained in any applicable requirement.

§ 70.7 Permit issuance, renewal, reopenings, and revisions.

(a) *Action on application.* (1) A permit, permit modification. or renewal may be issued only if all of the following condition have been met:

(i) The permitting authority has received a complete application for a permit, permit modification, or permit renewal, except that a complete application need not be received before issuance of a general permit under § 70.6(d) of this part;

(ii) Except for modifications qualifying for minor permit modification procedures under paragraphs (e) (2) and (3) of this section, the permitting authority has complied with the requirements for public participation under paragraph (h) of this section;

(iii) The permitting authority has complied with the requirements for notifying and responding to affected States under § 70.8(b) of this part;

(iv) The conditions of the permit provide for compliance with all applicable requirements and the requirements of this part; and

(v) The Administrator has received a copy of the proposed permit and any notices required under §§ 70.8(a) and 70.8(b) of this part, and has not objected to issuance of the permit under § 70.8(c) of this part within the time period specified therein.

(2) Except as provided under the initial transition plan provided for under § 70.4(b)(11) of this part or under regulations promulgated under title IV of title V of the Act for the permitting of affected sources under the acid rain program, the program shall provide that the permitting authority take final action on each permit application (including a request for permit modification or renewal) within 18 months, or such lesser time approved by the Administrator, after receiving a complete application.

(3) The program shall also contain reasonable procedures to ensure priority is given to taking action on applications for construction or modification under title I, parts C and D of the Act.

(4) The permitting authority shall promptly provide notice to the applicant of whether the application is complete. Unless the permitting authority requests additional information or otherwise notifies the applicant of incompleteness within 60 days of receipt of an application, the application shall be deemed complete. For modifications processed through minor permit modification procedures, such as those in paragraphs (e) (2) and (3) of this section, the State program need not require a completeness determination.

(5) The permitting authority shall provide a statement that sets forth the legal and factual basis for the draft permit conditions (including references to the applicable statutory or regulatory provisions). The permitting authority shall send

§ 70.7

this statement to EPA and to any other person who requests it.

(6) The submittal of a complete application shall not affect the requirement that any source have a preconstruction permit under title I of the Act.

(b) *Requirement for a permit.* Except as provided in the following sentence, § 70.4(b)(12)(i), and paragraphs (e) (2)(v) and (3)(v) of this section, no part 70 source may operate after the time that it is required to submit a timely and complete application under an approved permit program, except in compliance with a permit issued under a part 70 program. The program shall provide that, if a part 70 source submits a timely and complete application for permit issuance (including for renewal), the source's failure to have a part 70 permit is not a violation of this part until the permitting authority takes final action on the permit application, except as noted in this section. This protection shall cease to apply if, subsequent to the completeness determination made pursuant to paragraph (a)(4) of this section, and as required by § 70.5(a)(2) of this part, the applicant fails to submit by the deadline specified in writing by the permitting authority any additional information identified as being needed to process the application.

(c) *Permit renewal and expiration.* (1) The program shall provide that:

(i) Permits being renewed are subject to the same procedural requirements, including those for public participation, affected State and EPA review, that apply to initial permit issuance; and

(ii) Permit expiration terminates the source's right to operate unless a timely and complete renewal application has been submitted consistent with paragraph (b) of this section and § 70.5(a)(1)(iii) of this part.

(2) If the permitting authority fails to act in a timely way on a permit renewal, EPA may invoke its authority under section 505(e) of the Act to terminate or revoke and reissue the permit.

(d) *Administrative permit amendments.* (1) An "administrative permit amendment" is a permit revision that:

(i) Corrects typographical errors;

(ii) Identifies a change in the name, address, or phone number of any person identified in the permit, or provides a similar minor administrative change at the source;

(iii) Requires more frequent monitoring or reporting by the permittee;

(iv) Allows for a change in ownership or operational control of a source where the permitting authority determines that no other change in the permit is necessary, provided that a written agreement containing a specific date for transfer of permit responsibility, coverage, and liability between

the current and new permittee has been submitted to the permitting authority;

(v) Incorporates into the part 70 permit the requirements from preconstruction review permits authorized under an EPA-approved program, provided that such a program meets procedural requirements substantially equivalent to the requirements of §§ 70.7 and 70.8 of this part that would be applicable to the change if it were subject to review as a permit modification, and compliance requirements substantially equivalent to those contained in § 70.6 of this part; or

(vi) Incorporates any other type of change which the Administrator has determined as part of the approved part 70 program to be similar to those in paragraphs (d)(1) (i) through (iv) of this section.

(2) Administrative permit amendments for purposes of the acid rain portion of the permit shall be governed by regulations promulgated under title IV of the Act.

(3) *Administrative permit amendment procedures.* An administrative permit amendment may be made by the permitting authority consistent with the following:

(i) The permitting authority shall take no more than 60 days from receipt of a request for an administrative permit amendment to take final action on such request, and may incorporate such changes without providing notice to the public or affected States provided that it designates any such permit revisions as having been made pursuant to this paragraph.

(ii) The permitting authority shall submit a copy of the revised permit to the Administrator.

(iii) The source may implement the changes addressed in the request for an administrative amendment immediately upon submittal of the request.

(4) The permitting authority may, upon taking final action granting a request for an administrative permit amendment, allow coverage by the permit shield in § 70.6(f) for administrative permit amendments made pursuant to paragraph (d)(1)(v) of this section which meet the relevant requirements of §§ 70.6, 70.7, and 70.8 for significant permit modifications.

(e) *Permit modification.* A permit modification is any revision to a part 70 permit that cannot be accomplished under the program's provisions for administrative permit amendments under paragraph (d) of this section. A permit modification for purposes of the acid rain portion of the permit shall be governed by regulations promulgated under title IV of the Act.

(1) *Program description.* The State shall provide adequate, streamlined, and reasonable procedures for expeditiously processing permit modifications. The State may meet this obligation by

adopting the procedures set forth below or ones substantially equivalent. The State may also develop different procedures for different types of modifications depending on the significance and complexity of the requested modification, but EPA will not approve a part 70 program that has modification procedures that provide for less permitting authority, EPA, or affected State review or public participation than is provided for in this part.

(2) *Minor permit modification procedures*—(i) *Criteria.* (A) Minor permit modification procedures may be used only for those permit modifications that:

(1) Do not violate any applicable requirement;

(2) Do not involve significant changes to existing monitoring, reporting, or recordkeeping requirements in the permit;

(3) Do not require or change a case-by-case determination of an emission limitation or other standard, or a source-specific determination for temporary sources of ambient impacts, or a visibility or increment analysis;

(4) Do not seek to establish or change a permit term or condition for which there is no corresponding underlying applicable requirement and that the source has assumed to avoid an applicable requirement to which the source would otherwise be subject. Such terms and conditions include:

(A) A federally enforceable emissions cap assumed to avoid classification as a modification under any provision of title I; and

(B) An alternative emissions limit approved pursuant to regulations promulgated under section 112(i)(5) of the Act;

(5) Are not modifications under any provision of title I of the Act; and

(6) Are not required by the State program to be processed as a significant modification.

(B) Notwithstanding paragraphs (e)(2)(i)(A) and (e)(3)(i) of this section, minor permit modification procedures may be used for permit modifications involving the use of economic incentives, marketable permits, emissions trading, and other similar approaches, to the extent that such minor permit modification procedures are explicitly provided for in an applicable implementation plan or in applicable requirements promulgated by EPA.

(ii) *Application.* An application requesting the use of minor permit modification procedures shall meet the requirements of § 70.5(c) of this part and shall include the following:

(A) A description of the change, the emissions resulting from the change, and any new applicable requirements that will apply if the change occurs;

(B) The source's suggested draft permit;

(C) Certification by a responsible official, consistent with § 70.5(d), that the proposed modification meets the criteria for use of minor permit modification procedures and a request that such procedures be used; and

(D) Completed forms for the permitting authority to use to notify the Administrator and affected States as required under § 70.8.

(iii) *EPA and affected State notification.* Within 5 working days of receipt of a complete permit modification application, the permitting authority shall meet its obligation under § 70.8 (a)(1) and (b)(1) to notify the Administrator and affected States of the requested permit modification. The permitting authority promptly shall send any notice required under § 70.8(b)(2) to the Administrator.

(iv) *Timetable for issuance.* The permitting authority may not issue a final permit modification until after EPA's 45-day review period or until EPA has notified the permitting authority that EPA will not object to issuance of the permit modification, whichever is first, although the permitting authority can approve the permit modification prior to that time. Within 90 days of the permitting authority's receipt of an application under minor permit modification procedures or 15 days after the end of the Administrator's 45-day review period under § 70.8(c), whichever is later, the permitting authority shall:

(A) Issue the permit modification as proposed;

(B) Deny the permit modification application;

(C) Determine that the requested modification does not meet the minor permit modification criteria and should be reviewed under the significant modification procedures; or

(D) Revise the draft permit modification and transmit to the Administrator the new proposed permit modification as required by § 70.8(a) of this part.

(v) *Source's ability to make change.* The State program may allow the source to make the change proposed in its minor permit modification application immediately after it files such application. After the source makes the change allowed by the preceding sentence, and until the permitting authority takes any of the actions specified in paragraphs (e)(2)(v) (A) through (C) of this section, the source must comply with both the applicable requirements governing the change and the proposed permit terms and conditions. During this time period, the source need not comply with the existing permit terms and conditions it seeks to modify. However, if the source fails to comply with its proposed permit terms and conditions during this time period, the existing permit terms and conditions it seeks to modify may be enforced against it.

(vi) *Permit shield.* The permit shield under § 70.6(f) of this part may not extend to minor permit modifications.

§ 70.7

(3) *Group processing of minor permit modifications.* Consistent with this paragraph, the permitting authority may modify the procedure outlined in paragraph (e)(2) of this section to process groups of a source's applications for certain modifications eligible for minor permit modification processing.

(i) *Criteria.* Group processing of modifications may be used only for those permit modifications:

(A) That meet the criteria for minor permit modification procedures under paragraph (e)(2)(i)(A) of this section; and

(B) That collectively are below the threshold level approved by the Administrator as part of the approved program. Unless the State sets an alternative threshold consistent with the criteria set forth in paragraphs (e)(3)(i)(B) *(1)* and *(2)* of this section, this threshold shall be 10 percent of the emissions allowed by the permit for the emissions unit for which the change is requested, 20 percent of the applicable definition of major source in § 70.2 of this part, or 5 tons per year, whichever is least. In establishing any alternative threshold, the State shall consider:

(1) Whether group processing of amounts below the threshold levels reasonably alleviates severe administrative burdens that would be imposed by immediate permit modification review, and

(2) Whether individual processing of changes below the threshold levels would result in trivial environmental benefits.

(ii) *Application.* An application requesting the use of group processing procedures shall meet the requirements of § 70.5(c) of this part and shall include the following:

(A) A description of the change, the emissions resulting from the change, and any new applicable requirements that will apply if the change occurs.

(B) The source's suggested draft permit.

(C) Certification by a responsible official, consistent with § 70.5(d) of this part, that the proposed modification meets the criteria for use of group processing procedures and a request that such procedures be used.

(D) A list of the source's other pending applications awaiting group processing, and a determination of whether the requested modification, aggregated with these other applications, equals or exceeds the threshold set under paragraph (e)(3)(i)(B) of this section.

(E) Certification, consistent with § 70.5(d) of this part, that the source has notified EPA of the proposed modification. Such notification need only contain a brief description of the requested modification.

(F) Completed forms for the permitting authority to use to notify the Administrator and affected States as required under § 70.8 of this part.

(iii) *EPA and affected State notification.* On a quarterly basis or within 5 business days of receipt of an application demonstrating that the aggregate of a source's pending applications equals or exceeds the threshold level set under paragraph (e)(3)(i)(B) of this section, whichever is earlier, the permitting authority promptly shall meet its obligations under §§ 70.8 (a)(1) and (b)(1) to notify the Administrator and affected States of the requested permit modifications. The permitting authority shall send any notice required under § 70.8(b)(2) of this part to the Administrator.

(iv) *Timetable for issuance.* The provisions of paragraph (e)(2)(iv) of this section shall apply to modifications eligible for group processing, except that the permitting authority shall take one of the actions specified in paragraphs (e)(2)(iv) (A) through (D) of this section within 180 days of receipt of the application or 15 days after the end of the Administrator's 45-day review period under § 70.8(c) of this part, whichever is later.

(v) *Source's ability to make change.* The provisions of paragraph (e)(2)(v) of this section shall apply to modifications eligible for group processing.

(vi) *Permit shield.* The provisions of paragraph (e)(2)(vi) of this section shall also apply to modifications eligible for group processing.

(4) *Significant modification procedures*—(i) *Criteria.* Significant modification procedures shall be used for applications requesting permit modifications that do not qualify as minor permit modifications or as administrative amendments. The State program shall contain criteria for determining whether a change is significant. At a minimum, every significant change in existing monitoring permit terms or conditions and every relaxation of reporting or recordkeeping permit terms or conditions shall be considered significant. Nothing herein shall be construed to preclude the permittee from making changes consistent with this part that would render existing permit compliance terms and conditions irrelevant.

(ii) The State program shall provide that significant permit modifications shall meet all requirements of this part, including those for applications, public participation, review by affected States, and review by EPA, as they apply to permit issuance and permit renewal. The permitting authority shall design and implement this review process to complete review on the majority of significant permit modifications within 9 months after receipt of a complete application.

(f) *Reopening for cause.* (1) Each issued permit shall include provisions specifying the conditions under which the permit will be reopened prior to the expiration of the permit. A permit shall be reopened and revised under any of the following circumstances:

(i) Additional applicable requirements under the Act become applicable to a major part 70 source with a remaining permit term of 3 or more years. Such a reopening shall be completed not later than 18 months after promulgation of the applicable requirement. No such reopening is required if the effective date of the requirement is later than the date on which the permit is due to expire, unless the original permit or any of its terms and conditions has been extended pursuant to § 70.4(b)(10) (i) or (ii) of this part.

(ii) Additional requirements (including excess emissions requirements) become applicable to an affected source under the acid rain program. Upon approval by the Administrator, excess emissions offset plans shall be deemed to be incorporated into the permit.

(iii) The permitting authority or EPA determines that the permit contains a material mistake or that inaccurate statements were made in establishing the emissions standards or other terms or conditions of the permit.

(iv) The Administrator or the permitting authority determines that the permit must be revised or revoked to assure compliance with the applicable requirements.

(2) Proceedings to reopen and issue a permit shall follow the same procedures as apply to initial permit issuance and shall affect only those parts of the permit for which cause to reopen exists. Such reopening shall be made as expeditiously as practicable.

(3) Reopenings under paragraph (f)(1) of this section shall not be initiated before a notice of such intent is provided to the part 70 source by the permitting authority at least 30 days in advance of the date that the permit is to be reopened, except that the permitting authority may provide a shorter time period in the case of an emergency.

(g) *Reopenings for cause by EPA.* (1) If the Administrator finds that cause exists to terminate, modify, or revoke and reissue a permit pursuant to paragraph (f) of this section, the Administrator will notify the permitting authority and the permittee of such finding in writing.

(2) The permitting authority shall, within 90 days after receipt of such notification, forward to EPA a proposed determination of termination, modification, or revocation and reissuance, as appropriate. The Administrator may extend this 90-day period for an additional 90 days if he finds that a new or revised permit application is necessary or that the permitting authority must require the permittee to submit additional information.

(3) The Administrator will review the proposed determination from the permitting authority within 90 days of receipt.

(4) The permitting authority shall have 90 days from receipt of an EPA objection to resolve any objection that EPA makes and to terminate, modify, or revoke and reissue the permit in accordance with the Administrator's objection.

(5) If the permitting authority fails to submit a proposed determination pursuant to paragraph (g)(2) of this section or fails to resolve any objection pursuant to paragraph (g)(4) of this section, the Administrator will terminate, modify, or revoke and reissue the permit after taking the following actions:

(i) Providing at least 30 days' notice to the permittee in writing of the reasons for any such action. This notice may be given during the procedures in paragraphs (g) (1) through (4) of this section.

(ii) Providing the permittee an opportunity for comment on the Administrator's proposed action and an opportunity for a hearing.

(h) *Public participation.* Except for modifications qualifying for minor permit modification procedures, all permit proceedings, including initial permit issuance, significant modifications, and renewals, shall provide adequate procedures for public notice including offering an opportunity for public comment and a hearing on the draft permit. These procedures shall include the following:

(1) Notice shall be given: by publication in a newspaper of general circulation in the area where the source is located or in a State publication designed to give general public notice; to persons on a mailing list developed by the permitting authority, including those who request in writing to be on the list; and by other means if necessary to assure adequate notice to the affected public;

(2) The notice shall identify the affected facility; the name and address of the permittee; the name and address of the permitting authority processing the permit; the activity or activities involved in the permit action; the emissions change involved in any permit modification; the name, address, and telephone number of a person from whom interested persons may obtain additional information, including copies of the permit draft, the application, all relevant supporting materials, including those set forth in § 70.4(b)(3)(viii) of this part, and all other materials available to the permitting authority that are relevant to the permit decision; a brief description of the comment procedures required by this part; and the time and place of any hearing that may be held, including a statement of procedures to request a hearing (unless a hearing has already been scheduled);

(3) The permitting authority shall provide such notice and opportunity for participation by affected States as is provided for by § 70.8 of this part;

(4) *Timing.* The permitting authority shall provide at least 30 days for public comment and shall give notice of any public hearing at least 30 days in advance of the hearing.

§ 70.8

(5) The permitting authority shall keep a record of the commenters and also of the issues raised during the public participation process so that the Administrator may fulfill his obligation under section 505(b)(2) of the Act to determine whether a citizen petition may be granted, and such records shall be available to the public.

§ 70.8 Permit review by EPA and affected States.

(a) *Transmission of information to the Administrator.* (1) The permit program shall require that the permitting authority provide to the Administrator a copy of each permit application (including any application for permit modification), each proposed permit, and each final part 70 permit. The applicant may be required by the permitting authority to provide a copy of the permit application (including the compliance plan) directly to the Administrator. Upon agreement with the Administrator, the permitting authority may submit to the Administrator a permit application summary form and any relevant portion of the permit application and compliance plan, in place of the complete permit application and compliance plan. To the extent practicable, the preceding information shall be provided in computer-readable format compatible with EPA's national database management system.

(2) The Administrator may waive the requirements of paragraphs (a)(1) and (b)(1) of this section for any category of sources (including any class, type, or size within such category) other than major sources according to the following:

(i) By regulation for a category of sources nationwide, or

(ii) At the time of approval of a State program for a category of sources covered by an individual permitting program.

(3) Each State permitting authority shall keep for 5 years such records and submit to the Administrator such information as the Administrator may reasonably require to ascertain whether the State program complies with the requirements of the Act or of this part.

(b) *Review by affected States.* (1) The permit program shall provide that the permitting authority give notice of each draft permit to any affected State on or before the time that the permitting authority provides this notice to the public under § 70.7(h) of this part, except to the extent § 70.7(e) (2) or (3) of this part requires the timing of the notice to be different.

(2) The permit program shall provide that the permitting authority, as part of the submittal of the proposed permit to the Administrator [or as soon as possible after the submittal for minor permit modification procedures allowed under § 70.7(e) (2) or (3) of this part], shall notify the Administrator and any affected State in writing of any re-

fusal by the permitting authority to accept all recommendations for the proposed permit that the affected State submitted during the public or affected State review period. The notice shall include the permitting authority's reasons for not accepting any such recommendation. The permitting authority is not required to accept recommendations that are not based on applicable requirements or the requirements of this part.

(c) *EPA objection.* (1) The Administrator will object to the issuance of any proposed permit determined by the Administrator not to be in compliance with applicable requirements or requirements under this part. No permit for which an application must be transmitted to the Administrator under paragraph (a) of this section shall be issued if the Administrator objects to its issuance in writing within 45 days of receipt of the proposed permit and all necessary supporting information.

(2) Any EPA objection under paragraph (c)(1) of this section shall include a statement of the Administrator's reasons for objection and a description of the terms and conditions that the permit must include to respond to the objections. The Administrator will provide the permit applicant a copy of the objection.

(3) Failure of the permitting authority to do any of the following also shall constitute grounds for an objection:

(i) Comply with paragraphs (a) or (b) of this section;

(ii) Submit any information necessary to review adequately the proposed permit; or

(iii) Process the permit under the procedures approved to meet § 70.7(h) of this part except for minor permit modifications.

(4) If the permitting authority fails, within 90 days after the date of an objection under paragraph (c)(1) of this section, to revise and submit a proposed permit in response to the objection, the Administrator will issue or deny the permit in accordance with the requirements of the Federal program promulgated under title V of this Act.

(d) *Public petitions to the Administrator.* The program shall provide that, if the Administrator does not object in writing under paragraph (c) of this section, any person may petition the Administrator within 60 days after the expiration of the Administrator's 45-day review period to make such objection. Any such petition shall be based only on objections to the permit that were raised with reasonable specificity during the public comment period provided for in § 70.7(h) of this part, unless the petitioner demonstrates that it was impracticable to raise such objections within such period, or unless the grounds for such objection arose after such period. If the Administrator objects to the permit as a result of a petition filed under this paragraph, the permitting authority shall not issue the permit until EPA's objection has been resolved, except that a petition for review does not stay the effectiveness of a permit or its requirements if the permit was issued after the end of the 45-day review period and prior to an EPA objection. If the permitting authority has issued a permit prior to receipt of an EPA objection under this paragraph, the Administrator will modify, terminate, or revoke such permit, and shall do so consistent with the procedures in § 70.7(g) (4) or (5) (i) and (ii) of this part except in unusual circumstances, and the permitting authority may thereafter issue only a revised permit that satisfies EPA's objection. In any case, the source will not be in violation of the requirement to have submitted a timely and complete application.

(e) *Prohibition on default issuance.* Consistent with § 70.4(b)(3)(ix) of this part, for the purposes of Federal law and title V of the Act, no State program may provide that a part 70 permit (including a permit renewal or modification) will issue until affected States and EPA have had an opportunity to review the proposed permit as required under this section. When the program is submitted for EPA review, the State Attorney General or independent legal counsel shall certify that no applicable provision of State law requires that a part 70 permit or renewal be issued after a certain time if the permitting authority has failed to take action on the application (or includes any other similar provision providing for default issuance of a permit), unless EPA has waived such review for EPA and affected States.

§ 70.9 Fee determination and certification.

(a) *Fee Requirement.* The State program shall require that the owners or operators of part 70 sources pay annual fees, or the equivalent over some other period, that are sufficient to cover the permit program costs and shall ensure that any fee required by this section will be used solely for permit program costs.

(b) *Fee schedule adequacy.* (1) The State program shall establish a fee schedule that results in the collection and retention of revenues sufficient to cover the permit program costs. These costs include, but are not limited to, the costs of the following activities as they relate to the operating permit program for stationary sources:

(i) Preparing generally applicable regulations or guidance regarding the permit program or its implementation or enforcement;

(ii) Reviewing and acting on any application for a permit, permit revision, or permit renewal, including the development of an applicable requirement as part of the processing of a permit, or permit revision or renewal;

(iii) General administrative costs of running the permit program, including the supporting and tracking of permit applications, compliance certification, and related data entry;

(iv) Implementing and enforcing the terms of any part 70 permit (not including any court costs or other costs associated with an enforcement action), including adequate resources to determine which sources are subject to the program;

(v) Emissions and ambient monitoring;

(vi) Modeling, analyses, or demonstrations;

(vii) Preparing inventories and tracking emissions; and

(viii) Providing direct and indirect support to sources under the Small Business Stationary Source Technical and Environmental Compliance Assistance Program contained in section 507 of the Act in determining and meeting their obligations under this part.

(2)(i) The Administrator will presume that the fee schedule meets the requirements of paragraph (b)(1) of this section if it would result in the collection and retention of an amount not less than $25 per year [as adjusted pursuant to the criteria set forth in paragraph (b)(2)(iv) of this section] times the total tons of the actual emissions of each regulated pollutant (for presumptive fee calculation) emitted from part 70 sources.

(ii) The State may exclude from such calculation:

(A) The actual emissions of sources for which no fee is required under paragraph (b)(4) of this section;

(B) The amount of a part 70 source's actual emissions of each regulated pollutant (for presumptive fee calculation) that the source emits in excess of four thousand (4,000) tpy;

(C) A part 70 source's actual emissions of any regulated pollutant (for presumptive fee calculation), the emissions of which are already included in the minimum fees calculation; or

(D) The insignificant quantities of actual emissions not required in a permit application pursuant to § 70.5(c).

(iii) "Actual emissions" means the actual rate of emissions in tons per year of any regulated pollutant (for presumptive fee calculation) emitted from a part 70 source over the preceding calendar year or any other period determined by the permitting authority to be representative of normal source operation and consistent with the fee schedule approved pursuant to this section. Actual emissions shall be calculated using the unit's actual operating hours, production rates, and in-place control equipment, types of materials processed, stored, or combusted during the preceding calendar year or such other time period established by the permitting authority pursuant to the preceding sentence.

§ 70.10

(iv) The program shall provide that the $25 per ton per year used to calculate the presumptive minimum amount to be collected by the fee schedule, as described in paragraph (b)(2)(i) of this section, shall be increased each year by the percentage, if any, by which the Consumer Price Index for the most recent calendar year ending before the beginning of such year exceeds the Consumer Price Index for the calendar year 1989.

(A) The Consumer Price Index for any calendar year is the average of the Consumer Price Index for all-urban consumers published by the Department of Labor, as of the close of the 12-month period ending on August 31 of each calendar year.

(B) The revision of the Consumer Price Index which is most consistent with the Consumer Price Index for the calendar year 1989 shall be used.

(3) The State program's fee schedule may include emissions fees, application fees, service-based fees or other types of fees, or any combination thereof, to meet the requirements of paragraph (b)(1) or (b)(2) of this section. Nothing in the provisions of this section shall require a permitting authority to calculate fees on any particular basis or in the same manner for all part 70 sources, all classes or categories of part 70 sources, or all regulated air pollutants, provided that the permitting authority collects a total amount of fees sufficient to meet the program support requirements of paragraph (b)(1) of this section.

(4) Notwithstanding any other provision of this section, during the years 1995 through 1999 inclusive, no fee for purposes of title V shall be required to be paid with respect to emissions from any affected unit under section 404 of the Act.

(5) The State shall provide a detailed accounting that its fee schedule meets the requirements of paragraph (b)(1) of this section if:

(i) The State sets a fee schedule that would result in the collection and retention of an amount less than that presumed to be adequate under paragraph (b)(2) of this section; or

(ii) The Administrator determines, based on comments rebutting the presumption in paragraph (b)(2) of this section or on his own initiative, that there are serious questions regarding whether the fee schedule is sufficient to cover the permit program costs.

(c) *Fee demonstration.* The permitting authority shall provide a demonstration that the fee schedule selected will result in the collection and retention of fees in an amount sufficient to meet the requirements of this section.

(d) *Use of Required Fee Revenue.* The Administrator will not approve a demonstration as meeting the requirements of this section, unless it contains an initial accounting (and periodic updates as required by the Administrator) of how required fee revenues are used solely to cover the costs of

meeting the various functions of the permitting program.

§ 70.10 Federal oversight and sanctions.

(a) *Failure to submit an approvable program.* (1) If a State fails to submit a fully-approvable whole part 70 program, or a required revision thereto, in conformance with the provisions of § 70.4, or if an interim approval expires and the Administrator has not approved a whole part 70 program:

(i) At any time the Administrator may apply any one of the sanctions specified in section 179(b) of the Act; and

(ii) Eighteen months after the date required for submittal or the date of disapproval by the Administrator, the Administrator will apply such sanctions in the same manner and with the same conditions as are applicable in the case of a determination, disapproval, or finding under section 179(a) of the Act.

(2) If full approval of a whole part 70 program has not taken place within 2 years after the date required for such submission, the Administrator will promulgate, administer, and enforce a whole program or a partial program as appropriate for such State.

(b) *State failure to administer or enforce.* Any State program approved by the Administrator shall at all times be conducted in accordance with the requirements of this part and of any agreement between the State and the Administrator concerning operation of the program.

(1) Whenever the Administrator makes a determination that a permitting authority is not adequately administering or enforcing a part 70 program, or any portion thereof, the Administrator will notify the permitting authority of the determination and the reasons therefore. The Administrator will publish such notice in the FEDERAL REGISTER.

(2) If, 90 days after issuing the notice under paragraph (c)(1) of this section, the permitting authority fails to take significant action to assure adequate administration and enforcement of the program, the Administrator may take one or more of the following actions:

(i) Withdraw approval of the program or portion thereof using procedures consistent with § 70.4(e) of this part;

(ii) Apply any of the sanctions specified in section 179(b) of the Act;

(iii) Promulgate, administer, or enforce a Federal program under title V of the Act.

(3) Whenever the Administrator has made the finding and issued the notice under paragraph (c)(1) of this section, the Administrator will apply the sanctions under section 179(b) of the Act 18

months after that notice. These sanctions will be applied in the same manner and subject to the same deadlines and other conditions as are applicable in the case of a determination, disapproval, or finding under section 179(a) of the Act.

(4) Whenever the Administrator has made the finding and issued the notice under paragraph (c)(1) of this section, the Administrator will, unless the State has corrected such deficiency within 18 months after the date of such finding, promulgate, administer, and enforce, a whole or partial program 2 years after the date of such finding.

(5) Nothing in this section shall limit the Administrator's authority to take any enforcement action against a source for violations of the Act or of a permit issued under rules adopted pursuant to this section in a State that has been delegated responsibility by EPA to implement a Federal program promulgated under title V of the Act.

(6) Where a whole State program consists of an aggregate of partial programs, and one or more partial programs fails to be fully approved or implemented, the Administrator may apply sanctions only in those areas for which the State failed to submit or implement an approvable program.

(c) *Criteria for withdrawal of State programs.* (1) The Administrator may, in accordance with the procedures of paragraph (c) of this section, withdraw program approval in whole or in part whenever the approved program no longer complies with the requirements of this part, and the permitting authority fails to take corrective action. Such circumstances, in whole or in part, include any of the following:

(i) Where the permitting authority's legal authority no longer meets the requirements of this part, including the following:

(A) The permitting authority fails to promulgate or enact new authorities when necessary; or

(B) The State legislature or a court strikes down or limits State authorities to administer or enforce the State program.

(ii) Where the operation of the State program fails to comply with the requirements of this part, including the following:

(A) Failure to exercise control over activities required to be regulated under this part, including failure to issue permits;

(B) Repeated issuance of permits that do not conform to the requirements of this part;

(C) Failure to comply with the public participation requirements of § 70.7(h) of this part;

(D) Failure to collect, retain, or allocate fee revenue consistent with § 70.9 of this part; or

(E) Failure in a timely way to act on any applications for permits including renewals and revisions.

(iii) Where the State fails to enforce the part 70 program consistent with the requirements of this part, including the following:

(A) Failure to act on violations of permits or other program requirements;

(B) Failure to seek adequate enforcement penalties and fines and collect all assessed penalties and fines; or

(C) Failure to inspect and monitor activities subject to regulation.

(d) *Federal collection of fees.* If the Administrator determines that the fee provisions of a part 70 program do not meet the requirements of § 70.9 of this part, or if the Administrator makes a determination under paragraph (c)(1) of this section that the permitting authority is not adequately administering or enforcing an approved fee program, the Administrator may, in addition to taking any other action authorized under title V of the Act, collect reasonable fees to cover the Administrator's costs of administering the provisions of the permitting program promulgated by the Administrator, without regard to the requirements of § 70.9 of this part.

§ 70.11 Requirements for enforcement authority.

All programs to be approved under this part must contain the following provisions:

(a) *Enforcement authority.* Any agency administering a program shall have the following enforcement authority to address violations of program requirements by part 70 sources:

(1) To restrain or enjoin immediately and effectively any person by order or by suit in court from engaging in any activity in violation of a permit that is presenting an imminent and substantial endangerment to the public health or welfare, or the environment.

(2) To seek injunctive relief in court to enjoin any violation of any program requirement, including permit conditions, without the necessity of a prior revocation of the permit.

(3) To assess or sue to recover in court civil penalties and to seek criminal remedies, including fines, according to the following:

(i) Civil penalties shall be recoverable for the violation of any applicable requirement; any permit condition; any fee or filing requirement; any duty to allow or carry out inspection, entry or monitoring activities or, any regulation or orders issued by the permitting authority. These penalties shall be recoverable in a maximum amount of not less than $10,000 per day per violation. State law shall not include mental state as an element of proof for civil violations.

(ii) Criminal fines shall be recoverable against any person who knowingly violates any applicable requirement; any permit condition; or any fee or

Pt. 70, App. A

filing requirement. These fines shall be recoverable in a maximum amount of not less than $10,000 per day per violation.

(iii) Criminal fines shall be recoverable against any person who knowingly makes any false material statement, representation or certification in any form, in any notice or report required by a permit, or who knowingly renders inaccurate any required monitoring device or method. These fines shall be recoverable in a maximum amount of not less than $10,000 per day per violation.

(b) *Burden of proof.* The burden of proof and degree of knowledge or intent required under State law for establishing violations under paragraph (a)(3) of this section shall be no greater than the burden of proof or degree of knowledge or intent required under the Act.

(c) *Appropriateness of penalties and fines.* A civil penalty or criminal fine assessed, sought, or agreed upon by the permitting authority under paragraph (a)(3) of this section shall be appropriate to the violation.

APPENDIX A TO PART 70—APPROVAL STATUS OF STATE AND LOCAL OPERATING PERMITS PROGRAMS

This appendix provides information on the approval status of State and Local operating Permit Programs. An approved State part 70 program applies to all part 70 sources, as defined in that approved program, within such State, except for any source of air pollution over which a federally recognized Indian Tribe has jurisdiction.

Alabama

(a) Alabama Department of Environmental Management: submitted on December 15, 1993, and supplemented on March 3, 1994; March 18, 1994; June 5, 1995; July 14, 1995; and August 28, 1995: interim approval effective on December 15, 1995; interim approval expires December 15, 1997.

(b) City of Huntsville Department of Natural Resources and Environmental Management: submitted on November 15, 1993, and supplemented on July 20, 1995; interim approval effective on December 15, 1995; interim approval expires December 15, 1997.

(c) Jefferson County Department of Health: submitted on December 14, 1993, and supplemented on July 14, 1995: interim approval effective on December 15, 1995; interim approval expires December 15, 1997.

Arkansas

(a) The ADPCE submitted its Operating Permits program on November 9, 1993, for approval. Interim approval is effective on October 10, 1995. Interim approval will expire October 8, 1997.

(b) [Reserved]

California

The following district programs were submitted by the California Air Resources Board on behalf of:

(a) *Amador County Air Pollution Control District* (APCD) (complete submittal received on September 30, 1994): interim approval effective on June 2, 1995: interim approval expires June 3, 1997.

(b) *Bay Area Air Quality Management District:* Submitted on November 16, 1993, amended on October 27, 1994, and effective as an interim program on July 24, 1995. Revisions to interim program submitted on March 23, 1995, and effective on August 22, 1995, unless adverse or critical comments are received by July 24, 1995. Approval of interim program, including March 23, 1995, revisions, expires July 23, 1997.

(c) *Butte County APCD* (complete submittal received on December 16, 1993): interim approval effective on June 2, 1995: interim approval expires June 3, 1997.

(d) *Calaveras County APCD* (complete submittal received on October 31, 1994): interim approval effective on June 2, 1995: interim approval expires June 3, 1997.

(e) *Colusa County APCD* (complete submittal received on February 24, 1994): interim approval effective on June 2, 1995: interim approval expires June 3, 1997.

(f) *El Dorado County APCD* (complete submittal received on November 16, 1993): interim approval effective on June 2, 1995: interim approval expires June 3, 1997.

(g) *Feather River Air Quality Management District* (AQMD) (complete submittal received on December 27, 1993): interim approval effective on June 2, 1995: interim approval expires June 3, 1997.

(h) *Glenn County APCD* (complete submittal received on December 27, 1993): interim approval effective on August 14, 1995: interim approval expires August 13, 1997.

(i) *Great Basin Unified APCD* (complete submittal received on January 12, 1994): interim approval effective on June 2, 1995: interim approval expires June 3, 1997.

(j) *Imperial County APCD* (complete submittal received on March 24, 1994): interim approval effective on June 2, 1995: interim approval expires June 3, 1997.

(k) *Kern County APCD* (complete submittal received on November 16, 1993): interim approval effective on June 2, 1995: interim approval expires June 3, 1997.

(l) *Lake County AQMD* (complete submittal received on March 15, 1994): interim approval effective on August 14, 1995: interim approval expires August 13, 1997.

(m) *Lassen County APCD* (complete submittal received on January 12, 1994): interim approval effective on June 2, 1995: interim approval expires June 3, 1997.

(n) *Mariposa Air Pollution Control District:* submitted on March 8, 1995: approval effective on February 5, 1996 unless adverse or critical comments are received by January 8, 1996.

(o) *Mendocino County APCD* (complete submittal received on December 27, 1993): interim approval effective on June 2, 1995: interim approval expires June 3, 1997.

(p) *Modoc County APCD* (complete submittal received on December 27, 1993): interim approval effective on June 2, 1995: interim approval expires June 3, 1997.

(q) *Mojave Desert AQMD* (complete submittal received on March 10, 1995): interim approval effective on March 6, 1996: interim approval expires March 5, 1998.

(r) *Monterey Bay Unified Air Pollution Control District:* submitted on December 6, 1993, supplemented on February 2, 1994 and April 7, 1994, and revised by the submittal made on October 13, 1994: interim approval effective on November 6, 1995: interim approval expires November 6, 1997.

(s) *North Coast Unified AQMD* (complete submittal received on February 24, 1994): interim approval effective on June 2, 1995: interim approval expires June 3, 1997.

(t) *Northern Sierra AQMD* (complete submittal received on June 6, 1994): interim approval effective on June 2, 1995: interim approval expires June 3, 1997.

(u) *Northern Sonoma County APCD* (complete submittal received on January 12, 1994): interim approval effective on June 2, 1995: interim approval expires June 3, 1997.

(v) *Placer County APCD* (complete submittal received on December 27, 1993): interim approval effective on June 2, 1995: interim approval expires June 3, 1997.

(w) *the Sacramento Metropolitan Air Quality Management District:* (complete submittal received on August 1, 1994): interim approval effective on September 5, 1995: interim approval expires September 4, 1997.

(x) *San Diego Air Pollution Control District*: submitted on April 22, 1994 and amended on April 4, 1995 and October 10, 1995: approval effective on February 5, 1996, unless adverse or critical comments are received by January 8, 1996.

(y) *San Joaquin Valley Unified APCD* (complete submittal received on July 5 and August 18, 1995): interim approval effective on May 24, 1996: interim approval expires May 25, 1998.

(z) *San Luis Obispo County APCD* (complete submittal received on November 16, 1995): interim approval effective on December 1, 1995: interim approval expires December 1, 1997.

(aa) *Santa Barbara County Air Pollution Control District* (APCD) submitted on November 15, 1993, as amended March 2, 1994, August 8, 1994, December 8, 1994, and June 15, 1995: interim approval effective on December 1, 1995: interim approval expires December 1, 1997.

(bb) *Shasta County AQMD* (complete submittal received on November 16, 1993): interim approval effective on August 14, 1995: interim approval expires August 13, 1997.

(cc) *Siskiyou County APCD* (complete submittal received on December 6, 1993): interim approval effective on June 2, 1995: interim approval expires June 3, 997.

(dd) [Reserved]

(ee) *Tehama County APCD* (complete submittal received on December 6, 1993): interim approval effective on August 14, 1995: interim approval expires August 13, 1997.

(ff) *Tuolumne County APCD* (complete submittal received on November 16, 1993): interim approval effective on June 2, 1995: interim approval expires June 3, 1997.

(gg) *Ventura County Air Pollution Control District* (APCD) submitted on November 16, 1993, as amended December 6, 1993: interim approval effective on December 1, 1995: interim approval expires December 1, 1997.

(hh) *Yolo-Solano AQMD* (complete submittal received on October 14, 1994): interim approval effective on June 2, 1995: interim approval expires June 3, 1997.

Colorado

(a) Colorado Department Health–Air Pollution Control Division: submitted on November 5, 1993: effective on February 23, 1995: interim approval expires February 24, 1997.

(b) [Reserved]

Delaware

(a) Department of Natural Resources and Environmental Control: submitted on November 15, 1993 and amended on November 22, 1993, February 9, 1994, May 15, 1995 and September 5, 1995: interim approval effective on January 3, 1996: interim approval expires January 5, 1998.

(b) [Reserved]

District of Columbia

(a) Environmental Regulation Administration: submitted on January 13, 1994 and March 11, 1994: interim approval effective on September 6, 1995: interim approval expires September 8, 1997.

(b) [Reserved]

Florida

(a) Florida Department of Environmental Protection: submitted on November 16, 1993, and supplemented on July 8, 1994, November 28, 1994, December 21, 1994, December 22, 1994, and January 11, 1995: interim approval effective on October 25, 1995: interim approval expires October 25, 1997.

(b) [Reserved]

Georgia

(a) The Georgia Department of Natural Resources submitted on November 12, 1993, and supplemented on June 24, 1994: November 14, 1994: and June 5, 1995: interim approval effective on December 22, 1995: interim approval expires December 22, 1997.

(b) [Reserved]

Hawaii

(a) Department of Health: submitted on December 20, 1993: effective on December 1, 1994: interim approval expires December 1, 1996.

(b) [Reserved]

Illinois

(a) The Illinois Environmental Protection Agency: submitted on November 15, 1993: interim approval effective on March 7, 1995: interim approval expires March 7, 1997.

(b) [Reserved]

Indiana

(a) The Indiana Department of Environmental Management: submitted on August 10, 1994: interim approval effective on December 14, 1995: interim approval expires December 14, 1997.

(b) [Reserved]

Iowa

(a) The Iowa Department of Natural Resources submitted on November 15, 1993, and supplemented by correspondence dated March 15, 1994: August 8, 1994: October 5, 1994: December 6, 1994: December 15, 1994: February 6, 1995: March 1, 1995: March 23, 1995: and May 26, 1995. Interim approval effective on October 2, 1995: interim approval expires October 1, 1997.

Pt. 70, App. A

(b) [Reserved]

Kansas

(a) The Kansas Department of Health and Environment program submitted on December 12, 1994: April 7 and 17, 1995: November 14, 1995: and December 13, 1995. Full approval effective on February 29, 1996.

(b) [Reserved]

Kentucky

(a) Kentucky Natural Resources and Environmental Protection Cabinet: submitted on December 27, 1993, and supplemented on November 15, 1994, April 14, 1995, May 3, 1995 and May 22, 1995: interim approval effective on December 14, 1995: interim approval expires on December 14, 1997.

(b) Air Pollution Control District of Jefferson County, Kentucky: submitted on February 1, 1994, and supplemented on November 15, 1994, May 3, 1995, July 14, 1995 and February 16, 1996: full approval effective on April 22, 1996.

Louisiana

(a) The Louisiana Department of Environmental Quality, Air Quality Division submitted an Operating Permits program on November 15, 1993, which was revised November 10, 1994, and became effective on October 12, 1995.

(b) [Reserved]

Massachusetts

(a) Department of Environmental Protection: submitted on April 28, 1995: interim approval effective on May 15, 1996: interim approval expires May 15, 1998.

(b) [Reserved]

Minnesota

(a) Minnesota Pollution Control Agency: submitted on November 15, 1993: effective July 17, 1995: interim approval expires July 16, 1997.

(b) [Reserved]

Mississippi

(a) Department of Environmental Quality: submitted on November 15, 1993: full approval effective on January 27, 1995.

(b) [Reserved]

Missouri

(a) The Missouri Department of Natural Resources program submitted on January 13, 1995: August 14, 1995: September 19, 1995: and October 16, 1995. Interim approval effective on May 13, 1996.

(b) [Reserved]

Montana

(a) Montana Department of Health and Environmental Sciences—Air Quality Division: submitted on March 29, 1994: effective on June 12, 1995: interim approval expires June 11, 1997.

(b) [Reserved]

Nebraska; City of Omaha; Lincoln-Lancaster County Health Department

(a) The Nebraska Department of Environmental Quality submitted on November 15, 1993, supplemented by correspondence dated November 2, 1994, and August 29, 1995, and amended Title V rules submitted June 14, 1995.

(b) Omaha Public Works Department submitted on November 15, 1993, supplemented by correspondence dated April 18, 1994: April 19, 1994: May 13, 1994: August 12, 1994: and April 13, 1995. A delegation contract between the state and the city of Omaha became effective on June 6, 1995.

(c) Lincoln-Lancaster County Health Department submitted on November 12, 1993, supplemented by correspondence dated June 23, 1994. Full approval effective on November 17, 1995.

Nevada

The following district program was submitted by the Nevada Division of Environmental Protection on behalf of:

(a) *Nevada Division of Environmental Protection:* submitted on February 8, 1995: interim approval effective on January 11, 1996: interim approval expires January 12, 1998.

(b) Washoe County District Health Department: submitted on November 18, 1993: interim approval effective on March 6, 1995: interim approval expires February 5, 1997.

(c) *Clark County Air Quality Management District:* submitted on January 12, 1994 and amended on July 18 and September 21, 1994: interim approval effective on August 14, 1995: interim approval expires August 13, 1997.

New Jersey

(a) The New Jersey Department of Environmental Protection submitted an operating permit program on November 15, 1993, revised on August 10, 1995, with supplements on August 28, 1995, November 15, 1995, December 4, 1995, and December 6, 1995: interim approval effective on June 17, 1996: interim approval expires June 16, 1998.

(b) [Reserved]

New Mexico

(a) Environment Department: submitted on November 15, 1993: effective date on December 19, 1994: interim approval expires on December 19, 1996.

(b) City of Albuquerque Environmental Health Department, Air Pollution Control Division: submitted on April 4, 1994: effective on March 13, 1995: interim approval expires August 10, 1996.

North Carolina

(a) Department of Environment, Health and Natural Resources, Western North Carolina Regional Air Pollution Control Agency, Forsyth County Department of Environmental Affairs and the Mecklenburg County Department of Environmental Protection: submitted on November 12, 1993, and supplemented on December 17, 1993: February 28, 1994: May 31, 1994: and August 9, 1995: interim approval effective on December 15, 1995: interim approval expires December 15, 1997.

(b) [Reserved]

North Dakota

(a) North Dakota State Department of Health and Consolidated Laboratories—Environmental Health Section: submitted on May 11, 1994: effective on August 7, 1995: interim approval expires August 7, 1997.

(b) [Reserved]

Ohio

(a) The Ohio Environmental Protection Agency submitted on July 22, 1994; September 12, 1994; November 21, 1994; December 9, 1994; and January 5, 1995: full approval effective on October 1, 1995.

(b) [Reserved]

Oklahoma

(a) The Oklahoma Department of Environmental Quality submitted its operating permits program on January 12, 1994, for approval. Source category—limited interim approval is effective on March 6, 1996. Interim approval will expire March 5, 1998. The scope of the approval of the Oklahoma part 70 program excludes all sources of air pollution over which an Indian Tribe has jurisdiction.

(b) [Reserved]

Oregon

(a) Oregon Department of Environmental Quality: submitted on November 15, 1993, as amended on November 15, 1994, and June 30, 1995: full approval effective on November 27, 1995.

(b) Lane Regional Air Pollution Authority: submitted on November 15, 1993, as amended on November 15, 1994, and June 30, 1995: full approval effective on November 27, 1995.

Puerto Rico

(a) The Puerto Rico Environmental Quality Board submitted an operating permits program on November 15, 1993 with supplements on March 22, 1994 and April 11, 1994 and revised on September 29, 1995: full approval effective on March 27, 1996.

(b) [Reserved]

Rhode Island

(a) Department of Environmental Management: submitted on June 20, 1995: interim approval effective on July 5, 1996: interim approval expires July 6, 1998.

(b) [Reserved]

South Carolina

(a) Department of Health and Environmental Control: submitted on November 12, 1993: full approval effective on July 26, 1995.

(b) [Reserved]

South Dakota

(a) South Dakota Department of Environment and Natural Resources Division of Environmental Regulation: submitted on November 12, 1993: effective on April 21, 1995: interim approval expires April 22, 1997.

(b) [Reserved]

EDITORIAL NOTE: At 61 FR 2722, Jan. 29, 1996, appendix A to part 70 was amended by adding an entry for South Dakota. An entry already exists for South Dakota in the 1995 edition of this volume.

South Dakota

(a) South Dakota Department of Environment and Natural Resources—Division of Environmental Regulations: submitted on November 12, 1993: effective on February 28, 1996.

(b) [Reserved]

Tennessee

(a) [Reserved]

(b) Chattanooga-Hamilton County Air Pollution Control Bureau, Hamilton County, State of Tennessee: submitted on November 22, 1993, and supplemented on January 23, 1995, February 24, 1995, October 13, 1995, and March 14, 1996: full approval effective on April 25, 1996.

(c) The Knox County Department of Air Pollution Control: submitted on November 12, 1993, and supplemented on August 24, 1994: January 6 and 19, 1995: February 6, 1995: May 23, 1995: September 18 and 25, 1995: and March 6, 1996: full approval effective on May 30, 1996, in the FEDERAL REGISTER.

(d) The Metropolitan Health Department, Metropolitan Govenment of Nashville-Davidson County: submitted on November 13, 1993, and supplemented on April 19, 1994: September 27, 1994: December 28, 1994: and December 28, 1995: full approval effective on March 15, 1996.

Texas

(a) The TNRCC submitted its Operating Permits program on September 17, 1993, and supplemental submittals on October 28, 1993, and November 12, 1993, for approval. Source category-limited interim approval is effective on July 25, 1996. Interim approval will expire July 27, 1998. The scope of the approval of the Texas part 70 program excludes all sources of air pollution over which an Indian Tribe has jurisdiction.

(b) [Reserved]

Utah

(a) Utah Department of Environmental Quality—Division of Air Quality: submitted on April 14, 1994: effective on July 10, 1995.

(b) [Reserved]

Virginia

(a) Department of Environmental Quality: submitted on November 19, 1993: disapproval effective on January 4, 1995.

(b) [Reserved]

Pt. 70, App. A

Washington

(a) Department of Ecology (Ecology): submitted on November 1, 1993: effective on December 9, 1994: interim approval expires December 9, 1996.

(b) Energy Facility Site Evaluation Council (EFSEC): submitted on November 1, 1993: effective on December 9, 1994: interim approval expires December 9, 1996.

(c) Benton County Clean Air Authority (BCCAA): submitted on November 1, 1993 and amended on September 29, 1994 and April 12, 1995: effective on December 9, 1994: interim approval expires December 9, 1996.

(d) Northwest Air Pollution Authority (NWAPA): submitted on November 1, 1993: effective on December 9, 1994: interim approval expires December 9, 1996.

(e) Olympic Air Pollution Control Authority (OAPCA): submitted on November 1, 1993: effective on December 9, 1994: interim approval expires December 9, 1996.

(f) Puget Sound Air Pollution Control Agency (PSAPCA): submitted on November 1, 1993: effective on December 9, 1994: interim approval expires December 9, 1996.

(g) Southwest Air Pollution Control Authority (SWAPCA): submitted on November 1, 1993: effective on December 9, 1994: interim approval expires December 9, 1996.

(h) Spokane County Air Pollution Control Authority (SCAPCA): submitted on November 1, 1993: effective on December 9, 1994: interim approval expires December 9, 1996.

(i) Yakima County Clean Air Authority (YCCAA): submitted on November 1, 1993 and amended on September 29, 1994: effective on December 9, 1994: interim approval expires December 9, 1996.

West Virginia

(a) Department of Commerce, Labor and Environmental Resources: submitted on November 12, 1993, and supplemented by the Division of Environmental Protection on August 26 and September 29, 1994: interim approval effective on December 15, 1995: interim approval expires December 15, 1997.

(b) [Reserved]

Wisconsin

(a) Department of Natural Resources: submitted on January 27, 1994: interim approval effective on April 5, 1995: interim approval expires April 7, 1997.

(b) [Reserved]

Wyoming

(a) Department of Environmental Quality: submitted on November 19, 1993: effective on February 21, 1995: interim approval expires February 19, 1997.

(b) [Reserved]

[59 FR 55820, Nov. 9, 1994]

EDITORIAL NOTE: For FEDERAL REGISTER citations affecting appendix A to part 70, see the List of CFR Sections Affected in the Finding Aids section of this volume.

EFFECTIVE DATE NOTES: 1. At 61 FR 20155, May 6, 1996, appendix A to part 70 was amended by adding an entry for Rhode Island, effective July 5, 1996.

2. At 61 FR 32699, June 25, 1996, appendix A to part 70 was amended by adding an entry for Texas, effective July 25, 1996.

Major Source Operating Permit Application

F.2   Appendix F

MAJOR SOURCE OPERATING PERMIT APPLICATION - INDEX OF AIR POLLUTION PERMIT APPLICATION FORMS

1. ADMINISTRATION	
This application contains the following forms:	APC Form V.1, Facility Identification
	APC Form V.2, Operations and Flow Diagrams

2. EMISSIONS SOURCE DESCRIPTION		TOTAL NUMBER OF THIS FORM
This application contains the following forms (one form for each incinerator, printing operation, fuel burning installation, etc.):	APC Form V.3, Stack Identification	0
	APC Form V.4, Fuel Burning Non-Process Equipment	0
	APC Form V.5, Stationary Gas Turbines or Internal Combustion Engines	0
	APC Form V.6, Storage Tanks	0
	APC Form V.7, Incinerators	0
	APC Form V.8, Printing Operations	0
	APC Form V.9, Painting and Coating Operations	0
	APC Form V.10, Miscellaneous Processes	0
	APC Form V.33, Stage I and Stage II Vapor Recovery Equipment	0
	APC Form V.34, Open Burning	0

3. AIR POLLUTION CONTROL SYSTEM		TOTAL NUMBER OF THIS FORM
This application contains the following forms (one form for each control system in use at the facility):	APC Form V.11, Control Equipment - Miscellaneous	0
	APC Form V.12, Condensers	0
	APC Form V.13, Adsorbers	0
	APC Form V.14, Catalytic or Thermal Oxidation Equipment	0
	APC Form V.15, Cyclones/Settling Chambers	0
	APC Form V.16, Electrostatic Precipitators	0
	APC Form V.17, Wet Collection Systems	0
	APC Form V.18, Baghouse/Fabric Filters	0

MAJOR SOURCE OPERATING PERMIT APPLICATION - INDEX OF AIR POLLUTION PERMIT APPLICATION FORM

4. COMPLIANCE DEMONSTRATION		TOTAL NUMBER OF THIS FORM
This application contains the following forms (one form for each incinerator, printing operation, fuel burning installation, etc.):	APC Form V.19, Compliance Certification – Monitoring and Reporting - Description of Methods for Determining Compliance	0
	APC Form V.20, Continuous Emissions Monitoring	0
	APC Form V.21, Portable Monitors	0
	APC Form V.22, Control System Parameters or Operating Parameters of a Process	0
	APC Form V.23, Monitoring Maintenance Procedures	0
	APC Form V.24, Stack Testing	0
	APC Form V.25, Fuel Sampling and Analysis	0
	APC Form V.26, Record Keeping	0
	APC Form V.27, Other Methods	0
	APC Form V.28, Emissions from Process Emissions Sources/ Fuel Burning Installations/ Incinerators	0
	APC Form V.29, Emissions Summary for the Facility of for the Source Contained in This Application	0
	APC Form V.30, Current Emissions Requirements and Status	0
	APC Form V.31, Compliance Plan and Compliance Certification	0
	APC Form V.32, Air Monitoring Network	0

5.   STATEMENT OF COMPLETENESS AND CERTIFICATION OF COMPLIANCE

I have reviewed this application in its entirety and to the best of my knowledge, and based on information and belief formed after reasonable inquiry, the statements and information contained in this application are true, accurate, and complete. I have provided all the information that is necessary for compliance purposes an this

application consists of ___0___ pages and they are numbered from page __1__ to __0__. the status of this facility's compliance with all applicable air pollution control requirements, including the enhanced monitoring and compliance certification requirements of the Federal Clean Air Act, is reported in this application along with the methods to be used for compliance demonstration.

Name and Title of Responsible Official	Telephone Number with Area Code
Name, Title, and Phone Number	

Signature of Responsible Official	Date of Application

(FOR DEFINITION OF RESPONSIBLE OFFICIAL, SEE INSTRUCTIONS FOR APC FORM V.1)

CN-1007   RDA 1298

F.4　Appendix F

MAJOR SOURCE OPERATING PERMIT APPLICATION - APPLICATION COMPLETENESS CHECK LIST

I. IDENTIFICATION INFORMATION			
	_____ COMPLETE	_____ INCOMPLETE	_____ NOT APPLICABLE
A. FACILITY INFORMATION			
FACILITY NAME, LOCATION & MAILING ADDRESS	_____ YES	_____ NO	
PERMIT CONTACT PERSON	_____ YES	_____ NO	
RESPONSIBLE OFFICIAL	_____ YES	_____ NO	
PERMIT REQUESTED	_____ YES	_____ NO	
B. SOURCE DESCRIPTION			
1. OPERATIONAL INFORMATION:	_____ YES	_____ NO	
SIC CODE(S)	_____ YES	_____ NO	
LISTING AND DESCRIPTION OF EMISSION SOURCE(S)	_____ YES	_____ NO	
2. IDENTIFICATION AND DESCRIPTION OF ALTERNATIVE OPERATIVE SCENARIOS (IF APPLICABLE)	_____ YES	_____ NO	_____ N/A
C. PERMIT SHIELD REQUESTED	_____ YES	_____ NO	

II. EMISSIONS INFORMATION			
	_____ COMPLETE	_____ INCOMPLETE	_____ NOT APPLICABLE
A. QUANTIFICATION OF ALL EMISSIONS OF REGULATED AIR POLLUTANTS	_____ YES	_____ NO	
B. EMISSION SOURCES:			
IDENTIFICATION AND DESCRIPTION OF ALL EMISSION SOURCES IN SUFFICIENT DETAIL TO ESTABLISH THE BASIS FOR FEES AND APPLICABILITY OF REQUIREMENTS	_____ YES	_____ NO	
A LIST OF INSIGNIFICANT EMISSIONS UNITS OR ACTIVITIES EXEMPT BECAUSE OF SIZE OR PRODUCTION RATE	_____ YES	_____ NO	_____ N/A
C. PROCESS INFORMATION TO THE EXTENT IT IS NEEDED TO DETERMINE OR REGULATE EMISSIONS:			
FUELS	_____ YES	_____ NO	_____ N/A
RAW MATERIAL(S) / MATERIALS USED	_____ YES	_____ NO	_____ N/A
PRODUCTION RATES	_____ YES	_____ NO	_____ N/A
D. FOR REGULATED AIR POLLUTANTS, LIMITATIONS ON SOURCE OPERATIONS AFFECTING:			
EMISSIONS	_____ YES	_____ NO	_____ N/A
ANY WORK PRACTICE STANDARDS	_____ YES	_____ NO	_____ N/A
E. OTHER INFORMATION REQUIRED BY ANY APPLICABLE REQUIREMENTS FOR ALL REGULATED AIR POLLUTANTS SUCH AS:			
UTM COORDINATES OF EMISSION SOURCES	_____ YES	_____ NO	_____ N/A
FLOW RATES	_____ YES	_____ NO	_____ N/A
STACK PARAMETERS	_____ YES	_____ NO	_____ N/A
F. CALCULATIONS ON WHICH EMISSIONS RELATED INFORMATION ARE BASED	_____ YES	_____ NO	_____ N/A

MAJOR SOURCE OPERATING PERMIT APPLICATION - APPLICATION COMPLETENESS CHECK LIST

III. APPLICABILITY	_____ COMPLETE	_____ INCOMPLETE	_____ NOT APPLICABLE
A. CITATION AND DESCRIPTION OF ALL APPLICABLE REQUIREMENTS		_____ YES _____ NO	
B. OTHER SPECIFIC INFORMATION THAT MAY BE NECESSARY TO IMPLEMENT AND ENFORCE OTHER APPLICABLE REQUIREMENTS OF RULE 1200-3-9-.02(11) OF THE TENNESSEE AIR POLLUTION CONTROL REGULATIONS OR TO DETERMINE THE APPLICABILITY OF REQUIREMENTS		_____ YES _____ NO	
C. AN EXPLANATION OF ANY PROPOSED EXEMPTIONS FROM OTHERWISE APPLICABLE REQUIREMENTS		_____ YES _____ NO _____ N/A	

IV. COMPLIANCE	_____ COMPLETE	_____ INCOMPLETE	_____ NOT APPLICABLE
A. COMPLIANCE STATUS			
1. A DESCRIPTION OF THE COMPLIANCE STATUS OF THE SOURCE WITH RESPECT TO ALL APPLICABLE REQUIREMENTS		_____ YES _____ NO	
2. FOR APPLICABLE REQUIREMENTS WITH WHICH THE SOURCE IS IN COMPLIANCE, A STATEMENT THAT THE SOURCE WILL CONTINUE TO COMPLY WITH SUCH REQUIREMENTS		_____ YES _____ NO	
3. FOR APPLICABLE REQUIREMENTS THAT WILL BECOME EFFECTIVE DURING THE PERMIT TERM, A STATEMENT THAT THE SOURCE WILL MEET SUCH REQUIREMENTS ON A TIMELY BASIS		_____ YES _____ NO _____ N/A	
4. FOR REQUIREMENTS FOR WHICH THE SOURCE IS NOT IN COMPLIANCE AT THE TIME OF PERMIT ISSUANCE, A NARRATIVE DESCRIPTION OF HOW THE SOURCE WILL ACHIEVE COMPLIANCE WITH SUCH REQUIREMENTS		_____ YES _____ NO _____ N/A	
5. IDENTIFICATION AND DESCRIPTION OF AIR POLLUTION CONTROL EQUIPMENT AND COMPLIANCE MONITORING DEVICES OR ACTIVITIES		_____ YES _____ NO	
6. DESCRIPTION OF OR REFERENCE TO ANY APPLICABLE TEST METHOD FOR DETERMINING COMPLIANCE WITH EACH APPLICABLE REQUIREMENT		_____ YES _____ NO	
B. COMPLIANCE SCHEDULE	_____ COMPLETE	_____ INCOMPLETE	_____ NOT APPLICABLE
1. A SCHEDULE OF COMPLIANCE FOR SOURCES THAT ARE NOT IN COMPLIANCE WITH ALL APPLICABLE REQUIREMENTS AT THE TIME OF PERMIT ISSUES		_____ YES _____ NO _____ N/A	
2. A SCHEDULE FOR SUBMISSION OF CERTIFIED PROGRESS REPORTS NO LESS FREQUENTLY THAN EVERY SIX MOTHS FOR SOURCES REQUIRED TO HAVE A SCHEDULE OF COMPLIANCE TO REMEDY A VIOLATION		_____ YES _____ NO _____ N/A	
C. COMPLIANCE CERTIFICATION	_____ COMPLETE	_____ INCOMPLETE	_____ NOT APPLICABLE
1. CERTIFICATION OF COMPLIANCE WITH ALL APPLICABLE REQUIREMENTS BY A RESPONSIBLE OFFICIAL		_____ YES _____ NO	
2. A STATEMENT OF METHODS USED FOR DETERMINING COMPLIANCE, INCLUDING A DESCRIPTION OF MONITORING, RECORDKEEPING, AND REPORTING REQUIREMENTS AND TEST METHODS		_____ YES _____ NO	
3. A SCHEDULE FOR SUBMISSION OF COMPLIANCE CERTIFICATIONS		_____ YES _____ NO	
4. A STATEMENT INDICATING THE SOURCE'S COMPLIANCE STATUS WITH ANY APPLICABLE ENHANCED MONITORING AND COMPLIANCE CERTIFICATION REQUIREMENTS OF THE FEDERAL ACT		_____ YES _____ NO	

F.6   Appendix F

MAJOR SOURCE OPERATING PERMIT APPLICATION: FACILITY IDENTIFICATION

1. FACILITY NAME AND OWNER'S NAME IF DIFFERENT FROM THE FACILITY NAME:	FOR	APC COMPANY NO.
	APC	
MAILING ADDRESS (ST/RD/P.O. BOX):	USE	LOG/PERMIT NO.
CITY, STATE, ZIP CODE:	ONLY	

2. FACILITY LOCATION (ST/RD/HWY):	COUNTY NAME:
CITY OR DISTANCE TO NEAREST TOWN, ZIP CODE:	TELEPHONE NUMBER WITH AREA CODE:

3. FACILITY'S PRIMARY ACTIVITY AND THE FIRST TWO DIGITS OF THE FACILITY SIC CODE (S):

4. CONTACT PERSON'S NAME FOR THIS PERMIT: TITLE: TELEPHONE NUMBER WITH AREA CODE:

5. IF FACILITY IS LOCATED IN AN AREA DESIGNATED AS "NONATTAINMENT" OR "ADDITIONAL CONTROL", INDICATE THE POLLUTANT(S) FOR THE DESIGNATION.

6. LIST ALL VALID AIR POLLUTION PERMITS ISSUED TO THE SOURCES CONTAINED IN THIS APPLICATION [IDENTIFY ALL PERMITS WITH MOST RECENT PERMIT NUMBERS AND EMISSION SOURCE REFERENCE NUMBERS LISTED ON THE PERMIT(S)].

7. PERMIT REQUESTED FOR:

INITIAL APPLICATION TO OPERATE: _____   RELOCATION TO OPERATE: _____

MODIFICATION: _____   PERMIT RENEWAL TO OPERATE:_____

REVISION (ADMINISTRATIVE AMENDMENTS): _____

8. OWNER'S REGISTERED AGENT'S NAME & ADDRESS FOR SERVICE OF PROCESS   TELEPHONE NUMBER WITH AREA CODE

9. IS THIS FACILITY SUBJECT TO THE PROVISIONS GOVERNING PREVENTION OF ACCIDENTAL RELEASES OF HAZARDOUS AIR CONTAMINANTS CONTAINED IN CHAPTER 1200-3-32 OF THE TENNESSEE AIR POLLUTION CONTROL REGULATIONS?

_____ YES   _____ NO

IF THE ANSWER IS YES, ARE YOU IN COMPLIANCE WITH THE PROVISIONS OF CHAPTER 1200-3-32 OF THE TENNESSEE AIR POLLUTION CONTROL REGULATIONS?

_____ YES   _____ NO

10. PAGE NUMBER:	REVISION NUMBER:	DATE OF REVISION:

MAJOR SOURCE OPERATING PERMIT APPLICATION - OPERATION AND FLOW DIAGRAMS

1. PLEASE LIST, IDENTIFY AND DESCRIBE BRIEFLY PROCESS EMISSION SOURCES, FUEL BURNING INSTALLATION, AND INCINERATORS THAT ARE CONTAINED IN THIS APPLICATION. PLEASE ATTACH A FLOW DIAGRAM FOR THIS APPLICATION.

2. LIST ALL INSIGNIFICANT ACTIVITIES WHICH ARE EXEMPTED BECAUSE OF SIZE OR PRODUCTION RATE AND CITE THE APPLICABLE REGULATIONS.

3. ARE THERE ANY STORAGE PILES?

YES _____   NO _____

4. LIST THE STATES THAT ARE WITHIN 50 MILES OF YOUR FACILITY

5. PAGE NUMBER: REVISION NUMBER: DATE OF REVISION:

F.8　Appendix F

MAJOR SOURCE OPERATING PERMIT APPLICATION - STACK IDENTIFICATION

1. FACILITY NAME:	FOR APC USE ONLY	APC COMPANY NO.
2. STACK ID (OR FLOW DIAGRAM POINT IDENTIFICATION):		LOG/PERMIT NO.

3. EMISSION SOURCE (IDENTIFY):

4. STACK HEIGHT ABOVE GRADE IN FEET:

5. VELOCITY (DATA AT EXIT CONDITIONS): _____ (ACTUAL FEET PER SECOND)	6. INSIDE DIMENSIONS AT OUTLET IN FEET:
7. EXHAUST FLOW RATE AT EXIT CONDITIONS (ACFM):	8. FLOW RATE AT STANDARD CONDITIONS (DSCFM):
9. EXHAUST TEMPERATURE: _____ DEGREES FAHRENHEIT (^{O}F)	10. MOISTURE CONTENT (DATA AT EXIT CONDITIONS): GRAINS PER DRY STANDARD CUBIC _____ PERCENT _____ FOOT (gr/dscf)

11. EXHAUST TEMPERATURE THAT IS EQUALED OR EXCEEDED DURING NINETY (90) PERCENT OR MORE OF THE OPERATING TIME (FOR STACKS SUBJECT TO DIFFUSION EQUATION ONLY):

_____ (^{O}F)

12. IF THIS STACK IS EQUIPPED WITH CONTINUOUS POLLUTANT MONITORING EQUIPMENT REQUIRED FOR COMPLIANCE, WHAT POLLUTANT(S) DOES THIS EQUIPMENT MONITOR (e.g. OPACITY, SO_2, NO_x, etc.,)?

COMPLETE THE APPROPRIATE APC FORM(S) V.4, V.5, V.7, V.8, V.9, OR V.10 FOR EACH SOURCE EXHAUSTING THROUGH THIS STACK

13. DO YOU HAVE A BYPASS STACK:

_____ YES _____ NO

IF YES, DESCRIBE THE CONDITIONS WHICH REQUIRE ITS USE & COMPLETE APC FORM V.3 FOR THE BYPASS STACK. PLEASE IDENTIFY THE STACK NUMBER(S) OR FLOW DIAGRAM POINT NUMBER(S) EXHAUSTING THROUGH THIS BYPASS STACK.

14. PAGE NUMBER:	REVISION NUMBER:	DATE OF REVISION:

CN-1007

RDA 1298

MAJOR SOURCE OPERATING PERMIT APPLICATION - FUEL BURNING NON-PROCESS EQUIPMENT

1. FACILITY NAME:

2. LIST ALL FUEL-BURNING EQUIPMENT THAT IS AT THIS FUEL BURNING INSTALLATION (PLEASE COMPLETE AN APC V.4 FORM FOR EACH PIECE OF FUEL EQUIPMENT).

3. FUEL BURNING EQUIPMENT IDENTIFICATION NUMBER:	4. STACK ID OR FLOW DIAGRAM POINT IDENTIFICATION(S):

5. FUEL BURNING EQUIPMENT DESCRIPTION:

6. YEAR OF INSTALLATION OR LAST MODIFICATION OF FUEL BURNING EQUIPMENT:

7. FURNACE TYPE:	8. MANUFACTURER AND MODEL NUMBER (IF AVAILABLE):
9. MAXIMUM RATED HEAT INPUT CAPACITY (IN MILLION BTU/HOUR):	10. IF WOOD IS USED AS A FUEL, SPECIFY THE AMOUNT OF WOOD USED AS A FRACTION OF TOTAL HEAT INPUT.

11. FUELS:	PRIMARY FUEL	BACKUP FUEL #1	BACKUP FUEL #2	BACKUP FUEL #3
FUEL NAME				
ACTUAL YEARLY CONSUMPTION				

12. IF EMISSIONS FROM THIS FUEL BURNING EQUIPMENT ARE CONTROLLED FOR COMPLIANCE, PLEASE SPECIFY THE TYPE OF CONTROL:

13. LOCATION OF THIS FUEL BURNING EQUIPMENT ARE MONITORED FOR COMPLIANCE, PLEASE SPECIFY THE TYPE OF MONITORING:

14. LOCATION OF THIS FUEL BURNING INSTALLATION IN UTM COORDINATES:

UTM VERTICAL: _____ UTM HORIZONTAL:_____

15. NORMAL OPERATING SCHEDULE:

_____ HRS/DAY _____ DAYS/WK _____ DAYS/YR

16. DESCRIBE ANY FUGITIVE EMISSIONS ASSOCIATED WITH THIS PROCESS, SUCH AS OUTDOOR STORAGE PILES, OPEN CONVEYORS, MATERIAL HANDLING OPERATIONS. etc. (PLEASE ATTACH A SEPARATE SHEET IF NECESSARY).

17. PAGE NUMBER:	REVISION NUMBER:	DATE OF REVISION:

CN-1007   RDA 1298

MAJOR SOURCE OPERATING PERMIT APPLICATION - PRINTING OPERATIONS

1. FACILITY NAME:

2. PROCESS DESCRIPTION:

3. YEAR OF CONSTRUCTION OR LAST MODIFICATION:

4. STACK ID OR FLOW DIAGRAM POINT IDENTIFICATION(S):

IF THE EMISSIONS ARE CONTROLLED FOR COMPLIANCE, ATTACH AN APPROPRIATE AIR POLLUTION CONTROL SYSTEM FORM.
IF THIS PRINTING OPERATION IS MONITORED FOR COMPLIANCE, PLEASE ATTACH THE APPROPRIATE COMPLIANCE DEMONSTRATION FORM.

5. NORMAL OPERATING SCHEDULE:

_____ HRS/DAY _____ DAYS/WK _____ DAYS/YR

6. OPERATION TYPE: _____ FLEXOGRAPHIC _____ WEB-OFFSET _____ WEB-OFFSET (NON-HEATSET) _____ PACKAGING ROTOGRAVURE _____ PUBLICATION ROTOGRAVURE

_____ SCREEN PRINTING _____ OTHER (SPECIFY): _____

7. COMPLETE THE FOLLOWING TABLE - ATTACH ADDITIONAL TABLES AS NEEDED - FILL IN ONLY THE ITEMS NECESSARY FOR DETERMINATION OF COMPLIANCE WITH EMISSIONS STANDARD(S)

IDENTIFY INKS AND SOLVENTS: (NAME OR TYPE OF INK)	DAILY USAGE: GALLONS OR POUNDS		MONTHLY USAGE	COATING COMPOSITION: WEIGHT PERCENT AS APPLIED				INK DENSITY
	AVERAGE	MAXIMUM	MAXIMUM	SOLIDS WT (%)	SOLVENTS (VOCs) WT (%)	WATER WT (%)	EXEMPT SOLVENTS WT (%)	LBS/GAL
TOTAL INKS								

LIST THE THINNING SOLVENTS USED WITH THE INKS IDENTIFIED ABOVE:

(1):								
(2):								
CLEAN-UP SOLVENTS:								
OTHER (SPECIFY):								

8. LOCATION OF THIS OPERATION IN UTM COORDINATES:

UTM VERTICAL: _____ UTM HORIZONTAL: _____

9. PAGE NUMBER: _____ REVISION NUMBER: _____ DATE OF REVISION: _____

CN-1007

RDA 1298

F.10

MAJOR SOURCE OPERATING PERMIT APPLICATION - PAINTING AND COATING OPERATIONS

1. FACILITY NAME:

2. PROCESS DESCRIPTION:

3. YEAR OF CONSTRUCTION OR LAST MODIFICATION:

4. STACK ID OR FLOW DIAGRAM POINT IDENTIFICATION(S):

IF THE EMISSIONS ARE CONTROLLED FOR COMPLIANCE, ATTACH AN APPROPRIATE AIR POLLUTION CONTROL SYSTEM FORM.
IF THIS PRINTING OPERATION IS MONITORED FOR COMPLIANCE, PLEASE ATTACH THE APPROPRIATE COMPLIANCE DEMONSTRATION FORM.

5. NORMAL OPERATING SCHEDULE: _____ HRS/DAY _____ DAYS/WK _____ DAYS/YR

6. OVEN CURING (COMPLETE IF APPLICABLE):
NUMBER OF OVENS: _____ TEMPERATURE OF AIR CONTACTING COATED MATERIAL AS IT LEAVES THE OVEN (°F): _____

SPECIFY OVEN FUELS: _____ TOTAL MAXIMUM HEAT INPUT TO EACH OVEN: _____

7. APPLICATION TECHNIQUE AND TRANSFER EFFICIENCY (%):

8. COMPLETE THE FOLLOWING TABLE - ATTACH ADDITIONAL TABLES AS NEEDED - FILL IN ONLY THE ITEMS NECESSARY FOR DETERMINATION OF COMPLIANCE WITH EMISSIONS STANDARD(S)

IDENTIFY INKS AND SOLVENTS: (NAME OR TYPE OF INK)	MAXIMUM USAGE:			NORMAL USAGE	COATING COMPOSITION: WEIGHT PERCENT AS APPLIED								DENSITY OF SOLVENT FRACTION	COATING DENSITY
					SOLIDS		SOLVENTS (VOCs)		WATER		EXEMPT SOLVENTS			
	GAL/HR	GAL/MO	GAL/MO	GAL/MO	VOL (%)	WT (%)	WT (%)	VOL (%)	WT (%)	WT (%)	VOL (%)	WT (%)	LBS/GAL	LBS/GAL
TOTAL INKS														

LIST THE THINNING SOLVENTS USED WITH THE INKS IDENTIFIED ABOVE:

(1):

(2):

CLEAN-UP SOLVENTS:

OTHER (SPECIFY):

9. LOCATION OF THIS OPERATION IN UTM COORDINATES:

UTM VERTICAL: _____ UTM HORIZONTAL: _____

10. PAGE NUMBER: _____ REVISION NUMBER: _____ DATE OF REVISION: _____

CN-1007

RDA 1298

F.11

MAJOR SOURCE OPERATING PERMIT APPLICATION - MISCELLANEOUS PROCESSES

1. FACILITY NAME:	2. PROCESS IDENTIFICATION NUMBER:

3. STACK ID OR FLOW DIAGRAM POINT IDENTIFICATION(S):

IF EMISSIONS ARE CONTROLLED FOR COMPLIANCE, ATTACH THE APPROPRIATE AIR POLLUTION CONTROL SYSTEM FORM.

4. NORMAL OPERATING SCHEDULE: _____ HRS/DAY _____ DAYS/WK _____ DAYS/YR	5. YEAR OF CONSTRUCTION OR LAST MODIFICATION:

6. DESCRIBE THIS PROCESS (PLEASE ATTACH A FLOW DIAGRAM OF THIS PROCESS) AND CHECK ONE OF THE FOLLOWING:

_____ BATCH   _____ CONTINUOUS

7. LIST THE TYPES AND AMOUNTS OF RAW MATERIAL INPUT TO THIS PROCESS:

MATERIAL	STORAGE/MATERIAL HANDLING PROCESS	AVERAGE USAGE (UNITS)	MAXIMUM USAGE (UNITS)

8. LIST THE TYPES AND AMOUNTS OF PRIMARY PRODUCTS PRODUCED BY THIS PROCESS:

MATERIAL	STORAGE/MATERIAL HANDLING PROCESS	AVERAGE AMOUNT PRODUCED (UNITS)	MAXIMUM AMOUNT PRODUCED (UNITS)

9. PROCESS FUEL USAGE:

TYPE OF FUEL	MAX HEAT INPUT (10^6 BTU/HR)	AVERAGE USAGE (UNITS)	MAXIMUM USAGE (UNITS)

10. LIST ANY SOLVENTS, CLEANERS, etc., ASSOCIATED WITH THIS PROCESS:

IF THE EMISSIONS AND/OR OPERATIONS OF THIS PROCESS ARE MONITORED FOR COMPLIANCE, PLEASE ATTACH THE APPROPRIATE COMPLIANCE DEMONSTRATION FORM.

11. DESCRIBE ANY FUGITIVE EMISSIONS ASSOCIATED WITH THIS PROCESS, SUCH AS OUTDOOR STORAGE PILES, OPEN CONVEYORS, OPEN AIR SAND BLASTING, MATERIAL HANDLING OPERATIONS, etc. (PLEASE ATTACH A SEPARATE SHEET IF NECESSARY).

12. LOCATION OF THIS FUEL BURNING INSTALLATION IN UTM COORDINATES:

UTM VERTICAL: _____   UTM HORIZONTAL: _____

13. PAGE NUMBER:	REVISION NUMBER:	DATE OF REVISION:

MAJOR SOURCE OPERATING PERMIT APPLICATION
CONTROL EQUIPMENT - MISCELLANEOUS

1. FACILITY NAME:	2. EMISSION SOURCE (IDENTIFY):

3. STACK ID OR FLOW DIAGRAM POINT IDENTIFICATION(S):

4. DESCRIBE THE DEVICE IN USE. LIST THE KEY OPERATING PARAMETERS OF THIS DEVICE AND THEIR NORMAL OPERATING RANGE.
(e.g. PRESSURE DROP, GAS FLOW RATE, TEMPERATURE).

5. MANUFACTURER AND MODEL NUMBER (IF AVAILABLE):

6. YEAR OF INSTALLATION:

7. LIST OF POLLUTANT(S) TO BE CONTROLLED AND THE EXPECTED CONTROL EFFICIENCY FOR EACH POLLUTANT.

POLLUTANT	EFFICIENCY (%)	SOURCE OF DATA

8. DISCUSS HOW COLLECTED MATERIAL IS HANDLED FOR REUSE OR DISPOSAL.

9. IF THIS CONTROL EQUIPMENT IS IN SERIES WITH SOME OTHER CONTROL EQUIPMENT, STATE AND SPECIFY THE OVERALL EFFICIENCY.

10. PAGE NUMBER:	REVISION NUMBER:	DATE OF REVISION:

MAJOR SOURCE OPERATING PERMIT APPLICATION
CONTROL EQUIPMENT - MISCELLANEOUS

1. FACILITY NAME:	2. EMISSION SOURCE (IDENTIFY):

3. STACK ID OR FLOW DIAGRAM POINT IDENTIFICATION(S):

4. DESCRIBE THE DEVICE IN USE. LIST THE KEY OPERATING PARAMETERS OF THIS DEVICE AND THEIR NORMAL OPERATING RANGE. (e.g. PRESSURE DROP, GAS FLOW RATE, TEMPERATURE).

5. MANUFACTURER AND MODEL NUMBER (IF AVAILABLE):

6. YEAR OF INSTALLATION:

7. LIST OF POLLUTANT(S) TO BE CONTROLLED AND THE EXPECTED CONTROL EFFICIENCY FOR EACH POLLUTANT.

POLLUTANT	EFFICIENCY (%)	SOURCE OF DATA

8. DISCUSS HOW COLLECTED MATERIAL IS HANDLED FOR REUSE OR DISPOSAL.

9. IF THIS CONTROL EQUIPMENT IS IN SERIES WITH SOME OTHER CONTROL EQUIPMENT, STATE AND SPECIFY THE OVERALL EFFICIENCY.

10. PAGE NUMBER:	REVISION NUMBER:	DATE OF REVISION:

MAJOR SOURCE OPERATING PERMIT APPLICATION
CONTROL EQUIPMENT - CYCLONE/SETTLING CHAMBERS

1. FACILITY NAME:	2. PROCESS IDENTIFICATION NUMBER:

3. STACK ID OR FLOW DIAGRAM POINT IDENTIFICATION(S):

4. DESCRIBE THE DEVICE IN USE. LIST THE KEY OPERATING PARAMETERS OF THIS DEVICE AND THEIR NORMAL OPERATING RANGE.

5. LIST OF POLLUTANT(S) TO BE CONTROLLED AND THE EXPECTED CONTROL EFFICIENCY FOR EACH POLLUTANT.

POLLUTANT	EFFICIENCY (%)	SOURCE OF DATA

6. DISCUSS HOW COLLECTED MATERIAL IS HANDLED FOR REUSE OR DISPOSAL.

7. GAS FLOW RATE (ACFM):

8. IF THIS CONTROL EQUIPMENT IS IN SERIES WITH SOME OTHER CONTROL EQUIPMENT, STATE AND SPECIFY THE OVERALL EFFICIENCY.

9. PAGE NUMBER:	REVISION NUMBER:	DATE OF REVISION:

CN-1007

RDA 1298

MAJOR SOURCE OPERATING PERMIT APPLICATION
CONTROL EQUIPMENT - BAGHOUSES/FABRIC FILTERS

1. FACILITY NAME:	2. PROCESS IDENTIFICATION NUMBER:

3. STACK ID OR FLOW DIAGRAM POINT IDENTIFICATION(S):

4. DESCRIBE THE DEVICE IN USE. LIST THE KEY OPERATING PARAMETERS OF THIS DEVICE AND THEIR NORMAL OPERATING RANGE.

5. MANUFACTURER AND MODEL NUMBER (IF AVAILABLE):	6. YEAR OF INSTALLATION:

7. LIST OF POLLUTANT(S) TO BE CONTROLLED AND THE EXPECTED CONTROL EFFICIENCY FOR EACH POLLUTANT (SEE INSTRUCTIONS).

POLLUTANT	EFFICIENCY (%)	SOURCE OF DATA

8. DISCUSS HOW COLLECTED MATERIAL IS HANDLED FOR REUSE OR DISPOSAL.

9. IF THE BAGS ARE COATED, SPECIFY THE MATERIAL USED FOR COATING AND FREQUENCY OF COATING.

10. DOES THE BAGHOUSE COLLECT ASBESTOS CONTAINING MATERIAL?

YES _____   NO _____

IF "YES", PROVIDE DATA AS OUTLINED IN ITEM 10, INSTRUCTIONS FOR THIS FORM.

11. IF THIS CONTROL EQUIPMENT IS IN SERIES WITH SOME OTHER CONTROL EQUIPMENT, STATE AND SPECIFY THE OVERALL EFFICIENCY.

12. PAGE NUMBER:	REVISION NUMBER:	DATE OF REVISION:

COMPLIANCE CERTIFICATION - MONITORING AND REPORTING
DESCRIPTION OF METHODS USED TO DETERMINE COMPLIANCE

ALL SOURCES THAT ARE SUBJECT TO 1200-3-9-.02(11) OF TENNESSEE AIR POLLUTION CONTROL REGULATIONS ARE REQUIRED TO CERTIFY COMPLIANCE WITH ALL APPLICABLE REQUIREMENTS BY INCLUDING A STATEMENT WITHIN THE PERMIT APPLICATION OF THE METHODS USED FOR DETERMINING COMPLIANCE. THIS STATEMENT MUST INCLUDE A DESCRIPTION OF THE MONITORING, RECORDKEEPING, AND REPORTING REQUIREMENTS AND TEST METHODS. IN ADDITION, THE APPLICATION MUST INCLUDE A SCHEDULE FOR COMPLIANCE CERTIFICATION SUBMITTALS DURING THE PERMIT TERM. THESE SUBMITTALS MUST BE NO LESS FREQUENT THAN ANNUALLY AND MAY NEED TO BE MORE FREQUENT IF SPECIFIED BY THE UNDERLYING APPLICABLE REQUIREMENT OR THE TECHNICAL SECRETARY.

1. FACILITY NAME:

2. PROCESS EMISSION SOURCE, FUEL BURNING INSTALLATION, OR INCINERATOR (IDENTIFY):

3. STACK ID OR FLOW DIAGRAM POINT IDENTIFICATION(S):

4. THIS SOURCE AS DESCRIBED UNDER ITEM #2 OF THIS APPLICATION WILL USE THE FOLLOWING METHOD(S) FOR DETERMINING COMPLIANCE WITH APPLICABLE REQUIREMENTS (AND SPECIAL OPERATING CONDITIONS FROM AN EXISTING PERMIT). CHECK ALL THAT APPLY AND ATTACH THE APPROPRIATE FORM(S).

_____   CONTINUOUS EMISSIONS MONITORING (CEM) - APC FORM V.20
             POLLUTANT(S):

_____   EMISSION MONITORING USING PORTABLE MONITORS - APC FORM V.21
             POLLUTANT(S):

_____   MONITORING CONTROL SYSTEM PARAMETERS OR OPERATING PARAMETERS OF A PROCESS - APC FORM V.22
             POLLUTANT(S):

_____   MONITORING MAINTENANCE PROCEDURES - APC FORM V.23
             POLLUTANT(S):

_____   STACK TESTING - APC FORM V.24
             POLLUTANT(S):

_____   FUEL SAMPLING & ANALYSIS (FSA) - APC FORM V.25
             POLLUTANT(S):

_____   RECORDKEEPING - APC FORM V.26
             POLLUTANT(S):

_____   OTHER (PLEASE DESCRIBE) - APC FORM V.27
             POLLUTANT(S):

5. COMPLIANCE CERTIFICATION REPORTS WILL BE SUBMITTED TO THE DIVISION ACCORDING TO THE FOLLOWING SCHEDULE.

START DATE: _____

AND EVERY _____ DAYS THEREAFTER.

6. COMPLIANCE MONITORING REPORTS WILL BE SUBMITTED TO THE DIVISION ACCORDING TO THE FOLLOWING SCHEDULE.

START DATE: _____

AND EVERY _____ DAYS THEREAFTER.

7. PAGE NUMBER: REVISION NUMBER: DATE OF REVISION:

CN-1007   RDA 1298

**MAJOR SOURCE OPERATING PERMIT APPLICATION - COMPLIANCE DEMONSTRATION BY
MONITORING CONTROL SYSTEM PARAMETERS OR OPERATING PARAMETERS OF A PROCESS**

THE MONITORING OF A CONTROL SYSTEM PARAMETER OR A PROCESS PARAMETER SHALL BE ACCEPTABLE AS A COMPLIANCE
DEMONSTRATION METHOD PROVIDED THAT A CORRELATION BETWEEN THE PARAMETER VALUE AND THE EMISSION RATE OF A PARTICULAR
POLLUTANT IS ESTABLISHED.

1. FACILITY NAME:	2. STACK ID OR FLOW DIAGRAM POINT IDENTIFICATION(S):

3. EMISSION SOURCE (IDENTIFY):

4. POLLUTANT(S) OR PARAMETER BEING MONITORED:

5. DESCRIPTION OF THE METHOD OF MONITORING AND ESTABLISHMENT OF CORRELATION BETWEEN THE PARAMETER VALUE AND THE
EMISSION RATE OF A PARTICULAR POLLUTANT:

6. COMPLIANCE DEMONSTRATION FREQUENCY (SPECIFY THE FREQUENCY WITH WHICH COMPLIANCE WILL BE DEMONSTRATED):

7. PAGE NUMBER:	REVISION NUMBER:	DATE OF REVISION:

CN-1007

RDA 1298

MAJOR SOURCE OPERATING PERMIT APPLICATION
COMPLIANCE DEMONSTRATION BY MONITORING MAINTENANCE PROCEDURES

THE MONITORING OF A MAINTENANCE PROCEDURE SHALL BE ACCEPTABLE AS A COMPLIANCE DEMONSTRATION METHOD PROVIDED THAT A CORRELATION BETWEEN THE PROCEDURE AND THE EMISSION RATE OF A PARTICULAR POLLUTANT IS ESTABLISHED.

1. FACILITY NAME:

2. STACK ID OR FLOW DIAGRAM POINT IDENTIFICATION(S):

3. EMISSION SOURCE (IDENTIFY):

4. POLLUTANT(S) BEING MONITORED:

5. PROCEDURE BEING MONITORED:

6. DESCRIPTION OF THE METHOD OF MONITORING AND ESTABLISHMENT OF CORRELATION BETWEEN THE PROCEDURE AND THE EMISSION RATE OF A PARTICULAR POLLUTANT:

7. COMPLIANCE DEMONSTRATION FREQUENCY (SPECIFY THE FREQUENCY WITH WHICH COMPLIANCE WILL BE DEMONSTRATED):

8. PAGE NUMBER: REVISION NUMBER: DATE OF REVISION:

CN-1007   RDA 1298

MAJOR SOURCE OPERATING PERMIT APPLICATION
COMPLIANCE DEMONSTRATION BY FUEL SAMPLING AND ANALYSIS

1. FACILITY NAME:	2. STACK ID OR FLOW DIAGRAM POINT IDENTIFICATION(S):

3. EMISSION SOURCE (IDENTIFY):

4. POLLUTANT(S) BEING MONITORED:

5. FUEL BEING SAMPLED:

6. LIST THE FUEL SAMPLE COLLECTING AND ANALYZING METHODS USED (IF AN ASTM METHOD IS NOT APPLICABLE, PROPOSE A METHOD ACCEPTABLE TO THE TECHNICAL SECRETARY).

7. COMPLIANCE DEMONSTRATION FREQUENCY (SPECIFY THE FREQUENCY WITH WHICH COMPLIANCE WILL BE DEMONSTRATED):

8. PAGE NUMBER:	REVISION NUMBER:	DATE OF REVISION:

CN-1007

RDA 1298

MAJOR SOURCE OPERATING PERMIT APPLICATION
COMPLIANCE DEMONSTRATION BY RECORDKEEPING

RECORDKEEPING SHALL BE ACCEPTABLE AS A COMPLIANCE DEMONSTRATION METHOD PROVIDED THAT A CORRELATION BETWEEN THE PARAMETER VALUE RECORDED AND THE APPLICABLE REQUIREMENT IS ESTABLISHED.

1. FACILITY NAME:	2. STACK ID OR FLOW DIAGRAM POINT IDENTIFICATION(S):

3. EMISSION SOURCE (IDENTIFY):

4. POLLUTANT(S) OR PARAMETER BEING MONITORED:

5. MATERIAL OR PARAMETER BEING MONITORED AND RECORDED:

6. METHOD OF MONITORING AND RECORDING:

7. COMPLIANCE DEMONSTRATION FREQUENCY (SPECIFY THE FREQUENCY WITH WHICH COMPLIANCE WILL BE DEMONSTRATED):

8. PAGE NUMBER:	REVISION NUMBER:	DATE OF REVISION:

CN-1007   RDA 1298

MAJOR SOURCE OPERATING PERMIT APPLICATION
EMISSION FROM PROCESS EMISSION SOURCE / FUEL BURNING INSTALLATION / INCINERATOR

1. FACILITY NAME:	2. STACK ID OR FLOW DIAGRAM POINT IDENTIFICATION(S):

3. PROCESS EMISSION SOURCE/FUEL BURNING INSTALLATION/INCINERATOR (IDENTIFY):

4. COMPLETE THE FOLLOWING EMISSIONS SUMMARY FOR REGULATED AIR POLLUTANTS. FUGITIVE EMISSIONS SHALL BE INCLUDED. ATTACH CALCULATIONS AND EMISSION FACTOR REFERENCES.

AIR POLLUTANT	MAXIMUM ALLOWABLE EMISSIONS		ACTUAL EMISSIONS	
	TONS PER YEAR	RESERVED FOR STATE USE (POUNDS PER HOUR- ITEM 7, APC V.30)	TONS PER YEAR	RESERVED FOR STATE USE (POUNDS PER HOUR- ITEM 8. APC V.30)
PARTICULATE (TSP)				
(FUGITIVE EMISSIONS)				
SULFUR DIOXIDE				
(FUGITIVE EMISSIONS)				
VOLATILE ORGANIC COMPOUNDS				
(FUGITIVE EMISSIONS)				
CARBON MONOXIDE				
(FUGITIVE EMISSIONS)				
LEAD				
(FUGITIVE EMISSIONS)				
NITROGEN OXIDES				
(FUGITIVE EMISSIONS)				
TOTAL REDUCED SULFUR				
(FUGITIVE EMISSIONS)				
MERCURY				
(FUGITIVE EMISSIONS)				

(CONTINUED ON NEXT PAGE)

CN-1007   RDA 1298

	(CONTINUED FROM PREVIOUS PAGE)			
AIR POLLUTANT	**MAXIMUM ALLOWABLE EMISSIONS**		**ACTUAL EMISSIONS**	
	TONS PER YEAR	RESERVED FOR STATE USE (POUNDS PER HOUR- ITEM 7, APC V.30)	TONS PER YEAR	RESERVED FOR STATE USE (POUNDS PER HOUR- ITEM 8, APC V.30)
ASBESTOS				
(FUGITIVE EMISSIONS)				
BERYLLIUM				
(FUGITIVE EMISSIONS)				
VINYL CHLORIDES				
(FUGITIVE EMISSIONS)				
FLUORIDES				
(FUGITIVE EMISSIONS)				
GASEOUS FLOURIDES				
(FUGITIVE EMISSIONS)				

5. COMPLETE THE FOLLOWING <u>EMISSIONS SUMMARY FOR REGULATED AIR POLLUTANTS THAT ARE HAZARDOUS AIR POLLUTANT(S)</u>. FUGITIVE EMISSIONS SHALL BE INCLUDED. ATTACH CALCULATIONS AND EMISSION FACTOR REFERENCES.

AIR POLLUTANT AND CAS	**MAXIMUM ALLOWABLE EMISSIONS**		**ACTUAL EMISSIONS**	
	TONS PER YEAR	RESERVED FOR STATE USE (POUNDS PER HOUR- ITEM 7, APC V.30)	TONS PER YEAR	RESERVED FOR STATE USE (POUNDS PER HOUR- ITEM 8, APC V.30)

6. PAGE NUMBER: REVISION NUMBER: DATE OF REVISION:

CN-1007

MAJOR SOURCE OPERATING PERMIT APPLICATION

EMISSION SUMMARY FOR THE FACILITY OR FOR THE SOURCES CONTAINED IN THIS APPLICATION

1. FACILITY NAME:

2. COMPLETE THE FOLLOWING EMISSIONS SUMMARY FOR REGULATED AIR POLLUTANTS AT THIS FACILITY OR FOR THE SOURCES CONTAINED IN THIS APPLICATION.

AIR POLLUTANT	SUMMARY OF MAXIMUM ALLOWABLE EMISSIONS		SUMMARY OF ACTUAL EMISSIONS	
	TONS PER YEAR	RESERVED FOR STATE USE (POUNDS PER HOUR- ITEM 4, APC V.28)	TONS PER YEAR	RESERVED FOR STATE USE (POUNDS PER HOUR- ITEM 4. APC V.28)
PM-10				
PARTICULATE (TSP)				
SULFUR DIOXIDE				
VOLATILE ORGANIC COMPOUNDS				
CARBON MONOXIDE				
LEAD				
NITROGEN OXIDES				
TOTAL REDUCED SULFUR				
MERCURY				
ASBESTOS				
BERYLLIUM				
VINYL CHLORIDES				
FLUORIDES				
GASEOUS FLOURIDES				

(CONTINUED ON NEXT PAGE)

CN-1007   RDA 1298

(CONTINUED FROM PREVIOUS PAGE)				

3. COMPLETE THE FOLLOWING <u>EMISSIONS SUMMARY FOR REGULATED AIR POLLUTANTS THAT ARE HAZARDOUS AIR POLLUTANT(S) AT THIS</u> <u>FACILITY</u> OR FOR THE SOURCES CONTAINED IN THIS APPLICATION.

AIR POLLUTANT AND CAS	SUMMARY OF MAXIMUM ALLOWABLE EMISSIONS		SUMMARY OF ACTUAL EMISSIONS	
	TONS PER YEAR	RESERVED FOR STATE USE (POUNDS PER HOUR- ITEM 5, APC V.28)	TONS PER YEAR	RESERVED FOR STATE USE (POUNDS PER HOUR- ITEM 5. APC V.28)

4. PAGE NUMBER: REVISION NUMBER: DATE OF REVISION:

CN-1007   RDA 1298

MAJOR SOURCE OPERATING PERMIT APPLICATION: CURRENT EMISSIONS REQUIREMENT AND STATUS

1. FACILITY NAME:			2. EMISSION SOURCE NUMBER:			

3. DESCRIBE THE PROCESS EMISSION SOURCE/ FUEL BURNING INSTALLATION/ INCINERATOR.

4. IDENTIFY IF ONLY A PART OF THE SOURCE IS SUBJECT TO THIS REQUIREMENT	5. POLLUTANT	6. APPLICABLE REQUIREMENT(S): TN AIR POLLUTION CONTROL REGULATIONS, 40 CFR, PERMIT RESTRICTIONS, AIR QUALITY BASED STANDARDS	7. LIMITATION	8. MAXIMUM ACTUAL EMISSIONS	9. COMPLIANCE STATUS (IN/OUT)

10. OTHER APPLICABLE REQUIREMENTS (NEW REQUIREMENTS THAT APPLY TO THIS SOURCE DURING THE TERM OF THIS PERMIT)

11. PAGE NUMBER:	REVISION NUMBER:	DATE OF REVISION:

CN-1007

RDA 1298

F.26

MAJOR SOURCE OPERATING PERMIT APPLICATION
COMPLIANCE PLAN AND COMPLIANCE CERTIFICATION

1. FACILITY NAME:

2. LIST ALL THE PROCESS EMISSION SOURCE(S) OR FUEL BURNING INSTALLATION(S) OR INCINERATOR(S) THAT ARE PART OF THIS APPLICATION.

3. INDICATE THAT SOURCE(S) WHICH ARE CONTAINED IN THIS APPLICATION ARE PRESENTLY IN COMPLIANCE WITH ALL APPLICABLE REQUIREMENTS, BY CHECKING THE FOLLOWING.

_____ A. ATTACHED IS STATEMENT OF IDENTIFICATION OF THE SOURCE(S) CURRENTLY IN COMPLIANCE. WE WILL CONTINUE TO OPERATE AND MAINTAIN THE SOURCE(S) TO ASSURE COMPLIANCE WITH ALL THE APPLICABLE REQUIREMENTS FOR THE DURATION OF THE PERMIT.

_____ B. APC V.30 FORM(S) INCLUDES THE NEW REQUIREMENTS THAT APPLY OR WILL APPLY TO THE SOURCE(S) DURING THE TERM OF THE PERMIT. WE WILL MEET SUCH REQUIREMENTS ON A TIMELY BASIS.

4. INDICATE THAT THERE ARE SOURCE(S) THAT ARE CONTAINED IN THIS APPLICATION WHICH ARE NOT PRESENTLY IN FULL COMPLIANCE. BY CHECKING THE FOLLOWING.

_____ A. ATTACHED IS STATEMENT OF IDENTIFICATION OF THE SOURCE(S) NOT IN COMPLIANCE, NON-COMPLYING REQUIREMENT(S), BRIEF DESCRIPTION OF THE PROBLEM, AND THE PROPOSED SOLUTION.

_____ B. WE WILL ACHIEVE COMPLIANCE ACCORDING TO THE FOLLOWING SCHEDULE:

ACTION	DEADLINE

PROGRESS REPORTS WILL BE SUBMITTED:

START DATE: _____   AND EVERY 180 DAYS THEREAFTER UNTIL COMPLIANCE IS ACHIEVED.

5. STATE THE COMPLIANCE STATUS WITH ANY APPLICABLE ENHANCED MONITORING AND COMPLIANCE CERTIFICATION REQUIREMENTS THAT HAVE BEEN PROMULGATED UNDER SECTION 114(a)(3) OF THE CLEAN AIR ACT AS OF THIS DATE OF SUBMITTAL OF THIS APC FORM V.31.

6. PAGE NUMBER: REVISION NUMBER: DATE OF REVISION:

G

Hazardous Air Pollutants

Section 112: Hazardous Air Pollutants (7/8/96 update)

"This draft list includes current EPA staff recommendations for technical corrections and clarifications of the hazardous air pollutants (HAP) list in Section 112(b) (1) of the Clean Air Act. This draft has been distributed to apprise interested parties of potential future changes in the HAP list and is informational only. The recommended revisions of the current HAP list which are included in this draft do not themselves change the list as adopted by Congress and have no legal effect. EPA intends to propose specific revisions of the HAP list, including any technical corrections or clarifications of the list, only through notice and comment rule-making.

Chemical Abstracts service number	Pollutant
75-07-0	Acetaldehyde
60-35-5	Acetamide
75-05-8	Acetonitrile
98-86-2	Acetophenone
53-96-3	2-Acetylaminofluorene
107-02-8	Acrolein
79-06-1	Acrylamide
79-10-7	Acrylic acid
107-13-1	Acrylonitrile
107-05-1	Allyl chloride
92-67-1	4-Aminobiphenyl
62-53-3	Aniline
90-04-0	o-Anisidine
1332-21-4	Asbestos
71-43-2	Benzene (including benzene from gasoline)
92-87-5	Benzidine
98-07-7	Benzotrichloride
100-44-7	Benzyl chloride
92-52-4	Biphenyl
117-81-7	Bis (2-ethylhexyl) phthalate (DEHP)
542-88-1	Bis (chloromethyl) ether
75-25-2	Bromoform
106-99-0	1,3-Butadiene
156-62-7	Calcium cyanamide
105-60-2	Caprolactam (removed 6/18/96, 61FR30816)
133-06-2	Captan

Chemical Abstracts service number	Pollutant
63-25-2	Carbaryl
75-15-0	Carbon disulfide
56-23-5	Carbon tetrachloride
463-58-1	Carbonyl sulfide
120-80-9	Catechol
133-90-4	Chloramben
57-74-9	Chlordane
7782-50-5	Chlorine
79-11-8	Chloroacetic acid
532-27-4	2-Chloroacetophenone
108-90-7	Chlorobenzene
510-15-6	Chlorobenzilate
67-66-3	Chloroform
107-30-2	Chloromethyl methyl ether
126-99-8	Chloroprene
1319-77-3	Cresol/cresylic acid (mixed isomers)
95-48-7	o-Cresol
108-39-4	m-Cresol
106-44-5	p-Cresol
98-82-8	Cumene
N/A	2,4-D (2,4-Dichlorophenoxyacetic acid) (including salts and esters)
3547-04-4	DDE
334-88-3	Diazomethane
132-64-9	Dibenzofuran
96-12-8	1,2-Dibromo-3-chloropropane
84-74-2	Dibutyl phthalate
106-46-7	1,4-Dichlorobenzene
91-94-1	3,3'-Dichlorobenzidine
111-44-4	Dichloroethyl ether [Bis(2-chloroethyl) ether]
542-75-6	1,3-Dichloropropene
62-73-7	Dichlorvos
111-42-2	Diethanolamine
64-67-5	Diethyl sulfate
119-90-4	3,3'-Dimethoxybenzidine
60-11-7	4-Dimethylaminoazobenzene
121-69-7	N, N-Dimethylaniline
119-93-7	3,3'-Dimethylbenzidine
79-44-7	Dimethylcarbamoyl chloride
68-12-2	N,N-Dimethylformamide
57-14-7	1,1-Dimethylhydrazine
131-11-3	Dimethyl phthalate
77-78-1	Dimethyl sulfate
N/A	4,6-Dinitro-o-cresol (including salts)
51-28-5	2,4-Dinitrophenol
121-14-2	2,4-Dinitrotoluene
123-91-1	1,4-Dioxane (1,4-diethyleneoxide)
122-66-7	1,2-Diphenylhydrazine
106-89-8	Epichlorohydrin (1-chloro-2,3-epoxypropane)
106-88-7	1,2-Epoxybutane
140-88-5	Ethyl acrylate
100-41-4	Ethylbenzene

Chemical Abstracts service number	Pollutant
51-79-6	Ethyl carbamate (urethane)
75-00-3	Ethyl chloride (chloroethane)
106-93-4	Ethylene dibromide (dibromoethane)
107-06-2	Ethylene dichloride (1,2-dichloroethane)
107-21-1	Ethylene glycol
151-56-4	Ethyleneimine (aziridine)
75-21-8	Ethylene oxide
96-45-7	Ethylene thiourea
75-34-3	Ethylidene dichloride (1,1-dichloroethane)
50-00-0	Formaldehyde
76-44-8	Heptachlor
118-74-1	Hexachlorobenzene
87-68-3	Hexachlorobutadiene
N/A	1,2,3,4,5,6-Hexachlorocyclyhexane (all stereo isomers, including lindane)
77-47-4	Hexachlorocyclopentadiene
67-72-1	Hexachloroethane
822-06-0	Hexamethylene diisocyanate
680-31-9	Hexamethylphosphoramide
110-54-3	Hexane
302-01-2	Hydrazine
7647-01-0	Hydrochloric acid [hydrogen chloride (gas only)]
7664-39-3	Hydrogen fluoride (hydrofluoric acid)
123-31-9	Hydroquinone
78-59-1	Isophorone
108-31-6	Maleic anhydride
67-56-1	Methanol
72-43-5	Methoxychlor
74-83-9	Methyl bromide (bromomethane)
74-87-3	Methyl chloride (chloromethane)
71-55-6	Methyl chloroform (1,1,1-trichloroethane)
78-93-3	Methyl ethyl ketone (2-butanone)
60-34-4	Methylhydrazine
74-88-4	Methyl iodide (iodomethane)
108-10-1	Methyl isobutyl ketone (hexone)
624-83-9	Methyl isocyanate
80-62-6	Methyl methacrylate
1634-04-04	Methyl tert-butyl ether
101-14-4	4,4′-Methylenebis (2-chloroaniline)
75-09-2	Methylene chloride (dichloromethane)
101-68-8	4,4′-Methylenediphenyl diisocyanate (MDI)
101-77-9	4,4′-Methylenedianiline
91-20-3	Naphthalene
98-95-3	Nitrobenzene
92-93-3	4-Nitrobiphenyl
100-02-7	4-Nitrophenol
79-46-9	2-Nitropropane
684-93-5	N-Nitroso-N-methylurea
62-75-9	N-Nitrosodimethylamine
59-89-2	N-Nitrosomorpholine
56-38-2	Parathion
82-68-8	Pentachloronitrobenzene (quintobenzene)

Chemical Abstracts service number	Pollutant
87-86-5	Pentachlorophenol
108-95-2	Phenol
106-50-3	p-Phenylenediamine
75-44-5	Phosgene
7803-51-2	Phosphine
N/A	Phosphorus compounds
85-44-9	Phthalic anhydride
1336-36-3	Polychlorinated biphenyls (aroclors)
1120-71-4	1,3-Propane sultone
57-57-8	ß-Propiolactone
123-38-6	Propionaldehyde
114-26-1	Propoxur (baygon)
78-87-5	Propylene dichloride (1,2-dichloropropane)
75-56-9	Propylene oxide
75-55-8	1,2-Propylenimine (2-methylaziridine)
91-22-5	Quinoline
106-51-4	Quinone (p-benzoquinone)
100-42-5	Styrene
96-09-3	Styrene oxide
1746-01-6	2,3,7,8-Tetrachlorodibenzo-p-dioxin
79-34-5	1,1,2,2-Tetrachloroethane
127-18-4	Tetrachloroethylene (perchloroethylene)
7550-45-0	Titanium tetrachloride
108-88-3	Toluene
95-80-7	Toluene-2,4-diamine
584-84-9	2,4-Toluene diisocyanate
95-53-4	o-Toluidine
8001-35-2	Toxaphene (chlorinated camphene)
120-82-1	1,2,4-Trichlorobenzene
79-00-5	1,1,2-Trichloroethane
79-01-6	Trichloroethylene
95-95-4	2,4,5-Trichlorophenol
88-06-2	2,4,6-Trichlorophenol
121-44-8	Triethylamine
1582-09-8	Trifluralin
540-84-1	2,2,4-Trimethylpentane
108-05-4	Vinyl acetate
593-60-2	Vinyl bromide
75-01-4	Vinyl chloride
75-35-4	Vinylidene chloride (1,1-dichloroethylene)
1330-20-7	Xylenes (mixed isomers)
95-47-6	o-Xylene
108-38-3	m-Xylene
106-42-3	p-Xylene
	Antimony compounds
	Arsenic compounds (inorganic including arsine)
	Beryllium compounds
	Cadmium compounds
	Chromium compounds
	Cobalt compounds
	Coke oven emissions
	Cyanide compounds[a]
	Glycol ethers[b]

Chemical Abstracts service number	Pollutant
	Lead compounds
	Manganese compounds
	Mercury compounds
	Fine mineral fibers[c]
	Nickel compounds
	Polycyclic organic matter[d]
	Radionuclides (including radon)[e]
	Selenium compounds

NOTE: For all listings above which contain the word *compounds* and for glycol ethers, the following applies: Unless otherwise specified, these listings are defined as including any unique chemical substance that contains the named chemical (i.e., antimony, arsenic, etc.) as part of that chemical's infrastructure.

[a]X'CN where X = H' or any other group where a formal dissociation may occur, for example, KCN or $Ca(CN)_2$.

[b]Under review. Glycol ether definition draft options:

Possible correction to CAA 112(b)(1) footnote that would be consistent with OPPTS modified definition.

New OPPTS definition as published is

$R—(OCH_2CH_2)n—OR'$

where n = 1, 2, or 3
    R = alkyl C7 or less
    or R = phenyl or alkyl-substituted phenyl
    R' = H or alkyl C7 or less
    or OR' = carboxylic acid ester, sulfate, phosphate, nitrate, or sulfonate

CAA glycol ether definition exactly as in the statute (errors included):

"Includes mono- and di-ethers of ethylene glycol, diethylene glycol, and triethylene glycol $R—(OCH_2CH_2)_n—OR'$ where

n = 1, 2, or 3
R = alkyl or aryl groups
R' = R'H or groups which, when removed, yield glycol ethers with the structure $R—(OCH_2CH)_n—OH$. Polymers are excluded from the glycol category."

CAA glycol ether definition with technical correction made. (A 2 was left out of the last formula.)

"Includes mono- and di-ethers of ethylene glycol, diethylene glycol, and triethylene glycol $R—(OCH_2CH_2)_n—OR'$

where n = 1, 2, or 3
    R = alkyl or aryl groups
    R' = R, H, or groups which, when removed, yield glycol ethers with the structure $R—(OCH_2CH_2)_n—OH$. Polymers are excluded from the glycol category.

[c]Under review.
[d]Under review.
[e]A type of atom which spontaneously undergoes radioactive decay.

H

Categories of Source of HAPs and
Regulation Promulgation Schedule

Industry group	Source category[a]	Schedule date
Fuel Combustion	Engine test facilities	11/15/00
	Industrial boilers[b]	11/15/00
	Institutional/commercial boilers[b]	11/15/00
	Process heaters	11/15/00
	Stationary internal combustion engines[b]	11/15/00
	Stationary turbines[b]	11/15/00
Nonferrous Metals Processing	Primary aluminum production	11/15/97
	Secondary aluminum production	11/15/97
	Primary copper smelting	11/15/97
	Primary lead smelting	11/15/97
	Secondary lead smelting[j]	11/15/94
	Lead acid battery manufacturing	11/15/00
	Primary magnesium refining	11/15/00
Ferrous Metals Processing	Coke by-product plants	11/15/00
	Coke ovens: Charging, top side, and door leaks	12/31/92
	Coke ovens: Pushing, quenching, and battery stacks	11/15/00
	Ferroalloys production	11/15/97
	Integrated iron and steel manufacturing	11/15/00
	Non-stainless steel manufacturing—electric arc furnace (EAF) operation	11/15/97
	Stainless steel manufacturing—electric arc furnace (EAF) operation	11/15/97
	Iron foundries	11/15/00
	Steel foundries	11/15/00
	Steel pickling—HCl process	11/15/97

Industry group	Source category[a]	Schedule date
Mineral Products Processing	Alumina processing	11/15/00
	Asphalt concrete manufacturing	11/15/00
	Asphalt processing	11/15/00
	Asphalt roofing manufacturing	11/15/00
	Asphalt/coal tar application—metal pipes	11/15/00
	Chromium refractories production	11/15/00
	Clay products manufacturing	11/15/00
	Lime manufacturing	11/15/00
	Mineral wool production	11/15/97
	Portland cement manufacturing	11/15/97
	Taconite iron ore processing	11/15/00
	Wool fiberglass manufacturing	11/15/97
Petroleum and Natural Gas Production and Refining	Oil and natural gas production	11/15/97
	Petroleum refineries—catalytic cracking (fluid and other) units, catalytic reforming units, and sulfur plant units	11/15/97
	Petroleum refineries—other sources not distinctly listed[k]	11/15/94
Liquids Distribution	Gasoline distribution (stage 1)[g]	11/15/94
	Organic liquids distribution (nongasoline)	11/15/00
Surface Coating Processes	Aerospace industries[l]	11/15/94
	Auto and light-duty truck (surface coating)	11/15/00
	Flat wood paneling (surface coating)	11/15/00
	Large appliance (surface coating)	11/15/00
	Magnetic tapes (surface coating)[g]	11/15/94
	Manufacture of paints, coatings, and adhesives	11/15/00
	Metal can (surface coating)	11/15/00
	Metal coil (surface coating)	11/15/00
	Metal furniture (surface coating)	11/15/00
	Miscellaneous metal parts and products (surface coating)	11/15/00
	Paper and other webs (surface coating)	11/15/00
	Plastic parts and products (surface coating)	11/15/00
	Printing, coating, and dyeing of fabrics	11/15/00
	Printing/publishing (surface coating)	11/15/94
	Shipbuilding and ship repair (surface coating)	11/15/94
	Wood furniture (surface coating)	11/15/94

Industry group	Source category[a]	Schedule date
Waste Treatment and Disposal	Hazardous waste incineration	11/15/00
	Municipal landfills	11/15/00
	Publicly owned treatment works (POTW) emissions	11/15/95
	Sewage sludge incineration	11/15/00
	Site remediation	11/15/00
	Solid waste treatment, storage, and disposal facilities (TSDF)	11/15/94
Agricultural Chemicals Production	4-Chloro-2-methylphenoxyacetic acid production	11/15/97
	2,4-D Salts and esters production	11/15/97
	4,6-Dinitro-o-cresol production	11/15/97
	Captafol production[c]	11/15/97
	Captan production[c]	11/15/97
	Chloroneb production	11/15/97
	Chlorothalonil production[c]	11/15/97
	Dacthal production[c]	11/15/97
	Sodium pentachlorophenate production	11/15/97
	Tordon acid production[c]	11/15/97
Fibers Production Processes	Acrylic fibers/modacrylic fibers production	11/15/97
	Rayon production	11/15/00
	Spandex production	11/15/00
Food and Agriculture Processes	Baker's yeast manufacturing	11/15/00
	Cellulose food casing manufacturing	11/15/00
	Vegetable oil production	11/15/00
Pharmaceutical Production Processes	Pharmaceuticals production[c]	11/15/97
Polymers and Resins Production	Acetal resins production	11/15/97
	Acrylonitrile-butadiene-styrene production	11/15/94
	Alkyd/resins production	11/15/00
	Amino resins production	11/15/97
	Boat manufacturing	11/15/00
	Butadiene-furfural cotrimer (R-11)[c]	11/15/00
	Butyl rubber production	11/15/94
	Carboxymethylcellulose production	11/15/00
	Cellophane production	11/15/00
	Cellulose ethers production	11/15/00
	Epichlorohydrin elastomers production	11/15/94
	Epoxy resins production[h]	11/15/94
	Ethylene-propylene rubber production	11/15/94

Industry group	Source category[a]	Schedule date
Polymers and Resins Production	Flexible polyurethane foam production	11/15/97
	Hypalon production[c]	11/15/94
	Maleic anhydride copolymers production	11/15/00
	Methylcellulose production	11/15/00
	Methyl methacrylate-acrylonitrile-butadiene-styrene production[c]	11/15/94
	Methyl methacrylate-butadiene-styrene terpolymers production[c]	11/15/94
	Neoprene production	11/15/94
	Nitrile butadiene rubber production	11/15/94
	Nonnylon polyamides production[h]	11/15/94
	Nylon 6 production	11/15/97
	Phenolic resins production	11/15/97
	Polybutadiene rubber production[c]	11/15/94
	Polycarbonates production[c]	11/15/97
	Polyester resins production	11/15/00
	Polyethylene terephthalate production	11/15/94
	Polymerized vinylidene chloride production	11/15/00
	Polymethyl methacrylate resins production	11/15/00
	Polystyrene production	11/15/94
	Polysulfide rubber production[c]	11/15/94
	Polyvinyl acetate emulsions production	11/15/00
	Polyvinyl alcohol production	11/15/00
	Polyvinyl butyral production	11/15/00
	Polyvinyl chloride and copolymers production	11/15/00
	Reinforced-plastic composites production	11/15/97
	Styrene-acrylonitrile production	11/15/94
	Styrene-butadiene rubber and latex production[c]	11/15/94
Production of Inorganic Chemicals	Ammonium sulfate production—caprolactam by-product plants	11/15/00
	Antimony oxides manufacturing	11/15/00
	Chlorine production[c]	11/15/97
	Chromium chemicals manufacturing	11/15/97
	Cyanuric chloride production	11/15/97
	Fume silica production	11/15/00
	Hydrochloric acid production	11/15/00
	Hydrogen cyanide production	11/15/97
	Hydrogen fluoride production	11/15/00
	Phosphate fertilizers production	11/15/00
	Phosphoric acid manufacturing	11/15/00
	Quaternary ammonium compounds production	11/15/00
	Sodium cyanide production	11/15/97
	Uranium hexafluoride production	11/15/00

Industry group	Source category[a]	Schedule date
Production of Organic Chemicals	Synthetic organic chemical manufacturing[e]	11/15/92
Miscellaneous Processes	Aerosol can-filling facilities	11/15/00
	Benzyltrimethylammonium chloride production	11/15/00
	Butadiene dimers production	11/15/97
	Carbonyl sulfide production	11/15/00
	Chelating agents production	11/15/00
	Chlorinated paraffins production[c]	11/15/00
	Chromic acid anodizing[g]	11/15/94
	Commercial dry cleaning (perchloroethylene)—transfer machines	11/15/92
	Commercial sterilization facilities[g]	11/15/94
	Decorative chromium electroplating[g]	11/15/94
	Dodecanedioic acid production[c]	11/15/00
	Dry cleaning (petroleum solvent)	11/15/00
	Ethylidene norbornene production[c]	11/15/00
	Explosives production	11/15/00
	Halogenated solvent cleaners[g]	11/15/94
	Hard chromium electroplating[g]	11/15/94
	Hydrazine production	11/15/00
	Industrial cleaning (perchloroethylene)—dry-to-dry machines	11/15/92
	Industrial dry cleaning (perchloroethylene)—transfer machines	11/15/92
	Industrial process cooling towers[f]	11/15/94
	OBPA/1,3-diisocyanate production[c]	11/15/00
	Paint stripper users	11/15/00
	Photographic chemicals production	11/15/00
	Phthalate plasticizers production	11/15/00
	Plywood/particleboard manufacturing	11/15/00
	Polyether polyols production	11/15/97
	Pulp and paper production	11/15/97
	Rocket engine test firing	11/15/00
	Rubber chemicals manufacturing	11/15/00
	Semiconductor manufacturing	11/15/00
	Symmetric tetrachloropyridine production[c]	11/15/00
	Tire production	11/15/00
	Wood treatment	11/15/97
Categories of Area Sources[d]	Asbestos processing	11/15/94
	Chromic acid anodizing[g]	11/15/94
	Commercial dry cleaning (perchloroethylene)—dry-to-dry machines	11/15/92
	Commercial dry cleaning (perchloroethylene)—transfer machines	11/15/92
	Commercial sterilization facilities[g]	11/15/94
	Decorative chromium electroplating[g]	11/15/94
	Halogenated solvent cleaners[g]	11/15/94
	Hard chromium electroplating[g]	11/15/94

^{a}Only major sources within any category shall be subject to emission standards under Section 112 unless a finding is made of a threat of adverse effects to human health or the environment for the area sources in a category. All listed categories are exclusive of any specific operations or processes included under other categories that are listed separately.

^{b}Sources defined as electric utility steam generating units under Section 112(a)(8) shall not be subject to emission standards pending the findings of the study required under Section 112(n)(1).

^{c}Equipment handling specific chemicals for these categories or subsets of these categories is subject to a negotiated standard for equipment leaks contained in the HON, which was proposed on December 31, 1992. The HON includes a negotiated standard for equipment leaks from the SOCMI category and 20 non-SOCMI categories (or subsets of these categories). The specific processes affected within the categories are listed in Section XX.X0(c) of the March 6, 1991, *Federal Register* notice (56 FR 9315).

^{d}A finding of threat of adverse effects to human health or the environment was made for each category of area sources listed.

The following footnotes apply to source categories that are subject to court-ordered promulgation deadlines (differing from the above-listed regulatory deadlines) in accordance with a consent decree entered in *Sierra Club v. Browner,* case no. 93-0124 (and related cases) (D.C. Dist. Ct.).

^{e}Judicial deadline: 02/28/94
^{f}Judicial deadline: 07/31/94
^{g}Judicial deadline: 11/23/94
^{h}Judicial deadline: 02/28/95
^{i}Judicial deadline: 04/30/95
^{j}Judicial deadline: 05/31/95
^{k}Judicial deadline: 06/30/95
^{l}Judicial deadline: 07/31/95

MACT Rules—General Provisions

Subpart A—General Provisions

SOURCE: 59 FR 12430, Mar. 16, 1994, unless otherwise noted.

§ 63.1 Applicability.

(a) *General.* (1) Terms used throughout this part are defined in § 63.2 or in the Clean Air Act (Act) as amended in 1990, except that individual subparts of this part may include specific definitions in addition to or that supersede definitions in § 63.2.

(2) This part contains national emission standards for hazardous air pollutants (NESHAP) established pursuant to section 112 of the Act as amended November 15, 1990. These standards regulate specific categories of stationary sources that emit (or have the potential to emit) one or more hazardous air pollutants listed in this part pursuant to section 112(b) of the Act. This section explains the applicability of such standards to sources affected by them. The standards in this part are independent of NESHAP contained in 40 CFR part 61. The NESHAP in part 61 promulgated by signature of the Administrator before November 15, 1990 (i.e., the date of enactment of the Clean Air Act Amendments of 1990) remain in effect until they are amended, if appropriate, and added to this part.

(3) No emission standard or other requirement established under this part shall be interpreted, construed, or applied to diminish or replace the requirements of a more stringent emission limitation or other applicable requirement established by the Administrator pursuant to other authority of the Act (including those requirements in part 60 of this chapter), or a standard issued under State authority.

(4) The provisions of this subpart (i.e., subpart A of this part) apply to owners or operators who are subject to subsequent subparts of this part, except when otherwise specified in a particular subpart or in a relevant standard. The general provisions in subpart A eliminate the repetition of requirements applicable to all owners or operators affected by this part. The general provisions in subpart A do not apply to regulations developed pursuant to section 112(r) of the amended Act, unless otherwise specified in those regulations.

(5) [Reserved]

(6) To obtain the most current list of categories of sources to be regulated under section 112 of the Act, or to obtain the most recent regulation promulgation schedule established pursuant to section 112(e) of the Act, contact the Office of the Director, Emission Standards Division, Office of Air Quality Planning and Standards, U.S. EPA (MD-13), Research Triangle Park, North Carolina 27711.

(7) Subpart D of this part contains regulations that address procedures for an owner or operator to obtain an extension of compliance with a relevant standard through an early reduction of emissions of hazardous air pollutants pursuant to section 112(i)(5) of the Act.

(8) Subpart E of this part contains regulations that provide for the establishment of procedures consistent with section 112(l) of the Act for the approval of State rules or programs to implement and enforce applicable Federal rules promulgated under the authority of section 112. Subpart E also establishes procedures for the review and withdrawal of section 112 implementation and enforcement authorities granted through a section 112(l) approval.

(9) [Reserved]

(10) For the purposes of this part, time periods specified in days shall be measured in calendar days, even if the word "calendar" is absent, unless otherwise specified in an applicable requirement.

(11) For the purposes of this part, if an explicit postmark deadline is not specified in an applicable requirement for the submittal of a notification, application, test plan, report, or other written communication to the Administrator, the owner or operator shall postmark the submittal on or before the number of days specified in the applicable requirement. For example, if a notification must be submitted 15 days before a particular event is scheduled to take place, the notification shall be postmarked on or before 15 days preceding the event; likewise, if a notification must be submitted 15 days after a particular event takes place, the notification shall be postmarked on or before 15 days following the end of the event. The use of reliable non-Government mail carriers that provide indications of verifiable delivery of information required to be submitted to the Administrator, similar to the postmark provided by the U.S. Postal Service, or alternative means of delivery agreed to by the permitting authority, is acceptable.

(12) Notwithstanding time periods or postmark deadlines specified in this part for the submittal of information to the Administrator by an owner or operator, or the review of such information by the Administrator, such time periods or deadlines may be changed by mutual agreement between the owner or operator and the Administrator. Procedures governing the implementation of this provision are specified in § 63.9(i).

(13) Special provisions set forth under an applicable subpart of this part or in a relevant standard established under this part shall supersede any conflicting provisions of this subpart.

(14) Any standards, limitations, prohibitions, or other federally enforceable requirements established pursuant to procedural regulations in this

§ 63.2

part [including, but not limited to, equivalent emission limitations established pursuant to section 112(g) of the Act] shall have the force and effect of requirements promulgated in this part and shall be subject to the provisions of this subpart, except when explicitly specified otherwise.

(b) *Initial applicability determination for this part.* (1) The provisions of this part apply to the owner or operator of any stationary source that—

(i) Emits or has the potential to emit any hazardous air pollutant listed in or pursuant to section 112(b) of the Act; and

(ii) Is subject to any standard, limitation, prohibition, or other federally enforceable requirement established pursuant to this part.

(2) In addition to complying with the provisions of this part, the owner or operator of any such source may be required to obtain an operating permit issued to stationary sources by an authorized State air pollution control agency or by the Administrator of the U.S. Environmental Protection Agency (EPA) pursuant to title V of the Act (42 U.S.C. 7661). For more information about obtaining an operating permit, see part 70 of this chapter.

(3) An owner or operator of a stationary source that emits (or has the potential to emit, without considering controls) one or more hazardous air pollutants who determines that the source is not subject to a relevant standard or other requirement established under this part, shall keep a record of the applicability determination as specified in § 63.10(b)(3) of this subpart.

(c) *Applicability of this part after a relevant standard has been set under this part.* (1) If a relevant standard has been established under this part, the owner or operator of an affected source shall comply with the provisions of this subpart and the provisions of that standard, except as specified otherwise in this subpart or that standard.

(2) If a relevant standard has been established under this part, the owner or operator of an affected source may be required to obtain a title V permit from the permitting authority in the State in which the source is located. Emission standards promulgated in this part for area sources will specify whether—

(i) States will have the option to exclude area sources affected by that standard from the requirement to obtain a title V permit (i.e., the standard will exempt the category of area sources altogether from the permitting requirement);

(ii) States will have the option to defer permitting of area sources in that category until the Administrator takes rulemaking action to determine applicability of the permitting requirements; or

(iii) Area sources affected by that emission standard are immediately subject to the requirement to apply for and obtain a title V permit in

all States. If a standard fails to specify what the permitting requirements will be for area sources affected by that standard, then area sources that are subject to the standard will be subject to the requirement to obtain a title V permit without deferral. If the owner or operator is required to obtain a title V permit, he or she shall apply for such permit in accordance with part 70 of this chapter and applicable State regulations, or in accordance with the regulations contained in this chapter to implement the Federal title V permit program (42 U.S.C. 7661), whichever regulations are applicable.

(3) [Reserved]

(4) If the owner or operator of an existing source obtains an extension of compliance for such source in accordance with the provisions of subpart D of this part, the owner or operator shall comply with all requirements of this subpart except those requirements that are specifically overridden in the extension of compliance for that source.

(5) If an area source that otherwise would be subject to an emission standard or other requirement established under this part if it were a major source subsequently increases its emissions of hazardous air pollutants (or its potential to emit hazardous air pollutants) such that the source is a major source that is subject to the emission standard or other requirement, such source also shall be subject to the notification requirements of this subpart.

(d) [Reserved]

(e) *Applicability of permit program before a relevant standard has been set under this part.* After the effective date of an approved permit program in the State in which a stationary source is (or would be) located, the owner or operator of such source may be required to obtain a title V permit from the permitting authority in that State (or revise such a permit if one has already been issued to the source) before a relevant standard is established under this part. If the owner or operator is required to obtain (or revise) a title V permit, he/she shall apply to obtain (or revise) such permit in accordance with the regulations contained in part 70 of this chapter and applicable State regulations, or the regulations codified in this chapter to implement the Federal title V permit program (42 U.S.C. 7661), whichever regulations are applicable.

§ 63.2 Definitions.

The terms used in this part are defined in the Act or in this section as follows:

Act means the Clean Air Act (42 U.S.C. 7401 *et seq.,* as amended by Pub. L. 101–549, 104 Stat. 2399).

Actual emissions is defined in subpart D of this part for the purpose of granting a compliance extension for an early reduction of hazardous air pollutants.

Administrator means the Administrator of the United States Environmental Protection Agency or his or her authorized representative (e.g., a State that has been delegated the authority to implement the provisions of this part).

Affected source, for the purposes of this part, means the stationary source, the group of stationary sources, or the portion of a stationary source that is regulated by a relevant standard or other requirement established pursuant to section 112 of the Act. Each relevant standard will define the "affected source" for the purposes of that standard. The term "affected source," as used in this part, is separate and distinct from any other use of that term in EPA regulations such as those implementing title IV of the Act. Sources regulated under part 60 or part 61 of this chapter are not affected sources for the purposes of part 63.

Alternative emission limitation means conditions established pursuant to sections 112(i)(5) or 112(i)(6) of the Act by the Administrator or by a State with an approved permit program.

Alternative emission standard means an alternative means of emission limitation that, after notice and opportunity for public comment, has been demonstrated by an owner or operator to the Administrator's satisfaction to achieve a reduction in emissions of any air pollutant at least equivalent to the reduction in emissions of such pollutant achieved under a relevant design, equipment, work practice, or operational emission standard, or combination thereof, established under this part pursuant to section 112(h) of the Act.

Alternative test method means any method of sampling and analyzing for an air pollutant that is not a test method in this chapter and that has been demonstrated to the Administrator's satisfaction, using Method 301 in Appendix A of this part, to produce results adequate for the Administrator's determination that it may be used in place of a test method specified in this part.

Approved permit program means a State permit program approved by the Administrator as meeting the requirements of part 70 of this chapter or a Federal permit program established in this chapter pursuant to title V of the Act (42 U.S.C. 7661).

Area source means any stationary source of hazardous air pollutants that is not a major source as defined in this part.

Commenced means, with respect to construction or reconstruction of a stationary source, that an owner or operator has undertaken a continuous program of construction or reconstruction or that an owner or operator has entered into a contractual obligation to undertake and complete, within a rea-

sonable time, a continuous program of construction or reconstruction.

Compliance date means the date by which an affected source is required to be in compliance with a relevant standard, limitation, prohibition, or any federally enforceable requirement established by the Administrator (or a State with an approved permit program) pursuant to section 112 of the Act.

Compliance plan means a plan that contains all of the following:

(1) A description of the compliance status of the affected source with respect to all applicable requirements established under this part;

(2) A description as follows: (i) For applicable requirements for which the source is in compliance, a statement that the source will continue to comply with such requirements;

(ii) For applicable requirements that the source is required to comply with by a future date, a statement that the source will meet such requirements on a timely basis;

(iii) For applicable requirements for which the source is not in compliance, a narrative description of how the source will achieve compliance with such requirements on a timely basis;

(3) A compliance schedule, as defined in this section; and

(4) A schedule for the submission of certified progress reports no less frequently than every 6 months for affected sources required to have a schedule of compliance to remedy a violation.

Compliance schedule means: (1) In the case of an affected source that is in compliance with all applicable requirements established under this part, a statement that the source will continue to comply with such requirements; or

(2) In the case of an affected source that is required to comply with applicable requirements by a future date, a statement that the source will meet such requirements on a timely basis and, if required by an applicable requirement, a detailed schedule of the dates by which each step toward compliance will be reached; or

(3) In the case of an affected source not in compliance with all applicable requirements established under this part, a schedule of remedial measures, including an enforceable sequence of actions or operations with milestones and a schedule for the submission of certified progress reports, where applicable, leading to compliance with a relevant standard, limitation, prohibition, or any federally enforceable requirement established pursuant to section 112 of the Act for which the affected source is not in compliance. This compliance schedule shall resemble and be at least as stringent as that contained in any judicial consent decree or administrative order to which the source is subject. Any such schedule of compliance shall

§ 63.2

be supplemental to, and shall not sanction noncompliance with, the applicable requirements on which it is based.

Construction means the on-site fabrication, erection, or installation of an affected source.

Continuous emission monitoring system (CEMS) means the total equipment that may be required to meet the data acquisition and availability requirements of this part, used to sample, condition (if applicable), analyze, and provide a record of emissions.

Continuous monitoring system (CMS) is a comprehensive term that may include, but is not limited to, continuous emission monitoring systems, continuous opacity monitoring systems, continuous parameter monitoring systems, or other manual or automatic monitoring that is used for demonstrating compliance with an applicable regulation on a continuous basis as defined by the regulation.

Continuous opacity monitoring system (COMS) means a continuous monitoring system that measures the opacity of emissions.

Continuous parameter monitoring system means the total equipment that may be required to meet the data acquisition and availability requirements of this part, used to sample, condition (if applicable), analyze, and provide a record of process or control system parameters.

Effective date means: (1) With regard to an emission standard established under this part, the date of promulgation in the FEDERAL REGISTER of such standard; or

(2) With regard to an alternative emission limitation or equivalent emission limitation determined by the Administrator (or a State with an approved permit program), the date that the alternative emission limitation or equivalent emission limitation becomes effective according to the provisions of this part. The effective date of a permit program established under title V of the Act (42 U.S.C. 7661) is determined according to the regulations in this chapter establishing such programs.

Emission standard means a national standard, limitation, prohibition, or other regulation promulgated in a subpart of this part pursuant to sections 112(d), 112(h), or 112(f) of the Act.

Emissions averaging is a way to comply with the emission limitations specified in a relevant standard, whereby an affected source, if allowed under a subpart of this part, may create emission credits by reducing emissions from specific points to a level below that required by the relevant standard, and those credits are used to offset emissions from points that are not controlled to the level required by the relevant standard.

EPA means the United States Environmental Protection Agency.

Equivalent emission limitation means the maximum achievable control technology emission limitation (MACT emission limitation) for hazardous air pollutants that the Administrator (or a State with an approved permit program) determines on a case-by-case basis, pursuant to section 112(g) or section 112(j) of the Act, to be equivalent to the emission standard that would apply to an affected source if such standard had been promulgated by the Administrator under this part pursuant to section 112(d) or section 112(h) of the Act.

Excess emissions and continuous monitoring system performance report is a report that must be submitted periodically by an affected source in order to provide data on its compliance with relevant emission limits, operating parameters, and the performance of its continuous parameter monitoring systems.

Existing source means any affected source that is not a new source.

Federally enforceable means all limitations and conditions that are enforceable by the Administrator and citizens under the Act or that are enforceable under other statutes administered by the Administrator. Examples of federally enforceable limitations and conditions include, but are not limited to:

(1) Emission standards, alternative emission standards, alternative emission limitations, and equivalent emission limitations established pursuant to section 112 of the Act as amended in 1990;

(2) New source performance standards established pursuant to section 111 of the Act, and emission standards established pursuant to section 112 of the Act before it was amended in 1990;

(3) All terms and conditions in a title V permit, including any provisions that limit a source's potential to emit, unless expressly designated as not federally enforceable;

(4) Limitations and conditions that are part of an approved State Implementation Plan (SIP) or a Federal Implementation Plan (FIP);

(5) Limitations and conditions that are part of a Federal construction permit issued under 40 CFR 52.21 or any construction permit issued under regulations approved by the EPA in accordance with 40 CFR part 51;

(6) Limitations and conditions that are part of an operating permit issued pursuant to a program approved by the EPA into a SIP as meeting the EPA's minimum criteria for Federal enforceability, including adequate notice and opportunity for EPA and public comment prior to issuance of the final permit and practicable enforceability;

(7) Limitations and conditions in a State rule or program that has been approved by the EPA under subpart E of this part for the purposes of implementing and enforcing section 112; and

(8) Individual consent agreements that the EPA has legal authority to create.

Fixed capital cost means the capital needed to provide all the depreciable components of an existing source.

Fugitive emissions means those emissions from a stationary source that could not reasonably pass through a stack, chimney, vent, or other functionally equivalent opening. Under section 112 of the Act, all fugitive emissions are to be considered in determining whether a stationary source is a major source.

Hazardous air pollutant means any air pollutant listed in or pursuant to section 112(b) of the Act.

Issuance of a part 70 permit will occur, if the State is the permitting authority, in accordance with the requirements of part 70 of this chapter and the applicable, approved State permit program. When the EPA is the permitting authority, issuance of a title V permit occurs immediately after the EPA takes final action on the final permit.

Lesser quantity means a quantity of a hazardous air pollutant that is or may be emitted by a stationary source that the Administrator establishes in order to define a major source under an applicable subpart of this part.

Major source means any stationary source or group of stationary sources located within a contiguous area and under common control that emits or has the potential to emit considering controls, in the aggregate, 10 tons per year or more of any hazardous air pollutant or 25 tons per year or more of any combination of hazardous air pollutants, unless the Administrator establishes a lesser quantity, or in the case of radionuclides, different criteria from those specified in this sentence.

Malfunction means any sudden, infrequent, and not reasonably preventable failure of air pollution control equipment, process equipment, or a process to operate in a normal or usual manner. Failures that are caused in part by poor maintenance or careless operation are not malfunctions.

New source means any affected source the construction or reconstruction of which is commenced after the Administrator first proposes a relevant emission standard under this part.

One-hour period, unless otherwise defined in an applicable subpart, means any 60-minute period commencing on the hour.

Opacity means the degree to which emissions reduce the transmission of light and obscure the view of an object in the background. For continuous opacity monitoring systems, opacity means the fraction of incident light that is attenuated by an optical medium.

Owner or operator means any person who owns, leases, operates, controls, or supervises a stationary source.

Part 70 permit means any permit issued, renewed, or revised pursuant to part 70 of this chapter.

Performance audit means a procedure to analyze blind samples, the content of which is known by the Administrator, simultaneously with the analysis of performance test samples in order to provide a measure of test data quality.

Performance evaluation means the conduct of relative accuracy testing, calibration error testing, and other measurements used in validating the continuous monitoring system data.

Performance test means the collection of data resulting from the execution of a test method (usually three emission test runs) used to demonstrate compliance with a relevant emission standard as specified in the performance test section of the relevant standard.

Permit modification means a change to a title V permit as defined in regulations codified in this chapter to implement title V of the Act (42 U.S.C. 7661).

Permit program means a comprehensive State operating permit system established pursuant to title V of the Act (42 U.S.C. 7661) and regulations codified in part 70 of this chapter and applicable State regulations, or a comprehensive Federal operating permit system established pursuant to title V of the Act and regulations codified in this chapter.

Permit revision means any permit modification or administrative permit amendment to a title V permit as defined in regulations codified in this chapter to implement title V of the Act (42 U.S.C. 7661).

Permitting authority means: (1) The State air pollution control agency, local agency, other State agency, or other agency authorized by the Administrator to carry out a permit program under part 70 of this chapter; or

(2) The Administrator, in the case of EPA-implemented permit programs under title V of the Act (42 U.S.C. 7661).

Potential to emit means the maximum capacity of a stationary source to emit a pollutant under its physical and operational design. Any physical or operational limitation on the capacity of the stationary source to emit a pollutant, including air pollution control equipment and restrictions on hours of operation or on the type or amount of material combusted, stored, or processed, shall be treated as part of its design if the limitation or the effect it would have on emissions is federally enforceable.

Reconstruction means the replacement of components of an affected or a previously unaffected stationary source to such an extent that:

(1) The fixed capital cost of the new components exceeds 50 percent of the fixed capital cost that would be required to construct a comparable new source; and

§ 63.3

(2) It is technologically and economically feasible for the reconstructed source to meet the relevant standard(s) established by the Administrator (or a State) pursuant to section 112 of the Act. Upon reconstruction, an affected source, or a stationary source that becomes an affected source, is subject to relevant standards for new sources, including compliance dates, irrespective of any change in emissions of hazardous air pollutants from that source.

Regulation promulgation schedule means the schedule for the promulgation of emission standards under this part, established by the Administrator pursuant to section 112(e) of the Act and published in the FEDERAL REGISTER.

Relevant standard means:

(1) An emission standard;

(2) An alternative emission standard;

(3) An alternative emission limitation; or

(4) An equivalent emission limitation established pursuant to section 112 of the Act that applies to the stationary source, the group of stationary sources, or the portion of a stationary source regulated by such standard or limitation.

A relevant standard may include or consist of a design, equipment, work practice, or operational requirement, or other measure, process, method, system, or technique (including prohibition of emissions) that the Administrator (or a State) establishes for new or existing sources to which such standard or limitation applies. Every relevant standard established pursuant to section 112 of the Act includes subpart A of this part and all applicable appendices of this part or of other parts of this chapter that are referenced in that standard.

Responsible official means one of the following:

(1) For a corporation: A president, secretary, treasurer, or vice president of the corporation in charge of a principal business function, or any other person who performs similar policy or decision-making functions for the corporation, or a duly authorized representative of such person if the representative is responsible for the overall operation of one or more manufacturing, production, or operating facilities and either:

(i) The facilities employ more than 250 persons or have gross annual sales or expenditures exceeding $25 million (in second quarter 1980 dollars); or

(ii) The delegation of authority to such representative is approved in advance by the Administrator.

(2) For a partnership or sole proprietorship: a general partner or the proprietor, respectively.

(3) For a municipality, State, Federal, or other public agency: either a principal executive officer or ranking elected official. For the purposes of this part, a principal executive officer of a Federal agency includes the chief executive officer having responsibility for the overall operations of a principal geographic unit of the agency (e.g., a Regional Administrator of the EPA).

(4) For affected sources (as defined in this part) applying for or subject to a title V permit: "responsible official" shall have the same meaning as defined in part 70 or Federal title V regulations in this chapter (42 U.S.C. 7661), whichever is applicable.

Run means one of a series of emission or other measurements needed to determine emissions for a representative operating period or cycle as specified in this part.

Shutdown means the cessation of operation of an affected source for any purpose.

Six-minute period means, with respect to opacity determinations, any one of the 10 equal parts of a 1-hour period.

Standard conditions means a temperature of 293 K (68° F) and a pressure of 101.3 kilopascals (29.92 in. Hg).

Startup means the setting in operation of an affected source for any purpose.

State means all non-Federal authorities, including local agencies, interstate associations, and State-wide programs, that have delegated authority to implement: (1) The provisions of this part and/or (2) the permit program established under part 70 of this chapter. The term State shall have its conventional meaning where clear from the context.

Stationary source means any building, structure, facility, or installation which emits or may emit any air pollutant.

Test method means the validated procedure for sampling, preparing, and analyzing for an air pollutant specified in a relevant standard as the performance test procedure. The test method may include methods described in an appendix of this chapter, test methods incorporated by reference in this part, or methods validated for an application through procedures in Method 301 of appendix A of this part.

Title V permit means any permit issued, renewed, or revised pursuant to Federal or State regulations established to implement title V of the Act (42 U.S.C. 7661). A title V permit issued by a State permitting authority is called a part 70 permit in this part.

Visible emission means the observation of an emission of opacity or optical density above the threshold of vision.

§ 63.3 Units and abbreviations.

Used in this part are abbreviations and symbols of units of measure. These are defined as follows:

(a) *System International (SI) units of measure:*

A = ampere

g = gram

Hz = hertz
J = joule
°K = degree Kelvin
kg = kilogram
l = liter
m = meter
m³ = cubic meter
mg = milligram = 10^{-3} gram
ml = milliliter = 10^{-3} liter
mm = millimeter = 10^{-3} meter
Mg = megagram = 10^6 gram = metric ton
MJ = megajoule
mol = mole
N = newton
ng = nanogram = 10^{-9} gram
nm = nanometer = 10^{-9} meter
Pa = pascal
s = second
V = volt
W = watt
Ω = ohm
μg = microgram = 10^{-6} gram
μl = microliter = 10^{-6} liter

(b) *Other units of measure:*

Btu = British thermal unit
°C = degree Celsius (centigrade)
cal = calorie
cfm = cubic feet per minute
cc = cubic centimeter
cu ft = cubic feet
d = day
dcf = dry cubic feet
dcm = dry cubic meter
dscf = dry cubic feet at standard conditions
dscm = dry cubic meter at standard conditions
eq = equivalent
°F degree Fahrenheit
ft = feet
ft² = square feet
ft³ = cubic feet
gal = gallon
gr = grain
g-eq = gram equivalent
g-mole = gram mole
hr = hour
in. = inch
in. H_2O = inches of water
K = 1,000
kcal = kilocalorie
lb = pound
lpm = liter per minute
meq = milliequivalent
min = minute
MW = molecular weight
oz = ounces
ppb = parts per billion
ppbw = parts per billion by weight
ppbv = parts per billion by volume
ppm = parts per million

ppmw = parts per million by weight
ppmv = parts per million by volume
psia = pounds per square inch absolute
psig = pounds per square inch gage
°R = degree Rankine
scf = cubic feet at standard conditions
scfh = cubic feet at standard conditions per hour
scm = cubic meter at standard conditions
sec = second
sq ft = square feet
std = at standard conditions
v/v = volume per volume
yd² = square yards
yr = year

(c) *Miscellaneous:*

act = actual
avg = average
I.D. = inside diameter
M = molar
N = normal
O.D. = outside diameter
% = percent

§ 63.4 **Prohibited activities and circumvention.**

(a) *Prohibited activities.* (1) No owner or operator subject to the provisions of this part shall operate any affected source in violation of the requirements of this part except under—

(i) An extension of compliance granted by the Administrator under this part; or

(ii) An extension of compliance granted under this part by a State with an approved permit program; or

(iii) An exemption from compliance granted by the President under section 112(i)(4) of the Act.

(2) No owner or operator subject to the provisions of this part shall fail to keep records, notify, report, or revise reports as required under this part.

(3) After the effective date of an approved permit program in a State, no owner or operator of an affected source in that State who is required under this part to obtain a title V permit shall operate such source except in compliance with the provisions of this part and the applicable requirements of the permit program in that State.

(4) [Reserved]

(5) An owner or operator of an affected source who is subject to an emission standard promulgated under this part shall comply with the requirements of that standard by the date(s) established in the applicable subpart(s) of this part (including this subpart) regardless of whether—

(i) A title V permit has been issued to that source; or

(ii) If a title V permit has been issued to that source, whether such permit has been revised or modified to incorporate the emission standard.

§ 63.5

(b) *Circumvention.* No owner or operator subject to the provisions of this part shall build, erect, install, or use any article, machine, equipment, or process to conceal an emission that would otherwise constitute noncompliance with a relevant standard. Such concealment includes, but is not limited to—

(1) The use of diluents to achieve compliance with a relevant standard based on the concentration of a pollutant in the effluent discharged to the atmosphere;

(2) The use of gaseous diluents to achieve compliance with a relevant standard for visible emissions; and

(3) The fragmentation of an operation such that the operation avoids regulation by a relevant standard.

(c) *Severability.* Notwithstanding any requirement incorporated into a title V permit obtained by an owner or operator subject to the provisions of this part, the provisions of this part are federally enforceable.

§ 63.5 Construction and reconstruction.

(a) *Applicability.* (1) This section implements the preconstruction review requirements of section 112(i)(1) for sources subject to a relevant emission standard that has been promulgated in this part. In addition, this section includes other requirements for constructed and reconstructed stationary sources that are or become subject to a relevant promulgated emission standard.

(2) After the effective date of a relevant standard promulgated under this part, the requirements in this section apply to owners or operators who construct a new source or reconstruct a source after the proposal date of that standard. New or reconstructed sources that start up before the standard's effective date are not subject to the preconstruction review requirements specified in paragraphs (b)(3), (d), and (e) of this section.

(b) *Requirements for existing, newly constructed, and reconstructed sources.* (1) Upon construction an affected source is subject to relevant standards for new sources, including compliance dates. Upon reconstruction, an affected source is subject to relevant standards for new sources, including compliance dates, irrespective of any change in emissions of hazardous air pollutants from that source.

(2) [Reserved]

(3) After the effective date of any relevant standard promulgated by the Administrator under this part, whether or not an approved permit program is effective in the State in which an affected source is (or would be) located, no person may construct a new major affected source or reconstruct a major affected source subject to such

standard, or reconstruct a major source such that the source becomes a major affected source subject to the standard, without obtaining written approval, in advance, from the Administrator in accordance with the procedures specified in paragraphs (d) and (e) of this section.

(4) After the effective date of any relevant standard promulgated by the Administrator under this part, whether or not an approved permit program is effective in the State in which an affected source is (or would be) located, no person may construct a new affected source or reconstruct an affected source subject to such standard, or reconstruct a source such that the source becomes an affected source subject to the standard, without notifying the Administrator of the intended construction or reconstruction. The notification shall be submitted in accordance with the procedures in § 63.9(b) and shall include all the information required for an application for approval of construction or reconstruction as specified in paragraph (d) of this section. For major sources, the application for approval of construction or reconstruction may be used to fulfill the notification requirements of this paragraph.

(5) After the effective date of any relevant standard promulgated by the Administrator under this part, whether or not an approved permit program is effective in the State in which an affected source is located, no person may operate such source without complying with the provisions of this subpart and the relevant standard unless that person has received an extension of compliance or an exemption from compliance under § 63.6(i) or § 63.6(j) of this subpart.

(6) After the effective date of any relevant standard promulgated by the Administrator under this part, whether or not an approved permit program is effective in the State in which an affected source is located, equipment added (or a process change) to an affected source that is within the scope of the definition of affected source under the relevant standard shall be considered part of the affected source and subject to all provisions of the relevant standard established for that affected source. If a new affected source is added to the facility, the new affected source shall be subject to all provisions of the relevant standard that are established for new sources including compliance dates.

(c) [Reserved]

(d) *Application for approval of construction or reconstruction.* The provisions of this paragraph implement section 112(i)(1) of the Act.

(1) *General application requirements.* (i) An owner or operator who is subject to the requirements of paragraph (b)(3) of this section shall submit to the Administrator an application for approval of the construction of a new major affected

source, the reconstruction of a major affected source, or the reconstruction of a major source such that the source becomes a major affected source subject to the standard. The application shall be submitted as soon as practicable before the construction or reconstruction is planned to commence (but no sooner than the effective date of the relevant standard) if the construction or reconstruction commences after the effective date of a relevant standard promulgated in this part. The application shall be submitted as soon as practicable before startup but no later than 60 days after the effective date of a relevant standard promulgated in this part if the construction or reconstruction had commenced and initial startup had not occurred before the standard's effective date. The application for approval of construction or reconstruction may be used to fulfill the initial notification requirements of § 63.9(b)(5) of this subpart. The owner or operator may submit the application for approval well in advance of the date construction or reconstruction is planned to commence in order to ensure a timely review by the Administrator and that the planned commencement date will not be delayed.

(ii) A separate application shall be submitted for each construction or reconstruction. Each application for approval of construction or reconstruction shall include at a minimum:

(A) The applicant's name and address;

(B) A notification of intention to construct a new major affected source or make any physical or operational change to a major affected source that may meet or has been determined to meet the criteria for a reconstruction, as defined in § 63.2;

(C) The address (i.e., physical location) or proposed address of the source;

(D) An identification of the relevant standard that is the basis of the application;

(E) The expected commencement date of the construction or reconstruction;

(F) The expected completion date of the construction or reconstruction;

(G) The anticipated date of (initial) startup of the source;

(H) The type and quantity of hazardous air pollutants emitted by the source, reported in units and averaging times and in accordance with the test methods specified in the relevant standard, or if actual emissions data are not yet available, an estimate of the type and quantity of hazardous air pollutants expected to be emitted by the source reported in units and averaging times specified in the relevant standard. The owner or operator may submit percent reduction information if a relevant standard is established in terms of percent reduction. However, operating parameters, such as flow rate, shall be included in the submission to the ex-

tent that they demonstrate performance and compliance; and

(I) [Reserved]

(J) Other information as specified in paragraphs (d)(2) and (d)(3) of this section.

(iii) An owner or operator who submits estimates or preliminary information in place of the actual emissions data and analysis required in paragraphs (d)(1)(ii)(H) and (d)(2) of this section shall submit the actual, measured emissions data and other correct information as soon as available but no later than with the notification of compliance status required in § 63.9(h) (see § 63.9(h)(5)).

(2) *Application for approval of construction.* Each application for approval of construction shall include, in addition to the information required in paragraph (d)(1)(ii) of this section, technical information describing the proposed nature, size, design, operating design capacity, and method of operation of the source, including an identification of each point of emission for each hazardous air pollutant that is emitted (or could be emitted) and a description of the planned air pollution control system (equipment or method) for each emission point. The description of the equipment to be used for the control of emissions shall include each control device for each hazardous air pollutant and the estimated control efficiency (percent) for each control device. The description of the method to be used for the control of emissions shall include an estimated control efficiency (percent) for that method. Such technical information shall include calculations of emission estimates in sufficient detail to permit assessment of the validity of the calculations. An owner or operator who submits approximations of control efficiencies under this subparagraph shall submit the actual control efficiencies as specified in paragraph (d)(1)(iii) of this section.

(3) *Application for approval of reconstruction.* Each application for approval of reconstruction shall include, in addition to the information required in paragraph (d)(1)(ii) of this section—

(i) A brief description of the affected source and the components that are to be replaced;

(ii) A description of present and proposed emission control systems (i.e., equipment or methods). The description of the equipment to be used for the control of emissions shall include each control device for each hazardous air pollutant and the estimated control efficiency (percent) for each control device. The description of the method to be used for the control of emissions shall include an estimated control efficiency (percent) for that method. Such technical information shall include calculations of emission estimates in sufficient detail to permit assessment of the validity of the calculations;

§ 63.5

(iii) An estimate of the fixed capital cost of the replacements and of constructing a comparable entirely new source;

(iv) The estimated life of the affected source after the replacements; and

(v) A discussion of any economic or technical limitations the source may have in complying with relevant standards or other requirements after the proposed replacements. The discussion shall be sufficiently detailed to demonstrate to the Administrator's satisfaction that the technical or economic limitations affect the source's ability to comply with the relevant standard and how they do so.

(vi) If in the application for approval of reconstruction the owner or operator designates the affected source as a reconstructed source and declares that there are no economic or technical limitations to prevent the source from complying with all relevant standards or other requirements, the owner or operator need not submit the information required in subparagraphs (d)(3) (iii) through (v) of this section, above.

(4) *Additional information.* The Administrator may request additional relevant information after the submittal of an application for approval of construction or reconstruction.

(e) *Approval of construction or reconstruction.* (1)(i) If the Administrator determines that, if properly constructed, or reconstructed, and operated, a new or existing source for which an application under paragraph (d) of this section was submitted will not cause emissions in violation of the relevant standard(s) and any other federally enforceable requirements, the Administrator will approve the construction or reconstruction.

(ii) In addition, in the case of reconstruction, the Administrator's determination under this paragraph will be based on:

(A) The fixed capital cost of the replacements in comparison to the fixed capital cost that would be required to construct a comparable entirely new source;

(B) The estimated life of the source after the replacements compared to the life of a comparable entirely new source;

(C) The extent to which the components being replaced cause or contribute to the emissions from the source; and

(D) Any economic or technical limitations on compliance with relevant standards that are inherent in the proposed replacements.

(2)(i) The Administrator will notify the owner or operator in writing of approval or intention to deny approval of construction or reconstruction within 60 calendar days after receipt of sufficient information to evaluate an application submitted under paragraph (d) of this section. The 60-day approval or denial period will begin after the

owner or operator has been notified in writing that his/her application is complete. The Administrator will notify the owner or operator in writing of the status of his/her application, that is, whether the application contains sufficient information to make a determination, within 30 calendar days after receipt of the original application and within 30 calendar days after receipt of any supplementary information that is submitted.

(ii) When notifying the owner or operator that his/her application is not complete, the Administrator will specify the information needed to complete the application and provide notice of opportunity for the applicant to present, in writing, within 30 calendar days after he/she is notified of the incomplete application, additional information or arguments to the Administrator to enable further action on the application.

(3) Before denying any application for approval of construction or reconstruction, the Administrator will notify the applicant of the Administrator's intention to issue the denial together with—

(i) Notice of the information and findings on which the intended denial is based; and

(ii) Notice of opportunity for the applicant to present, in writing, within 30 calendar days after he/she is notified of the intended denial, additional information or arguments to the Administrator to enable further action on the application.

(4) A final determination to deny any application for approval will be in writing and will specify the grounds on which the denial is based. The final determination will be made within 60 calendar days of presentation of additional information or arguments (if the application is complete), or within 60 calendar days after the final date specified for presentation if no presentation is made.

(5) Neither the submission of an application for approval nor the Administrator's approval of construction or reconstruction shall—

(i) Relieve an owner or operator of legal responsibility for compliance with any applicable provisions of this part or with any other applicable Federal, State, or local requirement; or

(ii) Prevent the Administrator from implementing or enforcing this part or taking any other action under the Act.

(f) *Approval of construction or reconstruction based on prior State preconstruction review.* (1) The Administrator may approve an application for construction or reconstruction specified in paragraphs (b)(3) and (d) of this section if the owner or operator of a new or reconstructed source who is subject to such requirement demonstrates to the Administrator's satisfaction that the following conditions have been (or will be) met:

(i) The owner or operator of the new or reconstructed source has undergone a preconstruction review and approval process in the State in which the source is (or would be) located before the promulgation date of the relevant standard and has received a federally enforceable construction permit that contains a finding that the source will meet the relevant emission standard as proposed, if the source is properly built and operated;

(ii) In making its finding, the State has considered factors substantially equivalent to those specified in paragraph (e)(1) of this section; and either

(iii) The promulgated standard is no more stringent than the proposed standard in any relevant aspect that would affect the Administrator's decision to approve or disapprove an application for approval of construction or reconstruction under this section; or

(iv) The promulgated standard is more stringent than the proposed standard but the owner or operator will comply with the standard as proposed during the 3-year period immediately following the effective date of the standard as allowed for in § 63.6(b)(3) of this subpart.

(2) The owner or operator shall submit to the Administrator the request for approval of construction or reconstruction under this paragraph no later than the application deadline specified in paragraph (d)(1) of this section (see also § 63.9(b)(2) of this subpart). The owner or operator shall include in the request information sufficient for the Administrator's determination. The Administrator will evaluate the owner or operator's request in accordance with the procedures specified in paragraph (e) of this section. The Administrator may request additional relevant information after the submittal of a request for approval of construction or reconstruction under this paragraph.

§ 63.6 Compliance with standards and maintenance requirements.

(a) *Applicability.* (1) The requirements in this section apply to owners or operators of affected sources for which any relevant standard has been established pursuant to section 112 of the Act unless—

(i) The Administrator (or a State with an approved permit program) has granted an extension of compliance consistent with paragraph (i) of this section; or

(ii) The President has granted an exemption from compliance with any relevant standard in accordance with section 112(i)(4) of the Act.

(2) If an area source that otherwise would be subject to an emission standard or other requirement established under this part if it were a major source subsequently increases its emissions of hazardous air pollutants (or its potential to emit hazardous air pollutants) such that the source is a major source, such source shall be subject to the relevant emission standard or other requirement.

(b) *Compliance dates for new and reconstructed sources.* (1) Except as specified in paragraphs (b)(3) and (b)(4) of this section, the owner or operator of a new or reconstructed source that has an initial startup before the effective date of a relevant standard established under this part pursuant to section 112(d), 112(f), or 112(h) of the Act shall comply with such standard not later than the standard's effective date.

(2) Except as specified in paragraphs (b)(3) and (b)(4) of this section, the owner or operator of a new or reconstructed source that has an initial startup after the effective date of a relevant standard established under this part pursuant to section 112(d), 112(f), or 112(h) of the Act shall comply with such standard upon startup of the source.

(3) The owner or operator of an affected source for which construction or reconstruction is commenced after the proposal date of a relevant standard established under this part pursuant to section 112(d), 112(f), or 112(h) of the Act but before the effective date (that is, promulgation) of such standard shall comply with the relevant emission standard not later than the date 3 years after the effective date if:

(i) The promulgated standard (that is, the relevant standard) is more stringent than the proposed standard; and

(ii) The owner or operator complies with the standard as proposed during the 3-year period immediately after the effective date.

(4) The owner or operator of an affected source for which construction or reconstruction is commenced after the proposal date of a relevant standard established pursuant to section 112(d) of the Act but before the proposal date of a relevant standard established pursuant to section 112(f) shall comply with the emission standard under section 112(f) not later than the date 10 years after the date construction or reconstruction is commenced, except that, if the section 112(f) standard is promulgated more than 10 years after construction or reconstruction is commenced, the owner or operator shall comply with the standard as provided in paragraphs (b)(1) and (b)(2) of this section.

(5) The owner or operator of a new source that is subject to the compliance requirements of paragraph (b)(3) or paragraph (b)(4) of this section shall notify the Administrator in accordance with § 63.9(d) of this subpart.

(6) [Reserved]

(7) After the effective date of an emission standard promulgated under this part, the owner or operator of an unaffected new area source (i.e., an area source for which construction or reconstruction was commenced after the proposal date of the standard) that increases its emissions of (or its potential to emit) hazardous air pollutants such that

§ 63.6

the source becomes a major source that is subject to the emission standard, shall comply with the relevant emission standard immediately upon becoming a major source. This compliance date shall apply to new area sources that become affected major sources regardless of whether the new area source previously was affected by that standard. The new affected major source shall comply with all requirements of that standard that affect new sources.

(c) *Compliance dates for existing sources.* (1) After the effective date of a relevant standard established under this part pursuant to section 112(d) or 112(h) of the Act, the owner or operator of an existing source shall comply with such standard by the compliance date established by the Administrator in the applicable subpart(s) of this part. Except as otherwise provided for in section 112 of the Act, in no case will the compliance date established for an existing source in an applicable subpart of this part exceed 3 years after the effective date of such standard.

(2) After the effective date of a relevant standard established under this part pursuant to section 112(f) of the Act, the owner or operator of an existing source shall comply with such standard not later than 90 days after the standard's effective date unless the Administrator has granted an extension to the source under paragraph (i)(4)(ii) of this section.

(3)–(4) [Reserved]

(5) After the effective date of an emission standard promulgated under this part, the owner or operator of an unaffected existing area source that increases its emissions of (or its potential to emit) hazardous air pollutants such that the source becomes a major source that is subject to the emission standard shall comply by the date specified in the standard for existing area sources that become major sources. If no such compliance date is specified in the standard, the source shall have a period of time to comply with the relevant emission standard that is equivalent to the compliance period specified in that standard for other existing sources. This compliance period shall apply to existing area sources that become affected major sources regardless of whether the existing area source previously was affected by that standard. Notwithstanding the previous two sentences, however, if the existing area source becomes a major source by the addition of a new affected source or by reconstructing, the portion of the existing facility that is a new affected source or a reconstructed source shall comply with all requirements of that standard that affect new sources, including the compliance date for new sources.

(d) [Reserved]

(e) *Operation and maintenance requirements.* (1)(i) At all times, including periods of startup, shutdown, and malfunction, owners or operators shall operate and maintain any affected source, including associated air pollution control equipment, in a manner consistent with good air pollution control practices for minimizing emissions at least to the levels required by all relevant standards.

(ii) Malfunctions shall be corrected as soon as practicable after their occurrence in accordance with the startup, shutdown, and malfunction plan required in paragraph (e)(3) of this section.

(iii) Operation and maintenance requirements established pursuant to section 112 of the Act are enforceable independent of emissions limitations or other requirements in relevant standards.

(2) Determination of whether acceptable operation and maintenance procedures are being used will be based on information available to the Administrator which may include, but is not limited to, monitoring results, review of operation and maintenance procedures (including the startup, shutdown, and malfunction plan required in paragraph (e)(3) of this section), review of operation and maintenance records, and inspection of the source.

(3) *Startup, shutdown, and malfunction plan.* (i) The owner or operator of an affected source shall develop and implement a written startup, shutdown, and malfunction plan that describes, in detail, procedures for operating and maintaining the source during periods of startup, shutdown, and malfunction and a program of corrective action for malfunctioning process and air pollution control equipment used to comply with the relevant standard. As required under § 63.8(c)(1)(i), the plan shall identify all routine or otherwise predictable CMS malfunctions. This plan shall be developed by the owner or operator by the source's compliance date for that relevant standard. The plan shall be incorporated by reference into the source's title V permit. The purpose of the startup, shutdown, and malfunction plan is to—

(A) Ensure that, at all times, owners or operators operate and maintain affected sources, including associated air pollution control equipment, in a manner consistent with good air pollution control practices for minimizing emissions at least to the levels required by all relevant standards;

(B) Ensure that owners or operators are prepared to correct malfunctions as soon as practicable after their occurrence in order to minimize excess emissions of hazardous air pollutants; and

(C) Reduce the reporting burden associated with periods of startup, shutdown, and malfunction (including corrective action taken to restore malfunctioning process and air pollution control equipment to its normal or usual manner of operation).

(ii) During periods of startup, shutdown, and malfunction, the owner or operator of an affected source shall operate and maintain such source (including associated air pollution control equipment) in accordance with the procedures specified in the startup, shutdown, and malfunction plan developed under paragraph (e)(3)(i) of this section.

(iii) When actions taken by the owner or operator during a startup, shutdown, or malfunction (including actions taken to correct a malfunction) are consistent with the procedures specified in the affected source's startup, shutdown, and malfunction plan, the owner or operator shall keep records for that event that demonstrate that the procedures specified in the plan were followed. These records may take the form of a "checklist," or other effective form of recordkeeping, that confirms conformance with the startup, shutdown, and malfunction plan for that event. In addition, the owner or operator shall keep records of these events as specified in § 63.10(b) (and elsewhere in this part), including records of the occurrence and duration of each startup, shutdown, or malfunction of operation and each malfunction of the air pollution control equipment. Furthermore, the owner or operator shall confirm that actions taken during the relevant reporting period during periods of startup, shutdown, and malfunction were consistent with the affected source's startup, shutdown and malfunction plan in the semiannual (or more frequent) startup, shutdown, and malfunction report required in § 63.10(d)(5).

(iv) If an action taken by the owner or operator during a startup, shutdown, or malfunction (including an action taken to correct a malfunction) is not consistent with the procedures specified in the affected source's startup, shutdown, and malfunction plan, the owner or operator shall record the actions taken for that event and shall report such actions within 2 working days after commencing actions inconsistent with the plan, followed by a letter within 7 working days after the end of the event, in accordance with § 63.10(d)(5) (unless the owner or operator makes alternative reporting arrangements, in advance, with the Administrator (see § 63.10(d)(5)(ii))).

(v) The owner or operator shall keep the written startup, shutdown, and malfunction plan on record after it is developed to be made available for inspection, upon request, by the Administrator for the life of the affected source or until the affected source is no longer subject to the provisions of this part. In addition, if the startup, shutdown, and malfunction plan is revised, the owner or operator shall keep previous (i.e., superseded) versions of the startup, shutdown, and malfunction plan on record, to be made available for inspection, upon request, by the Administrator, for a period of 5 years after each revision to the plan.

(vi) To satisfy the requirements of this section to develop a startup, shutdown, and malfunction plan, the owner or operator may use the affected source's standard operating procedures (SOP) manual, or an Occupational Safety and Health Administration (OSHA) or other plan, provided the alternative plans meet all the requirements of this section and are made available for inspection when requested by the Administrator.

(vii) Based on the results of a determination made under paragraph (e)(2) of this section, the Administrator may require that an owner or operator of an affected source make changes to the startup, shutdown, and malfunction plan for that source. The Administrator may require reasonable revisions to a startup, shutdown, and malfunction plan, if the Administrator finds that the plan:

(A) Does not address a startup, shutdown, or malfunction event that has occurred;

(B) Fails to provide for the operation of the source (including associated air pollution control equipment) during a startup, shutdown, or malfunction event in a manner consistent with good air pollution control practices for minimizing emissions at least to the levels required by all relevant standards; or

(C) Does not provide adequate procedures for correcting malfunctioning process and/or air pollution control equipment as quickly as practicable.

(viii) If the startup, shutdown, and malfunction plan fails to address or inadequately addresses an event that meets the characteristics of a malfunction but was not included in the startup, shutdown, and malfunction plan at the time the owner or operator developed the plan, the owner or operator shall revise the startup, shutdown, and malfunction plan within 45 days after the event to include detailed procedures for operating and maintaining the source during similar malfunction events and a program of corrective action for similar malfunctions of process or air pollution control equipment.

(f) *Compliance with nonopacity emission standards*—(1) *Applicability.* The nonopacity emission standards set forth in this part shall apply at all times except during periods of startup, shutdown, and malfunction, and as otherwise specified in an applicable subpart.

(2) *Methods for determining compliance.* (i) The Administrator will determine compliance with nonopacity emission standards in this part based on the results of performance tests conducted according to the procedures in § 63.7, unless otherwise specified in an applicable subpart of this part.

(ii) The Administrator will determine compliance with nonopacity emission standards in this part by evaluation of an owner or operator's conformance with operation and maintenance requirements, including the evaluation of monitoring data, as specified in § 63.6(e) and applicable subparts of this part.

(iii) If an affected source conducts performance testing at startup to obtain an operating permit in

§ 63.6

the State in which the source is located, the results of such testing may be used to demonstrate compliance with a relevant standard if—

(A) The performance test was conducted within a reasonable amount of time before an initial performance test is required to be conducted under the relevant standard;

(B) The performance test was conducted under representative operating conditions for the source;

(C) The performance test was conducted and the resulting data were reduced using EPA-approved test methods and procedures, as specified in § 63.7(e) of this subpart; and

(D) The performance test was appropriately quality-assured, as specified in § 63.7(c) of this subpart.

(iv) The Administrator will determine compliance with design, equipment, work practice, or operational emission standards in this part by review of records, inspection of the source, and other procedures specified in applicable subparts of this part.

(v) The Administrator will determine compliance with design, equipment, work practice, or operational emission standards in this part by evaluation of an owner or operator's conformance with operation and maintenance requirements, as specified in paragraph (e) of this section and applicable subparts of this part.

(3) *Finding of compliance.* The Administrator will make a finding concerning an affected source's compliance with a nonopacity emission standard, as specified in paragraphs (f)(1) and (f)(2) of this section, upon obtaining all the compliance information required by the relevant standard (including the written reports of performance test results, monitoring results, and other information, if applicable) and any information available to the Administrator needed to determine whether proper operation and maintenance practices are being used.

(g) *Use of an alternative nonopacity emission standard.* (1) If, in the Administrator's judgment, an owner or operator of an affected source has established that an alternative means of emission limitation will achieve a reduction in emissions of a hazardous air pollutant from an affected source at least equivalent to the reduction in emissions of that pollutant from that source achieved under any design, equipment, work practice, or operational emission standard, or combination thereof, established under this part pursuant to section 112(h) of the Act, the Administrator will publish in the FEDERAL REGISTER a notice permitting the use of the alternative emission standard for purposes of compliance with the promulgated standard. Any FEDERAL REGISTER notice under this paragraph shall be published only after the public is notified and given the opportunity to comment. Such notice

will restrict the permission to the stationary source(s) or category(ies) of sources from which the alternative emission standard will achieve equivalent emission reductions. The Administrator will condition permission in such notice on requirements to assure the proper operation and maintenance of equipment and practices required for compliance with the alternative emission standard and other requirements, including appropriate quality assurance and quality control requirements, that are deemed necessary.

(2) An owner or operator requesting permission under this paragraph shall, unless otherwise specified in an applicable subpart, submit a proposed test plan or the results of testing and monitoring in accordance with § 63.7 and § 63.8, a description of the procedures followed in testing or monitoring, and a description of pertinent conditions during testing or monitoring. Any testing or monitoring conducted to request permission to use an alternative nonopacity emission standard shall be appropriately quality assured and quality controlled, as specified in § 63.7 and § 63.8.

(3) The Administrator may establish general procedures in an applicable subpart that accomplish the requirements of paragraphs (g)(1) and (g)(2) of this section.

(h) *Compliance with opacity and visible emission standards*—(1) *Applicability.* The opacity and visible emission standards set forth in this part shall apply at all times except during periods of startup, shutdown, and malfunction, and as otherwise specified in an applicable subpart.

(2) *Methods for determining compliance.* (i) The Administrator will determine compliance with opacity and visible emission standards in this part based on the results of the test method specified in an applicable subpart. Whenever a continuous opacity monitoring system (COMS) is required to be installed to determine compliance with numerical opacity emission standards in this part, compliance with opacity emission standards in this part shall be determined by using the results from the COMS. Whenever an opacity emission test method is not specified, compliance with opacity emission standards in this part shall be determined by conducting observations in accordance with Test Method 9 in appendix A of part 60 of this chapter or the method specified in paragraph (h)(7)(ii) of this section. Whenever a visible emission test method is not specified, compliance with visible emission standards in this part shall be determined by conducting observations in accordance with Test Method 22 in appendix A of part 60 of this chapter.

(ii) [Reserved]

(iii) If an affected source undergoes opacity or visible emission testing at startup to obtain an operating permit in the State in which the source is located, the results of such testing may be used to demonstrate compliance with a relevant standard if—

(A) The opacity or visible emission test was conducted within a reasonable amount of time before a performance test is required to be conducted under the relevant standard;

(B) The opacity or visible emission test was conducted under representative operating conditions for the source;

(C) The opacity or visible emission test was conducted and the resulting data were reduced using EPA-approved test methods and procedures, as specified in § 63.7(e) of this subpart; and

(D) The opacity or visible emission test was appropriately quality-assured, as specified in § 63.7(c) of this section.

(3) [Reserved]

(4) *Notification of opacity or visible emission observations.* The owner or operator of an affected source shall notify the Administrator in writing of the anticipated date for conducting opacity or visible emission observations in accordance with § 63.9(f), if such observations are required for the source by a relevant standard.

(5) *Conduct of opacity or visible emission observations.* When a relevant standard under this part includes an opacity or visible emission standard, the owner or operator of an affected source shall comply with the following:

(i) For the purpose of demonstrating initial compliance, opacity or visible emission observations shall be conducted concurrently with the initial performance test required in § 63.7 unless one of the following conditions applies:

(A) If no performance test under § 63.7 is required, opacity or visible emission observations shall be conducted within 60 days after achieving the maximum production rate at which a new or reconstructed source will be operated, but not later than 120 days after initial startup of the source, or within 120 days after the effective date of the relevant standard in the case of new sources that start up before the standard's effective date. If no performance test under § 63.7 is required, opacity or visible emission observations shall be conducted within 120 days after the compliance date for an existing or modified source; or

(B) If visibility or other conditions prevent the opacity or visible emission observations from being conducted concurrently with the initial performance test required under § 63.7, or within the time period specified in paragraph (h)(5)(i)(A) of this section, the source's owner or operator shall reschedule the opacity or visible emission observations as soon after the initial performance test, or time period, as possible, but not later than 30 days thereafter, and shall advise the Administrator of the rescheduled date. The rescheduled opacity or

visible emission observations shall be conducted (to the extent possible) under the same operating conditions that existed during the initial performance test conducted under § 63.7. The visible emissions observer shall determine whether visibility or other conditions prevent the opacity or visible emission observations from being made concurrently with the initial performance test in accordance with procedures contained in Test Method 9 or Test Method 22 in appendix A of part 60 of this chapter.

(ii) For the purpose of demonstrating initial compliance, the minimum total time of opacity observations shall be 3 hours (30 6-minute averages) for the performance test or other required set of observations (e.g., for fugitive-type emission sources subject only to an opacity emission standard).

(iii) The owner or operator of an affected source to which an opacity or visible emission standard in this part applies shall conduct opacity or visible emission observations in accordance with the provisions of this section, record the results of the evaluation of emissions, and report to the Administrator the opacity or visible emission results in accordance with the provisions of § 63.10(d).

(iv) [Reserved]

(v) Opacity readings of portions of plumes that contain condensed, uncombined water vapor shall not be used for purposes of determining compliance with opacity emission standards.

(6) *Availability of records.* The owner or operator of an affected source shall make available, upon request by the Administrator, such records that the Administrator deems necessary to determine the conditions under which the visual observations were made and shall provide evidence indicating proof of current visible observer emission certification.

(7) *Use of a continuous opacity monitoring system.* (i) The owner or operator of an affected source required to use a continuous opacity monitoring system (COMS) shall record the monitoring data produced during a performance test required under § 63.7 and shall furnish the Administrator a written report of the monitoring results in accordance with the provisions of § 63.10(e)(4).

(ii) Whenever an opacity emission test method has not been specified in an applicable subpart, or an owner or operator of an affected source is required to conduct Test Method 9 observations (see appendix A of part 60 of this chapter), the owner or operator may submit, for compliance purposes, COMS data results produced during any performance test required under § 63.7 in lieu of Method 9 data. If the owner or operator elects to submit COMS data for compliance with the opacity emission standard, he or she shall notify the Administrator of that decision, in writing, simultaneously

§ 63.6

with the notification under § 63.7(b) of the date the performance test is scheduled to begin. Once the owner or operator of an affected source has notified the Administrator to that effect, the COMS data results will be used to determine opacity compliance during subsequent performance tests required under § 63.7, unless the owner or operator notifies the Administrator in writing to the contrary not later than with the notification under § 63.7(b) of the date the subsequent performance test is scheduled to begin.

(iii) For the purposes of determining compliance with the opacity emission standard during a performance test required under § 63.7 using COMS data, the COMS data shall be reduced to 6-minute averages over the duration of the mass emission performance test.

(iv) The owner or operator of an affected source using a COMS for compliance purposes is responsible for demonstrating that he/she has complied with the performance evaluation requirements of § 63.8(e), that the COMS has been properly maintained, operated, and data quality-assured, as specified in § 63.8(c) and § 63.8(d), and that the resulting data have not been altered in any way.

(v) Except as provided in paragraph (h)(7)(ii) of this section, the results of continuous monitoring by a COMS that indicate that the opacity at the time visual observations were made was not in excess of the emission standard are probative but not conclusive evidence of the actual opacity of an emission, provided that the affected source proves that, at the time of the alleged violation, the instrument used was properly maintained, as specified in § 63.8(c), and met Performance Specification 1 in appendix B of part 60 of this chapter, and that the resulting data have not been altered in any way.

(8) *Finding of compliance.* The Administrator will make a finding concerning an affected source's compliance with an opacity or visible emission standard upon obtaining all the compliance information required by the relevant standard (including the written reports of the results of the performance tests required by § 63.7, the results of Test Method 9 or another required opacity or visible emission test method, the observer certification required by paragraph (h)(6) of this section, and the continuous opacity monitoring system results, whichever is/are applicable) and any information available to the Administrator needed to determine whether proper operation and maintenance practices are being used.

(9) *Adjustment to an opacity emission standard.* (i) If the Administrator finds under paragraph (h)(8) of this section that an affected source is in compliance with all relevant standards for which initial performance tests were conducted under § 63.7, but during the time such performance tests were conducted fails to meet any relevant opacity emission standard, the owner or operator of such source may petition the Administrator to make appropriate adjustment to the opacity emission standard for the affected source. Until the Administrator notifies the owner or operator of the appropriate adjustment, the relevant opacity emission standard remains applicable.

(ii) The Administrator may grant such a petition upon a demonstration by the owner or operator that—

(A) The affected source and its associated air pollution control equipment were operated and maintained in a manner to minimize the opacity of emissions during the performance tests;

(B) The performance tests were performed under the conditions established by the Administrator; and

(C) The affected source and its associated air pollution control equipment were incapable of being adjusted or operated to meet the relevant opacity emission standard.

(iii) The Administrator will establish an adjusted opacity emission standard for the affected source meeting the above requirements at a level at which the source will be able, as indicated by the performance and opacity tests, to meet the opacity emission standard at all times during which the source is meeting the mass or concentration emission standard. The Administrator will promulgate the new opacity emission standard in the FEDERAL REGISTER.

(iv) After the Administrator promulgates an adjusted opacity emission standard for an affected source, the owner or operator of such source shall be subject to the new opacity emission standard, and the new opacity emission standard shall apply to such source during any subsequent performance tests.

(i) *Extension of compliance with emission standards.* (1) Until an extension of compliance has been granted by the Administrator (or a State with an approved permit program) under this paragraph, the owner or operator of an affected source subject to the requirements of this section shall comply with all applicable requirements of this part.

(2) *Extension of compliance for early reductions and other reductions*—(i) *Early reductions.* Pursuant to section 112(i)(5) of the Act, if the owner or operator of an existing source demonstrates that the source has achieved a reduction in emissions of hazardous air pollutants in accordance with the provisions of subpart D of this part, the Administrator (or the State with an approved permit program) will grant the owner or operator an extension of compliance with specific requirements of this part, as specified in subpart D.

(ii) *Other reductions.* Pursuant to section 112(i)(6) of the Act, if the owner or operator of an existing source has installed best available control technology (BACT) (as defined in section 169(3) of the Act) or technology required to meet a lowest achievable emission rate (LAER) (as defined in section 171 of the Act) prior to the promulgation of an emission standard in this part applicable to such source and the same pollutant (or stream of pollutants) controlled pursuant to the BACT or LAER installation, the Administrator will grant the owner or operator an extension of compliance with such emission standard that will apply until the date 5 years after the date on which such installation was achieved, as determined by the Administrator.

(3) *Request for extension of compliance.* Paragraphs (i)(4) through (i)(7) of this section concern requests for an extension of compliance with a relevant standard under this part (except requests for an extension of compliance under paragraph (i)(2)(i) of this section will be handled through procedures specified in subpart D of this part).

(4)(i)(A) The owner or operator of an existing source who is unable to comply with a relevant standard established under this part pursuant to section 112(d) of the Act may request that the Administrator (or a State, when the State has an approved part 70 permit program and the source is required to obtain a part 70 permit under that program, or a State, when the State has been delegated the authority to implement and enforce the emission standard for that source) grant an extension allowing the source up to 1 additional year to comply with the standard, if such additional period is necessary for the installation of controls. An additional extension of up to 3 years may be added for mining waste operations, if the 1-year extension of compliance is insufficient to dry and cover mining waste in order to reduce emissions of any hazardous air pollutant. The owner or operator of an affected source who has requested an extension of compliance under this paragraph and who is otherwise required to obtain a title V permit shall apply for such permit or apply to have the source's title V permit revised to incorporate the conditions of the extension of compliance. The conditions of an extension of compliance granted under this paragraph will be incorporated into the affected source's title V permit according to the provisions of part 70 or Federal title V regulations in this chapter (42 U.S.C. 7661), whichever are applicable.

(B) Any request under this paragraph for an extension of compliance with a relevant standard shall be submitted in writing to the appropriate authority not later than 12 months before the affected source's compliance date (as specified in paragraphs (b) and (c) of this section) for sources that are not including emission points in an emissions average, or not later than 18 months before

the affected source's compliance date (as specified in paragraphs (b) and (c) of this section) for sources that are including emission points in an emissions average. Emission standards established under this part may specify alternative dates for the submittal of requests for an extension of compliance if alternatives are appropriate for the source categories affected by those standards, e.g., a compliance date specified by the standard is less than 12 (or 18) months after the standard's effective date.

(ii) The owner or operator of an existing source unable to comply with a relevant standard established under this part pursuant to section 112(f) of the Act may request that the Administrator grant an extension allowing the source up to 2 years after the standard's effective date to comply with the standard. The Administrator may grant such an extension if he/she finds that such additional period is necessary for the installation of controls and that steps will be taken during the period of the extension to assure that the health of persons will be protected from imminent endangerment. Any request for an extension of compliance with a relevant standard under this paragraph shall be submitted in writing to the Administrator not later than 15 calendar days after the effective date of the relevant standard.

(5) The owner or operator of an existing source that has installed BACT or technology required to meet LAER [as specified in paragraph (i)(2)(ii) of this section] prior to the promulgation of a relevant emission standard in this part may request that the Administrator grant an extension allowing the source 5 years from the date on which such installation was achieved, as determined by the Administrator, to comply with the standard. Any request for an extension of compliance with a relevant standard under this paragraph shall be submitted in writing to the Administrator not later than 120 days after the promulgation date of the standard. The Administrator may grant such an extension if he or she finds that the installation of BACT or technology to meet LAER controls the same pollutant (or stream of pollutants) that would be controlled at that source by the relevant emission standard.

(6)(i) The request for a compliance extension under paragraph (i)(4) of this section shall include the following information:

(A) A description of the controls to be installed to comply with the standard;

(B) A compliance schedule, including the date by which each step toward compliance will be reached. At a minimum, the list of dates shall include:

(*1*) The date by which contracts for emission control systems or process changes for emission control will be awarded, or the date by which or-

§ 63.6

ders will be issued for the purchase of component parts to accomplish emission control or process changes;

(2) The date by which on-site construction, installation of emission control equipment, or a process change is to be initiated;

(3) The date by which on-site construction, installation of emission control equipment, or a process change is to be completed; and

(4) The date by which final compliance is to be achieved;

(C) A description of interim emission control steps that will be taken during the extension period, including milestones to assure proper operation and maintenance of emission control and process equipment; and

(D) Whether the owner or operator is also requesting an extension of other applicable requirements (e.g., performance testing requirements).

(ii) The request for a compliance extension under paragraph (i)(5) of this section shall include all information needed to demonstrate to the Administrator's satisfaction that the installation of BACT or technology to meet LAER controls the same pollutant (or stream of pollutants) that would be controlled at that source by the relevant emission standard.

(7) Advice on requesting an extension of compliance may be obtained from the Administrator (or the State with an approved permit program).

(8) *Approval of request for extension of compliance.* Paragraphs (i)(9) through (i)(14) of this section concern approval of an extension of compliance requested under paragraphs (i)(4) through (i)(6) of this section.

(9) Based on the information provided in any request made under paragraphs (i)(4) through (i)(6) of this section, or other information, the Administrator (or the State with an approved permit program) may grant an extension of compliance with an emission standard, as specified in paragraphs (i)(4) and (i)(5) of this section.

(10) The extension will be in writing and will—

(i) Identify each affected source covered by the extension;

(ii) Specify the termination date of the extension;

(iii) Specify the dates by which steps toward compliance are to be taken, if appropriate;

(iv) Specify other applicable requirements to which the compliance extension applies (e.g., performance tests); and

(v)(A) Under paragraph (i)(4), specify any additional conditions that the Administrator (or the State) deems necessary to assure installation of the necessary controls and protection of the health of persons during the extension period; or

(B) Under paragraph (i)(5), specify any additional conditions that the Administrator deems

necessary to assure the proper operation and maintenance of the installed controls during the extension period.

(11) The owner or operator of an existing source that has been granted an extension of compliance under paragraph (i)(10) of this section may be required to submit to the Administrator (or the State with an approved permit program) progress reports indicating whether the steps toward compliance outlined in the compliance schedule have been reached. The contents of the progress reports and the dates by which they shall be submitted will be specified in the written extension of compliance granted under paragraph (i)(10) of this section.

(12)(i) The Administrator (or the State with an approved permit program) will notify the owner or operator in writing of approval or intention to deny approval of a request for an extension of compliance within 30 calendar days after receipt of sufficient information to evaluate a request submitted under paragraph (i)(4)(i) or (i)(5) of this section. The 30-day approval or denial period will begin after the owner or operator has been notified in writing that his/her application is complete. The Administrator (or the State) will notify the owner or operator in writing of the status of his/her application, that is, whether the application contains sufficient information to make a determination, within 30 calendar days after receipt of the original application and within 30 calendar days after receipt of any supplementary information that is submitted.

(ii) When notifying the owner or operator that his/her application is not complete, the Administrator will specify the information needed to complete the application and provide notice of opportunity for the applicant to present, in writing, within 30 calendar days after he/she is notified of the incomplete application, additional information or arguments to the Administrator to enable further action on the application.

(iii) Before denying any request for an extension of compliance, the Administrator (or the State with an approved permit program) will notify the owner or operator in writing of the Administrator's (or the State's) intention to issue the denial, together with—

(A) Notice of the information and findings on which the intended denial is based; and

(B) Notice of opportunity for the owner or operator to present in writing, within 15 calendar days after he/she is notified of the intended denial, additional information or arguments to the Administrator (or the State) before further action on the request.

(iv) The Administrator's final determination to deny any request for an extension will be in writing and will set forth the specific grounds on which the denial is based. The final determination will be made within 30 calendar days after presentation of additional information or argument (if the application is complete), or within 30 calendar days after the final date specified for the presentation if no presentation is made.

(13)(i) The Administrator will notify the owner or operator in writing of approval or intention to deny approval of a request for an extension of compliance within 30 calendar days after receipt of sufficient information to evaluate a request submitted under paragraph (i)(4)(ii) of this section. The 30-day approval or denial period will begin after the owner or operator has been notified in writing that his/her application is complete. The Administrator (or the State) will notify the owner or operator in writing of the status of his/her application, that is, whether the application contains sufficient information to make a determination, within 15 calendar days after receipt of the original application and within 15 calendar days after receipt of any supplementary information that is submitted.

(ii) When notifying the owner or operator that his/her application is not complete, the Administrator will specify the information needed to complete the application and provide notice of opportunity for the applicant to present, in writing, within 15 calendar days after he/she is notified of the incomplete application, additional information or arguments to the Administrator to enable further action on the application.

(iii) Before denying any request for an extension of compliance, the Administrator will notify the owner or operator in writing of the Administrator's intention to issue the denial, together with—

(A) Notice of the information and findings on which the intended denial is based; and

(B) Notice of opportunity for the owner or operator to present in writing, within 15 calendar days after he/she is notified of the intended denial, additional information or arguments to the Administrator before further action on the request.

(iv) A final determination to deny any request for an extension will be in writing and will set forth the specific grounds on which the denial is based. The final determination will be made within 30 calendar days after presentation of additional information or argument (if the application is complete), or within 30 calendar days after the final date specified for the presentation if no presentation is made.

(14) The Administrator (or the State with an approved permit program) may terminate an extension of compliance at an earlier date than specified if any specification under paragraphs (i)(10)(iii) or (i)(10)(iv) of this section is not met.

(15) [Reserved]

(16) The granting of an extension under this section shall not abrogate the Administrator's authority under section 114 of the Act.

(j) *Exemption from compliance with emission standards.* The President may exempt any stationary source from compliance with any relevant standard established pursuant to section 112 of the Act for a period of not more than 2 years if the President determines that the technology to implement such standard is not available and that it is in the national security interests of the United States to do so. An exemption under this paragraph may be extended for 1 or more additional periods, each period not to exceed 2 years.

§ 63.7 Performance testing requirements.

(a) *Applicability and performance test dates.* (1) Unless otherwise specified, this section applies to the owner or operator of an affected source required to do performance testing, or another form of compliance demonstration, under a relevant standard.

(2) If required to do performance testing by a relevant standard, and unless a waiver of performance testing is obtained under this section or the conditions of paragraph (c)(3)(ii)(B) of this section apply, the owner or operator of the affected source shall perform such tests as follows—

(i) Within 180 days after the effective date of a relevant standard for a new source that has an initial startup date before the effective date; or

(ii) Within 180 days after initial startup for a new source that has an initial startup date after the effective date of a relevant standard; or

(iii) Within 180 days after the compliance date specified in an applicable subpart of this part for an existing source subject to an emission standard established pursuant to section 112(d) of the Act, or within 180 days after startup of an existing source if the source begins operation after the effective date of the relevant emission standard; or

(iv) Within 180 days after the compliance date for an existing source subject to an emission standard established pursuant to section 112(f) of the Act; or

(v) Within 180 days after the termination date of the source's extension of compliance for an existing source that obtains an extension of compliance under § 63.6(i); or

(vi) Within 180 days after the compliance date for a new source, subject to an emission standard established pursuant to section 112(f) of the Act, for which construction or reconstruction is commenced after the proposal date of a relevant standard established pursuant to section 112(d) of the Act but before the proposal date of the relevant standard established pursuant to section 112(f) [see § 63.6(b)(4)]; or

§ 63.7

(vii) [Reserved]; or

(viii) [Reserved]; or

(ix) When an emission standard promulgated under this part is more stringent than the standard proposed (see § 63.6(b)(3)), the owner or operator of a new or reconstructed source subject to that standard for which construction or reconstruction is commenced between the proposal and promulgation dates of the standard shall comply with performance testing requirements within 180 days after the standard's effective date, or within 180 days after startup of the source, whichever is later. If the promulgated standard is more stringent than the proposed standard, the owner or operator may choose to demonstrate compliance with either the proposed or the promulgated standard. If the owner or operator chooses to comply with the proposed standard initially, the owner or operator shall conduct a second performance test within 3 years and 180 days after the effective date of the standard, or after startup of the source, whichever is later, to demonstrate compliance with the promulgated standard.

(3) The Administrator may require an owner or operator to conduct performance tests at the affected source at any other time when the action is authorized by section 114 of the Act.

(b) *Notification of performance test.* (1) The owner or operator of an affected source shall notify the Administrator in writing of his or her intention to conduct a performance test at least 60 calendar days before the performance test is scheduled to begin to allow the Administrator, upon request, to review and approve the site-specific test plan required under paragraph (c) of this section and to have an observer present during the test. Observation of the performance test by the Administrator is optional.

(2) In the event the owner or operator is unable to conduct the performance test on the date specified in the notification requirement specified in paragraph (b)(1) of this section, due to unforeseeable circumstances beyond his or her control, the owner or operator shall notify the Administrator within 5 days prior to the scheduled performance test date and specify the date when the performance test is rescheduled. This notification of delay in conducting the performance test shall not relieve the owner or operator of legal responsibility for compliance with any other applicable provisions of this part or with any other applicable Federal, State, or local requirement, nor will it prevent the Administrator from implementing or enforcing this part or taking any other action under the Act.

(c) *Quality assurance program.* (1) The results of the quality assurance program required in this paragraph will be considered by the Administrator when he/she determines the validity of a performance test.

(2)(i) *Submission of site-specific test plan.* Before conducting a required performance test, the owner or operator of an affected source shall develop and, if requested by the Administrator, shall submit a site-specific test plan to the Administrator for approval. The test plan shall include a test program summary, the test schedule, data quality objectives, and both an internal and external quality assurance (QA) program. Data quality objectives are the pretest expectations of precision, accuracy, and completeness of data.

(ii) The internal QA program shall include, at a minimum, the activities planned by routine operators and analysts to provide an assessment of test data precision; an example of internal QA is the sampling and analysis of replicate samples.

(iii) The external QA program shall include, at a minimum, application of plans for a test method performance audit (PA) during the performance test. The PA's consist of blind audit samples provided by the Administrator and analyzed during the performance test in order to provide a measure of test data bias. The external QA program may also include systems audits that include the opportunity for on-site evaluation by the Administrator of instrument calibration, data validation, sample logging, and documentation of quality control data and field maintenance activities.

(iv) The owner or operator of an affected source shall submit the site-specific test plan to the Administrator upon the Administrator's request at least 60 calendar days before the performance test is scheduled to take place, that is, simultaneously with the notification of intention to conduct a performance test required under paragraph (b) of this section, or on a mutually agreed upon date.

(v) The Administrator may request additional relevant information after the submittal of a site-specific test plan.

(3) *Approval of site-specific test plan.* (i) The Administrator will notify the owner or operator of approval or intention to deny approval of the site-specific test plan (if review of the site-specific test plan is requested) within 30 calendar days after receipt of the original plan and within 30 calendar days after receipt of any supplementary information that is submitted under paragraph (c)(3)(i)(B) of this section. Before disapproving any site-specific test plan, the Administrator will notify the applicant of the Administrator's intention to disapprove the plan together with—

(A) Notice of the information and findings on which the intended disapproval is based; and

(B) Notice of opportunity for the owner or operator to present, within 30 calendar days after he/she is notified of the intended disapproval, additional information to the Administrator before final action on the plan.

(ii) In the event that the Administrator fails to approve or disapprove the site-specific test plan within the time period specified in paragraph (c)(3)(i) of this section, the following conditions shall apply:

(A) If the owner or operator intends to demonstrate compliance using the test method(s) specified in the relevant standard, the owner or operator shall conduct the performance test within the time specified in this section using the specified method(s);

(B) If the owner or operator intends to demonstrate compliance by using an alternative to any test method specified in the relevant standard, the owner or operator shall refrain from conducting the performance test until the Administrator approves the use of the alternative method when the Administrator approves the site-specific test plan (if review of the site-specific test plan is requested) or until after the alternative method is approved (see paragraph (f) of this section). If the Administrator does not approve the site-specific test plan (if review is requested) or the use of the alternative method within 30 days before the test is scheduled to begin, the performance test dates specified in paragraph (a) of this section may be extended such that the owner or operator shall conduct the performance test within 60 calendar days after the Administrator approves the site-specific test plan or after use of the alternative method is approved. Notwithstanding the requirements in the preceding two sentences, the owner or operator may proceed to conduct the performance test as required in this section (without the Administrator's prior approval of the site-specific test plan) if he/she subsequently chooses to use the specified testing and monitoring methods instead of an alternative.

(iii) Neither the submission of a site-specific test plan for approval, nor the Administrator's approval or disapproval of a plan, nor the Administrator's failure to approve or disapprove a plan in a timely manner shall—

(A) Relieve an owner or operator of legal responsibility for compliance with any applicable provisions of this part or with any other applicable Federal, State, or local requirement; or

(B) Prevent the Administrator from implementing or enforcing this part or taking any other action under the Act.

(4)(i) *Performance test method audit program.* The owner or operator shall analyze performance audit (PA) samples during each performance test. The owner or operator shall request performance audit materials 45 days prior to the test date. Cylinder audit gases may be obtained by contacting the Cylinder Audit Coordinator, Quality Assurance Division (MD–77B), Atmospheric Research and Exposure Assessment Laboratory (AREAL), U.S.

EPA, Research Triangle Park, North Carolina 27711. All other audit materials may be obtained by contacting the Source Test Audit Coordinator, Quality Assurance Division (MD–77B), AREAL, U.S. EPA, Research Triangle Park, North Carolina 27711.

(ii) The Administrator will have sole discretion to require any subsequent remedial actions of the owner or operator based on the PA results.

(iii) If the Administrator fails to provide required PA materials to an owner or operator of an affected source in time to analyze the PA samples during a performance test, the requirement to conduct a PA under this paragraph shall be waived for such source for that performance test. Waiver under this paragraph of the requirement to conduct a PA for a particular performance test does not constitute a waiver of the requirement to conduct a PA for future required performance tests.

(d) *Performance testing facilities.* If required to do performance testing, the owner or operator of each new source and, at the request of the Administrator, the owner or operator of each existing source, shall provide performance testing facilities as follows:

(1) Sampling ports adequate for test methods applicable to such source. This includes:

(i) Constructing the air pollution control system such that volumetric flow rates and pollutant emission rates can be accurately determined by applicable test methods and procedures; and

(ii) Providing a stack or duct free of cyclonic flow during performance tests, as demonstrated by applicable test methods and procedures;

(2) Safe sampling platform(s);

(3) Safe access to sampling platform(s);

(4) Utilities for sampling and testing equipment; and

(5) Any other facilities that the Administrator deems necessary for safe and adequate testing of a source.

(e) *Conduct of performance tests.* (1) Performance tests shall be conducted under such conditions as the Administrator specifies to the owner or operator based on representative performance (i.e., performance based on normal operating conditions) of the affected source. Operations during periods of startup, shutdown, and malfunction shall not constitute representative conditions for the purpose of a performance test, nor shall emissions in excess of the level of the relevant standard during periods of startup, shutdown, and malfunction be considered a violation of the relevant standard unless otherwise specified in the relevant standard or a determination of noncompliance is made under § 63.6(e). Upon request, the owner or operator shall make available to the Administrator such records as may be necessary to determine the conditions of performance tests.

§ 63.7

(2) Performance tests shall be conducted and data shall be reduced in accordance with the test methods and procedures set forth in this section, in each relevant standard, and, if required, in applicable appendices of parts 51, 60, 61, and 63 of this chapter unless the Administrator—

(i) Specifies or approves, in specific cases, the use of a test method with minor changes in methodology; or

(ii) Approves the use of an alternative test method, the results of which the Administrator has determined to be adequate for indicating whether a specific affected source is in compliance; or

(iii) Approves shorter sampling times and smaller sample volumes when necessitated by process variables or other factors; or

(iv) Waives the requirement for performance tests because the owner or operator of an affected source has demonstrated by other means to the Administrator's satisfaction that the affected source is in compliance with the relevant standard.

(3) Unless otherwise specified in a relevant standard or test method, each performance test shall consist of three separate runs using the applicable test method. Each run shall be conducted for the time and under the conditions specified in the relevant standard. For the purpose of determining compliance with a relevant standard, the arithmetic mean of the results of the three runs shall apply. Upon receiving approval from the Administrator, results of a test run may be replaced with results of an additional test run in the event that—

(i) A sample is accidentally lost after the testing team leaves the site; or

(ii) Conditions occur in which one of the three runs must be discontinued because of forced shutdown; or

(iii) Extreme meteorological conditions occur; or

(iv) Other circumstances occur that are beyond the owner or operator's control.

(4) Nothing in paragraphs (e)(1) through (e)(3) of this section shall be construed to abrogate the Administrator's authority to require testing under section 114 of the Act.

(f) *Use of an alternative test method*—(1) *General.* Until permission to use an alternative test method has been granted by the Administrator under this paragraph, the owner or operator of an affected source remains subject to the requirements of this section and the relevant standard.

(2) The owner or operator of an affected source required to do performance testing by a relevant standard may use an alternative test method from that specified in the standard provided that the owner or operator—

(i) Notifies the Administrator of his or her intention to use an alternative test method not later than with the submittal of the site-specific test plan (if requested by the Administrator) or at least 60 days before the performance test is scheduled to begin if a site-specific test plan is not submitted;

(ii) Uses Method 301 in appendix A of this part to validate the alternative test method; and

(iii) Submits the results of the Method 301 validation process along with the notification of intention and the justification for not using the specified test method. The owner or operator may submit the information required in this paragraph well in advance of the deadline specified in paragraph (f)(2)(i) of this section to ensure a timely review by the Administrator in order to meet the performance test date specified in this section or the relevant standard.

(3) The Administrator will determine whether the owner or operator's validation of the proposed alternative test method is adequate when the Administrator approves or disapproves the site-specific test plan required under paragraph (c) of this section. If the Administrator finds reasonable grounds to dispute the results obtained by the Method 301 validation process, the Administrator may require the use of a test method specified in a relevant standard.

(4) If the Administrator finds reasonable grounds to dispute the results obtained by an alternative test method for the purposes of demonstrating compliance with a relevant standard, the Administrator may require the use of a test method specified in a relevant standard.

(5) If the owner or operator uses an alternative test method for an affected source during a required performance test, the owner or operator of such source shall continue to use the alternative test method for subsequent performance tests at that affected source until he or she receives approval from the Administrator to use another test method as allowed under § 63.7(f).

(6) Neither the validation and approval process nor the failure to validate an alternative test method shall abrogate the owner or operator's responsibility to comply with the requirements of this part.

(g) *Data analysis, recordkeeping, and reporting.* (1) Unless otherwise specified in a relevant standard or test method, or as otherwise approved by the Administrator in writing, results of a performance test shall include the analysis of samples, determination of emissions, and raw data. A performance test is "completed" when field sample collection is terminated. The owner or operator of an affected source shall report the results of the performance test to the Administrator before the close of business on the 60th day following the completion of the performance test, unless specified otherwise in a relevant standard or as approved otherwise in writing by the Administrator (see § 63.9(i)). The results of the performance test shall be submitted as part of the notification of compliance status required under § 63.9(h). Before a title V permit has been issued to the owner or operator of an affected source, the owner or operator shall send the results of the performance test to the Administrator. After a title V permit has been issued to the owner or operator of an affected source, the owner or operator shall send the results of the performance test to the appropriate permitting authority.

(2) [Reserved]

(3) For a minimum of 5 years after a performance test is conducted, the owner or operator shall retain and make available, upon request, for inspection by the Administrator the records or results of such performance test and other data needed to determine emissions from an affected source.

(h) *Waiver of performance tests.* (1) Until a waiver of a performance testing requirement has been granted by the Administrator under this paragraph, the owner or operator of an affected source remains subject to the requirements of this section.

(2) Individual performance tests may be waived upon written application to the Administrator if, in the Administrator's judgment, the source is meeting the relevant standard(s) on a continuous basis, or the source is being operated under an extension of compliance, or the owner or operator has requested an extension of compliance and the Administrator is still considering that request.

(3) *Request to waive a performance test.* (i) If a request is made for an extension of compliance under § 63.6(i), the application for a waiver of an initial performance test shall accompany the information required for the request for an extension of compliance. If no extension of compliance is requested or if the owner or operator has requested an extension of compliance and the Administrator is still considering that request, the application for a waiver of an initial performance test shall be submitted at least 60 days before the performance test if the site-specific test plan under paragraph (c) of this section is not submitted.

(ii) If an application for a waiver of a subsequent performance test is made, the application may accompany any required compliance progress report, compliance status report, or excess emissions and continuous monitoring system performance report [such as those required under § 63.6(i), § 63.9(h), and § 63.10(e) or specified in a relevant standard or in the source's title V permit], but it shall be submitted at least 60 days before the performance test if the site-specific test plan required under paragraph (c) of this section is not submitted.

(iii) Any application for a waiver of a performance test shall include information justifying the owner or operator's request for a waiver, such as

the technical or economic infeasibility, or the impracticality, of the affected source performing the required test.

(4) *Approval of request to waive performance test.* The Administrator will approve or deny a request for a waiver of a performance test made under paragraph (h)(3) of this section when he/she—

(i) Approves or denies an extension of compliance under § 63.6(i)(8); or

(ii) Approves or disapproves a site-specific test plan under § 63.7(c)(3); or

(iii) Makes a determination of compliance following the submission of a required compliance status report or excess emissions and continuous monitoring systems performance report; or

(iv) Makes a determination of suitable progress towards compliance following the submission of a compliance progress report, whichever is applicable.

(5) Approval of any waiver granted under this section shall not abrogate the Administrator's authority under the Act or in any way prohibit the Administrator from later canceling the waiver. The cancellation will be made only after notice is given to the owner or operator of the affected source.

§ 63.8 Monitoring requirements.

(a) *Applicability.* (1)(i) Unless otherwise specified in a relevant standard, this section applies to the owner or operator of an affected source required to do monitoring under that standard.

(ii) Relevant standards established under this part will specify monitoring systems, methods, or procedures, monitoring frequency, and other pertinent requirements for source(s) regulated by those standards. This section specifies general monitoring requirements such as those governing the conduct of monitoring and requests to use alternative monitoring methods. In addition, this section specifies detailed requirements that apply to affected sources required to use continuous monitoring systems (CMS) under a relevant standard.

(2) For the purposes of this part, all CMS required under relevant standards shall be subject to the provisions of this section upon promulgation of performance specifications for CMS as specified in the relevant standard or otherwise by the Administrator.

(3) [Reserved]

(4) Additional monitoring requirements for control devices used to comply with provisions in relevant standards of this part are specified in § 63.11.

(b) *Conduct of monitoring.* (1) Monitoring shall be conducted as set forth in this section and the relevant standard(s) unless the Administrator—

§ 63.8

(i) Specifies or approves the use of minor changes in methodology for the specified monitoring requirements and procedures; or

(ii) Approves the use of alternatives to any monitoring requirements or procedures.

(iii) Owners or operators with flares subject to § 63.11(b) are not subject to the requirements of this section unless otherwise specified in the relevant standard.

(2)(i) When the effluents from a single affected source, or from two or more affected sources, are combined before being released to the atmosphere, the owner or operator shall install an applicable CMS on each effluent.

(ii) If the relevant standard is a mass emission standard and the effluent from one affected source is released to the atmosphere through more than one point, the owner or operator shall install an applicable CMS at each emission point unless the installation of fewer systems is—

(A) Approved by the Administrator; or

(B) Provided for in a relevant standard (e.g., instead of requiring that a CMS be installed at each emission point before the effluents from those points are channeled to a common control device, the standard specifies that only one CMS is required to be installed at the vent of the control device).

(3) When more than one CMS is used to measure the emissions from one affected source (e.g., multiple breechings, multiple outlets), the owner or operator shall report the results as required for each CMS. However, when one CMS is used as a backup to another CMS, the owner or operator shall report the results from the CMS used to meet the monitoring requirements of this part. If both such CMS are used during a particular reporting period to meet the monitoring requirements of this part, then the owner or operator shall report the results from each CMS for the relevant compliance period.

(c) *Operation and maintenance of continuous monitoring systems.* (1) The owner or operator of an affected source shall maintain and operate each CMS as specified in this section, or in a relevant standard, and in a manner consistent with good air pollution control practices.

(i) The owner or operator of an affected source shall ensure the immediate repair or replacement of CMS parts to correct "routine" or otherwise predictable CMS malfunctions as defined in the source's startup, shutdown, and malfunction plan required by § 63.6(e)(3). The owner or operator shall keep the necessary parts for routine repairs of the affected equipment readily available. If the plan is followed and the CMS repaired immediately, this action shall be reported in the semiannual startup, shutdown, and malfunction report required under § 63.10(d)(5)(i).

(ii) For those malfunctions or other events that affect the CMS and are not addressed by the startup, shutdown, and malfunction plan, the owner or operator shall report actions that are not consistent with the startup, shutdown, and malfunction plan within 24 hours after commencing actions inconsistent with the plan. The owner or operator shall send a follow-up report within 2 weeks after commencing actions inconsistent with the plan that either certifies that corrections have been made or includes a corrective action plan and schedule. The owner or operator shall provide proof that repair parts have been ordered or any other records that would indicate that the delay in making repairs is beyond his or her control.

(iii) The Administrator's determination of whether acceptable operation and maintenance procedures are being used will be based on information that may include, but is not limited to, review of operation and maintenance procedures, operation and maintenance records, manufacturing recommendations and specifications, and inspection of the CMS. Operation and maintenance procedures written by the CMS manufacturer and other guidance also can be used to maintain and operate each CMS.

(2) All CMS shall be installed such that representative measurements of emissions or process parameters from the affected source are obtained. In addition, CEMS shall be located according to procedures contained in the applicable performance specification(s).

(3) All CMS shall be installed, operational, and the data verified as specified in the relevant standard either prior to or in conjunction with conducting performance tests under § 63.7. Verification of operational status shall, at a minimum, include completion of the manufacturer's written specifications or recommendations for installation, operation, and calibration of the system.

(4) Except for system breakdowns, out-of-control periods, repairs, maintenance periods, calibration checks, and zero (low-level) and high-level calibration drift adjustments, all CMS, including COMS and CEMS, shall be in continuous operation and shall meet minimum frequency of operation requirements as follows:

(i) All COMS shall complete a minimum of one cycle of sampling and analyzing for each successive 10-second period and one cycle of data recording for each successive 6-minute period.

(ii) All CEMS for measuring emissions other than opacity shall complete a minimum of one cycle of operation (sampling, analyzing, and data recording) for each successive 15-minute period.

(5) Unless otherwise approved by the Administrator, minimum procedures for COMS shall include a method for producing a simulated zero opacity condition and an upscale (high-level)

opacity condition using a certified neutral density filter or other related technique to produce a known obscuration of the light beam. Such procedures shall provide a system check of all the analyzer's internal optical surfaces and all electronic circuitry, including the lamp and photodetector assembly normally used in the measurement of opacity.

(6) The owner or operator of a CMS installed in accordance with the provisions of this part and the applicable CMS performance specification(s) shall check the zero (low-level) and high-level calibration drifts at least once daily in accordance with the written procedure specified in the performance evaluation plan developed under paragraphs (e)(3)(i) and (e)(3)(ii) of this section. The zero (low-level) and high-level calibration drifts shall be adjusted, at a minimum, whenever the 24-hour zero (low-level) drift exceeds two times the limits of the applicable performance specification(s) specified in the relevant standard. The system must allow the amount of excess zero (low-level) and high-level drift measured at the 24-hour interval checks to be recorded and quantified, whenever specified. For COMS, all optical and instrumental surfaces exposed to the effluent gases shall be cleaned prior to performing the zero (low-level) and high-level drift adjustments; the optical surfaces and instrumental surfaces shall be cleaned when the cumulative automatic zero compensation, if applicable, exceeds 4 percent opacity.

(7)(i) A CMS is out of control if—

(A) The zero (low-level), mid-level (if applicable), or high-level calibration drift (CD) exceeds two times the applicable CD specification in the applicable performance specification or in the relevant standard; or

(B) The CMS fails a performance test audit (e.g., cylinder gas audit), relative accuracy audit, relative accuracy test audit, or linearity test audit; or

(C) The COMS CD exceeds two times the limit in the applicable performance specification in the relevant standard.

(ii) When the CMS is out of control, the owner or operator of the affected source shall take the necessary corrective action and shall repeat all necessary tests which indicate that the system is out of control. The owner or operator shall take corrective action and conduct retesting until the performance requirements are below the applicable limits. The beginning of the out-of-control period is the hour the owner or operator conducts a performance check (e.g., calibration drift) that indicates an exceedance of the performance requirements established under this part. The end of the out-of-control period is the hour following the completion of corrective action and successful demonstration that the system is within the allow-

able limits. During the period the CMS is out of control, recorded data shall not be used in data averages and calculations, or to meet any data availability requirement established under this part.

(8) The owner or operator of a CMS that is out of control as defined in paragraph (c)(7) of this section shall submit all information concerning out-of-control periods, including start and end dates and hours and descriptions of corrective actions taken, in the excess emissions and continuous monitoring system performance report required in § 63.10(e)(3).

(d) *Quality control program.* (1) The results of the quality control program required in this paragraph will be considered by the Administrator when he/she determines the validity of monitoring data.

(2) The owner or operator of an affected source that is required to use a CMS and is subject to the monitoring requirements of this section and a relevant standard shall develop and implement a CMS quality control program. As part of the quality control program, the owner or operator shall develop and submit to the Administrator for approval upon request a site-specific performance evaluation test plan for the CMS performance evaluation required in paragraph (e)(3)(i) of this section, according to the procedures specified in paragraph (e). In addition, each quality control program shall include, at a minimum, a written protocol that describes procedures for each of the following operations:

(i) Initial and any subsequent calibration of the CMS;

(ii) Determination and adjustment of the calibration drift of the CMS;

(iii) Preventive maintenance of the CMS, including spare parts inventory;

(iv) Data recording, calculations, and reporting;

(v) Accuracy audit procedures, including sampling and analysis methods; and

(vi) Program of corrective action for a malfunctioning CMS.

(3) The owner or operator shall keep these written procedures on record for the life of the affected source or until the affected source is no longer subject to the provisions of this part, to be made available for inspection, upon request, by the Administrator. If the performance evaluation plan is revised, the owner or operator shall keep previous (i.e., superseded) versions of the performance evaluation plan on record to be made available for inspection, upon request, by the Administrator, for a period of 5 years after each revision to the plan. Where relevant, e.g., program of corrective action for a malfunctioning CMS, these written procedures may be incorporated as part of the affected source's startup, shutdown, and mal-

§ 63.8

function plan to avoid duplication of planning and recordkeeping efforts.

(e) *Performance evaluation of continuous monitoring systems—*(1) *General.* When required by a relevant standard, and at any other time the Administrator may require under section 114 of the Act, the owner or operator of an affected source being monitored shall conduct a performance evaluation of the CMS. Such performance evaluation shall be conducted according to the applicable specifications and procedures described in this section or in the relevant standard.

(2) *Notification of performance evaluation.* The owner or operator shall notify the Administrator in writing of the date of the performance evaluation simultaneously with the notification of the performance test date required under § 63.7(b) or at least 60 days prior to the date the performance evaluation is scheduled to begin if no performance test is required.

(3)(i) *Submission of site-specific performance evaluation test plan.* Before conducting a required CMS performance evaluation, the owner or operator of an affected source shall develop and submit a site-specific performance evaluation test plan to the Administrator for approval upon request. The performance evaluation test plan shall include the evaluation program objectives, an evaluation program summary, the performance evaluation schedule, data quality objectives, and both an internal and external QA program. Data quality objectives are the pre-evaluation expectations of precision, accuracy, and completeness of data.

(ii) The internal QA program shall include, at a minimum, the activities planned by routine operators and analysts to provide an assessment of CMS performance. The external QA program shall include, at a minimum, systems audits that include the opportunity for on-site evaluation by the Administrator of instrument calibration, data validation, sample logging, and documentation of quality control data and field maintenance activities.

(iii) The owner or operator of an affected source shall submit the site-specific performance evaluation test plan to the Administrator (if requested) at least 60 days before the performance test or performance evaluation is scheduled to begin, or on a mutually agreed upon date, and review and approval of the performance evaluation test plan by the Administrator will occur with the review and approval of the site-specific test plan (if review of the site-specific test plan is requested).

(iv) The Administrator may request additional relevant information after the submittal of a site-specific performance evaluation test plan.

(v) In the event that the Administrator fails to approve or disapprove the site-specific performance evaluation test plan within the time period specified in § 63.7(c)(3), the following conditions shall apply:

(A) If the owner or operator intends to demonstrate compliance using the monitoring method(s) specified in the relevant standard, the owner or operator shall conduct the performance evaluation within the time specified in this subpart using the specified method(s);

(B) If the owner or operator intends to demonstrate compliance by using an alternative to a monitoring method specified in the relevant standard, the owner or operator shall refrain from conducting the performance evaluation until the Administrator approves the use of the alternative method. If the Administrator does not approve the use of the alternative method within 30 days before the performance evaluation is scheduled to begin, the performance evaluation deadlines specified in paragraph (e)(4) of this section may be extended such that the owner or operator shall conduct the performance evaluation within 60 calendar days after the Administrator approves the use of the alternative method. Notwithstanding the requirements in the preceding two sentences, the owner or operator may proceed to conduct the performance evaluation as required in this section (without the Administrator's prior approval of the site-specific performance evaluation test plan) if he/she subsequently chooses to use the specified monitoring method(s) instead of an alternative.

(vi) Neither the submission of a site-specific performance evaluation test plan for approval, nor the Administrator's approval or disapproval of a plan, nor the Administrator' failure to approve or disapprove a plan in a timely manner shall—

(A) Relieve an owner or operator of legal responsibility for compliance with any applicable provisions of this part or with any other applicable Federal, State, or local requirement; or

(B) Prevent the Administrator from implementing or enforcing this part or taking any other action under the Act.

(4) *Conduct of performance evaluation and performance evaluation dates.* The owner or operator of an affected source shall conduct a performance evaluation of a required CMS during any performance test required under § 63.7 in accordance with the applicable performance specification as specified in the relevant standard. Notwithstanding the requirement in the previous sentence, if the owner or operator of an affected source elects to submit COMS data for compliance with a relevant opacity emission standard as provided under § 63.6(h)(7), he/she shall conduct a performance evaluation of the COMS as specified in the relevant standard, before the performance test required under § 63.7 is conducted in time to submit the results of the performance evaluation as specified in paragraph (e)(5)(ii) of this section. If a performance test is

not required, or the requirement for a performance test has been waived under § 63.7(h), the owner or operator of an affected source shall conduct the performance evaluation not later than 180 days after the appropriate compliance date for the affected source, as specified in § 63.7(a), or as otherwise specified in the relevant standard.

(5) *Reporting performance evaluation results.* (i) The owner or operator shall furnish the Administrator a copy of a written report of the results of the performance evaluation simultaneously with the results of the performance test required under § 63.7 or within 60 days of completion of the performance evaluation if no test is required, unless otherwise specified in a relevant standard. The Administrator may request that the owner or operator submit the raw data from a performance evaluation in the report of the performance evaluation results.

(ii) The owner or operator of an affected source using a COMS to determine opacity compliance during any performance test required under § 63.7 and described in § 63.6(d)(6) shall furnish the Administrator two or, upon request, three copies of a written report of the results of the COMS performance evaluation under this paragraph. The copies shall be provided at least 15 calendar days before the performance test required under § 63.7 is conducted.

(f) *Use of an alternative monitoring method—* (1) *General.* Until permission to use an alternative monitoring method has been granted by the Administrator under this paragraph, the owner or operator of an affected source remains subject to the requirements of this section and the relevant standard.

(2) After receipt and consideration of written application, the Administrator may approve alternatives to any monitoring methods or procedures of this part including, but not limited to, the following:

(i) Alternative monitoring requirements when installation of a CMS specified by a relevant standard would not provide accurate measurements due to liquid water or other interferences caused by substances within the effluent gases;

(ii) Alternative monitoring requirements when the affected source is infrequently operated;

(iii) Alternative monitoring requirements to accommodate CEMS that require additional measurements to correct for stack moisture conditions;

(iv) Alternative locations for installing CMS when the owner or operator can demonstrate that installation at alternate locations will enable accurate and representative measurements;

(v) Alternate methods for converting pollutant concentration measurements to units of the relevant standard;

(vi) Alternate procedures for performing daily checks of zero (low-level) and high-level drift that do not involve use of high-level gases or test cells;

(vii) Alternatives to the American Society for Testing and Materials (ASTM) test methods or sampling procedures specified by any relevant standard;

(viii) Alternative CMS that do not meet the design or performance requirements in this part, but adequately demonstrate a definite and consistent relationship between their measurements and the measurements of opacity by a system complying with the requirements as specified in the relevant standard. The Administrator may require that such demonstration be performed for each affected source; or

(ix) Alternative monitoring requirements when the effluent from a single affected source or the combined effluent from two or more affected sources is released to the atmosphere through more than one point.

(3) If the Administrator finds reasonable grounds to dispute the results obtained by an alternative monitoring method, requirement, or procedure, the Administrator may require the use of a method, requirement, or procedure specified in this section or in the relevant standard. If the results of the specified and alternative method, requirement, or procedure do not agree, the results obtained by the specified method, requirement, or procedure shall prevail.

(4)(i) *Request to use alternative monitoring method.* An owner or operator who wishes to use an alternative monitoring method shall submit an application to the Administrator as described in paragraph (f)(4)(ii) of this section, below. The application may be submitted at any time provided that the monitoring method is not used to demonstrate compliance with a relevant standard or other requirement. If the alternative monitoring method is to be used to demonstrate compliance with a relevant standard, the application shall be submitted not later than with the site-specific test plan required in § 63.7(c) (if requested) or with the site-specific performance evaluation plan (if requested) or at least 60 days before the performance evaluation is scheduled to begin.

(ii) The application shall contain a description of the proposed alternative monitoring system and a performance evaluation test plan, if required, as specified in paragraph (e)(3) of this section. In addition, the application shall include information justifying the owner or operator's request for an alternative monitoring method, such as the technical or economic infeasibility, or the impracticality, of the affected source using the required method.

(iii) The owner or operator may submit the information required in this paragraph well in ad-

§ 63.8

vance of the submittal dates specified in paragraph (f)(4)(i) above to ensure a timely review by the Administrator in order to meet the compliance demonstration date specified in this section or the relevant standard.

(5) *Approval of request to use alternative monitoring method.* (i) The Administrator will notify the owner or operator of approval or intention to deny approval of the request to use an alternative monitoring method within 30 calendar days after receipt of the original request and within 30 calendar days after receipt of any supplementary information that is submitted. Before disapproving any request to use an alternative monitoring method, the Administrator will notify the applicant of the Administrator's intention to disapprove the request together with—

(A) Notice of the information and findings on which the intended disapproval is based; and

(B) Notice of opportunity for the owner or operator to present additional information to the Administrator before final action on the request. At the time the Administrator notifies the applicant of his or her intention to disapprove the request, the Administrator will specify how much time the owner or operator will have after being notified of the intended disapproval to submit the additional information.

(ii) The Administrator may establish general procedures and criteria in a relevant standard to accomplish the requirements of paragraph (f)(5)(i) of this section.

(iii) If the Administrator approves the use of an alternative monitoring method for an affected source under paragraph (f)(5)(i) of this section, the owner or operator of such source shall continue to use the alternative monitoring method until he or she receives approval from the Administrator to use another monitoring method as allowed by § 63.8(f).

(6) *Alternative to the relative accuracy test.* An alternative to the relative accuracy test for CEMS specified in a relevant standard may be requested as follows:

(i) *Criteria for approval of alternative procedures.* An alternative to the test method for determining relative accuracy is available for affected sources with emission rates demonstrated to be less than 50 percent of the relevant standard. The owner or operator of an affected source may petition the Administrator under paragraph (f)(6)(ii) of this section to substitute the relative accuracy test in section 7 of Performance Specification 2 with the procedures in section 10 if the results of a performance test conducted according to the requirements in § 63.7, or other tests performed following the criteria in § 63.7, demonstrate that the emission rate of the pollutant of interest in the units of the relevant standard is less than 50 percent of the rel-

evant standard. For affected sources subject to emission limitations expressed as control efficiency levels, the owner or operator may petition the Administrator to substitute the relative accuracy test with the procedures in section 10 of Performance Specification 2 if the control device exhaust emission rate is less than 50 percent of the level needed to meet the control efficiency requirement. The alternative procedures do not apply if the CEMS is used continuously to determine compliance with the relevant standard.

(ii) *Petition to use alternative to relative accuracy test.* The petition to use an alternative to the relative accuracy test shall include a detailed description of the procedures to be applied, the location and the procedure for conducting the alternative, the concentration or response levels of the alternative relative accuracy materials, and the other equipment checks included in the alternative procedure(s). The Administrator will review the petition for completeness and applicability. The Administrator's determination to approve an alternative will depend on the intended use of the CEMS data and may require specifications more stringent than in Performance Specification 2.

(iii) *Rescission of approval to use alternative to relative accuracy test.* The Administrator will review the permission to use an alternative to the CEMS relative accuracy test and may rescind such permission if the CEMS data from a successful completion of the alternative relative accuracy procedure indicate that the affected source's emissions are approaching the level of the relevant standard. The criterion for reviewing the permission is that the collection of CEMS data shows that emissions have exceeded 70 percent of the relevant standard for any averaging period, as specified in the relevant standard. For affected sources subject to emission limitations expressed as control efficiency levels, the criterion for reviewing the permission is that the collection of CEMS data shows that exhaust emissions have exceeded 70 percent of the level needed to meet the control efficiency requirement for any averaging period, as specified in the relevant standard. The owner or operator of the affected source shall maintain records and determine the level of emissions relative to the criterion for permission to use an alternative for relative accuracy testing. If this criterion is exceeded, the owner or operator shall notify the Administrator within 10 days of such occurrence and include a description of the nature and cause of the increased emissions. The Administrator will review the notification and may rescind permission to use an alternative and require the owner or operator to conduct a relative accuracy test of the CEMS as specified in section 7 of Performance Specification 2.

(g) *Reduction of monitoring data.* (1) The owner or operator of each CMS shall reduce the monitoring data as specified in this paragraph. In addition, each relevant standard may contain additional requirements for reducing monitoring data. When additional requirements are specified in a relevant standard, the standard will identify any unnecessary or duplicated requirements in this paragraph that the owner or operator need not comply with.

(2) The owner or operator of each COMS shall reduce all data to 6-minute averages calculated from 36 or more data points equally spaced over each 6-minute period. Data from CEMS for measurement other than opacity, unless otherwise specified in the relevant standard, shall be reduced to 1-hour averages computed from four or more data points equally spaced over each 1-hour period, except during periods when calibration, quality assurance, or maintenance activities pursuant to provisions of this part are being performed. During these periods, a valid hourly average shall consist of at least two data points with each representing a 15-minute period. Alternatively, an arithmetic or integrated 1-hour average of CEMS data may be used. Time periods for averaging are defined in § 63.2.

(3) The data may be recorded in reduced or nonreduced form (e.g., ppm pollutant and percent O_2 or ng/J of pollutant).

(4) All emission data shall be converted into units of the relevant standard for reporting purposes using the conversion procedures specified in that standard. After conversion into units of the relevant standard, the data may be rounded to the same number of significant digits as used in that standard to specify the emission limit (e.g., rounded to the nearest 1 percent opacity).

(5) Monitoring data recorded during periods of unavoidable CMS breakdowns, out-of-control periods, repairs, maintenance periods, calibration checks, and zero (low-level) and high-level adjustments shall not be included in any data average computed under this part.

§ 63.9 Notification requirements.

(a) *Applicability and general information.* (1) The requirements in this section apply to owners and operators of affected sources that are subject to the provisions of this part, unless specified otherwise in a relevant standard.

(2) For affected sources that have been granted an extension of compliance under subpart D of this part, the requirements of this section do not apply to those sources while they are operating under such compliance extensions.

(3) If any State requires a notice that contains all the information required in a notification listed in this section, the owner or operator may send the Administrator a copy of the notice sent to the State to satisfy the requirements of this section for that notification.

(4)(i) Before a State has been delegated the authority to implement and enforce notification requirements established under this part, the owner or operator of an affected source in such State subject to such requirements shall submit notifications to the appropriate Regional Office of the EPA (to the attention of the Director of the Division indicated in the list of the EPA Regional Offices in § 63.13).

(ii) After a State has been delegated the authority to implement and enforce notification requirements established under this part, the owner or operator of an affected source in such State subject to such requirements shall submit notifications to the delegated State authority (which may be the same as the permitting authority). In addition, if the delegated (permitting) authority is the State, the owner or operator shall send a copy of each notification submitted to the State to the appropriate Regional Office of the EPA, as specified in paragraph (a)(4)(i) of this section. The Regional Office may waive this requirement for any notifications at its discretion.

(b) *Initial notifications.* (1)(i) The requirements of this paragraph apply to the owner or operator of an affected source when such source becomes subject to a relevant standard.

(ii) If an area source that otherwise would be subject to an emission standard or other requirement established under this part if it were a major source subsequently increases its emissions of hazardous air pollutants (or its potential to emit hazardous air pollutants) such that the source is a major source that is subject to the emission standard or other requirement, such source shall be subject to the notification requirements of this section.

(iii) Affected sources that are required under this paragraph to submit an initial notification may use the application for approval of construction or reconstruction under § 63.5(d) of this subpart, if relevant, to fulfill the initial notification requirements of this paragraph.

(2) The owner or operator of an affected source that has an initial startup before the effective date of a relevant standard under this part shall notify the Administrator in writing that the source is subject to the relevant standard. The notification, which shall be submitted not later than 120 calendar days after the effective date of the relevant standard (or within 120 calendar days after the source becomes subject to the relevant standard), shall provide the following information:

(i) The name and address of the owner or operator;

(ii) The address (i.e., physical location) of the affected source;

§ 63.9

(iii) An identification of the relevant standard, or other requirement, that is the basis of the notification and the source's compliance date;

(iv) A brief description of the nature, size, design, and method of operation of the source, including its operating design capacity and an identification of each point of emission for each hazardous air pollutant, or if a definitive identification is not yet possible, a preliminary identification of each point of emission for each hazardous air pollutant; and

(v) A statement of whether the affected source is a major source or an area source.

(3) The owner or operator of a new or reconstructed affected source, or a source that has been reconstructed such that it is an affected source, that has an initial startup after the effective date of a relevant standard under this part and for which an application for approval of construction or reconstruction is not required under § 63.5(d), shall notify the Administrator in writing that the source is subject to the relevant standard no later than 120 days after initial startup. The notification shall provide all the information required in paragraphs (b)(2)(i) through (b)(2)(v) of this section, delivered or postmarked with the notification required in paragraph (b)(5).

(4) The owner or operator of a new or reconstructed major affected source that has an initial startup after the effective date of a relevant standard under this part and for which an application for approval of construction or reconstruction is required under § 63.5(d) shall provide the following information in writing to the Administrator:

(i) A notification of intention to construct a new major affected source, reconstruct a major affected source, or reconstruct a major source such that the source becomes a major affected source with the application for approval of construction or reconstruction as specified in § 63.5(d)(1)(i);

(ii) A notification of the date when construction or reconstruction was commenced, submitted simultaneously with the application for approval of construction or reconstruction, if construction or reconstruction was commenced before the effective date of the relevant standard;

(iii) A notification of the date when construction or reconstruction was commenced, delivered or postmarked not later than 30 days after such date, if construction or reconstruction was commenced after the effective date of the relevant standard;

(iv) A notification of the anticipated date of startup of the source, delivered or postmarked not more than 60 days nor less than 30 days before such date; and

(v) A notification of the actual date of startup of the source, delivered or postmarked within 15 calendar days after that date.

(5) After the effective date of any relevant standard established by the Administrator under this part, whether or not an approved permit program is effective in the State in which an affected source is (or would be) located, an owner or operator who intends to construct a new affected source or reconstruct an affected source subject to such standard, or reconstruct a source such that it becomes an affected source subject to such standard, shall notify the Administrator, in writing, of the intended construction or reconstruction. The notification shall be submitted as soon as practicable before the construction or reconstruction is planned to commence (but no sooner than the effective date of the relevant standard) if the construction or reconstruction commences after the effective date of a relevant standard promulgated in this part. The notification shall be submitted as soon as practicable before startup but no later than 60 days after the effective date of a relevant standard promulgated in this part if the construction or reconstruction had commenced and initial startup had not occurred before the standard's effective date. The notification shall include all the information required for an application for approval of construction or reconstruction as specified in § 63.5(d). For major sources, the application for approval of construction or reconstruction may be used to fulfill the requirements of this paragraph.

(c) *Request for extension of compliance.* If the owner or operator of an affected source cannot comply with a relevant standard by the applicable compliance date for that source, or if the owner or operator has installed BACT or technology to meet LAER consistent with § 63.6(i)(5) of this subpart, he/she may submit to the Administrator (or the State with an approved permit program) a request for an extension of compliance as specified in § 63.6(i)(4) through § 63.6(i)(6).

(d) *Notification that source is subject to special compliance requirements.* An owner or operator of a new source that is subject to special compliance requirements as specified in § 63.6(b)(3) and § 63.6(b)(4) shall notify the Administrator of his/her compliance obligations not later than the notification dates established in paragraph (b) of this section for new sources that are not subject to the special provisions.

(e) *Notification of performance test.* The owner or operator of an affected source shall notify the Administrator in writing of his or her intention to conduct a performance test at least 60 calendar days before the performance test is scheduled to begin to allow the Administrator to review and approve the site-specific test plan required under § 63.7(c), if requested by the Administrator, and to have an observer present during the test.

(f) *Notification of opacity and visible emission observations.* The owner or operator of an affected source shall notify the Administrator in writing of the anticipated date for conducting the opacity or visible emission observations specified in § 63.6(h)(5), if such observations are required for the source by a relevant standard. The notification shall be submitted with the notification of the performance test date, as specified in paragraph (e) of this section, or if no performance test is required or visibility or other conditions prevent the opacity or visible emission observations from being conducted concurrently with the initial performance test required under § 63.7, the owner or operator shall deliver or postmark the notification not less than 30 days before the opacity or visible emission observations are scheduled to take place.

(g) *Additional notification requirements for sources with continuous monitoring systems.* The owner or operator of an affected source required to use a CMS by a relevant standard shall furnish the Administrator written notification as follows:

(1) A notification of the date the CMS performance evaluation under § 63.8(e) is scheduled to begin, submitted simultaneously with the notification of the performance test date required under § 63.7(b). If no performance test is required, or if the requirement to conduct a performance test has been waived for an affected source under § 63.7(h), the owner or operator shall notify the Administrator in writing of the date of the performance evaluation at least 60 calendar days before the evaluation is scheduled to begin;

(2) A notification that COMS data results will be used to determine compliance with the applicable opacity emission standard during a performance test required by § 63.7 in lieu of Method 9 or other opacity emissions test method data, as allowed by § 63.6(h)(7)(ii), if compliance with an opacity emission standard is required for the source by a relevant standard. The notification shall be submitted at least 60 calendar days before the performance test is scheduled to begin; and

(3) A notification that the criterion necessary to continue use of an alternative to relative accuracy testing, as provided by § 63.8(f)(6), has been exceeded. The notification shall be delivered or postmarked not later than 10 days after the occurrence of such exceedance, and it shall include a description of the nature and cause of the increased emissions.

(h) *Notification of compliance status.* (1) The requirements of paragraphs (h)(2) through (h)(4) of this section apply when an affected source becomes subject to a relevant standard.

(2)(i) Before a title V permit has been issued to the owner or operator of an affected source, and each time a notification of compliance status is required under this part, the owner or operator of such source shall submit to the Administrator a notification of compliance status, signed by the re-

sponsible official who shall certify its accuracy, attesting to whether the source has complied with the relevant standard. The notification shall list—

(A) The methods that were used to determine compliance;

(B) The results of any performance tests, opacity or visible emission observations, continuous monitoring system (CMS) performance evaluations, and/or other monitoring procedures or methods that were conducted;

(C) The methods that will be used for determining continuing compliance, including a description of monitoring and reporting requirements and test methods;

(D) The type and quantity of hazardous air pollutants emitted by the source (or surrogate pollutants if specified in the relevant standard), reported in units and averaging times and in accordance with the test methods specified in the relevant standard;

(E) An analysis demonstrating whether the affected source is a major source or an area source (using the emissions data generated for this notification);

(F) A description of the air pollution control equipment (or method) for each emission point, including each control device (or method) for each hazardous air pollutant and the control efficiency (percent) for each control device (or method); and

(G) A statement by the owner or operator of the affected existing, new, or reconstructed source as to whether the source has complied with the relevant standard or other requirements.

(ii) The notification shall be sent before the close of business on the 60th day following the completion of the relevant compliance demonstration activity specified in the relevant standard (unless a different reporting period is specified in a relevant standard, in which case the letter shall be sent before the close of business on the day the report of the relevant testing or monitoring results is required to be delivered or postmarked). For example, the notification shall be sent before close of business on the 60th (or other required) day following completion of the initial performance test and again before the close of business on the 60th (or other required) day following the completion of any subsequent required performance test. If no performance test is required but opacity or visible emission observations are required to demonstrate compliance with an opacity or visible emission standard under this part, the notification of compliance status shall be sent before close of business on the 30th day following the completion of opacity or visible emission observations.

(3) After a title V permit has been issued to the owner or operator of an affected source, the owner or operator of such source shall comply with all requirements for compliance status reports con-

§ 63.10

tained in the source's title V permit, including reports required under this part. After a title V permit has been issued to the owner or operator of an affected source, and each time a notification of compliance status is required under this part, the owner or operator of such source shall submit the notification of compliance status to the appropriate permitting authority following completion of the relevant compliance demonstration activity specified in the relevant standard.

(4) [Reserved]

(5) If an owner or operator of an affected source submits estimates or preliminary information in the application for approval of construction or reconstruction required in § 63.5(d) in place of the actual emissions data or control efficiencies required in paragraphs (d)(1)(ii)(H) and (d)(2) of § 63.5, the owner or operator shall submit the actual emissions data and other correct information as soon as available but no later than with the initial notification of compliance status required in this section.

(6) Advice on a notification of compliance status may be obtained from the Administrator.

(i) *Adjustment to time periods or postmark deadlines for submittal and review of required communications.* (1)(i) Until an adjustment of a time period or postmark deadline has been approved by the Administrator under paragraphs (i)(2) and (i)(3) of this section, the owner or operator of an affected source remains strictly subject to the requirements of this part.

(ii) An owner or operator shall request the adjustment provided for in paragraphs (i)(2) and (i)(3) of this section each time he or she wishes to change an applicable time period or postmark deadline specified in this part.

(2) Notwithstanding time periods or postmark deadlines specified in this part for the submittal of information to the Administrator by an owner or operator, or the review of such information by the Administrator, such time periods or deadlines may be changed by mutual agreement between the owner or operator and the Administrator. An owner or operator who wishes to request a change in a time period or postmark deadline for a particular requirement shall request the adjustment in writing as soon as practicable before the subject activity is required to take place. The owner or operator shall include in the request whatever information he or she considers useful to convince the Administrator that an adjustment is warranted.

(3) If, in the Administrator's judgment, an owner or operator's request for an adjustment to a particular time period or postmark deadline is warranted, the Administrator will approve the adjustment. The Administrator will notify the owner or operator in writing of approval or disapproval of the request for an adjustment within 15 cal-

endar days of receiving sufficient information to evaluate the request.

(4) If the Administrator is unable to meet a specified deadline, he or she will notify the owner or operator of any significant delay and inform the owner or operator of the amended schedule.

(j) *Change in information already provided.* Any change in the information already provided under this section shall be provided to the Administrator in writing within 15 calendar days after the change.

§ 63.10 Recordkeeping and reporting requirements.

(a) *Applicability and general information.* (1) The requirements of this section apply to owners or operators of affected sources who are subject to the provisions of this part, unless specified otherwise in a relevant standard.

(2) For affected sources that have been granted an extension of compliance under subpart D of this part, the requirements of this section do not apply to those sources while they are operating under such compliance extensions.

(3) If any State requires a report that contains all the information required in a report listed in this section, an owner or operator may send the Administrator a copy of the report sent to the State to satisfy the requirements of this section for that report.

(4)(i) Before a State has been delegated the authority to implement and enforce recordkeeping and reporting requirements established under this part, the owner or operator of an affected source in such State subject to such requirements shall submit reports to the appropriate Regional Office of the EPA (to the attention of the Director of the Division indicated in the list of the EPA Regional Offices in § 63.13).

(ii) After a State has been delegated the authority to implement and enforce recordkeeping and reporting requirements established under this part, the owner or operator of an affected source in such State subject to such requirements shall submit reports to the delegated State authority (which may be the same as the permitting authority). In addition, if the delegated (permitting) authority is the State, the owner or operator shall send a copy of each report submitted to the State to the appropriate Regional Office of the EPA, as specified in paragraph (a)(4)(i) of this section. The Regional Office may waive this requirement for any reports at its discretion.

(5) If an owner or operator of an affected source in a State with delegated authority is required to submit periodic reports under this part to the State, and if the State has an established timeline for the submission of periodic reports that is consistent with the reporting frequency(ies)

specified for such source under this part, the owner or operator may change the dates by which periodic reports under this part shall be submitted (without changing the frequency of reporting) to be consistent with the State's schedule by mutual agreement between the owner or operator and the State. For each relevant standard established pursuant to section 112 of the Act, the allowance in the previous sentence applies in each State beginning 1 year after the affected source's compliance date for that standard. Procedures governing the implementation of this provision are specified in § 63.9(i).

(6) If an owner or operator supervises one or more stationary sources affected by more than one standard established pursuant to section 112 of the Act, he/she may arrange by mutual agreement between the owner or operator and the Administrator (or the State permitting authority) a common schedule on which periodic reports required for each source shall be submitted throughout the year. The allowance in the previous sentence applies in each State beginning 1 year after the latest compliance date for any relevant standard established pursuant to section 112 of the Act for any such affected source(s). Procedures governing the implementation of this provision are specified in § 63.9(i).

(7) If an owner or operator supervises one or more stationary sources affected by standards established pursuant to section 112 of the Act (as amended November 15, 1990) and standards set under part 60, part 61, or both such parts of this chapter, he/she may arrange by mutual agreement between the owner or operator and the Administrator (or the State permitting authority) a common schedule on which periodic reports required by each relevant (i.e., applicable) standard shall be submitted throughout the year. The allowance in the previous sentence applies in each State beginning 1 year after the stationary source is required to be in compliance with the relevant section 112 standard, or 1 year after the stationary source is required to be in compliance with the applicable part 60 or part 61 standard, whichever is latest. Procedures governing the implementation of this provision are specified in § 63.9(i).

(b) *General recordkeeping requirements.* (1) The owner or operator of an affected source subject to the provisions of this part shall maintain files of all information (including all reports and notifications) required by this part recorded in a form suitable and readily available for expeditious inspection and review. The files shall be retained for at least 5 years following the date of each occurrence, measurement, maintenance, corrective action, report, or record. At a minimum, the most recent 2 years of data shall be retained on site. The remaining 3 years of data may be retained off

site. Such files may be maintained on microfilm, on a computer, on computer floppy disks, on magnetic tape disks, or on microfiche.

(2) The owner or operator of an affected source subject to the provisions of this part shall maintain relevant records for such source of—

(i) The occurrence and duration of each startup, shutdown, or malfunction of operation (i.e., process equipment);

(ii) The occurrence and duration of each malfunction of the air pollution control equipment;

(iii) All maintenance performed on the air pollution control equipment;

(iv) Actions taken during periods of startup, shutdown, and malfunction (including corrective actions to restore malfunctioning process and air pollution control equipment to its normal or usual manner of operation) when such actions are different from the procedures specified in the affected source's startup, shutdown, and malfunction plan (see § 63.6(e)(3));

(v) All information necessary to demonstrate conformance with the affected source's startup, shutdown, and malfunction plan (see § 63.6(e)(3)) when all actions taken during periods of startup, shutdown, and malfunction (including corrective actions to restore malfunctioning process and air pollution control equipment to its normal or usual manner of operation) are consistent with the procedures specified in such plan. (The information needed to demonstrate conformance with the startup, shutdown, and malfunction plan may be recorded using a "checklist," or some other effective form of recordkeeping, in order to minimize the recordkeeping burden for conforming events);

(vi) Each period during which a CMS is malfunctioning or inoperative (including out-of-control periods);

(vii) All required measurements needed to demonstrate compliance with a relevant standard (including, but not limited to, 15-minute averages of CMS data, raw performance testing measurements, and raw performance evaluation measurements, that support data that the source is required to report);

(viii) All results of performance tests, CMS performance evaluations, and opacity and visible emission observations;

(ix) All measurements as may be necessary to determine the conditions of performance tests and performance evaluations;

(x) All CMS calibration checks;

(xi) All adjustments and maintenance performed on CMS;

(xii) Any information demonstrating whether a source is meeting the requirements for a waiver of recordkeeping or reporting requirements under this part, if the source has been granted a waiver under paragraph (f) of this section;

§ 63.10

(xiii) All emission levels relative to the criterion for obtaining permission to use an alternative to the relative accuracy test, if the source has been granted such permission under § 63.8(f)(6); and

(xiv) All documentation supporting initial notifications and notifications of compliance status under § 63.9.

(3) Recordkeeping requirement for applicability determinations. If an owner or operator determines that his or her stationary source that emits (or has the potential to emit, without considering controls) one or more hazardous air pollutants is not subject to a relevant standard or other requirement established under this part, the owner or operator shall keep a record of the applicability determination on site at the source for a period of 5 years after the determination, or until the source changes its operations to become an affected source, whichever comes first. The record of the applicability determination shall include an analysis (or other information) that demonstrates why the owner or operator believes the source is unaffected (e.g., because the source is an area source). The analysis (or other information) shall be sufficiently detailed to allow the Administrator to make a finding about the source's applicability status with regard to the relevant standard or other requirement. If relevant, the analysis shall be performed in accordance with requirements established in subparts of this part for this purpose for particular categories of stationary sources. If relevant, the analysis should be performed in accordance with EPA guidance materials published to assist sources in making applicability determinations under section 112, if any.

(c) *Additional recordkeeping requirements for sources with continuous monitoring systems.* In addition to complying with the requirements specified in paragraphs (b)(1) and (b)(2) of this section, the owner or operator of an affected source required to install a CMS by a relevant standard shall maintain records for such source of—

(1) All required CMS measurements (including monitoring data recorded during unavoidable CMS breakdowns and out-of-control periods);

(2)–(4) [Reserved]

(5) The date and time identifying each period during which the CMS was inoperative except for zero (low-level) and high-level checks;

(6) The date and time identifying each period during which the CMS was out of control, as defined in § 63.8(c)(7);

(7) The specific identification (i.e., the date and time of commencement and completion) of each period of excess emissions and parameter monitoring exceedances, as defined in the relevant standard(s), that occurs during startups, shutdowns, and malfunctions of the affected source;

(8) The specific identification (i.e., the date and time of commencement and completion) of each

time period of excess emissions and parameter monitoring exceedances, as defined in the relevant standard(s), that occurs during periods other than startups, shutdowns, and malfunctions of the affected source;

(9) [Reserved]

(10) The nature and cause of any malfunction (if known);

(11) The corrective action taken or preventive measures adopted;

(12) The nature of the repairs or adjustments to the CMS that was inoperative or out of control;

(13) The total process operating time during the reporting period; and

(14) All procedures that are part of a quality control program developed and implemented for CMS under § 63.8(d).

(15) In order to satisfy the requirements of paragraphs (c)(10) through (c)(12) of this section and to avoid duplicative recordkeeping efforts, the owner or operator may use the affected source's startup, shutdown, and malfunction plan or records kept to satisfy the recordkeeping requirements of the startup, shutdown, and malfunction plan specified in § 63.6(e), provided that such plan and records adequately address the requirements of paragraphs (c)(10) through (c)(12).

(d) *General reporting requirements.* (1) Notwithstanding the requirements in this paragraph or paragraph (e) of this section, the owner or operator of an affected source subject to reporting requirements under this part shall submit reports to the Administrator in accordance with the reporting requirements in the relevant standard(s).

(2) *Reporting results of performance tests.* Before a title V permit has been issued to the owner or operator of an affected source, the owner or operator shall report the results of any performance test under § 63.7 to the Administrator. After a title V permit has been issued to the owner or operator of an affected source, the owner or operator shall report the results of a required performance test to the appropriate permitting authority. The owner or operator of an affected source shall report the results of the performance test to the Administrator (or the State with an approved permit program) before the close of business on the 60th day following the completion of the performance test, unless specified otherwise in a relevant standard or as approved otherwise in writing by the Administrator. The results of the performance test shall be submitted as part of the notification of compliance status required under § 63.9(h).

(3) *Reporting results of opacity or visible emission observations.* The owner or operator of an affected source required to conduct opacity or visible emission observations by a relevant standard shall report the opacity or visible emission results (produced using Test Method 9 or Test Method

22, or an alternative to these test methods) along with the results of the performance test required under § 63.7. If no performance test is required, or if visibility or other conditions prevent the opacity or visible emission observations from being conducted concurrently with the performance test required under § 63.7, the owner or operator shall report the opacity or visible emission results before the close of business on the 30th day following the completion of the opacity or visible emission observations.

(4) *Progress reports.* The owner or operator of an affected source who is required to submit progress reports as a condition of receiving an extension of compliance under § 63.6(i) shall submit such reports to the Administrator (or the State with an approved permit program) by the dates specified in the written extension of compliance.

(5)(i) *Periodic startup, shutdown, and malfunction reports.* If actions taken by an owner or operator during a startup, shutdown, or malfunction of an affected source (including actions taken to correct a malfunction) are consistent with the procedures specified in the source's startup, shutdown, and malfunction plan [see § 63.6(e)(3)], the owner or operator shall state such information in a startup, shutdown, and malfunction report. Reports shall only be required if a startup, shutdown, or malfunction occurred during the reporting period. The startup, shutdown, and malfunction report shall consist of a letter, containing the name, title, and signature of the owner or operator or other responsible official who is certifying its accuracy, that shall be submitted to the Administrator semiannually (or on a more frequent basis if specified otherwise in a relevant standard or as established otherwise by the permitting authority in the source's title V permit). The startup, shutdown, and malfunction report shall be delivered or postmarked by the 30th day following the end of each calendar half (or other calendar reporting period, as appropriate). If the owner or operator is required to submit excess emissions and continuous monitoring system performance (or other periodic) reports under this part, the startup, shutdown, and malfunction reports required under this paragraph may be submitted simultaneously with the excess emissions and continuous monitoring system performance (or other) reports. If startup, shutdown, and malfunction reports are submitted with excess emissions and continuous monitoring system performance (or other periodic) reports, and the owner or operator receives approval to reduce the frequency of reporting for the latter under paragraph (e) of this section, the frequency of reporting for the startup, shutdown, and malfunction reports also may be reduced if the Administrator does not object to the intended change. The procedures to implement the allowance in the preceding

sentence shall be the same as the procedures specified in paragraph (e)(3) of this section.

(ii) *Immediate startup, shutdown, and malfunction reports.* Notwithstanding the allowance to reduce the frequency of reporting for periodic startup, shutdown, and malfunction reports under paragraph (d)(5)(i) of this section, any time an action taken by an owner or operator during a startup, shutdown, or malfunction (including actions taken to correct a malfunction) is not consistent with the procedures specified in the affected source's startup, shutdown, and malfunction plan, the owner or operator shall report the actions taken for that event within 2 working days after commencing actions inconsistent with the plan followed by a letter within 7 working days after the end of the event. The immediate report required under this paragraph shall consist of a telephone call (or facsimile (FAX) transmission) to the Administrator within 2 working days after commencing actions inconsistent with the plan, and it shall be followed by a letter, delivered or postmarked within 7 working days after the end of the event, that contains the name, title, and signature of the owner or operator or other responsible official who is certifying its accuracy, explaining the circumstances of the event, the reasons for not following the startup, shutdown, and malfunction plan, and whether any excess emissions and/or parameter monitoring exceedances are believed to have occurred. Notwithstanding the requirements of the previous sentence, after the effective date of an approved permit program in the State in which an affected source is located, the owner or operator may make alternative reporting arrangements, in advance, with the permitting authority in that State. Procedures governing the arrangement of alternative reporting requirements under this paragraph are specified in § 63.9(i).

(e) *Additional reporting requirements for sources with continuous monitoring systems*—(1) *General.* When more than one CEMS is used to measure the emissions from one affected source (e.g., multiple breechings, multiple outlets), the owner or operator shall report the results as required for each CEMS.

(2) *Reporting results of continuous monitoring system performance evaluations.* (i) The owner or operator of an affected source required to install a CMS by a relevant standard shall furnish the Administrator a copy of a written report of the results of the CMS performance evaluation, as required under § 63.8(e), simultaneously with the results of the performance test required under § 63.7, unless otherwise specified in the relevant standard.

(ii) The owner or operator of an affected source using a COMS to determine opacity compliance during any performance test required under § 63.7 and described in § 63.6(d)(6) shall furnish the Ad-

§ 63.10

ministrator two or, upon request, three copies of a written report of the results of the COMS performance evaluation conducted under § 63.8(e). The copies shall be furnished at least 15 calendar days before the performance test required under § 63.7 is conducted.

(3) *Excess emissions and continuous monitoring system performance report and summary report.* (i) Excess emissions and parameter monitoring exceedances are defined in relevant standards. The owner or operator of an affected source required to install a CMS by a relevant standard shall submit an excess emissions and continuous monitoring system performance report and/or a summary report to the Administrator semiannually, except when—

(A) More frequent reporting is specifically required by a relevant standard;

(B) The Administrator determines on a case-by-case basis that more frequent reporting is necessary to accurately assess the compliance status of the source; or

(C) The CMS data are to be used directly for compliance determination and the source experienced excess emissions, in which case quarterly reports shall be submitted. Once a source reports excess emissions, the source shall follow a quarterly reporting format until a request to reduce reporting frequency under paragraph (e)(3)(ii) of this section is approved.

(ii) *Request to reduce frequency of excess emissions and continuous monitoring system performance reports.* Notwithstanding the frequency of reporting requirements specified in paragraph (e)(3)(i) of this section, an owner or operator who is required by a relevant standard to submit excess emissions and continuous monitoring system performance (and summary) reports on a quarterly (or more frequent) basis may reduce the frequency of reporting for that standard to semiannual if the following conditions are met:

(A) For 1 full year (e.g., 4 quarterly or 12 monthly reporting periods) the affected source's excess emissions and continuous monitoring system performance reports continually demonstrate that the source is in compliance with the relevant standard;

(B) The owner or operator continues to comply with all recordkeeping and monitoring requirements specified in this subpart and the relevant standard; and

(C) The Administrator does not object to a reduced frequency of reporting for the affected source, as provided in paragraph (e)(3)(iii) of this section.

(iii) The frequency of reporting of excess emissions and continuous monitoring system performance (and summary) reports required to comply with a relevant standard may be reduced only after

the owner or operator notifies the Administrator in writing of his or her intention to make such a change and the Administrator does not object to the intended change. In deciding whether to approve a reduced frequency of reporting, the Administrator may review information concerning the source's entire previous performance history during the 5-year recordkeeping period prior to the intended change, including performance test results, monitoring data, and evaluations of an owner or operator's conformance with operation and maintenance requirements. Such information may be used by the Administrator to make a judgment about the source's potential for noncompliance in the future. If the Administrator disapproves the owner or operator's request to reduce the frequency of reporting, the Administrator will notify the owner or operator in writing within 45 days after receiving notice of the owner or operator's intention. The notification from the Administrator to the owner or operator will specify the grounds on which the disapproval is based. In the absence of a notice of disapproval within 45 days, approval is automatically granted.

(iv) As soon as CMS data indicate that the source is not in compliance with any emission limitation or operating parameter specified in the relevant standard, the frequency of reporting shall revert to the frequency specified in the relevant standard, and the owner or operator shall submit an excess emissions and continuous monitoring system performance (and summary) report for the noncomplying emission points at the next appropriate reporting period following the noncomplying event. After demonstrating ongoing compliance with the relevant standard for another full year, the owner or operator may again request approval from the Administrator to reduce the frequency of reporting for that standard, as provided for in paragraphs (e)(3)(ii) and (e)(3)(iii) of this section.

(v) *Content and submittal dates for excess emissions and monitoring system performance reports.* All excess emissions and monitoring system performance reports and all summary reports, if required, shall be delivered or postmarked by the 30th day following the end of each calendar half or quarter, as appropriate. Written reports of excess emissions or exceedances of process or control system parameters shall include all the information required in paragraphs (c)(5) through (c)(13) of this section, in § 63.8(c)(7) and § 63.8(c)(8), and in the relevant standard, and they shall contain the name, title, and signature of the responsible official who is certifying the accuracy of the report. When no excess emissions or exceedances of a parameter have occurred, or a CMS has not been inoperative, out of control, repaired, or adjusted, such information shall be stated in the report.

(vi) *Summary report.* As required under paragraphs (e)(3)(vii) and (e)(3)(viii) of this section, one summary report shall be submitted for the hazardous air pollutants monitored at each affected source (unless the relevant standard specifies that more than one summary report is required, e.g., one summary report for each hazardous air pollutant monitored). The summary report shall be entitled "Summary Report—Gaseous and Opacity Excess Emission and Continuous Monitoring System Performance" and shall contain the following information:

(A) The company name and address of the affected source;

(B) An identification of each hazardous air pollutant monitored at the affected source;

(C) The beginning and ending dates of the reporting period;

(D) A brief description of the process units;

(E) The emission and operating parameter limitations specified in the relevant standard(s);

(F) The monitoring equipment manufacturer(s) and model number(s);

(G) The date of the latest CMS certification or audit;

(H) The total operating time of the affected source during the reporting period;

(I) An emission data summary (or similar summary if the owner or operator monitors control system parameters), including the total duration of excess emissions during the reporting period (recorded in minutes for opacity and hours for gases), the total duration of excess emissions expressed as a percent of the total source operating time during that reporting period, and a breakdown of the total duration of excess emissions during the reporting period into those that are due to startup/shutdown, control equipment problems, process problems, other known causes, and other unknown causes;

(J) A CMS performance summary (or similar summary if the owner or operator monitors control system parameters), including the total CMS downtime during the reporting period (recorded in minutes for opacity and hours for gases), the total duration of CMS downtime expressed as a percent of the total source operating time during that reporting period, and a breakdown of the total CMS downtime during the reporting period into periods that are due to monitoring equipment malfunctions, nonmonitoring equipment malfunctions, quality assurance/quality control calibrations, other known causes, and other unknown causes;

(K) A description of any changes in CMS, processes, or controls since the last reporting period;

(L) The name, title, and signature of the responsible official who is certifying the accuracy of the report; and

(M) The date of the report.

(vii) If the total duration of excess emissions or process or control system parameter exceedances for the reporting period is less than 1 percent of the total operating time for the reporting period, and CMS downtime for the reporting period is less than 5 percent of the total operating time for the reporting period, only the summary report shall be submitted, and the full excess emissions and continuous monitoring system performance report need not be submitted unless required by the Administrator.

(viii) If the total duration of excess emissions or process or control system parameter exceedances for the reporting period is 1 percent or greater of the total operating time for the reporting period, or the total CMS downtime for the reporting period is 5 percent or greater of the total operating time for the reporting period, both the summary report and the excess emissions and continuous monitoring system performance report shall be submitted.

(4) *Reporting continuous opacity monitoring system data produced during a performance test.* The owner or operator of an affected source required to use a COMS shall record the monitoring data produced during a performance test required under § 63.7 and shall furnish the Administrator a written report of the monitoring results. The report of COMS data shall be submitted simultaneously with the report of the performance test results required in paragraph (d)(2) of this section.

(f) *Waiver of recordkeeping or reporting requirements.* (1) Until a waiver of a recordkeeping or reporting requirement has been granted by the Administrator under this paragraph, the owner or operator of an affected source remains subject to the requirements of this section.

(2) Recordkeeping or reporting requirements may be waived upon written application to the Administrator if, in the Administrator's judgment, the affected source is achieving the relevant standard(s), or the source is operating under an extension of compliance, or the owner or operator has requested an extension of compliance and the Administrator is still considering that request.

(3) If an application for a waiver of recordkeeping or reporting is made, the application shall accompany the request for an extension of compliance under § 63.6(i), any required compliance progress report or compliance status report required under this part (such as under § 63.6(i) and § 63.9(h)) or in the source's title V permit, or an excess emissions and continuous monitoring system performance report required under paragraph (e) of this section, whichever is applicable. The application shall include whatever information the owner or operator considers useful to convince the Administrator that a waiver of recordkeeping or reporting is warranted.

§ 63.11

(4) The Administrator will approve or deny a request for a waiver of recordkeeping or reporting requirements under this paragraph when he/she—

(i) Approves or denies an extension of compliance; or

(ii) Makes a determination of compliance following the submission of a required compliance status report or excess emissions and continuous monitoring systems performance report; or

(iii) Makes a determination of suitable progress towards compliance following the submission of a compliance progress report, whichever is applicable.

(5) A waiver of any recordkeeping or reporting requirement granted under this paragraph may be conditioned on other recordkeeping or reporting requirements deemed necessary by the Administrator.

(6) Approval of any waiver granted under this section shall not abrogate the Administrator's authority under the Act or in any way prohibit the Administrator from later canceling the waiver. The cancellation will be made only after notice is given to the owner or operator of the affected source.

§ 63.11 Control device requirements.

(a) *Applicability.* This section contains requirements for control devices used to comply with provisions in relevant standards. These requirements apply only to affected sources covered by relevant standards referring directly or indirectly to this section.

(b) *Flares.* (1) Owners or operators using flares to comply with the provisions of this part shall monitor these control devices to assure that they are operated and maintained in conformance with their designs. Applicable subparts will provide provisions stating how owners or operators using flares shall monitor these control devices.

(2) Flares shall be steam-assisted, air-assisted, or non-assisted.

(3) Flares shall be operated at all times when emissions may be vented to them.

(4) Flares shall be designed for and operated with no visible emissions, except for periods not to exceed a total of 5 minutes during any 2 consecutive hours. Test Method 22 in appendix A of part 60 of this chapter shall be used to determine the compliance of flares with the visible emission provisions of this part. The observation period is 2 hours and shall be used according to Method 22.

(5) Flares shall be operated with a flame present at all times. The presence of a flare pilot flame shall be monitored using a thermocouple or any other equivalent device to detect the presence of a flame.

(6) Flares shall be used only with the net heating value of the gas being combusted at 11.2 MJ/ scm (300 Btu/scf) or greater if the flare is steam-assisted or air-assisted; or with the net heating value of the gas being combusted at 7.45 MJ/scm (200 Btu/scf) or greater if the flare is non-assisted. The net heating value of the gas being combusted in a flare shall be calculated using the following equation:

ER16MR94.000

Where:

H_T=Net heating value of the sample, MJ/scm; where the net enthalpy per mole of offgas is based on combustion at 25 °C and 760 mm Hg, but the standard temperature for determining the volume corresponding to one mole is 20 °C.

K=Constant =

ER16MR94.001

where the standard temperature for (g-mole/scm) is 20 °C.

C_i=Concentration of sample component i in ppmv on a wet basis, as measured for organics by Test Method 18 and measured for hydrogen and carbon monoxide by American Society for Testing and Materials (ASTM) D1946–77 (incorporated by reference as specified in § 63.14).

H_i=Net heat of combustion of sample component i, kcal/g-mole at 25 °C and 760 mm Hg. The heats of combustion may be determined using ASTM D2382–76 (incorporated by reference as specified in § 63.14) if published values are not available or cannot be calculated.

n=Number of sample components.

(7)(i) Steam-assisted and nonassisted flares shall be designed for and operated with an exit velocity less than 18.3 m/sec (60 ft/sec), except as provided in paragraphs (b)(7)(ii) and (b)(7)(iii) of this section. The actual exit velocity of a flare shall be determined by dividing by the volumetric flow rate of gas being combusted (in units of emission standard temperature and pressure), as determined by Test Method 2, 2A, 2C, or 2D in appendix A to 40 CFR part 60 of this chapter, as appropriate, by the unobstructed (free) cross-sectional area of the flare tip.

(ii) Steam-assisted and nonassisted flares designed for and operated with an exit velocity, as determined by the method specified in paragraph (b)(7)(i) of this section, equal to or greater than 18.3 m/sec (60 ft/sec) but less than 122 m/sec (400 ft/sec), are allowed if the net heating value of the gas being combusted is greater than 37.3 MJ/scm (1,000 Btu/scf).

(iii) Steam-assisted and nonassisted flares designed for and operated with an exit velocity, as determined by the method specified in paragraph (b)(7)(i) of this section, less than the velocity V_{max}, as determined by the method specified in this paragraph, but less than 122 m/sec (400 ft/sec) are allowed. The maximum permitted velocity, V_{max}, for flares complying with this paragraph shall be determined by the following equation:

$$Log_{10}(V_{max})=(H_T+28.8)/31.7$$

Where:

V_{max}=Maximum permitted velocity, m/sec.
28.8=Constant.
31.7=Constant.
H_T=The net heating value as determined in paragraph (b)(6) of this section.

(8) Air-assisted flares shall be designed and operated with an exit velocity less than the velocity V_{max}. The maximum permitted velocity, V_{max}, for air-assisted flares shall be determined by the following equation:

$$V_{max}=8.706+0.7084(H_T)$$

Where:

V_{max}=Maximum permitted velocity, m/sec.
8.706=Constant.
0.7084=Constant.
H_T=The net heating value as determined in paragraph (b)(6) of this section.

§ 63.12 State authority and delegations.

(a) The provisions of this part shall not be construed in any manner to preclude any State or political subdivision thereof from—

(1) Adopting and enforcing any standard, limitation, prohibition, or other regulation applicable to an affected source subject to the requirements of this part, provided that such standard, limitation, prohibition, or regulation is not less stringent than any requirement applicable to such source established under this part;

(2) Requiring the owner or operator of an affected source to obtain permits, licenses, or approvals prior to initiating construction, reconstruction, modification, or operation of such source; or

(3) Requiring emission reductions in excess of those specified in subpart D of this part as a condition for granting the extension of compliance authorized by section 112(i)(5) of the Act.

(b)(1) Section 112(l) of the Act directs the Administrator to delegate to each State, when appropriate, the authority to implement and enforce standards and other requirements pursuant to section 112 for stationary sources located in that State. Because of the unique nature of radioactive material, delegation of authority to implement and enforce standards that control radionuclides may require separate approval.

(2) Subpart E of this part establishes procedures consistent with section 112(l) for the approval of State rules or programs to implement and enforce applicable Federal rules promulgated under the authority of section 112. Subpart E also establishes

procedures for the review and withdrawal of section 112 implementation and enforcement authorities granted through a section 112(l) approval.

(c) All information required to be submitted to the EPA under this part also shall be submitted to the appropriate State agency of any State to which authority has been delegated under section 112(l) of the Act, provided that each specific delegation may exempt sources from a certain Federal or State reporting requirement. The Administrator may permit all or some of the information to be submitted to the appropriate State agency only, instead of to the EPA and the State agency.

§63.13 Addresses of State air pollution control agencies and EPA Regional Offices.

(a) All requests, reports, applications, submittals, and other communications to the Administrator pursuant to this part shall be submitted to the appropriate Regional Office of the U.S. Environmental Protection Agency indicated in the following list of EPA Regional Offices.

EPA Region I (Connecticut, Maine, Massachusetts, New Hampshire, Rhode Island, Vermont), Director, Air, Pesticides and Toxics Division, J.F.K. Federal Building, Boston, MA 02203–2211.

EPA Region II (New Jersey, New York, Puerto Rico, Virgin Islands), Director, Air and Waste Management Division, 26 Federal Plaza, New York, NY 10278.

EPA Region III (Delaware, District of Columbia, Maryland, Pennsylvania, Virginia, West Virginia), Director, Air, Radiation and Toxics Division, 841 Chestnut Street, Philadelphia, PA 19107.

EPA Region IV (Alabama, Florida, Georgia, Kentucky, Mississippi, North Carolina, South Carolina, Tennessee), Director, Air, Pesticides and Toxics, Management Division, 345 Courtland Street, NE., Atlanta, GA 30365.

EPA Region V (Illinois, Indiana, Michigan, Minnesota, Ohio, Wisconsin), Director, Air and Radiation Division, 77 West Jackson Blvd., Chicago, IL 60604–3507.

EPA Region VI (Arkansas, Louisiana, New Mexico, Oklahoma, Texas), Director, Air, Pesticides and Toxics, 1445 Ross Avenue, Dallas, TX 75202–2733.

EPA Region VII (Iowa, Kansas, Missouri, Nebraska), Director, Air and Toxics Division, 726 Minnesota Avenue, Kansas City, KS 66101.

EPA Region VIII (Colorado, Montana, North Dakota, South Dakota, Utah, Wyoming), Director, Air and Toxics Division, 999 18th Street, 1 Denver Place, Suite 500, Denver, CO 80202–2405.

EPA Region IX (Arizona, California, Hawaii, Nevada, American Samoa, Guam), Director, Air and Toxics Division, 75 Hawthorne Street, San Francisco, CA 94105.

EPA Region X (Alaska, Idaho, Oregon, Washington), Director, Air and Toxics Division, 1200 Sixth Avenue, Seattle, WA 98101.

(b) All information required to be submitted to the Administrator under this part also shall be submitted to the appropriate State agency of any State to which authority has been delegated under section 112(l) of the Act. The owner or operator of

§ 63.14

an affected source may contact the appropriate EPA Regional Office for the mailing addresses for those States whose delegation requests have been approved.

(c) If any State requires a submittal that contains all the information required in an application, notification, request, report, statement, or other communication required in this part, an owner or operator may send the appropriate Regional Office of the EPA a copy of that submittal to satisfy the requirements of this part for that communication.

§ 63.14 Incorporations by reference.

(a) The materials listed in this section are incorporated by reference in the corresponding sections noted. These incorporations by reference were approved by the Director of the Federal Register in accordance with 5 U.S.C. 552(a) and 1 CFR part 51. These materials are incorporated as they exist on the date of the approval, and notice of any change in these materials will be published in the FEDERAL REGISTER. The materials are available for purchase at the corresponding addresses noted below, and all are available for inspection at the Office of the Federal Register, 800 North Capitol Street, NW, suite 700, Washington, DC, at the Air and Radiation Docket and Information Center, U.S. EPA, 401 M Street, SW., Washington, DC, and at the EPA Library (MD–35), U.S. EPA, Research Triangle Park, North Carolina.

(b) The materials listed below are available for purchase from at least one of the following addresses: American Society for Testing and Materials (ASTM), 1916 Race Street, Philadelphia, Pennsylvania 19103; or University Microfilms International, 300 North Zeeb Road, Ann Arbor, Michigan 48106.

(1) ASTM D1946–77, Standard Method for Analysis of Reformed Gas by Gas Chromatography, IBR approved for § 63.11(b)(6).

(2) ASTM D2382–76, Heat of Combustion of Hydrocarbon Fuels by Bomb Calorimeter (High-Precision Method), IBR approved for § 63.11(b)(6).

(3) ASTM D2879–83, Standard Test Method for Vapor Pressure—Temperature Relationship and Initial Decomposition Temperature of Liquids by Isoteniscope, IBR approved for § 63.111 of subpart G of this part.

(4) ASTM D 3695–88, Standard Test Method for Volatile Alcohols in Water by Direct Aqueous-Injection Gas Chromatography, IBR approved for § 63.365(e)(1) of subpart O of this part.

(5) ASTM D 1193–77, Standard Specification for Reagent Water, IBR approved for Method 306, section 4.1.1 and section 4.4.2, of appendix A to part 63.

(6) ASTM D 1331–89, Standard Test Methods for Surface and Interfacial Tension of Solutions of Surface Active Agents, IBR approved for Method 306B, section 2.2, section 3.1, and section 4.2, of appendix A to part 63.

(7) ASTM E 260–91, Standard Practice for Packed Column Gas Chromatography, IBR approved for § 63.750(b)(2) of subpart GG of this part.

EDITORIAL NOTE: At 60 FR 64336, Dec. 15, 1995, in § 63.14, the following paragraphs (b)(4) through (b)(14) were added, although (b)(4) through (b)(7) already existed before this amendment.

(4) ASTM D523–89, Standard Test Method for Specular Gloss, IBR approved for § 63.782.

(5) ASTM D1475–90, Standard Test Method for Density of Paint, Varnish, Lacquer, and Related Products, IBR approved for § 63.788 appendix A.

(6) ASTM D2369–93, Standard Test Method for Volatile Content of Coatings, IBR approved for § 63.788 appendix A.

(7) ASTM D3912–80, Standard Test Method for Chemical Resistance of Coatings Used in Light-Water Nuclear Power Plants, IBR approved for § 63.782.

(8) ASTM D4017–90, Standard Test Method for Water and Paints and Paint Materials by Karl Fischer Method, IBR approved for § 63.788 appendix A.

(9) ASTM D4082–89, Standard Test Method for Effects of Gamma Radiation on Coatings for Use in Light-Water Nuclear Power Plants, IBR approved for § 63.782.

(10) ASTM D4256–89 [reapproved 1994], Standard Test Method for Determination of the Decontaminability of Coatings Used in Light-Water Nuclear Power Plants, IBR approved for § 63.782.

(11) ASTM D3792–91, Standard Test Method for Water Content of Water-Reducible Paints by Direct Injection into a Gas Chromatograph, IBR approved for § 63.788 appendix A.

(12) ASTM D3257–93, Standard Test Methods for Aromatics in Mineral Spirits by Gas Chromatography, IBR approved for § 63.786(b).

(13) ASTM E260–91, Standard Practice for Packed Column Gas Chromatography, IBR approved for § 63.786(b).

(14) ASTM E180–93, Standard Practice for Determining the Precision of ASTM Methods for Analysis and Testing of Industrial Chemicals, IBR approved for § 63.786(b).

(c) The materials listed below are available for purchase from the American Petroleum Institute (API), 1220 L Street, NW., Washington, DC 20005.

(1) API Publication 2517, Evaporative Loss from External Floating-Roof Tanks, Third Edition, February 1989, IBR approved for § 63.111 of subpart G of this part.

(2) API Publication 2518, Evaporative Loss from Fixed-roof Tanks, Second Edition, October 1991, IBR approved for § 63.150(g)(3)(i)(C) of subpart G of this part.

(d) *State and Local Requirements.* The materials listed below are available at the Air and Radiation Docket and Information Center, U.S. EPA, 401 M Street, SW., Washington, DC.

(1) *California Regulatory Requirements Applicable to the Air Toxics Program,* March 1, 1996, IBR approved for § 63.99(a)(5)(ii) of subpart E of this part.

(2) [Reserved]

[59 FR 12430, Mar. 16, 1994, as amended at 59 FR 19453, Apr. 22, 1994; 59 FR 62589, Dec. 6, 1994; 60 FR 4963, Jan. 25, 1995; 60 FR 33122, June 27, 1995; 60 FR 45980, Sept. 1, 1995; 61 FR 25399, May 21, 1996]

§ 63.15 Availability of information and confidentiality.

(a) *Availability of information.* (1) With the exception of information protected through part 2 of this chapter, all reports, records, and other information collected by the Administrator under this part are available to the public. In addition, a copy of each permit application, compliance plan (including the schedule of compliance), notification of compliance status, excess emissions and continuous monitoring systems performance report, and title V permit is available to the public, consistent with protections recognized in section 503(e) of the Act.

(2) The availability to the public of information provided to or otherwise obtained by the Administrator under this part shall be governed by part 2 of this chapter.

(b) *Confidentiality.* (1) If an owner or operator is required to submit information entitled to protection from disclosure under section 114(c) of the Act, the owner or operator may submit such information separately. The requirements of section 114(c) shall apply to such information.

(2) The contents of a title V permit shall not be entitled to protection under section 114(c) of the Act; however, information submitted as part of an application for a title V permit may be entitled to protection from disclosure.

Summary of MACT Compliance Dates

Summary of Compliance Dates for Promulgated Part 63 MACT Regulations

Preface

The purpose of the following tables is to provide, in a similar format, a summary of compliance dates for different MACT regulations. Each table contains a summary of implementation dates for a specific Part 63 MACT regulation. Although the information contained in each table has been reviewed for accuracy, it is not intended to replace the provisions in the actual MACT subpart. If questions arise regarding a specific requirement, please refer to the actual MACT regulation or call the OAQPS contact for that rule.

Index
Part 63 MACT Regulations and EPA Contacts

Phone number	OAQPS contact	Regulation	OC contact	Phone number
(919) 541-2452	Jim Szykman	MACT Subpart A: General Provisions	Belinda Breidenbach	(202) 564-7022
(919) 541-5254	Jan Meyer	MACT Subparts F–H: The HON	Jeff Kenknight	(202) 564-7033
(919) 541-5268	Amanda Agnew	MACT Subpart L: Coke Oven Batteries	Maria Malave	(202) 564-7027
(919) 541-1549	George Smith	MACT Subpart M: Perc Dry Cleaners	Joyce Chandler	(202) 564-7073
(919) 541-5420	Lalit Banker	MACT Subpart N:	Scott Throwe	(202) 564-7013
(919) 541-5289	Phil Mulrine	Chromium Electroplating		
(919) 541-0837	David Markwordt	MACT Subpart O: Ethylene Oxide Sterilizers	Karin Leff	(202) 564-7068
(919) 541-5289	Phil Mulrine	MACT Subpart Q: Ind. Process Cooling Towers	Mimi Guernica	(202) 564-2415
(919) 541-5397	Steve Shedd	MACT Subpart R: Gasoline Distribution	Julie Tankersley	(202) 564-7002
(919) 541-0283	Paul Almodovar	MACT Subpart T: Halogenated Solvent Cleaning	Tracy Back	(202) 564-7076
(919) 541-5608	Bob Rosensteel	MACT Subpart U: Polymers and Resins, Group I	Sally Sasnett	(202) 564-7074
(919) 541-5402	Randy McDonald	MACT Subpart W: Epoxy Resins and Nonnylon Polymides	Sally Sasnett	(202) 564-7074
(919) 541-2364	Kevin Cavender	MACT Subpart X: Secondary Lead Smelters	Jane Engert	(202) 564-5021
(919) 541-0837	David Markwordt	MACT Subpart Y: Marine Vessel Loading	Virginia Lathrop	(202) 564-7057
(919) 541-5672	Jim Durham	MACT Subpart CC: Petroleum Refineries	Tom Ripp	(202) 564-7003

Index
Part 63 MACT Regulations and EPA Contacts

Phone number	OAQPS contact	Regulation	OC contact	Phone number
(919) 541-2363	Michele Aston	MACT Subpart DD: Off-Site Waste and Recovery Operations	Ann Stephanos	(202) 564-7043
(919) 541-5261	Gail Lacy	MACT Subpart EE: Magnetic Tape Manufacture	Seth Heminway	(202) 564-7017
(919) 541-2452	Jim Szykman	MACT Subpart GG: Aerospace	Suzanne Childress	(202) 564-7018
(919) 541-2379	Mohammed Serageldin	MACT Subpart II: Shipbuilding and Repair	Suzanne Childress	(202) 564-7018
(919) 541-0283	Paul Almodovar	MACT Subpart JJ: Wood Furniture Manufacturing	Robert Marshall	(202) 564-7021
(919) 541-0859	Dave Salman	MACT Subpart KK: Printing and Publishing	Ginger Gotliffe	(202) 564-7072
(919) 541-5608	Bob Rosensteel	MACT Subpart JJJ: Polymers and Resins, Group IV	Sally Sasnett	(202) 564-7074

SUMMARY OF IMPLEMENTATION DATES
Existing-Source MACT Requirements

Subparts: F, G, and H
Source Category: Synthetic Organic Chemical Manufacturing Industry
OAQPS Contact: Jan Meyer, (919) 541-5254

Milestone	Comments on date	Date
Effective Date	Rule promulgation date	4/22/94; substantial revisions to wastewater to be published soon
O/O: Initial notification due	Typically within 120 days of the effective date	8/19/94
O/O: Submit special compliance monitoring or implementation plans	Not all rules have these plans	12/31/96
EPA/state: Review/approve special compliance monitoring or implementation plans	Not all rules have these plans	NA
O/O: Request for compliance extension (can be allowed up to 1 year for special circumstances)	Currently 12 months before compliance date	4/22/96
EPA/state: Approval of request for compliance extensions	Variable but generally within 3 months of request	
Compliance Date	No greater than 3 years after the effective date (not counting extensions)	Subparts F and G: 4/22/97 Subpart H: Variable, 10/24/94 through 10/23/95
O/O: Notice of performance test	Generally >30 days before performance test— §63.103(b)(2)	Obviously could be anytime before 4/22/97 to 9/19/97
EPA/state: Approval of site-specific test plan	<30 days after receipt	NA
Performance test*	Within 180 days after compliance date	Before 9/19/97
EPA/state: Attend performance test	Optional but recommended	
Compliance status reports* (before a Title V permit is functional for the source)	First report: <60 days from performance test/compliance date. Periodic—as required in standard	9/19/97 NCS; 5/98 periodic reports
EPA/state: **Review compliance status reports**	EPA/states need to report simple summaries of compliance status	No date assigned

O/O means the owner or operator performs the milestone.
EPA/state means EPA or a delegated state performs the milestone.
*The HON does not specify separate dates for completing the performance test and submitting the NCS report. The results of the performance test are to be included in the NCS, and that has to be submitted no later than 150 days from the compliance date for the subpart G provisions. For subpart H, the NCS report is to be submitted 90 days after the applicable compliance date, for example, 1/23/95 through 1/22/96.

SUMMARY OF IMPLEMENTATION DATES
New-Source MACT Requirements

Subparts: F, G, and H
Source Category: Synthetic Organic Chemical Manufacturing Industry
OAQPS Contact: Jan Meyer, (919) 541-5254

Milestone	Comments on date	Date*
Applicability Date for New Sources	Rule proposed date	12/31/92
O/O: Application for approval to construct/reconstruct a major emitting affected source. *Note:* Nonmajor emitting affected sources are not required to request approval to construct/reconstruct; they merely notify the EPA/state.	As soon as possible before commencing construction. Special allowance provided for sources commencing construction after proposal date but before effective date. [See §63.151(b)(2)(i) and (b)(2)(ii).]	No sooner than 7/21/94
EPA/state: Notice of complete information	Within 30 days from application by O/O [§63.5(e)(2)(i)]	
EPA/state: Approval/denial of construction/reconstruction	Within 60 days from notification of complete application [§63.5(e)(2)(i)]	
O/O: Notice of intended start-up Then follows existing source process...	30 to 60 days before start-up [§63.9(b)(4)(iv)]	
Compliance Date	Upon start-up unless rule provides special allowance	

O/O means the owner or operator performs the milestone.
EPA/state means EPA or a delegated state performs the milestone.
*All the dates depend on the action of the O/O; thus they cannot be filled in. This form may be useful for EPA/states working with specific O/O after an application for approval to construct/reconstruct has been submitted.

SUMMARY OF IMPLEMENTATION DATES
Existing-Source MACT Requirements

Subpart: L

Source Category: Coke Ovens NESHAP

OAQPS Contact: Amanda Agnew, (919) 541-5268

Note: Due to various criteria for determining implementation dates under the Coke Oven NESHAP, please contact Amanda Agnew at (919) 541-5268 for specific details.

Milestone	Comments on date	Date
Effective Date	Promulgation date	
O/O: Initial notification due	Typically within 120 days of effective date	
O/O: Request for compliance extension (up to 1 year for special circumstances)	Currently 12 months before compliance date	
EPA/state: Approval of request for compliance extensions	Variable but generally within 3 months of request	
Compliance Date	No greater than 3 years after effective date	
O/O: Submit special compliance-monitoring plans	Not all rules have these plans	
EPA/state: Review/approve special compliance-monitoring plans	Not all rules have these plans	
O/O: Notice of performance test	Generally >30 days before performance test	
EPA/state: Approval of site-specific test plan	<30 days of receipt	
Performance test	Within 180 days of compliance date	
EPA/state: Attend performance test		
Compliance status reports (before a Title V permit is functional for the source)	Initial: <60 days from performance test/compliance date Periodic: as required in standard	
EPA/state: **Review compliance status reports**	There is a need to report simple summaries of compliance status	

SUMMARY OF IMPLEMENTATION DATES
New-Source MACT Requirements

Subpart: L
Source Category: Coke Ovens NESHAP

OAQPS Contact: Amanda Agnew, (919) 541-5268

Note: Due to various criteria for determining implementation dates under the Coke
   Oven NESHAP, please contact Amanda Agnew at (919) 541-5268 for specific
   details.

Milestone	Comments on date	Date
Applicability Date for New Sources	Proposed rule date	
O/O: Application for approval to construct/reconstruct a major emitting affected source	As soon as practicable before commencing construction. Special allowance provided for sources commencing construction after proposal date but before effective date.	
EPA/state: Notice of complete information	+30 days from application	
EPA/state: Approval of construction/reconstruction	+60 days from application	
O/O: Notice of intended start-up	>60 days before start-up	
Then follows existing source process...		

SUMMARY OF IMPLEMENTATION DATES
Existing-Source MACT Requirements

Subpart: M
Source Category: Perchloroethylene Dry Cleaning Facilities
OAQPS Contact: George Smith, (919) 541-1549

Milestone	Comments on date	Date
Effective Date O/O: Initial notification due	Rule promulgation date Typically within 120 days of effective date	9/22/93 6/18/94
O/O: Submit special compliance-monitoring or implementation plans	Not all rules have these plans	N/A
EPA/state: Review/approve special compliance-monitoring or implementation plans	Not all rules have these plans	N/A
O/O: Request for compliance extension (can be allowed up to 1 year for special circumstances)	Currently 12 months before compliance date	No requests at present
EPA/state: Approval of request for compliance extensions	Variable but generally within 3 months of request	There are no requests pending approval
Compliance Date	No greater than 3 years after effective date (not counting extensions)	For work practices 12/20/93; for control hardware 9/23/96
O/O: Notice of performance test	Generally >30 days before performance test	N/A
EPA/state: Approval of site-specific test plan	<30 days after receipt	N/A
Performance test	Within 180 days after compliance date	N/A
EPA/state: Attend performance test	Optional but recommended	N/A
Compliance status reports (before a Title V permit is functional for the source)	First report: <60 days from performance test/compliance date Periodic: as required in standard	For work practices 6/18/94; for control hardware 10/23/96
EPA/state: **Review compliance status reports**	EPA/states need to report simple summaries of compliance status	No date assigned

O/O means the owner or operator performs the milestone.
EPA/state means EPA or a delegated state performs the milestone.

SUMMARY OF IMPLEMENTATION DATES
New-Source MACT Requirements

Subpart: M
Source Category: Perchloroethylene Dry Cleaning Facilities
OAQPS Contact: George Smith, (919) 541-1549

Milestone	Comments on date	Date*
Applicability Date for New Sources	Rule proposed date	12/18/91
O/O: Application for approval to construct/reconstruct a major emitting affected source. *Note:* Nonmajor emitting affected sources are not required to request approval to construct/reconstruct; they merely notify the EPA/state.	As soon as possible before commencing construction. Special allowance provided for sources commencing construction after proposal date but before effective date [§63.5(d)(1)(i)]	No instance of this has occurred—99% of sources are nonmajor
EPA/state: Notice of complete information	Within 30 days from application by O/O [§63.5(e)(2)(i)]	See above
EPA/state: Approval/denial of construction/reconstruction	Within 60 days from notification of complete application [§63.5(e)(2)(i)]	See above
O/O: Notice of intended start-up Then follows existing source process...	30 to 60 days before start-up [§63.9(b)(4)(iv)]	See above
Compliance Date	Upon start-up unless rule provides special allowance	Upon start-up

O/O means the owner or operator performs the milestone.
EPA/state means EPA or a delegated state performs the milestone.
*All the dates depend on the action of the O/O; thus they cannot be filled in. This form may be useful for EPA/states working with a specific O/O after an application for approval to construct/reconstruct has been submitted.

SUMMARY OF IMPLEMENTATION DATES
Existing-Source MACT Requirements

Subpart: N
Source Category: Chrome Electroplating
OAQPS Contact: Phil Mulrine, (919) 541-5289; or Lalit Banker, (919) 541-5420

Milestone	Comments on date	Date
Effective Date	Rule promulgation date	1/25/95
O/O: Initial notification due	Typically within 120 days of effective date	7/24/95
O/O: Submit special compliance-monitoring or implementation plans	Not all rules have these plans	N/A
EPA/state: Review/approve special compliance-monitoring or implementation plans	Not all rules have these plans	N/A
O/O: Request for compliance extension (can be allowed up to 1 year for special circumstances)	Currently 12 months before compliance date	Section 63.6(i)
EPA/state: Approval of request for compliance extensions	Variable but generally within 3 months of request	
Compliance Date	No greater than 3 years after effective date (not counting extensions)	1/25/96 for decorative, 1/25/97 for hard and anodizing
O/O: Notice of performance test	Generally >30 days before performance test	>60 days
EPA/state: Approval of site-specific test plan	<30 days after receipt	<30 days
Performance test	Within 180 days after compliance date	7/23/96
EPA/state: Attend performance test	Optional but recommended	7/24/97
Compliance status reports (before a Title V permit is functional for the source)	First report: <60 days from performance test/compliance date Periodic: as required in standard	<90 days
EPA/state: **Review compliance status reports**	EPA/states need to report simple summaries of compliance status	No date assigned

O/O means the owner or operator performs the milestone.
EPA/state means EPA or a delegated state performs the milestone.

SUMMARY OF IMPLEMENTATION DATES
New-Source MACT Requirements

Subpart: N
Source Category: Chrome Electroplating
OAQPS Contact: Phil Mulrine, (919) 541-5289; or Lalit Banker, (919) 541-5420

Milestone	Comments on date	Date*
Applicability Date for New Sources	Rule proposed date	12/16/93
O/O: Application for approval to construct/reconstruct a major emitting affected source. *Note:* Nonmajor emitting affected sources are not required to request approval to construct/reconstruct; they merely notify the EPA/state.	As soon as possible before commencing construction. Special allowance provided for sources commencing construction after proposal date but before effective date [§63.5(d)(1)(i)]	
EPA/state: Notice of complete information	Within 30 days from application by O/O [§63.5(e)(2)(i)]	
EPA/state: Approval/denial of construction/reconstruction	Within 60 days from notification of complete application [§63.5(e)(2)(i)]	
O/O: Notice of intended start-up	30 to 60 days before start-up [§63.9(b)(4)(iv)]	
Then follows existing source process…		
Compliance Date	Upon start-up unless rule provides special allowance	

O/O means the owner or operator performs the milestone.
EPA/state means EPA or a delegated state performs the milestone.
*All the dates depend on the action of the O/O; thus they cannot be filled in. This form may be useful for EPA/states working with a specific O/O after an application for approval to construct/reconstruct has been submitted.

SUMMARY OF IMPLEMENTATION DATES
Existing-Source MACT Requirements

Subpart: O
Source Category: Ethylene Oxide Commercial Sterilization Facilities
OAQPS Contact: Dave Markwordt, (919) 541-0837

Milestone	Comments on date	Date
Effective Date	Rule promulgation date	12/06/94
O/O: Initial notification due	Typically within 120 days of effective date	04/08/95
O/O: Submit special compliance-monitoring or implementation plans	Not all rules have these plans	N/A
EPA/state: Review/approve compliance-monitoring or implementation plans	Not all rules have these plans	N/A
O/O: Request for compliance extension (can be allowed up to 1 year for special circumstances)	Currently 12 months before compliance date	08/08/97
EPA/state: Approval of request for compliance extensions	Variable but generally within 3 months of request	By 11/08/97
Compliance Date	No greater than 3 years after effective date (not counting extensions)	08/08/98
O/O: Notice of performance test	Generally >30 days before performance test	60 days prior to test
EPA/state: Approval of site-specific test plan	<30 days after receipt	<30 days
Performance test	Within 180 days after compliance date	By 02/08/99
EPA/state: Attend performance test	Optional but recommended	
Compliance status reports (before a Title V permit is functional for the source)	First report: <60 days from performance test/compliance date Periodic: as required in standard	By 04/08/99
EPA/state: **Review compliance status reports**	EPA/states need to report simple summaries of compliance status	No date assigned

O/O means the owner or operator performs the milestone.
EPA/state means EPA or a delegated state performs the milestone.

SUMMARY OF IMPLEMENTATION DATES
New-Source MACT Requirements

Subpart: O
Source Category: Ethylene Oxide Commercial Sterilization Facilities
OAQPS Contact: Dave Markwordt, (919) 541-0837

Milestone	Comments on date	Date*
Applicability Date for New Sources	Rule proposed date	03/07/94
O/O: Application for approval to construct/reconstruct a major emitting affected source. *Note:* Nonmajor emitting affected sources are not required to request approval to construct/reconstruct; they merely notify the EPA/state.	As soon as possible before commencing construction. Special allowance provided for sources commencing construction after proposal date but before effective date [§63.5(d)(1)(i)]	As soon as possible
EPA/state: Notice of complete information	Within 30 days from application by O/O [§63.5(e)(2)(i)]	63.5(e)(2)(i) for >10 ton facilities
EPA/state: Approval/denial of construction/reconstruction	Within 60 days from notification of complete application [§63.5(e)(2)(i)]	63.5(e)(2)(i) for >10 ton facilities
O/O: Notice of intended start-up	30 to 60 days before start-up [§63.9(b)(4)(iv)]	30 to 60 days
Then follows existing source process...		
Compliance Date	Upon start-up unless rule provides special allowance	08/08/98 or upon start-up after 08/08/98

O/O means the owner or operator performs the milestone.
EPA/state means EPA or a delegated state performs the milestone.
*All the dates depend on the action of the O/O; thus they cannot be filled in. This form may be useful for EPA/states working with a specific O/O after an application for approval to construct/reconstruct has been submitted.

SUMMARY OF IMPLEMENTATION DATES
Existing-Source MACT Requirements

Subpart: Q
Source Category: Industrial Process Cooling Towers
OAQPS Contact: Phil Mulrine, (919) 541-5289

Milestone	Comments on date	Date
Effective Date	Promulgation date	09/08/94
O/O: Initial notification due	Typically within 120 days of effective date	09/08/95
O/O: Request for compliance extension (up to 1 year for special circumstances)	Currently 12 months before compliance date	N/A
EPA/state: Approval of request for compliance extensions	Variable but generally within 3 months of request	N/A
Compliance Date	No greater than 3 years after effective date	03/08/96
O/O: Submit special compliance-monitoring plans	Not all rules have these plans	N/A
EPA/state: Review/approve special compliance-monitoring plans	Not all rules have these plans	N/A
O/O: Notice of performance test	Generally >30 days before performance test	N/A
EPA/state: Approval of site-specific test plan	<30 days of receipt	N/A
Performance test	Within 180 days of compliance date	N/A
EPA/state: Attend performance test		N/A
Compliance status reports (before a Title V permit is functional for the source)	Initial: <60 days from performance test/compliance date Periodic: as required in standard	N/A
EPA/state: **Review compliance status reports**	There is a need to report simple summaries of compliance status	No date assigned

O/O means the owner or operator performs the milestone.
EPA/state means EPA or the delegated authority performs the milestone.

SUMMARY OF IMPLEMENTATION DATES
New-Source MACT Requirements

Subpart: Q
Source Category: Industrial Process Cooling Towers
OAQPS Contact: Phil Mulrine, (919) 541-5289

Milestone	Comments on date	Date*
Applicability Date for New Sources	Proposed rule date	08/12/93
O/O: Application for approval to construct/reconstruct a major emitting affected source	As soon as possible before commencing construction. Special allowance provided for sources commencing construction after proposal date but before effective date [§63.5(d)(1)(i)]	
EPA/state: Notice of complete information	Within 30 days from application by O/O [§63.5(e)(2)(i)]	
EPA/state: Approval of construction/reconstruction	Within 60 days from notification of complete application [§63.5(e)(2)(i)]	
O/O: Notice of intended start-up	30 to 60 days before start-up [§63.9(b)(4)(iv)]	
Then follows existing source process...		
Compliance Date	Upon start-up unless special allowance is provided	

O/O means the owner or operator performs the milestone.
EPA/state means EPA or the delegated authority performs the milestone.
*All the dates depend on the action of the O/O; thus they cannot be filled in. This would be identified after an application for approval to construct or reconstruct has been submitted by the O/O.

SUMMARY OF IMPLEMENTATION DATES
Existing-Source MACT Requirements

Subpart: R
Source Category: Gasoline Distribution MACT
OAQPS Contact: Steve Shedd, (919) 541-5397

Milestone	Comments on date	Date
Effective Date	Promulgation date	12/14/94
O/O: Initial notification due	Typically within 120 days of effective date	12/16/96
Notification date for area sources using screening equation	New for this MACT standard	12/16/96
O/O: Request for compliance extension (up to 1 year for special circumstances)	Currently 12 months before compliance date	*
EPA/state: Approval of request for compliance extensions	Variable but generally within 3 months of request	*
Compliance Date	No greater than 3 years after effective date	12/15/97
O/O: Submit special compliance-monitoring plans	Not all rules have these plans	*
EPA/state: Review/approve special compliance-monitoring plans	Not all rules have these plans	*
O/O: Notice of performance test	Generally >30 days before performance test	*
EPA/state: Approval of site-specific test plan	<30 days of receipt	*
Performance test	Within 180 days of compliance date	*
EPA/state: Attend performance test		*
Compliance status reports (before a Title V permit is functional for the source)	Initial: <60 days from performance test/compliance date Periodic: as required in standard	*
EPA/state: **Review compliance status reports**	There is a need to report simple summaries of compliance status	*

*Did not specify a date in rule; however, did cross-reference General Provisions for the appropriate dates.

SUMMARY OF IMPLEMENTATION DATES
New-Source MACT Requirements

Subpart: R
Source Category: Gasoline Distribution MACT
OAQPS Contact: Steve Shedd, (919) 541-5397

Milestone	Comments on date	Date
Applicability Date for New Sources	Proposed rule date	2/8/94
O/O: Application for approval to construct/reconstruct a major emitting affected source	As soon as practicable before commencing construction. Special allowance provided for sources commencing construction after proposal date but before effective date	*
EPA/state: Notice of complete information	+30 days from application	*
EPA/state: Approval of construction/reconstruction	+60 days from application	*
O/O: Notice of intended start-up	>60 days before start-up	*
Then follows existing source process. . .		*

*Did not specify a date in rule; however, did cross-reference General Provisions for the appropriate dates.

SUMMARY OF IMPLEMENTATION DATES
Existing-Source MACT Requirements

Subpart: T
Source Category: Halogenated Solvent Cleaning NESHAP
OAQPS Contact: Paul Almodovar, (919) 541-0283

Milestone	Comments on date	Date
Effective Date	Rule promulgation date	12/02/94
O/O: Initial notification due	Typically within 120 days of effective date	08/29/95
O/O: Submit special compliance-monitoring or implementation plans	Not all rules have these plans	N/A
EPA/state: Review/approve compliance-monitoring or implementation plans	Not all rules have these plans	N/A
O/O: Request for compliance extension (can be allowed up to 1 year for special circumstances)	Currently 12 months before compliance date	12/02/96
EPA/state: Approval of request for compliance extensions	Variable but generally within 3 months of request	03/02/97
Compliance Date	Within 3 years after effective date (not counting extensions)	12/02/97
O/O: Notice of performance test	Generally >30 days before performance test	
EPA/state: Approval of site-specific test plan	<30 days after receipt	
Performance test	Within 180 days after compliance date	3/2/98
EPA/state: Attend performance test	Optional but recommended	
Compliance status reports (before a Title V permit is functional for the source)	First report: <60 days from performance test/compliance date	2/2/97
	Periodic: as required in standard	
EPA/state: **Review compliance status reports**	EPA/states need to report simple summaries of compliance status	No date assigned

O/O means the owner or operator performs the milestone.
EPA/state means EPA or a delegated state performs the milestone.

SUMMARY OF IMPLEMENTATION DATES
New-Source MACT Requirements

Subpart: T
Source Category: Halogenated Solvent Cleaning NESHAP
OAQPS Contact: Paul Almodovar, (919) 541-0283

Milestone	Comments on date	Date*
Applicability Date for New Sources	Rule proposed date	11/29/93
O/O: Application for approval to construct/reconstruct a major emitting affected source. *Note:* Nonmajor emitting affected sources are not required to request approval to construct/reconstruct; they merely notify the EPA/state.	As soon as possible before commencing construction. Special allowance provided for sources commencing construction after proposal date but before effective date [§63.5(d)(1)(i)]	
EPA/state: Notice of complete information	Within 30 days from application by O/O [§63.5(e)(2)(i)]	
EPA/state: Approval/denial of construction/reconstruction	Within 60 days from notification of complete application [§63.5(e)(2)(i)]	
O/O: Notice of intended start-up	30 to 60 days before start-up [§63.9(b)(4)(iv)]	
Then follows existing source process...		
Compliance Date	Upon start-up unless rule provides special allowance	

O/O means the owner or operator performs the milestone.

EPA/state means EPA or a delegated state performs the milestone.

*All the dates depend on the action of the O/O; thus they cannot be filled in. This form may be useful for EPA/states working with a specific O/O after an application for approval to construct/reconstruct has been submitted.

SUMMARY OF IMPLEMENTATION DATES
Existing-Source MACT Requirements

Subpart: U
Source Category: Polymers and Resins, Group I
OAQPS Contact: Bob Rosensteel, (919) 541-5608

Milestone	Comments on date	Date
Effective Date	Promulgation date	09/05/96
O/O: Initial notification due	Typically within 120 days of effective date	N/A
O/O: Request for compliance extension (up to 1 year for special circumstances)	Currently 12 months before compliance date	9/5/98 (except eq. leaks/compressors)
EPA/state: Approval of request for compliance extensions	Variable but generally within 3 months of request	Within 30 days of receipt of information
Compliance Date	No greater than 3 years after effective date	09/05/99 for most emission points
O/O: Submit special compliance-monitoring plans	Not all rules have these plans	03/05/98* 09/05/98†
EPA/state: Review/approve special compliance-monitoring plans	Not all rules have these plans	07/05/98* 10/05/98†
O/O: Notice of performance test	Generally >30 days before performance test	At least 30 days before test
EPA/state: Approval of site-specific test plan	<30 days of receipt	N/A
Performance test	Within 150 days of compliance date	By 02/05/00
EPA/state: Attend performance test		
Compliance status reports (before a Title V permit is functional for the source)	Initial: <60 days from performance test/compliance date Periodic: as required in standard	02/05/00‡
EPA/state: **Review compliance status reports**	There is a need to report simple summaries of compliance status	Not assigned

O/O means the owner or operator performs the milestone.
EPA/state means EPA or a delegated state performs the milestone.
*Emissions averaging plan.
†Precompliance report.
‡Notification of compliance status.

SUMMARY OF IMPLEMENTATION DATES
New-Source MACT Requirements

Subpart: U
Source Category: Polymers and Resins, Group I
OAQPS Contact: Bob Rosensteel, (919) 541-5608

Milestone	Comments on date	Date*
Applicability Date for New Sources	Proposed rule date	06/12/95
O/O: Application for approval to construct/reconstruct a major emitting affected source	As soon as practicable before commencing construction. Special allowance provided for sources commencing construction after proposal date but before effective date	
EPA/state: Notice of complete information	+30 days from application	
EPA/state: Approval of construction/reconstruction	+60 days from application	
O/O: Notice of intended start-up	>60 days before start-up	
Then follows existing source process...		
Compliance Date	Upon start-up unless rule provides special allowance	

O/O means the owner or operator performs the milestone.
EPA/state means EPA or a delegated state performs the milestone.
*All the dates depend on the action of the O/O; thus they cannot be filled in at this time. Dates would be determined after an application for approval to construct or reconstruct has been submitted by the O/O.

SUMMARY OF IMPLEMENTATION DATES
Existing-Source MACT Requirements

Subpart: W
Source Category: Epoxy Resins Production and Nonnylon Polyamides Production
OAQPS Contact: Randy McDonald, (919) 541-5402

Milestone	Comments on date	Date
Effective Date	Rule promulgation date	March 8, 1995
O/O: Initial notification due	Typically within 120 days of effective date	June 6, 1995
O/O: Submit special compliance-monitoring or implementation plans	Not all rules have these plans	
EPA/state: Review/approve special compliance-monitoring or implementation plans	Not all rules have these plans	
O/O: Request for compliance extension (can be allowed up to 1 year for special circumstances)	Currently 12 months before compliance date	
EPA/state: Approval of request for compliance extensions	Variable but generally within 3 months of request	
Compliance Date	No greater than 3 years after effective date (not counting extensions)	March 8, 1998
O/O: Notice of performance test	Generally >30 days before performance test	
EPA/state: Approval of site-specific test plan	<30 days after receipt	
Performance test	Within 180 days after compliance date	
EPA/state: Attend performance test	Optional but recommended	
Compliance status reports (before a Title V permit is functional for the source)	First report: <60 days from performance test/compliance date Periodic: as required in standard	
EPA/state: **Review compliance status reports**	EPA/states need to report simple summaries of compliance status	No date assigned

O/O means the owner or operator performs the milestone.
EPA/state means EPA or a delegated state performs the milestone.

SUMMARY OF IMPLEMENTATION DATES
New-Source MACT Requirements

Subpart: W
Source Category: Epoxy Resins Production and Nonnylon Polyamides Production
OAQPS Contact: Randy McDonald, (919) 541-5402

Milestone	Comments on date	Date*
Applicability Date for New Sources	Rule proposed date	
O/O: Application for approval to construct/reconstruct a major emitting affected source. *Note:* Nonmajor emitting affected sources are not required to request approval to construct/reconstruct; they merely notify the EPA/state.	As soon as possible before commencing construction. Special allowance provided for sources commencing construction after proposal date but before effective date [§63.5(d)(1)(i)]	
EPA/state: Notice of complete information	Within 30 days from application by O/O [§63.5(e)(2)(i)]	
EPA/state: Approval/denial of construction/reconstruction	Within 60 days from notification of complete application [§63.5(e)(2)(i)]	
O/O: Notice of intended start-up	30 to 60 days before start-up [§63.9(b)(4)(iv)]	
Then follows existing source process...		
Compliance Date	Upon start-up unless rule provides special allowance	

O/O means the owner or operator performs the milestone.
EPA/state means EPA or a delegated state performs the milestone.
*All the dates depend on the action of the O/O; thus they cannot be filled in. This form may be useful for EPA/states working with a specific O/O after an application for approval to construct/reconstruct has been submitted.

SUMMARY OF IMPLEMENTATION DATES
Existing-Source MACT Requirements

Subpart: X
Source Category: Secondary Lead Smelting
OAQPS Contact: Kevin Cavender, (919) 541-2364

Milestone	Comments on date	Date
Effective Date	Promulgation date	06/23/95
O/O: Initial notification due	Typically within 120 days of effective date	10/23/95
O/O: Request for compliance extension (up to 1 year for special circumstances)	Currently 12 months before compliance date	06/23/96
EPA/state: Approval of request for compliance extensions	Variable but generally within 3 months of request	09/23/96
Compliance Date	No greater than 3 years after effective date	06/23/97
O/O: Submit special compliance-monitoring or implementation plans	Not all rules have these plans	12/23/96
EPA/state: Review/approve special compliance-monitoring plans	Not all rules have these plans	
O/O: Notice of performance test	Generally >30 days before performance test	>30 days before test
EPA/state: Approval of site-specific test plan	<30 days of receipt	<30 days after receipt
Performance test	Within 180 days of compliance date	By 12/23/97
EPA/state: Attend performance test		
Compliance status reports (before a Title V permit is functional for the source)	Initial: <60 days from performance test/compliance date Periodic: as required in standard	60 days from performance test
EPA/state: **Review compliance status reports**	There is a need to report simple summaries of compliance status	No date assigned

O/O means the owner or operator performs the milestone.
EPA/state means EPA or a delegated state performs the milestone.

SUMMARY OF IMPLEMENTATION DATES
New-Source MACT Requirements

Subpart: X
Source Category: Secondary Lead Smelting
OAQPS Contact: Kevin Cavender, (919) 541-2364

Milestone	Comments on date	Date*
Applicability Date for New Sources	Proposed rule date	06/09/94
O/O: Application for approval to construct/reconstruct a major emitting affected source	As soon as possible before commencing construction. Special allowance provided for sources commencing construction after proposal date but before effective date [§63.5(d)(1)(i)]	
EPA/state: Notice of complete information	Within 30 days from application by O/O [§63.5(e)(2)(i)]	
EPA/state: Approval of construction or reconstruction	Within 60 days from notification of complete application [§63.5(e)(2)(i)]	
O/O: Notice of intended start-up	At least 30 days before start-up [§63.9(b)(4)(iv)]	
Then follows existing source process...		
Compliance Date	Upon start-up unless rule provides special allowance	

O/O means the owner or operator performs the milestone.
EPA/state means EPA or a delegated state performs the milestone.
*All the dates depend on the action of the O/O; thus they cannot be filled in at this time. Dates would be determined after an application for approval to construct or reconstruct has been submitted by the O/O.

SUMMARY OF IMPLEMENTATION DATES
Existing-Source MACT Requirements

Subpart: Y
Source Category: Marine Tank Vessel Loading Operations
OAQPS Contact: Dave Markwordt, (919) 541-0837

Milestone	Comments on date	Date
Effective Date	Rule promulgation date	09/19/95
O/O: Initial notification due	Typically within 120 days of effective date	09/19/96
O/O: Submit special compliance-monitoring or implementation plans	Not all rules have these plans	
EPA/state: Review/approve special compliance-monitoring plans	Not all rules have these plans	
O/O: Request for compliance extension (up to 1 year for special circumstances)	Currently 12 months before compliance date	NA (extension built-in compliance date)
EPA/state: Approval of request for compliance extensions	Variable but generally within 3 months of request	NA
Compliance Date	Within 4 years after effective date (not counting extensions)	09/19/99
O/O: Notice of performance test	Generally >30 days before performance test	60 days prior to test
EPA/state: Approval of site-specific test plan	<30 days after receipt	<30 days after receipt
Performance test	Within 180 days after compliance date	Within 180 days
EPA/state: Attend performance test	Optional but recommended	
Compliance status reports (before a Title V permit is functional for the source)	First report: <60 days from performance test/compliance date Periodic: as required in standard	<60 days from test
EPA/state: **Review compliance status reports**	EPA/states need to report simple summaries of compliance status	No date assigned

O/O means the owner or operator performs the milestone.
EPA/state means EPA or a delegated state performs the milestone.

SUMMARY OF IMPLEMENTATION DATES
New-Source MACT Requirements

Subpart: Y
Source Category: Marine Tank Vessel Loading Operations
OAQPS Contact: Dave Markwordt, (919) 541-0837

Milestone	Comments on date	Date*
Applicability Date for New Sources	Rule proposed date	09/07/95
O/O: Application for approval to construct/reconstruct a major emitting affected source	As soon as possible before commencing construction. Special allowance provided for sources commencing construction after proposal date but before effective date [§63.5(d)(1)(i)]	As soon as practicable
EPA/state: Notice of complete information	Within 30 days from application by O/O [§63.5(e)(2)(i)]	63.5(e)(2)(i)
EPA/state: Approval or denial of construction/reconstruction	Within 60 days from notification of complete application [§63.5(e)(2)(i)]	63.5(e)(2)(i)
O/O: Notice of intended start-up	30 to 60 days before start-up [§63.9(b)(4)(iv)]	30 to 60 days before start-up
Then follows existing source process...		
Compliance Date	Upon start-up unless rule provides special allowance	09/19/99 or upon start-up if after 09/19/99

O/O means the owner or operator performs the milestone.
EPA/state means EPA or a delegated state performs the milestone.
*All the dates depend on the action of the O/O; thus they cannot be filled in at this time. Dates would be determined after an application for approval to construct or reconstruct has been submitted by the O/O.

SUMMARY OF IMPLEMENTATION DATES
Existing-Source MACT Requirements

Subpart: CC
Source Category: Petroleum Refineries: Other Sources Not Directly Listed
OAQPS Contact: Jim Durham, (919) 541-5672

Milestone	Comments on date	Date
Effective Date	Promulgation: 60 FR 43243	8/18/95
O/O: Request for compliance extension (up to 1 year for special circumstances)	1 year before compliance date [§63.6(i)(4)(i)(B)]	8/18/97
EPA/state: Approval of request for compliance extensions	Within 30 days notify O/O if application is complete or specify deficiencies [§63.6(i)(12)]	9/17/97
Compliance Date	3 years after effective date of 8/18/95 except for certain storage vessels [63.640(h)] [see note 5]	8/18/98
O/O: Notice of performance test	Provide 30 days' advance notice [63.642(d)(2)]	Variable
O/O: Implementation plan for compliance using emissions averaging [see note 6]	Submit 18 months before compliance date [63.653(d)(1)]	2/18/97
EPA/state: Notify O/O of status of plan to implement emissions averaging	Within 120 days notify O/O if plan is complete or specify deficiencies. Must then approve plan within 120 days of receiving sufficient information [63.653(d)(3)]	6/18/97
O/O: Notification of compliance status (NCS) report (before a Title V permit is functional for the source)	Submit 150 days after compliance date [63.654(f)]	1/15/99

1. O/O—Owner/Operator.
2. Initial notification not required (preamble p. 43255).
3. Special compliance monitoring plans not required.
4. Site-specific test plans not required (preamble p. 43254 and table 6 of rule).
5. Floating roof storage vessels must comply at first scheduled shutdown after August 18, 1998 [63.640(h)(4)].
6. Emissions averaging is an optional compliance alternative that few, if any, refineries are expected to choose.

SUMMARY OF IMPLEMENTATION DATES
New-Source MACT Requirements

Subpart: CC
Source Category: Petroleum Refineries: Other Sources Not Directly Listed
OAQPS Contact: Jim Durham, (919) 541-5672

Milestone	Comments on date	Date
Applicability Date for New Sources	Proposed rule: 59 FR 36130	7/15/94
O/O: Application for approval to construct or reconstruct a major affected source	As soon as practical before start of construction but no sooner than 90 days after promulgation [63.640(k)(2)(i)]	
EPA/state: Notice of complete information	30 days from receipt of application [63.5(e)(2)(i)]	
EPA/state: Approval of construction or reconstruction	60 days after notifying O/O that application is complete [63.5(e)(2)(i)]	
O/O: Notice of intended start-up	At least 30 but not more than 60 days' advance notice [63.9(b)(4)(iv)]	
Then follow existing source process...		
Compliance Date	Upon start-up	

SUMMARY OF IMPLEMENTATION DATES
Existing-Source MACT Requirements

Subpart: DD
Source Category: Off-Site Waste and Recovery Operations
OAQPS Contact: Michele Aston, (919) 541-2363

Milestone	Comments on date	Date
Effective Date	Rule promulgation date	07/01/96
O/O: Initial notification due	Typically within 120 days of effective date	10/28/96
O/O: Submit special compliance-monitoring or implementation plans	Not all rules have these plans	N/A
EPA/state: Review/approve special compliance-monitoring plans	Not all rules have these plans	N/A
O/O: Request for compliance extension (up to 1 year for special circumstances)	Currently 12 months before compliance date	07/01/98
EPA/state: Approval of request for compliance extensions	Variable but generally within 3 months of request	Specified in subpart A
Compliance Date	No greater than 3 years after effective date (not counting extensions)	07/01/99
O/O: Notice of performance test	Generally >30 days before performance test	Specified in subpart A
EPA/state: Approval of site-specific test plan	<30 days after receipt	Specified in subpart A
Performance test	Within 180 days after compliance date	12/27/99
EPA/state: Attend performance test	Optional but recommended	
Compliance status reports (before a Title V permit is functional for the source)	First report: <60 days from performance test/compliance date Periodic: as required in standard	Specified in subpart A
EPA/state: **Review compliance status reports**	EPA/states need to report simple summaries of compliance status	No date assigned

O/O means the owner or operator performs the milestone.
EPA/state means EPA or a delegated state performs the milestone.

SUMMARY OF IMPLEMENTATION DATES
New-Source MACT Requirements

Subpart: DD
Source Category: Off-Site Waste and Recovery Operations
OAQPS Contact: Michele Aston, (919) 541-2363

Milestone	Comments on date	Date*
Applicability Date for New Sources	Rule proposed date	10/13/94
O/O: Application for approval to construct/reconstruct a major emitting affected source. *Note:* Nonmajor emitting affected sources are not required to request approval to construct/reconstruct; they merely notify the EPA/state.	As soon as possible before commencing construction. Special allowance provided for sources commencing construction after proposal date but before effective date [§63.5(d)(1)(i)]	
EPA/state: Notice of complete information	Within 30 days from application by O/O [§63.5(e)(2)(i)]	
EPA/state: Approval/denial of construction/reconstruction	Within 60 days from notification of complete application [§63.5(e)(2)(i)]	
O/O: Notice of intended start-up	30 to 60 days before start-up [§63.9(b)(4)(iv)]	
Then follows existing source process...		
Compliance Date	Upon start-up unless rule provides special allowance	

O/O means the owner or operator performs the milestone.
EPA/state means EPA or a delegated state performs the milestone.
*All the dates depend on the action of the O/O; thus they cannot be filled in at this time. Dates would be determined after an application for approval to construct or reconstruct had been submitted by the O/O.

SUMMARY OF IMPLEMENTATION DATES
Existing-Source MACT Requirements

Subpart: EE
Source Category: Magnetic Tape Manufacturing
OAQPS Contact: Gail Lacy, (919) 541-5261

Milestone	Comments on date	Date
Effective Date	Promulgation date	12/15/94
O/O: Initial notification due	Typically within 120 days of effective date	04/14/95
O/O: Request for compliance extension (up to 1 year for special circumstances)	Currently 12 months before compliance date	12/15/95 or 12/15/96*
EPA/state: Approval of request for compliance extensions	Variable but generally within 3 months of request	3/15/96 or 3/15/97*
Compliance Date	Within 2 years after effective date if O/O does not need to install a new add-on air pollution control device to comply, or within 3 years after effective date if O/O needs to install a new add-on air pollution control device (not counting extensions)	12/15/96 without new control device *or* 12/15/97 with new control device
O/O: Submit special compliance-monitoring plans	Not all rules have these plans	N/A
EPA/state: Review/approve special compliance-monitoring plans	Not all rules have these plans	N/A
O/O: Notice of performance test	Generally >30 days before performance test	5/14/97 or 5/14/98*
EPA/state: Approval of site-specific test plan	<30 days of receipt	6/13/97 or 6/13/98*
Performance test	Within 180 days of compliance date	6/13/97 or 6/13/98*
EPA/state: Attend performance test		6/13/97 or 6/13/98*
O/O: Submit for approval an alternative limit when coating operation is not occurring	Optional for O/O. Submit no more than 180 days after performance test	12/10/97 or 12/10/98*

SUMMARY OF IMPLEMENTATION DATES
Existing-Source MACT Requirements (Continued)

Subpart: EE
Source Category: Magnetic Tape Manufacturing
OAQPS Contact: Gail Lacy, (919) 541-5261

Milestone	Comments on date	Date
EPA/state: Approve or disapprove alternative limit for when coating operation is not occurring	Within 60 days of receipt of application for alternative limit and other supplemental supporting information requested	2/8/98 or 2/8/99*
O/O: Submit plan identifying parameters to be monitored for capture efficiency	With the first compliance status report	8/12/97 or 8/12/98*
Compliance status reports (before a Title V permit is functional for the source)	First report: <60 days from performance test/compliance date. Periodic: as required in standard and general provisions. Facilities subject to HAP utilization limit have different schedules [63.707(j)]	8/12/97 or 8/12/98* semi-annual
EPA/state: **Review compliance status reports**	There is a need to report simple summaries of compliance status	No date assigned

O/O means the owner or operator performs the milestone.
EPA/state means EPA or a delegated state performs the milestone.
*Two dates are shown because there are two different compliance dates in the rule. The earlier date is tied with the earlier compliance date, and the later date is tied with the later compliance date. The differences in the compliance dates are explained in the compliance date section of the table.

SUMMARY OF IMPLEMENTATION DATES
New-Source MACT Requirements

Subpart: EE
Source Category: Magnetic Tape Manufacturing
OAQPS Contact: Gail Lacy, (919) 541-5261

Milestone	Comments on date	Date*
Applicability Date for New Sources	Proposed rule date	03/11/94
O/O: Application for approval to construct/reconstruct a major emitting affected source	As soon as possible before commencing construction. Special allowance provided for sources commencing construction after proposal date but before effective date [§63.5(d)(1)(i)]	
EPA/state: Notice of complete information	Within 30 days from application by O/O [§63.5(e)(2)(i)]	
EPA/state: Approval of construction/reconstruction	Within 60 days from notification of complete application [§63.5(e)(2)(i)]	
O/O: Notice of intended start-up	30 to 60 days before start-up [§63.9(b)(4)(iv)]	
Then follows existing source process...		
Compliance Date	Upon start-up unless rule provides special allowance	

O/O means the owner or operator performs the milestone.
EPA/state means EPA or a delegated state performs the milestone.
*All the dates depend on the action of the O/O; thus they cannot be filled in at this time. Dates would be determined after an application for approval to construct or reconstruct had been submitted by the O/O.

SUMMARY OF IMPLEMENTATION DATES
Existing-Source MACT Requirements

Subpart: GG
Source Category: Aerospace Manufacturing and Rework Facilities
OAQPS Contact: Jim Szykman, (919) 541-2452

Milestone	Comments on date	Date
Effective Date	Promulgation date	09/01/95
O/O: Initial notification due	Typically within 120 days of effective date	09/01/97
O/O: Request for compliance extension (up to 1 year for special circumstances)	Currently 12 months before compliance date	05/04/98
EPA/state: Approval of request for compliance extensions	Variable but generally within 3 months of request	60 days after receipt
Compliance Date	No greater than 3 years after effective date	06/01/98
O/O: Submit special compliance-monitoring plans	Not all rules have these plans	N/A
EPA/state: Review/approve special compliance-monitoring plans	Not all rules have these plans	N/A
O/O: Notice of performance test	Generally >30 days before performance test	60 days before performance test
EPA/state: Approval of site-specific test plan	<30 days of receipt	30 days after receipt
Performance test	Within 180 days of compliance date	02/28/99
EPA/state: Attend performance test		
Compliance status reports (before a Title V permit is functional for the source)	Initial: <60 days from performance test/compliance date Periodic: as required in standard	60 days after initial perf. test; then semi-annually [§63.753]
EPA/state: **Review compliance status reports**	There is a need to report simple summaries of compliance status	No date assigned

O/O means the owner or operator performs the milestone.
EPA/state means EPA or a delegated state performs the milestone.

SUMMARY OF IMPLEMENTATION DATES
New-Source MACT Requirements

Subpart: GG
Source Category: Aerospace Manufacturing and Rework Facilities
OAQPS Contact: Jim Szykman, (919) 541-2452

Milestone	Comments on date	Date*
Applicability Date for New Sources	Proposed rule date	05/25/94
O/O: Application for approval to construct/reconstruct a major emitting affected source	As soon as possible before commencing construction. Special allowance provided for sources commencing construction after proposal date but before effective date [§63.5(d)(1)(i)]	
EPA/state: Notice of complete information	Within 30 days from application by O/O [§63.5(e)(2)(i)]	
EPA/state: Approval of construction/reconstruction	Within 60 days from notification of complete application [§63.5(e)(2)(i)]	
O/O: Notice of intended start-up	30 to 60 days before start-up [§63.9(b)(4)(iv)]	
Then follows existing source process...		
Compliance Date	Upon start-up unless rule provides special allowance	

O/O means the owner or operator performs the milestone.
EPA/state means EPA or a delegated state performs the milestone.
*All the dates depend on the action of the O/O; thus they cannot be filled in at this time. Dates would be determined after an application for approval to construct or reconstruct had been submitted by the O/O.

SUMMARY OF IMPLEMENTATION DATES
Existing-Source MACT Requirements

Subpart: II
Source Category: Shipbuilding and Ship Repair Facilities (Coating Operations)
OAQPS Contact: Mohammed Serageldin, (919) 541-2379

Milestone	Comments on date	Date
Effective Date	Rule promulgation date	12/15/95
O/O: Initial notification due	Typically within 120 days of effective date (**extended to 180 days**)	06/13/96
O/O: Submit special compliance-monitoring or implementation plans	Not all rules have these plans	12/16/96
EPA/state: Review/approve special compliance-monitoring plans	Not all rules have these plans	12/16/97
O/O: Request for compliance extension (up to 1 year for special circumstances)	Currently 12 months before compliance date (**extended from 12/16/96 to 12/16/97**)	
EPA/state: Approval of request for compliance extensions	Variable but generally within 3 months of request	
Compliance Date	Within 3 years after effective date (not counting extensions)	12/16/97
O/O: Notice of performance test	>30 days before performance test	
EPA/state: Approval of site-specific test plan	<30 days after receipt	
Performance test	Within 180 days of compliance date	
EPA/state: Attend performance test	Optional but recommended	
Compliance status reports (before a Title V permit is functional for the source)	First report: <60 days from performance test/compliance date Periodic: as required in standard	
EPA/state: **Review compliance status reports**	EPA/states need to report simple summaries of compliance status	No date assigned
First reporting period ends		6/16/98
First compliance reporting due (and every 6 months thereafter)		8/16/98

O/O means the owner or operator performs the milestone.
EPA/state means EPA or a delegated state performs the milestone.

SUMMARY OF IMPLEMENTATION DATES
New-Source MACT Requirements

Subpart: II
Source Category: Shipbuilding and Ship Repair Facilities (Coating Operations)
OAQPS Contact: Mohammed Serageldin, (919) 541-2379

Milestone	Comments on date	Date*
Applicability Date for New Sources	Rule proposed date	12/6/94
O/O: Application for approval to construct/reconstruct a major emitting affected source. *Note:* Nonmajor emitting affected sources are not required to request approval to construct/reconstruct; they merely notify the EPA/state.	As soon as possible before commencing construction. Special allowance provided for sources commencing construction after proposal date but before effective date [§63.5(d)(1)(i)]	
EPA/state: Notice of complete information	Within 30 days from application by O/O [§63.5(e)(2)(i)]	
EPA/state: Approval/denial of construction/reconstruction	Within 60 days from notification of complete application [§63.5(e)(2)(i)]	
O/O: Notice of intended start-up	30 to 60 days before start-up [§63.9(b)(4)(iv)]	6 months prior to start-up
Then follows existing source process...		
Compliance Date	Upon start-up unless rule provides special allowance	Start-up date
First reporting period ends		6 months after start-up
First compliance report due (and every 6 months hereafter)		8 months after start-up

O/O means the owner or operator performs the milestone.
EPA/state means EPA or a delegated state performs the milestone.
*All the dates depend on the action of the O/O; thus they cannot be filled in at this time. Dates would be determined after an application for approval to construct or reconstruct had been submitted by the O/O.

SUMMARY OF IMPLEMENTATION DATES
Existing-Source MACT Requirements

Subpart: JJ
Source Category: Wood Furniture Manufacturing Operations NESHAP
OAQPS Contact: Paul Almodovar, (919) 541-0283

Milestone	Comments on date	Date
Effective Date	Rule promulgation date	12/7/95
O/O: Initial notification due	Typically within 120 days of effective date	9/3/96
O/O: Submit special compliance-monitoring or implementation plans	Not all rules have these plans	N/A
EPA/state: Review/approve special compliance-monitoring plans	Not all rules have these plans	N/A
O/O: Request for compliance extension (can be allowed up to 1 year for special circumstances)	Currently 12 months before compliance date	11/21/96 for sources emitting >50 tons of HAP in 1996; 12/7/97 for sources emitting <50 tons of HAP in 1996
EPA/state: Approval of request for compliance extensions	Variable but generally within 3 months of request	2/21/97; 3/7/98
Compliance Date	Within 3 years after effective date (not counting extensions)	11/21/97 for sources emitting >50 tons of HAP in 1996; 12/7/98 for sources emitting <50 tons of HAP in 1996
O/O: Notice of performance test	Generally >30 days before test	
EPA/state: Approval of site test plan	<30 days after receipt	
Performance test	Within 180 days of compliance date	5/21/98; 6/7/99
EPA/state: Attend performance test	Optional but recommended	
Compliance status reports (before a Title V permit is functional for the source)	First report: <60 days from performance test/compliance date Periodic: as required in standard	1/21/98; 2/7/99
EPA/state: **Review compliance status reports**	EPA/states need to report simple summaries of compliance status	No date assigned

SUMMARY OF IMPLEMENTATION DATES
New-Source MACT Requirements

Subpart: JJ
Source Category: Wood Furniture Manufacturing Operations NESHAP
OAQPS Contact: Paul Almodovar, (919) 541-0283

Milestone	Comments on date	Date*
Applicability Date for New Sources	Rule proposed date	12/6/94
O/O: Application for approval to construct/reconstruct a major emitting affected source. *Note:* Nonmajor emitting affected sources are not required to request approval to construct/reconstruct; they merely notify the EPA/state.	As soon as possible before commencing construction. Special allowance provided for sources commencing construction after proposal date but before effective date [§63.5(d)(1)(i)]	
EPA/state: Notice of complete information	Within 30 days from application by O/O [§63.5(e)(2)(i)]	
EPA/state: Approval/denial of construction/reconstruction	Within 60 days from notification of complete application [§63.5(e)(2)(i)]	
O/O: Notice of intended start-up	30 to 60 days before start-up [§63.9(b)(4)(iv)]	
Then follows existing source process...		
Compliance Date	Upon start-up unless rule provides special allowance	

O/O means the owner or operator performs the milestone.
EPA/state means EPA or a delegated state performs the milestone.
*All the dates depend on the action of the O/O; thus they cannot be filled in at this time. Dates would be determined after an application for approval to construct or reconstruct had been submitted by the O/O.

SUMMARY OF IMPLEMENTATION DATES
Existing-Source MACT Requirements

Subpart: KK
Source Category: Printing and Publishing
OAQPS Contact: Dave Salman, (919) 541-0859

Milestone	Comments on date	Date
Effective Date	Rule promulgation date	05/30/96
O/O: Initial notification due	Typically within 120 days of effective date	05/30/98
O/O: Submit special compliance-monitoring or implementation plans	Not all rules have these plans	
EPA/state: Review/approve special compliance-monitoring plans	Not all rules have these plans	
O/O: Request for compliance extension (up to 1 year for special circumstances)	Currently 12 months before compliance date	05/30/98
EPA/state: Approval of request for compliance extensions	Variable but generally within 3 months of request	<30 days of receipt of comp. app.
Compliance Date	No greater than 3 years after effective date (not counting extensions)	05/30/99
O/O: Notice of performance test	Generally >30 days before performance test	>60 days before test
EPA/state: Approval of site-specific test plan	<30 days after receipt	<30 days after receipt
Performance test	Within 180 days after compliance date	Within 180 days after comp. date
EPA/state: Attend performance test	Optional but recommended	
Compliance status reports (before a Title V permit is functional for the source)	First report: <60 days from performance test/compliance date Periodic: as required in standard	<60 days after complete initial perf. test
EPA/state: **Review compliance status reports**	EPA/states need to report simple summaries of compliance status	No date assigned

O/O means the owner or operator performs the milestone.
EPA/state means EPA or a delegated state performs the milestone.

SUMMARY OF IMPLEMENTATION DATES
New-Source MACT Requirements

Subpart: KK
Source Category: Printing and Publishing
OAQPS Contact: Dave Salman, (919) 541-0859

Milestone	Comments on date	Date*
Applicability Date for New Sources	Rule proposed date	03/14/95
O/O: Application for approval to construct/reconstruct a major emitting affected source. *Note:* Nonmajor emitting affected sources are not required to request approval to construct/reconstruct; they merely notify the EPA/state.	As soon as possible before commencing construction. Special allowance provided for sources commencing construction after proposal date but before effective date [§63.5(d)(1)(i)]	
EPA/state: Notice of complete information	Within 30 days from application by O/O [§63.5(e)(2)(i)]	
EPA/state: Approval/denial of construction/reconstruction	Within 60 days from notification of complete application [§63.5(e)(2)(i)]	
O/O: Notice of intended start-up	30 to 60 days before start-up [§63.9(b)(4)(iv)]	
Then follows existing source process...		
Compliance Date	Upon start-up unless rule provides special allowance	

O/O means the owner or operator performs the milestone.
EPA/state means EPA or a delegated state performs the milestone.
*All the dates depend on the action of the O/O; thus they cannot be filled in at this time. Dates would be determined after an application for approval to construct or reconstruct had been submitted by the O/O.

SUMMARY OF IMPLEMENTATION DATES
Existing-Source MACT Requirements

Subpart: JJJ
Source Category: Polymers and Resins, Group IV
OAQPS Contact: Bob Rosensteel, (919) 541-5608

Milestone	Comments on date	Date
Effective Date	Promulgation date	09/12/96
O/O: Initial notification due	Typically within 120 days of effective date	N/A
O/O: Request for compliance extension (up to 1 year for special circumstances)	Currently 12 months before compliance date	09/12/98
EPA/state: Approval of request for compliance extensions	Variable but generally within 3 months of request	12/12/98
Compliance Date	No greater than 3 years after effective date	09/12/99
O/O: Submit special compliance-monitoring plans	Not all rules have these plans	3/12/98* 9/12/98†
EPA/state: Review/approve special compliance-monitoring plans	Not all rules have these plans	07/12/98* 10/12/98†
O/O: Notice of performance test	Generally >30 days before performance test	At least 30 days before test
EPA/state: Approval of site-specific test plan	<30 days of receipt	N/A
Performance test	Within 150 days of compliance date	By 02/12/00 for most emission pts.
EPA/state: Attend performance test		
Compliance status reports (before a Title V permit is functional for the source)	Initial: <60 days from performance test/compliance date Periodic: as required in standard	02/12/00‡
EPA/state: **Review compliance status reports**	There is a need to report simple summaries of compliance status	Not assigned

O/O means the owner or operator performs the milestone.
EPA/state means EPA or a delegated state performs the milestone.
*Emissions averaging plan.
†Precompliance report.
‡Notification of compliance status.

SUMMARY OF IMPLEMENTATION DATES
New-Source MACT Requirements

Subpart: JJJ
Source Category: Polymers and Resins, Group IV
OAQPS Contact: Bob Rosensteel, (919) 541-5608

Milestone	Comments on date	Date*
Applicability Date for New Sources	Proposed rule date	03/29/95
O/O: Application for approval to construct/reconstruct a major emitting affected source	As soon as practicable before commencing construction. Special allowance provided for sources commencing construction after proposal date but before effective date	
EPA/state: Notice of complete information	+30 days from application	
EPA/state: Approval of construction/reconstruction	+60 days from application	
O/O: Notice of intended start-up	>60 days before start-up	
Then follows existing source process...		
Compliance Date	Upon start-up unless rule provides special allowance	Upon start-up

O/O means the owner or operator performs the milestone.
EPA/state means EPA or a delegated state performs the milestone.
*All the dates depend on the action of the O/O; thus they cannot be filled in at this time. Dates would be determined after an application for approval to construct or reconstruct had been submitted by the O/O.

K

De Minimis Levels

Case no.	Chemical name	De minimis level, tons/yr	Basis
57147	1,1-Dimethyl hydrazine	0.008	UR
79005	1,1,2-Trichloroethane	1	UR
79345	1,1,2,2-Tetrachloroethane	0.3	UR
96128	1,2-Dibromo-3-chloropropane	0.008	UR
122667	1,2-Diphenylhydrazine	0.09	UR
106887	1,2-Epoxybutane	Pending	
75558	1,2-Propylenimine (2-methyl aziridine)	Pending	
120821	1,2,4-Trichlorobenzene	10	CS
106990	1,3-Butadiene	0.2	UR
542756	1,3-Dichloropropene	1	DEF=1
1120714	1,3-Propane sultone	Pending	
106467	1,4-Dichlorobenzene (p)	3	UR
123911	1,4-Diozane (1,4-diethyleneoxide)	.6	UR
53963	2-Acetylaminofluorine	Pending	
532274	2-Chloroacetophenone	0.6	RfC
79469	2-Nitropropane	0.007	UR
640841	2,2,4-Trimethylpentane	5	DEF=5
1746016	2,3,7,8-Tetrachlorodibenzo-p-diozin	6E-07	UR
584849	2,4-Toluene diisocyanate	0.1	ACUTE
51285	2,4-Dinitrophenol	1	CS
121142	2,4-Dinitrotoluene	Pending	
94757	2,4-D, salts, esters(2,4-dichlorophenoxy acetic acid)	10	CS
95807	2,4-Toluene diamine	0.02	UR
88062	2,4,6-Trichlorophenol	6	UR
91941	3,3-Dichlorobenzidene	0.2	UR
119904	3,3'-Dimethoxybenzidine	5	UR
119937	3,3-Dimethyl benzidine	0.008	UR
92933	4-Nitrobiphenyl	5	DEF=5
100027	4-Nitrophenol	5	DEF=5
101144	4,4-Methylene bis(2-chloroaniline)	Pending	
534521	4,6-Dinitro-o-cresol, and salts	0.1	ACUTE
75070	Acetaldehyde	9	UR
75058	Acetonitrile	10	CS
98862	Acetophanone	1	CS
107028	Acrolein	0.04	RfC
79061	Acrylamide	0.02	UR

Case no.	Chemical name	De minimis level, tons/yr	Basis
79107	Acrylic acid	Pending	
107131	Acrylonitrile	0.3	UR
107051	Allyl chloride	1	DEF=1
62533	Aniline	1	UR
71432	Benzene	2	UR
92875	Benzidine	0.0003	UR
98077	Benzotrichloride	0.006	UR
100447	Benzyl chloride	0.1	ACUTE
97578	beta-Proprolactone	0.1	ACUTE
92524	Biphenyl	10	CS
117817	Bis(2-ethylhexyl)phthalate (DEHP)	5	UR
542981	Bis(chloromethyl)ether	0.0003	UR
75252	Bromoform	10	CAP-UR
158827	Calcium cyanamide	10	CS
105802	Caprolactam	10	CS
133062	Captan	10	CAP-UR
63252	Carbaryl	10	CS
75150	Carbon disulfide	1	CS
56235	Carbon tetrachloride	1	UR
463581	Carbonyl sulfide	5	DEF=5
120809	Catechol	5	DEF=5
57749	Chlordane	0.05	UR
7782505	Chlorine	0.1	ACUTE
79118	Chloroacetic acid	0.1	ACUTE
108907	Chlorobenzene	10	CS
510158	Chlorobenzilate	0.4	UR
67683	Chloroform	0.9	UR
107302	Chloromethyl methyl ether	0.1	ACUTE
126998	Chloroprene	10	CS
1319773	Cresols/cresylic acid (isomers and mixture)	1	DEF=1
95487	o-Cresol	1	DEF=1
108394	m-Cresol	1	DEF=1
106445	p-Cresol	1	DEF=1
98828	Cumene	10	CS
132649	Dibenzofuran	5	DEF=5
72559	DOE (p'p'-Dichlorodiphenyldichloroethylene)	0.2	UR
84742	Dibutylphthalate	10	CS
111444	Dichloroethyl ether (Bis(2-chloroethyl)ether)	0.06	UR
62737	Dichlorvos	0.2	UR
11422	Diethanolamine	5	DEF=5
60117	Dimethyl aminoazobenzene	1	DEF=1
79447	Dimethyl carbamoyl chloride	Pending	
68122	Dimethyl formamide	Pending	
131113	Dimethyl phthalate	10	CS
77781	Dimethyl sulfate	0.1	ACUTE
106898	Epichlorohydrin	2	RfC
140885	Ethyl acrylate	1	UR
100414	Ethyl benzene	10	CAP-RfC
51796	Ethyl carbamate (Urethane)	Pending	
75003	Ethyl chloride	10	CAP-RfC
106834	Ethylene dibromide (dibromoethane)	0.1	UR
107062	Ethylene dichloride (1,2-dichloroethane)	0.8	UR

Case no.	Chemical name	De minimis level, tons/yr	Basis
107211	Ethylene glycol	5	DEF=5
151564	Ethylene imine (aziridine)	Pending	
75218	Ethylene oxide	0.2	UR
96457	Ethylene thiourea	0.6	UR
75343	Ethylidene dichloride (1,1-dichloroethane)	1	DEF=1
62207765	Fluomine	1	CS
50000	Formaldehyde	2	UR
76448	Heptachlor	0.02	UR
118741	Hexachlorobenzene	0.04	UR
87683	Hexachorobutadiene	0.9	UR
77474	Hexachlorocyclopentadiene	0.1	ACUTE
67721	Hexachloroethane	5	UR
822060	Hexamethylene-1,6-diisocyanate	5	DEF=5
110543	Hexane	10	CAP-RfC
302012	Hydrazine	0.004	UR
7647010	Hydrochloric acid	10	CAP-RfC
7664393	Hydrogen fluoride	0.1	ACUTE
123319	Hydroquinone	10	CS
78591	Isophorone	10	CAP-UR
58899	Lindane (hexachlorcyclohexane, gamma)	0.05	UR
108316	Maleic anhydride	1	CS
67581	Methanol	10	CS
72435	Methoxychlor	10	CS
74839	Methyl bromide (bromomethane)	10	RfC
74873	Methyl chloride (chloromethane)	10	CAP-UR
71556	Methyl chloroform (1,1,1-trichloroethane)	10	CS
78933	Methyl ethyl ketone (2-butanone)	10	CAP-RfC
60344	Methyl hydrazine	0.06	UR
74884	Methyl iodide (iodomethane)	1	DEF=1
108101	Methyl isobutyl ketone	10	CS
624839	Methyl isocyanate	0.1	ACUTE
80626	Methyl methacrylate	10	CS
1634044	Methyl tert-butyl ether	10	CAP-RfC
12108133	Methylcyclopentadienyl manganese	0.1	ACUTE
75092	Methylene chloride (dichloromethane)	10	CAP-UR
101688	Methylene diphenyl diisocyanate	0.1	CS
91203	Naphthalene	10	CS
98953	Nitrobenzene	1	CS
62759	N-Nitrosodimethylamine	0.001	UR
684935	N-Nitroso-N-methylurea	Pending	
121697	N,N-Dimethylaniline	1	CS
95534	o-Toluidine	0.3	UR
56382	Parathion	0.1	ACUTE
82688	Pentachloronitrobenzene (quintobenzene)	0.3	UR
87865	Pentachlorophenol	0.7	UR
108952	Phenol	0.1	CS
75445	Phosgene	0.1	ACUTE
7803512	Phosphine	5	DEF=5
7723140	Phosphorous	0.1	ACUTE
85449	Phthalic anhydride	5	DEF=5
1336363	Polychlorinated biphenyls (aroclors)	0.009	UR

Case no.	Chemical name	De minimis level, tons/yr	Basis
106503	p-Phenylenediamine	10	CS
123386	Propionaldehyde	5	DEF=5
114261	Propoxur (baygone)	5	DEF=5
73875	Propylene dichloride (1,2-dichloropropane)	1	UR
75569	Propylene oxide	10	CAP-UR
91225	Quinoline	0.006	UR
106514	Quinone	5	DEF=5
100425	Styrene	1	DEF=1
127184	Tetrachloroethylene (perchloroethylene)	10	CAP-UR
7550450	Titanium tetrachloride	0.1	ACUTE
108883	Toluene	10	CAP-RfC
8001352	Toxaphene (chlorinated camphene)	0.06	UR
79016	Trichloroethylene	10	CAP-UR
121448	Triethylamine	10	CAP-RfC
1582098	Trifluralin	9	UR
108054	Vinyl acetate	Pending	
593602	Vinyl bromide (bromoethene)	1	UR
75014	Vinyl chloride	0.2	UR
75354	Vinylidene chloride (1,1-dichloroethylene)	0.4	UR
1330207	Xylenes (isomers and mixture)	10	CS
108383	m-Xylenes	10	CS
95476	o-Xylenes	10	CS
106423	p-Xylenes	10	CS
	Chemical compound classes		
—	Arsenic and inorganic arsenic compounds	0.005	UR
7784421	Arsine	5	DEF=5
—	Antimony compounds (except those specifically listed)*	5	DEF=5
1309644	Antimony trioxide	1	DEF=1
1345046	Antimony trisulfide	0.1	CS
28300745	Antimony potassium tartrate	1	CS
—	Beryllium compounds (except beryllium salts)	0.008	UR
—	Beryllium salts	Pending	
—	Cadmium compounds	0.01	UR
—	Chromium compounds (except hexavalent and trivalent)	5	DEF=5
—	Hexavalent chromium compounds	0.002	UR
—	Trivalent chromium compounds	5	DEF=5
—	Cobalt compounds (except those specifically listed)*	5	DEF=5
744084	Cobalt metal	0.1	ACUTE
10210681	Cobalt carbonyl	0.1	ACUTE
—	Coke oven emissions	0.03	UR
—	Cyanide compounds (except those specifically listed)*	5	DEF=5
143339	Sodium cyanide	10	CS
151508	Potassium cyanide	10	CS
—	Glycol ethers (except those specifically listed)	5	DEF=5
110805	2-Ethoxy ethanol	10	CAP-RfC
111762	Ethylene glycol monobutyl ether	10	CS
103364	2-Methoxy ethanol	1	CS
107992	Propylene glycol monomethyl ether	10	CAP-RfC
—	Inorganic lead compounds	0.6000	PSO
75741	Tetramethyl lead	0.1	ACUTE

Case no.	Chemical name	De minimis level, tons/yr	Basis
78002	Tetraethyl lead	0.1	ACUTE
7439965	Manganese and compounds	0.8	RfC
—	Mercury compounds (except those specifically listed)*	5	DEF=5
10045940	Mercuric nitrate	0.1	CS
748794	Mercuric chloride	1	CS
62384	Phenyl mercuric acetate	1	CS
—	Elemental mercury	0.6	IRIS
—	Mineral fiber compounds	Pending	
1332214	Asbestos	Pending	
—	Erionite	Pending	
—	Silica (crystalline)	Pending	
—	Talc (containing asbestos form fibers)	Pending	
—	Nickel compounds (except those specifically listed)*	1	DEF=1
13463393	Nickel carbonyl	0.1	ACUTE
12035722	Nickel refinery dust	0.08	UR
—	Nickel subsulfide	0.04	UR
—	Polycyclic organic matter (POM)		
58553	Benz(a)anthracene	1	DEF=1
50328	Benzo(a)pyrene	0.01	UR
205992	Benzo(b)fluoranthene	1	DEF=1
57976	7,12-Dimethylbenz(a)anthracene	1	DEF=1
225514	Benz(c)acridine	1	DEF=1
218019	Chrysene	1	DEF=1
53703	Dibenz(ah)anthracene	1	DEF=1
189559	1,2:7,8-Dibenzopyrene	1	DEF=1
193395	Indeno(1,2,3-cd)pyrene	1	DEF=1
—	Dioxins and Furans (TCDD equivalent)†	6E-07	UR
7782492	Selenium and compounds (except those specifically listed)*	1	DEF=1
7446346	Selenium sulfide	1	DEF=1
7488564	Selenium disuldfide	1	DEF=1
99999918	Radionuclides (including radon)	Pending	

Legend:
UR   Based on unit risk value
DEF=1   Used for carcinogens where no UR exists
RfC   Based on reference concentration in IRIS
CS   Used where no RfC is listed in IRIS CS=1–20: de minimis=10
            CS=21–40: de minimis=1; CS>40: de minimis=0.1
DEF=5   Used where no UR, RfC, or CS exists
CAP   UR, or RfC yielded a value>10 tons/yr.
        Thus a CAP of 10 tons/yr was used.
Pending   Awaiting data to assign a value.
IRIS   Integrated Risk Information System
*For this chemical group, specific compounds or subgroups are named specifically in this table. For the remainder of the chemicals of the chemical group, a single de minimis value is listed. This value applies to compounds which are not named specifically.
†The "toxic equivalent factor" method in EPA/625/3-89-016 should be used for PCDD/PCDF mixtures.

L

Accidental-Release Prevention Provisions

1. Part 68 is amended by redesignating subpart C as subpart F as follows:

   Subpart F Regulated Substances for Accidental Release Prevention

2. The table of contents of Part 68 is revised to read as follows:

Part 68—ACCIDENTAL RELEASE PREVENTION PROVISIONS

Subpart A General
68.1	Scope.
68.3	Definitions.
68.10	Applicability.
68.12	General requirements.
68.15	Management.

Subpart B Hazard Assessment
68.20	Applicability.
68.22	Off-site consequence analysis parameters.
68.25	Worst-case release scenario analysis.
68.28	Alternative release scenario analysis.
68.30	Defining off-site impacts—population.
68.33	Defining off-site impacts—environment.
68.36	Review and update.
68.39	Documentation.
68.42	Five-year accident history.

Subpart C Program 2 Prevention Program
68.48	Safety information.
68.50	Hazard review.
68.52	Operating procedures.
68.54	Training.
68.56	Maintenance.
68.58	Compliance audits.
68.60	Incident investigation.

Subpart D Program 3 Prevention Program
68.65	Process safety information.
68.67	Process hazard analysis.

3. The authority citation is revised to read as follows:

Authority: 42 U.S.C. 7412(r), 7601(a)(1), 7661-7661f.

4. Section 68.3 is amended to add the following definitions:

68.3   Definitions

Act means the Clean Air Act as amended (42 U.S.C. 7401 *et seq.*)
Administrative controls mean written procedural mechanisms used for hazard control.

AIChE/CCPS means the American Institute of Chemical Engineers/Center for Chemical Process Safety.

API means the American Petroleum Institute.

ASME means the American Society of Mechanical Engineers.

Catastrophic release means a major uncontrolled emission, fire, or explosion, involving one or more regulated substances that presents imminent and substantial endangerment to public health and the environment.

Classified information means "classified information" as defined in the Classified Information Procedures Act, 18 U.S.C. App. 3, section 1(a) as "any information or material that has been determined by the United States Government pursuant to an executive order, statute, or regulation, to require protection against unauthorized disclosure for reasons of national security."

Covered process means a process that has a regulated substance present in more than a threshold quantity as determined under §68.115 of this part.

Designated agency means the state, local, or Federal agency designated by the state under the provisions of §68.215(d) of this part.

Environmental receptor means natural areas such as national or state parks, forests, or monuments; officially designated wildlife sanctuaries, preserves, refuges, or areas; and Federal wilderness areas, that could be exposed at any time to toxic concentrations, radiant heat, or overpressure greater than or equal to the endpoints provided in §68.22(a) of this part, as a result of an accidental release and that can be identified on local U.S. Geological Survey maps.

Hot work means work involving electric or gas welding, cutting, brazing, or similar flame- or spark-producing operations.

Implementing agency means the state or local agency that obtains delegation for an accidental-release prevention program under subpart E, 40 CFR part 63. The implementing agency may, but is not required to, be the state or local air permitting agency. If no state or local agency is granted delegation, EPA will be the implementing agency for that state.

Injury means any effect on a human that results either from direct exposure to toxic concentrations; radiant heat; or overpressures from accidental releases or from the direct consequences of a vapor cloud explosion (such as flying glass, debris, and other projectiles) from an accidental release and that requires medical treatment or hospitalization.

Major change means introduction of a new process, process equipment, or regulated substance, an alteration of process chemistry that results in any change to safe operating limits, or other alteration that introduces a new hazard.

Mechanical integrity means the process of ensuring that process equipment is fabricated from the proper materials of construction and is properly installed, maintained, and replaced to prevent failures and accidental releases.

Medical treatment means treatment, other than first aid, administered by a physician or registered professional personnel under standing orders from a physician.

Mitigation or mitigation system means specific activities, technologies, or equipment designed or deployed to capture or control substances upon loss of containment to minimize exposure of the public or the environment. Passive mitigation means equipment, devices, or technologies that function without human, mechanical, or other energy input. Active mitigation means equipment, devices, or technologies that need human, mechanical, or other energy input to function.

NFPA means the National Fire Protection Association.

Offsite means areas beyond the property boundary of the stationary source, and areas within the property boundary to which the public has routine and unrestricted access during or outside business hours.

OSHA means the U.S. Occupational Safety and Health Administration.

Owner or operator means any person who owns, leases, operates, controls, or supervises a stationary source.

Population means the public.

Public means any person except employees or contractors at the stationary source.

Public receptor means offsite residences, institutions (e.g., schools, hospitals), industrial, commercial, and office buildings, parks, or recreational areas inhabited or occupied by the public at any time without restriction by the stationary source where members of the public could be exposed to toxic concentrations, radiant heat, or overpressure, as a result of an accidental release.

Replacement in kind means a replacement that satisfies the design specifications.

RMP means the risk management plan required under subpart G of this part.

SIC means Standard Industrial Classification.

Typical meteorological conditions means the temperature, wind speed, cloud cover, and atmospheric stability class, prevailing at the site based on data gathered at or near the site or from a local meteorological station.

Worst-case release means the release of the largest quantity of a regulated substance from a vessel or process line failure that results in the greatest distance to an endpoint defined in §68.22(a) of this part.

5. Section 68.10 is added to read as follows:

68.10   Applicability.

(a) An owner or operator of a stationary source that has more than a threshold quantity of a regulated substance in a process, as determined under §68.115 of this part, shall comply with the requirements of this part no later than the latest of the following dates:

(1) *[Insert date 3 years after the date of publication in the Federal Register]:*

(2) Three years after the date on which a regulated substance is first listed under §68.130 of this part; or

(3) The date on which a regulated substance is first present above a threshold quantity in a process.

(b) *Program 1 eligibility requirements.* A covered process is eligible for Program 1 requirements as provided in §68.12(b) of this part if it meets all of the following requirements:

(1) For the 5 years prior to the submission of an RMP, the process has not had an accidental release of a regulated substance where exposure to the substance, its reaction products, overpressure generated by an explosion involving the substance, or radiant heat generated by a fire involving the substance led to any of the following offsite:

(i) Death;

(ii) Injury; or

(iii) Response or restoration activities for an exposure of an environmental receptor;

(2) The distance to a toxic or flammable endpoint for a worst-case release assessment conducted under subpart B and §68.25 of this part is less than the distance to any public receptor, as defined in §68.30 of this part; and

(3) Emergency response procedures have been coordinated between the stationary source and local emergency planning and response organizations.

(c) *Program 2 eligibility requirements.* A covered process is subject to Program 2 requirements if it does not meet the eligibility requirements of either paragraph (b) or paragraph (d) of this section.

(d) *Program 3 eligibility requirements.* A covered process is subject to Program 3 requirements if the process does not meet the requirements of paragraph (b) of this section, and if either of the following conditions is met:

(1) The process is in SIC code 2611, 2812, 2819, 2821, 2865, 2869, 2873, 2879, or 2911; or

(2) The process is subject to the OSHA process safety management standard, 29 CFR 1910.119.

(e) If at any time a covered process no longer meets the eligibility criteria of its Program level, the owner or operator shall comply with the requirements of the new Program level that applies to the process and update the RMP as provided in §68.190 of this part.

6. Section 68.12 is added to read as follows:

68.12   General requirements.

(a) *General requirements.* The owner or operator of a stationary source subject to this part shall submit a single RMP, as provided in §§68.150 to

68.185 of this part. The RMP shall include a registration that reflects all covered processes.

(b) *Program 1 requirements.* In addition to meeting the requirements of paragraph (a) of this section, the owner or operator of a stationary source with a process eligible for Program 1, as provided in §68.10(b) of this part, shall:

(1) Analyze the worst-case release scenario for the process(es), as provided in §68.25 of this part; document that the nearest public receptor is beyond the distance to a toxic or flammable endpoint defined in §68.22(a) of this part; and submit in the RMP the worst-case release scenario as provided in §68.165 of this part;

(2) Complete the 5-year accident history for the process as provided in §68.42 of this part and submit it in the RMP as provided in §68.168 of this part;

(3) Ensure that response actions have been coordinated with local emergency planning and response agencies; and

(4) Certify in the RMP the following: "Based on the criteria in 40 CFR 68.10, the distance to the specified endpoint for the worst-case accidental release scenario for the following process(es) is less than the distance to the nearest public receptor: [list process(es)]. Within the past 5 years, the process(es) has (have) had no accidental release that caused offsite impacts provided in the risk management program rule [40 CFR 68.10(b)(1)]. No additional measures are necessary to prevent offsite impacts from accidental releases. In the event of fire, explosion, or a release of a regulated substance from the process(es), entry within the distance to the specified endpoints may pose a danger to public emergency responders. Therefore, public emergency responders should not enter this area except as arranged with the emergency contact indicated in the RMP. The undersigned certifies that, to the best of my knowledge, information, and belief, formed after reasonable inquiry, the information submitted is true, accurate, and complete. [Signature, title, date signed.]"

(c) *Program 2 requirements.* In addition to meeting the requirements of paragraph (a) of this section, the owner or operator of a stationary source with a process subject to Program 2, as provided in §68.10(c) of this part, shall:

(1) Develop and implement a management system as provided in §68.15 of this part;

(2) Conduct a hazard assessment as provided in §§68.20 through 68.42 of this part;

(3) Implement the Program 2 prevention steps provided in §§68.48 through 68.60 of this part or implement the Program 3 prevention steps provided in §§68.65 through 68.87 of this part;

(4) Develop and implement an emergency response program as provided in §§68.90 to 68.95 of this part; and

(5) Submit as part of the RMP the data on prevention program elements for Program 2 processes as provided in §68.170 of this part.

(d) *Program 3 requirements.* In addition to meeting the requirements of paragraph (a) of this section, the owner or operator of a stationary source with a process subject to Program 3, as provided in §68.10(d) of this part, shall:

(1) Develop and implement a management system as provided in §68.15 of this part;

(2) Conduct a hazard assessment as provided in §§68.20 through 68.42 of this part;

(3) Implement the prevention requirements of §§68.65 through 68.87 of this part;

(4) Develop and implement an emergency response program as provided in §§68.90 to 68.95 of this part; and

(5) Submit as part of the RMP the data on prevention program elements for Program 3 processes as provided in §68.175 of this part.

7. Section 68.15 is added to read as follows:

68.15   Management.

(a) The owner or operator of a stationary source with processes subject to Program 2 or Program 3 shall develop a management system to oversee the implementation of the risk management program elements.

(b) The owner or operator shall assign a qualified person or position that has the overall responsibility for the development, implementation, and integration of the risk management program elements.

(c) When responsibility for implementing individual requirements of this part is assigned to persons other than the person identified under paragraph (b) of this section, the names or positions of these people shall be documented and the lines of authority defined through an organization chart or similar document.

8. Subpart B is added to read as follows:

Subpart B Hazard Assessment

68.20   Applicability.
68.22   Offsite consequence analysis parameters.
68.25   Worst-case release scenario analysis.
68.28   Alternative release scenario analysis.
68.30   Defining offsite impacts—population.
68.33   Defining offsite impacts—environment.
68.36   Review and update.
68.39   Documentation.
68.42   Five-year accident history.

68.20   Applicability. This owner or operator of a stationary source subject to this part shall prepare a worst-case release scenario analysis as provided in §68.25 of this part and complete the 5-year accident history as provided in §68.42 of this part. The owner or operator of a Program 2 and 3 process must comply with all sections in this subpart for these processes.

68.22   Offsite consequence analysis parameters.

(a) Endpoints. For analyses of offsite consequences, the following endpoints shall be used:

(1) Toxics. The toxic endpoints provided in Appendix A of this part.

(2) Flammables. The endpoints for flammables vary according to the scenarios studied:

(i) Explosion. An overpressure of 1 psi.

(ii) Radiant heat/exposure time. A radiant heat of 5 kW/m^2 for 40 seconds.

(iii) Lower flammability limit. A lower flammability limit as provided in NFPA documents or other generally recognized sources.

(b) Wind speed/atmospheric stability class. For the worst-case release analysis, the owner or operator shall use a wind speed of 1.5 meters per second and F atmospheric stability class. If the owner or operator can demonstrate that local meteorological data applicable to the stationary source show a higher minimum wind speed or less stable atmosphere at all times during the previous 3 years, these minimums may be used. For analysis of alternative scenarios, the owner or operator may use the typical meteorological conditions for the stationary source.

(c) Ambient temperature/humidity. For worst-case release analysis of a regulated toxic substance, the owner or operator shall use the highest daily maximum temperature in the previous 3 years and average humidity for the site, based on temperature/humidity data gathered at the stationary source or at a local meteorological station; an owner or operator using the *RMP Offsite Consequence Analysis Guidance* may use 25°C and 50 percent humidity as values for these variables. For analysis of alternative scenarios, the owner or operator may use typical temperature/humidity data gathered at the stationary source or at a local meteorological station.

(d) Height of release. The worst-case release of a regulated toxic substance shall be analyzed assuming a ground level (0 feet) release. For an alternative scenario analysis of a regulated toxic substance, release height may be determined by the release scenario.

(e) Surface roughness. The owner or operator shall use either urban or rural topography, as appropriate. Urban means that there are many obstacles in the immediate area; obstacles include buildings or trees. Rural means there are no buildings in the immediate area and the terrain is generally flat and unobstructed.

(f) Dense or neutrally buoyant gases. The owner or operator shall ensure that tables or models used for dispersion analysis of regulated toxic substances appropriately account for gas density.

(g) Temperature of released substance. For worst case, liquids other than gases liquefied by refrigeration only shall be considered to be released at the highest daily maximum temperature, based on data for the previous 3 years appropriate for the stationary source, or at process temperature, whichever is higher. For alternative scenarios, substances may be considered to be released at a process or ambient temperature that is appropriate for the scenario.

68.25   Worst-case release scenario analysis.

(a) The owner or operator shall analyze and report in the RMP:

(1) For Program 1 processes, one worst-case release scenario for each Program 1 process;

(2) For Program 2 and 3 processes:

(i) One worst-case release scenario that is estimated to create the greatest distance in any direction to an endpoint provided in Appendix A of this part resulting from an accidental release of regulated toxic substances from covered processes under worst-case conditions defined in §68.22 of this part;

(ii) One worst-case release scenario that is estimated to create the greatest distance in any direction to an endpoint defined in §68.22(a) of this part resulting from an accidental release of regulated flammable substances from covered processes under worst-case conditions defined in §68.22 of this part; and

(iii) Additional worst-case release scenarios for a hazard class if a worst-case release from another covered process at the stationary source potentially affects public receptors different from those potentially affected by the worst-case release scenario developed under paragraphs (a)(2)(i) or (a)(2)(ii) of this section.

(b) Determination of worst-case release quantity. The worst-case release quantity shall be the greater of the following:

(1) For substances in a vessel, the greatest amount held in a single vessel, taking into account administrative controls that limit the maximum quantity; or

(2) For substances in pipes, the greatest amount in a pipe, taking into account administrative controls that limit the maximum quantity.

(c) Worst-case release scenario—toxic gases.

(1) For regulated toxic substances that are normally gases at ambient temperature and handled as a gas or as a liquid under pressure, the owner or operator shall assume that the quantity in the vessel or pipe, as determined under paragraph (b) of this section, is released as a gas over 10 minutes. The release rate shall be assumed to be the total quantity divided by 10 unless passive mitigation systems are in place.

(2) For gases handled as refrigerated liquids at ambient pressure:

(i) If the released substance is not contained by passive mitigation systems or if the contained pool would have a depth of 1 cm or less, the owner or operator shall assume that the substance is released as a gas in 10 minutes;

(ii) If the released substance is contained by passive mitigation systems in a pool with a depth greater than 1 cm, the owner or operator may assume that the quantity in the vessel or pipe, as determined under paragraph (b) of this section, is spilled instantaneously to form a liquid pool. The volatilization rate (release rate) shall be calculated at the boiling point of the substance and at the conditions specified in paragraph (d) of this section.

(d) Worst-case release scenario—toxic liquids.

(1) For regulated toxic substances that are normally liquids at ambi-

ent temperature, the owner or operator shall assume that the quantity in the vessel or pipe, as determined under paragraph (b) of this section, is spilled instantaneously to form a liquid pool.

(i)   The surface area of the pool shall be determined by assuming that the liquid spreads to 1 cm deep unless passive mitigation systems are in place that serve to contain the spill and limit the surface area. Where passive mitigation is in place, the surface area of the contained liquid shall be used to calculate the volatilization rate.

(ii) If the release would occur onto a surface that is not paved or smooth, the owner or operator may take into account the actual surface characteristics.

(2)   The volatilization rate shall account for the highest daily maximum temperature occurring in the past 3 years, the temperature of the substance in the vessel, and the concentration of the substance if the liquid spilled is a mixture or solution.

(3)   The rate of release to air shall be determined from the volatilization rate of the liquid pool. The owner or operator may use the methodology in the *RMP Offsite Consequence Analysis Guidance* or any other publicly available techniques that account for the modeling conditions and are recognized by industry as applicable as part of current practices. Proprietary models that account for the modeling conditions may be used provided the owner or operator allows the implementing agency access to the model and describes model features and differences from publicly available models to local emergency planners upon request.

(e)   Worst-case release scenario—flammables. The owner or operator shall assume that the quantity of the substance, as determined under paragraph (b) of this section, vaporizes resulting in a vapor cloud explosion. A yield factor of 10 percent of the available energy released in the explosion shall be used to determine the distance to the explosion endpoint if the model used is based on TNT-equivalent methods.

(f)   Parameters to be applied. The owner or operator shall use the parameters defined in §68.22 of this part to determine distance to the endpoints. The owner or operator may use the methodology provided in the *RMP Offsite Consequence Analysis Guidance* or any commercially or publicly available air dispersion modeling techniques, provided the techniques account for the modeling conditions and are recognized by industry as applicable as part of current practices. Proprietary models that account for the modeling conditions may be used provided the owner or operator allows the implementing agency access to the model and describes model features and differences from publicly available models to local emergency planners upon request.

(g)   Consideration of passive mitigation. Passive mitigation systems may be considered for the analysis of worst case provided that the mitigation system is capable of withstanding the release event triggering the scenario and would still function as intended.

(h)   Factors in selecting a worst-case scenario. Notwithstanding the provisions of paragraph (b) of this section, the owner or operator shall select as the worst case for flammable regulated substances or the worst case for regulated toxic substances, a scenario based on the following factors if such a scenario would result in a greater distance to an endpoint defined in §68.22(a) of this part beyond the stationary source boundary than the scenario provided under paragraph (b) of this section:

(1)   Smaller quantities handled at higher process temperature or pressure; and

(2)   Proximity to the boundary of the stationary source.

68.28   Alternative release scenario analysis.

(a)   The number of scenarios. The owner or operator shall identify and analyze at least one alternative release scenario for each regulated toxic substance held in a covered process(es) and at least one alternative release scenario to represent all flammable substances held in covered processes.

(b)   Scenarios to consider.

(1)   For each scenario required under paragraph (a) of this section, the owner or operator shall select a scenario:

(i)   That is more likely to occur than the worst-case release scenario under §68.25 of this part; and

(ii)   That will reach an endpoint offsite, unless no such scenario exists.

(2)   Release scenarios considered should include, but are not limited to, the following, where applicable:

(i)   Transfer hose releases due to splits or sudden hose uncoupling;

(ii)   Process piping releases from failures at flanges, joints, welds, valves and valve seals, and drains or bleeds;

(iii)   Process vessel or pump releases due to cracks, seal failure, or drain, bleed, or plug failure;

(iv)   Vessel overfilling and spill, or overpressurization and venting through relief valves or rupture disks; and

(v)   Shipping container mishandling and breakage or puncturing leading to a spill.

(c)   Parameters to be applied. The owner or operator shall use the appropriate parameters defined in §68.22 of this part to determine distance to the endpoints. The owner or operator may use either the methodology provided in the *RMP Offsite Consequence Analysis Guidance* or any commercially or publicly available air dispersion modeling techniques, provided the techniques account for the specified modeling conditions and are recognized by industry as applicable as part of current practices. Proprietary models that account for the modeling conditions may be used provided the owner or operator allows the implementing agency access to the model and describes model features and differences from publicly available models to local emergency planners upon request.

(d)   Consideration of mitigation. Active and passive mitigation systems may be considered provided they are capable of withstanding the event that triggered the release and would still be functional.

(e)   Factors in selecting scenarios. The owner or operator shall consider the following in selecting alternative release scenarios:

(1) The 5-year accident history provided in §68.42 of this part; and

(2) Failure scenarios identified under §68.50 or §68.67 of this part.

68.30   Defining offsite impacts—population.

(a) The owner or operator shall estimate in the RMP the population within a circle with its center at the point of the release and a radius determined by the distance to the endpoint defined in §68.22(a) of this part.

(b) Population to be defined. Population shall include residential population. The presence of institutions (schools, hospitals, prisons), parks and recreational areas, and major commercial, office, and industrial buildings shall be noted in the RMP.

(c) Data sources acceptable. The owner or operator may use the most recent Census data, or other updated information, to estimate the population potentially affected.

(d) Level of accuracy. Population shall be estimated to two significant digits.

68.33   Defining offsite impacts—environment.

(a) The owner or operator shall list in the RMP environmental receptors within a circle with its center at the point of the release and a radius determined by the distance to the endpoint defined in §68.22(a) of this part.

(b) Data sources acceptable. The owner or operator may rely on information provided on local U.S. Geological Survey maps or on any data source containing U.S.G.S. data to identify environmental receptors.

68.36   Review and update.

(a) The owner or operator shall review and update the offsite consequence analyses at least once every 5 years.

(b) If changes in processes, quantities stored or handled, or any other aspect of the stationary source might reasonably be expected to increase or decrease the distance to the endpoint by a factor of 2 or more, the owner or operator shall complete a revised analysis within 6 months of the change and submit a revised risk management plan as provided in §68.190 of this part.

68.39   Documentation. The owner or operator shall maintain the following records on the offsite consequence analyses:

(a) For worst-case scenarios, a description of the vessel or pipeline and substance selected as worst case, assumptions and parameters used, and the rationale for selection; assumptions shall include use of any administrative controls and any passive mitigation that were assumed to limit the quantity that could be released. Documentation shall include the anticipated effect of the controls and mitigation on the release quantity and rate.

(b) For alternative release scenarios, a description of the scenarios identified, assumptions and parameters used, and the rationale for the selection of specific scenarios; assumptions shall include use of any administrative controls and any mitigation that were assumed to limit the quantity that could be released. Documentation shall include the effect of the controls and mitigation on the release quantity and rate.

(c) Documentation of estimated quantity released, release rate, and duration of release.

(d) Methodology used to determine distance to endpoints.

(e) Data used to estimate population and environmental receptors potentially affected.

68.42   Five-year accident history.

(a) The owner or operator shall include in the 5-year accident history all accidental releases from covered processes that resulted in deaths, injuries, or significant property damage on-site, or known offsite deaths, injuries, evacuations, sheltering in place, property damage, or environmental damage.

(b) Data required. For each accidental release included, the owner or operator shall report the following information:

(1) Date, time, and approximate duration of the release;

(2) Chemical(s) released;

(3) Estimated quantity released in pounds;

(4) The type of release event and its source;

(5) Weather conditions, if known;

(6) On-site impacts;

(7) Known offsite impacts;

(8) Initiating event and contributing factors if known;

(9) Whether offsite responders were notified if known; and

(10) Operational or process changes that resulted from investigation of the release.

(c) Level of accuracy. Numerical estimates may be provided to two significant digits.

9. Subpart C is added to read as follows:

Subpart C Program 2 Prevention Program

68.48   Safety information.
68.50   Hazard review.
68.52   Operating procedures.
68.54   Training.
68.56   Maintenance.
68.58   Compliance audits.
68.60   Incident investigation.

68.48   Safety information.

(a) The owner or operator shall compile and maintain the following up-to-date safety information related to the regulated substances, processes, and equipment:

(1) Material Safety Data Sheets that meet the requirements of 29 CFR 1910.1200(g);

(2) Maximum intended inventory of equipment in which the regulated substances are stored or processed;

(3) Safe upper and lower temperatures, pressures, flows, and compositions;

(4) Equipment specifications; and

(5) Codes and standards used to design, build, and operate the process.

(b) The owner or operator shall ensure that the process is designed in compliance with recognized and generally accepted good engineering practices. Compliance with Federal or state regulations that address industry-specific safe design or with industry-specific design codes and standards may be used to demonstrate compliance with this paragraph.

(c) The owner or operator shall update the safety information if a major change occurs that makes the information inaccurate.

68.50 Hazard review.

(a) The owner or operator shall conduct a review of the hazards associated with the regulated substances, process, and procedures. The review shall identify the following:

(1) The hazards associated with the process and regulated substances;

(2) Opportunities for equipment malfunctions or human errors that could cause an accidental release;

(3) The safeguards used or needed to control the hazards or prevent equipment malfunction or human error; and

(4) Any steps used or needed to detect or monitor releases.

(b) The owner or operator may use checklists developed by persons or organizations knowledgeable about the process and equipment as a guide to conducting the review. For processes designed to meet industry standards or Federal or state design rules, the hazard review shall, by inspecting all equipment, determine whether the process is designed, fabricated, and operated in accordance with the applicable standards or rules.

(c) The owner or operator shall document the results of the review and ensure that problems identified are resolved in a timely manner.

(d) The review shall be updated at least once every 5 years. The owner or operator shall also conduct reviews whenever a major change in the process occurs; all issues identified in the review shall be resolved before start-up of the changed process.

68.52   Operating procedures.

(a) The owner or operator shall prepare written operating procedures that provide clear instructions or steps for safely conducting activities associated with each covered process consistent with the safety information for that process. Operating procedures or instructions provided by equipment manufacturers or developed by persons or organizations knowledgeable about the process and equipment may be used as a basis for a stationary source's operating procedures.

(b) The procedures shall address the following:

(1) Initial start-up;

(2) Normal operations;

(3) Temporary operations;

(4) Emergency shutdown and operations;

(5) Normal shutdown;

(6) Start-up following a normal or emergency shutdown or a major change that requires a hazard review;

(7) Consequences of deviations and steps required to correct or avoid deviations; and

(8) Equipment inspections.

(c) The owner or operator shall ensure that the operating procedures are updated, if necessary, whenever a major change occurs and prior to start-up of the changed process.

68.54   Training.

(a) The owner or operator shall ensure that each employee presently operating a process, and each employee newly assigned to a covered process have been trained or tested competent in the operating procedures provided in §68.52 of this part that pertain to their duties. For those employees already operating a process on *[insert date 3 years after the date of publication in the Federal Register],* the owner or operator may certify in writing that the employee has the required knowledge, skills, and abilities to safely carry out the duties and responsibilities as provided in the operating procedures.

(b) Refresher training. Refresher training shall be provided at least every 3 years, and more often if necessary, to each employee operating a process to ensure that the employee understands and adheres to the current operating procedures of the process. The owner or operator, in consultation with the employees operating the process, shall determine the appropriate frequency of refresher training.

(c) The owner or operator may use training conducted under Federal or state regulations or under industry-specific standards or codes or training conducted by covered process equipment vendors to demonstrate compliance with this section to the extent that the training meets the requirements of this section.

(d) The owner or operator shall ensure that operators are trained in any updated or new procedures prior to start-up of a process after a major change.

68.56   Maintenance.

(a) The owner or operator shall prepare and implement procedures to maintain the ongoing mechanical integrity of the process equipment. The owner or operator may use procedures or instructions provided by covered process equipment vendors or procedures in Federal or state regulations or industry codes as the basis for stationary source maintenance procedures.

(b) The owner or operator shall train or cause to be trained each employee involved in maintaining the ongoing mechanical integrity of the process. To ensure that the employee can perform the job tasks in a safe manner, each such employee shall be trained in the hazards of the process, in how to avoid or correct unsafe conditions, and in the procedures applicable to the employee's job tasks.

(c)   Any maintenance contractor shall ensure that each contract maintenance employee is trained to perform the maintenance procedures developed under paragraph (a) of this section.

(d)   The owner or operator shall perform or cause to be performed inspections and tests on process equipment. Inspection and testing procedures shall follow recognized and generally accepted good engineering practices. The frequency of inspections and tests of process equipment shall be consistent with applicable manufacturers' recommendations, industry standards or codes, good engineering practices, and prior operating experience.

68.58   Compliance audits.

(a)   The owners or operators shall certify that they have evaluated compliance with the provisions of this subpart at least every 3 years to verify that the procedures and practices developed under the rule are adequate and are being followed.

(b)   The compliance audit shall be conducted by at least one person knowledgeable in the process.

(c)   The owner or operator shall develop a report of the audit findings.

(d)   The owner or operator shall promptly determine and document an appropriate response to each of the findings of the compliance audit and document that deficiencies have been corrected.

(e)   The owner or operator shall retain the two (2) most recent compliance audit reports. This requirement does not apply to any compliance audit report that is more than 5 years old.

68.60   Incident investigation.

(a)   The owner or operator shall investigate each incident which resulted in, or could reasonably have resulted in, a catastrophic release.

(b)   An incident investigation shall be initiated as promptly as possible, but not later than 48 hours following the incident.

(c)   A summary shall be prepared at the conclusion of the investigation which includes at a minimum:

(1)   Date of incident;

(2)   Date investigation began;

(3)   A description of the incident;

(4)   The factors that contributed to the incident; and,

(5)   Any recommendations resulting from the investigation.

(d)   The owner or operator shall promptly address and resolve the investigation findings and recommendations. Resolutions and corrective actions shall be documented.

(e)   The findings shall be reviewed with all affected personnel whose job tasks are affected by the findings.

(f)   Investigation summaries shall be retained for 5 years.

10.   Subpart D is added to read as follows:

Subpart D Program 3 Prevention Program
68.65   Process safety information.
68.67   Process hazard analysis.

68.69   Operating procedures.
68.71   Training.
68.73   Mechanical integrity.
68.75   Management of change.
68.77   Pre-start-up review.
68.79   Compliance audits.
68.81   Incident investigation.
68.83   Employee participation.
68.85   Hot work permit.
68.87   Contractors.

68.65   Process safety information.

(a) In accordance with the schedule set forth in §68.67 of this part, the owner or operator shall complete a compilation of written process safety information before conducting any process hazard analysis required by the rule. The compilation of written process safety information is to enable the owner or operator and the employees involved in operating the process to identify and understand the hazards posed by those processes involving regulated substances. This process safety information shall include information pertaining to the hazards of the regulated substances used or produced by the process, information pertaining to the technology of the process, and information pertaining to the equipment in the process.

(b) Information pertaining to the hazards of the regulated substances in the process. This information shall consist of at least the following:

(1) Toxicity information;

(2) Permissible exposure limits;

(3) Physical data;

(4) Reactivity data;

(5) Corrosivity data;

(6) Thermal and chemical stability data; and

(7) Hazardous effects of inadvertent mixing of different materials that could foreseeably occur.

Note: Material Safety Data Sheets meeting the requirements of 29 CFR 1910.1200(g) may be used to comply with this requirement to the extent that they contain the information required by this subparagraph.

(c) Information pertaining to the technology of the process.

(1) Information concerning the technology of the process shall include at least the following:

(i) A block flow diagram or simplified process flow diagram;

(ii) Process chemistry;

(iii) Maximum intended inventory;

(iv) Safe upper and lower limits for such items as temperatures, pressures, flows or compositions; and,

(v) An evaluation of the consequences of deviations.

(2) Where the original technical information no longer exists, such information may be developed in conjunction with the process hazard analysis in sufficient detail to support the analysis.

(d) Information pertaining to the equipment in the process.

(1) Information pertaining to the equipment in the process shall include:

(i) Materials of construction;

(ii) Piping and instrument diagrams (P&IDs);

(iii) Electrical classification;

(iv) Relief system design and design basis;

(v) Ventilation system design;

(vi) Design codes and standards employed;

(vii) Material and energy balances for processes built after *[insert date 3 years after the date of publication in the Federal Register]*; and

(viii) Safety systems (e.g., interlocks, detection or suppression systems).

(2) The owner or operator shall document that equipment complies with recognized and generally accepted good engineering practices.

(3) For existing equipment designed and constructed in accordance with codes, standards, or practices that are no longer in general use, the owner or operator shall determine and document that the equipment is designed, maintained, inspected, tested, and operating in a safe manner.

68.67 Process hazard analysis.

(a) The owner or operator shall perform an initial process hazard analysis (hazard evaluation) on processes covered by this part. The process hazard analysis shall be appropriate to the complexity of the process and shall identify, evaluate, and control the hazards involved in the process. The owner or operator shall determine and document the priority order for conducting process hazard analyses based on a rationale which includes such considerations as extent of the process hazards, number of potentially affected employees, age of the process, and operating history of the process. The process hazard analysis shall be conducted as soon as possible, but not later than *[insert date 3 years after the date of publication in the Federal Register]*. Process hazards analyses completed to comply with 29 CFR 1910.119(e) are acceptable as initial process hazards analyses. These process hazard analyses shall be updated and revalidated, based on their completion date.

(b) The owner or operator shall use one or more of the following methodologies that are appropriate to determine and evaluate the hazards of the process being analyzed:

(1) What-If;

(2) Checklist;

(3) What-If/Checklist;

(4) Hazard and Operability Study (HAZOP);

(5) Failure Mode and Effects Analysis (FMEA);

(6) Fault tree analysis; or

(7) An appropriate equivalent methodology.

(c) The process hazard analysis shall address:

(1) The hazards of the process;

(2) The identification of any previous incident which had a likely potential for catastrophic consequences.

(3) Engineering and administrative controls applicable to the hazards and their interrelationships such as appropriate application of detection methodologies to provide early warning of releases. (Acceptable detection methods might include process monitoring and control instrumentation with alarms, and detection hardware such as hydrocarbon sensors.);

(4) Consequences of failure of engineering and administrative controls;

(5) Stationary source siting;

(6) Human factors; and

(7) A qualitative evaluation of a range of the possible safety and health effects of failure of controls.

(d) The process hazard analysis shall be performed by a team with expertise in engineering and process operations, and the team shall include at least one employee who has experience and knowledge specific to the process being evaluated. Also, one member of the team must be knowledgeable in the specific process hazard analysis methodology being used.

(e) The owner or operator shall establish a system to promptly address the team's findings and recommendations; assure that the recommendations are resolved in a timely manner and that the resolution is documented; document what actions are to be taken; complete actions as soon as possible; develop a written schedule of when these actions are to be completed; communicate the actions to operating, maintenance and other employees whose work assignments are in the process and who may be affected by the recommendations or actions.

(f) At least every five (5) years after the completion of the initial process hazard analysis, the process hazard analysis shall be updated and revalidated by a team meeting the requirements in paragraph (d) of this section, to assure that the process hazard analysis is consistent with the current process. Updated and revalidated process hazard analyses completed to comply with 29 CFR 1910.119(e) are acceptable to meet the requirements of this paragraph.

(g) The owner or operator shall retain process hazards analyses and updates or revalidations for each process covered by this section, as well as the documented resolution of recommendations described in paragraph (e) of this section for the life of the process.

68.69   Operating procedures.

(a) The owner or operator shall develop and implement written operating procedures that provide clear instructions for safely conducting activities involved in each covered process consistent with the process safety information and shall address at least the following elements:

(1) Steps for each operating phase:

(i) Initial start-up;

(ii) Normal operations;

(iii) Temporary operations;

(iv) Emergency shutdown including the conditions under which emergency shutdown is required, and the assignment of shutdown responsibility to qualified operators to ensure that emergency shutdown is executed in a safe and timely manner

(v) Emergency operations;

(vi) Normal shutdown; and

(vii) Start-up following a turnaround, or after an emergency shutdown.

(2) Operating limits:

(i) Consequences of deviation; and

(ii) Steps required to correct or avoid deviation.

(3) Safety and health considerations:

(i) Properties of, and hazards presented by, the chemicals used in the process;

(ii) Precautions necessary to prevent exposure, including engineering controls, administrative controls, and personal protective equipment;

(iii) Control measures to be taken if physical contact or airborne exposure occurs;

(iv) Quality control for raw materials and control of hazardous chemical inventory levels; and

(v) Any special or unique hazards.

(4) Safety systems and their functions.

(b) Operating procedures shall be readily accessible to employees who work in or maintain a process.

(c) The operating procedures shall be reviewed as often as necessary to assure that they reflect current operating practice, including changes that result from changes in process chemicals, technology, and equipment, and changes to stationary sources. The owner or operator shall certify annually that these operating procedures are current and accurate.

(d) The owner or operator shall develop and implement safe work practices to provide for the control of hazards during operations such as lockout/tagout; confined space entry; opening process equipment or piping; and control over entrance into a stationary source by maintenance, contractor, laboratory, or other support personnel. These safe work practices shall apply to employees and contractor employees.

68.71   Training.

(a) Initial training.

(1) Each employee presently involved in operating a process, and each employee before being involved in operating a newly assigned process, shall be trained in an overview of the process and in the operating procedures as specified in §68.69 of this part. The training shall include emphasis on the specific safety and health hazards, emergency operations including shutdown, and safe work practices applicable to the employee's job tasks.

(2) In lieu of initial training for those employees already involved in operating a process on [insert date 3 years after the date of publication in

the Federal Register] an owner or operator may certify in writing that the employee has the required knowledge, skills, and abilities to safely carry out the duties and responsibilities as specified in the operating procedures.

(b) Refresher training. Refresher training shall be provided at least every 3 years, and more often if necessary, to each employee involved in operating a process to assure that the employee understands and adheres to the current operating procedures of the process. The owner or operator, in consultation with the employees involved in operating the process, shall determine the appropriate frequency of refresher training.

(c) Training documentation. The owner or operator shall ascertain that each employee involved in operating a process has received and understood the training required by this paragraph. The owner or operator shall prepare a record which contains the identity of the employee, the date of training, and the means used to verify that the employee understood the training.

68.73   Mechanical integrity.

(a) Application. Paragraphs (b) through (f) of this section apply to the following process equipment:

(1) Pressure vessels and storage tanks;

(2) Piping systems (including piping components such as valves);

(3) Relief and vent systems and devices;

(4) Emergency shutdown systems;

(5) Controls (including monitoring devices and sensors, alarms, and interlocks); and

(6) Pumps.

(b) Written procedures. The owner or operator shall establish and implement written procedures to maintain the ongoing integrity of process equipment.

(c) Training for process maintenance activities. The owner or operator shall train each employee involved in maintaining the ongoing integrity of process equipment in an overview of that process and its hazards and in the procedures applicable to the employee's job tasks to assure that the employee can perform the job tasks in a safe manner.

(d) Inspection and testing.

(1) Inspections and tests shall be performed on process equipment.

(2) Inspection and testing procedures shall follow recognized and generally accepted good engineering practices.

(3) The frequency of inspections and tests of process equipment shall be consistent with applicable manufacturers' recommendations and good engineering practices. The frequency of inspections shall occur more often if determined to be necessary by prior operating experience.

(4) The owner or operator shall document each inspection and test that have been performed on process equipment. The documentation shall identify the date of the inspection or test, the name of the person who performed the inspection or test, the serial number or other identifier of the equipment on which the inspection or test was performed, a

description of the inspection or test performed, and the results of the inspection or test.

(e) Equipment deficiencies. The owner or operator shall correct deficiencies in equipment that are outside acceptable limits (defined by the process safety information in §68.65 of this part) before further use or in a safe and timely manner when necessary means are taken to assure safe operation.

(f) Quality assurance.

(1) In the construction of new plants and equipment, the owner or operator shall assure that equipment as it is fabricated is suitable for the process application for which it will be used.

(2) Appropriate checks and inspections shall be performed to assure that equipment is installed properly and consistent with design specifications and the manufacturer's instructions.

(3) The owner or operator shall assure that maintenance materials, spare parts and equipment are suitable for the process application for which they will be used.

68.75   Management of change.

(a) The owner or operator shall establish and implement written procedures to manage changes (except for "replacements in kind") to process chemicals, technology, equipment, and procedures; and changes to stationary sources that affect a covered process.

(b) The procedures shall assure that the following considerations are addressed prior to any change:

(1) The technical basis for the proposed change;

(2) Impact of change on safety and health;

(3) Modifications to operating procedures;

(4) Necessary time period for the change; and

(5) Authorization requirements for the proposed change.

(c) Employees involved in operating a process and maintenance and contract employees whose job tasks will be affected by a change in the process shall be informed of, and trained in, the change prior to start-up of the process or affected part of the process.

(d) If a change covered by this paragraph results in a change in the process safety information required by §68.65 of this part, such information shall be updated accordingly.

(e) If a change covered by this paragraph results in a change in the operating procedures or practices required by §68.69 of this part, such procedures or practices shall be updated accordingly.

68.77   Pre-start-up review.

(a) The owner or operator shall perform a pre-start-up safety review for new stationary sources and for modified stationary sources when the modification is significant enough to require a change in the process safety information.

(b) The pre-start-up safety review shall confirm that prior to the introduction of regulated substances to a process:

(1) Construction and equipment are in accordance with design specifications;

(2) Safety, operating, maintenance, and emergency procedures are in place and are adequate;

(3) For new stationary sources, a process hazard analysis has been performed and recommendations have been resolved or implemented before start-up; and modified stationary sources meet the requirements contained in management of change, §68.75 of this part.

(4) Training of each employee involved in operating a process has been completed.

68.79   Compliance audits.

(a) The owners or operators shall certify that they have evaluated compliance with the provisions of this section at least every 3 years to verify that the procedures and practices developed under the standard are adequate and are being followed.

(b) The compliance audit shall be conducted by at least one person knowledgeable in the process.

(c) A report of the findings of the audit shall be developed.

(d) The owner or operator shall promptly determine and document an appropriate response to each of the findings of the compliance audit, and document that deficiencies have been corrected.

(e) The owner or operator shall retain the two (2) most recent compliance audit reports.

68.81   Incident investigation.

(a) The owner or operator shall investigate each incident which resulted in, or could reasonably have resulted in, a catastrophic release of a regulated substance.

(b) An incident investigation shall be initiated as promptly as possible, but not later than 48 hours following the incident.

(c) An incident investigation team shall be established and shall consist of at least one person knowledgeable in the process involved, including a contract employee if the incident involved work of the contractor, and other persons with appropriate knowledge and experience to thoroughly investigate and analyze the incident.

(d) A report shall be prepared at the conclusion of the investigation which includes at a minimum:

(1) Date of incident;

(2) Date investigation began;

(3) A description of the incident;

(4) The factors that contributed to the incident; and

(5) Any recommendations resulting from the investigation.

(e) The owner or operator shall establish a system to promptly address and resolve the incident report findings and recommendations. Resolutions and corrective actions shall be documented.

(f) The report shall be reviewed with all affected personnel whose job tasks are relevant to the incident findings including contract employees where applicable.

(g) Incident investigation reports shall be retained for 5 years.

68.83   Employee participation.

(a) The owner or operator shall develop a written plan of action regarding the implementation of the employee participation required by this section.

(b) The owner or operator shall consult with employees and their representatives on the conduct and development of process hazards analyses and on the development of the other elements of process safety management in this rule.

(c) The owner or operator shall provide to employees and their representatives access to process hazard analyses and to all other information required to be developed under this rule.

68.85   Hot work permit.

(a) The owner or operator shall issue a hot work permit for hot work operations conducted on or near a covered process.

(b) The permit shall document that the fire prevention and protection requirements in 29 CFR 1910.22(a) have been implemented prior to beginning the hot work operations; it shall indicate the date(s) authorized for hot work and identify the object on which hot work is to be performed. The permit shall be kept on file until completion of the hot work operations.

68.87   Contractors.

(a) Application. This section applies to contractors performing maintenance or repair, turnaround, major renovation, or specialty work on or adjacent to a covered process. It does not apply to contractors providing incidental services which do not influence process safety, such as janitorial work, food and drink services, laundry, delivery or other supply services.

(b) Owner or operator responsibilities.

(1) The owner or operator, when selecting a contractor, shall obtain and evaluate information regarding the contract owner or operator's safety performance and programs.

(2) The owner or operator shall inform the contract owner or operator of the known potential fire, explosion, or toxic release hazards related to the contractor's work and the process.

(3) The owner or operator shall explain to the contract owner or operator the applicable provisions of subpart E of this part.

(4) The owner or operator shall develop and implement safe work practices consistent with §68.69(d) of this part, to control the entrance, presence, and exit of the contract owner or operator and contract employees in covered process areas.

(5) The owner or operator shall periodically evaluate the performance of the contract owner or operator in fulfilling his or her obligations as specified in paragraph (c) of this section.

(c) Contract owner or operator responsibilities.

(1) The contract owner or operator shall assure that each contract employee is trained in the work practices necessary to safely perform his/her job.

(2) The contract owner or operator shall assure that each contract employee is instructed in the known potential fire, explosion, or toxic release hazards related to his/her job and the process, and the applicable provisions of the emergency action plan.

(3) The contract owner or operator shall document that each contract employee has received and understood the training required by this section. The contract owner or operator shall prepare a record which contains the identity of the contract employee, the date of training, and the means used to verify that the employee understood the training.

(4) The contract owner or operator shall assure that each contract employee follows the safety rules of the stationary source including the safe work practices required by §68.69(d) of this part.

(5) The contract owner or operator shall advise the owner or operator of any unique hazards presented by the contract owner or operator's work, or of any hazards found by the contract owner or operator's work.

11. Subpart E is added to read as follows:

Subpart E Emergency Response

68.90   Applicability.
68.95   Emergency Response Program.

68.90   Applicability.

(a) Except as provided in paragraph (b) of this section, the owner or operator of a stationary source with Program 2 and Program 3 processes shall comply with the requirements of §68.95 of this part.

(b) The owner or operator of a stationary source whose employees will not respond to accidental releases of regulated substances need not comply with §68.95 of this part provided that they meet the following:

(1) For stationary sources with any regulated toxic substance held in a process above the threshold quantity, the stationary source is included in the community emergency response plan developed under 42 U.S.C. 11003;

(2) For stationary sources with only regulated flammable substances held in a process above the threshold quantity, the owner or operator has coordinated response actions with the local fire department; and

(3) Appropriate mechanisms are in place to notify emergency responders when there is a need for a response.

68.95   Emergency response program.

(a) The owner or operator shall develop and implement an emergency response program for the purpose of protecting public health and the environment. Such program shall include the following elements:

(1) An emergency response plan, which shall be maintained at the stationary source and shall contain at least the following elements:

(i) Procedures for informing the public and local emergency response agencies about accidental releases;

(ii) Documentation of proper first-aid and emergency medical treatment necessary to treat accidental human exposures; and

(iii) Procedures and measures for emergency response after an accidental release of a regulated substance;

(2) Procedures for the use of emergency response equipment and for its inspection, testing, and maintenance;

(3) Training for all employees in relevant procedures; and

(4) Procedures to review and update, as appropriate, the emergency response plan to reflect changes at the stationary source and ensure that employees are informed of changes.

(b) A written plan that complies with other Federal contingency plan regulations or is consistent with the approach in the National Response Team's *Integrated Contingency Plan Guidance* ("One Plan") and that, among other matters, includes the elements provided in paragraph (a) of this section, shall satisfy the requirements of this section if the owner or operator also complies with paragraph (c) of this section.

(c) The emergency response plan developed under paragraph (a)(1) of this section shall be coordinated with the community emergency response plan developed under 42 U.S.C. 11003. Upon request of the local emergency planning committee or emergency response officials, the owner or operator shall promptly provide to the local emergency response officials information necessary for developing and implementing the community emergency response plan.

12. Subpart G is added to read as follows:

Subpart G Risk Management Plan

68.150   Submission.
68.155   Executive summary.
68.160   Registration.
68.165   Offsite consequence analysis.
68.168   Five-year accident history.
68.170   Prevention program/Program 2.
68.175   Prevention program/Program 3.
68.180   Emergency response program.
68.185   · Certification.
68.190   Updates.

68.150   Submission.

(a) The owner or operator shall submit a single RMP that includes the information required by §§68.155 through 68.185 of this part for all covered processes. The RMP shall be submitted in a method and format to a central point as specified by EPA prior to *[insert date 3 years after the date of publication in the Federal Register]*.

(b) The owner or operator shall submit the first RMP no later than the latest of the following dates:

(1) *[insert date 3 years after the date of publication in the Federal Register]*;

(2) Three years after the date on which a regulated substance is first listed under §68.130 of this part; or

(3)   The date on which a regulated substance is first present above a threshold quantity in a process.

(c)   Subsequent submissions of RMPs shall be in accordance with §68.190 of this part.

(d)   Notwithstanding the provisions of §§68.155 to 68.190 of this part, the RMP shall exclude classified information. Subject to appropriate procedures to protect such information from public disclosure, classified data or information excluded from the RMP may be made available in a classified annex to the RMP for review by Federal and state representatives who have received the appropriate security clearances.

68.155   Executive summary.

The owner or operator shall provide in the RMP an executive summary that includes a brief description of the following elements:

(a)   The accidental release prevention and emergency response policies at the stationary source;

(b)   The stationary source and regulated substances handled;

(c)   The worst-case release scenario(s) and the alternative release scenario(s), including administrative controls and mitigation measures to limit the distances for each reported scenario;

(d)   The general accidental release prevention program and chemical-specific prevention steps;

(e)   The 5-year accident history;

(f)   The emergency response program; and

(g)   Planned changes to improve safety.

68.160   Registration.

(a)   The owner or operator shall complete a single registration form and include it in the RMP. The form shall cover all regulated substances handled in covered processes.

(b)   The registration shall include the following data:

(1) Stationary source name, street, city, county, state, Zip code, latitude, and longitude;

(2) The stationary source Dun and Bradstreet number;

(3) Name and Dun and Bradstreet number of the corporate parent company;

(4) The name, telephone number, and mailing address of the owner or operator;

(5) The name and title of the person or position with overall responsibility for RMP elements and implementation;

(6) The name, title, telephone number, and 24-hour telephone number of the emergency contact;

(7) For each covered process, the name and CAS number of each regulated substance held above the threshold quantity in the process, the maximum quantity of each regulated substance or mixture in the process (in pounds) to two significant digits, the SIC code, and the Program level of the process;

(8) The stationary source EPA identifier;

(9) The number of full-time employees at the stationary source;

(10) Whether the stationary source is subject to 29 CFR 1910.119;

(11) Whether the stationary source is subject to 40 CFR part 355;

(12) Whether the stationary source has a CAA Title V operating permit; and

(13) The date of the last safety inspection of the stationary source by a Federal, state, or local government agency and the identity of the inspecting entity.

68.165   Offsite consequence analysis.

(a) The owner or operator shall submit in the RMP information:

(1) One worst-case release scenario for each Program 1 process; and

(2) For Program 2 and 3 processes, one worst-case release scenario to represent all regulated toxic substances held above the threshold quantity and one worst-case release scenario to represent all regulated flammable substances held above the threshold quantity. If additional worst-case scenarios for toxics or flammables are required by §68.25(a)(2)(iii) of this part, the owner or operator shall submit the same information on the additional scenario(s). The owner or operator of Program 2 and Program 3 processes shall also submit information on one alternative release scenario for each regulated toxic substance held above the threshold quantity and one alternative release scenario to represent all regulated flammable substances held above the threshold quantity.

(b) The owner or operator shall submit the following data:

(1) Chemical name;

(2) Physical state (toxics only);

(3) Basis of results (give model name if used);

(4) Scenario (explosion, fire, toxic gas release, or liquid spill and vaporization);

(5) Quantity released in pounds;

(6) Release rate;

(7) Release duration;

(8) Wind speed and atmospheric stability class (toxics only);

(9) Topography (toxics only);

(10) Distance to endpoint;

(11) Public and environmental receptors within the distance;

(12) Passive mitigation considered; and

(13) Active mitigation considered (alternative releases only).

68.168   Five-year accident history.

The owner or operator shall submit in the RMP the information provided in §68.42(b) of this part on each accident covered by §68.42(a) of this part.

68.170   Prevention program/Program 2.

(a) For each Program 2 process, the owner or operator shall provide in the RMP the information indicated in paragraphs (b) through (k) of this section. If the same information applies to more than one covered process, the owner or operator may provide the information only once, but shall indicate to which processes the information applies.

(b)   The SIC code for the process.

(c)   The name(s) of the chemical(s) covered.

(d)   The date of the most recent review or revision of the safety information and a list of Federal or state regulations or industry-specific design codes and standards used to demonstrate compliance with the safety information requirement.

(e)   The date of completion of the most recent hazard review or update.

(1)   The expected date of completion of any changes resulting from the hazard review;

(2)   Major hazards identified;

(3)   Process controls in use;

(4)   Mitigation systems in use;

(5)   Monitoring and detection systems in use; and

(6)   Changes since the last hazard review.

(f)   The date of the most recent review or revision of operating procedures.

(g)   The date of the most recent review or revision of training programs;

(1)   The type of training provided—classroom, classroom plus on the job, on the job; and

(2)   The type of competency testing used.

(h)   The date of the most recent review or revision of maintenance procedures and the date of the most recent equipment inspection or test and the equipment inspected or tested.

(i)   The date of the most recent compliance audit and the expected date of completion of any changes resulting from the compliance audit.

(j)   The date of the most recent incident investigation and the expected date of completion of any changes resulting from the investigation.

(k)   The date of the most recent change that triggered a review or revision of safety information, the hazard review, operating or maintenance procedures, or training.

68.175   Prevention program/Program 3.

(a)   For each Program 3 process, the owner or operator shall provide the information indicated in paragraphs (b) through (p) of this section. If the same information applies to more than one covered process, the owner or operator may provide the information only once, but shall indicate to which processes the information applies.

(b)   The SIC code for the process.

(c)   The name(s) of the substance(s) covered.

(d)   The date on which the safety information was last reviewed or revised.

(e)   The date of completion of the most recent PHA or update and the technique used.

(1) The expected date of completion of any changes resulting from the PHA;

(2) Major hazards identified;

(3) Process controls in use;

(4) Mitigation systems in use;

(5) Monitoring and detection systems in use; and

(6) Changes since the last PHA.

(f) The date of the most recent review or revision of operating procedures.

(g) The date of the most recent review or revision of training programs.

(1) The type of training provided—classroom, classroom plus on the job, on the job; and

(2) The type of competency testing used.

(h) The date of the most recent review or revision of maintenance procedures and the date of the most recent equipment inspection or test and the equipment inspected or tested.

(i) The date of the most recent change that triggered management of change procedures and the date of the most recent review or revision of management of change procedures.

(j) The date of the most recent pre-start-up review.

(k) The date of the most recent compliance audit and the expected date of completion of any changes resulting from the compliance audit;

(l) The date of the most recent incident investigation and the expected date of completion of any changes resulting from the investigation;

(m) The date of the most recent review or revision of employee participation plans;

(n) The date of the most recent review or revision of hot work permit procedures;

(o) The date of the most recent review or revision of contractor safety procedures; and

(p) The date of the most recent evaluation of contractor safety performance.

68.180 Emergency response program.

(a) The owner or operator shall provide in the RMP the following information:

(1) Do you have a written emergency response plan?

(2) Does the plan include specific actions to be taken in response to accidental releases of a regulated substance?

(3) Does the plan include procedures for informing the public and local agencies responsible for responding to accidental releases?

(4) Does the plan include information on emergency health care?

(5) The date of the most recent review or update of the emergency response plan;

(6) The date of the most recent emergency response training for employees.

(b) The owner or operator shall provide the name and telephone number of the local agency with which the plan is coordinated.

(c) The owner or operator shall list other Federal or state emergency plan requirements to which the stationary source is subject.

68.185   Certification.

(a) For Program 1 processes, the owner or operator shall submit in the RMP the certification statement provided in §68.12(b)(4) of this part.

(b) For all other covered processes, the owner or operator shall submit in the RMP a single certification that, to the best of the signer's knowledge, information, and belief formed after reasonable inquiry, the information submitted is true, accurate, and complete.

68.190   Updates.

(a) The owner or operator shall review and update the RMP as specified in paragraph (b) of this section and submit it in a method and format to a central point specified by EPA prior to *[insert date 3 years after the date of publication in the Federal Register]*.

(b) The owner or operator of a stationary source shall revise and update the RMP submitted under §68.150 as follows:

(1) Within 5 years of its initial submission or most recent update required by paragraphs (b)(2)–(b)(7) of this section, whichever is later.

(2) No later than 3 years after a newly regulated substance is first listed by EPA;

(3) No later than the date on which a new regulated substance is first present in an already covered process above a threshold quantity;

(4) No later than the date on which a regulated substance is first present above a threshold quantity in a new process;

(5) Within 6 months of a change that requires a revised PHA or hazard review;

(6) Within 6 months of a change that requires a revised offsite consequence analysis as provided in §68.36 of this part; and

(7) Within 6 months of a change that alters the Program level that applied to any covered process.

(c) If a stationary source is no longer subject to this part, the owner or operator shall submit a revised registration to EPA within 6 months indicating that the stationary source is no longer covered.

13. Subpart H is added to read as follows:

Subpart H Other Requirements

68.200   Recordkeeping.
68.210   Availability of information to the public.
68.215   Permit content and air permitting authority or designated agency requirements.
68.220   Audits.

68.200   Recordkeeping.

The owner or operator shall maintain records supporting the implementation of this part for 5 years unless otherwise provided in subpart D of this part.

68.210   Availability of information to the public.

(a) The RMP required under subpart G of this part shall be available to the public under 42 U.S.C. 7414(c).

(b) The disclosure of classified information by the Department of Defense or other Federal agencies or contractors of such agencies shall be controlled by applicable laws, regulations, or executive orders concerning the release of classified information.

68.215   Permit content and air permitting authority or designated agency requirements.

(a) These requirements apply to any stationary source subject to part 68 and part 70 or 71 of this chapter. The 40 CFR part 70 or part 71 permit for the stationary source shall contain:

(1) A statement listing this part as an applicable requirement;

(2) Conditions that require the source owner or operator to submit:

(i) A compliance schedule for meeting the requirements of this part by the date provided in §68.10(a) of this part or;

(ii) As part of the compliance certification submitted under 40 CFR 70.6(c)(5), a certification statement that the source is in compliance with all requirements of this part, including the registration and submission of the RMP.

(b) The owner or operator shall submit any additional relevant information requested by the air permitting authority or designated agency.

(c) For 40 CFR part 70 or part 71 permits issued prior to the deadline for registering and submitting the RMP and which do not contain permit conditions described in paragraph (a) of this section, the owner or operator or air permitting authority shall initiate permit revision or reopening according to the procedures of 40 CFR 70.7 or 71.7 to incorporate the terms and conditions consistent with paragraph (a) of this section.

(d) The state may delegate the authority to implement and enforce the requirements of paragraph (e) of this section to a state or local agency or agencies other than the air permitting authority. An up-to-date copy of any delegation instrument shall be maintained by the air permitting authority. The state may enter a written agreement with the Administrator under which EPA will implement and enforce the requirements of paragraph (e) of this section.

(e) The air permitting authority or the agency designated by delegation or agreement under paragraph (d) of this section shall, at a minimum:

(1) Verify that the source owner or operator has registered and submitted an RMP or a revised plan when required by this part;

(2) Verify that the source owner or operator has submitted a source certification or in its absence has submitted a compliance schedule consistent with paragraph (a)(2) of this section;

(3) For some or all of the sources subject to this section, use one or more mechanisms such as, but not limited to, a completeness check, source audits, record reviews, or facility inspections to ensure that permitted sources are in compliance with the requirements of this part; and

(4) Initiate enforcement action based on paragraphs (e)(1) and (e)(2) of this section as appropriate.

68.220 Audits.

(a) In addition to inspections for the purpose of regulatory development and enforcement of the Act, the implementing agency shall periodically audit RMPs submitted under subpart G of this part to review the adequacy of such RMPs and require revisions of RMPs when necessary to ensure compliance with subpart G of this part.

(b) The implementing agency shall select stationary sources for audits based on any of the following criteria:

(1) Accident history of the stationary source;

(2) Accident history of other stationary sources in the same industry;

(3) Quantity of regulated substances present at the stationary source;

(4) Location of the stationary source and its proximity to the public and environmental receptors;

(5) The presence of specific regulated substances;

(6) The hazards identified in the RMP; and

(7) A plan providing for neutral, random oversight.

(c) Exemption from audits. A stationary source with a Star or Merit ranking under OSHA's voluntary protection program shall be exempt from audits under paragraph (b)(2) and (b)(7) of this section.

(d) The implementing agency shall have access to the stationary source, supporting documentation, and any area where an accidental release could occur.

(e) Based on the audit, the implementing agency may issue the owner or operator of a stationary source a written preliminary determination of necessary revisions to the stationary source's RMP to ensure that the RMP meets the criteria of subpart G of this part. The preliminary determination shall include an explanation for the basis for the revisions, reflecting industry standards and guidelines (such as AIChE/CCPS guidelines and ASME and API standards) to the extent that such standards and guidelines are applicable, and shall include a timetable for their implementation.

(f) Written response to a preliminary determination.

(1) The owner or operator shall respond in writing to a preliminary determination made in accordance with paragraph (e) of this section. The response shall state the owner or operator will implement the revisions contained in the preliminary determination in accordance with the timetable included in the preliminary determination or shall state that the owner or operator rejects the revisions in whole or in part. For each rejected revision, the owner or operator shall explain the basis for rejecting such revision. Such explanation may include substitute revisions.

(2) The written response under paragraph (f)(1) of this section shall be received by the implementing agency within 90 days of the issue of the preliminary determination or a shorter period of time as the implementing agency specifies in the preliminary determination as necessary to protect public health and the environment. Prior to the written response being due and upon written request from the owner or operator, the implementing agency may provide in writing additional time for the response to be received.

(g) After providing the owner or operator an opportunity to respond under paragraph (f) of this section, the implementing agency may issue the owner or operator a written final determination of necessary revisions to the stationary source's RMP. The final determination may adopt or modify the revisions contained in the preliminary determination under paragraph (e) of this section or may adopt or modify the substitute revisions provided in the response under paragraph (f) of this section. A final determination that adopts a revision rejected by the owner or operator shall include an explanation of the basis for the revision. A final determination that fails to adopt a substitute revision provided under paragraph (f) of this section shall include an explanation of the basis for finding such substitute revision unreasonable.

(h) Thirty days after completion of the actions detailed in the implementation schedule set in the final determination under paragraph (g) of this section, the owner or operator shall be in violation of subpart G of this part and this section unless the owner or operator revises the RMP prepared under subpart G of this part as required by the final determination, and submits the revised RMP as required under §68.150 of this part.

(i) The public shall have access to the preliminary determinations, responses, and final determinations under this section in a manner consistent with §68.210 of this part.

(j) Nothing in this section shall preclude, limit, or interfere in any way with the authority of EPA or the state to exercise its enforcement, investigatory, and information gathering authorities concerning this part under the Act.

14. Part 68 Appendix A is added to read as follows:

APPENDIX A: TABLE OF TOXIC ENDPOINTS
(as defined in §68.22 of this part)

CAS no	Chemical name	Toxic endpoint (mg/L)
107-02-8	Acrolein [2-Propenal]	0.0011
107-13-1	Acrylonitrile [2-Propenenitrile]	0.076
814-68-6	Acrylyl chloride [2-Propenoyl chloride]	0.00090
107-18-6	Allyl alcohol [2-Propen-1-ol]	0.036
107-11-9	Allylamine [2-Propen-1-amine]	0.0032
7664-41-7	Ammonia (anhydrous)	0.14
7664-41-7	Ammonia (conc. 20% or greater)	0.14
7784-34-1	Arsenous trichloride	0.010
7784-42-1	Arsine	0.0019
10294-34-5	Boron trichloride [Borane, trichloro-]	0.010
7637-07-2	Boron trifluoride [Borane, trifluoro-]	0.028
353-42-4	Boron trifluoride compound with methyl ether (1:1) [Boron, trifluoro[oxybis[methane]]-, T-4	0.023
7726-95-6	Bromine	0.0065
75-15-0	Carbon disulfide	0.16
7782-50-5	Chlorine	0.0087
10049-04-4	Chlorine dioxide [Chlorine oxide (ClO$_2$)]	0.0028
67-66-3	Chloroform [Methane, trichloro-]	0.49
542-88-1	Chloromethyl ether [Methane, oxybis[chloro-]	0.00025
107-30-2	Chloromethyl methyl ether [Methane, chloromethoxy-]	0.0018
4170-30-3	Crotonaldehyde [2-Butenal]	0.029
123-73-9	Crotonaldehyde, (E)- [2-Butenal, (E)-]	0.029
506-77-4	Cyanogen chloride	0.030
108-91-8	Cyclohexylamine [Cyclohexanamine]	0.16
19287-45-7	Diborane	0.0011
75-78-5	Dimethyldichlorosilane [Silane, dichlorodimethyl-]	0.026
57-14-7	1,1-Dimethylhydrazine [Hydrazine, 1,1-dimethyl-]	0.012
106-89-8	Epichlorohydrin [Oxirane, (chloromethyl)-]	0.076
107-15-3	Ethylenediamine [1,2-Ethanediamine]	0.49
151-56-4	Ethyleneimine [Aziridine]	0.018
75-21-8	Ethylene oxide [Oxirane]	0.090
7782-41-4	Fluorine	0.0039
50-00-0	Formaldehyde (solution)	0.012
110-00-9	Furan	0.0012
302-01-2	Hydrazine	0.011
7647-01-0	Hydrochloric acid (conc. 30% or greater)	0.030
74-90-8	Hydrocyanic acid	0.011
7647-01-0	Hydrogen chloride (anhydrous) [Hydrochloric acid]	0.030
7664-39-3	Hydrogen fluoride/Hydrofluoric acid (conc. 50% or greater) [Hydrofluoric acid]	0.016
7783-07-5	Hydrogen selenide	0.00066
7783-06-4	Hydrogen sulfide	0.042
13463-40-6	Iron, pentacarbonyl- [Iron carbonyl (Fe(CO)$_5$), (TB-5-11)-]	0.00044

APPENDIX A: TABLE OF TOXIC ENDPOINTS
(as defined in §68.22 of this part) (Continued)

CAS no	Chemical name	Toxic endpoint (mg/L)
78-82-0	Isobutyronitrile [Propanenitrile, 2-methyl-]	0.14
108-23-6	Isopropyl chloroformate [Carbonochloridic acid, 1-methylethyl ester]	0.10
126-98-7	Methacrylonitrile [2-Propenenitrile, 2-methyl-]	0.0027
74-87-3	Methyl chloride [Methane, chloro-]	0.82
79-22-1	Methyl chloroformate [Carbonochloridic acid, methylester]	0.0019
60-34-4	Methyl hydrazine [Hydrazine, methyl-]	0.0094
624-83-9	Methyl isocyanate [Methane, isocyanato-]	0.0012
74-93-1	Methyl mercaptan [Methanethiol]	0.049
556-64-9	Methyl thiocyanate [Thiocyanic acid, methyl ester]	0.085
75-79-6	Methyltrichlorosilane [Silane, trichloromethyl-]	0.018
13463-39-3	Nickel carbonyl	0.00067
7697-37-2	Nitric acid (conc. 80% or greater)	0.026
10102-43-9	Nitric oxide [Nitrogen oxide (NO)]	0.031
8014-95-7	Oleum (fuming sulfuric acid) [Sulfuric acid, mixture with sulfur trioxide]	0.010
79-21-0	Peracetic acid [Ethaneperoxoic acid]	0.0045
594-42-3	Perchloromethylmercaptan [Methanesulfenyl chloride, trichloro-]	0.0076
75-44-5	Phosgene [Carbonic dichloride]	0.00081
7803-51-2	Phosphine	0.0035
10025-87-3	Phosphorus oxychloride [Phosphoryl chloride]	0.0030
7719-12-2	Phosphorus trichloride [Phosphorous trichloride]	0.028
110-89-4	Piperidine	0.022
107-12-0	Propionitrile [Propanenitrile]	0.0037
109-61-5	Propyl chloroformate [Carbonochloridic acid, propylester]	0.010
75-55-8	Propyleneimine [Aziridine, 2-methyl-]	0.12
75-56-9	Propylene oxide [Oxirane, methyl-]	0.59
7446-09-5	Sulfur dioxide (anhydrous)	0.0078
7783-60-0	Sulfur tetrafluoride [Sulfur fluoride (SF_4), (T-4)-]	0.0092
7446-11-9	Sulfur trioxide	0.010
75-74-1	Tetramethyllead [Plumbane, tetramethyl-]	0.0040
509-14-8	Tetranitromethane [Methane, tetranitro-]	0.0040
7550-45-0	Titanium tetrachloride [Titanium chloride $(TiCl_4)$ (T-4)-]	0.020
584-84-9		0.0070
91-08-7	Toluene 2,4-diisocyanate [Benzene, 2,4-diisocyanato-1-methyl-]	0.0070
26471-62-5	Toluene 2,6-diisocyanate [Benzene, 1,3-diisocyanato-2-methyl-]	0.0070
75-77-4	Toluene diisocyanate (unspecified isomer) [Benzene, 1,3-diisocyanatomethyl-]	0.050
108-05-4	Trimethylchlorosilane [Silane, chlorotrimethyl-] Vinyl acetate monomer [Acetic acid ethenyl ester]	0.26

M

List of Regulated Toxic Substances and Threshold Quantities for Accidental-Release Prevention

Sec. 68.130: List of Substances

(a) Explosives are covered under section 112(r) of the Clean Air Act if they are classified by DOT in Division 1.1 (49 CFR part 172.102), defined as explosives that have a mass explosion hazard, and are present on site in quantities greater than 5000 pounds.

(b) Regulated toxic and flammable substances under section 112(r) of the Clean Air Act are the substances listed in Tables 1, 2, 3, and 4 of this section. Threshold quantities for listed toxic and flammable substances are specified in the tables.

(c) The bases for placing toxic and flammable substances on the list of regulated substances are explained in the notes to the list.

TABLE 1 TO SEC. 68.130   List of Regulated Toxic Substances and Threshold Quantities for Accidental-Release Prevention

[Alphabetical order—100 substances]

Chemical name	CAS no.	Threshold quantity (lb)	Basis for listing
Acetone cyanohydrin	75-86-5	5,000	(b)
Acrolein	107-02-8	1,000	(b)
Acrylonitrile	107-13-1	10,000	(b)
Acrylyl chloride	814-68-6	1,000	(b)
Allyl alcohol	107-18-6	5,000	(b)
Allylamine	107-11-9	1,000	(b)
Ammonia (anhydrous)	7664-41-7	1,000	(a,b)
Ammonia (aqueous solution, conc. 20% or greater)	7664-41-7	5,000	(a,b)
Aniline	62-53-3	5,000	(b)
Antimony pentafluoride	7783-70-2	1,000	(b)
Arsenous trichloride	7784-34-1	5,000	(b)
Arsine	7784-42-1	500	(b)
Benzal chloride	98-87-3	1,000	(b)
Benzenamine, 3-(trifluoromethyl)-	98-16-8	1,000	(b)
Benzotrichloride	98-07-7	500	(b)
Benzyl chloride	100-44-7	1,000	(b)
Benzyl cyanide	140-29-4	1,000	(b)
Boron trichloride	10294-34-5	1,000	(b)
Boron trifluoride	7637-07-2	1,000	(b)
Boron trifluoride compound with methyl ether (1:1)	353-42-4	5,000	(b)
Bromine	7726-95-6	1,000	(a,b)
Carbon disulfide	75-15-0	10,000	(b)
Chlorine	7782-50-5	1,000	(a,b)
Chlorine dioxide	10049-04-4	500	(d)
Chloroethanol	107-07-3	1,000	(b)
Chloroform	67-66-3	10,000	(b)
Chloromethyl ether	542-88-1	500	(b)
Chloromethyl methyl ether	107-30-2	1,000	(b)
Crotonaldehyde	4170-30-3	10,000	(b)
Crotonaldehyde, (E)-	123-73-9	10,000	(b)
Cyanogen chloride	506-77-4	1,000	(d)
Cyclohexylamine	108-91-8	5,000	(b)
Diborane	19287-45-7	500	(b)
Trans-1,4-dichlorobutene	110-57-6	1,000	(b)
Dichloroethyl ether	111-44-4	10,000	(b)
Dimethyldichlorosilane	75-78-5	1,000	(b)
Dimethylhydrazine	57-14-7	5,000	(b)
Dimethyl phosphorochloridothioate	2524-03-0	1,000	(b)
Epichlorohydrin	106-89-8	10,000	(b)
Ethylenediamine	107-15-3	10,000	(b)
Ethyleneimine	151-56-4	1,000	(b)
Ethylene oxide	75-21-8	5,000	(a,b)
Fluorine	7782-41-4	500	(b)
Formaldehyde	50-00-0	500	(b)
Formaldehyde cyanohydrin	107-16-4	5,000	(b)
Furan	110-00-9	1,000	(b)

TABLE 1 TO SEC. 68.130   List of Regulated Toxic Substances and Threshold Quantities for Accidental-Release Prevention (*Continued*)

[Alphabetical order—100 substances]

Chemical name	CAS no.	Threshold quantity (lb)	Basis for listing
Hydrazine	302-01-2	5,000	(b)
Hydrochloric acid (solution, conc. 25% or greater)	7647-01-0	5,000	(e)
Hydrocyanic acid	74-90-8	500	(a,b)
Hydrogen chloride (anhydrous)	7647-01-0	1,000	(a,b)
Hydrogen fluoride	7664-39-3	500	(a,b)
Hydrogen peroxide (conc. >52%)	7722-84-1	5,000	(b)
Hydrogen selenide	7783-07-5	500	(b)
Hydrogen sulfide	7783-06-4	1,000	(a,b)
Iron, pentacarbonyl-	13463-40-6	500	(b)
Isobutyronitrile	78-82-0	10,000	(b)
Isopropyl chloroformate	108-23-6	5,000	(b)
Lactonitrile	78-97-7	5,000	(b)
Methacrylonitrile	126-98-7	1,000	(b)
Methyl bromide	74-83-9	5,000	(b)
Methyl chloride	74-87-3	10,000	(a)
Methyl chloroformate	79-22-1	1,000	(b)
Methyl hydrazine	60-34-4	5,000	(b)
Methyl isocyanate	624-83-9	1,000	(a,b)
Methyl mercaptan	74-93-1	1,000	(b)
Methyl thiocyanate	556-64-9	10,000	(b)
Methyltrichlorosilane	75-79-6	1,000	(b)
Nickel carbonyl	13463-39-3	500	(b)
Nitric acid	7697-37-2	5,000	(b)
Nitric oxide	10102-43-9	1,000	(b)
Nitrobenzene	98-95-3	10,000	(c)
Parathion	56-38-2	1,000	(c)
Peracetic acid	79-21-0	1,000	(b)
Perchloromethylmercaptan	594-42-3	1,000	(b)
Phenol (liquid)	108-95-2	10,000	(c)
Phosgene	75-44-5	500	(a,b)
Phosphine	7803-51-2	1,000	(b)
Phosphorus oxychloride	10025-87-3	1,000	(b)
Phosphorus trichloride	7719-12-2	5,000	(b)
Piperidine	110-89-4	5,000	(b)
Propionitrile	107-12-0	1,000	(b)
Propyl chloroformate	109-61-5	5,000	(b)
Propyleneimine	75-55-8	10,000	(b)
Propylene oxide	75-56-9	10,000	(b)
Pyridine, 2-methyl-5-vinyl-	140-76-1	1,000	(b)
Sulfur dioxide	7446-09-5	1,000	(a,b)
Sulfuric acid	7664-93-9	5,000	(c)
Sulfur tetrafluoride	7783-60-0	1,000	(b)
Sulfur trioxide	7446-11-9	1,000	(a,b)
Tetramethyllead	75-74-1	1,000	(b)
Tetranitromethane	509-14-8	1,000	(b)
Thiophenol	108-98-5	1,000	(b)

TABLE 1 TO SEC. 68.130　List of Regulated Toxic Substances and Threshold Quantities for Accidental-Release Prevention

[Alphabetical order—100 substances]

Chemical name	CAS no.	Threshold quantity (lb)	Basis for listing
Titanium tetrachloride	7550-45-0	500	(b)
Toluene 2,4-diisocyanate	584-84-9	1,000	(a)
Toluene 2,6-diisocyanate	91-08-7	1,000	(a)
Toluene diisocyanate (unspecified isomer)	26471-62-5	1,000	(a)
Trichloroethylsilane	115-21-9	1,000	(b)
Trimethylchlorosilane	75-77-4	1,000	(b)
Vinyl acetate monomer	108-05-4	5,000	(b)
Vinyl chloride	75-01-4	10,000	(a)

Basis for listing:

a. Mandated for listing by Congress.

b. On EHS list, vapor pressure 0.5 mmHg or greater.

c. On EHS list, vapor pressure less than 0.5 mmHg, but has been involved in accidents resulting in death or injury.

d. Toxic gas.

e. Listed based on toxicity of hydrogen chloride, potential to release hydrogen chloride, and history of accidents.

TABLE 2 TO SEC. 68.130   List of Regulated Toxic Substances and Threshold Quantities for Accidental-Release Prevention

[CAS number order—100 substances]

Case no.	Chemical name	Threshold quantity (lb)	Basis for listing
50-00-0	Formaldehyde	500	(b)
56-38-2	Parathion	1,000	(c)
57-14-7	Dimethylhydrazine	5,000	(b)
60-34-4	Methyl hydrazine	5,000	(b)
62-53-3	Aniline	5,000	(b)
67-66-3	Chloroform	10,000	(b)
74-83-9	Methyl bromide	5,000	(b)
74-87-3	Methyl chloride	10,000	(a)
74-90-8	Hydrocyanic acid	500	(a,b)
74-93-1	Methyl mercaptan	1,000	(b)
75-01-4	Vinyl chloride	10,000	(a)
75-15-0	Carbon disulfide	10,000	(b)
75-21-8	Ethylene oxide	5,000	(a,b)
75-44-5	Phosgene	500	(a,b)
75-55-8	Propyleneimine	10,000	(b)
75-56-9	Propylene oxide	10,000	(b)
75-74-1	Tetramethyllead	1,000	(b)
75-77-4	Trimethylchlorosilane	1,000	(b)
75-78-5	Dimethyldichlorosilane	1,000	(b)
75-79-6	Methyltrichlorosilane	1,000	(b)
75-86-5	Acetone cyanohydrin	5,000	(b)
78-82-0	Isobutyronitrile	10,000	(b)
78-97-7	Lactonitrile	5,000	(b)
79-21-0	Peracetic acid	1,000	(b)
79-22-1	Methyl chloroformate	1,000	(b)
91-08-7	Toluene 2,6-diisocyanate	1,000	(a)
98-07-7	Benzotrichloride	500	(b)
98-16-8	Benzenamine, 3-(trifluoromethyl)	1,000	(b)
98-87-3	Benzal chloride	1,000	(b)
98-95-3	Nitrobenzene	10,000	(c)
100-44-7	Benzyl chloride	1,000	(b)
106-89-8	Epichlorohydrin	10,000	(b)
107-02-8	Acrolein	1,000	(b)
107-07-3	Chloroethanol	1,000	(b)
107-11-9	Allylamine	1,000	(b)
107-12-0	Propionitrile	1,000	(b)
107-13-1	Acrylonitrile	10,000	(b)
107-15-3	Ethylenediamine	10,000	(b)
107-16-4	Formaldehyde cyanohydrin	5,000	(b)
107-18-6	Allyl alcohol	5,000	(b)
107-30-2	Chloromethyl methyl ether	1,000	(b)
108-05-4	Vinyl acetate monomer	5,000	(b)
108-23-6	Isopropyl chloroformate	5,000	(b)
108-91-8	Cyclohexylamine	5,000	(b)
108-95-2	Phenol (liquid)	10,000	(c)
108-98-5	Thiophenol	1,000	(b)
109-61-5	Propyl chloroformate	5,000	(b)

TABLE 2 TO SEC. 68.130   List of Regulated Toxic Substances and Threshold Quantities for Accidental Release Prevention (*Continued*)

[CAS number order—100 substances]

CAS no.	Chemical name	Threshold quantity (lb)	Basis for listing
110-00-9	Furan	1,000	(b)
110-57-6	Trans-1,4-dichlorobutene	1,000	(b)
110-89-4	Piperidine	5,000	(b)
111-44-4	Dichloroethyl ether	10,000	(b)
115-21-9	Trichloroethylsilane	1,000	(b)
123-73-9	Crotonaldehyde, (E)	10,000	(b)
126-98-7	Methacrylonitrile	1,000	(b)
140-29-4	Benzyl cyanide	1,000	(b)
140-76-1	Pyridine, 2-methyl-5-vinyl	1,000	(b)
151-56-4	Ethyleneimine	1,000	(b)
302-01-2	Hydrazine	5,000	(b)
353-42-4	Boron trifluoride compound with methyl ether (1:1)	5,000	(b)
506-77-4	Cyanogen chloride	1,000	(d)
509-14-8	Tetranitromethane	1,000	(b)
542-88-1	Chloromethyl ether	500	(b)
556-64-9	Methyl thiocyanate	10,000	(b)
584-84-9	Toluene 2,4-diisocyanate	1,000	(a)
594-42-3	Perchloromethylmercaptan	1,000	(b)
624-83-9	Methyl isocyanate	1,000	(a,b)
814-68-6	Acrylyl chloride	1,000	(b)
2524-03-0	Dimethylphosphorochloridothioate	1,000	(b)
4170-30-3	Crotonaldehyde	10,000	(b)
7446-09-5	Sulfur dioxide	1,000	(a,b)
7446-11-9	Sulfur trioxide	1,000	(a,b)
7550-45-0	Titanium tetrachloride	500	(b)
7637-07-2	Boron trifluoride	1,000	(b)
7647-01-0	Hydrogen chloride (anhydrous)	1,000	(a,b)
7647-01-0	Hydrochloric acid (solution, conc. 25% or greater)	5,000	(e)
7664-39-3	Hydrogen fluoride	500	(a,b)
7664-41-7	Ammonia (anhydrous)	1,000	(a,b)
7664-41-7	Ammonia (aqueous solution, conc. 20% or greater)	5,000	(a,b)
7664-93-9	Sulfuric acid	5,000	(c)
7697-37-2	Nitric acid	5,000	(b)
7719-12-2	Phosphorus trichloride	5,000	(b)
7722-84-1	Hydrogen peroxide (conc. >52%)	5,000	(b)
7726-95-6	Bromine	1,000	(a,b)
7782-41-4	Fluorine	500	(b)
7782-50-5	Chlorine	1,000	(a,b)
7783-06-4	Hydrogen sulfide	1,000	(a,b)
7783-07-5	Hydrogen selenide	500	(b)
7783-60-0	Sulfur tetrafluoride	1,000	(b)

TABLE 2 TO SEC. 68.130   List of Regulated Toxic Substances and Threshold Quantities for Accidental Release Prevention (*Continued*)

[CAS number order—100 substances]

CAS no.	Chemical name	Threshold quantity (lb)	Basis for listing
7783-70-2	Antimony pentafluoride	1,000	(b)
7784-34-1	Arsenous trichloride	5,000	(b)
7784-42-1	Arsine	500	(b)
7803-51-2	Phosphine	1,000	(b)
10025-87-3	Phosphorus oxychloride	1,000	(b)
10049-04-4	Chlorine dioxide	500	(d)
10102-43-9	Nitric oxide	1,000	(b)
10294-34-5	Boron trichloride	1,000	(b)
13463-39-3	Nickel carbonyl	500	(b)
13463-40-6	Iron, pentacarbonyl	500	(b)
19287-45-7	Diborane	500	(b)
26471-62-5	Toluene diisocyanate (unspecified isomer)	1,000	(a)

Basis for listing:

a. Mandated for listing by Congress.

b. On EHS list, vapor pressure 0.5 mmHg or greater.

c. On EHS list, vapor pressure less than 0.5 mmHg, but has been involved in accidents resulting in death or injury.

d. Toxic gas.

e. Listed based on toxicity of hydrogen chloride, potential to release hydrogen chloride, and history of accidents.

TABLE 3 TO SEC. 68.130   List of Regulated Flammable Substances and Threshold Quantities for Accidental-Release Prevention

[Alphabetical order—62 substances]

Chemical name	CAS no.	Threshold quantity (lb)	Basis for listing
Acetaldehyde	75-07-0	10,000	(g)
Acetylene	74-86-2	10,000	(f)
Bromotrifluorethylene	598-73-2	10,000	(f)
1,3-Butadiene	106-99-0	10,000	(f)
Butane	106-97-8	10,000	(f)
1-Butene	106-98-9	10,000	(f)
2-Butene	107-01-7	10,000	(f)
Butene	25167-67-3	10,000	(f)
2-Butene-cis	590-18-1	10,000	(f)
2-Butene-trans	624-64-6	10,000	(f)
Carbon oxysulfide	463-58-1	10,000	(f)
Chlorine monoxide	7791-21-1	10,000	(f)
2-Chloropropylene	557-98-2	10,000	(g)
1-Chloropropylene	590-21-6	10,000	(g)
Cyanogen	460-19-5	10,000	(f)
Cyclopropane	75-19-4	10,000	(f)
Dichlorosilane	4109-96-0	10,000	(f)
Difluoroethane	75-37-6	10,000	(f)
Dimethylamine	124-40-3	10,000	(f)
2,2-Dimethylpropane	463-82-1	10,000	(f)
Ethane	74-84-0	10,000	(f)
Ethyl acetylene	107-00-6	10,000	(f)
Ethylamine	75-04-7	10,000	(f)
Ethyl chloride	75-00-3	10,000	(f)
Ethylene	74-85-1	10,000	(f)
Ethyl ether	60-29-7	10,000	(g)
Ethyl mercaptan	75-08-1	10,000	(g)
Ethyl nitrite	109-95-5	10,000	(f)
Hydrogen	1333-74-0	10,000	(f)
Isobutane	75-28-5	10,000	(f)
Isopentane	78-78-4	10,000	(g)
Isoprene	78-79-5	10,000	(g)
Isopropylamine	75-31-0	10,000	(g)
Isopropyl chloride	75-29-6	10,000	(g)
Methane	74-82-8	10,000	(f)
Methylamine	74-89-5	10,000	(f)
3-Methyl-1-butene	563-45-1	10,000	(f)
2-Methyl-1-butene	563-46-2	10,000	(g)
Methyl ether	115-10-6	10,000	(f)
Methyl formate	107-31-3	10,000	(g)
2-Methylpropene	115-11-7	10,000	(f)
1,3-Pentadiene	504-60-9	10,000	(f)

TABLE 3 TO SEC. 68.130 List of Regulated Flammable Substances and Threshold Quantities for Accidental-Release Prevention (*Continued*)

[Alphabetical order—62 substances]

Chemical name	CAS no.	Threshold quantity (lb)	Basis for listing
Pentane	109-66-0	10,000	(g)
1-Pentene	109-67-1	10,000	(g)
2-Pentene, (E)-	646-04-8	10,000	(g)
2-Pentene, (Z)-	627-20-3	10,000	(g)
Propadiene	463-49-0	10,000	(f)
Propane	74-98-6	10,000	(f)
Propylene	115-07-1	10,000	(f)
Propyne	74-99-7	10,000	(f)
Silane	7803-62-5	10,000	(f)
Tetrafluoroethylene	116-14-3	10,000	(f)
Tetramethylsilane	75-76-3	10,000	(g)
Trichlorosilane	10025-78-2	10,000	(g)
Trifluoro-chloroethylene	79-38-9	10,000	(f)
Trimethylamine	75-50-3	10,000	(f)
Vinyl acetylene	689-97-4	10,000	(f)
Vinyl ethyl ether	109-92-2	10,000	(g)
Vinyl fluoride	75-02-5	10,000	(f)
Vinylidene chloride	75-35-4	10,000	(g)
Vinylidene fluoride	75-38-7	10,000	(f)
Vinyl methyl ether	107-25-5	10,000	(f)

Basis for listing:
f. Flammable gas.
g. Volatile flammable liquid.

TABLE 4 TO SEC. 68.130   List of Regulated Flammable Substances and Threshold Quantities for Accidental-Release Prevention

[CAS number order—62 substances]

CAS no.	Chemical name	Threshold quantity (lb)	Basis for listing
60-29-7	Ethyl ether	10,000	(g)
74-82-8	Methane	10,000	(f)
74-84-0	Ethane	10,000	(f)
74-85-1	Ethylene	10,000	(f)
74-86-2	Acetylene	10,000	(f)
74-89-5	Methylamine	10,000	(f)
74-98-6	Propane	10,000	(f)
74-99-7	Propyne	10,000	(f)
75-00-3	Ethyl chloride	10,000	(f)
75-02-5	Vinyl fluoride	10,000	(f)
75-04-7	Ethylamine	10,000	(f)
75-07-0	Acetaldehyde	10,000	(g)
75-08-1	Ethyl mercaptan	10,000	(g)
75-19-4	Cyclopropane	10,000	(f)
75-28-5	Isobutane	10,000	(f)
75-29-6	Isopropyl chloride	10,000	(g)
75-31-0	Isopropylamine	10,000	(g)
75-35-4	Vinylidene chloride	10,000	(g)
75-37-6	Difluoroethane	10,000	(f)
75-38-7	Vinylidene fluoride	10,000	(f)
75-50-3	Trimethylamine	10,000	(f)
75-76-3	Tetramethylsilane	10,000	(g)
78-78-4	Isopentane	10,000	(g)
78-79-5	Isoprene	10,000	(g)
79-38-9	Trifluorochloroethylene	10,000	(f)
106-97-8	Butane	10,000	(f)
106-98-9	1-Butene	10,000	(f)
106-99-0	1,3-Butadiene	10,000	(f)
107-00-6	Ethyl acetylene	10,000	(f)
107-01-7	2-Butene	10,000	(f)
107-25-5	Vinyl methyl ether	10,000	(f)
107-31-3	Methyl formate	10,000	(g)
109-66-0	Pentane	10,000	(g)
109-67-1	1-Pentene	10,000	(g)
109-92-2	Vinyl ethyl ether	10,000	(g)
109-95-5	Ethyl nitrite	10,000	(f)
115-07-1	Propylene	10,000	(f)
115-10-6	Methyl ether	10,000	(f)
115-11-7	2-Methylpropene	10,000	(f)
161-14-3	Tetrafluoroethylene	10,000	(f)
124-40-3	Dimethylamine	10,000	(f)
460-19-5	Cyanogen	10,000	(f)
463-49-0	Propadiene	10,000	(f)
463-58-1	Carbon oxysulfide	10,000	(f)

TABLE 4 TO SEC. 68.130   List of Regulated Flammable Substances and Threshold Quantities for Accidental-Release Prevention

[CAS number order—62 substances]

CAS no.	Chemical name	Threshold quantity (lb)	Basis for listing
463-82-1	2,2-Dimethylpropane	10,000	(f)
504-60-9	1,3-Pentadiene	10,000	(f)
557-98-2	2-Chloropropylene	10,000	(g)
563-45-1	3-Methyl-1-butene	10,000	(f)
563-46-2	2-Methyl-1-butene	10,000	(g)
590-18-1	2-Butene-cis	10,000	(f)
590-21-6	1-Chloropropylene	10,000	(g)
598-73-2	Bromotrifluorethylene	10,000	(f)
624-64-6	2-Butene-trans	10,000	(f)
627-20-3	2-Pentene, (Z)	10,000	(g)
646-04-8	2-Pentene, (E)	10,000	(g)
689-97-4	Vinyl acetylene	10,000	(f)
1333-74-0	Hydrogen	10,000	(f)
4109-96-0	Dichlorosilane	10,000	(f)
7791-21-1	Chlorine monoxide	10,000	(f)
7803-62-5	Silane	10,000	(f)
10025-78-2	Trichlorosilane	10,000	(g)
25167-67-3	Butene	10,000	(f)

Basis for listing:
f. Flammable gas.
g. Volatile flammable liquid.

Clean Air Resources

General

EPA Public Information Center (PIC)

Contact:	PM-211B, 401 M Street, SW, Southeast Garage Level, Washington, DC 20460
Phone:	(202) 260-2080
Format:	Distributes numerous EPA publications. Refers questions to an appropriate contact in an EPA program office, clearinghouse, docket, or hotline.
Info. provided:	The PIC is one of EPA's primary contact points for general information and responds to inquiries on all major environmental topics. If you do not know where to start to look for information, this is a good first step.
Access/cost:	8:00 a.m.–5:30 p.m. Eastern, M–F; free

Hotlines

Acid Rain Hotline

Hotline:	(617) 674-7377 or (617) 641-5377
Format:	Hotline for main data file. Supplemental Data File (SDF), Adjunct Data File (ADF), and supporting materials.
Info. provided:	Fact sheets of proposed and final rules, reports and brochures regarding EPA's Acid Rain program, and copies of the National Allowance Data Base (NADB) on diskette. The NADB contains data for the purpose of allocating allowances to affected utility units during phase II of the Acid Rain program. The ADF contains emissions data on nontraditional utility units, which may or may not produce power for sale to the public. The SDF determines which of these nontraditional units qualify for the special allocation provisions of the CAAA. Database information is also available from the EPA Technology Transfer Network Bulletin Board System (see below) under

"CAAA," or from U.S. EPA, Acid Rain Division, or your regional EPA representative.

Access/cost:	9:00 a.m.–5:30 p.m. Eastern, M–F; free

AirRISC (Air Risk Information Support Center)

Contact:	Holly Reid, (919) 541-5344
Hotline:	(919) 541-0888
Format:	Hotline, reports/manuals, technical assistance projects
Info. provided:	Information and technical guidance pertaining to health, exposure, and risk assessment for toxic air pollutants and criteria pollutants.
Access/cost:	Free

CTC (Control Technology Center hotline)

Contact:	Bob Blaszczak, Control Technology Center, U.S. EPA (MD-13), Research Triangle Park, NC 27711
Phone:	(919) 541-0800 National
Format:	Hotline, Federal Small Business Assistance Program (FSBAP), Global Greenhouse Gases Technology Transfer Center (GGGTTC).
Info. provided:	No question is too simple! Technical support and guidance on air pollution emissions and control technology for stationary sources. Also the GGGTTC which includes information on greenhouse gas emissions, prevention, mitigation, and control strategies.
Access/cost:	7:30 a.m.–5:30 p.m. Eastern, M–F; free

Mobile sources

Hotline:	(800) 821-1237 New England states only
	(800) 631-2700 Massachusetts only
Info. provided:	Complaints regarding auto emission tampering, emission, auto warranty, recall, fuel issues, CFC recycling, auto air conditioning
Access/cost:	8:00 a.m.–8:00 p.m. Eastern, M–F; free

PPIC (Pollution Prevention Information Clearinghouse) Technical Support Hotline

Hotline:	(703) 821-4800 or Fax (703) 821-4775 or (703) 442-0584
Info. provided:	All PPIC documents are available through the hotline as well as the newsletter and information on how to access

their computer bulletin board system (PIES) through the SprintNews network.

Access/cost: 9:00 a.m.–5:00 p.m. Eastern, M–F; free.

Regional EPA office hotlines

Contacts:
    Region I   (CT, ME, MA, NH, RI, VT)
(617) 565-3420

Region II   (NY, NJ, PR, VI)   (212) 264-7834

Region III   (DE, DC, MD, PA, VA, WV)
(800) 438-2474

Region IV   (AL, FL, GA, KY, MS, NC, SC, TN)
(800) 241-1754

Region V   (IL, ID, MI, MN, OH, WI)
(800) 621-8431 or (800) 572-2515 in IL

Region VI   (AR, LA, NM, OK, TX)   (214) 655-7200

Region VII   (IA, KS, MO, NE)   (800) 223-0425

Region VIII   (CO, MT, ND, SD, UT, WY)
(800) 525-3022

Region IX   (AZ, CA, HI, NV)   (415) 744-1500

Region X   (AK, ID, OR, WA)   (800) 424-4EPA

Info. provided: General information on agency programs to the public and referrals as needed.

Stratospheric Ozone Information Hotline

Hotline: (800) 296-1996

Info. provided: Reports, fact sheets, brochures, and copies of proposed rules and regulations, all regarding EPA's Stratospheric Ozone Protection program

Access/cost: 10:00 a.m.–4:00 p.m. Eastern, M–F; free

TRI (Toxic Release Inventory)

Hotline: TRI User Support, (202) 260-1531

Format: Hotline, database, and fact sheets

Info. provided: Environmental releases of toxic chemicals to air, water, and in waste; includes transfers and publicly owned treatment works (POTWs). Annual fact sheets on releases in each state are also available.

Access/cost: TRI searches can be requested through TRI User Support; searches are free, and will take up to a few weeks to process depending upon the workload.

Computer Networks

OAQPS Technology Transfer Network (OAQPS TTN)

Contacts: TTN   (919) 541-5742

Jerry Mersch   (919) 541-5635

Herschel Rorex   (919) 541-5637

Format: On-line network

Info. provided: Multiple bulletin board network for access to many OAQPS technical and informational centers. It also contains calendars of public meetings, texts of Federal Register notices, testimony, regulatory schedules, etc.

Access/cost: Free access via modem, 24 h/day, 7 days/week, except 8:00 a.m.–noon Eastern on Mondays; set your communications software parameters to 8 data bits, a parity of N, and 1 stop bit. Call TTN at (919) 541-5742, go through the registration process, and you gain immediate access to the following electronic bulletin boards:

OAQPS—Office of Air Quality Planning and Standards
EMTIC—Emissions Measurements Technical Information Center
SCRAM—Support Center for Regulatory Air Models
CHIEF—Clearinghouse for Inventories and Emission Factors
CAAA—Clean Air Act Amendments
APTI—Air Pollution Training Institute
CTC—Control Technology Center
AMTIC—Ambient Monitoring Technology Information Center
AIRS—Aerometric Information Retrieval System
BLIS RACT/BACT/LAER Clearinghouse (RBLC)— Readily Available/Best Available Control Technology/ Lowest Achievable Emissions Rate
NATICH—National Air Toxics Information Clearinghouse
OMS—Office of Mobile Sources

AIRS (Aerometric Information Retrieval System)

Contacts: AIRS-Air Quality Hotline (1-800-333-7909)

For general questions/problems or specific questions/ problems about the operations of the database. Data retrieval request: Tim Link, (919) 541-5456 or Fax (919) 629-5663/2357.

Info. provided: A computer-based repository of information about airborne pollution in the United States. AIRS is comprised of four subsystems: Air Quality (AQS), Air Facility (AFS), Area/Mobile Source (AMS), and Geo-Common (GCS) subsystems.

Access/cost: Users may submit retrieval requests for AIRS data through the Freedom of Information Act from Tom Link. Users must agree to pay computer search costs in excess of $25. Remember to request a fee waiver for nonprofits.

CFR Title 40 Parts (Air-Related)

Title 40—Protection of the Environment

Part 50: National Primary and Secondary Ambient Air Quality Standards

Part 51: Requirements for Preparation, Adoption, and Submittal of Implementation Plans

Part 52: Approval and Promulgation of Implementation Plans

Part 53: Ambient Air Monitoring Reference and Equivalent Methods

Part 54: Prior Notice of Citizen Suits

Part 55: Outer Continental Shelf Air Regulations

Part 56: Regional Consistency

Part 57: Primary Nonferrous Smelter Orders

Part 58: Ambient Air Quality Surveillance

Part 60: Standards of Performance for New Stationary Sources

Part 61: National Emission Standards for Hazardous Air Pollutants

Part 62: Approval and Promulgation of State Plans for Designated Facilities and Pollutants

Part 63: National Emission Standards for Hazardous Air Pollutants for Source Categories

Part 65: Delayed Compliance Orders (Removed)

Part 66: Assessment and Collection of Noncompliance Penalties by EPA

Part 67: EPA Approval of State Noncompliance Penalty Program

Part 68: Chemical Accident Prevention Provisions

Part 69: Special Exemptions from Requirements of the Clean Air Act

Part 70: State Operating Permit Programs

Part 71: Federal Operating Permit Programs

Part 72: Permits

Part 73: Sulfur Dioxide Allowance System

Part 74: Sulfur Dioxide Opt-ins

Part 75: Continuous Emission Monitoring

Part 76: Acid Rain Nitrogen Oxides Emission Reduction Program

Part 77: Excess Emissions

Part 78: Appeal Procedures for Acid Rain Program

Part 79: Registration of Fuels and Fuel Additives

Part 80: Regulation of Fuels and Fuel Additives

Part 81: Designation of Areas for Air Quality Planning Purposes

Part 82: Protection of Stratospheric Ozone

Part 85: Control of Air Pollution from Motor Vehicles and Motor Vehicle Engines

Part 86: Control of Air Pollution from New and In-Use Motor Vehicles

Part 87: Control of Air Pollution from Aircraft and Aircraft Engines

Part 88: Clean-Fuel Vehicles

Part 89: Control of Emissions from New and In-Use Nonroad Engines

Part 90: Control of Emissions from Nonroad Spark-Ignition Engines

Part 91: Control of Emissions from Marine Spark-Ignition Engines

Part 93: Determining Conformity of Federal Actions to State or Federal Implementation Plans

P

Equivalent Lengths of Ductwork Fittings

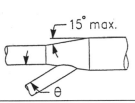
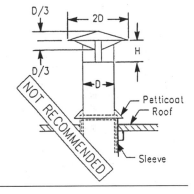

Pipe diam	90° Elbow* centerline radius			Angle of entry		H, no. of diameters		
D, in	1.5D	2.0D	2.5D	30°	45°	1.0D	0.75D	0.5D
3	5	3	3	2	3	2	2	9
4	6	4	4	3	5	2	3	12
5	9	6	5	4	6	2	4	16
6	12	7	6	5	7	3	5	20
7	13	9	7	6	9	3	6	23
8	15	10	8	7	11	4	7	26
10	20	14	11	9	14	5	9	36
12	25	17	14	11	17	6	11	44
14	30	21	17	13	21	7	13	53
16	36	24	20	16	25	9	15	62
18	41	28	23	18	28	10	18	71
20	46	32	26	20	32	11	20	80
24	57	40	32			13	24	92
30	74	51	41			17	31	126
36	93	64	52			22	39	159
40	105	72	59					
48	130	89	73					

*For 60° elbows—0.67 × loss for 90°
45° elbows—0.5 × loss for 90°
30° elbows—0.33 × loss for 90°

SOURCE: From ACGIH, 1995.

Appendix

Thermal Stability Rankings

Principal Hazardous Organic Constituent	Rank
ACETONITRILE {ETHANENITRILE} [2]	17-18
ACETONYLBENZYL-4-HYDROXYCOUMARIN (3-alpha-) {WARFARIN}	98-99
ACETOPHENONE {ETHANONE, 1-PHENYL-} [2]	85-88
ACETYL CHLORIDE {ETHANOYL CHLORIDE} [2]	92-97
ACETYL-2-THIOUREA (1-) {ACETAMIDE,N-[AMINOTHIOXOMETHYL]-}	286-290
ACETYLAMINOFLUORENE (2-) {ACETAMIDE.N-[9H-FLUOREN-2-YL]-}	69-77
ACROLEIN {2-PROPENAL}	106-107
ACRYLAMIDE {2-PROPENAMIDE}	60-64
ACRYLONITRILE {2-PROPENENITRILE} [2]	20
AFLATOXINS	200
ALDRIN	162-164
ALLYL ALCOHOL {2-PROPEN-1-OL}	116-118
AMINOBIPHENYL (4-) {[1,1' BIPHENYL]-4-AMINE}	51
AMITROLE {1H-1,2,4-TRIAZOL-3-AMINE}	208-209
ANILINE {BENZENAMINE}	46-50
ARAMITE	235-239
AURAMINE	180-181
AZASERINE {L-SERINE,DIAZOACETATE[ESTER]}	297
BENZAL CHLORIDE {ALPHA, ALPHA-DICHLOROTOLUENE} [2]	168-173
BENZANTHRACENE (1.2-) {BENZ[a]ANTHRACENE}	9
BENZENE [2]	3
BENZENETHIOL {THIOPHENOL} [2]	110
BENZIDINE {[1,1'-BIPHENYL]-4,4' DIAMINE}	60-64
BENZOQUINONE {1,4-CYCLOHEXADIENEDIONE}	89-91
BENZOTRICHLORIDE {TRICHLOROMETHYLBENZENE}	195-196
BENZO[a]PYRENE {1,2-BENZOPYRENE}	11
BENZO[b]FLUORANTHENE {2,3-BENZOFLUORANTHENE}	8
BENZO[j]FLUORANTHENE {7,8-BENZOFLUORANTHENE}	7
BENZYL CHLORIDE {CHLOROMETHYLBENZENE} [2]	127-130
BENZ[c]ACRIDINE {3,4-BENZACRIDINE}	85-88
bis(2-CHLOROETHOXY)METHANE	189-192
bis(2-CHLOROETHYL)ETHER [2]	183-186
bis(2-CHLOROISOPROPYL)ETHER	227-228
bis(2-ETHYLHEXYL)PHTHALATE	269-270
bis(CHLOROMETHYL)ETHER {METHANE-OXYbis[2-CHLORO-]}	222-223
BROMOACETONE {1-BROMO-2-PROPANONE}	136-140
BROMOFORM {TRIBROMOMETHANE} [2]	202-203
BROMOMETHANE {METHYL BROMIDE} [2]	31-33
BROMOPHENYL PHENYL ETHER (4-) {BENZENE,1-BROMO-4-PHENOXY-}	226
BRUCINE {STRYCHNIDIN-10-ONE,2,3-DIMETHOXY-}	245-246
BUTANONE PEROXIDE (2-) {METHYL ETHYL KETONE, PEROXIDE}	279
BUTYL-4,6-DINITROPHENOL (2-sec-) {DNBP}	187-188
CHLORAL {TRICHLOROACETALDEHYDE}	189-192
CHLORAMBUCIL	142
CHLORDANE (ALPHA AND GAMMA ISOMERS)	221
CHLORO-1,3-BUTADIENE (2-) {CHLOROPRENE}	69-77
CHLORO-2,3-EPOXYPROPANE (1-) {OXIRANE,2-CHLOROMETHYL-}	183-186
CHLOROACETALDEHYDE	166-167
CHLOROANILINE {CHLOROBENZENAMINE}	37
CHLOROBENZENE [2]	19
CHLOROBENZILATE	204-207
CHLOROCRESOL {4-CHLORO-3-METHYLPHENOL}	116-118

Principal Hazardous Organic Constituent	Rank
CHLORODIFLUOROMETHANE [2] [4]	151-153
CHLOROETHANE (ETHYL CHLORIDE) [4] [5]	126
CHLOROETHYLVINYLETHER (2-) {ETHENE,[2-CHLOROETHOXY]-} [2]	211-213
CHLOROMETHANE {METHYL CHLORIDE} [2]	29-30
CHLOROMETHYLMETHYL ETHER {CHLOROMETHOXYMETHANE}	218-220
CHLORONAPHTHALENE (1-) [2]	21-22
CHLOROPHENOL (2-)	102
CHLOROPHENYL THIOUREA (1-) {THIOUREA,[2-CHLOROPHENYL]-}	286-290
CHLOROPROPENE (3-) {ALLYL CHLORIDE} [2]	120
CHLOROPROPIONITRILE (3-) {3-CHLOROPROPANENITRILE} [2]	143-144
CHRYSENE {1,2-BENZPHENANTHRENE}	10
CITRUS RED No. 2 {2-NAPHTHOL,1-[(2,5-DIMETHOXYPHENYL)AZO]}	258-259
CRESOL (1,2-) {METHYLPHENOL}	104-105
CRESOL (1,3-) {METHYLPHENOL}	103
CRESOL (1,4-) {METHYLPHENOL} [2]	104-105
CROTONALDEHYDE {2-BUTENAL} [2]	113-115
CYANOGEN BROMIDE {BROMINE CYANIDE}	23-24
CYANOGEN CHLORIDE {CHLORINE CYANIDE}	17-18
CYANOGEN {ETHANEDINITRILE}	1
CYCASIN {beta-D-GLUCOPYRANOSIDE,[METHYL-ONN-AZOXY]METHYL-}	301
CYCLOHEXYL-4,6-DINITROPHENOL (2-)	187-188
CYCLOPHOSPHAMIDE	273-276
DAUNOMYCIN	291-292
DDD {DICHLORODIPHENYLDICHLOROETHANE}	145-146
DDE{1,1-DICHLORO-2,2-BIS(4-CHLOROPHENYLETHYLENE}	38
DDT {DICHLORODIPHENYLTRICHLOROETHANE}	175-178
Di-n-BUTYL PHTHALATE	261-265
Di-n-OCTYL PHTHALATE [2]	267
DI-n-PROPYLNITROSAMINE {N-NITROSO-DI-n-PROPYLAMINE}	303-318
DIALLATE {S-(2,3-DICHLOROALLYL)DIISOPROPYL THIOCARBAMATE}	235-239
DIBENZO[a,e]PYRENE {1,2,4,5-DIBENZOPYRENE}	16
DIBENZO[a,h]PYRENE {1,2,5,6-DIBENZOPYRENE}	14
DIBENZO[a,i]PYRENE {1,2,7,8-DIBENZOPYRENE}	15
DIBENZO[c,g]CARBAZOLE (7H-) {3,4,5,6-DIBENZCARBAZOLE}	100-101
DIBENZ[a,h]ACRIDINE {1,2,5,6-DIBENZACRIDINE}	92-97
DIBENZ[a,h]ANTHRACENE {1,2,5,6-DIBENZANTHRACENE}	12
DIBENZ[a,j]ACRIDINE {1,2,7,8-DIBENZACRIDINE}	92-97
DIBROMO-3-CHLOROPROPANE (1,2-)	214
DIBROMOETHANE (1,2-) {ETHYLENE DIBROMIDE}	199
DIBROMOMETHANE {METHYLENE BROMIDE} [2]	127-130
DICHLORO-1-PROPANOL (2,3-)	168-173
DICHLORO-2-BUTENE (1,4-)	136-140
DICHLORO-2-PROPANOL (1,1-)	145-146
DICHLORO-2-PROPANOL (1,3-)	147
DICHLOROBENZENE {1,2-DICHLOROBENZENE} [2]	23-24
DICHLOROBENZENE {1,3-DICHLOROBENZENE} [2]	25
DICHLOROBENZENE {1,4-DICHLOROBENZENE}	21-22
DICHLOROBENZIDINE (3,3'-)	67
DICHLORODIFLUOROMETHANE [2]	85-88
DICHLOROETHANE (1,1-) {ETHYLIDENE DICHLORIDE} [5]	175-178
DICHLOROETHANE (1,2-) [2]	131
DICHLOROETHENE (1,1-) [2]	42-44

Principal Hazardous Organic Constituent	Rank
DICHLOROETHENE (trans-1,2-) [2]	54
DICHLOROFLUOROMETHANE [2] [4]	154-157
DICHLOROMETHANE {METHYLENE CHLORIDE} [2]	65-66
DICHLOROPHENOL (2,4-)	113-115
DICHLOROPHENOL (2,6-)	113-115
DICHLOROPHENOXYACETIC ACID (2,4-) {2,4-D}	211-213
DICHLOROPROPANE (1,1-) [5]	182
DICHLOROPROPANE (1,2-) {PROPYLENE DICHLORIDE} [5]	179
DICHLOROPROPANE (1,3-) [5]	165
DICHLOROPROPANE (2,2-) [5]	224
DICHLOROPROPENE (1,1-) [2]	81-84
DICHLOROPROPENE (2,3-)	127-130
DICHLOROPROPENE (3,3-)	135
DICHLOROPROPENE (cis-1,3-)	121-125
DICHLOROPROPENE (trans-1,2-)	89-91
DICHLOROPROPENE (trans-1,3-)	121-125
DIELDRIN	161-163
DIEPOXYBUTANE (1,2,3,4-) {2,2'-BIOXIRANE}	194
DIETHYL PHTHALATE	256-257
DIETHYLSTILBESTEROL	108-109
DIHYDROSAFROLE {1,2-METHYLENEDIOXY-4-PROPYLBENZENE}	227-228
DIHYDROXY-ALPHA-[METHYLAMINO]METHYL BENZYL ALCOHOL (3,4-)	106-107
DIISOPROPYLFLUOROPHOSPHATE {DFP}	261-265
DIMETHOATE	235-239
DIMETHOXYBENZIDINE (3,3'-)	250
DIMETHYL PHTHALATE [2]	92-97
DIMETHYL-1-METHYLTHIO-2-BUTANONE,O-[(METHYLAMINO)-CARBONYL] OXIME (3,3-) {THIOFANOX}	218-220
DIMETHYLAMINOAZOBENZENE	255
DIMETHYLBENZIDINE (3,3'-)	78
DIMETHYLBENZ[a]ANTHRACENE (7,12-)	45
DIMETHYLCARBAMOYLCHLORIDE	175-178
DIMETHYLHYDRAZINE (1,1-) [5]	216-217
DIMETHYLHYDRAZINE (1,2-)	218-220
DIMETHYLPHENETHYLAMINE (alpha, alpha-)	60-64
DIMETHYLPHENOL (2,4-)	119
DINITROBENZENE (1,2-)	158-161
DINITROBENZENE (1,3-)	154-157
DINITROBENZENE (1,4-)	158-161
DINITROCRESOL (4,6-) {PHENOL,2,4-DINITRO-6-METHYL-}	189-192
DINITROPHENOL (2,4-)	183-186
DINITROTOLUENE (2,4-)	168-173
DINITROTOLUENE (2,6-)	168-173
DIOXANE (1,4-) {1,4-DIETHYLENE OXIDE} [2]	141
DIPHENYLAMINE {N-PHENYLBENZENAMINE}	42-44
DIPHENYLHYDRAZINE (1,2-)	251
DISULFOTON	261-265
DITHIOBIURET (2,4-) {THIOIMIDODICARBONIC DIAMIDE}	295-296
ENDOSULFAN	320
ENDRIN	278
ETHYL CARBAMATE {URETHAN} {CARBAMIC ACID, ETHYL ESTER}	204-207
ETHYL CYANIDE {PROPIONITRILE} [2]	89-91
ETHYL METHACRYLATE {2-PROPENOIC ACID, 2-METHYL-,ETHYL ESTER}	204-207

Principal Hazardous Organic Constituent	Rank
ETHYL METHANESULFONATE {METHANESULFONIC ACID, ETHYL ESTER}	261-265
ETHYLENE OXIDE {OXIRANE} [5]	174
ETHYLENE THIOUREA {2-IMIDAZOLIDINETHIONE}	291-292
ETHYLENEbisDITHIOCARBAMIC ACID	283
ETHYLENEIMINE {AZIRIDINE}	235-239
FLUORANTHENE {BENZO[j,k]FLUORENE}	6
FLUOROACETAMIDE (2-)	55-56
FLUOROACETIC ACID	42-44
FORMALDEHYDE {METHYLENE OXIDE}	46-50
FORMIC ACID {METHANOIC ACID}	39-40
GLYCIDYALDEHYDE {1-PROPANOL-2,3-EPOXY}	175-178
HEPTACHLOR	180-181
HEPTACHLOR EPOXIDE	193
HEXACHLOROBENZENE [2]	31-33
HEXACHLOROBUTADIENE (trans-1,3) [2]	92-97
HEXACHLOROCYCLOHEXANE {LINDANE} [2]	151-153
HEXACHLOROCYCLOPENTADIENE	168-173
HEXACHLOROETHANE [2]	202-203
HEXACHLOROPHENE {2,2'-METHYLENEbis[3,4,6-TRICHLOROPHENOL]}	136-140
HEXACHLOROPROPENE [2]	234
HEXAETHYL TETRAPHOSPHATE	298
HYDRAZINE (DIAMINE)	127-130
HYDROGEN CYANIDE {HYDROCYANIC ACID} [2]	2
INDENO(1,2,3-cd)PYRENE {1,10-(1,2-PHENYLENE)PYRENE}	13
IODOMETHANE {METHYL IODIDE}	210
ISOBUTYL ALCOHOL {2-METHYL-1-PROPANOL} [2]	112
ISODRIN	162-164
ISOSAFROLE {1,2-METHYLENEDIOXY-4-ALLYLBENZENE}	247-249
KEPONE	245-246
LASIOCARPINE	204-207
MALEIC ANHYDRIDE {2,5-FURANDIONE}	98-99
MALEIC HYDRAZIDE {1,2-DIHYDRO-3,6-PYRIDAZINEDIONE}	225
MALONONITRILE {PROPANEDINITRILE}	46-50
MELPHALAN {ALANINE,3-[p-bis(2-CHLOROETHYL)AMINO]PHENYL-,L-}	293-294
METHACRYLONITRILE {2-METHYL-2-PROPENENITRILE} [2]	65-66
METHAPYRILENE	195-196
METHOXYCHLOR	243-244
METHYL CHLOROCARBONATE {CARBONOCHLORIDIC ACID, METHYL ESTER}	46-50
METHYL ETHYL KETONE {2-BUTANONE} [2]	108-109
METHYL HYDRAZINE [5]	197-198
METHYL ISOCYANATE {METHYLCARBYLAMINE}	46-50
METHYL METHACRYLATE {2-PROPENOIC ACID, 2-METHYL-, METHYL ESTER}	60-64
METHYL METHANESULFONATE {METHANESULFONIC ACID, METHYL ESTER}	229
METHYL PARATHION	148-150
METHYL-2-METHYLTHIO-PROPIONALDEHYDE-O-(METHYLCARBONYL)OXIME(2-)	232-233
METHYLACTONITRILE (2-) {PROPANENITRILE,2-HYDROXY-2-METHYL}	116-118
METHYLAZIRIDINE (2-) {1,2-PROPYLENIMINE}	243-244
METHYLCHOLANTHRENE (3-)	68
METHYLENE BIS(2-CHLOROANILINE) (4,4-)	211-213
METHYLTHIOURACIL	269-270
METHYOMYL	232-233
MUSCIMOL {5-AMINOMETHYL-3-ISOAZOTOL}	208-209

Principal Hazardous Organic Constituent	Rank
MUSTARD GAS {bis[2-CHLOROETHYL]-SULFIDE}	132-134
N,N-BIS(2-CHLOROETHYL)2-NAPHTHYLAMINE {CHLORNAPHAZINE}	132-134
N,N-DIETHYLHYDRAZINE {1,2-DIETHYLHYDRAZINE}	216-217
n-BUTYLBENZYL PHTHALATE [2]	253
N-METHYL-N'-NITRO-N-NITROSOGUANIDINE	303-318
N-NITROSO-DI-ETHANOLAMINE {[2,2'-NITROSOIMINO]bisETHANOL}	303-318
N-NITROSO-DI-N-BUTYLAMINE {N-BUTYL-N-NITROSO-1-BUTANAMINE}	303-318
N-NITROSO-N-ETHYLUREA {N-ETHYL-N-NITROSOCARBAMIDE}	303-318
N-NITROSO-N-METHYLUREA {N-METHYL-N-NITROSOCARBAMIDE}	303-318
N-NITROSO-N-METHYLURETHANE	303-318
N-NITROSODIETHYLAMINE {N-ETHYL-N-NITROSOETHANAMINE}	303-318
N-NITROSODIMETHYLAMINE {DIMETHYLNITROSAMINE}	303-318
N-NITROSOMETHYLETHYLAMINE {N-METHYL-N-NITROSOETHANAMINE}	303-318
N-NITROSOMETHYLVINYLAMINE {N-METHYL-N-NITROSOETHENAMINE}	303-318
N-NITROSOMORPHOLINE	303-318
N-NITROSONORNICOTINE	303-318
N-NITROSOPIPERIDINE {HEXAHYDRO-N-NITROSOPYRIDINE}	303-318
N-NITROSOSARCOSINE	303-318
N-PHENYLTHIOUREA	286-290
n-PROPYLAMINE {1-PROPANAMINE}	79
NAPHTHALENE [2]	5
NAPHTHOQUINONE (1,4-) {1,4-NAPHTHALENEDIONE}	92-97
NAPHTHYL-2-THIOUREA (1-) {THIOUREA,1-NAPHTHALENYL-}	286-290
NAPHTHYLAMINE (1-)	52-53
NAPHTHYLAMINE (2-)	52-53
NICOTINE {(S)-3-[1-METHYL-2-PYRROLIDINYL]PYRIDINE}	273-276
NITROANILINE {4-NITROBENZENAMINE}	154-157
NITROBENZENE [2]	143
NITROGEN MUSTARD	132-134
NITROGEN MUSTARD N-OXIDE	299-300
NITROGLYCERINE {TRINITRATE-1,2,3-PROPANETRIOL} [5]	281
NITROPHENOL (4-)	148-150
NITROQUINOLINE-1-OXIDE (4-)	299-300
NITROSOPYRROLIDINE {N-NITROSOTETRAHYDROPYRROLE}	303-318
NITROTOLUIDINE (5-) {BENZENAMINE,2-METHYL-5-NITRO-}	166-167
O,O,O-TRIETHYL PHOSPHOROTHIOATE	261-265
O,O-DIETHYL S-[(ETHYLTHIO)METHYL]ESTER OF PHOSPHORODITHIOIC ACID	258-259
O,O-DIETHYL-O-2-PYRAZINYL PHOSPHOROTHIOATE	254
O,O-DIETHYL-S-METHYL ESTER OF PHOSPHORIC ACID	256-257
O,O-DIETHYLPHOSPHORIC ACID,O-p-NITROPHENYL ESTER	252
OCTAMETHYLPYROPHOSPHORAMIDE {OCTAMETHYLDIPHOSPHORAMIDE}	268
OXABICYCLO[2.2.1]HEPTANE-2,3-DICARBOXYLIC ACID (7-) {ENDOTHAL}	319
PARALDEHYDE {2,4,6-TRIMETHYL-1,3,5-TRIOXANE} [5]	266
PARATHION [5]	222-223
PENTACHLOROBENZENE [2]	31-33
PENTACHLOROETHANE [2]	154-157
PENTACHLORONITROBENZENE {PCNB}	235-239
PENTACHLOROPHENOL	151-153
PHENACETIN {N-[4-ETHOXYPHENYL]ACETAMIDE}	197-198
PHENOL {HYDROXYBENZENE}	100-101
PHENYLENEDIAMINE (1,2-) {BENZENEDIAMINE}	57-59
PHENYLENEDIAMINE (1,3-) {BENZENEDIAMINE}	57-59

Principal Hazardous Organic Constituent	Rank
PHENYLENEDIAMINE (1,4) {BENZENEDIAMINE}	57-59
PHOSGENE {CARBONYL CHLORIDE}	39-40
PHTHALIC ANHYDRIDE {1,2-BENZENEDICARBOXYLIC ACID ANHYDRIDE}	148-150
PICOLINE (2-) {PYRIDINE, 2-METHYL-}	81-84
PRONAMIDE {3,5-DICHLORO-N-[1,1-DIMETHYL-2-PROPYNYL] BENZAMIDE}	69-77
PROPANE SULFONE (1,3-) {1,2-OXATHIOLANE,2,2-DIOXIDE}	230
PROPYLTHIOURACIL	271
PROPYN-1-OL (2-) {PROPARGYL ALCOHOL}	55-56
PYRIDINE [2]	80
RESERPINE	273-276
RESORCINOL {1,3-BENZENEDIOL}	111
SACCHARIN {1,2-BENZOISOTHIAZOLIN-3-ONE,1,1-DIOXIDE}	231
SAFROLE {1,2-METHYLENE-4-ALLYLBENZENE}	247-249
STREPTOZOTOCIN	302
STRYCHNINE {STRYCHNIDIN-10-ONE}	272
SULFUR HEXAFLUORIDE [3]	4
TETRACHLOROBENZENE (1,2,3,5-TETRACHLOROBENZENE) [2] [4]	20
TETRACHLOROBENZENE (1,2,4,5-TETRACHLOROBENZENE)	29-30
TETRACHLORODIBENZO-p-DIOXIN (2,3,7,8-) {TCDD}	34
TETRACHLOROETHANE (1,1,1,2-) [2]	215
TETRACHLOROETHANE (1,1,2,2-) [2]	121-125
TETRACHLOROETHENE [2]	36
TETRACHLOROMETHANE {CARBONTETRACHLORIDE} [2]	136-140
TETRACHLOROPHENOL (2,3,4,6-)	136-140
TETRAETHYLDITHIOPYROPHOSPHATE	282
TETRAETHYLPYROPHOSPHATE	280
TETRANITROMETHANE [5]	284
THIOACETAMIDE {ETHANETHIOAMIDE}	81-84
THIOSEMICARBAZIDE {HYDRAZINECARBOTHIOAMIDE}	293-294
THIOUREA {THIOCARBAMIDE}	286-290
THIURAM {bis[DIMETHYLTHIOCARBAMOYL]DISULFIDE}	295-296
TOLUENE {METHYLBENZENE} [2]	35
TOLUENEDIAMINE (1,3-) {DIAMINOTOLUENE}	69-77
TOLUENEDIAMINE (1,4-) {DIAMINOTOLUENE}	69-77
TOLUENEDIAMINE (2,4-) {DIAMINOTOLUENE}	69-77
TOLUENEDIAMINE (2,6-) {DIAMINOTOLUENE}	69-77
TOLUENEDIAMINE (3,4-) {DIAMINOTOLUENE}	69-77
TOLUENEDIAMINE (3,5-) {DIAMINOTOLUENE}	69-77
TOLUIDINE HYDROCHLORIDE {2-METHYL-BENZENAMINE HYDROCHLORIDE}	273-276
TOLYLENE DIISOCYANATE {1,3-DIISOCYANATOMETHYLBENZENE}	277
TRICHLOROBENZENE (1,2,4-TRICHLOROBENZENE) [2]	26-27
TRICHLOROBENZENE (1,3,5-TRICHLOROBENZENE) [2] [4]	26-27
TRICHLOROETHANE (1,1,1-) {METHYL CHLOROFORM} [2]	201
TRICHLOROETHANE (1,1,2-) [2]	158-161
TRICHLOROETHENE [2]	41
TRICHLOROFLUOROMETHANE [2]	85-88
TRICHLOROMETHANE {CHLOROFORM} [2]	195-196
TRICHLOROMETHANETHIOL	189-192
TRICHLOROPHENOL (2,4,5-)	121-125
TRICHLOROPHENOL (2,4,6-)	121-125
TRICHLOROPHENOXYACETIC ACID (2,4,5-) {2,4,5-T}	240-241
TRICHLOROPHENOXYPROPIONIC ACID (2,4,5-) {2,4,5-TP} {SILVEX}	240-241

Principal Hazardous Organic Constituent	Rank
TRICHLOROPROPANE (1,2,3-) [2]	168-173
TRICHLORO-(1,2,2)-TRIFLUOROETHANE (1,1,2) [2] [3]	81-84
TRINITROBENZENE {1,3,5-TRINITROBENZENE}	183-186
tris(1-AZRIDINYL) PHOSPHINE SULFIDE	247-249
tris(2,3-DIBROMOPROPYL)PHOSPHATE	242
TRYPAN BLUE	260
URACIL MUSTARD {5-[bis(2-CHLOROETHYL)AMINO]URACIL}	285
VINYL CHLORIDE (CHLOROETHENE)	60-64

[1]Units of temperature are degrees Celsius.

[2]Boldface indicates compound thermal stability is "experimentally evaluated" (ranking based on UDRI experimental data coupled with reaction kinetic theory).

[3]Non-Appendix VIII compound.

[4]N.O.S. listing; ranking is presented based on either UDRI or literature experimental data coupled with reaction kinetic theory.

[5]*Italics* indicate compound thermal stability is ranked based on literature experimental data coupled with reaction kinetic theory.

SOURCE: From Taylor, 1990.

Index

ABOUT THE AUTHOR

E. Roberts Alley & Associates is an environmental engineering consulting firm based in Brentwood, Tennessee. The editor, Mr. E. Roberts Alley, P.E., R.L.S., D.E.E., has more than 35 years of experience in the design and teaching of industrial and domestic water, wastewater, hazardous waste, and air treatment facilities. He has taught courses in these subjects at Vanderbilt University and George Washington University.